"*A History of the English-Speaking Peoples Since 1900* is written with verve. . . . This history contains many good things, including a perceptive, revisionist account of Suez. It is beautifully written and will be widely read."

—*The Financial Times*

"With passion and scholarship, Andrew Roberts embraces the faith of Winston Churchill that the English speaking peoples are 'the last best hope of mankind.' . . . Here is a history that is as timely as it is valuable."

—Harold Evans, author of *The American Century*

"The Anglo-American century was not predestined. . . . But, as Roberts justly concludes, when the next empire rises, the essential fair-mindedness of the Anglo-Americans may be mourned." —Evan Thomas, *Newsweek*

"Roberts has interesting and perceptive things to say about the more exotic aspects of the Anglo-Saxon diaspora. . . . The finale of the story is the most sumptuous part." —Richard Overy, *The Sunday Telegraph*

"This is a very stimulating and original book that combines this well-established author's gift for careful research with his taste for rational contrariety and love of the comical and the obscure. . . . This is an excellent book. —*American Spectator*

"Elegantly written . . . Roberts shines a fresh light on today's geopolitical realities. A truly smart and important book."

—Douglas Brinkley, author of *The Great Deluge*

"This book is deeply researched, very well written and full of fresh thinking. . . . Andrew Roberts has written an extraordinarily wide-ranging, stimulating and necessary book." —Denis Judd, *History Today*

"Roberts has a cracking good tale to tell, and he tells it very well."

—*The New York Sun*

"This book is more entertaining than many novels. . . . It will have some readers purring in happy agreement, and others tearing their hair in fury."
—Allan Massie, *The Daily Telegraph*

"Thought-provoking, erudite, and opinionated in the best sense of the term, Andrew Roberts . . . is one of Britain's most talented young historians, and in these pages, American readers will delight in discovering him."
—Jay Winik, author of *The Great Upheaval*

"A feel-good history, a much needed shot in the arm for all English-speaking peoples who may be inclined at the present moment to feel a bit uncomfortable about their place in the scheme of things."
—*The Washington Times*

"An engagingly written and widely researched book. . . . Roberts has a flair for the apt quotation and the telling fact. His portrait of Churchill is refreshingly free of cant."
—Walter Russell Mead, *The American Interest*

"An exhilarating book. . . . A great achievement. . . . This book must rank as one of the great interventions in the culture wars of the last three decades."
—*The New Criterion*

"With much of the West engaged in self-flagellation and a baffling inability to recognise a mounting threat in its midst, Andrew Roberts steps into the fray with a trenchant, timely and powerfully Churchillian defence of the Anglo-American world."
—Justin Marozzi, *The Sunday Telegraph*

"A splendid new history. . . . This is a highly enjoyable book, to be dipped into at bedtime, as well as an important one to be studied and pondered."
—John O'Sullivan, *National Review*

Bolla Denehy

About the Author

ANDREW ROBERTS took a first in modern history at Gonville and Caius College, Cambridge. His books include *'The Holy Fox': The Life of Lord Halifax* (1991), *Eminent Churchillians* (1994), *The Aachen Memorandum* (1995), *Salisbury: Victorian Titan* (1999), *Napoleon and Wellington* (2001), *Hitler and Churchill: Secrets of Leadership* (2003), *What Might Have Been* (2004), and *Waterloo: Napoleon's Last Gamble* (2005). *Salisbury* won him the Wolfson History Prize and the James Stern Silver Pen Award for nonfiction. He writes regularly for *The Sunday Telegraph* and reviews widely, and he is a fellow of the Royal Society of Literature. His website can be found at www.andrew-roberts.net.

ALSO BY ANDREW ROBERTS

Waterloo: Napoleon's Last Gamble

What Might Have Been: Imaginary History from Twelve Leading Historians

'What Ifs' of History

Hitler and Churchill: Secrets of Leadership

Napoleon and Wellington

Salisbury: Victorian Titan

The Aachen Memorandum

Eminent Churchillians

'The Holy Fox': The Life of Lord Halifax

A History of the English-Speaking Peoples Since 1900

ANDREW ROBERTS

HARPER PERENNIAL

NEW YORK • LONDON • TORONTO • SYDNEY • NEW DELHI • AUCKLAND

HARPER ● PERENNIAL

First published in Great Britain in 2006 by Weidenfeld & Nicolson.

The first U.S. hardcover edition of this book was published in 2007 by HarperCollins Publishers.

A HISTORY OF THE ENGLISH-SPEAKING PEOPLES SINCE 1900. Copyright © 2006 by Andrew Roberts. All rights reserved. Printed in the United States of America. No part of this book may be used or reproduced in any manner whatsoever without written permission except in the case of brief quotations embodied in critical articles and reviews. For information address HarperCollins Publishers, 10 East 53rd Street, New York, NY 10022.

HarperCollins books may be purchased for educational, business, or sales promotional use. For information please write: Special Markets Department, HarperCollins Publishers, 10 East 53rd Street, New York, NY 10022.

FIRST HARPER PERENNIAL EDITION PUBLISHED 2008.

Library of Congress Cataloging-in-Publication Data is available upon request.

ISBN: 978-0-06-087599-2 (pbk.)

08 09 10 11 12   RRD   10 9 8 7 6 5 4 3 2 1

To Leonie Frieda

CONTENTS

LIST OF ILLUSTRATIONS

The author and publishers thank the following for permission to reproduce the illustrations in this book:

[1] Illustrate London News Picture Library
[2] Getty Images
[3] Bridgeman Art Library
[4] Bridgeman Art Library (Imperial War Museum)
[5] Imperial War Museum
[6] Weidenfeld and Nicholson Archive
[7] Topforo
[8] AKG London
[9] G.A. Duncan Photographs
[10] NASA
[11] Getty Images (Time/Life)
[12] REX Features

ACKNOWLEDGEMENTS

This book is emphatically not intended to be a comprehensive history of the English-speaking peoples, which would be impossible to write in one volume and anyhow probably rather dull to read. Instead of a textbook, this is a series of snapshots taken rather arbitrarily, episodically and idiosyncratically from the life of the English-speaking peoples since the dawn of the twentieth century, through whose shared experience I believe certain common themes emerge, almost unbidden.

For my purposes, the English-speaking peoples hail from those places where the majority of people speak English as their first language: the United States, the United Kingdom and her dependencies, Canada, Australia, New Zealand, the British West Indies and Ireland. The English diaspora has of course extended far further, such as to South Africa, Rhodesia (present-day Zimbabwe), Singapore, Hong Kong, and so forth, but I have confined myself to the places where the English language is spoken by the numerical majority as their first language today.

D-Day saw the supreme expression of the English-speaking peoples working together for the good of Civilisation, and I should like to thank the irrepressible Paul Woodadge of Battlebus Tours for conducting me on battlefield tours of Omaha beach, Beuzeville-au-Plain, La Fière, Utah beach, Les Mézières, Sainte-Marie-du-Mont, Bréville, Angoville-au-Plain, Merville Battery, Strongpoint Hillman, Sword beach, Pegasus Bridge, Juno beach, Sainte Mère Eglise, Lion-sur-Mer, Gold beach and Crépon, as well as to the Ryes Commonwealth War Cemetery at Bazenville and the Normandy American Cemetery at Colleville-sur-Mer. Paul's encyclopaedic knowledge of Normandy in 1944 has been invaluable in offering me insights into the campaign that represents the greatest single service made by the English-speaking peoples to the cause of Freedom.

My thanks also go to Mrs Esin Oktar for conducting me around several battlefields and cemeteries of the Gallipoli campaign, including Anzac Cove, 'V' Beach, The Nek, Lone Pine, Plugges's Plateau and Chunuk Bair, as well as the Kabatepe Museum. My understanding of that campaign has also benefited greatly from my contact with Prof. Ian Beckett of Northampton University, whom I should also like to thank.

This book has benefited from my conversations with people interested in the subject, and I would particularly like to thank the following for their thoughts, ideas and help: Carol Adelman, Rupert Allason, Joan Bright Astley, Jed Babbin, John Barnes, Michael Barone, Prof. Ian Beckett, Alan Bell, James C. Bennett, Mrs Benazir Bhutto, Robin Birley, Lord Black, Geoffrey Blainey, Philip Bobbitt, John Bolton, Sen. George Brandeis, Jung Chang and Jon Halliday, Robert Conquest, Alistair Cooke, Sir Zelman Cowen, Peter Day, Prof. David Dilks, Prof. Clement H. Dodd, Alexander Downer, Douglas Feith, Prof. Niall Ferguson, Alan Forward, Sir Martin Gilbert, Hon. Newt Gingrich, Dean Godson, Michael

Gove, Tom Gross, Katherine Harris, Simon Heffer, Prof. Peter Hennessy, Peter and Caroline Hopkins, Sir Alistair Horne, Prof. Sir Michael Howard, Colonel John Hughes-Wilson, General Sir Michael Jackson, Frank Johnson, Paul Johnson, General John K. Keane, Alan H. Kessel, Roger Kimball, Prof. Mervyn King, Dr Henry Kissinger, Irving Kristol, Lord Lambton, Lord Lamont, Richard Langworth, Bernard Lewis, Andrew Lownie, Hugh Lunghi, Kim McArthur, Sen. John McCain, John McCormack, Prof. Kenneth Minogue, John Montgomery, M. Ergün Olgun, John O'Sullivan, Allen Packwood, Stephen Parker, Hayden B. Peake, Richard Perle, Melanie Phillips, General Colin Powell, Lord Powell, David Pryce-Jones, James and Georgie Riley, Tony Ring, Patrick Robertson, Kenneth Rose, Donald Rumsfeld, Colin Russel, Mr Justice Antonin Scalia, Prof. Roger Scruton, Simon Sebag Montefiore, Victor Sebastyen, William Shawcross, James E. Shiele, Dr John Silber, Sen. Gordon H. Smith, Anthony Staunton, Irwin and Cita Stelzer, David Tang, Lady Thatcher, Chang Tsong-zung, Patrick Tyler, Prof. Claudio Véliz, Prof. Donald Cameron Watt, Lord Weidenfeld, the late Caspar Weinberger, David Wills, Keith Windschuttle, James Woolsey and Sir Peregrine Worsthorne.

Mr Ian Sayer has kindly allowed me privileged access to his Second World War archive of 60,000 unpublished documents, the largest such collection in private hands, and I would like to thank him for permission to quote from several of them.

This book has necessarily involved much archival research in some far-flung parts of the world, and I should like to thank the following people, in no particular order, for making that such an enjoyable experience over the past four years: Darrah McElroy at the (British) National Archives; Martin Collett of the Auckland War Memorial Museum Library; Jeff Flannery of the Department of Manuscripts in the Library of Congress in Washington; Campbell Bairstow of Trinity College, Melbourne; John Stinson of the Manuscripts Department of the New York Public Library; Allen Packwood, Director of the Churchill Archive Centre in Cambridge; the staff of the London Library and British Library; Fiona Kilby at the National Library of Australia in Canberra; Lara Webb of the library of the Royal Society of Arts; Sheila Markham of Brooks's Club library; the Librarian of the Knickerbocker Club, New York; Jenny Whaley of the Auckland City Library; Philip Reed, Director of the Cabinet War Rooms in London; Geoffrey Browne of the University of Melbourne; Sir John Lucas-Tooth, secretary of the Beefsteak Club; Colin Harris of the Bodleian Library Modern Manuscripts Room, Claire Scott of Christchurch City Library; Sean Noel of the Howard Gotlieb Archival Research Center, Boston University; Richard Palmer of the Lambeth Palace Library; Godfrey Waller, Superintendent of the Manuscript Reading Room at the Cambridge University Library; Janusz Janik of the State Library of Victoria in Melbourne; David Bell of the National Archives of Australia in Canberra; James Guest for my tour of the Victoria State Parliament; Peter Tuziak for my tour of the New South Wales State Parliament; Christina Young for my tour of the New Zealand Parliament; and the guides for my tours of the Old Parliament Building and the Australian Parliament in Canberra; Neil Porter at the Northern Ireland Assembly at Stormont; the librarian of the Metropolitan Club, Washington; Tim Lovell-Smith of the Alexander Turnbull Library of the National Library of New Zealand; Suzanne Mallon of the Mitchell-Dixon building of the State Library of New South Wales; Prof. Mark Francis of the University of Christchurch, and Heidi Kuglin of the Archives of New Zealand in Christchurch.

Characteristically generous have been my aunt and uncle, David and Susan Rowlands, for lending me their beautiful farmhouse in the Dordogne for five weeks in the summer of 2005, during which much of this book was written.

A large number of people have read parts of my manuscript for me and offered suggestions, and I would like to thank John Barnes, Prof. Paul Bew, Lord Black, Con Coughlin, Carlo D'Este, Jonathan Foreman, Susan Gilchrist, Dean Godson, Tom Gross, Frank Johnson, Lord Lamont, Patrick Maume, John McCormack, Simon Sebag Montefiore, Douglas Murray, Tim Newark, Lucy Pawle, David Pryce-Jones, William Shawcross, Andy Smith, Prof. Fred Smoler, Irwin Stelzer, Prof. Claudio Véliz, Lord Weidenfeld, David Wills, Keith Windschuttle, Paul Woodadge and Peter Wyllie. Eric Petersen, Roger Jenkin, Stephen Parker and my father Simon Roberts each read the entire manuscript, and I would like specially to acknowledge their invaluable help.

My wife, Susan Gilchrist, has helped with this book in a myriad of ways, principally through her ceaseless encouragement, and I cannot thank her enough. My editor Ion Trewin, literary agent Georgina Capel and copy-editor Linda Osband could not have been kinder and more enjoyable to work with.

This book is dedicated to Leonie Frieda, in recognition of our half-decade special relationship.

Andrew Roberts
www.andrew-roberts.net
January 2006

A Portrait of the English-Speaking Peoples at the Dawn of the Twentieth Century

'Propagate our language all over [the] world. ... Fraternal association with U.S. – this would let them in too. Harmonises with my ideas for future of the world. This will be the English speaking century.'

Winston Churchill's remarks to Cabinet, 12 July 1943[1]

'If one reflected on the most important events of the last millennium compared with the first, the ascent of the English-speaking peoples to predominance in the world surely ranked highest.'

Professor Deepak Lal, *In Praise of Empires*[2]

As the first rays of sunlight broke over the Chatham Islands, 360 miles east of New Zealand in the South Pacific, a little before 6 a.m. on Tuesday, 1 January 1901, the world entered a century that for all its warfare and perils would nonetheless mark the triumph of the English-speaking peoples. Few could have suspected it at the time, but the British Empire would wane to extinction during that period, while the American Republic would wax to such hegemony that it would become the sole global hyper-power. Assault after assault would be made upon the English-speaking peoples' primacy, each of which would be beaten off successfully, albeit sometimes at huge and tragic cost. Even as the twenty-first century dawned, they would be doughtily defending themselves still.

Just as we do not today differentiate between the Roman Republic and the imperial period of the Julio-Claudians when we think of the Roman Empire, so in the future no-one will bother to make a distinction between the British Empire-led and the American Republic-led periods of English-speaking dominance between the late-eighteenth and the twenty-first centuries. It will be recognised that in the majestic sweep of history they had so much in common – and enough that separated them from everyone else – that they ought to be regarded as a single historical entity, which only scholars and pedants will try to describe separately. A Martian landing on our planet might find linguistic or geographical factors more useful than ethnic ones when it came to analysing the differences between different groups of earthlings; the countries whose history this book covers are those where the majority of people speak English as their first language.

As the dominant world political culture since 1900, the English-speaking peoples would be constantly envied and often hated, which far from being anything perturbing has been the inescapable lot of all hegemonic powers since even before the days of Ancient Rome. Like the Romans, they would at times be ruthless, at times self-indulgent, and they too would sometimes find that the greatest danger to their continued *imperium* came not from their declared enemies without, but rather from vociferous critics within their own society.

Despite the harsh methods occasionally adopted to protect their status and safety from Wilhelmine Prussian militarism, then the Nazi-led Axis, then global Marxism-Leninism and presently from Islamic fundamentalism, the English-speaking peoples would remain the last, best hope for Mankind. The beliefs that they brought into the twentieth century largely actuate them yet; their values are still the best available in a troubled world; the institutions that made them great continue to inspire them today. Indeed, the beliefs, values and institutions of the English-speaking peoples are presently on the march.

In 1901, there was nothing inevitable about the domination that the English-speaking peoples' political culture would retain throughout the twentieth century and beyond. Wilhelmine Germany's burgeoning economic power was reflected in the massive High Seas Fleet that was being built specifically to challenge the Royal Navy. Third Republic France had a huge global empire and a thirst for revenge against Britain for slights, real and imagined, that she had received over the last century, culminating on the Upper Nile three years earlier. Tsarist Russia, the largest country in the world with a vast standing army, looked enviously at British India across the narrowing gap between them in Central Asia. Each would have liked to have seen the United States humiliated over her continued protection of Latin America through the Monroe Doctrine which excluded European imperialism from the American hemisphere. Within a decade, the German High Command had drawn up plans to shell Manhattan and land a 100,000-strong army in New England.

The world of 1901 was a multi-polar one of fiercely competing Great Powers. The idea that a century later the English-speaking peoples would hold unquestioned sway in the world, challenged only – and even then not mortally – by some disaffected fanatics from the rump of the Ottoman Empire, would have astounded Kaiser, Tsar and French president alike. Two global con-flagrations in the space of a generation, in which the English-speaking peoples escaped invasion – except those who lived in the Channel Islands – whereas no other Great Power did, explains much, but certainly not all.

*

There is something fitting about the first sunlight of the new century falling upon the Chatham Islands, since they eloquently illustrate the astounding reach of the English-speaking peoples. The forty isles and outcrops, only two of which – the Chatham and Pitt Islands – are inhabited, were first claimed in the name of King George III in 1791 by a Royal Navy lieutenant, William Broughton. The main island was named after his vessel, HM Brig *Chatham*, and the smaller, Pitt Island, eleven miles to the south-east, after William Pitt the Elder, 1st Earl of Chatham, a supporter of the American colonists.[3] The only way in which islands at almost precisely the other end of the world from the British Isles could have been successfully colonised was through the Royal Navy being the most powerful military machine in the world at the time, before the rise of Napoleon's *Grande Armée* half a decade later. Taking what the seventeenth-century diarist John Evelyn had called the 'command of the ocean', first achieved by the British and subsequently by the American Navy, was the first prerequisite for the English-speaking peoples' global dominance. The second became dominion over the skies.

As the new century dawned, both the British Empire and the American Republic were involved in protracted colonial wars, in South Africa and the Philippines respectively. War has been the almost constant lot of Mankind since the days of Rome, yet the English-speaking peoples have presided over a longer period of peace between the Great Powers than at any time since the Dark Ages. In 1901, neither Britain nor the United States saw herself as part of a greater entity, the English-speaking peoples. They were rivals, though newly friendly ones. It was not until the emergency of the early 1940s that the realisation finally dawned on both that they would be infinitely stronger together than the sum of their constituent parts. Even then, strong voices were raised in both countries against the formation of a 'Special Relationship'. Yet their reverses – Dunkirk, Pearl Harbor, Suez and Vietnam among them – have come when they were divided from one another. By contrast, their many victories – the 1918 summer offensive, North Africa 1942, Italy, the liberation of Europe 1944–5, the Berlin airlift, the Korean War, the Falklands, the collapse of Soviet communism, the Gulf War, the liberation of Kosovo and the overthrow of Saddam Hussein – all came when they were united.

In South Africa, the war was proving far more expensive in terms both of blood and treasure in January 1901 than anyone had predicted when it had broken out fifteen months earlier. An army of nearly half-a-million men had been fielded, larger than the enemy populations of the Transvaal and Orange Free State combined, yet on New Year's Day 1901 there was a serious armed incursion by Akrikaaner commandos into Queen Victoria's Cape Colony. Worse still for imperialists was the moral effect; the sheen of Empire that had seemed so bright at the Queen-Empress's Diamond Jubilee in 1897 seemed tarnished four years later as a result of the harsh measures employed to isolate and harry the Boer forces. Although the Boers' conventional forces were militarily defeated in the field, they nonetheless refused to surrender, resorting instead to a protracted insurgency campaign. The United States already had to deal with a popular revolt in the Philippines that was fought against her with very similar guerrilla and terror tactics. The problem of how to deal with asymmetric warfare being made upon them would be one that would perplex the English-speaking peoples several times over the next eleven decades.

'I was born at the end of the last century when Queen Victoria was still on the throne,' recalled the British soldier Major John Gordon-Duff in his auto-biography entitled *It Was Different Then*. 'The South African War was still being fought and the great British Empire was reaching its zenith, the prime minister was in the House of Lords and stable buckets were made of wood.'[4] From Gordon-Duff's perspective of 1975–6 – undoubtedly the worst peace-time twenty-four months in the history of the English-speaking peoples – the world of 1901 seemed like a glorious era of calm, stability and certainty for Britain. Quite apart from the reassuring composition of stable buckets, Great Britain boasted a Royal Navy of 330 ships, a City of London that was the world's financial hub and an admired Queen-Empress who had been on the throne for an unprecedented sixty-three years. Her empire comprised a quarter of the global population, covering one-fifth of the world's land surface. To most Britons their future prospects looked bright, the advance of Civil-isation seemed natural and assured, and there seemed no reason why Mankind should not be on an uninterrupted path towards what a young, newly elected British MP of the day named Winston Churchill was later to call 'the sunlit uplands'.

Yet for all its sprawling vastness, the leaders of the British Empire of 1901 were acutely conscious of the process by which, in the words of the Anglican hymnal, 'Earth's proud empires pass away'. Most of her senior decision-makers were keen to end her isolationist foreign policy, which they viewed as perilous rather than splendid. They feared that unless they contracted lasting alliances, they might become the object of a combination of hostile Great Powers and their *imperium* might thus soon, as Rudyard Kipling had dolefully predicted at the time of the Diamond Jubilee, be 'one with Nineveh and Tyre!' After all, as Churchill was also to put it, 'The shores of history are strewn with the wrecks of empires.'[5]

One Great Power from which Britain's leaders no longer suspected enmity was the United States. Across the Atlantic lay a republic that in 1898 had defeated in only eight months an ancient if ramshackle European power, Spain. America had healed her hitherto-debilitating Civil War wounds and saw herself as a single political entity with unlimited prospects; in the parlance of the day, a 'Manifest Destiny'. The time for Anglo-American amity in a dangerous world had come, and when just before his death in 1898 Otto von Bismarck was asked what was the decisive factor in modern history, he replied: 'The fact that the North Americans speak English.'

On Saturday, 6 September 1901, President William McKinley was shot in the breast and abdomen by an anarchist named Leon Czolgosz, whose hand he was shaking at the Pan-American stand of an exhibition in Buffalo. He died eight days later at 2.15 a.m. on Sunday the 14th. Large memorial services were held in London at Westminster Abbey and St Paul's Cathedral, testament to burgeoning Anglo-American friendship. The man sworn in as president at noon on 14 September, Theodore Roosevelt, was to be one of the most remarkable leaders of the English-speaking peoples. Just as Cecil Rhodes believed that the British Empire had a Manifest Destiny to govern an almost unlimited number of subject peoples, so too did his American counterparts trust that the United States had a Manifest Destiny in their continent and far beyond. For all its absurdity as a philosophical concept, since nothing in human affairs can be inevitable but death, this belief meant that America had already taken over from Spain an empire which her anti-imperial birthright forced her to refer to by any other euphemism.

In a sense the United States, despite her recently arrived immigrant populations who wanted no further entanglements with the Old Europe from which they had successfully escaped, had only one way to proceed once the Wild West had been tamed. Roosevelt understood that, and he came to power at precisely the right time for his country. Whatever else Czolgosz might have wanted to achieve, it is unlikely to have been an expansionist America, but such is the iron law of unintended consequences. Rancher, explorer, naturalist, reformer, historian, soldier, big-game hunter, sportsman, wit and patriot, Roosevelt believed in what he called 'the strenuous life' and personified America's new sense of boundless opportunities and interests, as well as the sheer robustness of his optimistic, thrusting, hardy country.

The ceremony held at Westminster Abbey, Britain's ancient church, on the last evening of the nineteenth century was 'thronged by standing worshippers and every seat taken', and the sermon was delivered by Canon Charles Gore. In many ways Gore personified the British ruling class, having been educated at Harrow School and Balliol College, Oxford, but his message was one of pessimism about the future, rather like those seemingly Olympian statesmen who were contemporaneously distancing themselves from isolationism. 'There was no doubt that the nineteenth century was closing with a certain widespread sense of disappointment and anxiety among many of those who care most for righteousness and truth in the world,' Gore intoned, before noting with melancholy the fact that since the deaths of the historian Thomas Carlyle and the poet Alfred, Lord Tennyson, 'There was no prophet for the people.'

Canon Gore was probably being deliberately controversial when he told his huge congregation: 'The dominant cry is the cry of Empire ... but ... it was poor in moral quality, and appeared, behind only too thin a veil, as the worship of our unregenerate British selves, without humility or fear of God.'[6] Criticism from churchmen, liberals and public thinkers of the mission of the English-speaking peoples was to be a recurring theme throughout the coming century. Later that year, the British Prime Minister, Lord Salisbury, was to complain that, 'England is, I believe, the only country in which, during a great war, eminent men write and speak as if they belonged to the enemy.'[7] In fact, the phenomenon was to recur throughout the English-speaking peoples over the coming decades, and in some engagements – such as at Suez and in Vietnam – opposition from a vociferous domestic minority was to doom their enterprises far more than foreign opponents.

Most Britons would not have agreed with Canon Gore. In the same copy of *The Times* newspaper that carried Gore's denunciation of the 'poor moral quality' of British imperialism, was printed the pastoral letter of the Bishop of Salisbury, which he had caused to be read aloud to all congregations in his diocese the previous Sunday. This sought to define the Empire's mission and did so in terms of which the vast majority of Britons would have heartily approved. He said that his flock

> seemed called upon to take their part in the establishment of a federated Empire which shall unite willing peoples in conditions of great local freedom, but with an attachment of loyal affection to a common centre in the British Crown, with a great mass of common laws and common interests, with the full voice of a worthy common language and a noble common literature, and, above all, bearing the decided impress of a common and enlightened Christianity. That seemed to be their duty, and they could not escape it.

Ever since the mid-1830s, the English-speaking peoples had considered it their civilising mission to apply – with varying degrees of force – their values and institutions to those areas of the world they believed would benefit from them. Although Britain was under no threat from them herself, Lord Palmerston imposed regime change in Spain, Portugal and Belgium, using the power of the Royal Navy to force liberal constitutions on countries that baulked at first but later came to value them. 'I hold that the real policy of England', he told the House of Commons in March 1848, 'is to be the champion of justice and right ... not becoming the Quixote of the world, but giving the weight of her moral sanction and support wherever she thinks justice is, and wherever she thinks that wrong has been done.' The neo-conservatives of President George W. Bush's Administration did not invent some brand new political philosophy in their desire to extend representative institutions to the Middle East.

In his 1833 speech on the renewal of the charter of the East India Company, which governed British India until 1858, the British statesman and historian Lord Macaulay argued that, 'by good government we may educate our subjects into a capacity for better government; that having become instructed in European knowledge they may, in some future age, demand European institutions'. Macaulay admitted that he could not say when such a time would come that such trusteeship could give way to independence, but when it did, 'it will be the proudest day in English history'.[8]

Much derided as merely an excuse for putting off indigenous self-government, in fact men like Macaulay believed profoundly in this sense of mission, just as today's neo-conservatives passionately believe in the advantages that might flow from instilling – through installing – democracy in such countries as Afghanistan and Iraq. Whether the Middle East proves too theocratic, obscurantist and in some places feudal to benefit from democracy remains to be seen, but neo-conservatism is certainly no new historical departure in the self-proclaimed mission of the English-speaking peoples.

The inequalities inherent in imperial British society at the turn of the century were obvious, even at the time. Free-market capitalism, resilient and untrammelled, was able in April 1901 to launch the new White Star liner the *Celtic*, built at Harland and Wolff's vast Belfast shipyard, at 700 feet in length and 20,800 tons in weight the largest ship in the world.[9] Yet in London that Christmas official returns showed that there were no fewer than 107,539 people receiving poor relief in the capital of the Empire, 39,409 of whom were termed 'outdoor paupers', or tramps. London also had a correspondingly large number of prostitutes, variously estimated at between 80,000 and 100,000. In one year alone – 1905 – when the Edwardians cracked down on brothel-keeping, no fewer than 944 women were charged with having sexual intercourse in the open air.

Partly as a response to social deprivation, the Labour Representation Committee was founded in the Memorial Hall, Farringdon Road, in London on 27 February 1900, dedicated to getting working men into Parliament. Always committed to employing constitutional means to achieve its ends, the following year it received its first great cause with the milestone Taff Vale judgment in the House of Lords, which reversed the Court of Appeal's decision that a trade union could not be sued in its registered name as a corporate body. The trade union in question, which went under the splendidly Victorian name of the Amalgamated Society of Railway Servants, therefore effectively lost its right to strike. A legitimate grievance was created that brought into being British parliamentary socialism, a force that was to remain highly influential for nearly a century, until Tony Blair defeated it within the Labour Party in May 1994.

*

When in March 1902 the American financier J.P. Morgan acquired a predominating influence in Cunard, White Star and other shipping lines, the
First Lord of the Admiralty, the 2nd Earl of Selborne, wrote to his father-in-
law, the Prime Minister, to ask what might be done. On the verge of retirement,
Lord Salisbury replied in a mood of resigned realism: 'It is very sad, but I'm
afraid America is bound to forge ahead and nothing can restore the equality
between us. If we had interfered in the Confederate War it was then possible
for us to reduce the power of the United States to manageable proportions.
But *two* such chances are not given to a nation in the course of its career.'[10] It
was certainly not the remark of a Yankophile, but was nonetheless uncannily
accurate when it came to predicting the future of the power-relationship
between the two leading nations of the English-speaking peoples over the
course of the coming century. Yet just as in science-fiction people are able to
live on through cryogenic freezing after their bodies die, so British post-
imperial greatness has been preserved and fostered through its incorporation
into the American world-historical project.

The Themistoclean foresight of Lord Salisbury's Unionist Government to
observe a benevolent neutrality towards the United States during the Spanish-
American War – while the rest of Europe sympathised openly with Spain –
ensured that the twentieth century dawned on the best state of Anglo-
American relations since the Revolution. Even as recently as 1896, the two
countries had almost gone to war over the absurd *casus belli* of a Venezuelan
border dispute, but five years later they were – at least at governmental level –
firm friends. The century was to see strains in the Special Relationship,
particularly in 1927, 1944, 1956, 1965 and 1994–5, but never the break for
which their rivals and opponents desperately hoped.

As the Spanish-American War broke out, the British Colonial Secretary,
Joseph Chamberlain, said publicly that however terrible a conflict, it would be
cheap 'if, in a great and noble cause, the Stars and Stripes and the Union Jack
should wave together over an Anglo-Saxon alliance'. He told the Anglophile
American Ambassador to London, John Hay, that he 'didn't care a hang what
they said on the Continent' and that, as Hay reported home, 'If we give up the
Philippines it will be a considerable disappointment to our English friends.'[11]
Meanwhile, the Empire's foremost poet, Rudyard Kipling, urged the United
States to 'Take up the White Man's Burden' in the Philippines. In March
1899, Britain and the United States collaborated together to frustrate German
ambitions in a struggle over who should control the Pacific island of Samoa,
then in September Hay, who by then had become US Secretary of State,
issued a Note on China which substantially supported Britain's Open Door
policy there. The scene was thus set for the Roosevelt Administration's pro-
British stance over the Boer War, one that was virulently opposed by many
Irish-Americans, German-Americans and, of course, Dutch-Americans.

★

One of the common threads uniting the wars of the English-speaking peoples in the twentieth century was that they have often suffered serious reverses in the first battle, or even the first campaign, before going on to ultimate victory. In the Boer War, the British were soundly defeated in 'Black Week' at Stromberg, Magersfontein and Colenso; in the Great War, they were forced to retreat from Mons; in the Second World War, they were humiliatingly flung off the continent at Dunkirk and the Americans were stricken at Pearl Harbor and the Philippines. Thereafter, in the Cold War, the initial years were characterised by Soviet expansion and provocation; North Korea attacked the South without warning in 1950, just as North Vietnam destabilised South Vietnam just over a decade later; after the surprise invasion of the Falklands, British marines were photographed lying prone on the ground before their Argentine captors; in the Gulf War, Saddam Hussein invaded Kuwait overnight; and in the War against Terror, the American people had to witness the horror of scores of their compatriots jumping to certain death from the upper storeys of Manhattan's Twin Towers sooner than be burnt alive.

This pattern of initial English-speaking humiliation, or even catastrophe, is too well-established to admit of any doubt about the recurring phenomenon. All were initial defeats, provocations or utter disasters early in the conflict, yet each served to rally the English-speaking peoples for the necessary sacrifices ahead. Nineteenth-century precursors further emphasise this historical ubiquity, such as the Alamo in the Texan War, the Charge of the Light Brigade in the Crimean War, Little Big Horn in the Great Sioux War, Maiwand in the Second Afghan War and Isandhlwana in the Zulu War. The English-speaking peoples rarely win the first battle, but they equally rarely lose the subsequent war.

The British Commander-in-Chief in South Africa, Lord Roberts, returned to a hero's welcome in Britain on 2 January 1901, when Queen Victoria presented him with the Order of the Garter. The celebrations turned out to be absurdly premature, however, because although the set-piece battle-fighting stage of the war was over, the Boers then fought a vicious insurgency campaign that lasted many months more. Roberts' Chief of Staff, Lord Kitchener, became Commander-in-Chief and it was he who took the difficult decisions that won the South African War. He illustrates another aspect of the experience of the English-speaking peoples in the twentieth century: the tendency for the right men to come to the fore in times of crises. Just as Wellington had leapfrogged the British High Command in the Peninsular War and Lord Palmerston had replaced Lord Aberdeen in the Crimean War, so in the next century the desperate need for the best people produced the political environments necessary for the most talented leaders to emerge.

Despite David Lloyd George being forced to escape from the Birmingham

Town Hall dressed as a police sergeant in December 1901 due to his opposition to the Boer War, fifteen years later he was the best person to replace the indecisive H.H. Asquith as prime minister and grasped the opportunity when it was offered. Similarly, Churchill became prime minister in 1940 under similar self-propulsion, on precisely the day that Hitler unleashed his *Blitzkrieg* in the West. In Franklin D. Roosevelt he found an ally of preternatural political talent, the greatest American president of the twentieth century. With the Soviet Union staggering under the weight of her own internal contradictions by the early 1980s, it required leaders of the English-speaking peoples with the staunch anti-Communist convictions of Ronald Reagan and Margaret Thatcher to go on to the offensive in a peaceful campaign of unswerving attrition. Similarly, George W. Bush and Tony Blair have been absolutely unwavering in their dedication to pursuing the War against Terror.

Nor have these leaders been churchmen or soldiers in uniform, as in many other countries. The separation of Church and State in the American constitution and the complete subjection of the armed forces to democratic control throughout the English-speaking peoples meant that they have been free of two undue influences that have time and again stunted other nations' opportunities in the twentieth century: theocracy and military dictatorship. While some of the people who made their names whilst soldiering have become successful politicians – including Theodore Roosevelt, Dwight D. Eisenhower, and (at a pinch) Winston Churchill and John F. Kennedy – they never had the threat of force at their back, as was the case with Hitler, Franco, Attatürk, Mao, Chiang Kai-shek, Lenin, de Gaulle, Pinochet, Mussolini, Perón, Zia, Somoza, Stalin, Horthy, Gadaffi, Saddam Hussein, Amin, Nasser, Mengistu, the Greek colonels and so many others from outside the English-speaking world.

Just as Bonapartism has been entirely foreign to the experience of the English-speaking peoples, so too have theocracy and political subservience to prelates, except for a short period in southern Ireland. Religion's direct involvement in politics has been generally kept to a minimum, except in areas such as education and abortion; and although there are signs that it might be on the increase in modern-day America, it is in many ways a cultural phenomenon and does not indicate that it will occupy the kind of central position that it historically has in the polities of Poland, Italy, Spain, parts of Latin America, the whole of the Middle East (except Israel) and significant portions of Africa and Asia. As Europe found in the sixteenth and seventeenth centuries, secular states are more successful than theocracies, and the English-speaking peoples have benefited enormously from making that discovery so early on.

An inevitable concomitant of power has been the envy of others. In the twentieth century, no less than in any other, the Great Powers excited the envy of lesser powers not necessarily because of how they behaved but simply because of what they were. In a speech on Monday, 13 May 1901, at the Hotel Metropole in London, Lord Salisbury used the opportunity of a banquet of the Nonconformist Unionist Association to warn foreign powers off even considering intervening in South Africa. After toasts to King Edward VII and the Royal Family, Salisbury was received with loud cheers as he rose to speak. What he told the Lord Chancellor, the Duke of Devonshire, a large group of peers and MPs, and the Association's members was uncompromising:

> When I was at the Foreign Office – I was there a long time – I used not infrequently to hear suggestions that our time had passed by, that our star was set, that we were living upon the benefit of the valour of those who had gone before us and upon successes in which we had no active share, and that if we meant to keep our place in the world new exertions were necessary. I need not say that I heard those suggestions with no kind of sympathy and with something of contempt. It is true that there had been spread around the world the impression that we should never fight again, and that every adversary had only to press hardly and boldly upon us to be certain that we should yield. It was a gross miscalculation on their part.

This was received with cheers, but none so loud as those that greeted the grave warning that followed:

> Make what deductions you will, lament as you will – and I heartily concur with you – as to the sacrifices we have been forced to make, still it is now a great achievement that there is no Great Power in the world but knows that if it defies the might of England, it defies one of the most formidable enemies it could possibly defy.[12]

It occasionally seemed as though a cabal of Britain's competitors, led by France, Germany and Russia, might try to intervene in the Boer War in order to impose a peace there, one not at all conducive to British interests in the region. Yet the might of the Royal Navy – which by a law of 1889 had to be larger than the fleets of the next two countries in the world combined – made this physically impossible, and the scheme came to nothing.

Resentment of the leading world power by its rivals in 1901 was simply a factor of the human condition and not the result of anything in particular that the British Empire had done in South Africa or anywhere else. This was superbly summed up by Britain's greatest-ever proconsul, Lord Curzon, in April 1900, when Viceroy of India. In a letter to his friend Lord Selborne he wrote:

> I never spend five minutes in inquiring if we are unpopular. The answer is written in red ink on the map of the globe. .... No, I would count everywhere on the

individual hostility of all the great Powers, but would endeavour to arrange things that they were not *united* against me. .... I would be as strong in small things as in big. This might be a counsel of perfection; but I should like to see the experiment tried.[13]

The vital importance of maintaining the authority and prestige of the English-speaking peoples – a duty which passed from Britain to America in the 1940s – was upheld throughout the century in every decade except the second half of the 1970s, when the US Congress prevented the Nixon and Ford Administrations from sustaining it, and when the Carter Administration wilfully attempted to abandon it. The period between the withdrawal from Vietnam in 1972 and the election of Ronald Reagan in 1980 thus constituted a perilous time for the English-speaking peoples, reminiscent of Churchill's characterisation of the Thirties with their 'long, dismal, drawling tides of drift and surrender'. Overall, however, the prestige of the English-speaking peoples, and pride in their reliability as allies and indefatigability as foes, has actuated their leaders, which is one explanation of their phenomenal global success since 1900. Prestige is a tangible benefit in the calculus of international relations, its loss a concomitant danger.

Although capitalism and global trade were making population a less and less important geopolitical factor, it is worth noting that the world's most populous countries in 1901 were China, with 350 million, British India with 294 million, Russia with 146 million, the United States with 75.9 million, Germany with 56.3 million, Japan with 45.4 million, Great Britain (including Ireland) with 41.4 million, France with 38.9 million, Italy with 32.4 million and Austria with 26.1 million.[14] A glance at the United States' key population and steel production figures show her already burgeoning power during this period. In 1880 there were around fifty million Americans, in 1901 nearly seventy-six and by 1905 no fewer than eighty-four. Meanwhile, in 1880 the US produced 3.84 million tons of steel (against Britain's 7.75 million and Germany's 2.69 million), whereas by 1900 the figures were 13.8 million, 9.0 million and 8.4 million respectively. The biggest leap came in 1907, when America's 25.8 million tons comprised more than Britain's 9.9 million and Germany's 12.7 million combined. Coal production figures for 1901 similarly mirror these other indicators of relative economic power, with the US producing 268 million tons, against Britain's 219 million and Germany's 112 million. It was calculated that at around the turn of the century, the United States could buy all the assets of Great Britain outright, with still enough left over to settle her national debt too.[15] Into this promising situation, with a determination to translate America's wealth into power, stepped the protean genius of Theodore Roosevelt.

In the debate over whether America was born great, achieved greatness or had greatness thrust upon her, the only possible conclusion must be: all three. That she was conscious of, at least, the possibility of her imperial greatness is evident from her public architecture. As one historian notes, the United States 'is ruled from a city that was built to replicate as far as possible parts of ancient Rome. No other modern nation is governed from a building called the Capitol.'[16] As if to echo in the financial and commercial worlds President Roosevelt's political and colonial expansionism, Andrew Carnegie founded US Steel in 1901, the world's first billion-dollar corporation, which proceeded to construct a town in Gary, Indiana, that could house no fewer than 200,000 workers. On 10 January 1901, oil had been discovered in Texas, and soon afterwards the internal combustion engine became integral to Western life, creating a vast, entirely new, world industry of automobiles.

These were the years of the great American 'robber-baron' businessmen – men like J.D. Rockefeller, J.P. Morgan, James H. Hill, Henry Ford and E.H. Harriman – the founders of modern tooth-and-claw capitalism, who tried to corner markets and establish cartels. Paradoxically, one of Theodore Roosevelt's major achievements was to use at the time unprecedented regulatory and legislative measures, collectively known as 'trust-busting', designed to foster the free-market competition that has more than any other single factor been the key to American greatness.

Staying at the forefront of all the major developments in automobiles, aeronautics, computers, finance, biotechnology and the information revolution – and of all their various key military applications – has enabled the English-speaking peoples to win and retain their global hegemony. That world leadership will only be ceded to whichever world power – possibly China or India – is capable of producing better products cheaper than they, in a similarly politically secure environment. It will happen, but hopefully this time it will not happen violently. It if does, however, the battle-lines are easily drawn. As Winston Churchill put it in May 1938, 'It is the English-speaking nations who, almost alone, keep alight the torch of Freedom. These things are a powerful incentive to collaboration. With nations, as with individuals, if you care deeply for the same things, and these things are threatened, it is natural to work together to preserve them.'

In 1901, Canada was a land of almost infinite possibility and opportunity. In the last two decades of the nineteenth century, her population had increased by a staggering 24%, yet from the 1901 census figure of 5,371,315 the population then grew by two million in the first decade and then a further one-and-a-half million in the next, registering an astonishing 64% increase over two decades. In 1891, only 350,000 people inhabited the vast territories between the eastern border of Manitoba and the Pacific Ocean; twenty years later, this had increased five-fold. Yet the most profound changes came in the central prairies, where in 1905 the provinces of Alberta and Saskatchewan were carved out of the North-West Territories, and where agriculture, manufacturing, timber, mining and finance flourished mightily. Although much of the increase in population came from immigration, there was a dramatic increase in the birth-rate too, concomitant with the 125% increase in the amount of farmed land in the two decades after 1901. No longer could Canada be thought of as 'a narrow and broken strip of land lying to the north of the American border'.

On 1 January 1901, Australians were celebrating from coast to coast. By imperial proclamation an 'indissoluble constitution' had come into being that day, creating a Commonwealth of Australia. The idea of federating Australia's six states into a fully fledged, self-governing nation within the Empire had been mooted since the mid-1850s, but it was not until the first day of the twentieth century that all the necessary political compromises were finally made and nationhood became a reality. That day a continent of four million people (and 100 million sheep) became a nation, from Perth in Western Australia – the most isolated large city on the planet – across the Red Continent to beautiful Hobart, capital of Tasmania.

The Governor-General the Earl of Hopetoun's procession through Sydney was the greatest imperial celebration since Queen Victoria's Diamond Jubilee. As well as Australian troops, there were detachments from twenty-one British units including the Life Guards, Royal Horse Artillery, Grenadier and Coldstream Guards, Black Watch, Rifle Brigade and Seaforth and Cameron Highlanders, as well as from twenty-four Indian Army regiments such as the 9th and 10th Bengal Lancers, Bombay Grenadiers and 1st and 5th Punjab Cavalry. A New South Wales bibliophile named Alfred Lee kept a huge book of invitations and ephemera from the Federation celebrations, which are notable for their heartfelt patriotic slogans such as 'Advance Australia', 'United Australia' and 'Birth of a Nation'.[17]

As the twentieth century dawned, New Zealand had every right to consider herself one of the most progressive and advanced nations on earth. She had obligatory conciliation and arbitration in all labour disputes; her state system of education was free, secular and compulsory; two-thirds of her 104,471 square miles were fit for agriculture or grazing; her legal system retained the best of English common law but added certain local addenda; and in 1898 Richard Seddon, the Lancashire-born Premier of New Zealand from May 1893 until his death in June 1906, had introduced an old-age pension bill providing pensions of £18 per annum for everyone over sixty-five, on the condition that they had not been imprisoned more than four times, had not deserted their spouse for more than six months, were 'of good moral character', did not have an income of more than £52 a year and did not have accumulated property of over £270.[18]

In 1893, New Zealand became the first country in the world to give women the vote; by 2005, she had a female prime minister, governor-general and speaker of her unicameral parliament. On New Year's Day 1901, New Zealand also introduced a universal penny postage rate 'to all important parts of the Empire'. Seddon saw no contradiction between being a progressive and a convinced imperialist. 'The flag that floats over us and protects us was expected to protect our kindred and countrymen who are in the Transvaal,' he said of the Boer War. 'We should take action [because] we are a portion of the dominant family of the world – we are of the English-speaking race.'

The beginning of the twentieth century saw the English-speaking Caribbean in a very under-developed state, with poverty widespread; in some places everyday life was little different than it had been in the mid-nineteenth century. Yet as the Jamaican-born US Secretary of State General Colin Powell was to recall in his autobiography, *My American Journey*,

> The British ended slavery in the Caribbean in 1833, well over a generation before the Americans. And after abolition, the lingering weight of servitude did not persist as long. The British were mostly absentee landlords, and West Indians were more or less left on their own. Their lives were hard, but they did not experience the crippling paternalism of the American plantation system, with white masters controlling every aspect of a slave's life. After the British ended slavery, they told my ancestors they were now British citizens with all the rights of any subject of the crown. That was an exaggeration; still, the British did produce good schools and made attendance mandatory. They filled the ranks of the lower civil service with blacks. Consequently, West Indians had an opportunity to develop attitudes of independence, self-responsibility, and self-worth.[19]

This was no mean legacy, but the twentieth century was to see the Caribbean remain by far the poorest part of the English-speaking world. The British West Indies raised regiments to fight for the Crown in both world wars, but otherwise the region slipped slowly but perceptibly from the British into the American zone of influence.

The visit of Queen Victoria to Dublin in April 1900 was, Palace officials insisted, an informal not a State visit to Ireland. Just short of her eighty-first birthday, she was prompted by a desire to recognise the gallantry of her Irish soldiers in South Africa, and she received a rousing reception, wearing the shamrock in her bonnet and jacket. The courtier Sir Frederick Ponsonby recalled in his autobiography how,

When we got into Dublin the mass of people wedged together in the street and every window, even on the roofs, was quite remarkable. Although I have seen many visits of this kind, nothing has ever approached the enthusiasm and even frenzy displayed by the people of Dublin. There were, however, two places where I heard ugly sounds like booing, but they only seemed like a bagpipe drone to the highly-pitched note of the cheering.

Although some Irish nationalists, such as Maud Gonne's Transvaal Committee and Arthur Griffiths' *United Irishman* newspaper, were supporting the Boers, several thousand Irishmen were fighting against them, and many had distinguished themselves in battle. With typical hyperbole, Griffiths' paper denounced Victoria as 'Queen of the famines, of the pestilences, of the emigrant ships, of the levelled homesteads, of the dungeons and gallows', but the nationalist movement was dismayed at the huge crowds that turned out to cheer her wherever she went.[20] A fortnight later it admitted: 'We have learnt a strange and bitter lesson; let it not be lost upon us. There is much to be done to absolve the land from the treachery of the last few weeks.' There had been some 'hostile cries' amongst the crowds, but these were, as the *Freeman's Journal* recorded, 'not many'. Otherwise full-throated loyalty was the leitmotif of the visit, and what James Joyce called 'the old pap of racial hatred' was entirely absent. When Edward VII visited Dublin two years later, the most serious act of protest came when Maud Gonne hung a black petticoat out of her window.[21]

Throughout the period covered by this book the experience of Ireland, or at least the southern twenty-six of the island's thirty-two counties, seems to run contrary to that of the rest of the English-speaking peoples. It provided the exception to every rule, disrupted every generalisation and pursued so different a route from the rest of the english-speaking peoples so often that it must be considered quite apart from the rest. Yet that was not the case in 1900, when the Queen received loyal accolades from ordinary people quite as fervent as any that would have been heard in Manchester, Glasgow, Adelaide, Toronto or Auckland. That said, 1905 saw the founding of Sinn Fein ('Ourselves Alone'), an anti-British revolutionary organisation that was to cause much misery over the coming century in its ultimately failed campaign to separate the whole island of Ireland from the United Kingdom.

Winston Churchill's fine four-volume book *A History of the English-Speaking Peoples* – 'their origins, their quarrels, their misfortunes and their reconciliation' – ends in January 1901, just before by far the more important and interesting part of their tale began. The Nobel Prizewinner for literature concluded his great work with the words:

The vast potentialities of America lay as a portent across the globe, as yet dimly recognised, save by the imaginative. But in the contracting world of better communications to remain detached from the pre-occupations of others was rapidly becoming impossible. The status of world-Power is inseparable from its responsibilities. ... The English-speaking peoples ... are now to become allies in terrible but victorious wars. Another phase looms before us, in which alliance will be once more tested and in which its formidable virtues may be to preserve Peace and Freedom. The future is unknowable, but the past should give us hope.

Churchill's unknowable future then is the English-speaking peoples' known past today. Here, therefore, is the next stage of their story.

ONE

Shouldering 'The White Man's Burden'

1900 – 4

The character of Theodore Roosevelt – America in the Philippines – The Boer War – Sport – The Pacific cable – Canadian patriotism – Robber-baron capitalism – The American West – Australian Federation – New Zealand's 'knight in icy armour' – Irish nationalism – The Hay-Pauncefoote Treaty – The Wright brothers' revolution – The Geographical Pivot of History *– A pogrom in Limerick –* The Entente Cordiale

'Dear Teddy, I came over here meaning to join the Boers, who I was told were Republicans fighting Monarchists; but when I got here I found the Boers talked Dutch while the British talked English, so I joined the latter.'
Letter from a Rough Rider veteran to Vice-President Theodore Roosevelt[1]

'If new nations come to power . . . the attitude of we who speak English should be one of ready recognition of the rights of the newcomers, of desire to avoid giving them just offense, and at the same time of preparedness in body and mind to hold our own if our interests are menaced.'
Theodore Roosevelt to Cecil Spring-Rice, 1904[2]

Theodore Roosevelt was brave, intelligent, well-travelled and had a photographic memory. He felt shame at his father's not having served in the Civil War, yet otherwise regarded him as 'the best man I ever knew', and he always 'strove for his father's posthumous blessing'.[3] Ever since shooting a crane at a lagoon near Thebes in adolescence, he loved slaughtering avifauna in vast quantities. He wanted to be a natural scientist while at Harvard – taking a 97 in zoology and graduating *magna cum laude* – but preferred the great outdoors to microscopes. An asthmatic, he was obsessed with the need to prove himself physically and was keen on boxing, rowing, riding, walking, skating, camping and sailing. He didn't smoke or gamble, drank sparingly and seems not to have been much interested in sex.[4] His time as US Civil Service Commissioner, head of the New York City Police Board, assistant secretary of the Navy, a dashing Rough Rider cavalry colonel in the Spanish-American War and a corruption-busting governor of New York won him fame early and – along with Czolgosz's fatal bullet in 1901 – helped make him at forty-two the youngest of all the presidents before or since. The sheer energy of the man – he leapt over chairs at the White House and once dragged an ambassador off to play tennis in a hailstorm – was part of his charm. On New Year's Day 1907, he shook the hands of no fewer than 8,513 people at a White House reception. The naturalist John Burroughs said that when Roosevelt entered a room, 'it was as if a strong wind had blown the door open'.

Within a few months of taking office, Roosevelt presented an awesome challenge to Congress and the nation. 'The American people must either build and maintain an adequate Navy,' he said, 'or else make up their minds definitely to accept a secondary position in international affairs, not only in politics but in commercial matters.'[5] As an early champion and friend of the incredibly influential, though little-known, American naval officer Alfred Thayer Mahan, author of the seminal work *The Influence of Sea Power on History 1666–1783*, Roosevelt understood international naval power politics like no other previous president. His huge expansion of the US Navy presaged the American eruption onto the global stage that was to be the single most important feature of world politics in what was to be dubbed 'the American Century'.

John F. Kennedy was puzzled that Americans rated Theodore Roosevelt so highly, considering that he never led the nation through any war (an estimation that might more profitably be extended to JFK himself). Roosevelt filled the White House like no other peacetime president; Mark Twain accorded the fact that he was 'the most popular human being that ever existed in the United States' to his 'joyous ebullitions of excited sincerity'. Yet there were solid achievements too: he won the Nobel Peace Prize for negotiating the Treaty of Portsmouth that ended the Russo-Japanese War in 1905 and began constructing the isthmian canal that linked his country's western ocean to its eastern, thus saving US warships from having to make the ninety-day journey around Cape Horn.

The process of splitting Panama from distant Colombia in order for the canal to be built has long been held against Roosevelt in Latin America; yet Panama had rebelled fifty times in fifty years – surely some kind of a record in international relations – and all he had to do in November 1903 was to let the fiftieth rebellion succeed. He sent the warship *Nashville* to Colon and refused Colombian troops permission to use a US-operated railway, something that international law permitted him to do.[6] The entire Panamanian coup was effected with the deaths of, according to the casualty report, 'one Chinaman and an ass', suffered when a stray shell hit Panama City. Senator Samuel

Hayakawa of California once said of the Panama Canal, 'We stole it, fair and square,' but the United States in fact paid vast sums for it. The higher direction of the feat of cutting the canal, which opened in 1914, was one of the greatest civil engineering achievements of the English-speaking peoples in the twentieth century, despite the manual work largely being undertaken by labourers from the British West Indies who suffered a high mortality rate.

In his foreign policy, Roosevelt fiercely defended the Monroe Doctrine, especially against German imprecations over Venezuela in 1902. When time after time during that crisis the war-games between the 'Blue Fleet' (American) and the 'Black Fleet' (German) undertaken at the US Naval War College resulted in Black Fleet victories, he forced on the pace of naval armament, which was ultimately to make the United States a world power by the time he left office in 1909. As one historian has perceptively put it, 'In terms of bloodshed and lives lost, America's rise to great power status could hardly have been more harmless.'[7]

Nor was Roosevelt's expansionism doctrinaire; he handed Cuba her independence in May 1902, 'after a brief period of military government that transformed the island from an abused, insanitary and poverty-stricken Spanish colony to a healthy new nation amply equipped to govern itself'.[8] Most of his interventions in Central America were undertaken reluctantly and at the urgent requests of the governments there, for, as he said about one crisis in the Dominican Republic, 'I have about the same desire to annex [islands] as a gorged boa constrictor might have to swallow a porcupine wrong end to.' In vigorously enforcing the Monroe Doctrine throughout the twentieth century, the United States deserves commendation for not allowing that continent to develop into a battleground between the Great Powers.

The Panama Canal was to bring the United States into West Indian and Latin American politics on a very regular basis, as a force for stability and the protection of property rights. It was under Theodore Roosevelt that the Caribbean gradually became an American lake. In December 1904, Roosevelt issued his Corollary to the Monroe Doctrine in which he reserved the United States' right to intervene in nations of the Western hemisphere that were plagued by 'wrongdoing or impotence'. His rebuke of Colombia over Panama in 1903 had been a case in point, and under the terms of the Corollary the US intervened in Cuba in 1906, Nicaragua in 1909 and 1912, Mexico in 1914, Haiti in 1915, the Dominican Republic in 1916, Guatemala in 1920, Honduras in 1924 and Panama in 1925, in the first quarter of the twentieth century alone. Usually these were very limited interventions for a specific purpose – often to overthrow corrupt, unpopular or undemocratic regimes – and did not last long, although in Haiti the occupation lasted nineteen years. For all the sarcasm directed towards her over this by liberal academics – 'The white man's burden in Latin America is heavy indeed,' wrote one from St John's College, Oxford, recently – the United States saved several of those states from revolution, civil war, expropriation, bloodshed and bankruptcy by her prompt willingness to act as a police power in her own backyard.[9]

Roosevelt read voraciously; when asked two years into his presidency which authors he had managed to study while at the White House, he provided a list of 114, including Thucydides, Aristotle, Gibbon, Tolstoy, Scott, Twain and Molière (in the original). His first action as president had been to invite a black man – Booker T. Washington – to dine with him at the White House, but it only happened once and that was pretty much the limit of his interest in black emancipation at a time when there were 100 lynchings taking place per year in the South.

It was Roosevelt's somewhat idiosyncratic campaign for 'Simplified Spelling' that abolished the 'u' in the American spellings of words like 'honour' and 'colour', but he never persuaded people to style him 'Rozevelt'. A fine phrase-maker, in his South American policies he adapted a West African adage: 'Speak softly and carry a big stick'; he described the railway magnate E.H. Harriman and other 'robber-barons' as 'malefactors of great wealth', and was tough in his anti-trust legislation against J.P. Morgan's Northern Securities Trust. He optimistically thought that a 'square deal' was possible between capital, labour and consumer.[10] Rather lyrically, he once said of the clash between them that, 'Envy and arrogance are the two opposite sides of the same black crystal.'

Roosevelt's progressive Republicanism was vital at a time when poverty was widespread, especially in the rural Midwest, where it is estimated that 'seven out of ten families lived below subsistence level, in the Negro South, and above all among the rapidly growing and shockingly underpaid immigrant populations of a thousand overcrowded towns and cities'.[11] In order to keep these Americans on the side of the social order that had accorded them so little materially, it was vital in that period to have a president who was committed to progressive reform.

The dichotomy between America's great net wealth and her large numbers of poor people is a constant throughout her history since 1900. It is probably one of the engines for her astounding economic success, in that the price for falling behind in American society has always been comparatively high and thus a constant inducement to hard work. What it has never been – thanks in part to the two Roosevelts – was so high that Americans have been tempted to try another system, any of which would be bound to have been worse and would have threatened her world-status as the engine of capitalism. Indeed, throughout the twentieth century and into the twenty-first it was the Anglo-American form of capitalism, of free enterprise, free trade and *laissez-faire* economics, that has consistently produced more prosperity than any other model.

Roosevelt was also the father of the conservation movement which saved the National Parks from development, and he backed both the Meat Inspection and the Pure Food and Drugs Acts. He fully deserves his place on Mount Rushmore, and it was both America's tragedy and his own that he refused a third presidency in 1908, which was his for the taking. More tragic still was his decision to run in 1912, after his appointed successor as president, William Howard Taft, had disappointed him, as anyone was bound to do. The result of Roosevelt's 'Bull Moose' candidacy was a split Republican vote and the election of the least impressive of the three candidates, Woodrow Wilson.

The statesmanship – and there can be no greater test of statesmanship than sticking to unpopular but correct policies in the face of a general election – of McKinley, Roosevelt and Hay laid the basis of the friendly co-operation of the English-speaking peoples in the coming century. As one historian has put it, 'By conducting a decidedly pro-British neutrality policy during the war, the United States Government bolstered the fledgling Anglo-American friendship and prepared the way for the emergence under President Theodore Roosevelt of the uniquely special relationship that would play such a crucial role in twentieth century international history.'[12] In retrospect, Lord Salisbury's decision to adopt a pro-American neutrality stance during the Spanish-American War was among the wisest of his long and sagacious career.

Under the terms of the Paris peace treaty that ended the Spanish-American War on 10 December 1898, the United States obtained Spain's colonies of Puerto Rico, Guam and the 7,107 islands of the Philippines. Cuba obtained her nominal independence, but was effectively a US protectorate. Although in 1897 President McKinley had stated that 'forcible annexation' of the Philippines 'cannot be thought of, by our code' and would amount to 'criminal aggression', twelve months later he was willing to pay Spain $20 million for their cession to the United States. Two major factors had influenced his change of mind. One was the concept of Manifest Destiny, or, as he put it, the responsibilities 'which we must meet and discharge as becomes a great nation on whose growth and career from the beginning the Ruler of Nations has plainly written the high command and pledge of civilisation'.[13] The other factor was 'the commercial opportunity to which American statesmanship cannot be indifferent. It is just to use every legitimate means for the enlargement of American trade.'

Trade has the virtue of generally benefiting colonial and colonist alike, and the total volume of Filipino overseas trade increased massively after 1898. In

1895, the Philippines – under the Spanish – undertook 62 million pesos' worth of trade with the rest of the world, which by 1909 – under the Americans – had more than doubled to 132 million; by 1913, it had more than trebled to 202 million and by 1920 was nearly ten times that amount, at 601 million pesos. The percentage of that vastly increased trade with the US grew from 13% in 1894 to 32% in 1909 to 43% in 1913 to 66% in 1920, despite the 10,000 miles distance.[14]

There were nonetheless many Americans who vehemently opposed the Treaty of Paris and the United States' inherent acceptance of responsibilities far beyond her shores. Populists, Democrats, a few Republicans led by Senator Samuel Hoare of Massachusetts, and the financier Andrew Carnegie all denounced it as contrary to the Declaration of Independence, the Constitution and especially George Washington's isolationist Farewell Address. 'Europe has a set of primary interests, which to us have none, or a very remote relation,' wrote Washington on leaving the presidency in the late summer of 1796. 'Hence she must be engaged in frequent controversies, the causes of which are foreign to our concerns.'[15]

Those words – wise for 1796 when it took seven weeks' sailing to reach America from Europe – made far less sense in the world that George Washington could never have foreseen, that of the railway, the telegraph, the aeroplane, the steamship, the submarine, the aircraft carrier, the jet, the internet, let alone the inter-continental ballistic missile. Under Washington the fastest a man could travel was on a galloping horse, yet by May 1869 the trans-continental railroad could take a passenger from New York to San Francisco in a few days, when before the journey would have taken six months. As the globe shrank so America's world role grew, and by the early twentieth century it had certainly outgrown its late-nineteenth-century mantras. (Perhaps surprisingly, George Washington himself did not shy away from imperialist connotations; indeed, he rarely used the word 'republic' to describe the United States, preferring the word 'empire'. When in October 1783 the Marquis de Lafayette proposed a visit by Washington to Europe, he replied that he would sooner tour what he called 'the New Empire' from Detroit via the Mississippi to the Carolinas.)

After furious debates over the Treaty of Paris, the McKinley Administration managed to secure the required two-thirds Senate majority in favour of the annexation programme, albeit by just one vote. Prominent supporters included Senator Orville H. Platt of Connecticut and Albert J. Beveridge of Indiana, the latter of whom, a Progressive senator from Indiana, argued that the Anglo-Saxons were the 'master organizers of the world' with a mission that took precedence over 'any question of the isolated policy of our country'.[16]

The next problem was that some Filipinos themselves, under their charismatic leader Emilio Aguinaldo, who had for two years led risings against the Spanish, declared an autonomous republic on 23 January 1899. Less than two weeks later, the first rebel soldier had been killed in a fight with American troops. Guerrilla tactics quickly replaced set-piece engagements, especially after General Arthur MacArthur, the US Military Governor of the Philippines, captured the rebel stronghold of Malolos on 31 March.[17] MacArthur estimated that there were no fewer than 1,026 'contacts' with the enemy between May 1900 and June 1901. Nor did MacArthur argue that the Filipinos were being intimidated into supporting Aguinaldo, since that could not, in his words, 'account for the united and spontaneous action of several millions of people'.[18]

The justification for the American presence in the Philippines has been assumed by some historians and economists to be almost solely exploitative, but this entirely fails to take into account the genuine sense of mission that actuated American policy-makers of the day. Throughout the twentieth century, the enlightened self-interest of American law-makers has been mistaken for money-grubbing greed, often because of the way that trade naturally followed the flag. The 1900 report of the Philippines Commission should be taken at face value, reflecting the sincerely held views of its members, when it stated that, 'The United States cannot withdraw from the Philippines. We are there and duty binds us to remain. There is no escape from our responsibility to the Filipinos and to Mankind for the government of the Archipelago and the amelioration of the condition of its inhabitants.'[19] Written off by Marxists and cynics as self-serving claptrap, in fact these sentiments accurately reflected the thinking of very many distinguished, intelligent and hard-working American public administrators, for whom duty was a watch-word and service a way of life. By increasing trade and commerce in the islands, they believed that the United States would help the Philippines towards prosperity in the region and eventual self-government. In the long run, they were proved right.

The Democrats denounced what they called 'a war of criminal aggression' that they alleged was entirely based on 'a greedy commercialism'.[20] Senator Hoare said that Filipinos had right on their side, and Senator Benjamin Tillman of South Carolina poured scorn on the Republicans, forcing McKinley to denounce as near-sacrilegious traitors those who equated Aguinaldo with George Washington. In the November 1900 elections, McKinley's 'Forward' policy won him a bigger majority than he had secured four years earlier. It seemed that the American people agreed with vice-presidential candidate Theodore Roosevelt's description of those who doubted America's Manifest Destiny as mere 'mollycoddles'.

As tends to happen in asymmetric, guerrilla wars, atrocities were committed on both sides. The insurgents punished pro-American Filipinos by wrapping the Stars and Stripes around their heads as turbans, pouring kerosene on them and then setting them alight. Suspected informers' lips were cut off and large numbers of *barrios* were razed to the ground in reprisal by the US, fifty-three in the Samar region of 5,276 square miles alone. Judge William Howard Taft, who in 1900 was appointed president of the commission of inquiry into conditions in the Philippines, described the rebels as 'a mafia on a large scale', as its daring godfather Aguinaldo continued to slip through US lines and evade capture by the 70,000 men of MacArthur's army. In all, the war cost the United States $175 million.

Aguinaldo's good fortune finally ran out on Saturday, 23 March 1901, when he fell for a brilliant American ruse, 'as desperate an undertaking as the heart and brain of a soldier ever carried to a successful conclusion'.[21] The following month he took an oath of allegiance to the United States, after which the war came to an official end, despite occasional minor flare-ups over the next couple of years.

Under the governorship of Judge Taft, the Philippines protectorate was guided towards the day when it could become an autonomous republic. The government health service was hugely expanded and a free primary school system introduced. Local elections were instituted, a bill of rights introduced (unsurprisingly excluding the right to bear arms), the peso was linked to the gold standard and a $3 million relief fund set up. Slavery, piracy, head-hunting and religious repression were all vigorously suppressed. An efficient Filipino constabulary and model prison system were introduced, and the hospitals that had been set up throughout the islands to fight smallpox, bubonic plague, cholera and malaria had in three decades brought the death-rate down almost to the level of the continental USA.[22] (The Spanish attitude to Filipino healthcare had long been along the lines of '*che sera sera*'.)

American administration of the Philippines was high-minded, but it was also practical. In January 1901, the Taft Commission reported in cipher to the Secretary of War, Elihu Root, in Washington on the question of drunkenness and prostitution in Manila. Drunkenness was no worse than in any American city of the same size, it explained, and schemes to check prostitutes for venereal disease, at between fifty cents and $2 a time, were 'better than a futile attempt at total suppression in an Oriental city of three hundred thousand', which they believed would only have the effect of 'pro-ducing greater evil'.[23]

In October 1907, Taft inaugurated the first popularly elected Assembly at Manila's opera house. A fifty-two-page report that year drawn up for Taft listed the number of areas in which the standard of living of ordinary Filipinos had improved exponentially since the Americans replaced the Spanish as their rulers. It almost constitutes a check-list of what a modern liberal democracy was able to bestow upon a developing country and included steel and concrete wharves at the newly renovated port of Manila; dredging the River Pasig; streamlining of the Insular Government; accurate, intelligible accounting; the construction of a telegraph and cable communications network; the establishment of a postal savings bank; large-scale road- and bridge-building; impartial and incorrupt policing; well-financed civil engineering; the conservation of old Spanish architecture; large public parks; a bidding process for the right to build railways; corporation law; and a coastal and geological survey of all the islands covering 115,000 square miles.[24]

The eight million Filipinos – roughly equal to the number of Japanese at that time – were about to enter upon the most prosperous chapter of their history. Of their demands for self-government, in July 1910 the Governor-General, W. Cameron Forbes, wrote to Taft, who had become president the previous year: 'It is more worth while to spend your life in trying to assist people who have enough ambition to want to manage their own affairs than it would be to work for people who had not that amount of initiative. I do not want in any manner or degree to discourage the desire for independence. I am adopting the policy of telling them that if they really want it to get busy and get those things without which it is impossible.' There is no reason for supposing that Forbes was misleading his president: the twentieth-century record of imperialism of the English-speaking peoples, be they American, British or Antipodean, was far superior to that of any of their rivals.

Language is an expression of power, and one area in which the United States imposed her will in the Philippines was to flood the islands with teachers, who educated the population in 'good citizenship and individual ambition', but most of all in English. By 1935, there were no fewer than 8,000 government schools teaching 1.23 million pupils, and the language they taught was the one that the English-speaking peoples have made their most valuable and longest-lasting export. On 4 July 1901, in a long letter to Taft about American aims in the Philippines, President McKinley wrote:

> In view of the great numbers of languages spoken by the different tribes, it is especially important to the prosperity of the Islands that a common medium of communication may be established, and it is obviously desirable that this medium should be the English language. Especial attention should be at once given to affording full opportunity to all the people of the Islands to acquire the use of the English language.

By 1925, the Education Survey's extensive tests of Filipino schoolchildren found that, 'In receiving dictation and in spelling in English they are almost the equal of American schoolchildren.'[25] Here was cultural imperialism at its most benign, since the twentieth century saw the massive expansion of English into the global *lingua franca*. (The very term illustrates how far English's once great rival, French, has slipped behind in the struggle for linguistic supremacy.) One of the reasons that the Philippines could be guided to true independence was that by the 1930s the multiracial population of the hundreds of islands, with their many and varying degrees of political development, could at last communicate easily with one another.

Equally benign was the protective umbrella that the United States threw over the Philippines, at least until it was thrust aside by the Japanese capture of Manila in January 1942. Had the islands not been an American protectorate, they might have stayed under the brutal Japanese occupation for far longer, since there would have been little strategic, economic or moral reason to have committed the huge amount of Allied blood and treasure under General Douglas MacArthur necessary to have liberated them. Just as the English-speaking peoples prevented the Aboriginal people of Australia and the Maoris of New Zealand from falling beneath the Japanese jackboot, so they freed the Filipinos, Malays, New Guineans, Burmese, Pacific Islanders, Koreans, Indo-Chinese, Hong Kong Chinese and many other Asian peoples from it too, services that are easily forgotten if one concentrates solely on so-called 'colonial exploitation'.

In the four years that the Philippines were part of Japan's so-called Greater East Asia Co-Prosperity Sphere, no fewer than five per cent of the entire Filipino population died.[26] Ultimately the English-speaking peoples' prosecution of the war against Japan in the Far East, including the deployment of the vital nuclear technology that had been pioneered and financed by them ever since the New Zealander Ernest Rutherford had split the atom, meant that the Far East could be liberated from the horrific depredations of Showa Japan.

Although the Boer War has long been denounced by historians as the British Empire's Vietnam, and characterised as being fought for gold and diamonds, and trumped up by greedy, jingoistic British politicians keen to bully the two small, brave South African republics, the truth was very different. Far from fighting for their own freedom, the Boers were really struggling for the right to oppress others, principally their black servant-slaves, but also the large non-Afrikaans white Uitlander ('foreigner') population of the Transvaal who worked their mines, paid 80% of the taxes and yet had no vote. The American colonists had fought under James Otis' cry that 'Taxation without representation is tyranny' in 1776, yet when Britain tried to apply that same rule to Britons in South Africa, she was accused of vicious interference.

The Transvaal was in no sense a democracy in 1899; no black, Briton, Catholic or Jew was allowed either to vote or to hold office. Every Boer was compelled to own a rifle, no non-Boer was authorised to. The business centre Johannesburg, with 50,000 mainly British inhabitants, was not even permitted a municipal council. The English language was specifically banned in all official proceedings. Judges were appointed by the Boer President, Paul Kruger, who controlled all the government monopolies from the manufacture of jam to that of dynamite. No open-air public meetings were permitted, newspapers were closed down arbitrarily and full citizenship was almost impossible to gain for non-Boers. Kruger ran a tight, tough, quasi-police state from his state capital, Pretoria.

Lord Salisbury and the British Colonial Secretary, Joseph Chamberlain, were genuinely outraged at the way that Pretoria treated the Britons who lived and worked in the Transvaal, and especially at how Kruger repeatedly raised the residency-period requirements for the franchise while ignoring the Uit-landers' petitions protesting at the way they were subjected to higher taxes, poorer school provision, police brutality, and the private and state monopolies that grossly inflated their cost of living. All this was humiliating for the British Government, which saw itself as their champion.

In March 1899, a British subject, Mr Edgar, was shot dead in a brawl by a drunken Boer policeman, who subsequently escaped punishment. No fewer than 21,684 Uitlanders signed a petition deploring this miscarriage of justice, except that this time they addressed it not to Kruger but to Queen Victoria. Chamberlain's Cabinet memorandum on the subject argued that it could not be ignored, otherwise 'British influence in South Africa will be severely shaken'. An ultimatum on the other hand would be likely to lead to war, and British forces in South Africa were totally unprepared for that (in a way they would surely not have been were the British leaders hoping and planning for a conflict). So Chamberlain enclosed a draft despatch 'intended as a protest, and still more as an appeal to public opinion'.

The real fear of the effect of loss of prestige – a genuine concern in regions where prestige truly mattered – was to recur again and again in the story of the English-speaking peoples. Empires run by tiny elites are to a great extent ruled as much through kudos as by deployable military power.[27] Although it cannot be quantified, prestige constitutes vital capital in international affairs, and the fear of losing it was always an authentic one.

Chamberlain's despatch to Pretoria acknowledged that the British Gov-ernment recognised the Transvaal's right to manage her own internal affairs, but then went into detail about how the Uitlanders were treated as second-class non-citizens, despite the enormous contribution they made to the country's prosperity. It mentioned in particular education costs, the liquor laws, the lack of political representation, arbitrary arrests, the partiality of the courts, press censorship, widespread corruption, the summary expulsion laws and the precedence given to Afrikaans over English, even in schools where almost all the children were British. The Boer War was thus partially fought over human rights, even though that concept, which was to bulk so large later in the century, was then in its cradle.

As Cecil Rhodes, the British-born Prime Minister of Cape Colony, had said about his visit to the Johannesburg Uitlanders in his trial after he had tried to overthrow Kruger in a coup in 1896, 'I saw a number of people many of whom had the feeling peculiar to our race, that they must have a share of the government of the country where they were paying taxes.'[28] In the same year as the failed coup, Roger Casement, the British Consul in Lourenço Marques, the port of Portuguese East Africa, had reported on the despair of the disenfranchised Uitlanders: 'What in my opinion an English minister has to fear in the Transvaal more than anything else is an alienation of the sympathies of the English-speaking peoples there.' Reports such as these spurred the British Government on to fresh efforts.

On 4 May 1899, the British High Commissioner in South Africa, Sir Alfred Milner, warned London that the Transvaal was arming quickly whilst propagating 'a ceaseless stream of malignant lies about the intentions of the British Government'. Milner protested that all he wanted was 'to obtain for the outlanders in the Transvaal a fair share of the Government of the country which owes everything to their exertion'. It was an insistence on representative institutions and civil rights, not a greedy desire to grab goldfields as has often been alleged, that led to the decision of the British Government to institute regime change in Pretoria. Kruger did try to haggle over the extent of Uitlander franchise, offering a seven-year residency qualification and a mere five seats in the thirty-five-seat Boer-dominated Volksraad. When Milner demanded more, Kruger told him, 'I am not ready to hand over my country to strangers. There is nothing else now to be done.' For his part, Milner privately called Kruger 'a frock-coated Neanderthal'.

Lord Salisbury found it hard to believe that 410,000 Boers could seriously be considering taking on the might of the British Empire at such a pitch of its fame and power. Intelligence sources nevertheless continued to suggest that the republics were arming to the maximum extent possible. Pretoria then suddenly declared war on Britain on 20 October 1899 and invaded the British colonies of Natal and Cape Colony, thereby deliberately starting a conflict which was to cost tens of thousands of lives, but which has ever since been perversely and unfairly blamed entirely on Britain.

Pomp and circumstance, esteem and conventions – smoke and mirrors – these things matter when running a Great Power, and the Transvaal and Orange Free State had rudely ripped them aside with their brave if suicidal declaration of war against an empire infinitely larger and more populous than their two tiny republics. London estimated – correctly – that only the most crushing response could re-establish the *status quo ante*, before other peoples across the Empire – starting in Cape Colony but quickly spreading across Africa and Asia – saw Britain's weakness and developed their own taste for revolt. The clash between the Boers and the British in South Africa was long in coming, but once it materialised it was a straightforward struggle for primacy of prestige, in order to see, in Lord Salisbury's candid words, 'who is Boss'.

A characteristic of the English-speaking peoples displayed both in South Africa and in the Philippines at the dawn of the twentieth century, and then fairly regularly ever since, was their tendency towards ruthlessness in warfare. For all that they are slow to anger, they have historically been very hard-nosed once the fighting was actually taking place, although this has often been tempered by a tendency to treat defeated enemies generously. Plenty of what in the luxury of peacetime have been called 'war crimes' have been laid at their door since 1901. In the Boer War, there was a scandal that centred on two Australian officers, Lieutenants H.H. 'Breaker' Morant and P.J. Handcock, who were executed for fighting a dirty war in the Zoutpansberg and Spelonken areas of the northern Transvaal in 1901, which involved the murder of eight Boer prisoners of war.[29] Even so, as with the prisoner-abuse scandal over Abu Ghraib prison in Iraq in 2004, once the relevant authorities were apprised of the facts of the case they acted decisively, through courts-martial.

No-one in history has done more for the concept of human beings having certain inalienable rights than the English-speaking peoples, and it is often solely because of their belief in the rule of law that abuses ever come to light and are punished. Every war has thrown up its dirty secret, such as Britain's Hola Camp in Kenya during the Mau Mau rebellion, or America's My Lai massacre in Vietnam, or at Abu Ghraib in the Iraq War. To expect anything different is to misunderstand the way humans behave in wartime, from whichever colour, creed or class they hail. What is needed is a legal device to correct abuses, and that is something that the English-speaking world has generally had in place, but which the Germans in 1900s' Angola, the Japanese in 1930s' China, the French in 1950s' Algeria and the Russians in 1980s' Afghanistan had not. The difference is not that the English-speaking peoples never commit crimes in wartime, but rather that their open societies and free press tend to ensure that these are punished, while many other societies' crimes rarely are, or are even acknowledged as such.

Despite the military reverses of Black Week back in December 1899, the Empire had been stalwart. 'It is the destiny of the British nation to spread good and just government over a large proportion of the Earth's surface,' was the opinion of New Zealand's *Waikato Argus* on 31 January 1900. 'There is only one sentiment throughout the Empire – we must win regardless of the cost in men and treasure!' The same attitude still pervaded imperial thinking a year later; an insurgency campaign, however vicious, was not about to dent trust in ultimate victory.

'Here was an enemy of a different type,' continued the official history, 'one who operated from no base and towards no objective, whose victories lay in escapes, and in the length of time during which he could remain untrapped; who could never be said to advance or retire, but merely to move, now this way, now that, his tactics rendered unfathomable either by utter lack or rapid change of purpose.'[30] Again and again throughout the twentieth and into the twenty-first century the English-speaking peoples have been harried by such unconventional military tactics, as practised by Subhas Chandra Bose's INA, the Mau Mau, EOKA, Malayan communist guerrillas, the Vietcong, the IRA, Al-Queda, the insurgency and others. Unlike their successor organisations, the Boers indulged in few atrocities, though there certainly were some, including the regularly reported abuse of the white flag of surrender.

The 'war crime' for which the British have been most commonly held responsible during the Boer War was the supposed ill-treatment of Afrikaans women and children in camps there. In fact, these 'concentration' camps – the term had no pejorative implication until the Nazi era – were set up for the Boers' protection off the veldt, and were run as efficiently and humanely as possible, given the Boer commandos' own constant disruption of rail-borne supplies into them. A civilian surgeon Dr Alec Kay, writing in 1901, gave a further reason why the death rates were so high:

> The Boers in the camps often depend on home remedies, with deplorable results. Inflammation of the lungs and enteric fever are frequently treated by the stomach of a sheep or goat, which has been killed at the bedside of the patient, being placed hot and bloody over the chest or abdomen; cow-dung poultices are a favourite remedy for many skin diseases; lice are given for jaundice; and crushed bugs for convulsions in children.[31]

The American public generally sided with the Boers, as did ex-President Benjamin Harrison, the Democrat leader William Jennings Bryan, Andrew Carnegie, the German-born Missouri senator General Carl Schurz, Henry Adams, the Chicago intellectual Clarence Darrow, the *New York Herald*, the *Washington Post*, the *Chicago Tribune*, the *Baltimore Sun*, the *Atlanta Constitution* and Joseph Pulitzer's *New York World* and *North American*. New York City Council and Boston Commonwealth Council passed unanimous votes of admiration for the Boers and a Boer delegation was fêted, staying at the Arlington Hotel in Washington where the whole street was illuminated by a reception committee. In the delegation's subsequent national tour, which took them as far west as San Francisco, audiences totalling hundreds of thousands turned out to hear their tales of British repression.

Three hundred Americans went out to fight for the Boers under a Captain O'Connor, who Kruger ordered to: 'Be good fellows, obey your commanders, look after your ponies.' Several, such as James 'Arizona Kid' Foster and J.H. 'Dynamite Dick' King, fought with distinction. Furthermore, 29,000 schoolboys from Pennsylvania, New York and Massachusetts signed an address of admiration to President Kruger, which was delivered by a messenger boy called James Smith. Arriving in the Boer capital on 28 May 1900, still dressed in his messenger's uniform, Smith gave Kruger the message, who tucked it in his pocket and the following day fled Pretoria before the advancing British.

The level of American support for the Boers only increased as the war progressed; Representative Fitzgerald of Manhattan proposed that the entire Afrikaans nation should decamp to the US, and the Governors of Arkansas and Colorado offered millions of acres for the scheme. In the Senate, the Republican William Mason of Illinois accused Britain of 'criminal aggression' and George Wellington of Maryland said of the Boers, 'Their foe has been our foe, and their battle for right has been a repetition of our own.' In fact, it was the Uitlanders who suffered taxation without representation, and the Boers who had invaded, but American public opinion was solidly behind the seeming underdog.

Yet none of this had any effect on the stalwart McKinley and Roosevelt Administrations, which maintained a strict policy of 'equal access' throughout the conflict. Since the Boer republics were landlocked and the Royal Navy controlled the ocean, that meant genuine access for only one side, leading historians to agree with the Democrats of the day that American neutrality was in effect 'thoroughly pro-British'.[32] The men at the top of the two Administrations – while admiring Boer courage – were tremendous Anglophiles, who also recognised that it was not in America's national interest for Britain to be humiliated in South Africa. This attitude only strengthened as the plight of British arms became acute during 'Black Week'.

On 24 September 1899, two weeks before the war broke out, the US Secretary of State John Hay had written to Henry White, the First Secretary at the London Embassy, 'The one indispensable feature of our foreign policy should be a friendly understanding with England.'[33] At the beginning of Black Week that December – where British forces were defeated thrice in six days – New York Governor Theodore Roosevelt told his great friend, the British diplomat Cecil Spring-Rice, 'It would be for the advantage of Mankind to have English spoken south of the Zambesi.' Anxiety was a factor in Roosevelt's calculations, for as he told the Spanish-American War hero,

Captain (later Admiral) Richard Wainwright, a disaster for the British Empire would place the United States 'in grave danger from the great European military and naval powers'.[34]

Such pessimism might seem astonishing from such a proponent of a power that was still fairly geographically isolated from European power politics, but it is an indication of how small the US Navy then still was and how globally American policy-makers were already thinking by the close of the nineteenth century. As Roosevelt wrote to another Englishman, Arthur (later Lord) Lee, in January 1900, 'I believe in five years it will mean a war between us and some of the great continental nations unless we are content to abandon our Monroe Doctrine for South America,' which the United States had shown over the Venezuelan crisis that she was very unwilling to do. Long seen as a proactive, even aggressive measure, in fact there was also a defensive element to Roosevelt's construction of the powerful White Fleet in the years before 1909.

Concern that a combination of France, Germany, Russia and perhaps other lesser powers might try to take advantage of Britain's travails during Black Week even led Roosevelt to tell the US Civil Service Commissioner, John R. Proctor, 'I should very strongly favour this country taking a hand in the game if the European continent selected this opportunity to try to smash the British Empire.' Later the next century, Roosevelt's distant cousin Franklin was to become deeply antipathetic towards that Empire, but Theodore certainly did not take that view in December 1899.

English-speaking fellow-feeling seems to have played as important a part in Administration thinking about the Boer War. Hay told Henry White in March 1900 how, 'The fight of England is the fight of civilisation and progress and all our interests are bound up in her success.'[35] That November, Hay even made the appalling diplomatic *faux pas*, after reading of a British victory, of telling the (pro-Boer) Netherlands Ambassador, 'At last we have had a success.' Despite his own Dutch blood, Roosevelt agreed with the sentiment. After Black Week, *Harper's Weekly* summed up this attitude by advising Americans not to 'lose sight of the stupendous fact that British prestige is in mortal danger; nor can we fail, if we have a proper pride of race, or a decent sense of gratitude, or a consciousness of what the English have accomplished in the homes of the savage races, to mourn over these disasters'. The reference to gratitude was reflected in Roosevelt's private remark that, 'I am keenly alive to the friendly countenance England gave us in 1898.'

On discovering that a Russian initiative was under way to try to 'mediate' in the war, Hay telegraphed to tell White to warn Lord Salisbury, who three days later made the public avowal that, 'Her Majesty's Government cannot accept the intervention of any other power.' In the event, the United States provided invaluable help for Britain; cartridges, hay, oats and preserved meats were shipped to South Africa in large quantities as well as 100,000 horses and 80,000 mules (comprising half the mules used by the British Army during the entire conflict). US banking loans underwrote one-fifth of the cost of the war, and her exports to South Africa rose from an average of $112 million in 1895–8 to $577 million during the war years of 1899–1902.

The Administration had to pay a political price, such as pressure at the 1900 Republican convention, but only the most innocuous plank on the war wound up in the final platform. Meanwhile, the Democratic candidate William Jennings Bryan denounced 'the ill-concealed Republican alliance with England' and extended support for 'the heroic burghers in their unequal struggle to maintain their liberty and independence', but to no avail, winning only 155 electoral college votes to McKinley's 292 in the November 1900 elections. Eight months after the election, Roosevelt reiterated to Spring-Rice how, 'I have always felt that by far the best possible result would be to have South Africa all united, with English as its common speech.'

Writing of the election to an English friend, Rev. Harry Wolryche-Whitmore of Quail Rectory, Bridgnorth, on 21 November, Roosevelt explained that,

> There is nothing in the office of Vice President, and I hated to leave the Governorship of New York, but I felt it was extremely important to beat Bryan who represented a compound of class hatred, semi-criminality, thoughtlessness, ignorance and sentimentality with a sprinkling of the sincere men who get wrong-headed on some particular point.

Clearly the vice-presidency was not 'quite enough' for Roosevelt, nor should it have been for a man of his talents and drive. The appearance not to desire office had long been a standard part of the Victorian statesman's repertoire, but it should seldom – if ever – be taken at face value.

In the realm of sport, America was certainly making her presence felt by 1901. Although at Wimbledon in June the Doherty brothers retained the Gentlemen's Doubles Championships against the Americans Davis and Ward, on Lower Killarney Lake in Ireland the University of Pennsylvania crew beat Dublin University's boat 'easily' over a three-mile course. In the Inter-University Athletic competition in New York that September, held between Yale, Harvard, Oxford and Cambridge, the American universities won six events to three and the following week the three fifteen-mile races of the America's Cup were won by the American yacht *Columbia*, beating Sir Thomas Lipton's *Shamrock II*.

Although the British invented the lion's share of the world's competitive sports and games – including soccer, rugby, cricket, golf, modern tennis, bobsleighing, bowls, croquet, racquets, table tennis, snooker, badminton and boxing – they also formulated the rules for many of those they hadn't invented, such as hockey, polo, ice-skating, canoeing, lacrosse and downhill skiing. 'If you can score points by hitting or kicking something, it was almost certainly invented by Britain's leisured classes, keen on exercise, team spirit and clear rules.'[36]

By the start of the twentieth century, however, British teams were regularly finding themselves being thrashed by other nations. Only twice, though – during the 1932/3 'bodyline' Ashes tour of Australia and the 2003 Rugby world cup – did sport lead to accusations of bad sportsmanship. Usually, it has proved a valuable social cement between the English-speaking peoples; no fewer than 30,000 rugby fans left the British Isles to follow the British Lions' 2005 tour of New Zealand, for example, and no two nations that play cricket with one another have gone to war. (India and Pakistan suspended play in 1960 prior to fighting each other five years later.)

Another effective cement was the Pacific cable, the legislation for which finally gained Royal Assent in 1901. There had been an Atlantic cable in operation since August 1858 and the idea for a trans-oceanic cable had been mooted as early as during Queen Victoria's Golden Jubilee in 1887, but the scheme kept foundering on the predictable issue of who was to pay for it. In July 1899, the British Chancellor of the Exchequer, Sir Michael Hicks Beach, finally agreed that government credit could be used to lay the cable, as the 'main factor in the situation' was 'the idea of co-operation between the Mother Country and the colonies'. Canada, too, offered financial help, not out of the prospect of material gain so much as being 'common partners in the scheme' for Imperial unity. Since Canada's average annual trade with Australia in the third quarter of the 1890s only amounted to £190,000, as against Britain's £53.17 million, the reason given for their commitment can be taken at face value. At that time Canadians sent 90,000 letters to Australia per annum, while receiving nearly seven million from the UK.[37]

Once started, the huge project was quickly completed, and on 8 December 1902 it became possible for Canadians to contact Australians and New Zealanders in real time, via Vancouver, the Fanning Islands, Fiji, the Norfolk Islands, Brisbane and Doubtless Bay. Canon Gore was wrong about there being 'no prophet for the people', since the poet and novelist Rudyard Kipling was just such a man, combining a seer's prescience with a poet's lyricism and a novelist's imagination. Kipling celebrated the great scheme with a three-stanza poem entitled *The Deep-Sea Cables*, whose last verse ran:

They have wakened the timeless Things; they have killed their father Time;
  Joining hands in the gloom, a league from the last of the sun.

Hush! Men talk to-day o'er the waste of the ultimate slime,
And a new Word runs between: whispering 'Let us be one!'[38]

The Pacific cable aided enormously what Disraeli had once described as 'a great policy of Imperial consolidation', and in July 1901 Lord Salisbury introduced a bill into the House of Lords changing the official title of the monarch to include references to the colonies, something that had not been considered necessary when Queen Victoria had ascended the Throne back in 1837.

If America's Administrations – as opposed to her people – were stalwart during the Boer War, the rest of the English-speaking peoples were magnificent. Even before the fighting broke out, the Canadian soldier Lieutenant-Colonel Sam Hughes said that his country should 'fulfil our part as the senior colony' and send troops to South Africa, and his call was heeded by 7,368 Canadians, eighty-nine of whom were to die in action there and 135 of whom succumbed to disease or accidents.[39] The enthusiastic celebrations for Queen Victoria's Diamond Jubilee of 1897 had raised Canadian patriotism to a high degree, and this volunteering for Crown service in a faraway war – every call was heavily over-subscribed – was a further manifestation of that.

The nationalism – indeed jingoism – engendered by the Boer War encouraged a profound desire amongst Canadians to differentiate themselves from Americans, whose population was as pro-Boer as her leaders were pro-British. Relations between the two countries had not been smooth, and the 1895–6 Venezuelan crisis had seriously perturbed Canada, not least when a Note by the then US Secretary of State, Richard Olney, seemed to deny Canada's right to stay in the British Empire, and Congress passed a $100 million appropriations measure for a 900,000-strong US army. In January 1896, both sides of the Canadian Parliament applauded the Government's announcement of the rearmament of the militia, and the remark of the Finance Minister, George E. Foster, that Canada would defend her British connections and heritage from the imprecations of the United States.[40]

At this distance of time it seems unimaginable that there might have been a US-Canadian war, not least because English-speaking democracies do not fight one another, but in 1896 forty-two out of America's forty-five state governors promised to enrol troops for one, and a senior American general, Nelson Miles, was quoted saying, 'Canada would fall into our hands as a matter of course.' (Be that as it may, the pounding that the eastern seaboard of the United States would have suffered from a Royal Navy that was hugely larger than the pre-Rooseveltian US Navy might not have made invasion worthwhile.)

Lord Salisbury's solving through arbitration of the Venezuelan crisis failed to end the resentment that Canadians felt towards America, which one historian has characterised as 'virtual Americanophobia' by 1899.[41] The Dingley Tariff Bill passed in 1897, instituting the highest US duties to date, caused outrage in Canada, which retaliated by granting preferential trading status to Britain, New South Wales and the British West Indies. Thereafter disputes abounded: there were lawsuits over lumber exports; Canadians were deported from the United States under an alien labour law; American strike-breakers operating in Canada left Canadian trade unionists incandescent; territory was vigorously argued over in Alaska and, in 1899, British Columbia passed legislation requiring all miners there to be British subjects. That year the Canadian Premier, Sir Wilfrid Laurier, even employed the noun 'War' in a discussion of the ways that a mining dispute in the Yukon might escalate, thereby creating an international sensation and understandable consternation in the British Foreign Office.

In 1901, Imperial unity served as a counterpoise to bad US-Canadian relations and afforded Canadians a means of preserving national identity, but also of standing up for what they believed to be their rights vis-à-vis their militarily, demographically and economically giant neighbour to the south. Complaining that their situation was like that of living next door to an elephant, Anglo-Canadians wholeheartedly embraced the alternative vision for which the British Empire stood. The twentieth century was to see ultra-loyal Canada brusquely rejected by Britain, yet not drawn into America's orbit as a result.

Back in 1897, Canada had commemorated the Queen-Empress's sixty years on the throne with coast-to-coast parades, military reviews, speeches, receptions, unveiling of monuments and statues, poems, newspaper articles, assemblies of schoolchildren, openings of parks, banquets and all the other paraphernalia of private and public celebration. In Winnipeg, it lasted for two weeks. Sir Wilfrid Laurier was welcomed to Britain 'like a visiting royal' and the Mounties drew loud cheers when they took part in the Queen's procession through London. In a speech in Liverpool, Laurier emphasised Canada's loyalty and her determination to maintain 'to the fullest extent the obligations and responsibilities as British subjects'.[42] Nor was the term 'British' a slip of Laurier's tongue, since most English Canadians were both nationalists and imperialists who saw no contradiction between the two 'since their objective was to create a stronger nation within the Empire, not to prepare for Canada's withdrawal from it'.[43] They thus genuinely saw themselves as simultaneously Canadians and Britons.

Why was this land of opportunity and growth so keen to become involved in a war 7,000 miles away, on behalf of an elderly Mother Country, in which Canada had no direct concern? In the 1960s, it was fashionable to explain this remarkable phenomenon in terms of conspiracy theories and 'manipulations

from Downing Street and British propaganda', yet the closer and more objectively the phenomenon has been examined the clearer has, in the words of one
historian, 'the salient fact emerged that English Canada had been eager, if not
anxious, to fight and had forced the Canadian Government to send troops'.
This was because of a deep-seated commitment to the concept of Imperial
unity and genuine Canadian identification with the Empire's cause. This was
so strong that even when the war was almost over in 1902, Laurier was
compelled to allow another Canadian contingent to go out there.

When in March 1901 Mr Bourassa's motion in the Canadian House of
Commons against participation was debated and voted upon, it was rejected
by 144 to 3, after which MPs rose to sing the national anthem. Of the thousands
of Protestant clergymen across Canada, only ten made public pronouncements
against the war. In Quebec Cathedral, the Rev. Frederick George Scott told
the Canadian military contingent that, 'We, a republic in monarchical form,
go out to crush a despotism in the form of a republic,' in order to extend to
the Transvaal 'light, liberty and religious toleration'. Even Irish-Canadian
MPs supported the war and, although many Quebecois were opposed to it
and the tricolour was seen more often during it, the Archbishop of Quebec
was in favour, as was Laurier himself, the first French-Canadian and Roman
Catholic Prime Minister of Canada.

The centre of loyalty to the Crown was Toronto, the headquarters of the
British Empire League. Burned by the Americans in the war of 1812, threatened in the rebellion of 1837, stalwart during the Fenian raids, Toronto had
demanded rearmament during the Venezuelan crisis of 1896. Of all the many
cities to give cash bonuses to volunteers for the Boer War, Toronto was the
most generous, although right across Canada provinces and municipalities
subscribed large sums to complement the $2 million government budget.
When the wounded started coming back to Canada from mid-July 1900, an
entire local community, including the militia and brass band, would turn out
to welcome home even a single soldier. The return of the Royal Canadian
Regiment that November brought huge crowds on to the streets in gratitude,
and two months later the Mounted Rifles and Artillery were greeted by civic
receptions and cheering multitudes at every single railway stop from Halifax to
Vancouver. Canadians contributed to plaques, fountains, statues, ornamental
gates and gold watches, in a manner they surely would not have done had the
war not been genuinely popular and perfectly in tune with their sense of
national identity.

As with the rest of the English-speaking peoples in this period, the fundamental
drive for self-improvement in Canada came from the genius of capitalism.
Although the idea for limited-liability joint-stock companies originated with
the Dutch in the late sixteenth century, they were brought to their peak by the

English-speaking peoples. The model for all subsequent chartered firms was the Vereenigde Oost-Indische Compagnie (Dutch East India Company), incorporated in 1602, which had limited liability and publicly traded shares at a proper stock exchange. It was between 1844 and 1862, however, that successive Company Acts passed by the British Parliament enshrined the basic principles which led to the exponential growth of market capitalism and created what a distinguished recent study has rightly described as, 'The most important organization in the world: the basis of the prosperity of the West and the best hope for the future of the rest of the world.'[44] By 2001, there were no fewer than five-and-a-half million companies registered in the United States of America.

Under those Victorian laws, companies no longer had to have strict specific purposes, and limited liability ensured that investors could only lose the amount that they had originally put into the firm. That, along with the public trading of shares of equal value, opened up the modern capitalist system that has brought prosperity to every society that has ever properly adopted it, while 'civilizations that once outstripped the West yet failed to develop private-sector companies – notably China and the Islamic world – fell farther and farther behind'. Nicholas Murray Butler, the President of Columbia University and Nobel Peace Prize laureate of 1931, equated the invention of the limited liability corporation with that of steam locomotion and electricity. Furthermore, they lived for ever: 'Companies have proved enormously powerful not just because they improve productivity, but also because they possess most of the legal rights of a human being, without the attendant disadvantages of biology: they are not condemned to die of old age and they can create progeny pretty much at will.'[45]

The way that capitalism, when allied to the right to own secure property and the rule of law, has unleashed the energy and ingenuity of Mankind has been remarkable and forms the basis of the English-speaking peoples' present global hegemony. So long as they retain the technological edge in the military field, the only way they can be replaced as the world-hegemon is through another Great Power adopting an even more effective form of capitalism. The way that the corporation has managed to harness human effort and render it hugely productive, in a manner that no other social invention has successfully achieved over time, proves how the idea of limited liability was one of genius:

> Companies increase the pool of capital available for productive investment. They allow investors to spread their risk by purchasing small and easily marketable shares in several enterprises. And they provide a way of imposing effective management structures on organizations. Of course, companies can simply ossify, but the fact that investors can simply put their money elsewhere is a powerful rejuvenator.[46]

Edmund Burke believed that since the cost of anything automatically decreases the more there is of it, and increases the fewer there is, capitalism was therefore a law of Nature and thus an invention of God. In fact it was the brainchild of some gifted Dutchmen, which was then brought to an ever-more productive pitch by the English-speaking peoples. The French, Swedish, social democratic, Japanese corporatist and various other models of capitalism have all failed dismally compared to the Anglo-Saxon version.

Although the way that the English-speaking peoples grasped and then perfected the idea of the corporation is the foremost key to their global success, it was by no means a foregone conclusion because the late-nineteenth and early-twentieth centuries saw the rise in America of a perversion of the essentially beneficial corporation, known as the Trust. With America producing 36% of the world's total industrial output by 1913, which was more than Germany's 16% and Britain's 14% combined, it was essential that monopoly capitalism did not replace the more efficient, genuinely competitive kind. By 1904, however, no less than two-fifths of manufacturing capital in the United States was contributed by trusts.[47] Something had to alter and, as so often throughout the story of modern capitalism, American flexibility won through, on this occasion in the shape of Theodore Roosevelt's 'trust-busting' campaigns using the hitherto-moribund Sherman Antitrust Act in over forty suits, the most famous of which was the dissolution of the Northern Securities Trust.

The economic growth of America at the turn of the century was astonishing: whereas America exported 40 million bushels of wheat in 1850, that figure had grown to 600 million by 1914. The total amount of capital in publicly traded manufacturing companies grew from $33 million in 1890 to over $7 billion by 1903, a 24,000% increase. Moreover, although Russia had 44,600 miles of railway in operation in 1906, Germany 36,000, India 29,800, France 29,700, Austria-Hungary 25,800, Great Britain 23,100 and Canada 22,400, the United States had a staggering 236,900 miles, more than all those other countries put together.

On 7 March 1901 it was announced that the arrangements had been concluded for the formation of the US Steel Corporation, with capital of $850 million, half in common stock and half in 7% cumulative preferred stock, and $304 million in bonds.[48] With men like Andrew Carnegie and Henry Clay Frick making steel according to the new theories of line production, Cornelius Vanderbilt and E.H. Harriman consolidating the railway industry and John D. Rockefeller's Standard Oil controlling the oil industry of America, it was clear that the scene was being set by the turn of the century for a titanic clash between the 'robber barons' of capitalism – vastly rich self-made pioneers and despots, whose cartels had cornered their respective markets – and the rest of American business.

Henry Ford, born in Greenfield, Michigan, in 1863, produced his first petrol-driven motor car in 1893. By 1899, he had founded his own company in Detroit and was designing his own cars, and in 1903 he incorporated the Ford Motor Company. Five years later he produced the Model 'T', the first car that the average American could afford. After he had developed the world's first assembly-line techniques, the price fell year-on-year so that by 1925 it cost $260; by the time it was replaced by the Model 'A' in 1928, he had sold fifteen million of them and there were more on the road than all other makes of car added together.

It is therefore perhaps too simplistic to see this struggle in the black-and-white colours favoured by some left-wing historians, since the 'robber-barons' do deserve credit for almost creating entire industries from virtually nothing. Men like Rockefeller and Morgan were giants because their ruthlessness and vision came at a time when American capitalism needed both. Yet in order to protect the competition ethic so vital for the system to work, their empires had to be split up. In that men of such vast wealth can be considered victims, they were victims of their own success. Roosevelt denounced the trust kings as 'the criminal rich, the most dangerous of criminal classes', and the fight was on. By the time that J.P. Morgan, 'the Napoleon of Wall Street', attended the House Banking and Currency Committee hearings under subpoena in December 1912, eighteen US financial institutions controlled aggregate capital resources of $25 billion, or two-thirds of the Gross Domestic Product (GDP).[49] Earlier that month, the Supreme Court had ordered the dissolution of the Union Pacific and Southern Pacific railways merger. In the clash between the 'robber-barons' and Congress's trust-busting legislation, the legislature won, with overall beneficial results for American – and world – capitalism.

Britain had led the world in the 'first wave' of industrial growth between 1790 and the mid-nineteenth century, especially in steam engines and the mass production of textiles. The 'second wave' from the mid-1840s until about 1890 had been driven by steel production and railroads, where the United States was starting to overtake Britain as the greatest economy on earth.[50] By the time the twentieth century dawned, Mankind had entered the third industrial wave – spearheaded by the chemical, electrical and automobile industries – which lasted until the Great Depression. During this period the United States established her undoubted economic dominance. It was of inestimable benefit to Britain that the power to which she was about to cede hegemony was her own younger cousin, which shared so many of her own political, moral, legal and linguistic mores and characteristics.

As well as a global power-shift towards America, the start of the twentieth century also saw the beginning of a shift in power geographically within the

USA, with political influence moving westwards. The addition of ten extra seats in the US Senate as Oklahoma (1907), New Mexico (1912), Arizona (1912), Hawaii (1959) and Alaska (1959) became states of the Union was only the start. Western representation in the House rose from sixty in 1900 to 127 in 1980. The first presidents with real western connections – Theodore Roosevelt had a ranch in North Dakota and Herbert Hoover was born in Iowa, raised in Oregon and educated in California – began to challenge the exclusively eastern and southern orientation of American foreign policy. Of the forty years between 1952 and 1992, 'genuine or honorary westerners' – Eisenhower, Johnson, Nixon, Reagan and Bush Snr – were in office for thirty-one of them.[51]

As well as an increased interest in the Pacific Rim region, which was probably inevitable anyhow given global economic trends, the rise of the West has meant that Western political issues moved higher up the national political agenda than they might otherwise have done. Rebellions that start there – such as the free enterprise rebellion that fears that local investment and employment is falling foul of federal government conservation policies, or the Proposition 13 tax revolt of the late Seventies, or the environmental campaign against unregulated Alaskan offshore oil-drilling – tend to gain national prominence.

Far from being, as it once was, the haven of individual enterprise, the West has been the recipient of vast amounts of federal funding during the twentieth century, so much so that by 1994 almost half of the total land area of the eleven westernmost states in the lower 48 – including 86% of Nevada – was owned or administered by federal agencies such as the Interior, Agriculture and Defense Departments. *The Oxford History of the American West* has gone so far as to state that, 'The effect has been the nationalization of Western America, the reduction of differences as sub-regions and cultures have been incorporated within national systems and found themselves participants in national programs.'

'The skies of the new century look down on no community set in happier conditions, or with the promise of a brighter future, than the new Australian Commonwealth,' enthused *The Argus*, the new Dominion's paper of record.[52] The *Melbourne Age* concurred, proclaiming that, 'Certainly never did a century dawn on a free people with a worthier heritage of political promise than that which has come to us this day.' Across the continent on that clear and sunny first day of the century the new song *Australia* was sung, church bells were rung, shops were covered in 'an abundance of foliage, eucalyptus boughs predominating', there were torchlight processions, firework displays, promenade concerts, patriotic poems and editorials, electric illuminations, brass bands, special midnight 'Commonwealth' church services, proclamations, children's monster-picnics, and 'wherever a pole of any description could be placed in position a flag of some sort was flying'. All to celebrate the day 'which makes Australians one people, with one home and one destiny'. A few Australians overdid it; according to the *Herald-Standard*, a Melbourne girl called Polly Miller was fined forty shillings for using obscene language in Little Lonsdale Street after having 'taken more than was good for her' by 7 a.m. (The paper concluded, not unreasonably, that she had 'commenced the celebrations of Commonwealth Day very early'.)

Australia soon found herself the richest of all the Dominions. As one historian has succinctly put it,

> The continent is prodigiously rich in things in the ground that you can dig up and sell at a massive profit. In the nineteenth century, it was gold; in the twentieth, it was iron ore, uranium, titanium and a series of other exotic materials that have generally been in international demand. More unpredictably there were fortunes to be made in agriculture: in particular in sheep in the rolling grasslands of Victoria and New South Wales and cattle (to provide roast dinners for Surrey Sunday lunches and, more recently, for Tokyo's hamburger-consuming youth).[53]

All this served to confirm the optimistic vein in which *The Argus'* leader-writer summed up the general view at the moment of Federation: 'We have a self-contained continent, the brightest, fairest and richest field on which a nation was ever planted. We are a section of one of the greatest races of history. We have a political constitution which we have framed for ourselves, the freest political science has yet evolved.' It was true; Australia had had the secret ballot and manhood suffrage since 1861, far earlier than the Mother Country which did not enjoy them till 1872 and 1918 respectively. Small wonder, therefore, that Australians looked to the future with confidence.

Australia's commitment to the Empire was reflected in the 16,000 troops she sent to fight in South Africa, 598 of whom did not come back; 10,000 were raised at Australia's own expense, the rest at Britain's.[54] A sense of patriotism and spirit of adventure were not the only spurs to recruitment; life was tough in Australia, large parts of which had suffered an extraordinary drought ever since the mid-1890s. Kimberley in Western Australia recorded the worst drought for sixteen years, and 1901 was the driest year on record for Victoria, too, with major bush fires, falling wheat yields and consequential rising bankruptcies.[55] Average life expectancy was 55 years 2 months for men, 58 years 10 months for women. The previous year, a bubonic plague had claimed the lives of 103 people in seven months in Sydney alone, a city where the night-cart carried away effluent, rather than efficient modern sewers. Nothing was allowed to dim the celebrations following Lord Hopetoun's proclamation of Federation, however, and on 1 January 1901 the streets of

Sydney were thronged with 'enthusiastic and excited sightseers' hours before the constitutional changeover took place.

Federation meant that the significant political differences between the states had to be addressed in a national context and many were deep-seated. As an Australian historian has recorded recently,

> New South Wales was free-market, Victoria protectionist; New South Wales had supported the North in the American Civil War, Victoria the South. The railway tracks had different gauges, duties were sometimes levied on goods moving from one state to another. There were quarantine restrictions between states, some of which remain to this day, and different public holidays – still not sorted out.[56]

Federation has been written off by some historians as 'a process in which bourgeois politicians sought to stitch up a business deal for their common economic advantage', and the celebrations of January 1901 as 'evidence merely of the colonists' pleasure at the prospect of a free party'. The Australian historian Charles Manning Clark, a Leninist, has called it a 'reactionary plot'. Yet in fact Australians were genuinely proud of and excited by the notion of a young country with an independent identity. Moreover, the concept of federation worked and was one of the many benefits that the English-speaking peoples were to bequeath Mankind. Initially tested successfully in the United States – despite the terrible bloodletting of 1861–5 – the means by which several states were able to form a unified polity while remaining largely self-determining in various important aspects of their internal affairs were brought to triumphant fruition in Australia and was later successfully adopted by the British West Indies.

The large, troublesome and often even semi-rebellious French-speaking population of Quebec has only been kept within Canada through generous federal arrangements with the rest of that country (along with the lack of any viable alternative). It is too early to say whether the devolution of powers from Westminster to Scotland and Wales in 1998 has been successful, but it was essentially an attempt at British federalism. Federalism is certainly no cure-all; there were later mistakes, as with the attempted fusion of Northern Rhodesia, Southern Rhodesia and Nyasaland into a 'Central African Federation' between 1953 and 1964, but overall the concept has solved many more problems than it has created and has been adopted by countries as far from the English-speaking tradition as Russia and Malaysia.

Patriotism, not plunder, was the primary motivation for Federation, and it seems to have been generally welcomed by the Aborigines. When Lord Brassey, the Governor of Victoria, visited Portland on 26 March 1899, he was welcomed by Mr Albert White of the Lake Condah mission station on behalf of the Aborigines of the Western District. White said that the Aborigines looked forward to the 'coming nation of Australia', which they hoped would 'enjoy peace from all foreign foes, happiness among themselves and prosperity in the age to come'.[57] Although the early years of nationhood were hard on the Aborigines of Victoria, they were hardly easy for any Australians in a land that was then being blighted by almost freak weather conditions.

The first elections under federal law were held on Wednesday, 16 December 1903, which saw a turn-out of nearly 47%. Since then continent-wide elections, which are no mean administrative feat, have been held every three years in Australia.[58] The task of moving Australia's Parliament, which was only intended to stay in Melbourne temporarily, did not see similar feats. Suggestions put to Australia's first prime minister, Sir Edmund Barton, for the name of the new federal capital included 'Yarramatta' and 'Australapolis', but it was not until 1927 that it finally moved to the specially designated site of Canberra.[59]

One of the first actions Australia took on becoming a fully self-governing nation was to protect herself as an English-speaking country by instituting tough immigration restrictions. In contrast to Canada, which had opened her doors to significant Russian, Chinese and other incoming populations, Australia passed an Immigration Restriction Act in 1901, which, once allied to the Quarantine Act of 1908, constituted the building blocks of her 'White Australia' policy that stayed in force until the 1960s. Under the 1901 Act, immigration was prohibited to anyone who could not write out at dictation fifty words of any European language chosen by the immigration officer. Immigration officers could thus ask a Greek or Italian to take down fifty words in Serbo-Croat.

Of course, Australia was right to restrict immigration on some other grounds. A senior public health official was to argue for 'the strict prohibition against the entrance into our country of certain races of aliens whose uncleanly customs and absolute lack of sanitary conscience form a standing menace to the health of any community'.[60] While those words sound harsh and deeply politically incorrect today, huge epidemics in China were killing hundreds of thousands of people at the time, and Australia had the right (and duty) to protect herself from similar outbreaks.

Most Australians at that stage of their national development wished their country to stay recognisably British and were proud of the way 'they did not have large French – or Dutch – speaking groups in their country, like Canada and South Africa; they played cricket; named their suburbs after English places – Brighton, Sandringham, Ramsgate, Windsor; cherished British-made goods; and fawned on the royal family'.[61] This was to change over time, especially after the Gallipoli débâcle of 1915, but in 1900 Australia was proud of her very British identity. It was only much later that the famous line from Hobbes was to be bastardised, and 'the Poms' were characterised as 'solitary, poor, nasty, British and short'.

*

The invention of refrigeration turned New Zealand into something more than just 'the paradise of the Pacific'. Owing to the pioneering work done by the New Zealanders Thomas Brydone and William Davidson, by 1882 it was possible for 7,500 frozen lamb carcasses to make the three-month journey to London on the modified chartered ship *Dunedin*, with all but one arriving in good condition. The cargo fetched twice the price it would have at home, and by 1933 New Zealand provided half of all British lamb, mutton, cheese and butter imports, all through the miracle of refrigeration.[62] Before the 1880s, the decade when New Zealand experienced the depression that ravaged much of the world, sheep meat was almost a waste product, boiled to make candles or often simply thrown away. After Brydone and Davidson, it became – along with wool – a mainstay of the economy of that beautiful, robust but remote outpost of the English-speaking world. The historian James Belich has described refrigeration in his book *Paradise Reforged* as 'the knight in icy armour that rode to the rescue of the New Zealand economy'.

One nation that stayed relatively backward throughout the early years of the twentieth century was Ireland, where political, religious and racial differences – despite Queen Victoria's happy visit to Dublin – were never far from the surface. By early 1900, the United Irish League (UIL), which had only been launched two years earlier, had spread its tentacles across almost the whole island of Ireland, the latest in a long line of organisations devoted to agrarian agitation. In the previous quarter-century there had been the Land League, the Irish National League and the Plan of Campaign, but the UIL was to be more radical in its method of anti-British and anti-landlord protest than any of its previous incarnations. Drawing its strength from the rural Catholic peasantry, the UIL was to become a very significant nationalist and republican organisation by 1903, when it split as a result of the Irish Land Act of that year.[63] Irish agitation followed by British concession followed by further Irish agitation followed by further British concession was to be the pattern of Anglo-Irish relations until the Easter Rising of 1916.

Undoubtedly there was real, grinding poverty in Ireland, especially in the south-west, in the early part of the twentieth century. As well as attempting, in the words of its recent historian, Philip Bull, 'to revive a popular, grassroots, extra-parliamentary movement in the countryside in the hope of reinvigorating a decaying parliamentary nationalism', the UIL also hoped to transfer the ownership of agricultural land from landlords to tenant farmers. Founded by the former MPs William O'Brien and Michael Davitt, long-time nationalist agitators, it was described by the latter as 'a fighting combination of the people', although it was always more radical rhetorically than in actuality.

Starting in West Mayo, the UIL organised mass demonstrations in favour of tenants who were evicted from their landlords' properties, formenting boycotts and intimidating shopkeepers who served the class enemy. They would organise marches to the houses of landlords, in which hundreds of demonstrators would jeer at the terrified inhabitants. Although the Government constantly considered prosecuting the leadership for incitement, they believed (probably wrongly) that it wanted to be imprisoned. The authorities' position was, as Bull has pointed out, 'an impossible one. Prosecution might have given the aura of martyrdom to O'Brien, but failure to prosecute saved the UIL from the setback of his imprisonment and humiliated the Government and police by showing up their apparent weakness.'[64]

The local Roman Catholic clergy provided some of the UIL's most aggressive activists, while clothing it with a respectability that it largely did not deserve. (This was to become a serious problem; no fewer than twenty priests attended Sinn Fein's first public meeting.)[65] Only two months after the UIL's founding, 122 landlords needed and were receiving police protection. Rent strikes and monster demonstrations were its forte, and many priests feared that, 'as the League extended its influence, their status and authority were being undermined. Many also recognised that only by supporting and participating in the League could they exercise a moderating influence upon its actions. Consequently, there was a gradual but steady accretion of clerical support.' Archbishop McEvilly of Tuam even had a pastoral letter read out in churches in February 1898 that endorsed the UIL's redistributive agrarian policies. The following September, he wrote to several of his clergy instructing them to end their opposition to the UIL, on the grounds that there must be no friction between the people and their clergy.

Constitutional Irish nationalism had been completely split in 1890 by the resignation of Charles Stewart Parnell from its leadership in Westminster, over a sensational divorce case in which he was cited as a co-respondent. The UIL was still attempting to repair the rift in 1898, seven years after Parnell's death, and by 1900 it had largely been successful. That reunion, along with a revival of political consciousness and some successes over agrarian grievances, made the UIL a success. Irish flags were flown over the courthouses of those county councils where nationalists were in the majority, and the UIL even formed its own 'National Convention' that it described as the 'Parliament of the Irish People', which aped the institution by which true power was legitimately authorised in Ireland, namely at Westminster.

The dawn of the twentieth century saw Anglo-American amity solidified in the treaty signed on Monday, 18 November 1901, by the Secretary of State John Hay and the British Ambassador to Washington, Lord Pauncefoote. The Hay-Pauncefoote Treaty covered the proposed isthmian canal that was due to be excavated in Nicaragua, even though it was eventually cut in Panama instead. Once the Senate ratified it by 72 votes to 6 on 16 December, the 'Special Relationship' had been inaugurated. Since this Relationship has been the single most important geopolitical factor of the twentieth century and beyond, the Hay-Pauncefoote Treaty can be regarded as a great act of statesmanship by Lord Salisbury and Theodore Roosevelt. It was also a clear indication that the United States' international stature had hugely increased in the twenty years since 1881, when Britain had refused to grant America the right to construct, operate and fortify an isthmian canal.

The Treaty inaugurated a string of Anglo-American agreements between 1901 and 1909. Britain applied pressure to Canada to resolve a border dispute between Alaska and British Columbia in the US's favour in 1903; the long-running Newfoundland fisheries' dispute was then settled to mutual satisfaction, as was a dispute over Jamaica in 1907. As one historian has put it, 'Roosevelt constantly kept his primary objective, Anglo-American unity, sharply in focus.' Although the FDR-Churchill, Reagan-Thatcher and Bush-Blair friendships were to be a major theme of post-1900 Anglo-American amity, the impersonal but excellent working relationship between Salisbury and Roosevelt was in a sense the predicator of all three. The Salisbury ministry took the wise and long-sighted view that American maritime expansion posed no threat to the British Empire, and a *de facto* naval alliance grew between Britain and the United States up until 1927.

It had been under the Royal Navy in the nineteenth century that Britain had originally established what the distinguished Indian political scientist Professor Deepak Lal calls a 'Liberal International Economic Order', whose major attributes were free trade, free mobility of capital, sound money due to the gold standard, property rights guaranteed by law, piracy-free transportation, political stability, low domestic taxation and spending, and 'gentlemanly' capitalism run from the City of London. 'Despite Marxist and nationalist cant,' he writes, 'the British Empire delivered astonishing growth rates, at least to those places fortunate enough to be coloured pink on the globe.' The United States was to inherit the duty of protecting, promoting and expanding this Liberal International Economic Order in the coming century, and the Hay-Pauncefoote Special Relationship and *de facto* naval understanding were to be important stepping stones along that path.

Even more important than ruling the waves in the twentieth century has been the English-speaking peoples' dominion over the skies. The first successful heavier-than-air powered flight, undertaken by the brothers Orville and Wilbur Wright on the dunes at Kill Devil Hill near the fishing village of Kittyhawk, North Carolina, at 10.35 a.m. on Thursday, 17 December 1903, constituted a seminal event in the history of the English-speaking peoples. Although the first flight itself, in a freezing 24 mph wind with Orville lying prone at the controls to reduce resistance and Wilbur running alongside, only covered 120 feet and lasted twelve seconds, the world changed irrevocably from that moment. During that day the brothers made a number of ascents in the 12-horsepower gasoline-powered plane, optimistically dubbed *Flyer*, the longest covering 852 feet and lasting fifty-nine seconds. 'I found control of the front rudder quite difficult,' Orville noted in his diary. 'As a result the machine would rise suddenly, then as suddenly dart for the ground.'

Yet on that day a manned, power-driven, heavier-than-air machine had flown. Not only had the world changed, but so too had the ability of the English-speaking peoples to maintain their hegemony over it. By staying at the forefront of almost every advance in civil and military aeronautics throughout the twentieth century and into the twenty-first, the English-speaking peoples were able to bring decisive power to bear on their many and varied opponents. Air power was to become a central part of the reason why the English-speaking peoples have survived and prospered so successfully since 1900.

Only three years were to pass between the Wright brothers flying and the British Army officer, Lieutenant J.W. Dunne, designing the first military aeroplane in 1906. (An enterprising fellow, he later tried to invent a method of demonstrating that time was only relative, by recording his dreams.) Dunne undertook much of his early work of invention, founded in part by the War Office, at least in the beginning, at Blair Atholl in Scotland, his prototypes protected from prying eyes by the Duke of Atholl's estate workers. By the time of the outbreak of war in 1914, Dunne's prototype had been bought by an American boat-builder called Stirling Burgess and the Canadian Army had bought the Dunne-Burgess mark II plane in order to undertake aerial photo-reconnaissance of the Western Front. Sadly the 47-foot-wingspan aircraft was too badly damaged on its journey over to Europe on board *SS Athenia*, but it stands as an example of what the co-operation of the English-speaking peoples might have achieved. (A New Zealander called Richard Pearse was also a very early aviator, who, although he got airborne with a two-cylinder engine earlier than the Wright brothers, was not able to exercise enough control of his machine to usurp their claim to have invented heavier-than-air flight.)

Several people very nearly flew before the Wright brothers, including John Stringfellow in a tri-plane with an exceptionally light steam engine in 1868, and the American-born Sir Hiram Maxim with two 180-horsepower steam engines in 1894. Had Percy Sinclair Pilcher, a British marine engineer, not died of injuries sustained in a gliding accident in October 1899, he might well have also beaten the Wrights, since he had patented an aircraft powered by a petrol engine, which he had designed in an engineering works in Great Peter Street in London.[66] Pilcher's glider had been inadvertently left out in the rain, and was sodden and heavy, yet he decided to demonstrate its powers to Lord

Braye and his guests at Stanford Hall, Leicestershire, and died in the attempt. Four years later, the Wright brothers took to the skies, and by October 1905 Wilbur was airborne for more than half an hour, flying as far as twenty-four-and-a-half miles.

Warfare had long been the mother of aerial invention; ballooning thrived before the Napoleonic Wars as a means of observation and during the Franco-Prussian War as a means of communication, and it was to be the Great War that gave the spur to exploit the Wright brothers' breakthrough to the full. Since then and up to the present day, the race has been on for the English-speaking peoples always to develop new military aircraft that can hold sway in the skies above battlefields. The Spitfire's superiority over the Messerschmitt 109 and 110, the P-51 Mustang's superiority over both the German and Japanese fighters and interceptors, and more recently the F-16 and F-18's superiority over the MIG-29 all contributed decisively. (The P-51 was emblematic of Anglo-American co-operation, as it was originally designed in America for the RAF; powered by a Rolls-Royce 'Merlin' engine and built by Packard, it shot down almost 4,000 German planes alone.) Recently, in the Gulf War, Kosovo War and Iraq War, air power proved decisive very early on. Indeed, the day that the English-speaking peoples fall behind in the contest to build the world's best fighter and bomber aircraft will be the one when their primacy is doomed.

Someone who quickly appreciated the strategic importance of air power, but who applied it to dangerous ill-use, was the distinguished geopolitical theorist Halford Mackinder, Director of the London School of Economics and Reader in Geography at Oxford University. On Monday, 25 January 1904, Mackinder delivered a lecture to the Royal Geographical Society in Exhibition Road, London, the reverberations of which were thirty-seven years later to cause the world to hold its breath. Mackinder was the first person to ascend Mount Kenya, the founder of modern British geography as an academic discipline, and, after 1904, he was to become a Liberal Unionist MP (1910–22) and British High Commissioner for South Russia (1919–20) during the Russian Civil War. A member of any number of boards, committees and royal commissions, as well as the Privy Council, Mackinder was used to being listened to with respect. Those who gave their considered responses to Mackinder's address, which was entitled *The Geographical Pivot of History*, included the future Chichele Professor of Military History Spencer Wilkinson, the geographer Sir Thomas Holdich and the future First Lord of the Admiralty and Secretary of State for India, Leo Amery.[67] It was thus to a very high-powered gathering that Mackinder delivered his thoughts and ideas, yet they were to spread far further than the immediate audience in Exhibition Road.

The central theses of his lecture were that because of the modern development of steam navigation the world was shrinking, and that the fulcrum, or 'pivot area', of the world lay in Eastern Europe and southern Russia. The 'heart-land' of the globe, he argued with five maps, was in that vital region over which so many armies had fought in the past, which he called 'Euro-Asia'. The maps alone give an indication of the sweep of Mackinder's theorising and are entitled *The Natural Seats of Power; Continental and Arctic drainage; Political divisions of Eastern Europe at the accession of Charles I; Political Divisions of Eastern Europe at the Time of the Third Crusade* and *Eastern Europe before the nineteenth century.* For any number of historical, economic, geographical and strategic reasons, Mackinder argued, control over Eastern Europe and southern Russia held the key to global domination. He got an enthusiastic hearing, although Wilkinson complained that Mackinder's choice of the Mercator projection for his world map tended to exaggerate the size of the British Empire, whose naval duty it would be 'to hold the balance between the divided forces which work on the continental area'.

It was not until 1919 that Mackinder expanded his thesis into a book, *Democratic Ideals and Reality.* The intervening world war had done little to cause Mackinder to question his own thesis, and the book contained the following statement, clearly addressed to the members of the peace conference then assembled at Versailles:

> When our Statesmen are in conversation with the defeated enemy, some airy cherub should whisper to them from time to time saying: 'Who rules East Europe commands the Heartland: Who commands the Heartland commands the World-Island: Who rules the World-Island commands the World.'[68]

Of course the experience of the Great War and the Brest-Litovsk Treaty, under which the Germans had ruled Eastern Europe and commanded the Heartland, but failed to command the World-Island let alone the World, should have caused Mackinder to review or jettison his 1904 thoughts, but it didn't. Like many a polemical academic, he tried to fit the facts into his theory rather than *vice versa.* He was knighted in 1920.

Far from it being the victorious Allied statesmen who listened to Mackinder, however, in fact it was the defeated enemy who did. In Britain, his book went virtually un-reviewed, whereas it was closely studied in Germany, where it became an article of faith with the *Geopolitik* school of German thinkers. The 'airy cherub' who did the whispering was General Karl Haushofer, who reproduced Mackinder's *Natural Seats of Power* map no fewer than four times in his periodical *Zeitschrift für Geopolitik.* Among his many paeans to Mackinder in the inter-war years, in 1937 Haushofer described the 1904 paper as 'the greatest of all geographical views', adding that he had never 'seen anything greater than these few pages of a geopolitical masterpiece'. (For all his admiration of Mackinder, Haushofer still loathed the British; in his review of *Democratic Ideals and Reality*, he reminded his readers of Ovid's maxim to learn from one's opponent, and described Mackinder as the 'hateful enemy'.)

Religious toleration has been a mainstay of the English-speaking peoples since 1900; powerful emotions that have been channelled elsewhere in the world into suppressing minorities because of the way they choose to worship – or not to worship – particular deities have been generally absent from the secular societies of the English-speaking world, with corresponding advantages both for social unity and the ability of those minorities to contribute to the greater good. As David Landes has pointed out in *The Wealth and Poverty of Nations*, religious intolerance 'proved great for purity but bad for business, knowledge and know-how'.[69]

In the twentieth century, the best gauge of a society's attitude towards religious toleration has been its treatment of the Jews, and although they have undoubtedly been socially discriminated against – especially in the pre-war period – they have never been persecuted in the English-speaking world except in 1904 in Ireland, a country whose special historical development makes its experience very different from the rest of the English-speaking peoples since 1900.

At the turn of the century, there were fewer than 4,000 Jews in Ireland, too few to be considered any kind of a cultural threat to the overwhelming Roman Catholic majority. Ireland had certainly not had any history of anti-Semitism, not least, as Mr Deasy says in James Joyce's *Ulysses*, because it didn't let any in. (There were only 472 Irish Jews in 1881.) Yet in Limerick in spring 1904, there was a pogrom against the few who had managed to emigrate there from the institutionally anti-Semitic nations of Eastern Europe. Whipped up by the preachings of Father John Creagh, it started in January as a boycott of Jewish businesses, and soon Jews were hissed at by crowds in the street and mud was thrown at them. They were then physically attacked, with cries of 'Down with the Jews!', 'Death to the Jews!' and 'We must hunt them out!'[70]

Although the Irish parliamentary leader, John Redmond, and Michael Davitt both condemned the Limerick boycott, the local Irish Party MP supported it. When Rabbi Levin of Limerick asked the Catholic bishop to denounce what was happening, he made no public statement. Arthur Griffith, the founder of Sinn Fein, supported the boycott in the *United Irishman* newspaper, although it was opposed by the Unionist *Irish Times*. Soon Jews in Limerick were being refused service in shops. By April, twenty of the city's thirty-five Jewish families had been put out of business. Assaults on them continued and the boycott went on into the autumn. By 1905, not surprisingly, 'virtually the entire Jewish community in the city joined the exodus from Limerick'.[71] When the following year Creagh left Ireland for the Philippines, he was thanked by three local newspapers for what he had said about the Jews.

★

The *Entente Cordiale* – concluded on Friday, 8 April 1904, between the British Foreign Secretary, Lord Lansdowne, and the French Ambassador to London, Paul Cambon – proved to be one of the world's longest-lasting alliances, and is still in (at least nominal) existence. It was henceforth the unspoken assumption that in any war central to their continued existence, Britain and France would fight on the same side. Of course the *Entente* has worked in France's interests more than in Britain's, for the inescapable geographical fact that any country capable of threatening Britain's independence was likely to have attacked France beforehand.

There were secret clauses added to the public understanding that constituted the *Entente*, by which the Royal Navy would make certain maritime dispositions in the North Sea in wartime capable of allowing France to counter a threat in the Mediterranean. These gave rise, in the Commons debate on the Anglo-French Agreement on 11 August 1904, to the splendid answer given by Earl Percy in response to a question about possible secret clauses from a Scottish Liberal MP James Weir: 'Speculation and conjecture as to the existence or non-existence of secret clauses in international treaties is a public privilege, the maintenance of which depends on official reticence.' In pure *realpolitik* terms, Britain did herself few favours by concluding the *Entente Cordiale* in 1904, thereby shackling herself to the fortunes of a nation that was even then in faster imperial decline than she, and which stayed so ever since. The *Entente Cordiale* was the geopolitical equivalent of handcuffing Britain to a drowning man, yet short of making an alliance with aggressive and unpredictable Wilhelmine Germany, there was little alternative.

It started off well enough on 8 April 1904, under the avuncular eye of King Edward VII, when a series of difficult Anglo-French problems were solved at the stroke of a pen, involving Moroccan territory, Egyptian finance, Newfoundland fisheries, Madagascan sovereignty and the Suez Canal. For all its romantic title, the *Entente Cordiale* was in fact a hard-headed multi-faceted business deal that was treated in the French foreign office, the Quai d'Orsay, as just that, but which in the British Foreign Office was accorded greater significance than it deserved.

Thenceforth, from 1904 until 1940 Britain's fate was intimately linked to that of France, even though the subsequent story was one of having to fight two world wars, primarily because France was incapable of defending herself alone. Of course the gargantuan ambitions of both Kaiser Wilhelm II and Adolf Hitler had to be halted, but the *Entente Cordiale* failed to prevent either of their attacks on France in 1914 or 1940. Britain instead sacrificed her freedom to manoeuvre by being connected to France and gained little from it thereby. Militarily, she was forced to send two expeditionary forces to continental Europe, the first of which was painfully bogged down in trenches for four years and the second of which was routed and forced to evacuate. Nonetheless, there was little alternative in 1904, and the *Entente Cordiale* was responsible for preventing the whole of Europe coming under the domination of a violently aggressive Imperial Germany ten years later.

TWO

America Arrives

1905–14

The murder of Harry Galt – Lords Curzon and Cromer as proconsuls – General von Schlieffen has a Plan – HMS Dreadnought *– The San Francisco earthquake –* The Struggle for Foreign Trade *– Winston Churchill provokes a wager –* The Federal Reserve Bank *– The Great White Fleet – The birth of the British Secret Service – War scares – Suicide of a Suffragette – Anglophobia – The outbreak of the Great War*

'The same causes which have raised Great Britain to her present exalted position will (probably in the course of the next century) raise the United States of America to a degree of industry, wealth and power which will surpass the position in which England stands as at present England excels little Holland. The naval power of the Western world will surpass that of Great Britain as greatly as its coasts and rivers exceed those of Britain in extent and magnitude.'   The German political economist Friedrich List in 1844

'Germany had been preparing every resource, perfecting every skill, developing every invention, which would enable her to master the European world; and, after mastering the European world, to dominate the rest of the world. Everybody had been looking on. Everybody had known. ... Yet we were all living in a fool's paradise.'

Woodrow Wilson, Sioux Falls, South Dakota, 8 September 1919

The British Empire with which Wilhelm II had such a strange love-hate relationship saw the murder of one of its junior administrators in the summer of 1905, which is only a minute footnote in the history of that institution but is nonetheless illuminating for the light it shines on the way that it was run. At 6.30 p.m. on the evening of Friday, 19 May 1905, Harry St George Galt, the thirty-three-year-old acting sub-commissioner of the Western Province of Uganda, was murdered while sitting on the veranda of a rest house near Ibanda in the Ankole district, roughly 150 miles west of the capital Kampala. Someone entered the compound and flung a spear into his chest, which pierced his lung. 'Look, cook,' he called to a servant, 'a savage has speared me!' He then fell down dead. Despite very exhaustive inquiries being made at the time, the assassin's identity was no clearer than his motive.

It might have arisen out of a rivalry between the tribal leaders of Toro and Ankole, whose mutual antipathies went back decades, and not have been primarily about Galt at all.[1] The suspected murderer called Rutaraka, whose corpse was found in circumstances that implied he had not committed suicide as locals initially suggested, might have killed Galt on behalf of one tribe in order for the other to be blamed, but very soon there were 'produced such a bewildering series of accusations, retractions, and contradictions' that the British investigating officer was almost reduced to despair. After a trial two men were convicted, who were subsequently freed on appeal, and the mystery still baffles people to this day, despite a huge amount of evidence in the secretariat archives in Entebbe. It is a complicated story that incorporates witchcraft, cover-ups, buffalo-poaching, at least two double-crosses, exile, a drinking party, smallpox-infected milk and a hunch-backed dwarf ('one of the most terrible cases of distorted mind and body I have ever witnessed').[2]

A white-painted cone of stones about 15 feet high records where the murder took place, and a Galt Memorial Hall was erected in the district headquarters at Mbarara, which was used as a magistrates' court. Galt is remarkable not only for the whodunit that still surrounds his murder. Britain had declared a Protectorate over Uganda during Lord Rosebery's premiership in June 1894 and granted her independence on 9 October 1962, and in all those sixty-eight years – over two-thirds of a century – he was the only British administrative officer ever to have been assassinated there.[3] An African country of nearly 94,000 square miles with a population in 1955 of just over five million, including 48,000 Asians and 5,600 Europeans, might have been considered hugely difficult to police; those who seek to portray the British Empire as a tyranny need to explain why places like Uganda produced so little popular insurgency against British rule.

It is certainly not enough to argue that the native people lived in fear of reprisals from the single battalion of the King's African Rifles, led by some two dozen British officers, ten British senior NCOs and 850 Ugandan soldiers and NCOs stationed there. Far more likely is that they recognised the benefits that British rule brought. As early as December 1901, the Great Ugandan Railway was built; this huge four-and-a-half-year project involved constructing a railroad 550 miles into the heart of Africa, from Mombasa on the coast to the source of the Nile itself, at Lake Victoria. The last spike was driven into the earth by Florence Preston, the wife of the railway's engineer, at the town that used to bear her name, Port Florence.

In 1962, the incoming Prime Minister, Dr Milton Obote, asked Sir Walter Coutts, the last governor, to stay on as Governor-General after independence, eloquent testament to the friendliness of the handover. The Spanish-born American philosopher and Harvard professor, George Santayana, wrote of the 'sweet, just, boyish masters' who ruled the British Empire in its final phase. In Uganda, men like Coutts and his private secretary Alan Forward tried their best to rule some twenty different peoples, derived from three racial groupings speaking some twenty different languages, who lived in four kingdoms and ten districts. That they managed to achieve this without more Britons than poor Harry Galt being killed is an astonishing tribute to their incorrupt, beneficial and just ideals.

Since Uganda became independent from Britain in 1962, it has not enjoyed one single peaceful transfer of power. Even so much as a glance at the disastrous post-independence history of Uganda – Obote's self-appointment as president, the military coup and subsequent dictatorship of Idi Amin, the border war against Kenya, Amin's expulsion of the Ugandan Asians, the vicious fifteen-year civil war between 1971 and 1986, the Tanzanian invasion, the economic collapse, the insurrection of the Lord's Resistance Army terrorists, and so on and so horrifically on – will convince any objective person that the brief period of British rule constituted a far happier time for ordinary Ugandans than any before or since. Fortunate were the Africans who were colonised by Britain, as opposed to the Germans, Portuguese, Spanish, Italians or, worst of all, the Belgians. When Algeria finally won her independence from France in 1962 – the same year as Uganda – the death toll in her 'savage war of peace' there was over one million. In the first half-century of France's rule from 1830, the native population of Algeria had fallen from 4 million to 2.5 million.

The experience of Sudan was not unlike Uganda. Between her conquest by General Kitchener in 1898 and independence on New Year's Day 1956, she was governed by a tiny elite of British administrators, called the Sudan Political Service. As a reporter wrote from Omdurman, the capital of Sudan, in April 2005,

Many Sudanese have affectionate memories of their colonial past. ... The men from the Sudan Political Service were chosen from Oxbridge colleges for their sporting and academic prowess, prompting the quip that Sudan was a nation of 'blacks ruled by Blues'. In the 1930s there were only 130 of them – governing one million square miles of Africa's largest country. Sudan still depends on their achievements. As early as 1916 the country had one of Africa's best railway networks, stretching from Port Sudan on the Red Sea to El Obeid in the deserts of the Kordofan.[4]

According to Gordon Obat, a columnist of the independent daily newspaper, the *Khartoum Monitor*, 'Wherever you go, people still remember the name of the British district commissioner in their area. They were seen as working for the good of the people.' Jibril Abdullah Ali, a local historian, points out the buildings in the capital of north Darfur, al-Fasher, that were originally built by British army engineers; they include the first school, a hospital, law courts, army barracks, an airport and other government institutions that use the same buildings half a century later. 'When Wilfred Thesiger, the explorer, arrived as an assistant district commissioner in northern Darfur in the mid-1930s,' recalled another foreign correspondent in 2004, 'he and a handful of British officers administered an area the size of France. He was fresh from university and was equipped with only a rifle, a camel and a uniform. His only luxuries were a few books and a Christmas hamper sent by his mother from Fortnum and Mason.[5] 'Other nations might have built a modern unified world,' concludes the historian Arthur Herman, 'but they probably would not have done it as quickly, efficiently, elegantly – or as humanely.'

By complete contrast, Kaiser Wilhelm II's army in German South-West Africa (modern-day Namibia) killed around 75,000 members of the Herero, Nama and Damara peoples between 1904 and 1907.[6] Of course, between 1914 and 1945 the Europeans proved themselves more than capable of slaughtering each other too. Armageddon knew no racial discrimination between 1914 and 1918. Nor did the destruction of tribal peoples end with the collapse of European colonialism. The Khmer Rouge massacred up to two million people in Cambodia long after the French had left. The history of Africa shows one million dead in the Biafran War, 1.3 million killed in the long-running post-independence Sudanese civil war, and another million dead in Mozambique since 1975. Hutus and Tutsis used machetes for massive ethnic cleansing in Rwanda and Burundi in the 1990s, and massacres in the Darfur region of the Sudan in the early twenty-first century cost 70,000 lives; none of which can reasonably be blamed on Europeans.

The Mau Mau revolt against British rule in Kenya in the 1950s cost the lives of between 2,000 and 3,000 loyal Kenyans. In all, 1,090 terrorists were hanged and at times as many as 71,000 were held in detention. On occasion the pro-British black Kenyan loyalists committed serious abuses against the mainly Kikuyu tribe Mau Mau terrorists. In 1959, the revelation that twelve terrorists had been beaten to death in the Hola Camp provoked an outcry in the House of Commons, and within four years Kenya was granted her independence. In all, between 12,000 and 20,000 Mau Mau were killed during the fourteen years of troubles, or below the average of post-war conflicts. Even if the figure was as high as 20,000 – as one recent study by historian David Anderson suggests – it was certainly nothing like the 300,000 recently claimed in a book entitled *Britain's Gulag*, or the 450,000 claimed by an egregious BBC TV documentary equally provocatively entitled *Kenya: White Terror.*[7] Such figures are truly absurd.

As Viceroy of India from 1899 to 1905, George Curzon ruled over the region that is today spanned by Burma, Bangladesh, Sri Lanka, Pakistan and of course India itself. The foreign policy he formulated from Calcutta further covered imperial relations with China, Afghanistan, Iran and Iraq. He was the most talented colonial administrator of an empire teeming with them, actuated by altruism of motive that seems almost other-worldly to us today. He brought peace and prosperity to a continent, yet his viceroyalty ended in dejection and bitterness.

As a prize-winning scholar at both Eton and Oxford, Curzon seemed destined for glory. In his successes – a parliamentary seat at twenty-four, ministerial rank at thirty-four, the viceroyalty at thirty-nine – and his grandeur as heir to Kedleston, one of the finest stately homes in Britain, the very British bacillus of envy was excited amongst lesser intellects. Curzon did nothing to placate this with ingénue remarks such as, when watching some troops bathing in a river, 'I never knew the lower classes had such white skins,' or his dictum that 'Gentleman do not take soup at luncheon.' He even once suggested – probably apocryphally or in self-mockery – that Big Ben should be turned off at night since its chimes disturbed his sleep.

Yet for all his failings, Curzon had a talent for building and administering empires that amounted to genius. His foresight was prodigious; in his 1894 book *Problems of the Far East*, he drew attention to the strategic importance of a tiny hamlet in Indo-China named Dien Bien Phu. His understanding of and sympathy for the scores of races that made up the teeming 300 million of British India was probably unparalleled in imperial annals going back to Rome (which was geographically a small empire by British imperial standards, only covering the Mediterranean seaboard, northern Egypt and parts of Western Europe, with relatively few inhabitants by comparison with Queen Victoria's vast and populous domains).

Curzon's energy in creating the legal, financial and physical infrastructures through which civic society could prosper was generally recognised as the work of 'a superior person'. Yet his very success bred almost universal resentment. Field-Marshal Lord Kitchener, the Commander-in-Chief of the British Army in India, possessed an ego fully equal to his, and even a sub-continent proved too small for both men. There were rows stirred up by Lord Salisbury's nephew and successor as prime minister, Arthur Balfour, as well as by supporters of Kitchener back home, during one of which Curzon ill-temperedly resigned. He held high rank thereafter – in the Cabinet during the Great War and as foreign secretary from 1919 to 1924 – but when his chance came for the premiership in 1923, King George v chose Stanley Baldwin instead.

Curzon burst into tears and was predictably dismissive of his rival: 'Not even a public figure. A man of no experience. And of the utmost insignificance.' Two years later, aged only sixty-six, he seems simply to have lost the will to live.

Curzon's reputation as the greatest of all the British Empire's proconsuls rests securely upon the six years of his viceroyalty of India, with its financial and currency reforms, his work for Indian historical monuments and, above all, his reconciling of colonial peoples to the enlightened despotism that characterised British rule in the subcontinent. He was a difficult man for colleagues to warm to – one contemporary described his speeches in the House of Commons as like 'a divinity addressing black beetles' – but an easy one to admire.

The other great imperial administrator of the period was Sir Evelyn Baring, 1st Earl of Cromer, who ruled Egypt from almost the moment he stepped ashore at Alexandria as the British agent, consul-general and plenipotentiary in September 1883 until his resignation nearly a quarter of a century later, in 1907. Yet despite all the multifarious benefits he bestowed during his time there, he is cordially loathed in Egypt today; as recently as 1998 a group of Egyptian students asked the local archivist in the small Norfolk town of Cromer whereabouts he was buried, so that they could go to spit on his grave.

Cromer was a big enough man to take such posthumous unpopularity in his stride. As Kipling wrote in *The White Man's Burden*, the reward of spending a lifetime bringing peace and prosperity to the late-Victorian Empire was merely 'The blame of those ye better/The hate of those ye guard.' Nor did what his biographer describes as Cromer's 'bulky, imperious, overly self-confident' character, let alone his reputation for brusqueness, help to make him loved. Yet there was a privately delightful personality behind the man who in public always looked 'as though he was modelling for his own statue'. What is often forgotten about men like Curzon and Cromer was that they were always intensely conscious of the importance of bearing themselves with almost exaggerated dignity in public, because they represented the Crown and the Empire for millions of subjects, who expected it of them and would despise anything less. The youthful Cromer might have been a 'hedonistic, spendthrift young army officer', but by the age of forty-two he was ready for huge responsibilities.

Throughout the Roman Empire, Egypt was looked upon solely as a means by which proconsuls taxed the peasantry to the utmost, prior to returning to Rome rich enough to retire in luxury or to pursue further political ends. It was the ultimate imperial cash-cow, milked to exhaustion for centuries. Under Britain, by contrast, men like Cromer gave their lives to the country, returning to Britain no richer than when they went out. The fact that the Roman *imperium* lasted in Egypt for 650 years, from the battle of Actium in 31 BC until the Persian invasion of 619 AD, while Britain's lasted only the seventy-two years between 1882 and 1954 only underlines the first law of modern imperialism: that no good deed goes unpunished.

The modern echoes of one of Cromer's central achievements – keeping Egypt free from Islamic fundamentalism – ought not to be lost on today's decision-makers among the English-speaking peoples. Cromer was constantly exploring ways to undermine the attraction for Egyptians of what he termed 'nationalist demagogues and religious fanatics', about as good a shorthand description for Saddam Hussein and Osama bin Laden respectively as one is likely to receive from the 1880s. 'The political regeneration of Mohammedism' was kept at bay by ceaseless progressive projects in the fields of irrigation, education, taxation and fiscal practices, as well as by acute military Intelligence. Cromer also tended to favour the interests of the Egyptian *fellahin* rural peasantry over European holders of Egyptian bonds, much to the bond-holders' ire.

Part of Cromer's problem lay in the fact that he was always, as his biographer puts it, 'the unofficial ruler of a country of ambiguous status – part Ottoman, part [British] colony, part independent nation with imperial ambitions of its own', and so he had to step warily, at least in his early days there, so as not to allow any combination of powerful interests to threaten British *de facto* rule on the Nile. Lord Salisbury believed that Cromer succeeded because of 'the natural superiority which a good Englishman in such a position is pretty sure to show'. Perhaps a more modern view is that he was tough-minded, had a strong physical constitution, could be a wily political operator, was a workaholic, had an instinctive feel for balance sheets and profit-and-loss accounts, and vigorously opposed Westminster interference with 'the man on the spot'. He also had British regiments stationed around Egypt ready to crush nascent nationalist movements, such as the attempted Abbas insurrection of 1894.

The story of Cromer's near-quarter-century proconsulship in Egypt was one of constant striving to improve the lot of the rural Egyptian, if not necessarily also the urban factory worker. He was a Whig in politics, but one who believed in low taxes and balanced budgets, and wiping out the appalling financial mismanagement and corruption of many of the khedives. The complicated constitutional penumbra in which Cromer was forced to operate – Egypt officially belonged to the Sultan of Turkey and was never formally part of the British Empire – meant that he must be placed 'in a category of one, somewhere between longstanding viceroy, a provincial governor, an international banker, and an ambassador, and yet with a different relationship to those he governed than any of these'.[8] He imposed his will by constantly outmanoeuvring French investors, Cairo journalists, British Radical politicians, Egyptian premiers, gung-ho generals and Turkish sultans, and one hopes that back in 1998 those Egyptian students were misdirected by the Norfolk archivist.

The progressive transfer of power on to Egyptians after Cromer left Egypt led to an outbreak of political violence there, whose highlights were the shooting in February 1910 of the Egyptian Prime Minister; the attempted assassination in 1915 of the new head of state, Sultan Husayn Kamil; a similar attempt on the life of the Prime Minister, Abd al-Khaliq Tharwat, in 1933; and the murder of Sir Lee Stack, Governor of the Sudan, in 1924. (It fell to a schoolteacher called Najib al-Hilbawi to try to kill Sultan Husayn Kamil, by dropping a bomb from a hired upper room on to his procession as it left the palace in Alexandria. He lit the fuse with his cigarette and threw the bomb, which failed to explode. Although he escaped, he left his ashtray containing the ends of cigarettes that he had smoked waiting for the procession to begin, and his tobacconist identified them as his. The tobacconist was certain because he had cheated his customer, using a cheaper variety of tobacco than Hilbawi had ordered.)[9]

In December 1905, the retiring Chief of Staff of the German Army, General Alfred von Schlieffen, drew up a memorandum for his successor General Helmuth von Moltke – known to history as the Schlieffen Plan – that made some suggestions about how Germany might crush France in any future conflict. Put simply, it envisaged the German Army executing a massive right-flanking movement through Belgium to sweep down between the French Armies and the Channel and envelop western France. While the left flank stayed on the defensive, the flanking manoeuvre would result in a decisive battle being fought to the east of Paris, which would result in the capture of the French capital. After the destruction of France as a great power, the German Army could be transferred east to fight the Russians.[10]

Although Germany did invade France via Belgium in 1914, it did so in a very different way, and scholars have conclusively proved that the Schlieffen Plan was not really a plan at all in the conventional military sense. Indeed, one has recently convincingly argued that the Plan did not actually exist.[11] The sketchiness of Schlieffen's memorandum, and the fact that it was followed only in its single main essential – the path through Belgium – does not, however, absolve Germany of war guilt, since in 1911 Moltke reviewed Schlieffen's memorandum and indicated in his notes that Belgium should indeed be the future invasion route. This was to be as politically suicidal as it was strategically sound, for it drew Germany into a war it was hard to win, against the British Empire as well as France, one for which British and French staff officers had been planning their responses since the *Entente Cordiale* of 1904. Although he cannot be blamed for failing to predict it, the whole English-speaking peoples were united in fighting Germany within six years of Moltke's fateful decision.

As Chief of Staff, Moltke was not helped by Kaiser Wilhelm II, who made several erratic interventions in the plans in the belief that he was an inspired strategist. German generals despaired when the Army's annual military manoeuvres were turned into a farce by the Supreme War Lord insisting on having plenty of cavalry charges, although he was reluctant to stage them in the rain.[12]

Only a few months after the Schlieffen Plan was drawn up, on Saturday, 10 February 1906, King Edward VII, resplendent in his full-dress uniform as an admiral of the fleet, launched the British battleship HMS *Dreadnought* at Portsmouth. Suddenly the world changed almost as much as it had when the Wright brothers had taken to the air twenty-seven months earlier. *Dreadnought*'s ten 12-inch and twenty-four 12-pounder guns made the rest of the world's fighting ships obsolete overnight, and the race was on to see whether Germany or Britain could build more of them. On 5 June, the Third German Naval Bill provided for large increases in the numbers and size of battleships. The technological race soon became so rapid that by 1913 the *Dreadnought* was herself no longer regarded as a ship of the first line.[13]

In 1903, the relative naval strengths of the Great Powers had been, in terms of numbers of battleships in service: Great Britain 67, France 39, United States 27, Germany 27, Italy 18, Russia 18 and Japan 5.[14] Suddenly those figures meant next to nothing as dreadnoughts alone formed the measure of naval greatness. British defence estimates for the year 1910 amounted to £68 million, more than Germany at £64 million, Russia at £63 million, France at £52 million, Italy at £24 million and Austria-Hungary at £17 million. Yet after 1910, while Britain's spending stayed static, Germany's expanded exponentially, with potentially disastrous consequences.

By the outbreak of the Great War, Britain had slipped badly behind Germany in defence spending, although the Royal Navy's Grand Fleet was still superior to the High Seas Fleet in size and quality. That year Germany's defence estimates stood at £110.8 million, Russia's at £88.2 million, Britain's at £68 million, France's at £57.4 million, Austria-Hungary's at £36.4 million and Italy's at £28.2 million. In the all-important number of dreadnoughts, however, Britain had 19, Germany 13, the United States 8, France, Russia and Italy 6 each and Japan 3. Of course that did not mean that German U-boats could not make major incursions on British merchant shipping, of which the British Empire could boast 21 million tons in 1914, in contrast to Germany's 5.5 million, the United States' 5.4 million and France's 2.3 million. Britain also had far greater foreign investments upon which she could count as the conflict progressed, totalling £3.6 billion in 1914 against Germany's £1.08 billion and France's £1.74 billion.

These figures comprehensively undermine the accusations, made both at the time and subsequently, that it was the British rather than the German Empire that actively sought the conflict. (HMS *Dreadnought* herself became the flagship of the Home Fleet from 1907 until 1912 and remained part of that fleet thereafter. She then served with the 4th battle squadron in the North Sea during the first two years of the Great War and on 18 March 1915 she rammed and sank the German submarine *U-29*. Placed on reserve in 1919, she was finally sold for scrap in 1922.)

At the time of the Silver Jubilee Review of the Fleet at Spithead in 1935, Churchill proudly pointed out to the Liberal MP Robert Bernays all the vessels that he had been responsible for commissioning at the Admiralty before the Great War, representing all but three of the fifteen capital vessels present.[15] (No fewer than 160 Royal Navy ships were present on that occasion; seventy years later, the Royal Navy could only muster twenty-one vessels for the bicentenary celebrations for the battle of Trafalgar.)

For all the British naval superiority before the Great War, on land she was still only an insignificant military power, with tiny numbers in her armed forces compared with her global rivals. With no history of invasion, at least for several centuries, and thus with no culture of conscription, her entire Empire only had a maximum of 800,000 men under arms in 1912, over half of whom were colonial forces. Russia, meanwhile, could boast an army of 5.5 million, Germany 4.1 million, France 3.9 million, Austria-Hungary 2.3 million and Italy 1.2 million. Only the United States Army, numbering 100,000 men, was smaller than the British Army. The English-speaking peoples clearly cannot be accused of pursuing an aggressive policy in the immediate pre-war period; all they hoped for was to protect the status quo.

At 5.12 a.m. on Wednesday, 18 April 1906, a massive earthquake hit San Francisco. A section of the San Andreas Fault had 'dislodged itself by several metres, triggering a minute-long earthquake'.[16] The worst in American history, it is estimated to have measured between 7.9 and 8.3 on the yet-to-be-invented Richter scale. Wooden buildings that had been thrown up hastily in previous decades – such as during the 1849 Gold Rush – collapsed, especially in the working-class South of Market community, causing firestorms that went on to ravage two-thirds of the city over three days. The destruction was huge: 28,000 buildings were destroyed over 2,600 acres, causing $400,000 worth of damage. The loss of life was immense: over 3,000 people were killed, 9,000 injured and up to a quarter of a million left homeless. Newsreels of the day show smashed buildings reminiscent of Ground Zero, and men and women forming orderly lines waiting for free meals being distributed from tents.

The city of San Francisco and then the state of California were the primary and secondary organisers of relief, and did the job well, and the War Secretary William Taft acted with commendable efficiency, federalising the National

Guard by 7 a.m. on the morning of the disaster, thus allowing them to report to the city's mayor. The police force responded magnificently and there was virtually no rioting or looting.

The San Andreas Fault, a 750-mile-long gap between the North Atlantic tectonic plate (which is stationary) and the Pacific plate (which is moving north), makes San Francisco a fantastically dangerous place to live over the very long term. Just as New Orleans is a city twenty feet below sea level, protected from the ocean and a huge lake by levees in a part of the world where hurricanes strike regularly, so too was San Francisco a natural catastrophe waiting to happen. As one recent historian of the 1906 earthquake describes it, the Fault is 'a living, breathing, ever-evolving giant that slumbers lightly under the surface of the earth'.[17] In a sense it is a tribute to Californians' optimism that they choose to live somewhere that 'the Big One', as they call it, could happen at any time.

The vital necessity to New Zealand of foreign trade in the early part of the century is demonstrated by the fact that although the foreign trade of the United States, which had a population of 82.9 million, was only £7 per capita in 1904, in New Zealand, whose population was only 845,000, it was £33. Britain, with 44 million inhabitants, did nearly £21 of foreign trade per capita, which was far more than Germany (£10), France (£9) and certainly Italy (£4). By contrast Australia, with a population of a little over 4 million, enjoyed trade of over £29 per capita; Canada, with 5.41 million, had £17; but Russia, with 143 million, had only a little over £1. (Australia contained huge differentials; Western Australia, for example, had a population of 236,500 and trade per capita of £71/12/10, about as high as anywhere in the world, whereas Tasmania, with a population of 178,826, only enjoyed £31 per capita foreign trade.)[18] Natural traders by history, geography, inclination and necessity, the English-speaking peoples led in terms of international trade and were to continue to do so throughout the coming century. As the United States' vast internal market matured, she too moved up the global tables.

The importance to New Zealand of her exports was starkly outlined by a paper read to the Auckland Institute by the mathematician Professor H.W. Segar on 21 October 1907. Entitled *The Struggle for Foreign Trade*, Segar delivered a wide-ranging analysis of the major issues facing world commerce, including the likely effects of the coming canal through Panama, the rise in the population of Germany of over four million between 1900 and 1905, the emergence of Japan as a Great Power and even the abolition of Chinese footbinding. 'And with German trade will grow naturally and inevitably the German navy' was one acute warning, whereas he was a full century out when he said that 'China has been going to awaken for fifty years past, but it would appear that at last we are now in the presence of the realisation.'[19]

In his analysis of the opportunities and dangers of the future, Segar urged New Zealanders to concentrate on making their agriculture more intensive and productive, extending the cultivation of land and limiting population growth, otherwise it would 'lead to a rapid approximation in the condition of her workers to that of the old countries'. It was wise advice, very largely followed, although New Zealand's embrace of manufacturing also allowed her to increase her population from 885,000 in 1904 to 4.1 million by 2005.

On 12 April 1908, Winston Churchill entered the British Cabinet for the first time, as President of the Board of Trade. As a result, a member of the Beefsteak Club in London, Mr H.A. Newton, paid Mr W.G. Elliot £2, since back in February 1903 he had wagered that Churchill would not achieve that ambition within ten years. Whereas the expression of political views is cheap, those opinions that are backed by hard cash deserve more notice, even if club wagers are usually the result of post-prandial disagreements. For decades, wagers were registered in the Beefsteak's betting book by club members such as Harold Macmillan, Alfred Duff Cooper, Prince Francis of Teck, the 11th Duke of Devonshire and scores of others, and it thus provides an interesting social and political commentary on those issues on which members were willing to put their money where their mouths were. 'H[is] E[xcellency] Count Benckendorff lays Mr [Maurice] Baring £50 to £1 that the Pope does not drive through the streets of Rome within a month of his election' was a typical bet. (Baring paid up in August 1903.)

The issues over which Beefsteak members wagered during the century included Shakespeare quotations, the outcome of the Russo-Japanese War, golf championships, bloodstock pedigrees, female suffrage, whether the murderer Dr Crippen would be caught ('If he commits suicide before capture the bet is to be off'), the Varsity boat race, the distance between the Athenæum and Reform Clubs, the outbreak and course of various Balkan wars, whether Sir Roger Casement would be hanged, the date of the next Armenian massacre in Constantinople, the longevity of various French Republics, Irish Home Rule, legal cases (particularly capital ones), the origin of the term 'the weasel' in the nursery rhyme about twopenny rice, characters from Dickens, consols futures prices, Irish newspaper circulations, The Ashes test matches, whether compulsory military service would be introduced during the Great War, future tax rates, the height of the secretary of state for war, by-elections and general elections, which was the princely House of Svanetia, whether a certain (unnamed) club member would be imprisoned, whether air marshals' full-dress uniform included spurs, Gordon Brown's love life, whether more hardback books have been published about Beethoven or cricket, and whether Queen Elizabeth I had a water closet installed at Hampton Court.

Bets – usually of champagne, vintage port, 'a good lunch' or fairly small amounts of money, but occasionally up to £1,000 – were also laid on who would be the next Pope, Warden of All Souls, Commander-in-Chief, Archbishop of Canterbury, Chancellor of Oxford and headmaster of Eton.[20]

Over at Brooks's, the Whig club in St James's, July 1908 saw a ten-guinea wager between Churchill's private secretary, Edward Marsh, and Henry Somerset 'that there would not be a war between any two great European powers within twenty years', which must in the seemingly settled state of the world have looked like a reasonable bet, however horrifically it was mocked only six years later. (Somerset went on to be mentioned twice in despatches.) Brooks's members tended to wager more than at the Beefsteak, but on similarly recondite issues, including the profit of the 1910 Liverpool Agricultural Show, whose house at Eton the maître d'hotel of the Royal Spithead Hotel attended, whether Chou En-lai would appear at the official saluting base at Peking's 1968 May Day parade, whether Black Velvet was drunk before the Franco-Prussian War, whether 'a certain lady was a Roman Catholic', and in May 1941 'His Grace the Duke of St Albans bets Sir Mark Grant Sturgis £5 that Sir [Jock] Delves Broughton will be hanged for the murder of the Earl of Erroll'. (He wasn't, but did commit suicide in the Adelphi Hotel in Liverpool in December 1942.) Perhaps the strangest wager was Mr Oliver Knox's bet against Mr Roger Lubbock of a case of 1949 Pol Roger against a bottle, 'that on 1 Feb 1960 Mr Lubbock will not be prepared to exchange his eldest child (his wife notwithstanding) for a Rolls-Royce or the most desirable car on the market'. (Lubbock paid up.)

In 1900, only eighteen countries had a central bank; by 2005, the number was 174.[21] The oldest continuing central bank is the Bank of England, founded in 1694, which, since the Swedish Riksbank (founded 1668) did not take on central bank functions until considerably later, therefore allows the English-speaking peoples to claim the invention of yet another of the key concepts of the global economy.

It was the banking crisis of 1907 that persuaded Congress that the United States needed a central bank. The panic that threatened the American banking system for a few nail-biting days in late October 1907 was ended by a powerful Wall Street consortium headed by J.P. Morgan which stabilised the situation before financial meltdown could occur. Nonetheless, afterwards politicians of both parties agreed that fundamental reform was necessary, not least because the 1907 panic had been the fifth such serious crisis since 1873.

Paradoxically enough, one of the effects of the 1907 banking crisis was to prove the importance of the US economy, even when it was showing weakness. The ramifications of the event outside America were such that no-one could be in any doubt as to America's new power. Thus even in her weakness she advertised her strength. The National Monetary Commission was set up by Congress in 1908 to report on 'what changes are necessary or desirable in the monetary system of the United States' in order to prevent crises like the 1907 near-disaster from recurring. It wound up recommending 'essentially an American system, scientific in its methods, and democratic in its control', but not before producing no fewer than twenty-two incredibly detailed volumes reporting on other monetary and banking systems around the world.

The hard-working Commission decided not to adopt the Bank of England's essentially evolutionary model, despite commending 'the wisdom of the men who have controlled its operations', but instead preferred to institute 'legislative enactments' in order to set up the new Federal Reserve Bank. On 23 December 1913, therefore, Congress passed the Glass-Owen Currency Act, known as the Federal Reserve Bank Act, which established a Board with power over monetary policy and the twelve district Federal Reserve Banks, thus creating the first US national central banking system since the 1830s. The very considerable weaknesses inherent in the Federal Reserve System were not to become apparent until the Wall Street Crash of October 1929.

(The power of the Governor of the Bank of England in British society was ably demonstrated by Lord Cunliffe, who held the post between 1913 and 1918. Reluctantly giving evidence before a royal commission at the special request of the Chancellor of the Exchequer, Cunliffe was asked what were the reserves of the Bank. He answered that they were 'very, very considerable'. When pressed to give an even approximate figure, he replied that he would be 'very, very reluctant to add to what he had said'. Thereupon questioning ceased.)[22]

George Washington's Farewell Address of 1796, in which the first president had warned against long-term entangling alliances, had long been quoted as a warning against America founding her own empire along the lines of those of the Old World. Yet a close analysis of Washington's etymology reveals that he did not shy away from describing the United States as an empire on occasion. In 1783, he called them a 'new empire' and a 'rising empire', and in 1786 he wrote that, 'However unimportant America may be considered at present ... there will assuredly come a day when this country will have some weight in the scale of empires.'[23] The Address was read out in its entirety in Congress every February until the mid-1970s, when fortunately the tradition lapsed. By then Washington's description of America – 'With slight shades of difference, you have the same Religion, Manners, Habits and Political Principles' – was as obsolete as his anti-internationalist message.

The actual day 'when this country will have some weight in the scale of empires' dawned on Monday, 22 February 1909, when Theodore Roosevelt visited Hampton Roads, Virginia, to witness the return of the Great White Fleet after a fourteen-month, 45,000-mile circumnavigation of the world. On board the presidential yacht *Mayflower*, Roosevelt watched seven miles of bright white ships – they were painted battle-grey soon afterwards – simultaneously fire him a twenty-one-gun salute. 'We have definitely taken our place among the world Great Powers,' he said, and he was right. The plumes of smoke from the tall funnels of the sixteen battleships signalled to the globe that America had arrived. The Fleet's commander, Rear-Admiral Robley D. Evans, said his ships were ready for 'a feast, a frolic or a fight'.

The places that the Fleet had visited subtly underlined this important new factor of global geopolitics. Leaving Hampton Roads on 16 December 1907 – waved goodbye by Roosevelt with the words, 'Did you ever see such a fleet and such a day?' – the ships had steamed to the Caribbean, past the new possessions of Cuba and Puerto Rico, then down the east coast and up the west coast of South America, protected as never before by the Monroe Doctrine. Each country of the Latin American portion of the world cruise at which the Fleet stopped, including Brazil, Argentina, Chile, Peru and Mexico, could have been under no illusions about what this massive new force meant. When the Spanish-American War broke out in April 1898, the American Navy was tiny by Great Power standards, consisting of only 4 battleships, 7 small battleships, 19 cruisers and 13 torpedo-boats.[24] A decade later, she had the Great White Fleet, which, unlike the Turkish pre-war Ottoman Fleet, the German High Seas Fleet and the Soviet Russian Fleet, was not destined to become a great white elephant.

After Mexico, the Fleet visited Hawaii (annexed to the US in July 1898), New Zealand and Australia, China, the Philippines and Japan. It then sailed across the Indian Ocean, through the Suez Canal and the Mediterranean, and then across the Atlantic. As an historian of America's explosion onto the world scene records, 'The cruise not only impressed the world with America's new-found military strength but excited the imagination of Americans as well. A million people had turned out in San Francisco to welcome the ships before their voyage across the Pacific.'[25]

In the Tsushima Straits on 27 May 1904, the fleet of Japan, which was not even a Great Power by the standards of the day, had put all eight battleships of the Russian Imperial Fleet out of action in the course of a battle lasting only three-quarters of an hour. The implications of this victory had reverberated around every admiralty and chancellery of the world, and had already led to the 1905 Revolution in Russia. Roosevelt had mediated the Treaty of Portsmouth between Japan and Russia the following September, and was fully cognisant of Japan's dramatic emergence on the world scene as a naval power. It was partly to impress Japan with America's new-found naval power that the Great White Fleet had set sail. By the time it returned, no sentient strategist could doubt that – even before the Panama Canal opened in 1914 – the United States had a two-ocean navy that gave her the right to take her new place amongst the Great Powers of the earth.

The successful thirty-six-minute flight of the French engineer Louis Blériot across the English Channel from Les Barraques near Calais to Dover Castle on Sunday, 25 July 1909, had profound implications for Britain. As the immodestly named aircraft *Blériot XI* landed to claim the *Daily Mail*'s £1,000 prize for the feat, H.G. Wells peered into the future and warned: 'Never was a slacking, dull people so liberally served with warnings of what is in store for them. This is no longer an inaccessible island.' No longer isolated, splendidly or otherwise, the British had to put behind them political concepts that had served them well between the Congress of Berlin in 1878 and the Anglo-French *Entente* of 1904.

Henceforth, British cities were a mere thirty-six-minute bombing flight from the Channel and whichever hegemonic power controlled the Atlantic's eastern seaboard. By 1935, Winston Churchill was so conscious of the destructive power of the bomber that he spoke to the Liberal MP Robert Bernays about how he would like to abolish all civil and military aircraft, saying, 'I would make it as punishable to own an aeroplane as to commit an unnatural sexual offence.' The Atlantic itself being over 100 times wider than the Channel, the United States took rather longer to learn the impossibility of splendid isolation in the twentieth century.

One of the ways in which Britain hoped to keep abreast of any prospective enemy's power in the air was through the hugely increased use of espionage, and in setting up modern intelligence services in 1909 she chose her top spymaster well, in the shape of Sir Mansfield Cumming. His toughness was demonstrated five years later, at 9 p.m. on 2 October 1914, when his son Alistair Cumming, a twenty-four-year-old subaltern in the Seaforth Highlanders, was driving him in a Rolls-Royce 'somewhere in France'. For reasons unknown the car crashed into a tree and overturned, crushing the father's leg and trapping him, while mortally injuring the son. Hearing Alistair slowly dying, Sir Mansfield cut the last strips of skin off his own leg with his pocket knife and crawled over to his boy to cover him with his coat. Nine hours later he was found unconscious beside his son's corpse. In his diary he recorded laconically, 'Poor old Ally died.' Rarely were imperial upper lips stiffer than Sir Mansfield Cumming's.

Cumming was a fifty-year-old, semi-retired naval commander living on a Southampton houseboat when the call came from the Admiralty in August 1909 to set up the progenitor of both MI5 and MI6. With a tiny staff on a shoestring budget, and despite constant inter-departmental squabbles with the Foreign Office, the War Office and the Admiralty, he had built up an impressive organisation by the time the Great War broke out five years later. By 1915 he was running no fewer than 1,024 agents and scoring a number of significant coups, not least the discovery – through a Belgian woman's use of the sexual 'honeytrap' – of the whole German spy network in Paris.

The terror of sudden secret invasion which gripped pre-Great War Britain – as exemplified in Erskine Childers' classic thriller *The Riddle of the Sands* – was largely just irrational xenophobia. (At one point Parliament had to be reassured that there were not 66,000 German troops in disguise in London, ready to collect their arms from a depot near Charing Cross.) But the scares did at least lead to the creation of a service which, for all its amateurishness and occasional absurdities, did Britain proud. Cumming was able to provide the armed services and the relevant government departments with information about the Central Powers' troop movements, morale, strategy and trade during the four years of conflict. When the Germans managed to break the trade embargo on condoms by trading through the Swiss, for example, Cumming soon found out about it.

Cumming's use of people from all social backgrounds – unlike some other spook chiefs of the day, he was no snob – was part of his success. Amongst their number was Colonel Joe 'Klondyke' Boyle, a Canadian who ran away to sea, became amateur heavyweight boxing champion of America, made a million dollars in the Gold Rush and equipped a Canadian machine-gun detachment out of his own pocket. (He might not sound like a natural under-cover agent, but Cumming employed him nonetheless.) When Boyle first met Cumming's agent in Russia, George Hill, Hill had just used a swordstick on an enemy agent, noting how the weapon left 'only a slight film of blood halfway up the blade and a dark stain at the tip'. The story of the early MI6 is gratifyingly full of murders, disguises, squeaking hinges, codebooks, green ink and agents willing to commit suicide sooner than be captured.[26] Rather like the fictional character 'Q' in the James Bond novels, Cumming delighted in gadgetry and was on the look-out for new inventions, or, as in the following case, a new use for a very old one. Frank Stagg, who was seconded to Cumming from the Admiralty in 1915, years later recalled how,

> Secret inks were our stock-in-trade – and all were anxious to obtain some which came from a natural source of supply. I shall never forget 'C''s delight when the chief censor, Worthington, came one day with the announcement that one of his staff had found out that semen would not respond to iodine vapour [commonly used for developing secret writing], and told the Old Man that he had to remove the discoverer from the office immediately, as his colleagues were making life intolerable by accusations of masturbation.[27]

(One hopes the children of the official in question didn't ask him the famous question: 'What did you do in the Great War, Daddy?')

★

The British Chancellor of the Exchequer, David Lloyd George, had not involved himself much in foreign affairs before Friday, 21 July 1911, when he spoke at the Mansion House in the City of London. The highly provocative arrival of a German gunboat, the *Panther*, in the Moroccan port of Agadir three weeks earlier had ignited Franco-German antipathy, because Morocco was considered within the French zone of influence, and a statement on the situation was felt to be needed from the British Government. Britain, said Lloyd George, could not be 'treated where her interests were vitally affected as if she were of no account', the implication being that she would support France, who had interests in North Africa, against Germany, which did not. Prestige, not immediate interests or even direct threat, was the key factor, one that his listeners recognised as naturally as Lloyd George.

Within six months, British prestige was brought to perhaps its highest pitch. The proudest day of the long story of the English-speaking peoples' mission in India fell on Tuesday, 12 December 1911, when King George V and Queen Mary held the spectacular Delhi Durbar. All the princes of India paid fealty to their sovereign in a scene of unimaginable splendour, martial show and haughty pageantry. 'Enthroned on high beneath a golden dome,' enthused *The Times*, 'their Majesties the King-Emperor and Queen-Empress were acclaimed by over one hundred thousand of their subjects. The ceremony at its culminating moment exactly defined the Oriental conception of the ultimate repository of imperial power.'

Their Majesties were in a very different humour when on Wednesday, 4 June 1913, Emily Wilding Davison, a forty-one-year-old English literature graduate and veteran suffragette from Longhorsley, near Morpeth in Northumberland, committed suicide by flinging herself in front of the King's bay colt, the 50-1 outsider Anmer, as it rounded the Tattenham Corner in the Derby horse-race at Epsom. 'The horse struck the woman with its chest,' recorded the next day's *Daily Mirror*, 'knocking her down among the flying hoofs ... and she was desperately injured. ... Blood running from her mouth and nose.' An obsessed single-issue fanatic, Davison seemed not to care whether she injured the innocent jockey, Herbert 'Diamond' Jones, who was flung violently to the ground when she grabbed the reins. 'Anmer turned a complete somersault and fell upon his jockey,' reported the *Mirror*. Bleeding badly, both people were taken to hospital with their injuries, which Jones fortunately survived with only concussion, a fractured rib, cuts and bruises. (Anmer finished the race with bruised shins.)

Today, twenty feet of silver nitrate newsreel footage show Miss Davison rushing out in front of the horse and the terrible impact when it hit her. She suffered a fractured skull and never regained consciousness, dying four days later at Epsom Cottage Hospital, her bed hung with the purple, green and white bunting of the suffragette movement. She was given a heroine's funeral by the Women's Social and Political Union; 6,000 women attended her funeral in Bloomsbury, with ten bands and twelve clergymen accompanying her coffin from Victoria Station to King's Cross. References were made to the 'noble sacrifice' made by 'a fallen warrior and crusader'.

Queen Mary was less impressed, sending Jones a telegram commiserating with him upon his 'sad accident caused through the abominable conduct of a brutal lunatic woman'.[28] Davison's mother was equally unsympathetic: 'I cannot believe that you could have done such a dreadful act,' she wrote in a letter the dying woman was never to read. As for genuine 'fallen warriors' who made truly 'noble sacrifices', Herbert Jones' three brothers Reggie, Percy and Jack were all killed fighting on the Western Front in the Great War. He himself later committed suicide, 'while the balance of his mind was disturbed'.

It is hard to escape the conclusion – despite the fact that she had bought a return ticket from Victoria Station to Epsom Downs – that martyrdom was precisely what Emily Davison had sought. She had attempted suicide twice the previous year in Strangeways prison on the same day, once when she flung herself thirty feet down a wrought-iron staircase. She had earlier thrown iron balls labelled 'bomb' through windows and had been imprisoned for constantly setting fire to pillar-boxes and for throwing stones at Lloyd George, which had the text 'Rebellion against tyrants is obedience to God' written on paper wrapped around them. When in prison, she had barricaded herself into her prison cell and was only flushed out by the use of water cannon. 'Davison might not have shared the religious conviction of today's suicide bombers,' commented the *New Statesman* on the ninety-second anniversary of her stunt, 'but she had an equal disregard for her individual status within the struggle.'[29] In fact, her status was assured by her suicide, just as the suicide bombers' is by theirs in the warped hierarchy of Islamic 'martyrdom'.

The hunger strike, in which Emily Davison was greatly experienced, was to become a powerful weapon in the English-speaking world in the twentieth century. Before taking up hunger-striking, the suffragettes had padlocked themselves to railings, invaded No. 10 Downing Street, cut telegraph wires, committed arson, disrupted political meetings, slashed paintings in art galleries, burned pillar-boxes, vandalised Kew Gardens' orchid-house and harangued the Houses of Parliament through a loudhailer from a small boat on the Thames; to little avail. Yet when in June 1909 the suffragettes went on hunger strike and were force-fed using a method hitherto only employed in lunatic asylums, the conscience and chivalry of Edwardian society revolted.

The Liberal Home Secretary Reginald McKenna passed what was nick-named the 'Cat and Mouse Act', which allowed him to release hunger-strikers so that they did not die in government hands and to re-arrest them afterwards at will. McKenna recognised the enormous public-relations coup that a death in custody would have for the suffragette cause. Force-feeding was used along with sudden release and subsequent re-arrest, hence the Act's nickname. Yet even hunger-striking did not gain the suffragettes what was later described in another context as 'the oxygen of publicity' so successfully as the suicide of Emily Davison.

The suffragette movement, indeed the feminist movement in general, has strengthened the English-speaking peoples in a way that only the most patriotic of its original leaders could have intended. By allowing, in fact by encouraging, the mass recruitment of a female workforce into the economy – initially during the First World War – they effected a revolution that has massively assisted the development of Western capitalism. The eruption of a potential upper figure of 50% of the population into remunerative work has unleashed enormous human potential, with the concomitant advantage of keeping wages low relative to that which they might have been without the widespread hiring of female labour.

The way that the English-speaking peoples led the world in embracing female suffrage is a sign of its political maturity and liberalism, but also of its enlightened self-interest. Pitcairn Island gave women the vote in 1838; Britain extended it to unmarried women in local elections in 1869, the same year that Wyoming adopted equal suffrage at state level. Canada granted the right in local elections in 1883. New Zealand became the first country in the world to allow women to vote in national elections on 19 September 1893, followed by Australia in 1902. Only after the English-speaking peoples experimented successfully with female franchise, progressively dropping property and age restrictions and extending it to married women, did the rest of the world begin to adopt it. First was Scandinavia, with Finland, Norway, Denmark and Iceland going down that route between 1906 and 1915. (Sweden had allowed unmarried women to vote in local elections since 1869, but did not institute general female suffrage until 1921.)

After the Great War, between 1918 and 1920, Britain, Ireland, Canada and the United States all adopted female suffrage for national elections, although not all on precisely the same age-basis as for men. The Representation of the People Act passed the British Parliament easily in 1918, thereby losing Lord Howick his £1 bet against Lord Osborne Beauclerk at Brooks's that women would not be enfranchised before 1920. Quebec was the last Canadian province to enact female suffrage, in 1940. It was some time before other major countries followed down the English-speaking and Scandinavian trail.

Turkey, Portugal, Spain and India instituted female suffrage in the 1930s, the last because of British rule. Bulgaria and various South American states adopted it during the Second World War, but it was not until after the war that France, Hungary, Italy, Japan (with restrictions), Romania and Yugoslavia felt able to trust their female populations with the vote. The remainder of the 1940s saw much of the rest of the non-Arab world adopt it, especially after the Universal Declaration of Human Rights included a commitment to it in Article 21. Yet still it was not until the 1950s, a full half-century after the Australasian trail-blazers, that countries like Greece, Mexico, Colombia, Egypt, Pakistan and Tanzania adopted it on a federal footing. The 1960s saw Cyprus, Paraguay, much of the rest of Africa, Afghanistan (until the Taliban), Ecuador and Yemen take the plunge. Switzerland waited until 1971. Countries such as Samoa and Kazakhstan did not enfranchise women until the 1990s, and in 2005 restricted or blatantly unequal suffrage still existed in Bhutan, Kuwait, Lebanon, Oman and Saudi Arabia.

Partly as a result of the early enfranchisement of women in the English-speaking world, and the realisation that in their voting intentions they proved if anything a conservative rather than radical – let alone revolutionary – electoral force, other opportunities opened up. 'German universities had ceased to be all-male preserves only in 1900,' notes one historian, 'twenty years later than their Anglo-Saxon counterparts.'[30] The first female undergraduates did not attend the University of Berlin until 1908, whereas they had arrived at Washington University in 1869, Cambridge in 1870, Toronto and Adelaide in 1877, Oxford in 1878, Harvard in 1879 and Brown in 1891.

The net effect of the English-speaking peoples leading the sexual revolution has been that they were able to take advantage of the greater legitimacy that a fully enfranchised female population gives to harness their potential to the economy. Women competing with men has increased productivity, driven down wage-inflation and unleashed creativity. The hugely increased spending power of women in the twentieth century had an incalculable effect on modernising the development, marketing and sales of almost every imaginable brand and product, with infinite advantages for Western capitalism over its African, Asian, Latin American and certainly Arab rivals. Co-opting the female half of the population into the English-speaking peoples' consumer revolution proved a secret weapon of genius.

The Second World War saw women operate as a not-so-secret weapon, too. Although a higher proportion of women worked in private service in Germany during the war than in Britain or the United States, the Nazi regime did not like conscripting either older women or women with children for war service, who were often the most productive. The English-speaking peoples had no such qualms in harnessing them for their war effort.[31] Furthermore, the emphasis on the importance of domesticity in Nazi political philosophy meant that the German *hausfrau* was often allowed to retain her maidservant, in contrast to the experience of the middle classes of the English-speaking peoples during the war.

Between July 1942 and May 1945, the number of working women in the United States increased by no less than 50%.[32] In further contrast to Germany, women were allowed limited military involvement too; on 14 May 1942, the US Congress founded the Women's Auxiliary Army Corps, based on the British equivalent, the Queen Mary's Army Auxiliary Corps of 1917–21. In December 1941, as well as calling up men over the age of eighteen-and-a-half, the National Service Bill rendered single women between the ages of twenty and thirty liable to military service. In Total War, it was the English-speaking peoples who mobilised the entire population far more efficiently than the Axis.

The effect of American economic power was apparent to her leaders before the Great War brought it to the forefront of the world stage. The North Carolinian Walter Hines Page became American Ambassador to London in May 1913, having been successively editor of *Forum*, *Atlantic Monthly* and *World's Work*. That October he wrote to President Wilson in almost hubristic terms, but with great prescience:

> The future of the world belongs to us. A man needs to live here, with two economic eyes in his head, a very little time to be sure of this. Everybody will see it presently. These English are spending their capital, and it is their capital that continues to give them their vast power. Now what are we going to do with the leadership of the world presently when it clearly falls into our hands? And how can we use the English for the *highest* uses of democracy?[33]

Page's analysis was that Britain was on the brink of class warfare, with hatred of Lloyd George endemic among the upper classes with whom he liked to spend his time. He predicted loss of confidence and financial disaster, and, as a result, he told Wilson, 'The great economic tide of the century flows our way. We shall have the big world questions to decide presently.' He appreciated the way that although Britons saw the American Government as foreign, they did not tend to see American people in the same way. He had the membership list of a London club on his desk, 'wherein the members are classified as British, Colonial, American and Foreign – quite unconsciously'.

The longer Page stayed in London, the more Anglophile he became, even before the outbreak of war. Writing in January 1914 to Wilson's amanuensis, Colonel Edward M. House, who was an Anglophobe, he put forward his scheme for 'the tightest sort of alliance, offensive and defensive, between all Britain, colonies and all, and the United States', which would achieve disarmament, arbitration and 'dozens of good things'. This visionary went on: 'As I come to think of it, turning this way and that, there always comes to me just as I am falling asleep this reflection: the English-speaking peoples now rule the world in all essential facts. They alone and Switzerland have permanent free government. In France there's freedom, but for how long? In Germany and Austria – hardly.' After dismissing Scandinavia and the Benelux countries as 'small and exposed', and democracy in South America and Japan as merely developing, he stated, 'Only the British lands and the United States have the most treasure, the best fighters, the most land, the most ships – the future in fact.' The only problem, he felt, other than Irish-Americans such as the New York Senator James O'Gorman, was that, 'We choose to be ruled by an obsolete remark made by George Washington.'[34]

German Anglophobia grew exponentially in the years just prior to the Great War. An exemplar of it can be seen in the title of Oscar Schmitz's 1914 book *The Land Without Music*, in which Britain was depicted as a cultural wasteland. Yet only the year before its publication, Ralph Vaughan Williams had completed *A London Symphony*, George Butterworth his symphonic rhapsody *A Shropshire Lad*, Arnold Bax his orchestral work *The Garden of Fand*, Ivor Gurney his choral settings *Five Elizabethan Songs*, while Gustav Holst, who despite his Swedish name was an Englishman born in Cheltenham, completed his orchestral work *St Paul's Suite*. Hardly a wasteland.

The Kaiser himself, in an interview with the *Daily Telegraph* published on 28 October 1908, claimed to be 'a friend of England, but you make things very difficult for me … the prevailing sentiment among large sections of my own people is not friendly to England'. This was almost a genuine clash of civilisations because, as a modern historian has pointed out, radical German nationalists in the early twentieth century also attacked what they called

> Amerikanismus, or more quaintly, Komfortismus, the Western bourgeois addiction to physical comfort, to security, to money, to individual privacy, to the pursuit, in short, of personal happiness, enshrined in the American constitution. This was contrasted by these same thinkers with German heroism, cultural authenticity, spirit, pure blood and native soil, and above all to the will of every German hero to sacrifice himself to the greater cause of the Fatherland. The First World War was described by these radicals literally as a war against the West.[35]

The German sociologist Werner Sombart, in his 1915 book *Merchants and Heroes*, sought to contrast the racially mongrel merchant nations of Britain, America and France with his pure and heroic Fatherland. Much of the same kind of thinking is evident in modern Islamic fundamentalist views on the West, which is presented as 'soulless, decadent, rationalistic, rootless, money-grubbing and corrupting'.

In reply to the German characterisation of Britons as music-less mercantile cowards, the British came up with the spy-thriller. The 1915 classic of the genre, *The Thirty-Nine Steps*, was written by John Buchan, a Scot who served as Governor-General of Canada from 1935 until his death in 1940. He employed a subtle English-speaking-peoples' theme for the personnel who together foil the diabolical plot of the German espionage organisation the Black Stone, which has discovered the Royal Navy's fleet dispositions. Richard Hannay is a Scot who emigrated to South Africa as a child and fought for the Crown in the Boer War. The second hero is an American, Franklin P. Scudder, who uncovers Berlin's dastardly plan. Written at a time of German spy hysteria, *The Thirty-Nine Steps* sold over 25,000 copies within three weeks of publication.[36] The character of Scudder, whose bravery does not save him from being skewered to the ground by a long-bladed knife at the end of chapter two, did not need to have been an American. Yet Buchan chose Scudder's nationality carefully; defeating Germany was going to be the task of all the English-speaking peoples.

The press on both sides of the Atlantic tended to fuel nationalist feeling. American politics was described by Sir Cecil Spring-Rice as 'dullness, occasionally relieved by rascality'. Yet American journalism, especially the downmarket 'yellow press', was worse. The artist Sir Edward Burne-Jones was delighted by a correction paragraph in one notoriously inaccurate American newspaper that read: 'Instead of being arrested, as we stated, for kicking his wife down a flight of stairs and hurling a kerosene lamp after her, the Revd James P. Wellman died unmarried four years ago.'

Of works attempting to explain the outbreak of the First World War there are any number. They include books written immediately afterwards blaming 'the Hun'; books written a few years later by writers embarrassed by the Versailles guilt clauses that took a revisionist line and exculpated Germany and Austria-Hungary; books written by Nazi hacks blaming the Allies' lust for world conquest; books written by wartime and post-war British and French historians blaming the Central Powers again, and so on. This is as it should be; as the great Dutch historian Pieter Geyl said, 'History is an argument without end.' For our own time, a fine analysis was written by the American historian Professor David Fromkin, whose conclusions at the end of his book *Europe's Last Summer: Why the World Went to War in 1914* are as uncompromising as they are definitive. 'The international conflict in the summer of 1914 consisted of two wars, not one,' he explains. 'Both were started deliberately. They were started by rival empires that were bound together by mutual need.' In both the cases of the Habsburg and the Hohenzollern Empires, the decision to go to war 'was made by a few individuals at the top, whose peoples were unaware that such decisions were being considered, let alone made'.[37] The lack of proper democracy in either country therefore clearly exacerbated the situation.

'The wars were about power,' continues Fromkin. 'Specifically, they were about the ranking among the great European powers that at the time ruled most of the world. Both Germany and Austria-Hungary believed themselves to be on the way down. Each started a war to stay where it was.' There was a double irony here, in that although it was true of Austria-Hungary, Germany had the world's second largest economy after America and was burgeoning in 1914. After the war, with her High Seas Fleet lying under the Scottish waves at Scapa Flow and her economy prone to pneumonia whenever America caught a cold, she undoubtedly was on the way down, but not in 1914.

It was Austria-Hungary's little local war against Serbia – relatively unimportant in itself given the history of Balkan conflicts since 1875 – that, in Fromkin's estimation, 'provided the German generals with the conditions they needed in order to start a war of their own: a European conflict, which grew into a global conflict'. Far from being the pointless, unnecessary war that poets, playwrights and screenplay-writers have continually depicted since the mid-Twenties, in fact 'It was about the most important issue in politics: who should rule the world.'[38] As spring turned into summer in 1914, all Europe needed for a general conflagration was a spark, and it didn't have long to wait.

On Sunday, 28 June 1914, Gavrilo Princip, a nineteen-year-old postman's son, shot Archduke Franz Ferdinand of Austria-Hungary, the nephew of the Habsburg Emperor, in Sarajevo, the capital of Bosnia-Herzegovina. This was the 9/11 of the early twentieth century, a gross act of terrorism that was to trigger a long and painful war. (The gun Princip used was discovered in a Jesuit community house in Austria in 2004 and is now on display in the Vienna museum of military history. It looks rather like a starting pistol, which in a sense it was.)

Why, though, should the murder have led directly to the death sixteen months later of the twenty-five-year-old Private W. Tamarapa of the New Zealand Maori Battalion, who is buried in the Portianou Military Cemetery on the island of Lemnos? Why should a Maori New Zealander have died in Turkey and been buried in Greece because an Austrian had been shot by a Serb in Bosnia? As the notes for Churchill's *A History of the English-Speaking Peoples* show, his original intention had been to take his book up to the outbreak of war in 1914. The last chapter was to be called 'The Relations of the English-Speaking Races Before the Great War.' The explanation for the service (and sadly the subsequent death) of Private Tamarapa is that the relations between the peoples of the British Empire were so close in 1914 that an assault on the independence of one was automatically considered an assault on the independence of all, and Germany's march through Belgium was certainly such an assault.

Belgium was a British creation intended to ensure the Channel ports stayed out of the hands of an hegemonic power, and the threat to incorporate Belgium into Germany was a direct threat to Britain. In a sense it was even more of a direct threat than that posed by Napoleon at Boulogne in 1804, since the inferior French Fleet had at least had to wait for a following wind before they could invade Britain, whereas that was no longer necessary for the powerful German High Seas Fleet in the era of oil-powered engines.

The Germans were not going to let so perfect an opportunity as the Archduke's assassination go by without taking full advantage of it. Recent work by the German historian Professor John Röhl demonstrates that the Kaiser's court and government carefully orchestrated the Austrian response to the assassination in order to ensure that the summer crisis ended in full-scale war, and Professor Michael Howard has shown how 'Lutheran Germany saw its *Tugend* (virtue) threatened by the godless materialism of the Anglo-Saxons and the empty rationalism of France', and deluded themselves into the Teutonically self-righteous belief that somehow they themselves had been attacked. It was this 'accusation of conspiracy that the Germans trumpeted in the first days of August 1914'.[39]

'These wretched colonies will all be independent, too, in a few years,' Disraeli wrote to Lord Malmesbury in August 1852, 'and are millstones round our necks.' Few political predictions could have been more comprehensively disproved by events than that one, when in August 1914 the colonies instinctively rallied to the British Crown. Even before the Great War broke out, the Dominions and colonies of the British Empire left the Central Powers in no doubt as to what their reaction would be if the Mother Country were forced to go to war. Distance was absolutely no factor. As Lord Elton in his 1945 work *Imperial Commonwealth* recalled, 'The Dominions entered the conflict instantly and without hesitation. The recent loosening of formal ties had not, as German observers supposed, relaxed the subtler bonds of kinship, sympathy and a common way of life.'[40]

The vote on 31 July 1914 in the New Zealand Parliament to send an expeditionary force was unanimous. Australia similarly offered 20,000 men before war was even declared. With Canada, all three Dominions equipped, trained and paid for their own forces throughout the conflict, which had not been the case in the Boer War. Even South Africa, which had been an enemy in 1902, declared for Britain in 1914, and former enemies like the Boer commando Jan Christian Smuts went on to command operations against German East Africa.

Sir Robert Borden, Prime Minister of Canada, cabled London with his country's support on 1 August, three days before war was declared, while the leader of the opposition, Sir Wilfrid Laurier, told the Canadian House of Commons that it was Canada's 'duty ... at once ... to let the friends and foes of Great Britain know, that there is in Canada but one mind and one heart'. That month a Canadian expeditionary force was founded and by September over 40,000 Canadians had volunteered to fight 'to maintain the honour of the Empire'. (The raising of the Princess Patricia's Canadian Light Infantry Regiment was authorised on 10 August and its ranks had been filled in eight days.)

A powerful literary genre argues that the Great War was somehow 'unnecessary', that Britain ought not to have got involved. Sometimes stated overtly, at others implied elliptically, the assumption seems to be that Britain was engaged in a senseless waste of blood and treasure. So were the 908,371 life-sacrifices made by the British Empire between 1914 and 1918 really unnecessary? Were the deaths in the trenches that stretched from the Channel to the Swiss frontier an avoidable error? When President Roosevelt asked Churchill for an adjective to describe the Second World War, he chose 'unnecessary'; but was its predecessor unnecessary also?

For over three centuries, entirely out of *realpolitik* reasons of self-preservation, Britain had pursued a policy of supporting a balance of power in Europe, attempting to ensure that no single nation dominated the continent. Be it Philip II's Spain or Louis XIV and Napoleon's France, total hegemony could not be allowed to go to any single Great Power. 'England has ever watched the Channel ports with especial jealousy,' Lord Robert Cecil wrote of William Pitt's foreign policy during the Napoleonic Wars. 'It has always been one of the cardinal maxims of our foreign policy that they should not fall into the hands of any power whom she had need to fear.'

As early as 1898, Kaiser Wilhelm II's Germany had established through her construction of the High Seas Fleet, which was clearly intended to challenge the Royal Navy, that she needed to be feared. The Fleet was too large for self-defence or even to patrol Germany's relatively small colonial empire; it was an invasion fleet. 'I believe a war is unavoidable and the sooner the better,' said Helmuth von Moltke, Chief of the German General Staff, at a War Council in December 1912, and the Kaiser agreed. A hungry would-be hegemon, which her own Chancellor described as 'parvenu', Germany pushed Austria-Hungary ever closer towards quarrelling with Serbia and thus also her Slavic protector, Russia. At that vital War Council meeting, the Kaiser made it perfectly clear that he understood that a German attack on France would entail war with Britain also. The unequivocal message had come from the British Foreign Secretary, Sir Edward Grey, via the German Ambassador in London and, according to the record of the meeting, the Kaiser 'greeted this information as a desirable clarification of the situation'.

Should the German High Command put into operation their notorious Schlieffen Plan, by which France was attacked with a wide, sweeping right-flanking manoeuvre through Belgium, British participation in the war would be automatically triggered by the British guarantee of Belgian neutrality, formally given when that country was created in 1839. Because this is precisely what happened, it is clear that Imperial Germany – and she alone – bears the full responsibility for the outbreak of that terrible war. Far from being a futile, unnecessary conflict, Britain went to war in 1914 for the noblest possible ideal and best possible reason: her honour and self-defence. To have attempted to renege upon, or legalistically to wriggle out of, the 1839 Treaty of London would not only have been unacceptable to British public opinion at that time, but, as Grey later put it, 'We should have been isolated; we should have had no friend in the world; no one would have hoped or feared anything from us, or thought our friendship worth having. We should have been discredited and held to have played an inglorious part.' It was not the British way.

Furthermore, without British intervention there can be little doubt that France would have been overwhelmed, probably in only a matter of weeks, as in 1870 and 1940. But unlike in 1870, the Germans had no intention of merely annexing a province, exacting relatively light reparations and then withdrawing after three years. The Germany of Kaiser Wilhelm II had ambitions far closer to the Third Reich's than to his grandfather's limited desire for Alsace-Lorraine. As Röhl has demonstrated, Wilhelm II wanted nothing short of a German-dominated continent, what he later tellingly described as 'a United States of Europe under German leadership'. German control of such an entity from Brest to the Polish border would ultimately have posed a mortal danger to Britain's continued existence as an independent power.

Even fighting on two fronts, Germany and Austria-Hungary were able to defeat Russia by early 1918. Had they been relieved of a two-front war by a quick victory on the Western Front due to the absence of the British Expeditionary Force, the victory over Russia would have taken place far sooner. By November 1918, instead of celebrating the victory over the Central Powers as she did that month, Britain would have faced the bleak prospect of being isolated, dishonoured and with an implacable foe's huge battle fleet in the Channel ports. 'Jews and mosquitoes are a nuisance that humanity must get rid of in some way or manner,' wrote Kaiser Wilhelm II, Germany's All-Highest. 'I believe the best would be gas!' To have allowed such a man to rule Europe would have been a crime against Western Civilisation to an almost equal degree as to have failed to challenge Hitler's bid for the control of Europe twenty years later.

No-one would ever seek to doubt or deny the horrors of the Great War, with its gas, mud, machine-guns and mass slaughter, but questioning the merits of the manner in which a war was fought is quite different from questioning its motive. Its tactics might be doubted, its necessity cannot. For Britain and her Empire to have stood aside in 1914, looking only to her defences and colonies while Europe was ravaged, would only have postponed the day of reckoning whilst divesting her of her Russian, French, Belgian,

Italian, Japanese and ultimately American allies. Stern and unchanging commandments must dictate the foreign policy of a small island lying only twenty-two miles off the continental littoral. The foremost is that no power, least of all a great naval one, can be allowed to establish continental hegemony and control of the adjacent ports. It was as true for the Soviet Union during the Cold War as for Philip II at the time of the Spanish Armada. In the First World War no less than in the Second, Britons did not die for a vain cause.

Yet the fact remains that whilst France and Russia – like Serbia – were obliged to fight the Great War because they were invaded by the Central Powers, Britain was not herself directly attacked. In his speech to the House of Commons on Monday, 3 August 1914, Sir Edward Grey brought together all the reasons that Britain nonetheless ought to fight, and persuaded many Britons who, until he spoke, were uncertain about participation in the conflict. By that time Belgium had been invaded by Germany, prompting demands to punish it for its aggression and for Britain to stand up for the rights of small countries.

Grey did make the national security case in his masterly speech in the House of Commons on 3 August, but he made several other important points too. 'We have consistently worked with a single mind, with all the earnestness in our power, to preserve peace,' he told a packed chamber that anxious Monday afternoon, and there is no reason to disbelieve him.[41] War was not in the national interests of a 'satisfied' imperial Great Power like Britain that had nothing to gain but everything to lose from a global conflict. It is true that she had looked for such conflicts in the past – she was at war for over half of the entire 'long' eighteenth century of 1689–1815 and had won an empire thereby – but she certainly did not need a conflict at any time in the twentieth century. 'I should like the House to approach this crisis', Grey continued, 'from the point of view of British interests, British honour and British obligations, free from all passions as to why peace has not been preserved.'

To get a snapshot picture of just how weighty Britain's imperial obligations were in 1914, and what she had on her mind over the two days in which Europe slipped into war, here is a sample of the written and oral parliamentary answers that ministers were giving on 3 and 4 August 1914, none of them to do with the Balkan crisis. They were busy answering questions from MPs on tribal customs in Assam; the governorship of Tasmania; the Bengal Military Orphan Society; an outbreak of foot-and-mouth disease in Tipperary; the South African Native Land Act; Coptic newspapers in Egypt; press freedom in Lahore; the University of Calcutta; the abdication of the Raja of Cochin; Indian police pensions; 'suicide by burning among girl wives in Bengal'; cocaine possession in India; Masai cattle in British East Africa; taxation in the Straits Settlements and Malay States; sanitation in the Punjab; and the Anglo-Persian Oil Company Bill. Such was the width of British responsibilities.

In his speech, Grey was at pains to emphasise how, of the two rival European power blocs of the Triple Alliance (Germany, Austria-Hungary and Italy) and the *Triple Entente* (Britain, France and Russia), the latter 'was not an alliance – it was a diplomatic group', to which 'up till yesterday, we have also given no promise of anything more than diplomatic support'. Despite the *Entente Cordiale*, if the House wished to leave France in the lurch, Grey was effectively saying, it was perfectly free to do so. 'We are not parties to the Franco-Russian alliance,' he said later in the debate and, therefore, 'the House of Commons remains perfectly free' not to go to war if it did not wish to.

This raised the question of the extent of the secret Staff talks that had been going on between the naval and military High Commands of Britain and France, which Grey made clear had been authorised by the Government ever since the 1906 Moroccan crisis, but which were only conducted 'on the undertaking that nothing which passed between military or naval experts should bind either Government or restrict in any way their freedom to make a decision as to whether or not they would give that support when the time arose'. He read out a letter he had written to the French Ambassador on 22 November 1912, which simply agreed that in the event of 'grave reason to expect an unprovoked attack by a third power', Britain and France 'should immediately discuss with the other whether both Governments should act together to prevent aggression and to preserve peace'.

The decision in 1906 to authorise these secret Anglo-French Staff talks was taken by a tiny group of Cabinet ministers. Other than Grey himself, only the Prime Minister Sir Henry Campbell-Bannerman, the Chancellor of the Exchequer Herbert Asquith and the Secretary for War Richard Haldane knew about them; the rest of the Cabinet was kept in the dark. The absolute necessity to shroud issues of this importance in utter secrecy made it impractical to tell something as notoriously leaky as a Cabinet of over a dozen people. This decision to keep the decision on a ministerial 'need-to-know' basis predated that of the six members of the Attlee Government in 1946 who decided to build the hydrogen bomb, the equally small number in the Wilson Government who in 1967 decided to buy the Chevaline nuclear deterrent, and the half-dozen members of the Blair Government in 2003 who were permitted to see the Attorney-General's report on the legality of the Iraq War.

When vital national interests are at stake, the English-speaking peoples have rightly tended to put security and operational efficiency before collective Cabinet responsibility. Campbell-Bannerman's ministry took the right decision, although since he died in office in 1908 it fell to his successor Asquith to explain it. (His last words were, 'This is not the end of me,' which can either be taken as a profound statement of faith or just another politician's broken promise.)

'We are in the presence of a European conflagration,' Grey told a sombre Commons, asking rhetorically, 'Can anybody set limits to the consequences that may arise out of it?'[42] He went on to mention the danger of 'a combination of other Fleets in the Mediterranean' (i.e. Austria-Hungary and the then-still-neutral Italy) threatening vital British trade routes. He told the House that he had the previous day informed the French Ambassador that, 'If the German fleet comes into the Channel or through the North Sea to undertake hostile operations against the French coast or shipping, the British Fleet will give all the protection in its power', subject, of course, to the sanction that Grey was hoping to receive from Parliament. 'The French fleet is now in the Mediterranean, and the northern and western coasts of France are absolutely undefended ... because of the feeling of confidence and friendship which has existed between the two countries.'

The Treaty of 1839 was 'a Treaty with a history', said Grey, who mentioned Bismarck's promise in 1870 to respect Belgian neutrality when he attacked France during the Franco-Prussian War. Another vital difference from 1870 was that Napoleon III had hubristically precipitated the Franco-Prussian War by sending the Kaiser's grandfather, Wilhelm I, an insulting telegram. Grey continued:

> My own feeling is that if a foreign fleet engaged in a war with France, which France had not sought and in which she had not been the aggressor, came down the English Channel and bombarded and battered the undefended coasts of France, we could not stand aside and see this going on practically within sight of our eyes, with our arms folded, looking on dispassionately, doing nothing!

This was especially true if France's coast had only been left unprotected due to a secret naval agreement made by British ministers, as he had just admitted was the case. Rarely in those days did a *Hansard* reporter insert an exclamation mark in the official report of a parliamentary debate; Grey's plea was the heartfelt one of a man who had written a blank cheque that he now needed someone else to honour. With this eloquent combination of *realpolitik*, support for 'undefended' France, indignation over breaking of the 1839 Treaty and appeals to British honour, Grey was able, in his peroration, to call upon 'the determination, the resolution, the courage, and the endurance of the whole country'.[43] All four were going to be sorely needed over the next four years.

The United States was simply too geographically far removed and politically isolationist to help prevent war breaking out in 1914, but, in the opinion of Professor Deepak Lal,

> If the Americans had joined the British in creating an Anglo-American *imperium* to maintain the Pax, the terrible events of the last century could perhaps have

been avoided. The joint industrial and military might of an Anglo-American *imperium* ... could have prevented the Kaiser's gamble to achieve mastery in Europe.

THREE

The First Assault: Prussian Militarism

1914 – 17

The myth of war enthusiasm – The battle of Mons – The Zeppelin terror – Colonel Edward M. House – Catastrophe at Gallipoli – The U-boat threat – The sinking of the Lusitania *– A League of Nations – The effect of the machine-gun – Private 12768's war – The British West Indian Regiment – The battle of Jutland – Armageddon for the aristocracy – The Easter Rising – General John Pershing – The Somme offensive – Lord Lansdowne proposes peace – David Lloyd George – America arms herself – Room 40 cracks the German naval codes – The battle of Vimy Ridge – French mutinies – The Royal House of Windsor – The fall of Jerusalem – Rutherford splits the atom*

'You sunburn'd sicklemen, of August weary.'

William Shakespeare, *The Tempest*

'We shelled the Turks from 9 to 11: and then, being Sunday, had Divine Service.' Royal Navy report to the Admiralty, Gallipoli, 1915[1]

Although it is true that large crowds congregated in London on the 2nd, 3rd and 4th of August 1914, the British did not celebrate the outbreak of war in the jingoistic manner often alleged. People will throng the streets on what are clearly going to be historic occasions for all sorts of non-political reasons: to say they had been there, to watch what is going on, to exchange the latest information, to tell their grandchildren, to see famous people, to avoid feeling excluded, to be first with the news, even to experience the unusual sensation (particularly in those days) of speaking unbidden and un-introduced to strangers. The presence of large numbers of spectators should not imply popular support for whatever the Government was deciding at the time. Added to that, central London was traditionally where large crowds gathered on Bank Holiday Monday, even without the prospect of 'catching a glimpse of ministers', as the Tory-leaning *Globe* newspaper reported that people were trying to do on 3 August 1914.

Lloyd George wrote in his memoirs how, 'I shall never forget the warlike crowds that thronged Whitehall and poured into Downing Street, whilst the cabinet was deliberating on the alternative of peace or war', yet it was not the crowds' decision, but the Cabinet's, and the essence of Cabinet government is not to be swayed by the passing sentiment of the streets.[2] There has long been an assumption that large, naïve, enthusiastic crowds helped impel the nation towards catastrophe, yet in fact plenty of evidence exists to suggest otherwise.

It is possible that the crowds outside Buckingham Palace and along the Mall, estimated variously at between 6,000 and 10,000, came to see the Changing of the Guard and listen to the bands playing. In a city of eight million inhabitants, it represents only 0.1% of the population, not including tourists and day-trippers. The *South London Observer* recorded that London County Council's tramway receipts for Bank Holiday Monday, 3 August, were £9,622, only a fractional increase on the takings for the previous year. Several newspapers reported how the looming prospect of war wrecked the enjoyment of the Bank Holiday. 'Nowhere was there the slightest sign of "Mafficking"', reported the *Hampstead Record*, 'and it was obvious to the observer that the idea of war was distasteful to all.'[3]

A large anti-war demonstration in Trafalgar Square on 2 August was only heckled by 'a negligible contingent of youths in front of the southern plinth'. The South Wales Miners' Federation refused a request from the Government on 3 August to cut short the annual holiday on patriotic grounds, and only 100 of their 11,000 members actually did so. The Welsh-language newspapers, especially the Baptist and Methodist ones, wrote of 'Civilization breaking down' and of how 'the remains of this war will be left behind for generations, in hate, jealousy, misery and poverty'. The Bishop of Lincoln prophesied that 'a continental war could be nothing short of disastrous when one thought of the militarism of Europe, of the hell of the battlefields, of the miseries of the wounded, of ruined peasants'. On 2 August, a young trade unionist called Ernest Bevin called for a general strike against the war at a meeting on Bristol Downs.

C.P. Scott's *Manchester Guardian* and A.G. Gardiner's *Daily News* were both strongly anti-war until it had actually been declared, and much of the provincial press – including papers such as the *Cambridge Daily News*, the *Northern Daily Mail*, and the *Oxford Chronicle* – believed that Britain should declare her neutrality. Newspaper letters columns reflected a widespread desire for neutrality, at least until the die had been cast on 4 August. Nor was there a general feeling, as is so often stated about the Great War with so little evidence, that it would 'be all over by Christmas'. The true mood of the country was one of sombre, concerned realism. Our forefathers were thus not the naïve, bloodthirsty chauvinists of the popular imagination; they needed persuading to go to war against such Great Powers as Germany and Austria-Hungary, and what persuaded them was the cold-blooded execution of the Schlieffen Plan.

After Belgium's ill-treatment by German forces, a tsunami of moral outrage was unleashed from one end of the English-speaking peoples to the other. Henley Henson, the Dean of Durham, took part in Sunday church parades with the local regiment, the Durham Light Infantry, and even participated in their recruitment drives along with lords-lieutenant and local politicians. He later recalled how,

> I was profoundly impressed by the fact that the argument which seemed to be most effective was genuinely altruistic. The Germans never realised the effect in Great Britain of their perfidy in attacking Belgium, and their atrocious method of attack. The miners were little moved by the dangers to Great Britain, for they were comfortably assured that Great Britain was impregnable, but the treatment of Belgium stirred a flame of moral indignation in their minds, and created a determination to come to the rescue which I can only describe as chivalrous.[4]

It was not true that Britain was impregnable, for, as Churchill correctly observed, Admiral Lord Jellicoe could 'lose the war in an afternoon' if the German High Seas Fleet had defeated the Royal Navy in a battle in the North Sea, but it was to be nearly two years before a general naval engagement on that scale was to take place.

The German invasion of Belgium involved nearly a million Germans marching through that country. Under the (mistaken) belief that they would be opposed by mass guerrilla resistance, they killed between 5,000 and 6,000 civilians, including women and children, committed arson and destroyed the ancient library at the University of Louvain.[5] They therefore behaved precisely in the way that Allied propagandists needed in order to make Belgium's 'martyrdom' symbolise the struggle against 'Hunnish' militarism. The occupation of Belgium involved Germany deporting no fewer than 120,000 workers to Germany and occupied northern France. The Belgian economy was systematically destroyed 'through requisitioning, pillage, dismantling, currency manipulation and forced unemployment, and set Flemings against Walloons with the aim of permanently controlling the country'.[6] What kept Belgium fed from 1915 was the Commission for Relief of Belgium, a massive philanthropic organisation that was led by Americans, who thereby saw at first hand what German domination of Europe would most likely look like.

The commander of the American Expeditionary Force, General John Pershing, saw Germany's invasion of neutral Belgium as a wasted opportunity for the United States, even though she had not been a signatory to the 1839 Treaty of London that had guaranteed Belgian independence. 'I cannot escape the conviction', he wrote in his war memoirs in 1931, that in view of this defiance of neutral rights the United States made a grievous error in not immediately entering a vigorous protest. . . . If our people had grasped its meaning they would have at least insisted upon preparation to meet effectively the later cumulative offences of Germany against the law of nations. . . . The fact is that the world knew only too well that we had for years neglected to make adequate preparations for defense, and Germany therefore dared to go considerably further than she would have gone if we had been even partially ready to support our demands by force.

The argument in favour of strong defences and against appeasement is one that became common after the Second World War with regard to the rise of Hitler, but here General Pershing was making it *before* that conflict and, moreover, in relation to the Kaiser's Germany. In the crisis of August 1914, he contended,

Thus we presented the spectacle of the most powerful nation in the world sitting on the sidelines, almost idly watching the enactment of the greatest tragedy of all time, in which it might be compelled at any moment to take an important part. It is almost inconceivable that there should have been such an apparent lack of foresight in Administration circles regarding the probable necessity for an increase in our military forces. . . . The inaction played into the hands of Germany, for she knew how long it would take us to put an army in the field, and governed her actions accordingly.[7]

Pershing was right; both Field-Marshal Paul von Hindenburg and Field-Marshal Erich von Ludendorff felt something approaching contempt for the size and likely effectiveness of the American armed forces, which were until the spring of 1918 thousands of miles away from the theatres of war that they knew mattered. It was a contempt that was not to survive familiarity, but very soon after the war began, those two Generals took over effective control of the German State. 'Who were our politicians during the war?' Admiral Georg von Müller, chief of the German naval cabinet between 1914 and 1918, rhetorically asked afterwards. 'Hindenburg, Ludendorff and the political branch of the Great General Staff.'

It was said that Helmuth von Moltke, the Chief of the German General Staff after 1906, only laughed twice in his life: once when a certain fortress was declared to be impregnable and once when his mother-in-law died. If the fortress was Liège, his mirth was justified, since it surrendered to Ludendorff without a fight on 7 August 1914. Yet his plans for victory went awry the following month when the Germans were halted at the battle of the Marne, with French troops at one point being rushed to the front in Paris buses and taxi-cabs, and the advance ended in the stable entrenchments and tactical stalemates that characterised the next four years of war. To those of the 'Schlieffen School', this was all the fault of Moltke for not following Schlieffen's precept: 'Strengthen the right.'[8] After the Marne, Moltke was sacked.

It was not the outbreak of war itself that encouraged vast numbers of Britons to flock to the colours – only 100,000 had enlisted by 22 August – so much as the publication of the Mons Despatch of Tuesday, 25 August 1914. The defeat in the battle of Mons was presented in stark though heroic terms. The headlines in *The Times* that day: 'Namur Lost', 'German Success in Belgium', 'British Army's Stern Fight', 'German Readiness for War: Paris First, then London', 'German Forces in Lunéville', 'British Force Fighting Well', 'Liège Forts Holding Out', 'The Germans Strike Hard', 'Civilians as Screens For [German] Troops' and 'The British at Mons' could not fail to stir the anger and patriotism of Britons. 'If, then, the fortune of war compels us to make a general retirement,' *The Times* military correspondent reported from the Front, 'we must do so step by step, keeping our armies together ... and making every step of the enemy in advance as costly as possible.'[9]

On the same page as the Despatch lay an editorial entitled 'England's Call', which was a masterpiece of recruitment literature and bears repetition because in the week between 30 August and 5 September no fewer than 174,901 Britons applied to join the colours. However hyperbolic, even corny, it might sound to cynical modern ears, it was answered triumphantly. 'It is our English boast that the spirit of the country ever rises with ill-fortune,' the editorial began. 'The unexpected downfall of Namur is a spur which, if we mistake not, will excite our countrymen to redoubled efforts.' After a brief reference to 'the grim realities before us', it went on:

> We are committed to a life-and-death struggle for all that we hold dearest with the mightiest military Monarchy in the world. ... Now is the time for all men of British blood to bethink them what they may do to safeguard their national inheritance. ... Lord Kitchener has told them the first steps they must take. Now is the moment to answer his call. The best way to serve the country is the simplest way. It is to give him the men he demands for its defence.

The article concluded: 'When our troops are under fire and our Allies have met with reverses, there is no place amongst us for the idler or the loafer. England needs all her sons.'[10] As a recent historian has written of the analysis of the patterns of recruitment in the early part of the Great War, 'Far from signing up in a burst of enthusiasm at the outbreak of the war, the largest single component of volunteers enlisted at exactly the moment when the war turned serious. Men did not join the British Army expecting a picnic stroll to Berlin but in the expectation of a desperate fight for national defence.'[11]

★

On Tuesday, 19 January 1915, world history entered a new and horrifying phase when the English-speaking peoples sustained the first bombing attack of civilians from the air. The invention that was applied by the Kaiser's High Command to attack Britain owed more to the frères Montgolfier, however, than to the Wright brothers. In an attack on the East Anglian ports of Great Yarmouth, Sheringham and King's Lynn, two Zeppelins dropped twenty-four 50-kilogram high-explosive bombs and 3-kilogram incendiaries. Only four people were killed, sixteen injured and £7,740 worth of damage was done. Yet a line had been crossed that was never to be re-established. The destruction caused in Coventry and Dresden, Hiroshima and Hanoi can in a sense be traced back to those twenty East Anglian casualties.

The year 1915 saw only a further nineteen Zeppelin raids, during which 37 tons of bombs were dropped, killing 181 people and injuring 455. By February 1916, the Army had introduced searchlights and converted a variety of sub 4-inch calibre guns to anti-aircraft use, but the best way to bring Zeppelins down was by dropping bombs on them from above. The first pilot to achieve this was R.A.J. Warneford of the Royal Naval Air Service, flying a Morane Parasol on 7 June 1915, who dropped six 9-kilogram bombs on *LZ 37* over Ghent, winning the Victoria Cross in the process.

Twenty-three separate Zeppelin raids in 1916 saw a total of 125 tons of bombs being dropped, killing 293 people and wounding 691. By mid-1916, the British had developed forward-firing fighter aircraft, and although improvement in Zeppelin design meant that the Germans could fly at twice the altitude – over 10,000 feet – the first night-fighter victory came on 2 September 1916, when Lieutenant William Leefe Robinson, flying a BE2c of No. 39 (Home Defence) Squadron, shot down the German Army Schütte-Lanz airship *SL11*, for which he too was awarded a Victoria Cross. (He was later himself shot down in April 1917, but survived and made several escape attempts from prison, dying in December 1918.)

In 1917 and 1918, there were only eleven Zeppelin raids against Britain, the last of which occurred on 5 August 1918 resulting in the death of the German naval airship director, Peter Strasser. Of the eighty-eight Zeppelins built during the Great War, over sixty were destroyed, half by Allied action and half in accidents. In a total of fifty-one raids, in which 5,806 bombs were dropped, 557 people were killed and 1,358 wounded. It was hardly worth the cost, lives and sheer effort involved, although wartime production was disrupted, twelve fighter squadrons were diverted from the Western Front and up to 10,000 men were placed on ground-based air defences. The significance was not in the effect, so much as the precedent. The targeting of civilians – 162 people died in one London school in 1917 – was a specific violation of the 1907 Hague Convention, yet the policy had been personally authorised by the Kaiser. Long before the attack on Guernica in 1937, Mankind had thus entered a period of hitherto-unimaginable horror. A top priority for the English-speaking peoples was henceforth to develop fighter aircraft that could minimise the danger to their own cities, while simultaneously building bomber aircraft that would maximise the threat to their enemies'.

Soon after the Zeppelin assault, the Germans employed another new but far more lethal terror-weapon. Back in 1901, Winston Churchill had predicted that, 'The wars of peoples will be more terrible than the wars of kings.' On 22 April 1915, during the second battle of Ypres, poison gas was deployed for the first time in the West, emitted from some 5,000 cylinders. As Lord Allenby's biographer recalls, 'A greenish-yellow cloud of chlorine that initially puzzled onlookers drove a French Senegalese battalion from its trenches and exposed a gap that was hastily filled by newly-arrived troops from the Canadian Division.'[12] The Germans had fortunately made no preparations for exploiting the wide gap that had been opened up in the Allied line, and the Canadians and other units in the British 2nd Army finally stemmed their advance in a bitterly fought battle that lasted three days. Although the German losses were 35,000 the British, French and Canadians lost twice that number. Yet the line had held. 'After two thousand years of Mass,' rhymed the poet Thomas Hardy, 'we've got as far as poison gas.'

Writing from Germany on 15 February 1915, the American Ambassador to Berlin, James W. Gerard, told Colonel Edward M. House of the Germans: 'Make no mistake, they will win on land and probably get a separate peace from Russia, then get the same from France or overwhelm it, and put a large force in Egypt, and perhaps completely blockade England. Germany will make no peace proposals, but I am sure if a reasonable peace is proposed *now* (in a matter of days, even hours), it would be accepted.'[13] He asked that any Allied peace offer be sent to him '*verbally and secretly here*', but not to bother with one that involved Germany paying any indemnity 'to Belgium or anyone else'. A further measure of Gerard's naïveté can be gained from his remark: 'I do not think the Kaiser ever actually wanted the war.'

House, who was in London on a European tour, replied on 1 March: he was favourable to the idea of peace talks, which he thought would go well if begun, but of the British he wrote, 'These are slow-moving people.' As for Gerard's presumption of a British defeat, House wrote, 'Your prediction as to the final outcome is not shared by anyone here, from the highest to the lowest. If the war lasts six months longer, England will have a navy that is more than equal to the combined navies of the world. That is something for us Americans to think of; in fact, it is something for everyone to think of. . . .' House's fear of the power that a strengthened Royal Navy would afford Britain was a subject he was to return to regularly. Writing to President Wilson from Berlin on 20 March, he said that he had told the German Foreign Minister, Count Alfred Zimmermann, that, 'We recognized England had a perfect right to have a navy sufficient to prevent invasion, but further than that she should not go. He was exceedingly sympathetic to this thought and I think it will have a tendency to put us on a good footing here.'

House went even further a week later when he suggested telling the German Chancellor, Theobald von Bethmann-Hollweg, that, 'through the good offices of the United States, England might be brought to concede at the final settlement the Freedom of the Seas', and that he had already told the Chancellor 'that the United States would be justified in bringing pressure upon England in this direction, for our people had a common interest with Germany in that question'. The Freedom of the Seas was shorthand for allowing neutral countries to trade with Germany, which the British naval blockade prevented; to have conceded it would have severely reduced Britain's chances of victory, as House well knew. House's primary objective was to prevent American intervention in the war, which he recognised was most likely to happen as a result of Germany's unrestricted submarine warfare. He hoped that if Germany agreed to suspend such attacks, Britain might allow the Freedom of the Seas, but on 15 March the British Cabinet – with only Grey dissenting – refused this compromise, 'even if it meant braving the threat of the submarine'. Seven weeks later, House's worst fears were realised when the Atlantic liner, the *Lusitania*, was sunk by a U-boat.

At 4.30 a.m., just before dawn on Sunday, 25 April 1915, thirty-six boats landed at a cove at Ari Burnu, on the western side of the Gallipoli peninsula. Launched from three Royal Navy battleships, the *Queen*, *Prince of Wales* and *London*, they contained 1,500 men from the Australian and New Zealand Army Corps (ANZACs), who quickly moved up the steep slopes towards Chunuk Bair, a high point that dominates the peninsula and with it the Narrows of the Dardanelles beyond. Simultaneously, fifteen miles to the south at the southern tip of the peninsula at Cape Helles, the 29th British Division landed at five beaches – rather unimaginatively codenamed, from east to west, 'S', 'V', 'W', 'X' and 'Y' – and made its way towards the dominant hill mass of Achi Baba. In the fiercely contested fighting on 'W' beach, two officers and four men of the Lancashire Fusiliers won, in a phrase that became immortal, 'six VCs before breakfast'.

The British had generally experienced success with amphibious operations against enemy coastlines. The landings at Aboukir in 1801, Blouberg near Cape Town in 1806, Port Natal in 1842, the Crimea in 1854, Canton in 1858 and Alexandria in 1882 had all gone well, usually because the firepower of the Royal Navy had cleared the landing beaches of the enemy. At the Dagu Forts in China on one afternoon in June 1859, however, a British force of eleven gunboats and 700 marines had come to grief attacking across wide mudflats. Gunboats ran aground, there was a ferocious barrage, men were landed in waist-deep mud and were beheaded by the Chinese defenders.[14] Tragically, Gallipoli was to follow the Dagu rather than the earlier models.

In a sense, the whole Gallipoli campaign was a sign of the desperation that senior British strategists felt at the static nature of the Western Front eight months into the war, and their keenness to try any other way to break the stalemate. The Russian defender of Sevastopol in the Crimean War, General Todleden, was once asked what was the best defensive ground in the world. 'An exercise ground which is perfectly flat,' he replied.[15] Flanders was just such a theatre of operations. As Prince Philip put it in a speech about Gallipoli in 1987, 'Unlike any previous wars, the front lines of the antagonists had become a continuous line of entrenchments from the coast of the North Sea to the Alps. Out-flanking, in the accepted military understanding of that term, was therefore impossible.' With absolutely no historical precedent to guide them, policy-makers had to look at completely fresh ways of going beyond the Western European land-mass to attempt to find an out-flanking manoeuvre, and 'the nearest thing to a left flank was Turkey'.[16]

The Dardanelles represented the windpipe of Russia's trade. Some 90% of her grain and half of all her exports had passed through the Bosphorus before the war, which was of course blocked off to her as soon as Russia declared war on Turkey on 2 November 1914. During the long winter months her other principal ports, Murmansk and Vladivostok, were completely ice-bound. Thus her economy suffered considerably by her inability to export her wheat surplus, whereas clearing the Straits would have allowed the Russian Black Sea Fleet to operate in support of the Allies in the Mediterranean, as well as allowing her to trade.

Many of the assumptions made by most of the participants in the War Council discussions about the benefits of the Gallipoli campaign were wildly optimistic, however. Just because troops were landed there, it was assumed that the peninsula would fall. Then it was assumed that the Turkish mines in the Dardanelles could all be swept. Afterwards it was assumed that the fleet would be able to withstand bombardment from the Asiatic side of the Straits and get to Constantinople. (Vice-Admiral Sir John Duckworth had forced the Straits in 1807, but got no closer to Constantinople than eight miles distant, and Turkish gunnery had improved exponentially in the intervening century.) Just because an Allied Fleet appeared off the Golden Horn, it was assumed that Turkey would drop out of the war. This, it was assumed, would bring Italy, Romania, Bulgaria and probably Greece into the war on the Allied side and open up a third front against the Central Powers. That, it was hoped, might well knock Austria-Hungary out of the war, which would in turn take the pressure off Russia, as the Grand Duke Nicholas had begged for in a telegram on New Year's Day 1915. There were, therefore, a large number of dominoes that had to fall between the heights of Ari Burnu and the Chancellery in Berlin.

Somehow, sedulously, the conviction arose that an amphibious landing in a rocky peninsula in the northern Aegean was the key to winning a war against a Great Power whose only coastline was on the Baltic. It is an example of wishful thinking – today known as 'groupthink' – infecting otherwise intelligent (even cynical) policy-makers. By 12 April, the Cabinet Secretary Sir Maurice Hankey had become doubtful and wrote a memorandum to the Prime Minister, which concluded: 'The military operation appears, therefore, to be to a certain extent a *gamble* upon the supposed shortage of supplies and inferior fighting qualities of the Turkish Armies.'[17]

This last turned out to be the most erroneous assumption of them all, since the well-supplied Turkish soldiers both fought valiantly and were superbly commanded by Mustapha Kemal (later Kemal Atatürk). It might have seemed like simple racism on behalf of the War Council, such assumptions of superiority being quite normal for those days, but in fact it was perfectly understandable considering Turkey's recent performance in wars. Turkish soldiers had been defeated by Mehmet Ali in the late 1830s, fled the battlefield of Balaclava in the Crimean War, been flung out of Bulgaria in 1877–8 and lost the rest of Turkey-in-Asia in the First Balkan War. Who could have predicted that 'the sick man of Europe' would have leapt from his bathchair quite so energetically the moment Allied troops landed at Gallipoli?

The blame for the catastrophe has come to rest squarely on the shoulders of Winston Churchill, however hard he might have tried to apportion it differently at the Committee of Inquiry and subsequently in his book, *The World Crisis*. With Hankey, the War Minister Herbert Kitchener, the Chancellor of the Exchequer Lloyd George, the First Sea Lord Jackie Fisher and Prime Minister Herbert Asquith all advocates of it – albeit at different times and at different tempos – history has treated Churchill harshly. Although, being Churchill, he was always the most eloquent and public advocate of the sextet. Because decision-making for the higher strategy of the war was in the hands of a War Council that had no agenda, regular meetings or even minute-taking, there were endless opportunities for misunderstandings, both accidental and deliberate. Many of the senior generals were in France, and the Chief of the Imperial General Staff Sir James Wolfe Murray appears to have been overawed by Churchill, who (presumably behind his back) nicknamed him 'Sheep'. Kitchener was a particularly difficult man to oppose, let alone try to overrule. The British Empire's avenger – he avenged the 1885 defeat at Khartoum thirteen years later in his Sudan campaign and the 1881 defeat at

Majuba Hill nineteen years later during his South African command – he turned up to Cabinet meetings in his field-marshal's uniform.

The naval commitment made by the Empire to the operation was astonishing and a tribute to its ability to project power over the oceans. The imposing Royal Navy memorial at Cape Helles bears the names of the ships involved, including 25 battleships, 1 battle-cruiser (*Inflexible*), 8 cruisers, 13 light cruisers, 32 torpedo-boat destroyers, 6 torpedo-boats, 21 monitors, 7 sloops, 4 gunboats, 17 submarines, 27 minesweepers, 7 auxiliary minesweepers, 1 minelayer, 3 kite balloon ships, 2 seaplane carriers, 8 armed steamers and 20 hospital ships. It is thus very hard to accept Churchill's explanation in *The World Crisis* that 'a dozen old ships, half a dozen divisions, or a few hundred thousand shells were allowed to stand between them and success'.[18] Three things that the campaign was not kept short of were ships, soldiers and ordnance; in all, the Allies committed 180,000 men and 110 guns to the expedition, as well as the veritable armada listed above.

The Allied naval assault on the Dardanelles began on 19 February 1915, but although the Turkish outer forts on the Gallipoli peninsula were bombarded, they were not silenced. This meant that the minefields were not cleared before Admiral Sir Sackville Carden, the Commander-in-Chief in the Mediterranean, ordered the Fleet up the Straits on 18 March. Three battleships were sunk that day and another one damaged, for the loss of sixty British and 600 French seamen killed. The operation was called off by Carden, who suspected – rightly as it much later turned out – that there were many more mines further up the Straits.

It was only after this failure of the naval operation that the campaign became a land-based one, with the landings of 25 April. With no experience of combined operations to speak of, and only three months' training in Egypt which had a very different climate from Gallipoli, troops of the 29th Division, ANZACs and Royal Naval Division were put ashore on S, V, W, X and Y beaches and at Anzac Cove on that cold and foggy Sunday morning. Meanwhile, a French colonial division landed at Kum Kale on the Asiatic side. (It is an oft-forgotten fact that even more French than Australian soldiers died in the Gallipoli campaign.)

The commander of the operation, General Sir Ian Hamilton, had held successful commands on the North-West Frontier of India and in the Boer War; he knew the value of Intelligence, and the Navy had mapped the coast from the sea while spotter aircraft photographed it from above. 'Even so, the true nature of the peninsula's broken topography would only become apparent once the landings had taken place.'[19] Once it had been, the entire operation should have been called off, but by then the Allies were completely committed.

The density of fire was so heavy on Chocolate Hill – one of the many outcrops on the peninsula – that in the Kabatepe Museum in Gallipoli there are eleven bullets that hit each other *in mid-air*. It is a reflection of the intensity of the fighting that of the 18,166 New Zealanders who lost their lives in the Great War, no fewer than one-third have no known graves so complete was the carnage at Gallipoli. A letter written from Sedel Bahr on Tuesday, 27 April, by the Royal Naval Air Service Lieutenant-Commander Josiah Wedgwood to his fellow MP Winston Churchill, described the landing on V beach and related how,

> In ten minutes there were some four hundred dead and wounded on the beach and in the water. Not more than 10% got safe to shore and took shelter under the sand hedge. .... Thereafter the wounded cried out all day and for 36 hours – in every boat, lighter, hopper, and all along the shore. It was horrible, and all within two hundred yards of our guns trying to find and shoot the shooters. ... The wounded were still crying and drowning on that awful spit.[20]

Today, hundreds of young Australians and New Zealanders visit Anzac Cove at dawn every 25 April, sleeping on the beaches in the cold the night beforehand, in order to pay their respects to the sacrifices made by their forefathers. Over 100,000 New Zealanders served during the Great War, from a country with a total population of only 1.1 million in 1914. The sacrifices of Gallipoli were the crucible in which the national identity of both Australia and New Zealand were forged, but sadly and unnecessarily this has taken on an unwarrantedly anti-British character. It was a fine thing, fifteen years after Australian Federation, for that country to step away from the Mother Country's apron strings, but need it have been done with such bitter criticism over a campaign that had after all led to huge British losses also?

Equally, a sterile debate began between Australian, New Zealander and British military historians about which armies fought bravest, who was to blame for certain incidents, and so on. This was emphasised by those who wish to divide the English-speaking peoples. In the Peter Weir movie *Gallipoli*, for example, in which Mel Gibson starred in 1981, British officers are depicted drinking tea on the beach and refusing to come to the aid of the hard-pressed Australians. In fact, all three armies conceived a high appreciation of each other's gallantry at the time, as is expressed in countless letters, diaries and papers of the participants. The Gallipoli campaign was a disaster, but the idea that cynical and stupid upper-class British officers casually threw away the lives of idealistic young Australians and New Zealanders is a vicious travesty of the truth. Far better Rudyard Kipling, who in his sublime ode to the Melbourne Shrine of Remembrance, wrote, with regard to Gallipoli, with admiration

of certain men who strove to reach
Through the red surf the crest no man might hold,
And gave their name for ever to a beach
Which shall outlive Troy's tale when time is old.

As well as at Gallipoli, Australian troops served in an astonishing number of theatres of the Great War: in France and Belgium (especially at the battles of the Somme, Bullecourt, Messines, Ypres, Amiens, Passchendaele and the Hindenburg Line); in New Guinea and Malaysia; on the high seas off the Cocos Islands, where HMAS *Sydney* sank the German light cruiser *Emden* on 9 November 1914; in the West Atlantic, North Sea and the Otranto Barrage; and also in the Middle East (especially at Romani, Gaza and Beersheba, the Jordan Valley, Megiddo and Damascus). A young and numerically small country, with a 1914 population of four million, she lost no fewer than 58,961 killed in the Great War and 166,811 wounded, an enormous and awesome contribution to victory. In all, 416,809 Australians enlisted for service in the First World War, representing 38.7% of the total male population aged between eighteen and forty-four. As the historian of Australia, Geoffrey Blainey, has put it, 'The nation, reluctant to accept its convict past, greeted Gallipoli as the sign that it had redeemed its beginnings and come of age.'[21]

Similarly, New Zealand troops took part in some of the hardest-fought battles of the war. Around the panelled walls of her parliament chamber are carved the place-names of Hebuterne, La Vacquerie, Messines, Egypt, Samoa, Gallipoli, Somme, La Basséville, Passchendaele, Bapaume, Le Quesnoy and Palestine, testament to the ubiquity of her forces in important theatres of the conflict. The beautiful memorial chapel of St Michael and St George in Christchurch Cathedral, on New Zealand's South Island, was opened in 1949 by General Sir Bernard Freyberg, the altar of which was carved by Dean Carrington in memory of his son Christopher, who was killed in France in 1916. Nine regimental colours hang above the names of local people who died in the two world wars (including no fewer than sixteen nurses of the NZANS). To pick one entirely at random, on the guidon of the 1st (Canterbury Yeomanry Cavalry) Mounted Rifle Regiment are battle honours for Gallipoli, Syria and Palestine in the Great War alone.

In London's Embankment Gardens stands a delicate and handsome small bronze sculpture of a soldier mounted on a camel, a monument dedicated to 'the glorious and immortal memory of the officers, NCOs and men of the Imperial Camel Corps who fell in action or died of wounds and disease in Egypt, Syria and Palestine' between 1916 and 1918. There were four battalions, and the names of 197 Australians, 98 Britons, 41 New Zealanders and 9 men from the Hong Kong and Singapore Battery of the Royal Garrison Artillery are inscribed upon it. What is remarkable, apart from the pan-imperial nature of the recruitment, was the number of places where the Corps saw action, including Romani, Baharia, Mazar, Dakhla, El Arish and Maghdaba in 1916, Rafa, Hassana, Gaza, Sana Redoubt, Beersheba, Bir Khu Weilfe and Hill 265 in 1917, and Amman, Jordan Valley and Mudawara in 1918. Victory in the Middle Eastern theatre would have been impossible were Britain to have been forced to rely on forces from the metropolitan United Kingdom alone.

A key indicator for Great Britain's chances of survival between 1914 and 1918 would be the quarterly figures for the tonnage of her merchant shipping that the Germans managed to sink. A small overcrowded island that could not survive on the food she grew herself, in both world wars the prospect of starving Britain into surrender through submarine action opened up Germany's best chances of victory. Although of course every nerve was strained to replace the losses as quickly as possible, the shipping depletions were severe. Between August and December 1914, nearly 700,000 tons of shipping were lost by Britain, but this dropped to 215,000 the next quarter, not rising to more than 365,000 tons in any quarter until September 1916. Then, however, the figures increased catastrophically: in October-December 1916, they were 617,000; in January-March 1917, 912,000; and the worst quarter of the war came in April-June 1917, when no fewer than 1,361,870 tons were lost, followed by the second-worst the next quarter when a further 953,000 were lost. These dangerously high figures continued up in the 600,000 to 700,000 range until almost the end of the war, with 512,000 sunk between July and September 1918.

In total, the world losses of shipping during the Great War amounted to fifteen million tons, out of which no fewer than nine million were British. This, out of a pre-war figure of twenty-one million tons for the whole British Empire merchant shipping fleet, shows quite how important victory at sea was to the survival of Britain. Nonetheless, the shipbuilding programme had been brought by the Government to such a pitch that by 1919 the world's biggest merchant fleets were those of the British Empire at 18.6 million tons, the United States at 13.1 million, Germany at 3.5 million, Japan at 2.3 million and France at 2.2 million. The English-speaking peoples thus together had nearly four times the tonnage of the next three nations combined.

On Friday, 7 May 1915, the Atlantic passenger liner *Lusitania* was sunk by a U-boat off the coast of Ireland with the loss of 1,198 lives, 128 of them American. Germany, which had been conducting unrestricted submarine warfare against merchant shipping, was forced to suspend it in the face of national outrage that brought the United States to the brink of a declaration of war against it. Less than a month earlier, German forces had initiated the use of poison gas on the Western Front, and five days after the *Lusitania* was sunk the Bryce Report detailed German atrocities against Belgian civilians. With all three coming in such quick succession, a spontaneous outburst against the 'barbaric' and 'inhumane' Germans broke out in Britain, despite the fact that many immigrants were naturalised in all but name.[22] Before 1891, Germans constituted the largest group of immigrants in Britain after the Irish, and in the 1911 census there were 53,324 recorded as resident there. Nonetheless, after the *Lusitania* was sunk, riots broke out in Liverpool, with thousands of people attacking 550 German-owned shops in Liverpool over five days. Subsequent riots in London and Manchester damaged another 2,000 properties owned by Germans or often people with German-sounding names.

Meanwhile, petitions bearing half-a-million signatures were presented to the House of Commons demanding the wholesale internment of the entire German population for the duration of the war. On 13 May, Asquith announced that all enemy aliens of military age were to be interned while their dependants were repatriated. Anti-German hysteria was whipped up by newspapers, and only a few people, such as Sylvia Pankhurst, complained. After an air raid on the East End, she wrote: 'Prominent newspapers fill their columns with articles which consume ignorant, nervous and excitable people with a suspicious terror that transforms for them the poor Hoxton baker and his old mother into powerful spies, able at will to summon fleets of Zeppelins.'

According to the recently published diaries of Alfred Duff Cooper, a British diplomat working in the Foreign Office at the time of the *Lusitania*'s sinking, it was not taken for granted that an American declaration of war against Germany would automatically help the Allies. 'We should of course get no more ammunition from her and her army and navy would be of little value,' he wrote on the day of the attack.[23] Duff Cooper assumed that Germany had deliberately wanted to provoke the United States into a declaration of war that could therefore not be in the Allies' interests. Five days later, he came to a surprising conclusion:

> The feeling in America is so strong that they may be forced to go to war. If they do they will be simply playing into Germany's hands and we are most anxious to prevent it. It is a pity that the mass of the people both here and in America don't realize this, but our press is very restrained and sensible for once. ... The disadvantage to us if America joins in will be immediate and military, but I cannot help feeling that in the long run neutral nations and even the more thoughtful of Germans could not fail to be impressed by the spectacle of all the most civilized nations in the world joined in alliance against one enemy.[24]

Duff Cooper therefore hoped for American involvement for morale rather than matériel reasons, so small were the United States' Army and Navy, which would want to keep their ammunition production that was presently being exported to Britain. It was a short-sighted view that two years of losses on the Western Front, and a severe dearth of able-bodied young men in Britain by 1917, was radically to alter. (Duff Cooper himself took advantage of this shortage to leave the Foreign Office in June 1917 and join the Grenadier Guards, with whom he won the DSO after capturing eighteen Germans during 'the Battle of the Mist' on the Albert Canal in August 1918.)[25]

Winston Churchill fully concurred with Pershing's rather than Duff Cooper's view. Had the United States entered the war in 1915, he wrote later, 'what abridgement of the slaughter, what sparing of the agony; what ruin, what catastrophes would have been prevented; in how many homes would an empty chair be occupied today; how different would be the shattered world in which victors and vanquished alike are condemned to live!'

Within a fortnight of the sinking, Colonel House returned to the subject of Britain giving up her blockade of Germany, which the Cabinet had resolutely refused to do at its meeting back on 15 March. House felt that if a deal could be reached between Britain and Germany in which Britain's blockade and Germany's unrestricted submarine warfare were both suspended, the United States could both stay out of the war and resume exporting foodstuffs to Germany. He instructed Walter Page to speak to Grey, but the Ambassador prevaricated, saying that he knew that the British Government 'would not consider for a moment the proposal to lift the embargo'. The vegetarian Page had recently spent 'two cold, wet, miserable nights' at Walmer Castle near Deal with Asquith – being given nothing to eat but meat – and he had left under the distinct impression that any such submarine-blockade deal would be bad for the British, who wouldn't accept it anyhow.[26]

By-passing Page altogether, House went to see Grey himself, who he found 'even more receptive of the suggestion than when I last saw him', only this time Grey wanted a German promise to discontinue the use of asphyxiating or poison gasses to be appended to any agreement over the blockade and submarine policies. Grey made it clear that he could only speak for himself at that stage, and not the Government, adding that 'in ordinary times if the cabinet refused to acquiesce in his view, he would resign; but that he did not feel justified in doing this at time of war'. Nonetheless, Grey dictated a draft understanding, which House immediately cabled to Gerard in Berlin on 19 May.

The British Foreign Secretary was therefore in effect going behind the Cabinet's back in negotiating with the Germans – via the Americans – over blockade, submarine and poison-gas policy, although, as House told the President, 'I assumed the entire responsibility, so if things go wrong, you and Sir Edward can disclaim any connexion with it.' Such a negotiation had the advantage from House's point of view of placing Wilson 'in the position of doing everything possible to avert war between the United States and Germany'. In the event, Germany rejected the offer brusquely, determined to make full military use of the U-boat. This diplomatic démarche did not become public knowledge until House's memoirs were published in 1926, when they caused a storm among those British policy-makers who regarded the whole concept of 'Freedom of the Seas' with undisguised anathema. They felt that the Royal Navy provided for perfect freedom of the seas, at least for friendly nations.

With an irony that could only be appreciated years later, the movement to establish a League of Nations was begun in America, at a meeting in Philadelphia to found the American League to Enforce Peace on 17 June 1915. The prime movers on that occasion were William H. Taft and the former US Ambassador to Belgium, Theodore Marburg. The meeting declared: 'It is desirable for the United States to join a league of nations binding the signatories to ... jointly use forthwith both their economic and military forces against any one of their number that goes to war, or commits acts of hostility, against another of the signatories before any question shall be committed [to a Council of Conciliation].' Similar leagues were soon afterwards set up in France, Italy and Britain, which had two in the League of Nations Society and the League of Free Nations Association.[27]

On 22 October 1915, the order was issued for the British Army to set up a machine-gun corps separate from individual regiments. The British had hitherto fought with only two machine-guns per battalion, but the success of the weapon – on both sides – had made it clear that a new organisational structure was badly needed. Over the following months the various divisional machine-gun units were re-designated in a new corps that by the end of the war had 6,432 officers and 124,920 men. To replace the heavy Vickers machine-guns, the battalions were issued with Lewis light machine-guns. By December 1916, there was one Lewis gun for every four platoons, which had been increased to one for every two by 1918, not including some used for anti-aircraft duties.[28]

A German machine-gunner in 1916 reported how, 'We were very surprised to see them walking, we had never seen that before. ... They went down in their hundreds. You didn't have to aim, we just fired into them.' Although it is a myth that the Germans started the war with a higher proportion of machine-guns than the Allies, they organised them better regimentally.[29] In 1916, independent machine-gun companies were brought together in groups of three, called Maschinengewehr-Scarfschützen-Abteilungen, by which time each division contained seventy-two heavy machine-guns, which increased to 350 by 1918. The Germans also made great use of light machine-guns and automatic rifles. As late as May 1916, a French infantry officer saw a regiment of lancers forming up for the attack, and a fellow officer joked blackly: 'They're holding back all these fellows for the breakthrough, the famous breakthrough that we've been waiting for for two years. ... You know there's nothing like a lance against machine-guns.'[30]

Although the concept of a machine-gun had been around for almost as long as muskets themselves, there was no metal until the nineteenth century that could withstand the pressures of sustained mass firing, nor manufacturing techniques capable of creating the necessary fractional tolerances for every part of such a complex weapon. Indeed, when the machine-gun idea was mooted as a commercial proposition in 1718, one wag rhymed: 'Fear not, my friends, this terrible machine, They're only wounded that have shares therein.' Everything had changed in 1862 when the North Carolina-born inventor, Richard Jordan Gatling, produced a crank-operated gun capable of firing 200 rounds per minute. This was advanced upon twenty-two years later when Maine-born Hiram Maxim perfected a weapon that fired automatically at the press of a trigger until released. Maxim, who lived in Kent in England for much of his life, also took out patents for gas apparatus and electric lamps; he was knighted in 1901. In his intriguingly entitled 1975 book *The Social History of the Machine Gun*, John Ellis explored the profound military, political, social and even moral effects that the machine-gun was to have upon modern society, for never in the field of human conflict could so many be killed so quickly by so few.

The first and most obvious effect came in the American Civil War and then in African conflicts. At the battles of Ulundi in the Zulu War in 1879, Tel-el-Kebir in the Egyptian campaign in 1882 and especially Omdurman in the Sudan campaign in 1898, machine-guns played a major part. After Omdurman, where General Kitchener lost forty-eight killed and the Dervishes over 11,000, it was remarked that, 'In most of our wars it has been the dash, the skill, and the bravery of our officers and men that have won the day, but in this case the battle was won by a quiet scientific gentleman living in Kent.'[31] While the English-speaking peoples have kept the technological edge on the invention and development of weaponry, they have managed to retain the hegemony of the world. The machine-gun, the tank, the Spitfire, the Lancaster, B-29 and B-52 bombers, the H-bomb and A-bomb, Agent Orange, the F-16 fighter, stealth aircraft, the 'daisy-cutter' bomb – each were ground-breaking weapons and all were first deployed by the English-speaking peoples, and to devastating effect. The primary reason that peace reigned between the Great Powers for over sixty years after 1945 – an unprecedented length of time in modern history – was that the English-speaking peoples possessed weaponry of such power and sophistication that no rival power could possibly defeat them in a general tactical war.

(The understandable enthusiasm to abolish certain types of weapons –

such as the anti-personnel mine – needs to be set against the fact that some have been needed by the English-speaking peoples. The British used minefields to cover their retreat across the Western Desert in the Second World War and to narrow Rommel's fields of attack. The best use of mines in North Africa came before the second battle of El Alamein, at Alam Halfa, towards the end of August 1942, when Rommel's major offensive against Egypt was blunted after the Afrika Korps ran into a deep minefield protecting the British positions. After this German attack petered out, the British then took up the offensive, leading to the second battle of El Alamein in October, which was a turning point of the war for the English-speaking peoples. A blanket ban of the kind advocated by Diana, Princess of Wales, in the 1990s would, had it been in place in the Thirties, have spelt disaster for Montgomery at the fulcrum moment of the war in North Africa.)

John Jackson was a private in the 1st battalion of the 79th Regiment, Cameron Highlanders, who fought on the Western Front between 1915 and the end of the war, being present at the battles of Loos in 1915, the Somme in 1916 and in Flanders in 1917, where he won the Military Medal at the battle of Passchendaele. He faced the great German spring offensive of 1918 and was present at the breaking of the Hindenburg Line at the end of September in that extraordinary year. As a regimental signaller, he crawled over open ground to maintain telephone links, despite shell- and rifle fire.[32] Jackson was not a Highlander himself, hailing from Cumberland and working in Glasgow on the Caledonian Railway when the war broke out. He chose the Camerons because, as a skilled working man in a secure job, he wished to serve alongside men with similar backgrounds, his choice determined by 'the class of men joining the various units'. He was invalided back to 'Blighty' in 1916, re-trained, took part in the advance on the Hindenburg Line, was gassed, but was there on the morning of Monday, 16 December 1918, when, with their colours brought out of storage from Edinburgh, the Camerons marched over the German frontier with bayonets fixed, kilts swirling, drums crashing and the regimental pipers playing 'the 79ths Farewell'.

Jackson's short memoir of the Great War, *Private 12768*, was written in 1926, just before the flood of 'generational-betrayal' literature, such as the poems of Wilfred Owen and Siegfried Sassoon, Erich Maria Remarque's *All Quiet on the Western Front*, and numerous other books, plays and films began to distort history's view of the war, emphasising the supposed futility and waste far above aspects of strategic necessity, high morale and good comradeship. What is evident from Jackson's memoir is the lack of cynicism that one now associates with the Great War. 'Again and again throughout his testimony,' the Chichele Professor of War History at Oxford University, Hew Strachan, has commented, 'Jackson bears testimony, both direct and indirect, to the fellow-feeling and mutual respect that existed between officers and other ranks, and between the soldiers and their non-commissioned officers.'[33]

When on Sunday, 26 September 1915, Jackson's commanding officer, Lieutenant-Colonel A.F. Douglas-Hamilton, was shot leading an extra-ordinarily brave assault on 'Hill 70' at the battle of Loos, he was awarded a posthumous Victoria Cross. 'I well remember the pride we all felt when the news became known,' recalled Jackson. 'Had he survived I feel sure he'd have been as proud of his regiment, and the fight it made, as we were of his honour.' In all, the First Division won no fewer than six Victoria Crosses that day. 'The losses of the division ran into thousands and our own battalion had lost 700 of the 950 who went into action,' recalled Jackson. 'Of the whole forces engaged, the casualties for the week-end totalled over sixty-nine thousand officers and men.' Jackson never doubted that the cause had been just and that Britain had had to fight to prevent German domination of the European continent. Furthermore, only three weeks after Loos, 'In spite of the horrors we had passed through in the great battle, we began to pick up again our jaunty devil-may-care ways.' They were a very remarkable generation.

On Wednesday, 2 June 1915, General Sir William Manning, the Governor of Jamaica, wrote to the Colonial Secretary Lewis Harcourt in London suggesting that a regiment be raised from volunteers from the British West Indies.[34] By November 1918, no fewer than 15,204 men had served in the eleven battalions of the British West Indies Regiment (BWIR), which saw service in Palestine, Egypt, Mesopotamia, East Africa, India, France, Italy, Belgium and England.[35] The contribution from each colony ranged from the very large (10,280 Jamaicans out of a population of 331,552) to the relatively small – 229 from the Leeward Isles out of a population of 127,189. Nor was service confined to the land; in November 1914, a number of stokers from St Lucia went down in HMS *Good Hope* at the battle of Coronel. The same month the German light-cruiser *Karlsruhe* blew up accidentally on the way to attack Barbados, an event that was kept secret from the British until wreckage began washing up on St Vincent and Grenada six months later. The war saw restrictions and shortages, but these were largely offset by a large rise in the price of sugar. There was also heavy patriotic buying of non-interest-bearing war loans. 'Throughout the war the loyalty of West Indians was impressive,' concluded an historian of the region in that period.[36]

While the War Office in London was happy enough to see black soldiers in non-combatant roles, such as labour battalions behind the lines in France, the concept of them fighting in the trenches of the Western Front was far more controversial. Officers who had commanded black troops in Africa argued strenuously with General Callwell, the Director of Military Operations, that blacks' 'natural fighting qualities' and good physiques meant that Britain should not waste the opportunity of throwing His Majesty's West Indian subjects into the fray. The fate of the 24th Regiment at the hands of the Zulus at the battle of Isandhlwana in 1879 was invoked.[37]

On the side of the argument to employ blacks were their large numbers, their relatively low wages and their keenness to serve, as well as the fact that the Empire was frankly running low on whites. Against that were ranged a very wide but demonstrably thin series of arguments, most of them surprisingly prejudiced even for that period and milieu. Some in the War Office believed that blacks were too ignorant to be able to throw bombs properly; others thought it was 'just the kind of thing the French would resort to'; others were concerned about the reaction of South African whites; it was even feared that the Germans might object![38] Sir Leslie Probyn, Governor of Barbados, made the excuse that, 'Their colour would make them dangerously conspicuous to the Germans.' West Indians in Britain who tried to join up were often rejected by officials, even though Indians had served in the British Army for centuries and were to make a huge contribution to the war effort. (Several units were prepared to accept them, however.)

The War Office rejected the Colonial Office's offer of an overseas contingent, saying that blacks needed to stay in the West Indies for local defence, as though the Kaiser were planning a surprise attack on St Kitts and Nevis. Even if it was truly the case, as a letter to the West India Committee Circular pointed out, local riflemen would hardly be able to withstand the guns of a German cruiser. Fearing the bad political effect of refusing the West Indies' offer altogether, Harcourt suggested that they be sent to fight the Turks in Egypt. This was turned down by the War Office, whose offer of using Caribbean troops as peace-keepers in captured West African territories was in turn refused by the Colonial Office. The effect of the impasse began to be felt in the West Indies, and the liberal *Federalist* newspaper of Grenada was not alone in concluding that the reason was 'the nasty cowardly skin prejudice characteristic of the empire'.[39] The War Office's reluctance was all the more surprising considering that the West Indies Regiment had been in existence since the Napoleonic Wars and had seen plenty of service since then, albeit always in the Caribbean theatre.

The person who intervened to break the impasse was George V, who on 17 April 1915 told the Colonial Office that he could not help 'thinking that it would be very politic to gratify the wish of the West Indies to send a Regiment to the front'. He then saw Lord Kitchener, the Secretary for War, who (quite disingenuously) claimed to have supported the idea all along against Colonial Office obstruction. In the opinion of a recent historian of the West Indies during the Great War, 'The decision of the King to intervene effectively ensured that the West Indian involvement in the war would no longer be frustrated by the bureaucratic and racial attitudes of the War Office and the Colonial Office.'[40]

The news that a regiment was to be raised was greeted with enormous zeal among the public, especially when it was made clear that the troops would be paid the same rates as British soldiers, namely a shilling a day for privates and two shillings and four-pence for non-commissioned officers. 'This is history,' trumpeted the radical newspaper, the *West Indian*. 'We have before us today a blank page on which to write our glorious record. This is an hour that will not sound again for centuries. ... West Indians, most of whom are descendants of slaves, fighting for human liberty together with immediate sons of the Motherland in Europe's classic field of war made famous from ancient Grecian days to the days of Marlborough and Wellington.'

Newspapers were instrumental in encouraging West Indians in joining the colours, as were the major religious denominations such as the Baptists, Anglicans, Catholics and Wesleyans. Posters and carefully staged films were produced for those whose illiteracy cut them off from written propaganda, and in British Honduras the Governor even reminded would-be recruits of their bravery in fighting against the British as proving their martial valour. Using ancient precedents, delinquents were given the alternative of enlistment by magistrates who were otherwise going to send them to gaol; bounties were paid by patriotic employers; tax-exemptions were instituted for volunteers; the good rates of pay were emphasised and everything was done to encourage that strange combination of adventure, duty, patriotism and rise in status that donning the King's uniform was meant to engender.

On 21 January 1916, the first battalion of the British West Indies Regiment embarked on HMS *Marathon* for Alexandria, where they joined the Egyptian Expeditionary Force. The second battalion sailed to Crown Hill Barracks, Plymouth. They did well in training, and, as a third and fourth battalion arrived, they were then sent out to any number of tasks, most of which did not involve actual fighting. They built defences for the Suez Canal, joined the East Africa Expeditionary Force in Kenya and the Indian Expeditionary Force in Mesopotamia, carried ammunition, built trenches, mended roads, carried stretchers, loaded and unloaded ships, undertook guard duty at inland water transport camps, served as clerks, carpenters, blacksmiths and motor-boat drivers, and those who in civilian life had been electricians, fitters or engine drivers were transferred to the Royal Engineers. The hopes of their officers that they would be deployed together as a fighting regiment on a single battle front were dashed against the continued prejudice of the War Office, where it was assumed that they would only function well in warm climates. The tasks they undertook were important, but not the martial ones they had hoped for when they joined up.

Following terrific losses at the battle of the Somme, the War Office finally began to employ large numbers of non-white labourers on the Western Front, some 193,500 in all from China, India, South Africa, Egypt, Malta, the Seychelles and the West Indies after 1916.[41] During a conference in Cairo in November 1916, it was also decided to allow some members of the BWIR to be tested in battle, partly because further recruitment was being badly hampered by the fact that the regiment had seen so little action. Whenever the BWIR did come under fire, it behaved with exemplary courage, which was reflected in the large numbers of decorations per capita that they were to receive. 'They were subjected to enemy artillery bombardment, sniper fire, exploding ammunition dumps and aerial attacks,' records one historian. 'In France, life was also made uncomfortable by the prevalence of fleas, lice and rats, while in Egypt there were problems with scorpions, lizards, snakes and especially flies. Nevertheless, in every theatre the West Indians consistently displayed courage and discipline.'[42] Their decorations included no fewer than 19 Military Crosses, 11 Military Crosses with Bar, 37 Military Medals, 11 Military Medals with Bar, 49 mentions in despatches, 11 *Médailles d'Honneur* and 14 Royal Humane Society's Medals, a proud total for any unit. The Commander-in-chief, General Sir Douglas Haig, praised their contribution highly.

Losses were relatively low: six killed and sixteen wounded in the Arras offensive of April to June 1917; sixteen killed and sixty wounded in the Messines offensive of June to July 1917; fifty-seven killed and 377 wounded at Passchendaele between July and December 1917. In one sense, the racism of the War Office saved the BWIR from the horrors of front-line service in the Great War and thus undoubtedly saved the lives of several thousand West Indians. They might have been aching to serve, and would undoubtedly have been very brave, but had they been given a section of the Western Front to themselves they would certainly have endured the same terrible losses as the French, Canadians, British, Australians, New Zealanders and Germans. As it was, the West Indians took their honoured place in the English-speaking peoples' defence of Civilisation as prominently as they were permitted between 1914 and 1918. As with Australia and New Zealand, Great War service had a profound long-term effect. Ian Beckett of the US Marine Corps University has concluded, 'West Indians generally identified themselves as British, and enjoyed a generally warm welcome in wartime Britain, but that war service did lead to the beginning of a greater collective consciousness about what it meant to be West Indian.'[43]

Back at Brooks's Club in 1915, Sir John Fuller had bet Captain Percy Creed a guinea that the German High Seas Fleet would 'come out' before the cessation of hostilities, and he won his wager a year later when, on 30 May 1916, Vice-Admiral Reinhard Scheer put to sea and cruised north towards the Skagerrak, with Franz von Hipper's five battle-cruisers leading the way. They were on their way towards a major clash with the Royal Navy's Grand Fleet, which took place later that day off Jutland.

The battle of Jutland was a huge engagement, involving no fewer than 151 Royal Navy vessels – including twenty-eight dreadnought-class battleships and seventy-three destroyers versus ninety-nine German ones, among them sixteen dreadnought battleships and sixty-one destroyers. (Each fleet had about forty-five submarines, yet none were used.) Although tactically the Royal Navy came off worse, losing fourteen ships and suffering 6,784 casualties against the Germans' eleven ships and 3,039 casualties, it was a strategically successful engagement because the High Seas Fleet returned to harbour and never left it for the rest of the war, and the blockade of Germany could be resumed. The Germans required a decisive victory at Jutland, and by failing to win they had effectively shot the single most powerful bolt in their formidable armoury. They had been building up their capabilities for eighteen years, but Jutland was not the blow that they needed to wrest control of the North Sea from the Royal Navy.

For all the hellishness of the battle of Jutland itself, combat at sea did attract some chivalry during the Great War; when Vice-Admiral Sir Frederick Sturdee sunk the German fleet off the Falkland Islands in December 1914, he invited the *Gneisenau*'s captain, Commander Pochhammer, to dinner on his flagship and placed him at his right hand. After dinner, the Admiral warned the Commander that since the King's health was about to be drunk, perhaps he would prefer to abstain from the toast. On the contrary, replied Pochhammer, he had accepted the Royal Navy's dinner invitation so of course he would conform to her traditions.

Fighting at the battle of Jutland was the HMS *New Zealand*, a battle-cruiser that had been launched in 1911 as a gift to the Royal Navy from the people of that Dominion. When she had sailed to New Zealand in 1913, over half-a-million people went to the various ports around the country to view her, out of a total population of 1.1 million. A Maori chieftain had proclaimed her to be a lucky ship, and so it proved. Although she took part in three major engagements in the Great War, including Jutland, she suffered no casualties. It was a spokesman of the Maoris who, in 1918, spoke a moving truth about why his people had fought so enthusiastically for the British Crown:

We know of the Samoans, our kin: we know of the Eastern and Western natives of German Africa; and we know of the extermination of the Hereros, and that is enough for us. For seventy-eight years we have been, not under the rule of the British, but taking part in the ruling of ourselves, and we know by experience that the foundations of British sovereignty are based upon the eternal principles of liberty, equity and justice.[44]

One of those killed at the battle of Jutland on 31 May 1916 was Bernard Bailey, a seventeen-year-old midshipman in *HMS Defence*. The third son of the 2nd Baron Glanusk, Bailey was the youngest of the 270 peers or sons of peers to perish in the Great War. (Another casualty had been his elder brother Gerald, a lieutenant in the Grenadier Guards who had been killed in action in August 1915.) Although of course all classes in Britain suffered grievously in the Great War, it was the upper classes who suffered the most per capita, with profound political and social implications for the future of their caste and the nation. Many of the factors that led to the decline of the power and authority of the British aristocracy in the twentieth century had been in place in the late nineteenth, such as the continuing agricultural depression since the 1880s, the County Councils Act of 1888 and the imposition of death duties from 1894, but the scything down of such a large proportion of the bravest and the best of the British nobility between 1914 and 1918 played the most prominent part.

Political and social leadership was ingrained in the upper classes; 1914 began four years of their slaughter. As the historian Hugh Cecil has put it,

> When Armageddon came in 1914, as they expected of themselves, the gentry and aristocracy took a lead in joining up to serve their country. Challenged already by the emerging democracy, they felt the obligation to prove themselves in their historic role as warriors. Those who could, pulled strings to join the regular regiments of the BEF and get out to the Front within a few months of the war's beginning. Even before the Somme, the death toll among the sons of the landed classes was devastating.[45]

From the 685 families listed in *Debrett's* and the *Complete Peerage for the British Isles* of 1914, no fewer than 1,500 men served in the armed forces over the next four years. The number of peers or peers' sons who died in the Great War numbered even more than during the mass culling of the aristocracy during the Wars of the Roses.[46] Almost one hundred of them had been born in the 1890s and were thus only between seventeen and twenty-eight at the time of their deaths. The same general figures extended to the rest of the governing class. Of the 5,650 Etonians who served in the Great War, 1,157 lost their lives; the death toll amongst the 838 Balliol men was 183, or 22%. Families such as the Grenfells, Listers, Horners, Asquiths and Charterises paid an inordinately high price for the privilege of leadership that they had long enjoyed but were shortly to lose. Thirty-two peerages and thirty-eight baronetcies became extinct between 1914 and 1918, as the losses, particularly in the Guards Division, Royal Artillery and Rifle Brigade, mounted.

The multiplication of war-related death duties meant that one-quarter of all the land in England and Wales changed hands in the three years after 1918. Another consequence of the war was that British politics became the preserve of the old. Whereas before 1914 ambitious and talented young aristocrats entered the House of Commons and rose relatively quickly – the average age of a minister entering Asquith's Cabinet was fifty-four – after 1918 the field was left to the older survivors. The average age of a minister entering Baldwin's 1935 Cabinet, for example, was seventy. This made for the kind of cautious politics typified by the Respectable Tendency who dominated the National Governments of 1931 to 1940.

On Easter Monday, 24 April 1916, as the battle of Verdun was reaching its height, between 1,000 and 1,500 volunteers from Sinn Fein and the Irish Republican Brotherhood staged a rebellion against British rule in Dublin. Although the rebels captured the General Post Office in Sackville Street and the Four Courts on Inns Quay, the Rising was put down by the Royal Irish Constabulary with support from some regular forces. It had not sparked a general Irish uprising against Britain, as was hoped by the nationalists, and by 1 May it had been suppressed after some fairly heavy fighting (although, of course, nothing like so heavy as was taking place in France at the time).

The help that Irish republicans received from and gave to Germany at Easter 1916, even as tens of thousands of both Protestant and Catholic Irishmen were fighting against the Kaiser on the Western Front in order to protect the British Isles from invasion, led the English writer Roy Kerridge to conclude that,

> Irish 'patriotism' is something of a sham. As of old, the Irish have a warrior class, now known as the IRA. . . . The warrior class do not mind destroying Ireland if by such means they can spite England. Throughout their history they have invited would-be rulers from France, Spain and Germany. Far from being proudly independent patriots, they are willing to be ruled by any foreigner as long as he's an enemy of England. The pre-provisional IRA toyed with the idea of an alliance with Soviet Russia. . . . We English are the unwitting enemy of the warrior Irish no matter what we do, whether we like Ireland or not.[47]

The idea that the Kaiser, if successful in a war against Britain and thus master of the European continent for decades to come, would have allowed southern Ireland to be free and independent is a geopolitical absurdity, making Kerridge's analysis all the more credible.

The execution after summary trials of sixteen of the rebels, including their leader Patrick Pearse and the wheelchair-bound James Connolly, however legally correct and well-deserved considering the wartime circumstances, was a disastrous political error on behalf of the Government solely because it led to their 'martyrdom' by the republican movement. Many romantic songs and paeans were written in Gaelic and English glorifying the Easter 'heroes', and as W.B. Yeats' biographer and the historian of modern Ireland, Roy Foster, has pointed out, 'Initially seen as a lunatic gesture by an unrepresentative minority, the Rising came to be canonised as the founding event of the independent Irish state.'[48]

Sir Roger Casement had landed with a consignment of arms from a German submarine on 20 April to lead the rebellion, but was captured four days later and executed for high treason on 3 August. For twenty years a member of the British consular service, an admirer of Queen Victoria and a humanitarian campaigner, he had nonetheless thrown in his lot with the Germans. A movement to spare his life was stymied by the circulation by the British Government of his 'Black Diaries' relating to his very promiscuous homo-sexuality, which meticulously detailed the shape and length of every penis with which he had come into contact over many years. (For eighty-six years, the republican movement sought to argue that these had been forged by British Intelligence, even after a symposium funded by the Irish Government and hosted by the Royal Irish Academy found otherwise. In August 2002, an independently commissioned, careful, professional forensic examination of the diaries concluded that they were without doubt genuine. Even now, however, some republicans still deny it, because, in the words of one historian, the argument had by then become 'essentially theological, in the end a matter of faith rather than reason'.)[49]

Irish sympathies can be deduced from the fact that when in 1966 Casement's corpse was disinterred from Pentonville prison and flown to Ireland, he was given a full state funeral, whereas only in the twenty-first century have Irish officials attended the interments of those who had fought against the Kaiser. On 'Rebel tours' of Dublin, tourists are shown bullet-holes on the pillars of the General Post Office, which are said to have been caused by the British, but which were in fact probably created by nothing more violent than weather erosion.[50] (Another explanation is that they were caused during the Irish Civil War, and thus had nothing to do with the British either.)

Overall, Irishmen gave their lives for the Empire in the Great War at a high rate, even compared to Australians and New Zealanders: 50,000 Irishmen died out of a population of 4.376 million, or 1.14%, which was only slightly lower than Australia's 1.25% and New Zealand's 1.66%, but higher than Canada's 0.76%.[51] It is a sign of how loyal most Irishmen were to the Crown, even while a tiny minority of them raised the flag of rebellion over Dublin at Easter 1916.

The torpedoing without warning of the unarmed French Channel ferry steamer, the *Sussex*, on 24 March 1916, in which fifty people died and several Americans were injured, helped Woodrow Wilson to demand from the German Government the immediate cessation of unrestricted U-boat warfare. Germany made what was called the Sussex Pledge, suspending her campaign against passenger shipping and declaring her intention to prosecute the campaign only according to the rules of the International Prize Court. This meant that from 4 May, U-boats would not sink vessels without warning and would make provision for the rescuing of crews.[52] This did not affect the British Government's decision on 7 June to denounce, by order-in-council, the 1909 Declaration of London, which had codified the rules of blockade, defined contraband and generally benefited neutral powers. Britain was now fighting for her life and would not be hamstrung by agreements made under very different circumstances in peacetime. As Field-Marshal Ludendorff expostulated in his war memoirs, 'The violation of International Law was to be made legal and valid! We in the East also felt the effects of England's continued violation of International Law.'[53] Although complaints about such violations need to be seen in perspective when coming from a senior member of the High Command that invaded Belgium, Ludendorff was right that Britain was in breach of her commitments, but at least she announced the fact publicly.

Although the US Congress passed an Act to reorganise the army in June, and appropriated $300 million for naval expansion in August, in General John Pershing's view 'scarcely a move was made' to put either plan into operation before America's entrance into the war.[54] John Joseph Pershing was born in Linn County, Missouri, in 1860, graduating from West Point in 1886 as senior cadet captain, the highest honour the Military Academy conferred. He took part in the campaigns against the Apache Indians in New Mexico and Arizona as a second lieutenant in the 6th Cavalry between 1886 and 1890, and was chief officer of volunteers in Cuba during the Spanish-American War, before serving in the Philippines until 1903. After that he was on the General Staff, then a military attaché during the Manchurian Wars, then governor of the Moro province in the Philippines. For all his own many close brushes with death on campaign, the worst tragedy of his life struck in 1915 when his wife Helen and their three daughters perished in a fire at their home in Presidio, San Francisco.

A true soldier's soldier, Pershing commanded the troops sent to Mexico to pursue Pancho Villa, and it was while he was there that he was appointed to command the American Expeditionary Forces in Europe in 1917. Pershing believed that the Administration and Congress missed a fine opportunity in the spring of 1916 'to organise and equip half-a-million combat troops and request numbers of support troops, i.e. bring up Army and National Guard to war strength'. This, he firmly believed, could have made all the difference. As he saw it, albeit in retrospect, with Russia being 'a factor to be reckoned with' in the spring of 1916, French morale still high after their victory at Verdun and British forces reaching 'their maximum power', with the eruption of half-a-million American troops onto the Western Front, 'the war could have been brought to a victorious conclusion before the end of that year'.[55] Like all counterfactuals it is unproveable, but the point of view of America's commander in the European theatre must be accorded considerable weight.

The losses sustained by the British Army on the first day of the Somme offensive – Saturday, 1 July 1916 – not only completely dwarfed all earlier battles in British history, but exceeded many other entire wars. The 57,470 casualties – including 19,240 dead – of that single day bear equivalence to the 16,000 British soldiers killed in the Boer War and the 20,000 killed in the Crimean War, and were many times more than those killed in the Zulu War or the Indian Mutiny. As an extreme example, the Charge of the Light Brigade cost 110 dead in seven minutes, whereas the first day of the Somme killed 175 times that number. In all, the 142-day battle between 1 July and 19 November 1916 left 1.22 million men dead or wounded, of whom 398,671 were British.

A novel of 1929, written by an expatriate Australian aesthete-turned-soldier called Frederic Manning entitled *Her Privates We*, goes some way towards explaining what it must have been like to have fought in that battle. In the vast canon of First World War literature, along with such masterpieces as Robert Graves' *Goodbye to All That*, Alexander Solzhenitsyn's *August 1914*, Vera Brittain's *Testament of Youth* and Erich Maria Remarque's *All Quiet on the Western Front*, Manning's haunting semi-autobiographical tale became an international bestseller. The story of Private Bourne and his ever-dwindling band of comrades was, according to Ernest Hemingway, 'the best and noblest book of men in war that I have ever read. I read it over once each year to remember how things really were so that I will never lie to myself nor to anyone else about them.'

Manning wrote under a pseudonym 'Private 19022', produced very little else afterwards and died early, in 1935, aged only fifty-two. T.S. Eliot and the artist Sir William Rothenstein attended his funeral, but few others. Australia, usually quick to lionise her literary figures, largely turned her back on the expatriate. Despite his ability perfectly to reproduce the authentic patois of his comrades in the King's Own Shropshire Rifles, Manning was as utterly unlike the archetypal squaddie as it was possible to be. His biographer explores the manifold contradictions in his personality:

> A Catholic who was an Epicurean philosopher, a sceptic who was a believer, a Tory who was a democrat, a recluse who had a gift for friendship, a soldier in the worst of modern battles who was afraid of crossing Piccadilly Circus before the war, a bachelor with affectionate relationships with both men and women, Frederic Manning was a complex, puzzling and intriguing personality as well as a fine writer.[56]

(The book's title is a lewd pun on the lines from *Hamlet*: 'On fortune's cap we are not the very button . . . Then you live about her waist, or in the middle of her favours? . . . 'Faith, her privates we.' Such a low gag might have been expected from Private 19022, but hardly from the immensely cultivated senior book reviewer of the *Spectator*.)

It was not until 1977 that an unexpurgated edition of *Her Privates We* was published, in which the swear words of the original manuscript were reinstated, and 'beggar', 'cow' and 'muckin'' were returned to their baser originals. The obscenities did not seem at all out of place in the context of rats eating corpses in shell-holes, bodies being ripped apart and all the other true obscenities of the Somme offensive. One possible reason why it was never placed in the first rank of Great War novels, despite dealing with the same subject as did Wilfred Owen, Siegfried Sassoon and Joan Littlewood's 1960s' stage-play *Oh! What a Lovely War*, is that it is not anti-war. Despite being written at the height of the political movement that sought to portray the Great War as a great crime, Manning instead presented it as merely a horrible but unavoidable part of the human condition. 'To call it a crime against Mankind is to miss at least half its significance,' he wrote in the Preface, 'it is also the punishment of a crime.' Fatalism was a strong theme running through the book, but not cynicism, and the blood-stained experience of the rest of the century rendered his analysis more realistic than that of many of the more fashionable pacifist authors of the Twenties and Thirties.

Nor did the book contain a denunciation of the British officer class. Some of the most sympathetic characters are the harassed but honourable captains and lieutenants trying to do their best for their men in the face of Armageddon. Part of the book's raw power came from the fact that characters were drawn not to emphasise political points so much as to let the reader understand what it was really like to have been on the Western Front. Although there were some gripping moments of action in the book, much of it was set in billets behind the lines, as the troops chat and scrounge, booze and drill and occasionally try to fornicate. It is an antidote to the outrageous lampooning of *Oh! What a Lovely War*, which was 'motivated not only by a fierce class hatred of the establishment, but also by the contemporary international situation . . . of the Cuban missile crisis and developing American involvement in Vietnam'.[57]

The Great War myth insists that the British generals in that conflict were mainly 'aristocratic cavalry officers who lived in chateaux well behind the front line and sent millions of young men, either duped working-class heroes or sensitive middle-class poets, to their deaths'.[58] In fact, as the historian Robin Neillands has pointed out, of the forty-seven men who commanded Western Front divisions in the war, only nine were cavalrymen.[59] Most were middle-class professional soldiers, not upper-class twits, and the two generals who were genuinely upper class, Julian Byng and Henry Rawlinson, were amongst the best. The Great War generals tended to be infantrymen in their late forties, who had seen plenty of active service. Instead of living in chateaux behind the lines, no fewer than ninety-seven were killed in the war and a further 146 were wounded or taken prisoner. The same number of major-generals – three – were killed in action at the battle of Loos, for example, as in the whole of the Second World War. They were also personally brave; ten held VCs and 126 DSOs. Staff officers – 'red tabs' – worked fourteen-hour days and knew the front lines well. The military historian Richard Holmes considers them to have been 'honest, brave and hardworking' soldiers, who were 'all too well aware of the consequences of their mistakes'. It was, after all, their own brothers, cousins and friends who were dying in the trenches day after day, month after month. To mistake their stiff upper lips, which were necessary for maintaining morale, for insensate heartlessness is a very common mistake.

Since the armies of the English-speaking peoples were the only ones still capable of mounting any kind of offensive in 1918, with morale still high, and considering that they demonstrably won the war with a series of impressive victories that summer and autumn, Field-Marshal Sir Douglas Haig could not have been the blunderer depicted in the play and film *Oh! What a Lovely War* or in Alan Clark's book *The Donkeys*. In fact, far from being the 'butcher and bungler' of popular mythology, Haig learnt fastest and best of the Allied generals how to defeat our most formidable and efficient foreign enemy since the Napoleonic Wars. They took twenty-two years to win; in 1914 the High Command managed it in four. If there had been a way of fighting the war differently once trenches stretched 400 miles from the Channel to the Swiss border, the Germans or French might have discovered one, but neither did. Modern critics of the High Command need to explain how they would have fought the Great War differently.

Trench warfare had been a major feature in the Crimean War, the American Civil War, the Boer War and the Russo-Japanese War, so it was hardly new. Yet far from constantly going over the top, or 'over the bags' as it was called at the time, it was perfectly possible for an infantryman to see two years' service in the trenches without ever taking part in a major offensive. When they did, the casualty rate was horrifically high, although death by disease was very low, for the first time in history. Discipline was not simply kept up through fear. Only 346 of the 3,080 death sentences for desertion and cowardice were actually carried out – out of an army of over five million men – and usually to reinforce discipline in a particular unit or at a particular time.[60] All but a handful of those shot were justly convicted by the law as it stood at the time. Of course they were hardier men than today, often brought up in tough environments devoid of any kind of luxury. Yet the reason most Britons fought – apart from the sense of comradeship and regimental pride fostered in all armies – was that they rightly believed that Britain's safety depended upon victory. Some 704,208 Britons died in the war – 1.53% of the population – but they did not consider that they were dying in vain.

The losses on the Somme and elsewhere led the US Ambassador to Britain, Walter Hines Page, to recall how, before the United States' entry into the war,

> I suppose a thousand English women have been to see me – as a last hope – to ask me to have enquiries made in Germany about their 'missing' sons or husbands, generally sons. They are of every class and rank, from marchioness to scrub-woman. Every one tells her story with the same dignity of grief, the same marvellous self-restraint, the same courtesy and deference and sorrowful pride. Not one has whimpered. . . . It's the breed.[61]

Public displays of emotion were thought, rightly, to be bad for civilian morale; there was genuine, heartfelt lamentation for loved-ones, of course, but it was patriotically kept private. Elsewhere Page wrote:

> They never weep; their voices do not falter. Not a tear have I seen yet. They take it as part of the price of greatness and of empire. You guess at their grief only by their reticence. They use as few words as possible and then courteously take themselves away. It isn't an accident that these people own a fifth of the world. Utterly unwarlike, they outlast anybody else when war comes.

Few statesmen in the English-speaking world in November 1916 had the weighty authority – what the Romans called *auctoritas* – of Henry Charles Keith Petty-Fitzmaurice, 5th Marquess of Lansdowne. Seventy years old, a large landowner and Unionist grandee, a former governor-general of Canada, viceroy of India, foreign secretary and leader of the Conservative Party in the House of Lords, he was a figure of huge influence. Back in 1905, Mr Mowbray Morris had bet Mr Spencer Gore a sovereign at the Beefsteak Club that Lansdowne would be the next Conservative prime minister, but since there were no Conservative prime ministers for the next eighteen years, the opportunity passed. Lansdowne had been an enthusiastic supporter of the war at its outbreak, joining other Unionist Party leaders in pressing Asquith to send the 100,000-strong British Expeditionary Force to France. His youngest and favourite son Charlie, who was already in the Army, was killed at the battle of Ypres two months later. In 1915, he turned over his vast and beautiful stately home, Bowood, to the authorities for use as a hospital for the war-wounded, who were nursed by Charlie's widow Violet.[62]

By the autumn of 1916, Lansdowne, who was a minister without portfolio and a member of the War Cabinet, found that his staunch support for the war had yielded to doubt, and on 13 November 1916 he produced a memorandum for his colleagues that called for a negotiated peace. In it he pointed to reports from the president of the Board of Trade, Walter Runciman, saying that 'our shipbuilding was not keeping pace with our losses' and predicting 'a complete breakdown in shipping ... much sooner than June 1917'.[63] He then cited the President of the Board of Agriculture's report on 'Food Prospects', which predicted bread price rises and stated that in some parts of the country 'it was no longer a question of maintaining a moderate standard of cultivation, but whether cultivation will cease'. A report from the First Lord of the Admiralty of 14 October 1916 had stated that the enemy were increasing the numbers, size and strength of their U-boats and 'the submarine difficulty is becoming acute, and that, in spite of all our efforts, it seems impossible to provide an effective rejoinder to it'.

With acute manpower problems – especially in the light of Ireland's unwillingness to supply the 150,000 men that could be expected from that country – and casualties of over 1.1 million, including 15,000 officers killed, Lansdowne thought it right to ask,

> what our plight and the plight of the civilised world will be after another year, or, as we are sometimes told, two or three more years of a struggle as exhausting as that in which we are engaged. No one for a moment believes we are going to lose the War; but what is our chance of winning it in such a manner, and within such limits of time, as will enable us to beat our enemy to the ground and impose upon him the kind of terms which we so freely discuss?

Lansdowne continued: 'We are slowly but surely killing off the best of the male population of these islands'; meanwhile the financial burden came to more than £5 million per day and 'the responsibility of those who needlessly prolong such a War is not less than that of those who needlessly provoke it'. After noting how small were the gains won by Allied gallantry in both 1915 and 1916, Lansdowne asked: 'Can we afford to go on paying the same kind of price for the same sort of gains?' He then surveyed the uninspiring prospects of the Italian offensive, the Russian and Romanian fronts, Salonika, problems with neutral powers like America and the 'war weariness' of France. Most prescient of all he wrote: 'The domestic situation in Russia is far from reassuring. There have been alarming disorders both at Moscow and in Petrograd. Russia has had five ministers of the interior in twelve months, and the fifth is described as being by no means secure in his seat.'[64]

Sir Douglas Haig's diary entry for Saturday, 25 November, recorded how when the War Secretary, Lloyd George, came to bid him goodbye as he was leaving for France, 'He told me that he considered the political situation serious. Lord Lansdowne had written a terrible paper urging that we should make peace now, if the naval, military, financial and other heads of department could not be certain of victory by next autumn.'[65] Lloyd George appreciated the weight of Lansdowne's influence. 'Coming from a statesman of Lansdowne's position and antecedents,' he wrote later, the memorandum 'made a deep impression. No one could accuse him of being a mere "pacifist".

He was the author of the Entente Cordiale of 1904.' By 'antecedents', Lloyd George meant that Lansdowne hailed from one of the grandest Whig families in Britain; he only joined the Unionists after Gladstone had promoted Irish Home Rule in 1886.

There was something innately Whiggish about the way that Lansdowne was capable of stepping back from the passions of the war to look at the likely state of Europe after its conclusion. Although the Whigs were ardent in their love-lives, they were rarely so in their politics. Since the Glorious Revolution of 1688–9, which they had engineered, they had proved masters of political compromises, and they had often negotiated the end of wars to Britain's advantage, rarely advocating fighting to the bitter end. The Napoleonic Wars were only fought to the finish of Napoleon because the Tories were in government under Lord Liverpool, for example, while the peace of Paris that ended the Crimean War was negotiated by the Whig Government of Lord Palmerston. Lansdowne was thus acting securely in the Whig tradition. (The minister most interested in peace negotiations in 1940, Lord Halifax, also hailed from the Whig tradition in politics and only became a Conservative in 1909; his grandfather had married the daughter of the Whig Premier, Earl Grey, and had served in all the Whig Governments from 1832 until 1874.)

Suitably forewarned by Lloyd George, Haig delivered a particularly upbeat assessment to the Cabinet of the prospects for the British Army for 1917. In it he concluded that after the victory of the Somme, 'an appreciable proportion of the German soldiers are now practically beaten men, ready to surrender if they could find opportunity, thoroughly tired of the War, and hopeless of eventual success'. Furthermore, stated the Commander-in-Chief, 'It is true that the amount of ground gained is not great. That is nothing. Our proved ability to get the enemy to move at all from his defensive positions was the valuable result of the fighting.'[66] Unsurprisingly, after his paper was respectfully rejected by his colleagues, Lansdowne declined to join the Lloyd George Cabinet when Asquith was overthrown the following month, since it was committed to an even more vigorous prosecution of the war.

Although co-ordinating the peace approach with the other Allies such as France, Russia and Italy would have been hard, it would not have been impossible in November 1916. Similarly, ensuring that the Germans fully withdrew from France and Belgium and all captured territories would have required tough negotiation. Yet an armistice in late 1916 when Lord Lansdowne first advocated it would probably have saved the twentieth century from the horrors of the Bolshevik and Nazi revolutions and their aftermaths. Lansdowne's reference to 'the plight of the civilised' was even more apposite than he could ever have guessed.

It took twelve months for Lord Lansdowne to go public with his views in favour of a negotiated peace, which he did with the true aristocrat's insouciant disregard for public reaction. In the year since he had put forward his proposal to the Cabinet, the first Russian Revolution had broken out; as he wrote of his stately home to his daughter Evie that July, 'I am glad Bowood is not in Russia – but you may live to see trouble in this country if things go less well than people want.' By the time he made his public call for a negotiated peace, Russia had undergone the second revolution of 1917, and the Bolsheviks controlled Moscow and Petrograd. After *The Times* rejected his 'peace' letter for publication, Lansdowne sent it to the *Daily Telegraph*, where it was published on Thursday, 29 November, under the fantastically innocuous (and indeed disingenuous) headline: 'Co-ordination of Allied War Aims'.

It was fortunate that Lansdowne was so utterly indifferent to what others thought of him, because he was immediately deluged with obloquy, especially from the Northcliffe press. 'I am bomb-proof against all newspaper abuse,' he had said during a crisis when he was foreign secretary twelve years earlier, and that served him well. A lesser man might have buckled under the weight of the attacks, the friends who cut him, the fact that the Censor began opening his correspondence, his 'official excommunication' from the Unionist Party, but above all the way that the press 'slammed into him as if he were either a traitor or a doddering representative of the privileged classes out to save their wealth and power'.[67] He could take solace from the avalanche of letters of support he received from soldiers at the Front and the small but vocal British peace movement.

Of course by November 1917 the right time for pursuing peace – during Emperor Karl of Austria's overtures of a year earlier – had passed. Yet Lansdowne had tried at the right time and in the right way – through a secret memorandum in the War Cabinet – and so his second attempt was better than nothing. He followed it up in early March 1918 with another letter to the *Telegraph*, which was published only a fortnight before the German offensive that pushed the Allies back forty miles on the Western Front. By then the time for negotiations had gone, and only a smashing defeat could punish Germany for the war she had unleashed four years earlier.

The month after Lansdowne submitted his initial memorandum, Lloyd George entered Downing Street at the sprint and never let up. Taking the premiership from Herbert Asquith in December 1916, he utterly revolutionised the administration of government and the war, creating new ministries – those of labour, pensions, shipping and food – and slimming down the ultimate decision-making body to a five-man War Cabinet. He also established a Cabinet Secretariat answerable directly to him, with Maurice Hankey in charge. Without fixed-term premierships, the British have long been ruthless in sacking wartime prime ministers who are not perceived as up to the task: Aberdeen was replaced by Palmerston during the Crimean War, Asquith by

Lloyd George in the Great War and Chamberlain by Churchill in the Second World War. One admirer of 'the Welsh wizard', the historian Hywel Williams, has gone so far as to say that Lloyd George's period in office gave Britain her most bracing experience of executive government since the Protectorate of Oliver Cromwell.

After December 1916, Lloyd George was forced to fight the Great War depending on Conservative support in the Commons and the fine leadership qualities he displayed thus needed to include acute coalition diplomacy. His relations with the 'patient, loyal and diligent' Tory leader Andrew Bonar Law, his 'indispensable anchor-man', were a central feature of his ministry. The other crucial relationship was with the military High Command, approximately 90% of whom, in the estimation of his biographer John Grigg, cordially despised him. Lloyd George had on grounds of conscience opposed the Boer War in which many of the generals had served, and he had also been the man who introduced the class-based 1909 budget that led to the aristocracy losing its veto rights over legislation. The friction between the Prime Minister and the brass-hats, especially the Chief of the Imperial General Staff Sir William Robertson and the Commander-in-Chief Sir Douglas Haig, had many dangerous consequences.

Lloyd George did not believe that the appalling level of casualties on the Western Front was necessary for victory, but such was the power of the Service Chiefs that he could not impose his views upon them until almost the very end. Despite this, he held no fewer than 308 meetings of the War Cabinet in a twelve-month period in order to try.[68] According to the generally sympathetic Grigg, Lloyd George was 'seriously deficient' in physical courage (especially when it came to being anywhere near high explosive shells), as well as being vain, a hypochondriac, desperate for adulation and of course a quasi-bigamist. Yet for all that he was a great war leader, who was able to explain to the British people what the sacrifice of a generation of their sons and brothers had ultimately been for.

General Pershing had the measure of the German High Command when he commented in his war memoirs that, 'The date of resuming indiscriminate submarine warfare, 1 February 1917, was timed with the idea that the greater part of neutral and British shipping could be destroyed before we could be ready, should we by any chance enter the war.'[69] As it was, even though the United States declared war against Germany in April 1917, it was not until January 1918 that the American Expeditionary Force 1st Division went into the line in France, and over a year before American troops actually fought their first distinct battle, at Château-Thierry on 3 June 1918. Yet when they did, such was their commitment that they were able to place almost two million enthusiastic young American recruits – nicknamed 'doughboys' – in France and no fewer than 430,000 troops on one battlefield against Germany.

Getting the US Army ready for active service in Europe was an astonishing achievement. As a recent review of Gary Mead's book, *The Doughboys*, succinctly explained, in April 1917

> America's army of one hundred thousand ranked seventeenth in the world, together with Denmark. It had seen no major action since the Civil War, half a century before. It was equipped with rifles, sawn-off shotguns, and a few hundred machine-guns. The army had no tanks or aircraft, and only enough artillery and shells for a single, nine-hour barrage. To the US War Department, which had made no attempt to study methods of warfare developed since 1914, America's most important arm was the cavalry.[70]

Much of the credit for getting the US Army ready for such a major commitment overseas should go to Pershing himself. Once the seemingly endless number of American troops started arriving at the Western Front – and by September 1918 there were two million more in training in America – the German will to resist was sapped to breaking point.

Germany's resumption of unrestricted submarine warfare in September 1917 – suspended after the *Sussex* sinking – was a sign that the war was entering a new and most dangerous phase. Having failed to win the war in an afternoon, as they might have at the battle of Jutland, the Germans were now committed to starving Britain into surrender, come what may.[71] Yet in their struggle to cut off Britain's maritime windpipe, the Germans came up against a largely unsung genius in the shape of Professor Sir Alfred Ewing, one of those fortunate people who are flung up by circumstance and ability to be able to serve their country just when they are most needed. In late 1902 Ewing, the Professor of Applied Mechanics at Cambridge, was lured from his post at King's College by the famously energetic Second Sea Lord, Sir John 'Jackie' Fisher, to become Director of Naval Education. 'I believe a syndicate of Naval Rip van Winkles has been formed to attack the scheme,' Fisher warned Ewing that December. 'They went to ask [Admiral] Lord Charles Beresford to join them. He told them they ought to be stuffed and put in a museum!'[72] A month later Fisher added a rider to another letter, saying, 'You may always rely on me backing you up through thick and thin, through fire and water!' Fisher was a formidable supporter to have in the Whitehall jungle, and Ewing prospered, setting up both the Dartmouth and Osborne naval academies. By the outbreak of the Great War he had a high reputation in the Admiralty and could be entrusted with one of the most important roles of the war.

Born and educated in Dundee, Ewing qualified as an engineer and became a specialist in earthquake measurement when teaching at Tokyo University in the late 1870s. He was a dynamic professor at Cambridge, was married to a great-

great-great niece of George Washington from Claymont, West Virginia, and, according to *The Times'* announcement of his appointment, 'has exhibited a savoir faire which ought to stand him in good stead in a Government department'. Yet he had also always been interested in ciphers and codes, ever since as a small boy he had won a newspaper prize for solving an acrostic puzzle.[73]

Walking with Ewing to the United Services Club for lunch on the day war broke out, 4 August 1914, Rear-Admiral Henry 'Dummy' Oliver, the Director of Naval Intelligence, remarked that some of the Admiralty's Marconi wireless-telegraphy stations dotted around the British coast were passing on to him coded signals that were piling up on his desk, but he had no idea what they represented. After lunch Ewing collected them and took them to his room. He soon realised that they were German naval signals and that 'they would be of incalculable value if they could be interpreted'.[74] What followed next was an example of superb, inspired amateurism in the finest traditions of the English-speaking peoples, fortunately supported to the hilt by the powers in the Admiralty, who early on grasped the inestimable advantages that would flow from being able to read the German High Seas Fleet's codes. The success of the operation even led to the accusation after the war that Britain's success in code-breaking proved that she had been preparing for the war for years.[75]

In centuries past, frigates had been the eyes of the Fleet and used for scouting patrols, but by 1914 mines and submarines had put an end to that method of intelligence-gathering, especially when, a month after war was declared, three patrolling cruisers, *Hogue*, *Cressy* and *Aboukir*, were sunk by a single German U-boat with great loss of life. Ewing, researching among the code-books at Lloyd's of London, at the British Museum and in the General Post Office, was working against time. 'It was a time of enthusiasm, of eagerness to be of service,' wrote his son and biographer. 'Official barriers were already broken down. That he had come from the Admiralty on War business was an open sesame to the most secluded places.'

The operation started out small; he secured the help of a few discreet friends with a good knowledge of German, and they sat around the table in his office labouring on the documents, but 'finding it rather like hammering on an iron safe with the object of knowing what was inside'.[76] The stream of telegraphed intercepts, addressed 'Ewing, Admiralty', grew and grew until 2,000 were being received every day. The first Sunday that Ewing took off work was not until late October 1914, otherwise every available moment was spent trying to crack the codes.

It was to be Ewing who set up the legendary 'Room 40', the department of the Admiralty that was to crack the German naval codes.[77] The ninety code-breakers of Room 40 included five professors, four schoolmasters from Osborne and Dartmouth naval colleges, a City tycoon, a dress designer, a priest, the music critic of the *Morning Post*, wounded soldiers, linguists, publishers, lawyers, the ancient historian Frank Adcock and the son of Gordon Selfridge's butler.

Ewing, who was something of a ladies' man, did not discriminate against women; he found that they were particularly good on the political side. Room 40 saw, as Ewing proudly pointed out ten years later, 'a wide diversity in age, temperament, habits, state'. So vital was the national emergency in cracking the German codes that Edwardian prejudices were not allowed to interpose themselves; here was a meritocracy. The King's College, Cambridge, classicist Alfred Dillwyn 'Dilly' Knox, perhaps Room 40's most brilliant member, was allowed to work from a bath down the corridor in Room 53, since he claimed that codes were best cracked 'in an atmosphere of soap and steam'.[78] Another fellow of King's, the historian Frank Birch, appeared as Widow Twankey at the London Palladium.[79] Unconventional and eccentric personalities, many of whom would have found it difficult to fit in with military or civil service discipline and routine, found that Room 40 appreciated their talents. If the rest of the Admiralty found the organisation somewhat bizarre, they were nevertheless delighted with what their oddballs, 'boffins' and occasional misfits began to produce.

Ewing recalled the way that some tasks needed merely assiduity, patience and care – no mean attributes – whereas others required rare aptitudes amounting to genius. 'Fortunately a few members of staff possessed that incommunicable faculty for inspired guessing. To a good many there would come flashes, sometimes brilliant flashes; but two or three, when they had acquired experience, seemed to live in an atmosphere of continuous light. They would leap or fly to conclusions with an agility incomprehensible to my pedestrian wits.'[80] In fact Ewing's wits were anything but pedestrian, as acknowledged by the post-war press who delighted in coining the following soubriquets for him: 'Sherlock Holmes of Whitehall', 'The Cipher King', 'The U-boat Trapper', 'Eavesdropper Ewing', 'Admiralty Sleuth-Hound' and 'The Navy's Nosey Parker'.

A short, thick-set man with keen blue eyes and shaggy eyebrows, and a disarmingly quiet Scottish voice, Ewing presided over the expansion of Room 40, the name of the room in the old Admiralty building into which the operation spilled from Ewing's office, but which also 'suggested nothing and stirred no-one's curiosity'. What Ewing badly needed was a lucky break, and one came only a few weeks into the war when two Russian vessels sank the German light-cruiser *Magdeburg* in the Gulf of Finland and found the corpse of a drowned signalman still grasping a German Naval signal book, which the Russian Naval Attaché, Captain Wolcoff, passed on to Ewing on 17 October 1914.

The codebook from the *Magdeburg* continued in use until May 1917. As Ewing recalled,

Its code consisted of groups of letters, commonly three letters in a group. For this it used 31 letters, namely the 26 original ones and five more. The usual process of keying, as was presently discovered, consisted of substituting for each letter another letter, according to a prescribed plan. Occasionally, however, for communications which were intended to be ultra-secret and confined to particular correspondents, recourse was had to more complicated methods, especially in the later stages of the war. These gave my staff of solvers some interesting nuts to crack. The staff with much practice became so skilful that the nuts, however hard, were cracked with scarcely an exception.[81]

At roughly the same time two amateur radio hams, a barrister called Russell Clarke and a Somerset squire named H. Hippisley, came to Whitehall to tell Ewing that they were sure they could intercept German naval signals on the 400-metre wave, which was used for all traffic in the Heligoland Bight and for signalling to U-boats. These two men were granted permission to set up a listening station at Hunstanton, which passed messages to Ewing by direct line, a further five being set up later. Soon messages were being intercepted and read, such as signals to open the gate in the boom at the great German naval base at Wilhelmshaven, clear indications when battleships or battle-cruisers were setting sail into the Baltic.[82]

It took some time before the cipher was cracked by Ewing's boffins. It turned out to be a transposition cipher of the same type used by Julius Cæsar. In the early days, the Germans changed the cipher key relatively infrequently, although later in the war they did so daily because they suspected that their codes had been broken. However, by then Room 40 had built up such an expertise in German wireless customs that they usually cracked the new cipher by noon each day. When new signal books were introduced by the Germans, they too were captured from Zeppelins and submarines. The British admirals at the battles of Dogger Bank, Jutland and the Atlantic all benefited from Room 40 information, but they did not always make best use of it, particularly not at Jutland.[83]

Nor was it only naval traffic that Room 40 cracked. When a German consul called Wassmann planned a raid on the Persian oil pipeline at Abadan, Room 40 warned a British man-of-war, which attacked the raiding party first. The Germans fled and Wassmann's baggage wound up in a cellar in the India Office, which fortunately a Room 40 staff member was sent to check. It produced the German diplomatic cipher. Very soon, the heavy diplomatic traffic between Berlin and Vienna, Berlin and Madrid (for onward passage to North and South America), and Berlin and Bulgaria and Turkey was being intercepted. Soon German plans to sabotage American ammunition factories, the Canadian Pacific Railway and the Trans-Siberian Railway were uncovered and foiled. Nothing, however, was as useful as the deciphering of the communication that became known to history as 'The Zimmermann Telegram'.

On 16 January 1917, Count Arthur Zimmermann, the German Foreign Minister, despatched a telegram to Count von Eckhardt, the German Minister in Mexico City, about what to do in the event of the United States and Germany going to war as a result of the coming U-boat campaign. The telegram instructed von Eckhardt to offer the President of Mexico an alliance – ideally with Japan too – against the United States, which would restore the 'lost territory of Texas, New Mexico and Arizona' to Mexico. In order to ensure that the telegram reached Mexico City, the Germans sent it by several routes. It was telegraphed from Nauen to Seyville in Long Island, to be sent to Count von Bernstorff, the German Ambassador to Washington, to pass on to Mexico, but one was also handed to the Swedish Minister in Berlin, to be sent to Washington via Stockholm and Buenos Aires (where it went via British cables). Cheekily enough a third was sent as an enciphered attachment through the American Ambassador in Berlin. 'Was it cynical humour that led the Germans to employ the American State Department and its Embassy in Berlin as a channel for such a communication?' wondered Ewing a decade later.[84]

Once the publisher Nigel de Grey and the Presbyterian minister Revd. William Montgomery in Room 40 had decoded it, and with America hovering on the brink of war, it was imperative for Britain that the Zimmermann Telegram should be made public. Yet it was almost equally important that she should not be seen as the perpetrator, and the secret of the existence of Room 40 needed to be protected. This explosive material had to be detonated, and soon, in order for the English-speaking peoples to be able to stand in the same line of battle, but it must be done in a way that would not allow the Germans to blame British Intelligence. (Nigel de Grey worked for the Medici Society fine art publishers between the wars and then returned to cryptography at Bletchley Park in the Second World War.)

Arthur Balfour, the Foreign Secretary, told Ambassador Page the essential parts of the Zimmermann Telegram, for communication to President Wilson. Wilson verified it and gave it to the American press, which published it on 1 March 1917, with the implication that the American Intelligence services had discovered it. There was 'a German tumult of enquiry and recrimination as to responsibility for the leakage', but it was as nothing compared to the tumult in America about the German perfidy implied by the telegram itself. As Ewing recalled, 'The curtain which hid Room 40 remained wholly undisturbed. From behind it we listened to the storm with serenity, especially to the caustic comments of some newspapers on the superiority of the American Intelligence service over ours. The cryptographic machine continued to function as silently as before.'[85]

Although the Zimmermann Telegram was genuine enough, the schemes contained within it were demonstrably absurd. The idea that Mexico could have wrested the three southern states from America in 1917 was laughable, and one historian has even likened Count Zimmermann to his fellow-Teuton, Baron Munchausen.[86] Yet Germany's willingness to indulge in such fantasies with Mexico italicised her fundamental hostility to the United States, which was very soon reciprocated by the American press and public. Within six weeks of the publication of the telegram, Wilson had declared war.

(As a footnote to the story, as late as July 1979, Sir Brooks Richards of the Cabinet Office attempted to persuade Professor R.V. Jones to cut three passages from a reprint in the Notes and Records of the Royal Society of London of Ewing's speech to the Edinburgh Philosophical Institution in December 1927. Considering that the speech, entitled 'Some Special War Work' and concerning Room 40's role in the Great War, related to events that had taken place over sixty years before, it is an indication of the elephantine memory and great secretiveness of the British state.[87] The passages that the Cabinet Office wished to have excised from Ewing's speech – which itself had been delivered to 1,500 people fifty years previously – all related to the Zimmermann Telegram. All three of the desired excisions referred to the fact that British Intelligence was routinely intercepting and deciphering the diplomatic codes of neutral countries such as Sweden and Mexico.)[88]

If Australia and New Zealand defined themselves as independent nations partly through the sacrifice of Gallipoli, Canada's apotheosis moment came on Easter Monday 1917 with her victory at Vimy Ridge. On the outbreak of war Canada's Premier, Sir Robert Borden, had promised a Canadian army of half-a-million men. Different parts of Canada responded to the call differently. Only 4.7% of men eligible from Quebec volunteered by 1917, for example, compared with 15.5% of those from western Canada and 14.4% from Ontario. (This despite the fact that the war was being fought largely to protect metropolitan France.) After mobilisation in August 1914, the brand new First Canadian Infantry Division had been organised, equipped, trained, shipped to France and deployed in action by February 1915, taking only six months.

The Division had fought valiantly in the intervening two years, but nothing could have prepared it for the assault on Vimy Ridge on 9 April 1917. At 5.30 a.m. that day, the Canadian Corps, consisting of four Canadian infantry divisions, stormed and captured one of the best-defended German positions on the Western Front, exactly the spot from which the French had been hurled back in May and June 1915 with an estimated 150,000 casualties. The battle of Arras was the Allied preliminary to the great Nivelle offensive, with heavy gas and bombardment attacks before the British First Army under General H.S. Horne and Third Army under Sir Edmund Allenby assaulted the positions of General von Falkenhausen's Sixth Army, aided by air support.

The weather was sleety and wet, and the mud beneath the Canadians' feet was slippery. All four Canadian divisions attacked simultaneously, over 30,000 men in all. With over four miles of tunnelling on four different levels – some of which has been preserved – dug by the Royal Engineers and 7th Canadian infantry brigade, some troops (but by no means all) were able to avoid the murderous frontal attack through no-man's-land. The Ridge is an escarpment overlooking the Artois – or Douai – plain, a strategically commanding position near the town of Arras. Sections of it were sixty yards high and a warren of artillery-proof trenches, bunkers and natural caves defended it. No fewer than three rows of German trenches heavily protected by barbed wire and machine-gun nests separated the Canadians from the heights, yet as the Australian General Sir John Monash was later to write, 'It is impossible to overrate the advantages that accrued to the Canadian Corps from the close and constant association of all four divisions with the others. That was the prime factor in achieving the brilliant conquest of Vimy Ridge.'

The main planner of the operation was the Canadian Major-General Arthur Currie, who took over the command of the Corps from General Byng after the battle. His preparations included a mock replica of the terrain, detailed aerial reconnaissance, tunnel-digging, a light railway to move in heavy artillery and a daily 2,500 tons of explosives' bombardment of the Ridge for a fortnight before the assault. It was Currie who invented the concept of a 'creeping barrage', which moved closer and closer forward towards the enemy lines. At dawn on Easter Monday, over a four-mile front, Currie's 1st Canadian Infantry Division, Major-General Harry Burstall's 2nd, Major-General L.J. Lipsett's 3rd, and Major-General David Watson's 4th all attacked the Ridge.

By the afternoon the Canadians had captured most of it, and three days later they also took the high points of Hill 135 and Hill 145. Although the Germans attempted a counter-attack, they were beaten off and withdrew under cover of darkness. The cost to the Canadians was high: 10,602 casualties, of whom nearly 3,602 were killed in action, but these were a fraction of the Allied losses suffered trying to take the Ridge two years earlier. Although the offensive failed to break through as hoped, Arras was relieved, the Germans were demoralized and the Ridge was to prove 'a firm anchor' for the major Allied offensives which were to win the war the following year. Furthermore, as Brigadier-General Alexander Ross DSO, Commander of the 28th (North West) Canadian battalion at the battle, said later, 'It was Canada from the Atlantic to the Pacific on parade. I thought then in those few minutes I witnessed the birth of a nation.'

*

In the debate over American entry into the Great War, the words of George Washington's Farewell Address – "'Tis our true policy to steer clear of permanent alliances, with any portion of the foreign world' – were quoted against Woodrow Wilson, just as they were used later against Franklin Roosevelt in the Second World War and indeed every president who has tried to promote close political engagement with Europe and the outside world. After the war was over, in his speech in Minneapolis on 8 September 1919, Woodrow Wilson fully accepted that the United States ought to have entered the war earlier than she did. 'Her military men published books and told us what they were going to do,' he said of Germany, 'but we dismissed them. We said "The thing is a nightmare. The man is a crank. It could not be that he speaks for a great Government. The thing is inconceivable and can not happen." Very well, could it not happen? Did it not happen? ... The great nations of the world have been asleep.'

Between 25 May and 10 June 1917, serious mutinies took place in almost every unit of the French Army along the line of the Chemin des Dames on the Western Front. The horrors of the second battle of Verdun, which between 21 February and 18 December 1916 cost France 542,000 casualties and Germany 434,000, had broken the will of the French Army to return to the offensive, not just for the rest of the war, but effectively for the rest of the century. In the village church at Tocane Saint-Apré in the Dordogne, to take a small local *église* entirely at random, the war memorial lists the names of sixty-six men killed between 1914 and 1918 – including nine *Croix de guerre* and two *Légion d'honneurs* – but those of only four people who were killed between 1939 and 1945, a not atypical proportion to be found on such memorials all across France.

The mutinies had profound implications for the war effort required from France's English-speaking allies, but for nearly three weeks in the early summer of 1917 the upheavals might even have taken France out of the war altogether, handing Germany victory almost overnight. Occasionally British exasperation with France was expressed privately, as when Balfour said in 1917, 'It will scarcely be contended that our position as guarantor of French credit is one of special privilege or advantage.' Such delicate understatements aside, however, the alliance held up well, even in that year of widespread mutiny in French ranks.

As a biographer of Philippe Pétain, the victor of Verdun, has noted about the period as May turned into June 1917,

The eruptions of indiscipline were taking on a much more sinister form. In short, what could hitherto possibly be described as 'military strikes', in the sense of a simple refusal to obey orders, had become what can reasonably be described as 'mutinies', in the sense of taking over direct leadership of military units. The protests of May, which were largely about the conditions under which soldiers were required to live, assumed in June clear overtones of social revolution. 'Down with the War!' 'Throw down your arms!' were slogans heard more and more frequently as the mutineers gained confidence in their numbers. Whole companies were disappearing into the forests.[89]

For reasons of national pride, the French attempted to hide the extent of the problem from their British and American allies, with some success, at least at the beginning.

On 17 July 1917, the British Privy Council proclaimed that henceforth the name of the Royal Family would be changed from Saxe-Coburg-Gotha to Windsor, having divested itself of its previous surname as well as 'all other German degrees, styles, titles, dignitaries, honours and appellations'. After a large number of alternative names were considered – including Plantagenet, York, England, Lancaster, D'Este and Fitzroy – Lord Stamfordham's suggestion of Windsor was adopted, after a minor title once held by Edward III and the castle in Berkshire in which the family spent much of the year.

This anti-German gesture, made at a critical juncture of the First World War, produced one of Kaiser Wilhelm II's few jokes, when he remarked with rather heavy Teutonic humour that he looked forward to attending a performance of *The Merry Wives of Saxe-Coburg-Gotha*. A more serious and altogether grander criticism came from the Bavarian Count Albrecht von Montgelas, who observed that, 'The true royal tradition died on that day in 1917 when, for a mere war, King George V changed his name.'[90] In fact, the College of Arms could not even determine whether the royal surname had originally been Guelph, Wettin or Saxe-Coburg-Gotha in the first place.

The effect in Britain was instantaneous and overwhelmingly positive. With the whole family swapping Germanic-sounding for overtly British names – the Teck family became the Cambridges and took the Earldom of Athlone, the Battenbergs were transformed into Mountbattens with the marquisate of Milford Haven – the Royal Family proclaimed itself thoroughly British, to national applause. King George V was himself quintessentially British, finding German 'a rotten language'. When H.G. Wells criticised his 'alien and uninspiring court', the King retorted: 'I may be uninspiring, but I'll be damned if I'm an alien.'

On 2 September 1917 Admiral Alfred von Tirpitz, who had only retired from commanding the German Imperial Fleet the previous year, founded the Fatherland Party, an anti-democracy and anti-peace organisation, telling its inaugural meeting:

The war has developed into a life and death struggle between two world philo-

sophies: the German and the Anglo-American. The question today is whether
we can hold our own against Anglo-Americanism or whether we must sink down
and become mere manure for others (Völkerdünger). The colossal struggle that
Germany is waging is therefore not one for Germany alone; what is really at issue
is the liberty of the continent of Europe and its peoples against the all-devouring
tyranny of Anglo-Americanism.

Except for the rot about manure and all-devouring tyranny, Tirpitz was
substantially correct. The Fatherland Party was aiming for what the German
historian Fritz Fischer half a century later described as 'the anti-parliamentary,
dictatorial, one-party state', one that came into being in 1933, and the Anglo-
American concept of pluralist democracy was indeed an opposing 'world
philosophy'. Within a month of Tirpitz's speech, hundreds of miles to the east
another great nation fell into the hands of those who wanted to create another
anti-parliamentary, dictatorial, one-party state. Tirpitz's proto-fascism, Hitler's fascism, Soviet Red fascism, today's Islamo-fascism: all are similar world
philosophies profoundly antagonistic to that which actuates the English-speaking peoples.

When Karl Marx died in 1883, the British Prime Minister, the incredibly
well-read William Gladstone, had never heard of him. Yet thirty-four years
later in the Bolshevik Revolution of October 1917 his philosophy became the
declared creed of the largest country in the world. The honourable efforts of
the English-speaking peoples to try to strangle Bolshevism in its cradle are
often forgotten, but 7,500 Americans, 4,000 Canadians and 1,600 Britons
served as ground troops in the 1918–19 campaigns to intervene in the Russian
Civil War on the side of the Tsarist White Russian opposition, attempting to
crush Lenin's revolution before it was able to infect Mankind. They failed, of
course, but it was a most noble cause. A wide-ranging French report entitled
The Black Book of Communism, published in 1999, put the numbers killed by
Soviet, Chinese, Cambodian and all other communist regimes in the twentieth
century at 100 million people. From October 1917, the English-speaking
peoples were faced with a new and ruthless foe, one whose ideology demanded
the unanimity of Marxism-Leninism across the planet as its goal. Far more
than Russia, Lenin himself wanted Germany and Britain to fall to communism
because 'they, for him, as a revolutionary, were the key countries'.[91] Having
failed to destroy Bolshevism at birth, the English-speaking peoples had to
spend the next three-quarters of a century trying to survive its imprecations.

Meanwhile, on the Middle Eastern Front, Lloyd George had asked General
Sir Edmund Allenby to capture Jerusalem 'as a Christmas present' for the
British Empire, which he proceeded to do with a fortnight to spare, on 9
December 1917. (The white flag used to surrender the city can be seen at the

Imperial War Museum in London.) With Mecca and Baghdad lost by the Ottomans, this was their third great holy city to fall to the enemy.

Allenby's campaign had been an example of inter-English-speaking soldiery at its best. On 31 October 1917, the 4th Australian Horse conducted the last large-scale cavalry charge of the British Empire at the battle of Beersheba. The Turks, commanded by the German Marshal Erich von Falkenhayn, had been taken in by the ruse of a British staff officer, seemingly in reconnaissance, who on being chased from a Turkish outpost dropped a haversack containing papers suggesting the attack on the inland Beersheba front was a mere feint, with the real attack taking place against Gaza on the coast. Falkenhayn diverted forces to an attack that never came, fatally weakening himself at Beersheba. (There were many elaborate deceptions employed by Allenby in that campaign, which quite belied his nickname 'The Bull' and included 'the taking-over and preparation of a hotel in Jerusalem as a false GHQ, the throwing of new bridges over the Jordan, the pitching of new camps in its valley and the manning of them with fifteen thousand dummy horses made of canvas, for which mule-drawn sleighs raised clouds of dust at intervals to suggest that they were trotting out to drink in the Jordan'.)[92]

The contribution of disparate parts of the English-speaking peoples was important for overall victory in Palestine. Not only were the Australian Light Horse key to the campaign, but the Australian Flying Corps also proved vital for the September 1918 offensive against Damascus. The capture of Jerusalem hugely undermined the prestige of the Ottoman Empire, that sprawling edifice that had been in relative decline since it failed to capture Vienna in 1683, yet which still managed to retain the allegiance – if not necessarily the affection or regard – of millions of Middle Eastern Muslims 234 years later. Disastrously for the future of the Middle East, the victorious Western Allies were about to help consign that ramshackle, unstable but essentially co-operative and moderate empire to history, in one of the worst of Woodrow Wilson's well-meaning blunders.

As well as the martial emergence of the United States on the world scene, and the Bolshevik Revolution, the year 1917 was to witness an invention of the English-speaking peoples that might well prove more important and long-lasting than either phenomenon. Ernest Rutherford was born at Spring Grove near Nelson, on the South Island of New Zealand, in August 1871, the fourth child of twelve. His father was a wheelwright turned small-scale farmer, his mother a schoolteacher, and both believed in the value of education, sacrificing much to ensure that 'Ern' got the best education possible. One of the first acts of the New Zealand colony after establishing itself – as with so many of the Antipodean colonies – had been to set up universities. Along with the Anglican cathedral, gentleman's club, cricket ground, statue of the Queen-Empress, freemason's lodge, place-names from the Old Country and gothic parliament house, the early foundation of an Oxbridge-modelled university is an infallible sign of British Victorian colonial development. New Zealand's oldest university college was founded in 1850, only a decade after the Treaty of Waitangi that brought a (fragile) peace to the country.

In 1892, Rutherford won the only senior scholarship available for mathematics in New Zealand. After taking three degrees from Canterbury College, Christchurch – founded as early as 1869 – he became one of the first 'foreign' students to be admitted to Cambridge University in England. 'That's the last potato I will ever dig,' he said on receiving the news on his parents' farm.[93] While at Cambridge, Rutherford divined the electrical properties of solids and then used wireless waves as a method of signalling, the commercial possibilities of which were then picked up by the Italian, Guglielmo Marconi. In 1898, aged only twenty-seven, Rutherford took up the post of Professor of Physics (Macdonald Chair of Physics) at McGill University in Montreal. It was there he made the first of his three major studies as he unravelled the truth about radioactive atoms and the natural transmutation of elements. He was elected to the Royal Society of Canada in 1900 and later of London in 1903. In 1908 he was awarded the Nobel Prize in Chemistry, which amused him since he considered himself a physicist.

At McGill, Rutherford discovered the nuclear model of the atom. (An atom is so small that one million placed side by side would only stretch across the full-stop at the end of this sentence. Furthermore, the nucleus is 1,000 times smaller than an atom.) Rutherford then gave papers at scientific meetings in Australia and New Zealand, and during the Great War Sir Ernest – as he by then was – worked in Britain on acoustic methods for detecting submarines, leading a delegation to the United States to share the technology after America entered the war. Born in New Zealand, researching in Britain and Canada, lecturing in Australia and New Zealand and helping the United States' war effort, he was thus a one-man exemplar of the English-speaking peoples' capacity for spectacular mutual assistance.

Rutherford's greatest achievement came in 1917, when he split the atom whose nuclear model he had discovered. As his entry in *The Dictionary of New Zealand Biography* explains,

> While bombarding lightweight atoms with alpha rays, he observed outgoing protons of energy larger than that of the incoming alpha particles. From this observation he correctly deduced that the bombardment had converted oxygen atoms into nitrogen atoms. He thus became the world's first successful alchemist and the first person to split the atom, his third great claim to lasting scientific fame.[94]

Two years later, he became Director of the Cavendish Laboratory in Cambridge, where the neutron was discovered by James Chadwick in 1932 and where John Cockcroft and Ernest Walton later split the atom by entirely artificial means using protons. Rutherford died at Cambridge on 19 October 1937, disgracefully enough as a result of delays in operating on his partially strangulated umbilical cord. His ashes were interred in Westminster Abbey.

It was Ernest Rutherford who recognised that the energy involved in the radioactive decay of an atom was millions of times that of a chemical bond, and that this was internal to all atoms. He hoped that methods for extracting the energy would not be discovered until Mankind was living at peace, and for once he was wrong, thankfully. For his discovery during the First World War was eventually – given the commitment and drive of the English-speaking peoples – to bring to an end the Second. By 1945, a world war could be ended by nuclear means and Western Europe had a means of being protected from Stalin's vast Red Army.

FOUR

Peace Guilt

1918–19

'The age of nationalities will not be of short duration, or of a very tranquil character.'   Benjamin Disraeli to Mrs Brydges Willyams, April 1860

'This is not peace; it is an armistice for twenty years.'   Marshal Foch in 1919

Thursday, 21 March 1918, witnessed the start of Ludendorff's huge and terrible spring offensive on the Western Front, the most dangerous moment of the Great War for the Allied cause, excepting perhaps the day of the battle of Jutland. Recognising that America's manpower was going to be brought to bear relentlessly after the spring, the German High Command hoped to end the war quickly, using vast numbers of men transferred from the Eastern Front once peace was signed with Bolshevik Russia at Brest-Litovsk on 3 March. The first attacks opened on a sixty-mile front between Arras and La Fère following a five-hour bombardment by more than 6,000 cannon, when three German armies, the 17th, 2nd and 18th, struck the right flank of the British sector, defended by the Third Army under Byng and the Fifth under General Sir Hubert Gough.

As well as indomitable heroism, the offensive saw a great wrong perpetrated on a distinguished British commander that was not righted for many years. Gough's Fifth Army had been spread thin on a forty-two-mile front lately taken over from the exhausted and demoralised French. The reason why the Germans did not break through to Paris, as by all the laws of strategy they ought to have done, was the heroism of the Fifth Army and its utter refusal to break. They fought a thirty-eight-mile rearguard action, contesting every village, field and, on occasion, yard. 'Many of them went for forty-eight hours without food or drink,' records one of Lloyd George's biographers, Donald McCormick. 'Some battalions fought two German divisions in a single day, while others were reduced to mere skeleton forces of anything from two hundred to twenty men. On one occasion only two machine guns out of forty-eight were left in action.'[1]

With no reserves and no strongly defended line to its rear, and with eighty German divisions against fifteen British, the Fifth Army fought the Somme offensive to a standstill on the Ancre, not retreating beyond Villers-Bretonneux. As Colonel Wilfrith Elstob VC of the Manchester Regiment told his men in almost his last words, 'There is only one degree of resistance and that is to the last round and the last man.' It has been rightly described as 'an epic in military history and human endurance; never in the annals of war, ancient or modern, had the human spirit endured so much as in that spring of 1918'.[2] Finding Pétain more concerned with protecting Paris than with assisting Haig – and the French capital *was* bombarded with long-range cannon between 23 March and 7 August – the British appealed to have Marshal Foch take over from Pétain as supreme commander of the Western Front. At a meeting of the Allied War Council at Beauvais on 3 April, this was agreed. Pershing had already generously offered his eight divisions in France to Foch due to the emergency.

The first German offensive of the spring came to an end at Mondidier on 5 April, after Foch shifted the reserves in a manner that Pétain had refused to. The British had lost 163,000 casualties, the French 77,000 and German figures were almost as high as both combined. The Germans also now had to contend with a unified Allied High Command, something that had not really existed up until then. In that sense the Beauvais meeting had been critical. Yet it was also there that Lloyd George, casting about for a scapegoat for the fact that the British Army had been forced to retreat so far, demanded that Gough be cashiered. 'Gough is unworthy of future employment,' he told Douglas Haig. The British Commander-in-Chief meanwhile wrote in his diary: 'The Prime Minister looked as if he had been thoroughly frightened and he seemed still in a funk. ... And he appears to me to be a thorough impostor.'

Of course Gough took it like the officer and gentleman he was, telling an admiring Haig, 'Very well, Douglas, you'll have a busy time. I'll say no more. Goodbye and good luck.' Aged forty-eight and after a lifetime in the Army, Gough was disgraced, ordered home on half-pay and only given the consolation that there would be an official inquiry (upon which Lloyd George subsequently reneged). Writing in the *Illustrated Sunday Herald* seven months afterwards, the reporter Robert Blatchford sought to tell the truth about the way that Gough had forced the spring offensive to peter out without the feared

German breakthrough. 'I may at last venture to attempt some account of a glorious and tragic epoch of the war hitherto obscured by an invidious fog of official mystery,' he began.

> The story of this retreat has never been given to this country. Instead we have had vague speeches, dubious hints and ominous silences, and it is not too much to say that there has accumulated in the public mind an indefinite but dark suspicion that our generals or our armies failed before the German attacks in March and that our Fifth Army, under General Gough, was badly and ingloriously defeated. The March retreat, so far from being discreditable to our soldiers, was more arduous and more brilliant than the famous retreat from Mons. It was a retreat during which the Fifth Army contested every bit of ground against almost overwhelming odds and in which the bulk of our regiments fought without rest or sleep for seven days and nights.[3]

Lloyd George, who had criticised Gough in the House of Commons, only finally admitted his error on 30 April 1936 – eighteen years after the sacking – when in researching his war memoirs he was forced to admit that he had been wrong. 'I need hardly say that the facts which have come to my knowledge since the war have completely changed my mind as to the responsibility for this defeat. You were completely let down and no general could have won the battle under the conditions in which you were placed,' he subsequently wrote to Gough.[4] It was a handsome enough apology, except that it had been Lloyd George himself who let Gough down, not least by not ascertaining the facts at the time. The next year at a reunion of the Fifth Army, Lloyd George had a letter read in which he admitted that, 'The refusal of the Fifth Army to run away even when it was broken was the direct cause of the failure of the great German offensive of 1918.' In 1937, a full nineteen years after he had protected Amiens – and thus Paris – from the greatest German assault of the war, Gough was awarded the Grand Cross of the Order of the Bath. Handing it to him King George VI said, 'I suppose you can take this as a recognition of the gratitude of your country.'

The military historian Sir Max Hastings has accurately summed up the central point about the Great War:

> There never was a quick path to victory on the Western Front, because the 1914–18 technology of destruction vastly outstripped that of communication and mobility. Defenders could always reinforce a threatened sector more quickly than the attackers could advance across it, until the German army had been worn down by four years of loss and Allied blockade. Haig's commitment to the doctrine of attrition seems repugnant, because its human cost was unspeakable. Yet he was correct that victory was unattainable without it.[5]

Early June 1918 saw the eruption on to the Western Front of the American Army, which was comprehensively 'blooded' at the battles of Château-Thierry from 3 to 4 June and then at Belleau Wood from 6 to 26 June. Along with the French Tenth Colonial Division, the US Second and Third Divisions succeeded in pushing the German Seventh Army back across the River Marne. Following this success, James Harbord's Marine Corps was given the extremely difficult task of capturing Belleau Wood. The initial assault on 6 June across an open wheat field that was commanded by German machine-guns led to the highest single-day casualties in the Corps' history, until the capture of Tarawa in the Gilbert Islands from the Japanese in November 1943. In the twenty-day battle, the Wood was captured and recaptured no fewer than six times before the Germans were finally forced to cede it. By the end, the Americans suffered 9,777 casualties, of whom 1,811 were fatal. The Germans lost 9,500 men and more than 1,600 prisoners. (Today the Bois Belleau is officially called the Bois de la Brigade de Marine.)

During the second battle of the Marne, the US Third Division again blunted Ludendorff's assault, which turned out to be the last great German offensive of the war. The arrival of over 400,000 fresh and enthusiastic troops in the shape of General Pershing's American Expeditionary Force was demoralising for the Germans, whose great hopes for the Aisne offensive had resulted in a serious dent in the Allied front – a salient thirty miles wide and twenty deep – but not the required breakthrough. By mid-July 1918, the time had come for the Allied counter-offensives that were eventually to win the war, in which the US First and Second Divisions spearheaded the Tenth Army's attack, and where the six other American Divisions also took part.

If any single day can be thought of as being decisive in breaking the Germans' will to resist further, it was probably Thursday, 8 August 1918, the first day of the Amiens offensive. Haig threw Rawlinson's British Fourth Army and Eugène Debeny's French First Army against the unsuspecting German Eighteenth and Second Armies, who were expecting an attack further north in Flanders. The Canadian and Anzac corps went 'over the bags' without preliminary bombardment, preceded by tanks, and thrust deep through a dense fog catching the Germans off-guard, supported after the fog cleared by a conglomerate Allied air force. More than 15,000 prisoners and 400 guns were captured and the Germans were flung back ten miles. Ludendorff admitted that 8 August had been the German Army's 'Black Day' and privately remarked flatly, 'The war must be ended!'[6] Nonetheless, it took another three months, in which the Allies won a series of very significant and impressive victories, before the Germans bowed. In the last four days of September 1918 alone, no fewer than 45,000 American 'doughboys' were killed and wounded in battle.

The tough Allied blockade, which Germany claimed had led to the deaths of over three-quarters-of-a-million people through starvation, also contributed to the collapse of German domestic morale by the autumn of 1918. It was so harsh that the steward of Admiral Beatty's flagship at the time of the official surrender noted that the five German officers present finished off a whole leg of mutton down to the bone, and one of them tried to secrete an entire cheddar cheese under his greatcoat on leaving.

The casualties sustained by the British Empire during the Great War numbered 908,371 soldiers and 30,633 civilians killed, as well as 2,090,212 wounded, out of a total force mobilised of 8.9 million. France, meanwhile, with a slightly smaller number mobilised at 8.4 million, sustained no fewer than 1.383 million soldiers and 40,000 civilians killed and 4.266 million wounded. French losses were therefore three times the number of Americans who have died in every foreign war from 1776 to the present day. The United States came through the ordeal with 50,585 killed and 205,690 wounded. The greatest losses were sustained by Germany, which lost 1.8 million soldiers and 760,000 civilians killed, and 4.2 million wounded, and Russia, which lost 1.7 million soldiers and 2 million civilians killed and 4.95 million wounded.[7] Austria-Hungary suffered heavily too, with over 900,000 killed. Overall the war is estimated to have mobilised over 63 million soldiers (42.2 million Allied, 22.8 million Central Powers) and cost the lives in battle of over 8 million soldiers (4.88 million Allied, 3.13 million Central Powers) and 6.6 million civilians (3.15 million Allies, 3.45 million Central Powers). The total number of wounded numbered no fewer than 21,228,813 people. Such was the appalling human cost of the Central Powers' grasp for global hegemony in August 1914; the most incredible phenomenon of the century was that only one generation later they were ready for a sequel performance.

The scale on which the British Empire's colonies responded to London's call was astounding. Between 1914 and 1918, Australia, a country of only five million, sent 300,000 men overseas to fight for the Empire, of whom 60,000 never returned. New Zealand, with a population of only a little over one million, sent a staggering 11% of her total population, of whom 17,000 died. Canada, with eight million, sent no fewer than 600,000, of whom she saw 60,000 killed. The butcher's bill for preventing the Kaiser dominating Europe was thus almost – but not quite – prohibitive.

On Saturday, 22 March 1919, in fine sunny weather, the Guards Division and Household Cavalry paraded through the streets of London on their return from Germany. The route took eight hours to complete, winding through tumultuous crowds lining the streets of central London on both sides of the Thames. Watching from a balcony of Devonshire House in Piccadilly, opposite the Ritz Hotel, was the eight-year-old Simon Fraser – later the D-Day hero the 15th Lord Lovat – who almost sixty years later recalled the way that 'the battalions were welcomed with a roar of continuous cheering that carried with it all the surge and thunder of the *Odyssey*. ... I had never heard such a sound: it left me breathless and seemed to shake the air.' Old soldiers in the crowds bared their heads as the colours went by, and the regimental band of the Irish Guards struck up *Saint Patrick's Day* as they marched past the saluting base. 'The jaunty tune delighted me,' Lovat recalled, 'but, looking round, I was astonished to see the tears running down grown-up cheeks on every side. I was in Lady Kenmare's party. They had lost their eldest son in France. *Lacrimae rerum* – a child is lucky not to understand the tragedy of war.'[8]

'At every crisis he crumpled,' wrote Winston Churchill of Kaiser Wilhelm II. 'In defeat, he fled; in revolution, he abdicated; in exile, he remarried.' While remarriage might not quite be conclusive proof of moral pusillanimity, the Kaiser did indeed flee Germany to Doorn in Holland, his country having become a republic on 9 November. Since there was an active 'Hang the Kaiser' movement in Britain and France at the time, it was probably for the best as far as he was concerned, but it did mean the German Empire ended there and then, only forty-seven years after it had begun so hubristically in the Hall of Mirrors at Versailles with his grandfather's proclamation as Kaiser in 1871. Back in December 1914 at Brooks's, Cecil Beck had bet Captain Neil Primrose ten guineas against one 'that the present German Crown Prince is never crowned German Emperor'. (Primrose was absolved of his debt; he had died fighting in the Buckinghamshire Yeomanry near Gaza on 17 November 1917.)[9]

Whilst it is easy, and indeed understandable, to mourn the extinction of the Habsburg, Romanov and certainly the Ottoman Empires, no such sympathy can really be extended to the Hohenzollern one. The true nature of the Kaiser can be discerned in the letters and especially the proto-Nazi postcards he sent from his exile in Doorn to his school-friend, the American author, traveller and socialite Poultney Bigelow, during the 1920s and 1930s, and especially scrawled purple pencil postscripts such as 'Democracy = Bolshevism!' In March 1927, he wrote to Bigelow about how, 'The Hebrew race ... are the most inveterate enemies at home and abroad; they remain what they are and always were, the forgers of lies and the masterminds governing unrest, revolution, upheaval by spreading infamy with the help of their poisoned caustic, satyrical spirit. If the world wakes up it should mete out to them the punishment in store for them, which they deserve.' Seven months later he added, 'For such infernal criminals a sound, regular, international, all world's Pogrom à la Russe would be the best cure.'[10]

A *Dolchstosslegende* (stab-in-the-back myth) was promoted by the Nazis and other extreme nationalists that insisted that Germany had been *Im Felde unbesiegt* (undefeated in the field), but only betrayed by 'the criminals of November', such as socialists, capitalists, Jews, pacifists and, in Ludendorff's belief, 'the secret forces of freemasonry'. In fact, however, the German Army had sustained a series of overwhelming straightforward defeats during the summer and autumn of 1918 at the hands of the Allies. Yet such had been their earlier successes in 1914 that the Germans had surrendered before the fighting reached their borders, so their own territories were not ravaged by marching armies. Back on 1 August 1914, Oswald Partington MP had wagered Lord Murray £100 to £10 'that in the event of a General European War, Berlin will not be occupied by a hostile Army', and he was right. (It might have been better if Berlin had been occupied by the Allies, since the Weimar capital was the scene of the abortive Sparticist uprising in January 1919 and much unrest thereafter.) The *Dolchstosslegende* was allowed to germinate precisely because when they surrendered in November 1918, Germany's armies were all on foreign soil. Despite the terrible ravages of the Allied blockade, the country had not suffered the kind of domestic physical destruction that 'Bomber' Harris was to visit upon it in 1940–5. As the historian Margaret Macmillan put it in her book *Peacemakers*, 'Of course things might have been very different if Germany had been more thoroughly defeated.'[11]

'What is our task?' asked Lloyd George in a speech at Wolverhampton on 24 November 1918. 'To make Britain a fit country for heroes to live in.' The election three weeks later, on 14 December, resulted in a staggering 520 seats for Lloyd George's coalition – nicknamed the 'coupon' election because receipt of a supportive letter from the Prime Minister virtually guaranteed a candidate's success – and was an overwhelming personal endorsement for the Welsh Wizard, and a feat that even Winston Churchill was unable to replicate in the post-war election of 1945. It reinforced Lloyd George's authority, and full Cabinet meetings, as opposed to the meetings of ad hoc smaller bodies, were not re-introduced until October 1919.

When Woodrow Wilson visited London on Boxing Day 1918, he found, in the words of *The Times* reporter, 'a mighty roar of acclamation ... a wonderful, spontaneous tribute to a man and the nation he typifies'. Lord Sandhurst likened it to the 1911 Coronation, which was significant since he was the Lord Chamberlain. Ever since noon, men, women and children had been streaming towards Buckingham Palace, many of them carrying or wearing flags, 'and at the moment when the carriages rolled by, flags and handkerchiefs fluttered vigorously'. The King and Queen had gone to Victoria Station to greet the President, and they returned in five carriages with postilions wearing their Ascot liveries, which Sandhurst thought 'splendid, nothing could have been more effective'.

The Victoria Memorial had been festooned with Union Jacks and Stars and Stripes, which did not prevent large numbers of sailors climbing up it; around its base and on its steps were women in WRAF khaki and WREN blue. People packed against the railings and at 3.15 p.m. they started chanting: 'We want Wilson!' An American sailor and airman climbed up one of the stone pillars of the Palace railings, waving the Stars and Stripes, and in 'a voice that made a megaphone unnecessary', the sailor called for three cheers for the King that met with a deafening response from 10,000 throats. After that he called for cheers for President Wilson, General Pershing, Sir Douglas Haig, Lloyd George and Marshal Foch; each were 'given with rousing heartiness'.

Then a Scottish soldier with a Union Jack also climbed up the pillar 'and the British and American flags were waved together' to new outbursts of cheering. 'Next the Dominions joined in. First an Australian soldier in hospital blue scrambled up the railings – the foothold on the pillar was by now fully occupied – and then a New Zealand man. . . . The line might have grown, but the window behind the Palace balcony was opened, and this diverted the attention of the crowd. For a few seconds there was silence, and then a tremendous cheer went up. . . .'[12]

Lord Sandhurst recorded in his diary what it felt like to walk, along with the Lord Steward Lord Farquhar, on to the balcony at that moment:

> I went out first, a tremendous cheer met me and the King and President following me, the noise became deafening, then the Queen and Mrs Wilson. The page brought the Queen the little Union Jack she had waved on the occasion of the Armistice, and the Queen gave it to Mrs Wilson, who waved it, the President waving his hat, the noise deafening. The President made a short speech of which I don't suppose anyone heard a word, at any rate I didn't and I was close to him; however the reporter did, which was the main thing.[13]

That reporter noted how the Royal Family, which included Princess Mary and the Duke of Connaught, 'by standing a little in the background tactfully indicated their recognition that the enthusiasm of the people was intended for the President'. The greatest of the cheers came when Mrs Wilson waved her little Union Jack above her head.

Once the wounded men and their nurses, who were in the Palace forecourt, began to call for a speech, the cheering 'rose to a tumult', and Wilson made a three-sentence speech 'to say how much your splendid tribute to my own dear country today is appreciated. I hope you may live long enough to enjoy the fruits of the victory which you have achieved.' All the while aeroplanes flew overhead, catching, in the words of *The Times* reporter, 'the golden glint of the setting sun'.

Not just in Britain but across Europe, America held an almost mythical status in the minds of millions by the end of the Great War. In his 1924 novel *Hotel Savoy*, one his several elegies to the Austro-Hungarian Empire, Joseph Roth wrote of one of his characters, the revolutionary ex-soldier Zwonimir:

> He loved America. When a billet was good he said 'America'. When a position had been well fortified he said 'America'. Of a 'fine' lieutenant he would say 'America', and because I was a good shot he would say 'America' when I scored bulls-eyes.

As the comic writers R.C. Sellar and W.B. Yeatman wrote about the post-Great War period in their book *1066 And All That*, 'America was thus clearly top nation, and History came to a.' It was because of this tremendous prestige enjoyed by America that President Wilson's 'Fourteen Points' formed the basis for the reorganisation of the world. As he travelled from Brest to Paris in December 1918, peasants knelt in prayer for the success of his mission alongside the railway track.[14]

Back on 8 January 1918, Wilson had enumerated fourteen war aims for which the Allies were fighting, which he hoped would be central to any peace treaty. They included: 'Open covenants of peace, openly arrived at'; 'Absolute freedom of navigation upon the seas'; 'The removal, so far as possible, of all economic barriers'; 'Adequate guarantees given and taken that national armaments will be reduced to the lowest point consistent with domestic safety'; 'A free, open-minded, and absolutely impartial adjustment of all colonial claims'; 'The evacuation of all Russian territory'; the evacuation and res-toration of Belgium; Alsace-Lorraine to be returned to France; a readjustment of the frontiers of Italy along lines of nationality; autonomous development for the peoples of Austria-Hungary; the evacuation of Romania, Serbia and Montenegro; a sovereign Turkey shorn of her empire; an independent Poland; and, finally, 'A general association of nations must be formed under specific covenants for the purpose of affording mutual guarantees of political inde-pendence and territorial integrity to great and small states alike' (i.e. the League of Nations). It was an inspiring-sounding checklist at the time, but it also included perils, tripwires and at the very least severe problems for the future.

(Of course, one of the reasons that America was able to provide moral leadership for the rest of the world was that she had already got all her domestic massacring over and done with by the dawn of the twentieth century. Rudyard Kipling admitted in his autobiography how he 'never got over the wonder of a people who, having extirpated the aboriginals of their continent more completely than any modern race had ever done, honestly believed they were a godly New England community, setting examples to brutal Mankind. This wonder I used to explain to Theodore Roosevelt, who made the glass cases of Indian relics [in the Smithsonian Institution] shake with his rebuttals.')

★

Just as Britain was fundamentally opposed to the so-called Freedom of the Seas, seeing it as a way to circumscribe future naval campaigns, so too did several of the Dominions suspect that 'an impartial and open-ended adjustment of colonial claims' meant that they would be prevented from administering those colonies that they had shed blood in capturing. On 23 January 1919, the Empire's Prime Ministers – Lloyd George of Britain, Robert Borden of Canada, William Massey of New Zealand, William Hughes of Australia and Louis Botha of South Africa (assisted by Smuts) – met the Council of Ten of the League of Nations, including Woodrow Wilson, to tell them that they expected to be allowed to continue to rule over those territories captured from the defeated powers, principally Germany and Turkey. British armies were in control of Palestine, Mesopotamia, the Cameroons and German East Africa. Australia had taken New Guinea, New Zealand had taken Samoa and South Africa had taken German South-West Africa (roughly modern-day Namibia), and they did not mean to give them up. Although Canada had no territorial acquisitions herself, she fully supported the right of conquest of the others.

As Churchill, then Colonial Secretary, later recalled of the meeting, 'A jagged debate ensued.'[15] At one point Wilson said: 'And do you mean, Mr Hughes, that in certain circumstances Australia would place herself in opposition to the opinion of the whole civilized world?' Hughes replied simply: 'That's about it, Mr President.' In the end, Wilson had to accept the compromise formula hammered out by Borden and Botha that sovereignty over the colonies would be veiled through the use of the term 'League of Nations Mandate'. Under that form of words, they continued to rule the territories (and generally to rule them with exceptional efficiency, decency and fairness). Instead of winding up in a worse position as a result of the war, the Empire grew to its greatest-ever territorial extent. Asquith and Grey had no desire for further expansion in 1914, but five years later German colonies in Africa and Asia came under British trusteeship and a million-strong British army occupied much of the Middle East.[16] It was at Versailles that the British Empire reached its vertex. So vast was the British Empire as a result of the treaty that by 1922 a coal-burning round-the-world steamship could, with the single exception of the three-day stint across the South Atlantic from Ascension Island to Trinidad, be sure that every nightfall it could 'berth safely under the lee shore of a British dependency'.[17]

The way in which the Dominions signed the Versailles Treaty independently and joined the League of Nations separately under their own names marked an important stage in their national stories. During the Great War, the British Empire raised armies totalling nearly nine million men, of whom three million came from outside the United Kingdom. The contributions of

Canada, Australia and New Zealand, who between them raised over one million troops despite their small populations, needed to be recognised and rewarded. By 1919, 'It was now impossible for them to feel that foreign policy and its military consequences could be left to the United Kingdom. After the First World War the implicit autonomy of the Dominions in these remaining spheres of policy had to be made explicit.'[18] Over trade and tariffs, the Dominions had long exercised fiscal autonomy, but in 1921 Ottawa appointed an ambassador to the United States, and at the Washington Conference that year to discuss naval arms limitations, Canada, Australia and New Zealand were each represented separately. Back in 1902, during the negotiations over the Anglo-Japanese alliance, for example, Britain had spoken for all of them. The consequences for imperial unity that these new autonomies threw up were dealt with by Arthur Balfour in the Imperial Conference of 1926. Guiltless of having caused the war, the English-speaking peoples as 'top-dog' powers had had nothing to gain from a general separation, and everything to lose.

Ever since Macaulay's Education Minute of 1835, it had been envisaged that the British would quit their Empire as soon as they had educated their successors. It was thus the first empire in human history to be manufactured with a sell-by date attached. Now, however, it attained the greatest span in its history, if one includes the various Protectorates, such as Egypt, awarded it under the terms of Versailles. Yet just as its extent reached its widest, so the cracks in its edifice were beginning to become very apparent. The most obvious was the loss of 908,371 of its brightest, bravest and best young men over the previous four blood-spattered years.

The first draft of the proposed Covenant of the League of Nations sparked intense debate across the world, but especially in the United States where opinion was deeply divided about the extent to which a burgeoning world power like theirs should have its hands tied by a powerful supranational body. One of the most considered responses came from Elihu Root, a man of tremendous distinction whose voice carried enormous weight. Born in Clinton, New York, in February 1845, Root had been McKinley's and Roosevelt's Secretary of War, then Chairman of the Republican National Convention, Secretary of State, President of the American Society of International Law and American Bar Association, and Senator for New York. He had also won the Nobel Peace Prize in 1912. He personified the American Establishment, yet he was unhappy about several key areas of the League's constitution, and was just the kind of man that Wilson needed to appease in order to win approval for his brainchild.

On 29 March 1919, Root published his comments on the League's proposed Covenant in an open letter to the Chairman of the Republican National

Committee, saying that the scheme had great value but correspondingly 'very serious faults', which needed to be addressed. These were that arbitration before conflict was not made obligatory on all signatories; although an international court was mentioned, no details were given about how it would operate; the United States must preserve the Monroe Doctrine separately from the League, and in undertaking to preserve the territorial integrity of all League members, the world's borders would be effectively set in aspic for evermore. To Root, 'It would not only be futile; it would be mischievous. Change and growth are the law of life, and no generation can impose its will in regard to the growth of nations and the distribution of power upon succeeding generations.'[19] Root had led a thirty-three-man diplomatic mission to Russia after the March 1917 revolution and was well aware of the danger of anarchy sweeping through Russia, Germany and Eastern Europe. He wanted a League of Nations, just not one that proved counter-productive to peace in the long run. There were other issues concerning arms limitation and immigration that he considered also needed amendment, even so he concluded, 'I think it will be the clear duty of the United States to enter into the agreement.' Yet it was not to be.

In an article published in *Metropolitan* in March 1919, Theodore Roosevelt – whose untimely death at the age of sixty-one had taken place on 6 January – seemed to strike an ultra-isolationist note, one that was certainly made great use of by organisations that were sprouting up around this time, such as the League for the Preservation of American Independence. 'We have finished the great war with Germany,' Roosevelt wrote. 'I do not believe in keeping our men on the other side to patrol the Rhine, or police Russia, or interfere in Central Europe or the Balkan peninsula. ... Mexico is our Balkan peninsula.' Sadly, America listened to Roosevelt's posthumous advice and turned in upon herself in 1919, but it is worth considering a world in which for the next two decades after March 1919 the United States had indeed kept her men 'on the other side' of the Atlantic, as she had to do for the half-century after 1945.

Those troops could have 'patrolled the Rhine' and thereby prevented Adolf Hitler from remilitarising the land adjacent to it in 1936, and they could have 'interfered in Central Europe' to stop the Nazis marching into Prague, which happened twenty years to the week after Roosevelt's article. They did both after 1945, for considerably longer than twenty years. In short, they could have saved the world – and themselves – untold misery and bloodshed, rendering Civilisation far better service than was possible in Mexico. Active American involvement in the general world settlement of 1919 was a crucial prerequisite for its success. 'The whole Treaty had been deliberately, and ingeniously, framed by Mr Wilson himself to render American cooperation essential,' thought Harold Nicolson, part of the British delegation to Versailles.[20] Yet as the Conference progressed, in Nicolson's words, it slowly dawned upon the delegates that America might not ratify the Treaty, and although it 'was never mentioned between us, it became the ghost at all our feasts'.

Less than two weeks before he died, Roosevelt wrote to Rudyard Kipling, stating that Woodrow Wilson's parents were born in England and Scotland, adding, 'I have always insisted that the really good understanding [between] the British Empire and the United States would not come except insofar as we developed a thoroughly American type, separate from every European type and free alike from mean antipathy and mean cringeing.' He denied the claims of 'the Wilson adherents and the Sinn Feiners and pro-Germans and Socialists and Pacifists' that he was pro-British. However, he did add that, 'Because of the almost identity of the written (as opposed to the spoken) language and from other reasons I think that on the whole, and when there isn't too much gush and effusion and too much effort to bring them together, the people of our two countries are naturally closer than those of any others.'[21] Roosevelt intensely deprecated the 'good, mushy, well-meaning creatures who are always striving to bring masses of Englishmen and Americans together,' likening them to a philanthropist he once knew who was saddened by the historic antipathy between New York's police and fire departments. In order for them to 'get together', the rich man had hired the Yankees' stadium for a friendly game of baseball. The moment the umpire's decision was disputed in the middle of the first innings, Roosevelt recalled, both sides 'got together' in a vast brawl, with thousands of 'stalwart ... men in uniform' exchanging blows.

At 4.30 p.m. on the afternoon of Sunday, 13 April 1919, a seventy-five-strong detachment of native troops – only fifty of whom were armed with rifles – were ordered by Brigadier-General Reginald 'Rex' Dyer to march to the enclosed space called the Jallianwala Bagh in Amritsar in the Punjab, to break up a prohibited political meeting of between 15,000 and 20,000 people. Without giving any warning to the unarmed crowd, the fifty (mainly Nepalese Gurkha and Baluch from Sind) riflemen opened fire and continued shooting for ten minutes. Only after each man had fired thirty-three rounds was the cease-fire order given and the detachment marched away, leaving 379 killed and around 1,000 injured.[22]

Although there were those who argued that Dyer's actions had prevented a second Indian Mutiny, he was deprived of his command and sent home to Britain, and it became clear that although he was never cashiered he would not be given any future employment. Army officers complained that since Dyer was acting in support of the civil power, he had been let down by both the British and Indian Governments in fulfilling his undeniably unpleasant, but necessary duty. For the Indian Home Rule movement, of course, the Amritsar massacre was a propaganda god-send.

The massacre needs to be seen in its political context, in particular the launch by Mohandas Gandhi in February of the Satyagraha mass movement to win Indian Home Rule through passive disobedience. Very soon events got out of hand, and the moderate, law-abiding campaign of Gandhi and his immediate supporters was comprehensively suborned by more violent elements amongst the Indian independence movement. On 30 March, anarchy hit Delhi. Within a week local disturbances across the Punjab led to a situation resembling a general rebellion there. Nor were matters calmed by Gandhi's arrest on 9 April.

It was the events of the next day, 10 April, that led directly to Dyer's actions. Amritsar had significant Sikh, Muslim and Hindu populations, and was scheduled to be the place where the All-India Congress Party would meet later in the year to demand Indian independence. Two local agitators – Dr Kitchlew, a Muslim barrister, and Dr Satya Pal, a Hindu assistant surgeon – organised a strike there for Sunday, 6 April, despite official orders against both. On 8 April, Miles Irving, the Deputy Commissioner, requested reinforcements from Lahore, since he only had seventy-five armed and 100 unarmed constables with which to control a city of 150,000 inhabitants.

More controversially, on the morning of 10 April, Irving had Kitchlew and Pal arrested. On hearing of this, a mob stoned a small picket of British and Indian troops and police at Amritsar's railway crossing (the city being an important entrepôt as well as trading centre). The order was given to fire in order to prevent the civil lines from being rushed, which led to twenty casualties amongst the rioters. Elsewhere in Amritsar, another mob went on an anti-European rampage, which was not prevented by the native police present, who remained passive. Three bank officials were lynched in their offices, their corpses burnt in the street. Two other officials were murdered near the goods yard. An elderly female missionary, Miss Sherwood, was hauled off her bicycle, beaten to the ground repeatedly and left for dead in a gutter. Posters appeared on walls which alleged that the British raped Indian girls in Amritsar and urged Indians to 'dishonour' and 'clear the country of all English ladies'.[23] Furthermore, buildings connected with the British were defaced or damaged.

In an empire based entirely on the prestige of the ruling power – an empire of mirrors in which tiny numbers of British troops had to control the teeming millions of a multi-ethnic sub-continent – this kind of behaviour could not be allowed to go unpunished. Once again the reason was the protection of prestige, without which the British Empire in India would have simply evaporated overnight. (There had always been a dearth of British troops in India, which was held through native co-operation rather than by force. In 1885, the ratio of native inhabitants to British soldiers was 4,219 to 1.)

By sunset on 10 April, the reinforcements, including 260 Gurkhas, had arrived from Lahore and Irving handed over official responsibility to restore order to Brigadier-General Dyer, commander of the Jullundur Brigade. There can be no doubt that had the kind of murderous unrest seen in Amritsar on 10 April spread throughout the Punjab, and perhaps thereafter northern India, the bloodshed would have been horrific. Furthermore, the distances were as vast as the soldiers available to the authorities were few. In order to deal with the unrest in an area the size of Yorkshire, Dyer had fewer than 1,200 British and native troops. Taking control at 9 p.m., he made an immediate public declaration that all processions, gatherings and demonstrations were pro-hibited and 'All gatherings will be fired upon.' Such an unequivocal command could only be backed up by force, otherwise it would have had the opposite effect to that intended and demonstrated British lack of resolve.

Partly to restore order, and partly because, as he later put it, 'We look upon women as sacred, or ought to', Dyer commanded that any Indian men who wished to use the lane in which Miss Sherwood had been assaulted must proceed down it on bellies and elbows between the hours of 6 a.m. and 8 p.m. The order only lasted for five days and it was hoped thereby to prevent any gloating about what had happened there, but the 'crawling order', as it soon came to be called, was later used to question Dyer's motives. In fact 'fancy punishments', as they were called, were common in India at the time. Skipping exercises and suchlike were used as alternatives to hard labour at Kasur, for example, and it was not unusual for poets to escape physical punishment if they wrote poems praising British martial law that they then read in the market-place. Some independence activists obeyed the crawling order as a way of mocking Dyer; indeed, one had to be stopped after crawling there three times.[24]

Much has been made of Dyer's personality, as though that might explain the repressive measure he took. Had his father's settlement in India before the Mutiny instilled in him a terror of the Indian mob? Was his alleged unsociability at Staff College a factor in his psychological make-up? What if anything can the fact that Dyer spoke fluent Persian, Punjabi and other Indian dialects tell us? In fact, as the historian Brian Bond has pointed out, Dyer's long military career 'provides no sinister indications of latent irresponsibility or a latent bloodlust. Happily married, popular with his colleagues and men and above average in military efficiency, Reginald Dyer was certainly not the monster soon to be portrayed by Indian propagandists.'[25]

Although he had Amritsar reconnoitred from the air, and marched through it with troops and two armoured cars on 12 April, he only faced some insults and catcalls from the crowds and he only arrested two people. The next morning he reissued his proclamation against public meetings at no fewer than nineteen prominent places in the city, with beating drums and much ceremony. The warning that assemblies would be dispersed by force of arms was then repeated several times in both Urdu and Punjabi, languages understood by over 90% of the inhabitants of Amritsar. No sentient inhabitant of Amritsar could have possibly been under any doubt about the possibly fatal consequences of attending a political rally that day.

When Dyer heard at 1 p.m. that a meeting protesting against his proclamation was due to be held at 4.30 p.m. that very afternoon, he considered it a perfect opportunity to crush the incipient rebellion. When shortly after 4 p.m. the General heard that the meeting had begun early, he led his small force to the Bagh, the entrance to which was so narrow that his two armoured cars had to be left outside. He saw the large crowd about 100 yards away, in a dusty open space, about 200 yards long, which had houses on all sides. Despite Indian propagandists' subsequent claims, there were three or four narrow passages running off the Bagh and in places the boundary walls were low enough for men to climb over without difficulty. Although the crowd had no firearms, many did have *lathis* – metal-tipped sticks – and a concerted rush, which Dyer's staff officer Captain Briggs feared might happen, could have spelt disaster. Dyer therefore directed the fire into the centre of the crowd, and kept firing until his men only had enough ammunition to cover their return to base. Because they feared retribution if they moved forward individually, the wounded were not taken care of by the authorities and the curfew meant that the dead had to be left all night where they fell.

Although much about that terrible day is disputed, the fact that it pacified the Punjab virtually overnight is not. After 18 April, it was not necessary for another shot to be fired throughout the entire region. A deputation of Indian merchants and shopkeepers soon afterwards thanked the General for preventing looting and destruction, and he received many other such tokens of gratitude; the guardians of the Golden Temple – the central shrine of the Sikh faith – invested him there as an honorary Sikh. The readers of the *Morning Post* in Britain raised over £26,000 for his subsequent legal costs. Sir Michael O'Dwyer, the Lieutenant-Governor of the Punjab, also gave official approval of Dyer's actions on 16 April, and in his autobiography he wrote that 'not only did Dyer's action kill the rebellion at Amritsar, but as the news got around, would prevent its spreading elsewhere'.

Further support came from General William Benyon, Dyer's Divisional commander, who said that his 'strong measure' had 'prevented any further trouble in the Lahore Divisional area'. The Adjutant-General of India, Sir Henry Havelock Hudson, stressed to the Legislative Council of India that in situations such as that prevailing in Amritsar in early April, moderation was taken advantage of as weakness while severity was denounced as murder. 'When a rebellion has been started against the Government,' he said in conclusion, 'it is tantamount to a declaration of war. War cannot be conducted in accordance with standards of humanity to which we are accustomed in peace.' The Commander-in-Chief in India agreed, saying, 'the semi-educated native ... takes clemency as proof of weakness'. Even the Roman Catholic Archbishop of Simla wrote that Dyer had 'saved the Punjab, and, in the opinion of many, saved India'.

For all this support from the men on the spot, the people who knew India best and who had the responsibility of administering her, Dyer's professional reputation was soon comprehensively blackened. A committee, named after its chairman Lord Hunter, was appointed, the hearings of which turned into a *de facto* trial of Dyer. It had no members sitting on it with any Indian administrative experience, but three Indian lawyers, two of whom were personally hostile to the Punjab Government. The investigation was restricted solely to events in Bombay, Delhi and the Punjab and so did not even cover Dyer's claim that he had also nipped in the bud a threatened attack from Afghanistan. Native witnesses hostile to him were not put under oath or cross-examined, nor was he allowed any legal counsel. 'I had a choice of carrying out a very distasteful and horrible duty', he told the committee, 'of suppressing disorder or of becoming responsible for all future bloodshed.' Although the Hunter Committee's majority criticised Dyer in its report for firing before giving the crowd a chance to disperse and for continuing to fire afterwards, the three Indians also submitted a minority report that was vitriolic.

Even though the Government of India wrote to the India Office on 3 May 1920 to say that there had been 'a dangerous rising which might have had widespread and serious effects on the rest of India', the reply from the Secretary for India, Edwin Montagu, who had long been sympathetic to Indian Home Rule, took the point of view of the Hunter Committee and especially its minority report. Dyer was relieved of his command and recalled to Britain, although he kept his Army pension. He suffered a stroke in November 1921 and died in 1927.

Today's reactions to Dyer's deed are of course uniformly damning. In 2005, in reviews of a biography entitled *The Butcher of Amritsar*, it was variously described as 'an unforgivable atrocity', 'state terrorism', 'a heinous crime', 'the biggest and bloodiest blot on the generally benign record of British rule in India', and so on.[26] As well, of course, as tarnishing the prestige of the British Empire, and presenting it to world opinion as cruelly oppressive, one of the results was that senior British soldiers were convinced that, in the words of one historian, 'in the last resort the British would hesitate to repress disorders by the use of force: the imperial grasp was slackening. If necessary the British could be prized from power inch by inch by threats or by calculated outbreaks of violence.[27] Gandhi spotted how this unwillingness to employ force would shame the British into making concessions.

In the Commons, Churchill, then Secretary of State for War, described Amritsar as 'an extraordinary event, a monstrous event, an event which stands in sinister isolation', and criticised Dyer for his 'intention of terrorising not merely the rest of the crowd, but the whole district or country'. That was indeed what had happened, but if the Amritsar district, Punjab region or northern India generally had carried on in revolt, many more than 379 people would have lost their lives. (As a postscript, it is worth recording that, on 6 June 1984, the Government of India sent tanks against Sikh extremists who were inside the Golden Temple in Amritsar, massacring over 250 people. The orders were given by Indira Gandhi, who largely escaped global criticism since she was not a British imperialist like Reginald Dyer.)

On 14 and 15 June 1919, Captain John Alcock and his navigator, Arthur Whitten-Brown, flew non-stop across the Atlantic from Lester's Field, near St John's, Newfoundland, to crash-land in a bog in Clifden, Ireland. Five months later, the South Australian brothers Ross and Keith Smith flew from Hounslow near London to Darwin in Australia's Northern Territories in twenty-eight days. All four were knighted by King George V for their daring. Both teams won prizes of £10,000 from the *Daily Mail* and the Australian Government respectively, which both insisted on sharing with their mechanics. *The New York Times* headline, 'Alcock and Brown Fly Across Atlantic; Make 1,890 Miles in 16 Hours, 12 Minutes; Sometimes Upside Down in Dense, Icy Fog', captures the essence of their adventure. The flight was threatened by engine trouble, thick fog and snow, and Brown had repeatedly to climb onto the aircraft's wings to remove ice from the engines. They averaged 118 mph in a modified Vickers Vimy bomber. Alcock was killed just months later, aged only twenty-seven, flying to the Paris air show. Brown lived until 1948. Both men are commemorated in a memorial statue built at Heathrow Airport in 1954. The world was shrinking, bringing the English-speaking peoples closer than ever before, and the heroism of men like Alcock and Brown was in the forefront of that process.

One week before the signing of the Treaty of Versailles, an event took place which highlighted how recalcitrant, indeed revanchist, Germany was likely to become in the years ahead. Under Article 23 of the Armistice agreement of 11 November 1918, seventy-four named warships of the German High Seas Fleet had to be handed over for internment in an Allied or neutral port. It was decided that the best place for this would be the Royal Navy's base at Scapa Flow in the Orkney Isles. This powerful and undefeated fleet and its 20,000-strong crew therefore set sail for the Firth of Forth, where they were met by an Allied force of no fewer than 250 vessels under Admiral Beatty, including most of the Grand Fleet and an American battleship squadron, with a total of forty-four capital ships. There the German imperial flag was hauled down and the ammunition removed, before sailing on to confinement at Scapa Flow. Once there the crews numbered only 4,565 seamen plus 250 officers and warrant officers, which was later reduced further to just seventeen hundred.

When the naval clauses of the Treaty were made public a few days before its signature on 28 June 1919, it was revealed that the German Fleet was to be surrendered to the Allies and not sailed back to Germany. 'For a couple of days,' their commander Rear-Admiral Ludwig von Reuter recalled, 'it lay like lead on the minds of the men.' Then, on the bright, clear morning of Saturday, 21 June 1919, the sailors of the Imperial High Seas Fleet decided to take their revenge.

Waiting for the British first battle squadron to leave harbour for morning torpedo practice, at 10.30 a.m. Reuter signalled his Fleet: 'Paragraph 11. Confirm.' It was an hour before all the ships acknowledged the signal. Von Reuter's ships then opened the sea-cocks, 'letting the grey-green waters of the North Sea rush in'.[28] The first to be scuttled was Admiral Reinhardt Scheer's flagship at the battle of Jutland, the *Friedrich der Grosse*, at 12.10 p.m., and over the next five hours she was followed to the bottom of Scapa Flow by no fewer than fifty-one other vessels, the last being the battle-cruiser, the *Hindenburg*, at 5 p.m.

A party of pupils from the local Stromness Higher Grade School was on a tour of the base on a local steamer at noon that day and thus had a perfect view of these vast black warships as, 'suddenly and without warning and almost simultaneously, those huge vessels began to list to port or to starboard'. Years later one of them recalled how, 'Some keeled over and plunged headlong ... others were rapidly settling down in the ocean, with little more showing than their masts and funnels.' As steam billowed through the ships' vents, 'a dreadful roaring hiss' could be heard. For miles around the base, the sea was covered with the detritus of a great sunken fleet – boats, hammocks, life-belts, oil, chests, spars and endless flotsam. (Today, the radioactive-free metal from the ships is used for certain types of sensitive scientific instruments, since it was already underwater at the time that the Hiroshima blast contaminated much of the rest of the world's metal.)

The only British warships present at the scuttling were the destroyers *Vespa* and *Vega*, which signalled to the rest of the battle squadron to return to base at full speed. The Royal Navy succeeded in beaching the 15-inch-gun *Baden* and the three cruisers *Nürnberg*, *Emden* and *Frankfurt*, but all the other major ships sank. Nine German sailors were shot during the operation, but no-one drowned. All that the British Admiral, Sydney Fremantle, could do when he returned to port was to summon von Reuter and his staff on to his flagship *Revenge* and rant against this 'breach of naval honour'. He later recalled how 'They stood with expressionless faces, clicked their heels and descended the accommodation ladder without a word.' At least one Naval Intelligence officer, Francis Toye, thought Fremantle to have been inexcusably hypocritical in his reaction, believing that the Admiral, 'in his heart of hearts, like everyone else, must have been delighted at this solution of a dangerous and knotty problem ... in similar circumstances an English admiral would have done exactly the same. Of course he would.'[29]

With the sudden disappearance between the waves of Scapa Flow of over 400,000 tons of enemy warships, not for a century had the English-speaking peoples been so powerful vis-à-vis the rest of the world as at 5 p.m. on 21 June 1919. The Royal Navy now had nearly half of the world's stock of capital ships, an even better position than when Lord Salisbury had instituted the 'Two-Power Standard' in 1889, ordaining that she should be as large as the next two Great Powers' navies combined. The last time the Royal Navy had been in such utterly undisputed naval dominance was immediately after the battle of Trafalgar 114 years before, when she had a tonnage of 570,000 over the 360,000 tons of France, Spain and the Netherlands combined.[30]

The shadows gathered later that Saturday than on any other evening of the year, but they gathered over an empire that – even as it seemed at its most powerful – had in fact begun its long, slow and painful decline. In a sense what *The Times* reporter had seen on the aeroplanes circling Buckingham Palace during President Wilson's speech on Boxing Day 1918 – 'the golden glint of the setting sun' – might be taken as a metaphor for the British Empire as it reached its zenith.

On Saturday, 28 June 1919, five years to the day after the assassination of the Archduke Franz Ferdinand, a peace treaty was signed at the palace of Versailles. Junior officers and diplomats stood on tables and chairs outside the Hall of Mirrors, craning their necks to watch David Lloyd George, Georges Clemenceau, Woodrow Wilson and Vittorio Orlando signing the document. 'To bed, sick of life,' wrote Harold Nicolson that night. When later he asked Arthur Balfour why in the Orpen painting of the German delegation signing the Treaty he was depicted averting his gaze, Balfour answered, 'I make it a rule never to stare at someone who is in obvious distress.' As a result of the Treaty, seventy million Germans would be surrounded by a *cordon sanitaire* of small and medium-sized states, which, as one historian has pointed out, 'were domestically unstable, economically increasingly dependent on Germany, and which had to rely for their independence on the goodwill and assistance of faraway Great Powers'.[31]

The Austro-Hungarian and Ottoman Empires have been likened to old, beautiful vases that no-one appreciates until they have been smashed into hundreds of pieces, impossible to restore. The tragedy of the Trianon and Sèvres Treaties was that the vases weren't broken by accident, but flung to the ground by the hubristic vandals of Versailles. Although the dissolution of the Hohenzollern Empire had already taken place with the Kaiser's flight, there was no reason why the Habsburg and Ottoman Empires needed to be dissolved also, and plenty of good ones why they ought not to have been, since Trianon and Sèvres threw up quite as many problems as they solved. In his novel *Embers*, Sándor Márai wrote of how

> Vienna and the monarchy made up one enormous family of Hungarians, Germans, Moravians, Czechs, Serbs, Croats, and Italians, all of whom secretly understood that the only person who could keep order among this fantastical welter of longings, impulses, and emotions was the Emperor, in his capacity of sergeant-major and imperial majesty, government clerk in shirtsleeve protectors and grand seigneur, unmannerly clod and absolute ruler.

By breaking up that family and forcing its emperor into exile, President Wilson created a maelstrom of inherently unstable, competing nationalities increasingly prone, as the years went by, to the lure of fascism.

Back on 1 August 1914, Lord Murray had bet his brother Captain Arthur Murray ten guineas at Brooks's 'that if, arising out of the Austro-Serbian crisis, there is a general European War, there will be no crowned head in Europe except the King of England ten years from the date upon which the war breaks out'. In fact, there were still plenty left, in Scandinavia, Italy and the Benelux countries, but in the main his pessimism had been justified.

The post of caliph had been the supreme leadership of the Islamic faith since the seventh-century successors of the Prophet Muhammed, yet on 3 March 1924 the Turkish Nationalist Government abolished it, sending Abdul Mejid into exile in Paris, where he died in 1944. It was, wrote the *Daily Telegraph* at the time, 'one of the most astonishing acts of suicidal recklessness of modern or ancient times', prompting 'the inevitable stirring up of the Muslim world'.[32] Radical Islam has indeed called for the restoration of the caliphate to rule the billion-strong 'Islamic nation', or *Umma*, under *Sharia* law. In his first video broadcast after the 9/11 atrocities, in a videotape released on 7 October 2001, Osama bin Laden said, 'What the United States tastes today is insignificant compared to what we have tasted for decades. Our nation has been tasting this humiliation and contempt for more than eighty years.' The fact that it was not the United States but his fellow Muslims of the Turkish Grand National Assembly in Ankara that in fact abolished both the Ottoman sultanate in 1922 and the caliphate two years later seems not to have registered with the leader of Al-Queda.

The economist John Maynard Keynes described Lloyd George, with pardonable exaggeration (if unpardonable anti-Welshness), as

This extraordinary figure of our time, this siren, this goat-footed bard, this half-human visitor to our time from the hag-ridden magic and enchanted woods of Celtic antiquity. One catches in his company that flavour of final purposelessness, inner irresponsibility, existence outside or away from our Saxon good and evil, mixed with cunning, remorselessness, love of power, that lend fascinating enchantment and tenor to the fair-seeming magicians of North European folk lore.[33]

An appreciation of the personality of the sage of Llanystumdwy, David Lloyd George, is a vital prerequisite to an understanding of what happened at Versailles. In 1886, he had warned his future wife Margaret Owen that 'My supreme idea is to get on. To this idea I shall sacrifice everything – except I trust honesty. I am prepared to thrust love itself under the wheels of my juggernaut if it obstructs the way.' In fact, Lloyd George thrust honesty under the wheels of his juggernaut even before love. Clementine Churchill believed Lloyd George to be 'a descendant of Judas Iscariot', and she was not alone. He was also the first prime minister since Sir Robert Walpole to leave Number 10 much richer than when he entered it, as shameful a comment on him as it is a source for pride for the other fifty premiers who turned down the multifarious opportunities for making money from the post. At one point he even set up an office at Number 10 for the sale of honours.

The Hall of Mirrors, with its slightly distorting, disorientating aspect, was the ideal place to stage the signing of the Versailles Treaty. The major problem with recognising separate ethnic elements in Southern and Eastern Europe is that they can so easily bifurcate into smaller and smaller groupuscles. A classic example of this came in January 2005, when 2,000 Hungarian descendants of Attila the Hun demanded official recognition by the EU as a separate ethnic minority. Under Hungarian law, ethnic groups of 1,000 people or more can qualify for the status of an official minority, thereby winning a selection of generous subsidies and special privileges. 'As a member of the European Union,' said the group's self-appointed leader, Joshua Imre Novak, 'Hungary should not be suppressing a minority.' Many experts dismiss the claims as bogus, arguing that Attila has no surviving descendants, yet the very absurdity of the situation highlights how the issue of self-determination can be taken to ridiculous lengths.[34]

At the time, Winston Churchill enthused over the Treaty's provisions as it affected the small nations of Eastern Europe. 'Probably less than three percent of the European population are living under Governments whose nationality they repudiate,' he wrote in 1929, 'and the map of Europe has for the first time been drawn in general harmony with the wishes of its peoples.'[35] This was an exaggeration; the figure was far higher. The incredibly detailed maps used by Woodrow Wilson, preserved in the Library of Congress, went into fantastic, sometimes village-by-village detail about racial groupings, yet borders had to be drawn somewhere for the hyphenated new countries like Czecho-slovakia and Yugo-slavia.[36] In order to protect these new artificial borders, long-term commitments from all the victors needed to be entered into in good faith. It was the world's tragedy that this did not happen. To give Wilson his due, he did tell the Senate on his return from France about 'the difficulty of laying down straight lines of settlement anywhere on a field on which the old lines of international relationship, and the new alike, followed so intricate a pattern and were for the most part cut so deep by historical circumstances which dominated action even where it would have been best to ignore or reverse them'.

In 1920, the Oxford don Geoffrey Madan noted in his commonplace book that Lieutenant-Colonel Charles Repington's book *The First World War* had 'a shocking title', since it presupposed that there would be a second. The central accusation made against the Versailles Treaty is that its supposed harshness on Germany made *revanchism*, and thus future conflict, more rather than less likely. 'The final crime', opined *The Economist* summing up the twentieth century in its December 1999 issue, 'was the treaty of Versailles, whose harsh terms would ensure a Second World War.' This is pretty much the standard verdict on the Conference and Treaty which redrew Europe's frontiers until Hitler violated them two decades later. The received wisdom is that because another war broke out twenty years after the Treaty was signed, *ergo* it must have been flawed. Yet Adolf Hitler had plans of conquest and dreams of scourging the Bolsheviks and Jews that would have led him far beyond the frontiers that any peacemakers could possibly have agreed for Germany at Versailles. To blame Versailles for Hitler's war is, as Margaret Macmillan puts it in her book *Peacemakers*, 'to ignore the actions of everyone – political leaders, diplomats, soldiers, ordinary voters – for twenty years between 1919 and 1939'.

Had the Treaty actually been *harsher* on Germany – specifically if it had divided the country in two (as happened in 1945) or more (as was the case before 1871) separate entities – then there might have been no *via dolorosa* of Rhineland-Anschluss-Sudetenland-Danzig for Europe to walk between 1936 and 1939. The problem with Versailles was not that it was 'Carthaginian', as Keynes so eloquently argued in his influential philippic *The Economic Consequences of the Peace*, but that it left Germany in a physical position to launch her fifth war of territorial aggrandisement in three-quarters of a century to 1939. After the battle of Cannae in 216 BC, where Hannibal killed 55,000 Romans, he sent their gold wedding rings back to be strewn on the floor of the Carthaginian Senate in an impassioned plea for reinforcements with which he could besiege and capture Rome. Their refusal led to their destruction seventy years later, and explains why there is today only one identifiable Carthaginian monument on the site of the city, appropriately enough a tomb. Carthaginian peaces can work, but only if competently applied.

A peace that partitioned Germany in 1919, perhaps even returning her to the pre-1870 status of many small states, might well have prevented a second world war. The problem with the peacemakers of Versailles was that they were willing to wound but afraid to strike, although admittedly it did not look that way at the time. It was not the Versailles Treaty itself, so much as the United States and others' refusal to stand by its measures to curb German rearmament, come what may, that exposed the weakness of the security it was designed to instill. Although there was a clause relating to Germany's 'War Guilt' in the Treaty, it was not long before the Western powers were suffering from something almost as bad: 'Peace Guilt'. This was largely because of the attack on the Treaty by Keynes, who did more than anyone (other than Hitler) to undermine European security in the 1930s because of his sensationalist book *The Economic Consequences of the Peace*, published in December 1919. In it, he accused the Treaty's authors of every conceivable sin. Because exactly twenty years later Europe was indeed plunged once more into war and destruction, the assumption is made that Keynes was a great prophet and his prognosis regarding Versailles was correct. In fact, it was partly because of Keynes' book – which achieved sales of 140,000 copies in eleven translations – that the Versailles settlement broke down, which only then led to the resurgence of war and destruction in Europe. Sir William Beveridge believed that the book was read 'by – at a moderate computation – half a million people who never read an economic work before and probably will not read one again'.

For all their rather arid titles, *The Economic Consequences of the Peace* and its 1922 successor, *A Revision of the Treaty*, were blistering political tracts and had relatively little to do with economics. Keynes, who had attended the Versailles Conference as a representative of the British Treasury, described it as 'a nightmare, and everyone there was morbid', with a 'hot and poisoned' atmosphere, 'treacherous' halls, a 'morass' full of statesmen who were 'subtle and dangerous spellbinders' as well as 'the subtlest sophisters and most hypocritical draughtsmen', who were guilty of 'debauchery of thought and speech' because they were inspired by 'greed, sentiment, prejudice and deception'. The Treaty that these supposed moral and intellectual inadequates finally concocted, after their 'empty and arid intrigue', was 'contorted, miserable, utterly unsatisfactory to all parties'.[37]

Keynes' hyperbole ought to have given his vast readership pause for thought, not least when he described Woodrow Wilson as 'a blind and deaf Don Quixote', who was (not unreasonably therefore) 'playing blind man's buff' at Versailles, before his 'collapse' there and his subsequent 'extraordinary betrayal'. Keynes found it hard to decide whether 'the dreams of designing diplomats' were better or worse than 'the unveracities of politicians' in the creation of a treaty so marred by 'insincerity', 'an apparatus of self-deception' and 'a web of Jesuitical exegesis'. Here are the epithets and adjectives he used on various occasions to describe the Versailles Treaty, which became part of the accepted, received wisdom of the West over the next two decades: 'dishonourable', 'ridiculous and injurious', 'abhorrent and detestable', 'imbecile greed', 'oppression and rapine', 'sow the decay of the whole civilized life of Europe' and 'one of the most outrageous acts of a cruel victor in civilized history'.[38]

Keynes' book was published in America in January 1920 to huge applause. Its passages ridiculing Wilson and arguing that he had been bamboozled by Lloyd George were repeated in every newspaper. On 10 February, Senator William Borah quoted from it at length in the Senate, concluding that the Versailles Treaty was 'a crime born of blind revenge and insatiable greed'.[39] As General Smuts later stated, 'These few pages about Wilson in Keynes's book made an Aunt Sally of the noblest figure – perhaps the only noble figure – in the history of the War, and they led a fashion against Wilson that was adopted by the intelligentsia of the day.'[40] The refusal of the US Senate to ratify the Versailles Treaty – perhaps America's most fateful and worst decision in the history of the twentieth century – was itself ratified by Keynes' book published that same month.

On 29 April 1945 – only nine days before the end of the Second World War – a brave young Free French soldier called Étienne Mantoux was killed near a Bavarian village in the Danube Valley. A graduate of Oxford, the London School of Economics and a Rockefeller fellow at the Institute for Advanced Study at Princeton, Mantoux served in the French air force and as an observation officer attached to General Leclerc's Second Division. He refused to leave the fighting forces for an administrative post in the French provisional government until the war was won, something he sadly did not live to see. Nor did he therefore live to see the publication in October 1945 of his book *The Carthaginian Peace, or The Economic Consequences of Mr Keynes*, a brilliant and persuasive exposé of the myriad faults in Keynes' critique of Versailles. Mantoux's father, Captain Paul Mantoux, an economic historian, had served as an interpreter at some of the private meetings of the Conference and profoundly disagreed with Keynes' descriptions of their atmosphere, but the son's criticisms were based on the case Keynes had made against the Allies, especially over territory and reparations.

'Never before could a Peace Treaty have met with such vehement and indiscriminate abuse,' Mantoux argued,

not on the part of the vanquished merely, but on the part of the victors. That resentment should have expressed itself on behalf of the defeated party was inevitable. Had a solution existed that could satisfy all parties at once, no war would have been fought. However generous the Allies might show themselves, defeat could never be pleasant. Merely to forgo her ambitions of European conquest must have enraged Germany; to lose even her former hegemony over Europe must have infuriated her more.[41]

Mantoux diagnosed the Western democracies of the Thirties as suffering from what the French journalist Raoul de Roussy de Sales had dubbed in his 1941 book, *The Making of Tomorrow*, as a 'guilt-complex'. Clemenceau said, 'We do not have to beg pardon for our victory', but after Keynes that is exactly what the West felt it had to do. The high priest of what Mantoux dubbed 'Meaculpism', Keynes ensured that Britain and France considered the Treaty to be a 'breach of engagements and of international morality' that was equally as bad as Germany's invasion of Belgium.[42]

Mantoux did not attempt to deny that in order to enforce the Treaty there would have to be established a permanent 'military occupation of the Rhine bridgeheads, in which the principal Allies would all participate'. Although the French asked for this 'physical guarantee', both Lloyd George and Wilson rejected it, tragically as it turned out when in 1936 Hitler re-militarised the Rhineland. Instead of standing by his brainchild after the war, in 1938 Lloyd George wrote a 735-page book entitled *The Truth About the Peace Treaties*, in which he attempted to distance himself from many of the provisions of Versailles, implicitly blaming them on the French, and principally on Georges Clemenceau and Raymond Poincaré, both of whom were safely dead by then. In particular, Lloyd George wrote of:

> Clemenceau, with a powerful head and the square brow of the logician – the head conspicuously flat-topped, with no upper storey in which to lodge the humanities, the ever vigilant and fierce eye of an animal who has hunted and been hunted all his life. The idealist amused him so long as he did not insist on incorporating his dreams in a Treaty which Clemenceau had to sign. . . . He listened with eyes and ears lest Wilson should by a phrase commit the Conference to some proposition which weakened the settlement from a French standpoint.[43]

Yet if Clemenceau was entirely lacking in the humanities, Poincaré was worse, according to Lloyd George's *ex post facto* analysis. The French President had 'a dull and sterile mind. He had no wit or imagination or play of fancy. . . . His was the triumph of commonplace qualities well-proportioned, well-trained and consistently well-displayed. . . . In business he was a fussy little man who mistook bustle for energy.' Together, Lloyd George more than implied, the mistrustful Clemenceau, pedantic Poincaré and over-idealistic Wilson had somehow forced the British Premier into an over-harsh peace, which, in his preface to *The Truth About the Peace Treaties*, he said was then exacerbated by subsequent statesmen who took 'a discreditable advantage of their temporary superiority to deny justice to those who, for the time being, were helpless to exact it' (i.e. the Germans). Such was the climate of opposition to Versailles created by Keynes that Lloyd George did not do more than the bare minimum to attempt to defend and justify the Treaty that bore his own signature. Furthermore, the month in which he published his book was September 1938, just as Chamberlain flew to Munich, and thus absolutely the worst time to propagate the line that Germany had been maltreated by the Allies.

Mantoux also blamed Keynes for the way that Versailles was held to prove to Americans that Europeans were each as bad as one another, 'that they were all equally revengeful, equally Machiavellian, equally imperialistic; that the entry of America in the last war had been a ghastly mistake; and that the issue of any new one would be to her a matter of indifference, for an Allied victory would probably be no better than Versailles and a German victory could certainly be no worse'.[44] These were disastrous sentiments for many Americans to profess in the inter-war years, blaming Europe's ills on the supposed 'blind revenge and insatiable greed' that were held to have actuated the peacemakers of 1919. Keynes once famously remarked in his *General Theory* that 'Practical men, who believe themselves to be quite exempt from any intellectual influences, are usually the slaves of some defunct economist.' Yet for over eight decades after the publication of *The Economic Consequences of the Peace*, our world-historical view has been influenced by the same defunct economist who promulgated the assertion.

On his return from Versailles, Woodrow Wilson commended the Treaty to the US Senate on 10 July 1919, describing it in his opening remarks as 'nothing less than a world settlement'.[45] He established the United States' moral superiority from the outset, stating that, 'We entered the war as the disinterested champions of right and we interested ourselves in the peace in no other capacity.' It was perfectly true; America asked for nothing for herself in 1919. The Fourteen Points were devoid of selfish demands for the United States, which has received little enough credit for the fact over the decades. 'Two great empires had been forced into political bankruptcy,' said Wilson of Germany and Austria-Hungary, 'and we were the receivers.' Yet speaking of a third, Wilson claimed that the Ottoman Empire 'never had any real unity. It had been held together only by pitiless, inhuman force. Its peoples cried aloud for release, for succour from unspeakable distress, for all that the new day of hope seemed at last to bring within its dawn.' This messianic attitude was as absurd as it was hyperbolic.

Of Germany, far more reasonably, Wilson said that, 'The monster that had resorted to arms must be put in chains that could not be broken.' This was to be achieved by the League of Nations, set up under Article Ten of the Treaty, 'the only hope for mankind'. He then asked: 'Shall we or any other free people hesitate to accept this great destiny? Dare we reject it and break the heart of the world? ... There can be no question of our ceasing to be a world power. The only question is whether we can effuse the moral leadership this offered us, whether we shall accept or reject the confidence of the world.'

Listening to him was a senator who did indeed intend to break the world's heart, Henry Cabot Lodge of Massachusetts, who Wilson had utterly failed to persuade. Often portrayed as a know-nothing isolationist, Lodge was in fact a far more intelligent and subtle man than that, someone who Wilson signally failed to cultivate as he should have over such an important foreign policy initiative. A stalwart supporter of the war, Lodge had written to the former British Ambassador to Washington, Lord Bryce, on 3 June 1918, the day that American forces first clashed with the Germans: 'I have the profoundest faith that we shall win and by winning I mean putting Germany in such a position that she never again can repeat the horrors which she has precipitated upon the world.'[46]

After the war Lodge, who was Chairman of the Senate Foreign Relations Committee, supported the foundation of Czechoslovakia and a barrier of Slavic states against German aggression, wanted the US to be the leading power in the Allied force in Constantinople and believed that, 'We must protect those stricken people of Asia Minor.' But on the issue of a League of Nations he told Bryce on 14 October 1918: 'We have got a good league now – the Allies and the US. As Roosevelt said the other day it is a going concern. Why look for anything else at the moment?' Yet Lodge was not adamantly or ideologically opposed to 'a League of Nations as a means of maintaining the peace of the world, but it is all too vague. Nobody states the conditions. ... As Mr Wilson has not confided to the Senate, or anybody else so far as I am aware, just what he means by a league of Nations or what he expects us to agree to, it is impossible to discuss it effectively in the Senate.' Fearful of 'any arrangement which would sabotage our independence and sovereignty', Lodge warned Bryce on 16 January 1919 that he was unconvinced.

In March 1919, after a dinner for senators at the White House at which President Wilson 'did not realise how loosely the [League] draft is drawn, how many opportunities for difference and dissension it offers', Lodge persuaded thirty-nine senators to sign a resolution hoping that Versailles 'may revise this draft in such a way as to make it acceptable to the Senate'.[47] He was thus not the sheer wrecker of later Democratic demonology. The problem was Wilson, who failed to send the Senate his draft proposals for the League, and later, in Lodge's view, 'gave the Senate and the country to understand that it was to be that draft or nothing'. All Lodge wanted to safeguard was the Monroe Doctrine, the fact that there would be no international army or navy, and the right of Congress to prevent the US going to war. He also thought that questions of tariffs and immigration ought to be left to member-states, otherwise the League would look too much like a world-government.

The arrangements for a country to leave the League – requiring it to get the unanimous support of the nine countries on the Council and a majority of the other member-states – was likened by Lodge to a situation whereby 'our constitution had required for amendment the unanimous consent of the thirteen original states and the majority of the other states'. Lodge felt that Article Ten setting up the League should not be attached to the peace treaty but debated separately, since it was too important to be hurried through under pressure. All these were reasonable demands, yet the vain, over-confident and surprisingly impulsive President failed to address any of them, refusing the amendments and reservations of senators who had read the small print of what was proposed and baulked at a situation in which the United States could, in theory, be required to furnish an army of 200,000 men to take over Anatolia and Armenia.[48]

Because Wilson insisted on keeping Article Ten tacked on to the Versailles Treaty, the US Senate rejected it by 55 votes to 39 on 19 November 1919. Two months later, it also specifically voted against joining the League of Nations. (The United States made a separate peace with Germany, Austria and Hungary in 1921.) 'If by any mysterious influence of error,' Woodrow Wilson had warned two months earlier in a speech at Sioux Falls, South Dakota, 'America should not take the leading part in this enterprise of concerted power, the world would experience one of those reversals of sentiment, one of those penetrating chills of reaction ... for if America goes back upon mankind, mankind has no other place to turn.' Yet it was largely Wilson's own fault that it happened, and an historic opportunity was missed.

American Energy

1920–9

'Interest is not necessarily amoral; moral consequences can spring from interested acts. Britain did not contribute any less to international order for having
a clear-cut concept of its interest which required it to prevent the domination
of the Continent by a single power (no matter in what way it was threatened)
and the control of the seas by anybody (even if the immediate intentions were
not hostile).'   Henry Kissinger, 'Central Issues of American Foreign Policy'[1]

'My sad conviction is that people can only agree about what they're not really
interested in.'
                                                          Bertrand Russell[2]

On 16 January 1919 – the same day that the US Senate voted against
joining the League of Nations – the Eighteenth Amendment to the US
Constitution, prohibiting the manufacture, sale and distribution of alcoholic
beverages, was ratified by the last of the states. Prohibition had already been
in operation in much of Canada during the Great War. Some provinces, such
as Prince Edward Island, had had it since 1901, and Alberta passed her
prohibition law in 1916. Quebec, showing the positive side of her French
influence, was the last in 1919. The provinces repealed these illiberal and
often-unenforceable laws during the Twenties, Quebec being the first in 1920,
Alberta and Saskatchewan following in 1924, with Prince Edward Island the
last in 1948. New Zealand sensibly voted against it in a referendum in April
1919.

Although Woodrow Wilson vetoed the Volstead Prohibition Enforcement

Bill in late October 1919 – the same month that Norway adopted Prohibition – the House and the Senate passed it in early November and it came into force on 16 January 1920, one year after the states' ratification. At this distance of time it seems incredible that a civilised, open, democratic country like the United States – which has the 'pursuit of happiness' as one of its stated goals – could possibly have banned alcohol altogether for over a decade. It is usually explained – not least by anti-Americans – by reference to the puritanism of the Pilgrim Fathers, but this is clearly insufficient, not least because three centuries separated the *Mayflower* docking from the Volstead Act. Illiberal bossiness, as evidenced by modern political correctness, is also put forward. Whatever the reasons for it, liberation came on 5 December 1933, when the Twenty-first Amendment to the Constitution repealed the Eighteenth, and Americans were finally allowed to drink alcohol again.

The partition of the island of Ireland into two separate political entities happened under the Government of Ireland Act of 1920, which set up a Northern Irish parliament in Belfast and a Southern Irish one in Dublin. Sinn Fein forcibly resisted this and attacked not only the British forces but also those Irishmen in the south who did not support their campaign for a united Ireland entirely shorn of any British presence. Two years later, the Irish Free State Act repealed the earlier one and set up the self-governing Dominion of the Irish Free State in the southern twenty-six counties.

The new arrangements led to disaster for the Protestants of the south, just as the Unionists had been predicting ever since Home Rule had first erupted onto the British political agenda half a century earlier. Protestants in the south were 'menaced, boycotted, frightened, plundered or deprived of their land'.[3] Houses, churches and public buildings were burnt down, as were many of the country's most beautiful stately homes, such as Palmerston in County Kildare, Castleboro in County Wexford and Desart Court in County Kilkenny. There were massacres too; fourteen Protestants were killed on a single day in West Cork in April 1922, for example. Loyal and conscientious Catholic officers of the Royal Irish Constabulary and their families were also forced to flee north to the United Kingdom, for fear of retribution from their nationalist co-religionists. It was the first and only mass displacement of any native group in the British Isles since the seventeenth century.

The demographic statistics since 1922 are compelling; at the time of partition Catholics in the North made up approximately 34% of the six counties, whereas by the end of the century this had grown to 44% of the population, a proportion that is believed to be rising. By stark contrast, Protestants made up more than 12% of the population in the southern part of the island in 1920, yet by 2000 they were less than 2%.[4] The English-speaking peoples have not experienced ethnic cleansing and large-scale population transfers in their post-1900 history, except in Ireland, which again proved its exception to the general experience of the English-speaking peoples regarding religious toleration.

Although partition has proved remarkably successful and durable in Ireland since 1922, it inaugurated a series of other partitions by the British that were to have far less sanguine consequences. The partition of India in 1947 led to hundreds of thousands of deaths and is still disputed in Kashmir six decades later; the partition of Palestine the following year scarcely fared much better as a long-term solution, as did the partition of Cyprus after the Turkish invasion of July 1974.

'The business of America is business,' said President Calvin Coolidge, and by 1920 there were no fewer than 220 sales-management books in the catalogue of the Library of Congress. Books such as *The Science of Successful Salesmanship* (1904) and *Successful Sales Management: A Practical Application of Principles of Scientific Sales Management to Selling* (1913) heralded a new understanding of the importance of salesmanship, and by the 1950s salesmen had been turned into 'well-respected heralds of free enterprise'. Yet for all their importance in extending choice to the consumer, the salesman has long attracted only sneering condescension and occasional outright hostility from academic, literary and intellectual circles throughout the English-speaking world, which was epitomised by both Sinclair Lewis' 1922 novel *Babbit*, arguably his most significant literary work, with its scathing portrait of American middle-class life, and Arthur Miller's 1949 play *Death of a Salesman*, with its 'womanizing, self-pitying, worn-out' anti-hero Willy Loman.[5]

A recent book, *A Century of Icons*, has listed 100 logos, trade marks, slogans and jingles that have emanated from American companies to enter the global consciousness during the twentieth century.[6] For all the anti-capitalist prejudice of much of academia against the commercial and marketing successes of American industry, it is a remarkable achievement for a single country to have created such phenomenally successful brands as 'Coca-Cola, Heinz Baked Beans, Campbell's Soup, Kellogg's Cornflakes, Kodak cameras, Marlboro Man with his dangling cigarette, Cadillac luxury cars, Ronald McDonald, Gillette Razors, Levi's Jeans, down to Absolut vodka, Victoria's Secret lingerie, the AppleMac and the Yahoo Internet search engine'.[7]

By 1920, America had completely outstripped the rest of the world in many of the key economic indicators; she was producing 645.5 million tons of coal per annum, compared with Britain's 229.5 million and Germany's 107.5 million. She was also producing a massive 443 million barrels of petroleum per annum, against Mexico's 163 million, Russia's 25 million and the Dutch East Indies' 17 million. This fuelled the 8.88 million cars licensed in the US, against Britain's mere 0.66 million. Despite having a vastly larger workforce than both Britain and France, the United States had fewer workers involved in strikes – at 1,463,054 – than either of them.[8]

Warren G. Harding was inaugurated on 4 March 1921 in a ceremony that had to be paid for by the private sector, so parsimonious was Congress about his plans for a grand parade. This was all the more surprising since the previous November he had beaten the Democrat James M. Cox by 404 electoral college votes to 127, with 16.1 million popular votes against 9.1 million. The Republicans also retained their majorities in the House by 301 to 131 and the Senate by 59 to 37. Harding was no orator; indeed, a contemporary once said of him, 'His speeches left the impression of an army of pompous phrases moving over the landscape in search of an idea; sometimes these meandering words would actually capture a straggling thought and bear it triumphantly a prisoner in their midst, until it died of servitude and overwork.'

In his inaugural address, Harding advocated 'a return to normalcy'. The word did not then exist – he presumably meant normality – but inaugural addresses by American presidents are an ideal time to invent words, and ever since then it has. But Harding could not re-invent the normalcy of the pre-Great War world, and one of the primary reasons for that was the enormous war debts that the rest of the Great Powers had built up with the United States. The economic and financial costs of the war are estimated – including war expenditure, property and merchant shipping losses – to have been in the order of: Germany $58.07 billion, the British Empire $51.97 billion, France $49.8 billion, the United States $32.32 billion, Austria $23.70 billion and Italy $18.14 billion. In all, the Central Powers spent $86.23 billion against more than twice that by the Allies at $193.89 billion. Much of it had been lent by the United States, and she expected to be repaid.

In 1921 Winston Churchill was elected president of the English-Speaking Union. (Like the Rhodes Scholarships and the Pilgrims Society, the English-Speaking Union was and remains an invaluable contributor to the 'network commonwealth' of the Anglosphere.) Only the previous year, in 1920, Churchill had pronounced that, 'The consciousness of a common purpose in great matters between Britain and the United States is the only sure guarantee of the future peace of the world.' Yet after he delivered the presidential address, he wrote to his wife Clementine from the War Office to say:

> It was uphill work to make an enthusiastic speech about the United States at a time when so many hard things are said about us over there and when they are wringing the last penny out of their unfortunate allies. All the same there is only one road for us to tread, and that is to keep as friendly with them as possible, to be overwhelmingly patient and to wait for the growth of better feelings.[9]

Although the questions of the repayment of war loans and German reparations were complex ones, it was too much to expect that debtor and creditor should view them in the same way, and Britain and the United States certainly did not in the inter-war period. They were a constant source of antagonism, and not the least of the reasons why Neville Chamberlain, who was Chancellor of the Exchequer in 1923–4 and 1931–7, came to dislike and even to distrust the Americans. Similarly, Americans could not understand why money that had been borrowed in good faith could not be repaid according to the terms specified at the time. As Calvin Coolidge said of Britain and France, 'They hired the money, didn't they?'

Most Americans tended to view the international war loans in the same way as private or commercial indebtedness. Fortunately many leaders appreciated the strain that repayment put on what Churchill called 'the delicate machinery of international exchange'.[10] In considering Britain's ability to repay, however, ordinary Americans contrasted the balanced budget in Britain with their own deficits and, as a result, in Churchill's words, 'They brand Britain as a defaulter – a dishonest debtor.' Yet apart from shipments of gold, which was not available in sufficient quantities and which America anyhow did not need, the debt could only be discharged by selling more goods and services to the United States – which would damage the American economy and increase unemployment – or by reducing purchases from the US and taxing American commodities to create dollar credits from which payments might be made. The second route would injure both countries and create new frictions. 'While this ugly and irritating business of the War Debt remains in suspense it is a real barrier to Anglo-American friendship,' wrote Churchill in May 1938, yet the debt and its interest was not finally paid off until the end of the year 2005. The United States is rarely commended for her patience over this; but she deserves to be.

Other than in reparations and loan matters, the United States tended not to involve herself in international matters in the Twenties, sticking to the spirit of her vote not to join the League. Joe Martin, a veteran Massachusetts congressman and twice speaker of the House, recalled in his memoirs how,

> Foreign affairs were an inconsequential problem in Congress in the 1920s. For one week the House Foreign Affairs Committee debated to the exclusion of all other matters the question of authorizing a $20,000 appropriation for an international poultry show in Tulsa. This item, which we finally approved, was about the most important issue that came before the committee in the whole session.[11]

Nonetheless in mid-1931, with the Great Depression being felt acutely throughout the world, President Hoover came up with a plan over war debts and reparations that exhibited once again his country's capacity for altruism.

On 20 June, he proposed a twelve-month moratorium on all repayments, both principal and interest. It was immediately criticised in the US, but also in France, which wanted to exact the maximum in reparations from Germany. Nonetheless Hoover, using the new transatlantic telephone line, won support from no fewer than fifteen nations by 6 July. Congress did not concur until December, and even when it did it added that war debts could not be restructured entirely, as the President had hoped.

The moratorium failed to stimulate economic growth in Europe; Germany was soon gripped by a major banking crisis, Britain left the gold standard and France insisted that after the year was up the issues needed to be re-addressed, but it nonetheless stands as a noble American gesture. Before the moratorium expired, another effort was made to solve the issues of the war debt and reparations at Lausanne in 1932, but this failed also. At the expiration of the moratorium, some of the former Allies continued to make small downpayments, but only Finland managed to discharge her debt entirely.

Feliks Edmundovich Dzerzhinsky had spent half his adult life in Tsarist prisons before, at the age of forty in 1917, he was appointed by Lenin to command the Bolshevik secret police, the Cheka, with a seat on the Central Committee of the Communist Party. He had fought in the Russian Revolution of 1905 and had been one of the key organisers of the October 1917 *coup d'état*. The story of his time in the Cheka, then its successors the GPU and OGPU until his death in 1926, is one long tale of sadism and mass slaughter. His zealotry in the post even shocked other hardened Bolsheviks. He set up the concentration camps, gagged the press, organised the first show trials, purged the churches, wiped out all internal political opposition, sent tens of thousands of writers, academics and intellectuals to Siberia, and was largely responsible for turning Russia from an inefficiently run autocracy into a ruthlessly totalitarian regime.

With a staff of 143,000 by 1921, the Cheka employed terror and torture to destroy what it called 'anti-Soviet subversion', but which was often just innocent life. As Orlando Figes has recorded in *A People's Tragedy*, each of Dzerzhinsky's regional bureaus had their own favoured torture methods:

In Kharkov they burned the victim's hands in boiling water until the blistered skin could be peeled off. The Tsaritsyn Cheka sawed its victims' bones in half. In Voronezh they rolled their naked victims in nail-studded barrels. In Armivir they crushed their skulls by tightening a leather strap with an iron bolt around the head. In Kiev they affixed a cage with rats to the victim's torso and heated it so the enraged rats ate their way through the victim's guts in an effort to escape. In Odessa they chained their victims to planks and pushed them slowly into a furnace or a tank of boiling water. A favourite winter torture was to pour water on the naked victims until they became living ice statues. ... Another had the victims buried alive, or kept in a coffin with a corpse. Some chekas forced their victims to watch their loved ones being tortured, raped or killed.[12]

This was the way that the Bolsheviks built their workers' utopia.

A very small but heroic band of anti-communist writers sought to expose the true nature of Soviet tyranny to the English-speaking peoples from the 1920s onwards. Honourable mentions should be given to Eugene Lyons, Max Eastman, Isaac Don Levine, Robert Rindl, Malcolm Muggeridge, Arthur Koestler, Richard Pipes and Robert Conquest, amongst others.[13] All too often their message was met by disbelief, quibbling over facts, accusations of partiality or worse, and occasionally outright ridicule. When finally the Berlin Wall fell and the archives became available to Western scholars, it was discovered that many of these writers had if anything underestimated the true scale of the horrors being perpetrated by the Bolshevik regime against their own peoples.

'It is fitting', joked Herbert Asquith of the state funeral of Andrew Bonar Law in November 1923, 'that we should have buried the unknown prime minister by the side of the Unknown Soldier.' What wells of bitterness were plumbed in Asquith's cruel quip about the man who, in December 1916, had put his country before both his Party and his own ambition to allow Lloyd George to take the post of wartime premier. Bonar Law, who had briefly been prime minister between the fall of Lloyd George in October 1922 and his terminal illness seven months later, might have made the riposte to Asquith that he gave to Clemençeau: 'All great men are humbugs.'

Bonar Law might not have been a great man, but he had qualities invaluable in a statesman and much admired by the English-speaking peoples, including nobility of character, modesty, frankness, intelligence and decency. He might have been prime minister for only 211 days after October 1922 – the virtually blank page between the thick volumes of Lloyd George and Stanley Baldwin – but he has left legacies quite as hardy as those of his detractor Asquith.

Bonar Law was a man who hailed from far outside the grand *cursus honorum* of Tory leaders; his father was an Ulster-Scots Presbyterian clergyman from New Brunswick, Canada, his mother a well-connected Glaswegian who helped to find him an apprenticeship in the Scottish steel trade. Brains and application – he had read Gibbon's *Decline and Fall of the Roman Empire* thrice by the age of twenty-one – when applied to business won him the financial independence that was then a prerequisite for a career in Tory politics.

Unexpectedly winning a seat in the 1900 Unionist landslide, he was given office after only two years. It was the split over imperial preference – the future trading arrangements of the British Empire – that allowed Bonar Law to emerge as the compromise candidate between Austen Chamberlain and the Tory stalwart Walter Long. The two sides so loathed one another that Bonar Law deftly slipped between them to win the Party leadership in November 1911.

There were soon to be complaints about the extent to which Bonar Law was being advised by the very controversial Maxwell Aitken (later Lord Beaverbrook), a fellow Canadian and son of the manse, but insiders recognised that he was quite bright enough to steer his own path. Over issues such as Irish Home Rule, the Marconi Scandal and the outbreak of the Great War, Bonar Law established a fine reputation as a statesman of resolve. After the Unionists joined the Government in December 1916, he became indispensable to Lloyd George whilst, as he put it, 'hanging on to the coat tails of the Little Man and holding him back'. For his part, Lloyd George found his new ally 'apprehensive and fearless', meaning that he was cautious before agreeing to a policy and afterwards dauntless in seeing it through.

The deaths in battle of his two sons crushed Bonar Law emotionally, but they did not weaken his resolution, and he broke with Lord Lansdowne and those few Unionists who wanted an armistice before victory was won. It was not until Bonar Law had to resign as Lord Privy Seal and Leader of the Commons due to high blood pressure in March 1921 that the Lloyd George coalition started to career wildly off-course, nearly landing Britain in a war against Turkey in the autumn of 1922, which led to the complete rupturing of her relations with France.

By then Lloyd George was attracting much criticism over the way that he conducted British foreign policy: in effect, by-passing the Foreign Office and running the whole operation through an elite phalanx of advisors whom he had installed in a private office built in the garden of Number 10 nicknamed 'the garden suburb'. This mushroomed during his premiership, much to the chagrin of the rest of Whitehall, and especially the diplomats who had traditionally helped run British foreign policy.

'We cannot alone act as the policeman of the world,' Bonar Law wrote to *The Times* on 6 October 1922. It was as well that he announced that in the pages of the newspaper of record, because it came as news to an empire that was well used to acting in precisely that way, particularly in areas of the world adjacent to an ocean. Yet in a sense Bonar Law was right; without the United States being closely involved and without the League of Nations having teeth, there could be no world policeman, at least not a British Empire that was exhausted by the Great War, doubtful of its future and no longer the pugnacious force it once was.

Somewhat reluctantly, but with a natural sense of political timing, Bonar Law allowed himself to be persuaded to return from retirement, and by little more than his very presence at a Party meeting at the Carlton Club on 19 October 1922, he seized Lloyd George's premiership. The British journalist Malcolm Muggeridge wrote in *The Infernal Grove* that, 'To succeed pre-eminently in British public life it is necessary to conform either to the popular image of a bookie or of a clergyman.' After the Conservative Party voted to withdraw from the Lloyd George coalition and govern by itself, the Cabinet saw a wholesale swapping of bookies for clergymen. Out went Churchill, F.E. Smith and Lloyd George himself; in came Bonar Law, Stanley Baldwin, Neville Chamberlain, Sir William Joynson-Hicks and Edward Wood (later Lord Halifax), about as restrained and grave a parliamentary grouping as it was possible to find anywhere other than the bench of bishops.

A year later Bonar Law was dead from throat cancer, hugely mourned by the House of Commons over which he had established a mastery through intellect and character and by a country that trusted and admired him. Lacking in wit and that most elusive political quality, charisma, Bonar Law was utterly straight and deserved a much better soubriquet than the mocking one that Asquith pinned on him. He was not so much the Unknown Prime Minister as the Unappreciated one.

(Asquith's ire against the men who had brought him down in 1916 was more than matched by that of his 'true Cordelia' daughter, Lady Violet Bonham Carter. When pregnant with her first child, Cressida, she worried lest the fact that Lloyd George's 'face and personality has been branded deeper on my soul than any other human being' would mean that the baby would be born looking like him. She thought the historian Lewis Namier was 'king-leech, a lethal bloodsucking bore', Stafford Cripps was 'without one glint of humour, originality, warmth, humanity', Kitchener 'culpably apathetic', RAB Butler a 'slimy hypocrite', Curzon a 'swollen-headed mischief-maker', Baldwin 'full of pig-charm', Beaverbrook simply unspeakable and Neville Chamberlain guilty of 'blind, smug inertia'. She also resented having to attend her stepmother Margot Asquith's funeral because of the time it took from her canvassing in the 1945 general election. Clementine Churchill she thought – in a rather curious simile – 'stupid as an owl'. When forty years later Roy Jenkins broke the news of her father's long affair with her best friend, she said she found it hard to believe, as 'Venetia was so *plain*.')[14]

On the January day in 1924 that Ramsay MacDonald was sworn in as the first Labour Prime Minister, King George V wrote in his diary: 'Today twenty-three years ago dear Grandmama died. I wonder what she would have thought of a Labour government?' If the palpitations that Gladstone gave her were anything to go by, Queen Victoria would probably have had apoplexy.

Yet overall in the twentieth century British monarchs have tended to work just as well with Labour and Liberal Governments as with Conservative ones. King George V sponsored the constitutional conference of July 1910, which eventually led to the Parliament Act, as well as (albeit reluctantly) pledging to create the necessary number of Liberal peers to push the measure through against the last-ditcher Tories. In July 1914, he sponsored the Buckingham Palace conference on Irish Home Rule, again to the chagrin of Unionist diehards. As well as asking MacDonald to form the 1924 Administration, he agreed to his request for a dissolution of Parliament that October when not constitutionally obliged to, and two years later blocked the Baldwin Government's proposal to embargo trade-union funds during the General Strike. In a reign which covered the Great War, the General Strike and the Great Crash, George V acted with admirable constitutional propriety and fairness to Labour. Indeed, he perhaps had too exaggerated a respect for the forces of socialism, turning down his cousin Tsar Nicholas II's hopes of asylum in Britain for fear of exciting left-wing republicanism at home.

George VI similarly enjoyed a good working relationship with the Labour Party. Clement Attlee's biographer stated that the King 'trusted Attlee as a bulwark against demagogic change'. The Home Secretary Herbert Morrison was so royalist that it half-amused and half-disgusted the Chancellor of the Exchequer, Hugh Dalton, who remarked about how the King 'accepted calmly and willingly the changes of political outlook and of personality in the kind of minister he has known throughout his reign'. He could hardly realistically have done otherwise but as adaptability is the Windsors' secret weapon, he even accepted the stripping of his role as Emperor of India without a murmur of regret.

The weekly Buckingham Palace meetings between the famously taciturn Attlee and the shy, stuttering, diffident King were best summed up by George VI's remark to Ernest Bevin in 1948: 'He sits opposite me, but I can't get him to talk.' Dalton feared the King would leak this fact to 'Tory and court circles', but of course he never did. When the King died in 1952, Attlee cried. His successor Hugh Gaitskell had found the King generally sympathetic, but not over the issue of making dentures and spectacles free on the National Health Service. 'You might as well give them free shoes,' the seated King remarked, pointing to his own perfectly shod feet. Queen Elizabeth II's relationship with her politicians was summed up by Sir Godfrey Agnew, the former Clerk of the Privy Council, who told the Labour Cabinet Minister Richard Crossman that far from preferring the Tories, 'The Queen doesn't make fine distinctions between politicians of different parties. They all roughly belong to the same social category in her view.' He, perhaps fortunately, forbore to elucidate where that category stood in relation to other callings.

*

The Northern Territory of Australia, described by the journalist and poet Andrew 'Banjo' Paterson as 'the great lone land by the grey Gulf Water', has had an astonishing variety of projects for settlement. Addressing the seventeenth meeting of the Australasian Association for the Advancement of Science in Adelaide in August 1924, Stephen H. Roberts of Alphington, Victoria, waxed almost poetic about the romantic way that throughout Australia's history the Territory had drawn 'those adventurous spirits who ever seek the untamed frontier':

> The same men who pushed their sheep into the saltbush interior of New South Wales or who raised cattle in the Fijian uplands resorted again and again to the Northern Territory. Flock-masters from South Australia and Queensland led the way; British investors followed; and then came a confused medley of projects the world over. Malagasy tribesmen and Bessarabian communists, Patagonian Welshmen, and Californian dry-farmers were all attracted. On one occasion there was to be a reproduction of Japanese feudal society; on another, colonisation by an expedition of Italian redshirts under Garibaldi himself. Shanghai coolies and Pathan camel-drivers were imported, while, ever and anon, the bankers of the south and the tropical experts of the East dabbled in the hope that the potentialities might be realized.[15]

Roberts identified a period between 1911 and 1924 when, after any number of imaginative agricultural experiments had been tried and had failed as a result of the climate, Australians had learnt that cattle-grazing was the best use of two-thirds of the 355 million acres of the Territory, or at least the luxuriant grasslands from the coast to the foot of the mountain ranges in the north. The Great War saw a rise in meat prices and large-scale construction of freezing facilities, but even by the early 1920s government reports were bemoaning the fact that there was only one head of cattle per square mile. Neither those who had pronounced that Australia had unbounded tropical possibilities for developing the interior, nor those who persisted in seeing the continent as having 'a dead heart' of mere desert, helped the Northern Territory, since they were both wrong. Gradual growth, taking advantage of more artesian bore-holes, more and better stock routes and railways, more shipping facilities and freezing works, and more stock inspection were the proven way forward for the Territory, but were ones that came to a shuddering halt during the Great Depression, which hit Australia particularly hard.

By 1938, the discrepancy between the heavily populated fascist powers and the under-populated lands of the English-speaking peoples was impossible to miss. Germany had 366 people to the square mile, Italy 358 and Japan 352. Britain had 468, it was true, but New Zealand had only 15, Canada 3 and Australia 2.[16] As Churchill warned in May 1938,

The wide, open spaces of the British Empire are a standing temptation to imperialist adventure by foreign Powers. . . . Canada and Australia cannot be safe unless they add greatly to their British stock. New Zealand and South Africa must also increase their British populations to be secure from attack. In strengthening these great dominions we would strengthen the whole Empire – and all the peace-loving nations of the earth.

Germany and Japan's quest for *Lebensraum* ('living space') was set to be the great issue of the 1940s, with profound implications for national security. From where, though, did its leading exponent get the idea?

The period that Hitler spent in the Landsberg Prison between April and December 1924 was not wasted; his political testament *Mein Kampf* was dictated there to his close friend and future deputy Führer, Rudolf Hess. Hess had been a student of Professor Karl Haushofer and, according to Hitler's biographer Ian Kershaw, his writings were most probably 'one significant source' for Hitler's notions about *Lebensraum*; indeed, Haushofer's influence 'was probably greater than the Munich professor was later prepared to acknowledge'.[17]

At the Nuremberg Trials, Haushofer naturally argued that Hitler had misunderstood his works, but the historian Karl Lange has cast serious doubt on this assertion, while another German historian, Werner Maser, has analysed *Mein Kampf* closely enough to assert that Hitler was familiar with the theories of both Haushofer and Sir Halford Mackinder, from whom Haushofer received the 'World Pivot' theory referred to in Chapter Two. Haushofer visited both Hess and Hitler in prison, just as Hitler was dictating *Mein Kampf,* and gave Hess books and articles on geopolitics, including works by Mackinder.[18] A direct connection can therefore be made between Mackinder's academic address to the Royal Geographical Society in London in 1904 and Hitler's lunatic lunge for *Lebensraum*, Operation Barbarossa, thirty-seven years later.

Even as late as 1943, with the Gog and Magog of totalitarianism locked in combat in 'the heart-land' of 'the World-Island', the eighty-two-year-old Mackinder pronounced in the journal *Foreign Affairs* that,

> All things considered, the conclusion is unavoidable that if the Soviet Union emerges from this war as conqueror of Germany she must rank as the greatest land Power on the globe. . . . The Heartland is the greatest natural fortress on earth. For the first time in history it is manned by a garrison sufficient both in number and quality.[19]

Not surprisingly, Mackinder himself resented being accused of laying the foundations for Nazi militarism and *Lebensraum*. On receiving the Charles P. Daly Medal from the American Geographical Society in 1944, he said, 'Whatever Haushofer adopted from me he took from an address I gave before the Royal Geographical Society just forty years ago, long before there was any question of a Nazi party.' This was literally true, but rather beside the point. Just as J.M. Keynes wrote that people who think of themselves as having no economic preconceptions are often in thrall to a defunct economist, so the strategic views of an English geographer might well have spawned the *Lebensraum* philosophy that lay behind the greatest and most destructive invasion in the history of Mankind.

It seems almost otiose to point out that Mackinder's theory was actually wrong. Eastern Europe only became the pivot of world history in 1941–5 because Hitler invaded it, which he was not compelled to do because it was the Heartland of the World-Island. Since 1945 it has been a relative backwater, both under communism and even since its liberation in 1989–91. The Soviets, although they won in 1945, did not emerge as the greatest land power on the globe, except in terms of sheer amount of territory, and if the two English-speaking countries of North America are combined, not even then. Finally, 'Who commands the Heartland' did not even 'command the World-Island', as the western side of Europe stayed free of Soviet domination. Sir Halford Mackinder died in March 1947, just before his theory was revealed as utter codswallop.

There is certainly some evidence from the mid-1920s that Japan was considering pursuing *Lebensraum* at the expense of the English-speaking peoples. In the four hours between 2 a.m. and dawn on Sunday, 2 August 1925, a large ship, well-lit but unidentified, was seen to sail off Nobby's Head, near Newcastle, New South Wales in Australia. She moved about ten miles east to south-east of the Head, which forms the opening to the mouth of the Hunter River's south channel and the northern limit of the Newcastle coalfield. 'The strange craft,' recorded the historian R.D. Walton, 'moving in and out of a cloud of haze from the Broken Hill Proprietary steelworks, changed course several times, occasionally coming to a complete halt.'[20] She was spotted by several unrelated observers changing course suddenly whenever it looked as though other vessels were likely to come close to her. At dawn she sailed off towards the east.

On its own such a phenomenon might be unremarkable, but coming as it did in an extended pattern of strange behaviour by ships around Newcastle between 1919 and 1927, it is highly likely that the Japanese Imperial Navy was carrying out fairly sophisticated espionage activities against Australian economic targets around Newcastle, possibly with the final object of crippling or capturing them in the event of war. Unfortunately, almost all of Japan's naval Intelligence files were destroyed before the end of the Second World War, but work that has been done piecing together various incidents in the post-Great War period certainly tends to suggest that Japan had aggressive intent towards mainland Australia in the early-to-mid 1920s. With a population several times that of Australia's and a yen for conquest that became obvious to all by late 1941, Japan posed an evident threat to the sovereign independence of the English-speaking peoples in the Antipodes. As early as 1920, the Australian Chiefs of Staff concluded that Japan had the shipping necessary to land 100,000 troops in Australia in a single convoy.[21]

If this invasion had taken place anywhere on the east coast, it would have been at Newcastle, the heart of the northern New South Wales coalfield and a lynchpin of the Australian economy. In 1927, 56% of all the coal used in New South Wales was mined there and 93% of all coal exports from the state came from there. Tasmania, for example, depended on it for over half her coal supply. Broken Hill Proprietary produced more than 98% of the energy used by Australian industrial plant that came from coal, and early in the decade it also dominated in transport too. If Japan were able to deny Australia the use of that coalfield during wartime, it would have dealt her a devastating blow.

In 1924, a party of Japanese were observed photographing the beaches near Fort Wallace, north of Newcastle. Even more seriously, the assistant harbour-master at Newcastle caught sight of a map being used by a Japanese merchant naval officer, which was quickly concealed but which certainly corresponded to no published map then in existence in Australia.[22] Espionage stories are notoriously hard to prove or disprove, but the suspicion mounted in the minds of the Australian security officials that a surprise invasion was at least being considered in Tokyo.

In one month alone – March 1925 – both the large steamer *Chofuku Maru* and three days later another Japanese vessel, the cargo ship *Meiko Maru*, were observed by the signalling station at Nobby's Head passing close to Merewether Beach five miles south of Newcastle, the latter as close as one mile out to sea but several miles off any recognised or safe sea-route. Further such incidents took place in May and June 1926 involving different vessels with much the same *modus operandi*. Local customs officials judged 'the time, place and general conditions' as 'totally opposed' to smuggling operations; furthermore, the explanations given for the strange behaviour of the ships were completely unbelievable. One agent claimed that the skipper of the *Havo* needed to drop off mutinous Chinese firemen at Merewether, even though Newcastle harbour itself was only a quarter of an hour away and no firemen were unloaded.[23]

Early 1927 also saw traverses of the coast north of Newcastle in similarly weird circumstances. At this distance of time and without the relevant Japanese documentation we cannot know for certain that the Japanese Navy was photographing and taking soundings of the best beaches to use in order to invade

Australia at her most economically vulnerable place, but the evidence suggests that Newcastle, New South Wales, was being sized up as the Pearl Harbor of the 1920s. There can be no doubt from the horrors that Japan later visited on what it (with sinister Orwellian overtones) called the 'Greater Asian Co-Prosperity Sphere' that a successful invasion of Australia would have led to massacre and misery on a scale not experienced by the English-speaking peoples on their own soil since the American Civil War.

In 1924, the great wave that had brought one-and-a-half-million immigrants, mainly Eastern European Jews and southern Italians, to New York since 1880 was finally ended by radical changes in immigration policy. By 1910, 41% of all New Yorkers were foreign-born. 'The immigrants settled in suffocating number in the "Little Italies" and "Jew Towns", as their ghettos in lower Manhattan and Brooklyn were called. ... Some families remained frozen in their ethnic slums for generations. According to prevailing myth, they did not have the rugged entrepreneurial individualism to make it.'[24]

It was true; yet America deserves credit for taking in this huge influx of foreigners, who for the most part were escaping far worse poverty, prejudice and despotism than they suffered from the worst rack-renting landlord in New York. The naturalisation process most underwent on Ellis Island was humiliating for many immigrants – especially in the ascribing to them of new, Anglicised names – but compared to the knouts and pogroms of Russia and Poland it was hugely benign. Less than two decades after the gates temporarily closed in 1924, those Eastern European Jews who had not left faced genocide.

According to the census and other data taken at the time, the average amount of money that Italian immigrants arrived with in 1900 was $4.81, and in the century's first decade 47% of the arriving Italians and 27% of the Jews could neither read nor write.[25] Only 16% of the Italians had a trade; most had been day-labourers or peasants. Of course America did not open the flood-gates out of altruism or out of concern for human rights, but largely because of the booming economy's ravenous need for labour, and neither did the immigrants all participate fully or immediately in the American Dream. Yet the programme did allow millions to escape violence and to create for themselves a better life, and that deserves to be recognised as a fine achievement of the English-speaking peoples.

One of the foremost engines of American commercial energy in the United States in the 1920s was its Jewish community, especially that of New York City. The first Jews to arrive in America were twenty-three refugees from Brazil, who stepped ashore at New Amsterdam in 1654. They were imme-diately distrusted by the colony's governor, Peter Stuyvesant, who suspected that they would live by 'their customary usury and deceitful trading with the Christians'.[26] Yet by the time of the American Revolution, five Jewish communities were thriving in New York, Philadelphia, Rhode Island, Georgia and South Carolina.

Large-scale Jewish immigration into America began in the 1820s from Germany and then continued from Eastern Europe through the rest of the nineteenth century. Europe's loss in terms of Jewish commerce, energy, intellect and culture was America's gain, and by the 1920s the community – based mainly on the cities of the east coast – was strong and thriving, contributing to almost every aspect of American life. *Forverts* (later *Forward*), a Yiddish newspaper edited from New York's Lower East side, sold almost 200,000 copies daily. As a review of Hasia R. Diner's recent book *The Jews of the United States* put it,

> Jews became highly successful in cinema, literature and journalism, but they were also keen to assimilate; Oscar Hammerstein wrote *Oklahoma*, a patriotic musical about the American frontier, and Betty Persky became Lauren Bacall. . . . By the end of the Second World War, the American Jewish community had become the biggest and most powerful in the world.[27]

Anti-Semitism abounded in the 1920s and 1930s, and the imposition of immigration quotas condemned possibly tens of thousands or more to the grasping maw of the Holocaust, but compared to the persecution from which so many Jews escaped to America from Russia, Poland, the Ukraine and latterly Nazi Germany, the native English-speaking peoples on both sides of the Atlantic treated the Jews relatively well. Alienation, disruption and exclusion were all too often their lot in American society of the 1920s and 1930s – even occasionally lynching in Southern states – yet overall they thrived better in the English-speaking world than anywhere else before the creation of the State of Israel in 1948.

Meanwhile, the twentieth-century Jewish contribution to finance, science, the arts, academe, commerce and industry, literature, charity and politics in the English-speaking world has been astonishing, relative to their tiny numbers. Although they make up less than half of 1% of the world's population, between 1901 and 1950 Jews won 14% of all the Nobel Prizes awarded for Literature and Science, which increased to 29% between 1951 and 2000.[28]

'Nations on the gold standard are ships whose gangways are joined,' said Churchill, employing a strange analogy since nautically that can work for both good and ill. His re-joining of the gold standard at the wrong rate in 1925 and Calvin Coolidge's conservative economic policies have both been blamed for the Wall Street Crash of 1929. Because Coolidge came into office on Harding's death in August 1923 and left office on 4 March 1929, only seven months before the Crash, it is generally assumed that his Administration must have been primarily responsible. Both Coolidge's supporters and detractors are hamstrung by this most secretive of American presidents having destroyed almost all his personal papers before delivering the rest to the Library of Congress and Forbes Library in Northampton, Massachusetts.[29]

Genuinely shy people rarely enter politics, but Coolidge seems to have been an exception. He was reserved and taciturn to an extraordinary degree, and, according to his 1940 biographer Claude M. Fuess, 'his secretiveness is almost unparalleled among American statesmen'. However, it is perfectly possible that even if Coolidge had not performed an *auto da fé* on his personal correspondence, there might not be too much to justify or condemn the President anyhow, since he rarely felt it incumbent upon himself to explain his Administration through letters to friends. He is apocryphally credited with having explained his celebrated taciturnity with the words: 'I have never been hurt by what I have not said,' and that probably extended to what he had not written, too. Nonetheless, Coolidge was the last president to write all his own speeches.

Coolidge's Vice-President, General Charles G. Dawes, was awarded the Nobel Prize in 1925 for the Plan which bore his name that attempted to redraw the payment schedule for German reparations in an equitable and commonsensical manner. Coolidge's support for the Dawes Plan was unequivocal and denoted a return, if sadly short-lived, of American interest in the affairs of continental Europe.[30]

In October 1926, Britain's Lord President of the Council, Earl Balfour, agreed to chair the Inter-Imperial Relations Committee of the Imperial Conference. As Arthur James Balfour he had been Prime Minister over twenty years earlier, and this task was to be, as his biographer attests, 'his last great achievement' of a very long life in politics.[31] As British Foreign Secretary at the Versailles Conference, Balfour had dealt with the Dominions' representatives and he well understood their demands for greater autonomy from Britain. The Treaties of Lausanne and Locarno had seen the Dominions refusing blindly to follow the direction of British foreign policy, so how were the conflicting demands of imperial unity and Dominion autonomy to be reconciled? Balfour's way of squaring the circle was to formulate the principle of absolute Dominion equality, for, as he had told the House of Lords, 'None of us conceive that of this conglomeration of free states one is above the other.'[32] As Balfour frankly stated in his Declaration to the Imperial Conference:

> The tendency of equality of status was both right and inevitable. Geographical and other conditions made this impossible of attainment by way of federation. The only alternative was by way of autonomy; and along this road it has steadily been sought. Every self-governing member of the Empire is now the master of its destiny.

Lord Vansittart said of Arthur Balfour that, 'He viewed events with the detachment of a choirboy at a funeral service', but in the two declarations bearing his name – his 1917 one that committed Britain to 'establish in Palestine a national home for the Jews' and the 1926 one above – that famously detached statesman profoundly affected the course of history.

The deliberate course that Britain took to secure her colonies' loyalty was that of continual appeasement. Almost every stirring of nascent nationalism was met by Westminster's willingness to accord greater and greater self-government and local responsibility. Ever-mindful of the result of the mismanagement of American affairs between 1775 and 1783, successive British governments worked hard to avoid an Australian Bunker Hill or a Canadian Yorktown. Throughout the nineteenth and twentieth centuries the policy worked, and only in South Africa did a significant white colonial population try to oppose London by force. And when the semi-autonomous Boer colonies of Transvaal and Orange Free State did rebel, they found 30,000 colonials from the rest of the Empire volunteering to fight against them, something that would not have happened if Canadians, New Zealanders and Australians had felt themselves oppressed by Britain. As the American philosopher George Santayana wrote of the British Empire in 1922, 'Never since the heroic days of Greece has the world had such a sweet, such a boyish master. It will be a black day for the human race when scientific blackguards, conspirators, churls and fanatics manage to supplant him.'

The history of the constitutional development of the 'Old' Commonwealth is the story of step-by-step voluntary renunciation of control. Canadian self-government was presaged in the Durham Report in 1839 and its confederation was founded in 1867, New Zealand was granted her first constitution in 1852 and Australia federated in 1901. Meanwhile, successive imperial conferences between 1887 and 1907 only ever pulled delicately on the bit, with London refusing to pressurise the Empire into adopting full-scale imperial preference before its time.

This dignified, good-natured abdication of control contrasts starkly with France's 'savage war of peace' in Algeria and Indo-China, Germany's atrocities in Namibia and Belgium's genocidal exploitation of the Congo. Although there are of course significant Commonwealth minorities who resent Britain's close links with their countries – the Quebecois who still regret Wolfe's victory in 1759, for example, or Irish-Australian republicans – they must acknowledge that subjects of the British Crown since 1900 were never the victims of the kind of genocidal treatment meted out to the citizens of the continental European empires. Similarly, the Dominions themselves behaved well towards their own colonies. The distinguished constitutionalist lawyer Professor Colin Aikman, contributing to a book about New Zealand's record in the Pacific Islands during the twentieth century, pointed out how, in the field of constitutional development, 'New Zealand may be able to feel some satisfaction at the way in which she has guided Western Samoa and the Cook Islands to autonomy.'

One of the problems the English-speaking peoples have long encountered in their diplomacy has been their insistence on precise language and verifiable outcomes. This was well expressed on 16 September 1927 by the British Foreign Secretary Austen Chamberlain, who, as head of the British delegation to the Geneva Disarmament Conference, reported to the Prime Minister Stanley Baldwin how,

> In a number of smaller matters our preference for the real and practical, and the cold douche of common sense which we administer, are repugnant to the races who express themselves in a much more rhetorical form, who love broad generalisations and noble sentiments, and are less careful about the precise meaning of words they use and the undertakings they give than is compatible with our sense of what we owe to ourselves and others.[33]

He was principally referring to the Latin races, whose statesmen might sound more committed to disarmament because of their grandiloquent phraseology, but who were far more difficult to pin down on actualities than the Americans or Germans.

Yet it was finally over disagreements with the Americans in Geneva in 1927 that the conference broke down, producing the most dangerous moment for Anglo-American amity since 1896. In the words of the historian Professor Donald Cameron Watt, this was

> accompanied by the most virulent Anglophobic press campaigns [in America]. ... The Committee of Imperial Defence in London was worried enough to institute a secret inquiry into what would be involved in a war with America. The doctrinaires on the General Board of the US Navy did not abandon war planning against Britain until the mid-1930s. Admiral Sir Roger Keyes wrote to Churchill expressing his desire that when the Royal Navy stormed the principal US naval base he might be in command. The European press was full of speculation as to when, not if, war would break out between Britain and America.[34]

As in 1896, sanity prevailed, but it would be quite wrong to see the twentieth as a century of unbroken Anglo-American friendship. Churchill had been on the receiving end of American hostility to Britain a quarter of a century earlier, on a speaking tour of the Midwest in December 1900, when he had been hissed and booed by a pro-Boer crowd at the University of Michigan at Ann Arbor and only just managed to charm a rowdy and angry Irish audience in Chicago through extravagant praise of the courage of the Dublin Fusiliers.[35]

Only the previous year, in September 1899, Churchill had written dismissively, even cruelly, to his mother about a magazine she was proposing to publish, using phrases very different indeed from the ones that he was to adopt regarding Anglo-American amity between 1941 and 1945:

> Your title *The Anglo-Saxon* ... only needs the Union Jack and the Star Spangled Banner crossed on the cover to be suited to one of [Alfred, later Lord] Harmsworth's cheap Imperialist productions. ... As for the motto 'Blood is thicker than water' I thought that that had long ago been relegated to the pot house music hall. ... Your apparent conception of a hearty production frothing with patriotism and a popular idea of the Anglo-American alliance – that wild impossibility – will find no room among the literary ventures of the day.[36]

Churchill, who was no stranger to the pot-house music hall and had once instigated a riot in support of the right of prostitutes to frequent the Leicester Square Empire, added a sarcastic cartoon of the Union Jack enfurled with the Stars and Stripes. His strictures went unheeded by his American mother; *The Anglo-Saxon Review* was published in June 1899, but only survived for ten issues, closing in September 1901.

On the evening of 21 September 1928, entertaining guests at Chartwell, Churchill 'talked very freely about the USA'. The diary of one of them, a Tory MP called James Scrymgeour-Wedderburn, records how, 'He thinks they are arrogant, fundamentally hostile to us, and that they wish to dominate world politics. He thinks their "big Navy" talk is a bluff which we ought to call. He considers we ought to say firmly that we must decide for ourselves how large a navy we require, and that America must do the same.'[37] Churchill, who was Chancellor of the Exchequer at the time, was only slightly more guarded in his Cabinet memoranda, complaining on 19 November that over 'great events' such as the Irish Question, the Washington Treaty and Anglo-American debt settlement, 'Whatever may have been done at enormous cost and sacrifice to keep up friendship is apparently swept away by the smallest little tiff or misunderstanding, and you have to start again and placate the Americans by another batch of substantial or even vital concessions.'[38]

Someone else who understood the implications of the emergent power of American capitalism at the time – and its capability of dominating world politics – was Adolf Hitler. In his little-known 1928 sequel to *Mein Kampf*, a book on Nazi foreign policy that remained unpublished until 1961 and was not published in English until 2003, he aerated his views on 'the Anglo-Saxon peoples' at length. It was not hard to see why a political pundit writing in 1928 might opine that, 'The pride of the English today is no different from the pride of the Ancient Romans', but it took an impressively prophetic analysis to state that before too long the USA would be the sole hegemonic power in the world. Ironically enough, it was Hitler's own lunatic declaration of war in December 1941 that helped thrust America into that position.

Here was one of Hitler's more acute statements and predictions from 1928: 'The size of the internal American market and its wealth of buying power ... guarantee the American automobile industry internal sales figures that alone permit production methods that would simply be impossible in Europe. At issue is the general motorization of the world – a matter of immeasurable significance.' And another: 'There is a movement of devout adherents that wishes to counter the union of the American states with a European one, in order to prevent the hegemony of the North American continent.'

'In the future,' the prospective Führer predicted, 'the only state that will be able to stand up to North America will be the state that has understood how ... to raise the racial value of its people and bring it into the most practical national form for its purpose.' That form was, of course, a single European super-state under the ultimate political direction of the largest, most populous, hardest-working and best geographically placed power: Germany. It represents Hitler's first mention of the United States as a future enemy.[39]

'With the American Union a new power factor has emerged on a scale that threatens to nullify all the previous state power relationships and hierarchies,' Hitler wrote, arguing that emigration from Europe to America was a form of Darwinian survival of the fittest, and the 'menacing American hegemonic position' was 'determined primarily by the quality of the American people and only secondarily by its *Lebensraum*'. The United States' eighteenth- and nineteenth-century *Lebensraum* over the American continent had, in Hitler's opinion, created an internal market that increased production levels 'and thus production facilities that decrease the cost of the product to such a degree that, despite the enormous wages, underselling no longer seems at all possible'.[40] These were no more than the economic commonplaces of the day, but they prove that Hitler's interest in the United States was not confined to the cheap Westerns he read and movies he watched.

Hitler chose the classic example to describe the American industrial threat, writing of the development of the automobile industry:

> It is not only that we Germans, despite our ludicrous wages, are not in a position to export successfully against the American competition even to a small degree; we must watch how American vehicles are proliferating even in our own country. [41]

Protectionism has long been the default position of those anti-Americans who recognise that their own societies are unwilling or simply incapable of competing against sophisticated American capitalism.

Neither was Hitler, at least in 1928, deriding the United States' racial stock in the way that he regularly did during the war. In his second book, he even went so far as to tell his readers (leaving gaps for the figures that his fact-checkers would fill in later), 'That the American union is able to rise to such a threatening height is not based on the fact that ... million people form a state there, but on the fact that ... million square kilometres of the most fertile and richest soil are inhabited by ... million people of the highest racial quality.'[42]

One of the reasons why the English-speaking peoples could spread across continents and live and work in relative comfort, even in desert conditions by the 1920s, was air-conditioning, whose social and economic impact was unquantifiable, but enormous. America led the world in the invention, development and installation of what was to become a multi-billion-dollar industry, but Australia and Canada soon followed, allowing them to master the elements and establish cities in places that for millennia had been either too hot or too cold for human habitation. With its advent, pioneered by the English-speaking peoples, cities like Perth in Western Australia (average temperature in January 76 degrees fahrenheit) and Anchorage in Alaska (average January temperature –9 degrees) could not only survive but thrive.

The cooling of government buildings in Washington began in 1928, when arguments that it was a waste of taxpayers' money were overruled on the health grounds that warm air carried more microbes.[43] Hitherto politicians had sought to escape the boiling District of Columbia in summertime, but after 1928 they tended to stay on for longer. Gore Vidal, in one of his less fatuous rodomontades, has argued that the air-conditioning of Washington offices led to American imperial ambitions being promoted by workaholics who would otherwise have gone to their country estates to enjoy the cool breezes Washington is denied in summertime. Although Franklin D. Roosevelt refused to have air-conditioning in the White House, preferring to work in his shirt-sleeves, Richard Nixon would have it put at its coldest possible setting and then have fires lit in the grates.

It was not long before class distinctions intruded, even in the sphere of air-conditioning. In order to allow ladies to show off their fur coats, the lobby of the Fontainebleau Hotel in Miami 'was kept famously chilly, and had a showy staircase so that people could quit the lift at first-floor level and make a grand entrance'. Richer people had central air-conditioning, while the less well off lodged theirs in window openings, 'with the take-up moving block-by-block, since there was peer-pressure against being the first to install it (showing off) but also against being among the last (letting the neighbourhood down).'[44]

On Saturday, 4 February 1928 the electrical engineer John Logie Baird gave the first demonstration of colour television at the Dominion Theatre, London, on a 9-foot by 12-foot screen. He had produced televised images in outline back in 1924, and two years later had transmitted images of moving thirty-line silhouettes over telephone lines between London and Glasgow. A fascinating if obviously occasionally infuriating character, Baird was

> a scientific nuisance from the moment he was given pocket money. Aged thirteen, he bought an old oil engine, wrapped lead plates in flannel, packed them into jam jars filled with sulphuric acid, and produced the first electrically lit house in Helensburgh, the town where he was born, thirty miles from Glasgow. The experiment ended when, because the light bulbs in the corridor were rather dim, his father fell downstairs, and wee Baird contracted lead poisoning.[45]

There was definitely something of the con-man about Baird; before he invented television he had tried to manufacture artificial diamonds by connecting the power-grid of the Clyde Valley Electrical Power Company – of which he was assistant manager – through a carbon rod, plunging the whole valley into darkness. Afterwards he marketed a supposed cure for haemorrhoids, then a special 'Baird Undersock' which absorbed sweat, and later he attempted to export mango jam from Trinidad. After that he tried to sell 'Baird's Speedy Cleaner' (basically caustic soda), then artificial flowers, and finally a couple of tons of Australian honey that had somehow come his way off the docks. Baird was therefore hardly another Alexander Fleming or Ernest Rutherford, yet the last money-making contraption he devised has given instruction and pleasure to billions.

It was in the spring of 1923 that the thirty-four-year-old Baird invented television, from a bed-sit in Hastings, with 'tin, cardboard, darning needles, cotton reels, a powerful electric lamp, a bull's eye lens from the local cycle shop, and an electric fan motor'.[46] At one point he tried to make a cell from a human eye that he had managed to remove from the Charing Cross Ophthalmic Hospital, wrapped in cotton wool, but after he failed to dissect it with a razor he 'gave it up and threw the whole mess into the canal'. This irascible, suspicious, difficult, egotistical man – part-genius, part-charlatan – was somehow also the same person who invented not only television but also 'the first television that could see at night, the first video recording, the first television by telephone line, the first stereoscopic television, the first large-screen presentation, the first commercial television sets' as well as the first air-to-ground reconnaissance system, which could detect water ripples on reservoirs. Yet he never became rich, and in 1959 the BBC – with which he had fought throughout his career – refused to place his bust in the foyer of their new Television Centre in White City. Before his death in Bexhill-on-Sea aged only fifty-seven in June 1946, Baird had given demonstrations of higher-definition television systems which he called 'Telechrome' and had pioneered an invention that has utterly changed the world, probably for the better. Overall.

The record of British inventiveness has been one of the staggering success stories of the modern world, yet a recent survey by the Science Museum in London showed that 58% of Britons did not know that their countrymen had invented trains and 77% did not know they had invented the jet engine.[47] The first steam engine was invented by a Briton in 1698, the electric motor in 1821, the telephone in 1875, the internet in 1989. Other British inventions include the traffic light, electromagnet, underground train, light bulb, the pneumatic tyre, radar, vaccination, penicillin, cloning and the steel-ribbed umbrella, yet Britons stubbornly fail to give themselves credit for their inventiveness. Furthermore, as the journalist Anthony Browne adds,

> Britain's scientists have done more to unravel the mysteries of nature than any others. Of the four main forces of nature, Brits unravelled the mysteries of two – Newton with gravity and James Clerk Maxwell with electromagnetic radiation. Of the three planets unknown to the ancients, two [Uranus and Neptune] were discovered by the British. Britain is second only to the US in the number of Nobel prizes it has won – twice as many as France and seven times as many as Italy and Japan.[48]

In 1934, Percy Shaw, a forty-four-year-old road-mender from Yorkshire, was driving down a road on a foggy night near his birthplace, Halifax, when he saw a cat's eyes reflecting the light from his headlights. This – assuming this possibly apocryphal tale is true – set him to thinking about an invention that might use small roadside mirrors employing car headlights to mark out roads and thus prevent accidents. In 1935, he patented his invention of two glass and metal beads which, when pushed against their moulded rubber casing as cars drove over them, automatically cleaned themselves in rainwater that they could collect in a specially designed depression. Trade-marked 'Cat's-eyes', the invention made a fortune for Shaw once it was adopted globally and has undoubtedly saved untold numbers of lives.

As well as Baird's colour TV, the year 1928 also witnessed another Scot producing one of the greatest boons to Mankind ever delivered by the English-speaking peoples, when in September a forty-seven-year-old Scottish bacteriologist, Alexander ('Alec') Fleming, discovered penicillin. It has been described as 'possibly the greatest advance ever made in the entire history of medicine', and with good reason; penicillin in particular and antibiotics in general opened the way to almost universal relief from pain.[49]

As so often in important scientific discoveries, there was indeed a single definable 'Eureka' moment made by a brilliant individual, but only after the general area had been explored by very hard-working scientists, many more of whom later widened the potential made possible by the Eureka moment.

Fleming did not discover penicillin without a penumbra of fellow researchers' contributions in related fields, and it was ten years before the discovery was properly exploited, yet it was undoubtedly his.

Fleming's father was an Ayrshire sheep farmer who died when Fleming was seven years old. The strong emphasis always placed on education in Scotland meant that the local school at Loudoun Moor, his next school at Darvel and later Kilmarnock Academy equipped him well for later life. At fourteen he moved down to London with his two brothers to live with a fourth brother and continue his education at the Polytechnic Institute in Regent Street. Having spent four years as a clerk in a City shipping office, in 1901 a small legacy allowed him to become a student at St Mary's Hospital medical school in Paddington. Never was a legacy better bequeathed.

After winning the senior entrance scholarship in natural science, Fleming went on to take virtually every prize and scholarship available to him throughout his years as a medical student, including the London University Gold Medal of 1908. Yet nor was he simply a swat; as an undergraduate and post-graduate he took an active part in the swimming, shooting and golf clubs, as well as acting in student theatrical entertainments.[50] After graduating he became the apprentice to the formidable Sir Almroth Wright in St Mary's inoculation department, following Wright to Boulogne on the outbreak of the Great War to study the treatment of war wounds for the Medical Research Committee as a lieutenant (later a captain) in the Royal Army Medical Corps.

During the war, Fleming devised numerous ingenious experiments that were to make 'outstanding contributions to knowledge of the bacteriology and treatment of septic wounds'. He published a paper in the medical journal *The Lancet* in 1915 on the malign significance of *Streptococcus pyogenes*, advocated the early removal of necrotic (dead) tissue in wounds and made major advances in knowledge about gas gangrene. Even if he had not discovered penicillin, Fleming would deserve an honourable place in the annals of the English-speaking peoples' contribution to pain relief and cure. He spent the early and mid-1920s discovering lysozyme, an anti-microbial substance produced by the body's tissues, which he called the body's natural antibiotic, and in developing new techniques to demonstrate its diffusion he created methods that were to serve him well in his studies of penicillin.[51]

Yet it was in September 1928 that Fleming made the discovery that was to bring untold benefits to Mankind. That month he returned from holiday to his small laboratory at St Mary's to find a pile of petri dishes on which he had been growing colonies of bacteria. They were waiting to be cleaned. On one of them he noticed a mould had grown that had inhibited the growth of a colony of staphylococcus germs. This mould, commonly found on bread and called *Penicillium notatum*, Fleming turned into a liquid he called Penicillin. It was some time before it was realised that the unusually cold snap during Fleming's absence had allowed the *penicillium* to flourish in the petri dish, but to this day it remains a mystery how it entered the lab at all, although it is claimed that a window had been left open. (The Fountains Abbey public house across the road in Praed Street proudly claimed to have been the source, although modern researchers think infiltration from another lab more likely.)

Fleming's June 1929 paper on the phenomenon in the *British Journal of Experimental Pathology* was ground-breaking. In it he described most of penicillin's properties that became universally known: its readily filterable active agent, its marked action on *pyogenic cocci* and the diphtheria group of bacilli, its non-toxicity and non-irritant nature on animals even in enormous doses, its efficacy as an antiseptic, its capacity for being injected into infected areas. Even in dilutions of one in one thousand, penicillin worked against *pyogenic cocci*.

Yet because there were no trained chemists in the St Mary's lab, and because Fleming himself moved on to other areas such as sulphonamides, the discovery of penicillin took another decade to be fully appreciated for the incredible breakthrough that it was. 'Neither the time when the discovery was made,' wrote the leading pharmacologist Sir Henry Dale, 'nor, perhaps, the scientific atmosphere of the laboratory in which [Fleming] worked, was propitious to such further enterprise as its development would have needed.' For nearly a decade, therefore, little was done to promote this genuine 'wonder drug'.

Penicillin's incredible, indeed world-changing, properties were properly developed at Oxford by an Australian pathology professor called Howard (later Lord) Florey, a German Jewish refugee named Ernst (later Sir Ernst) Chain and an Englishman Norman Heatley. They produced penicillin in large quantities, proving its qualities just in time for its use during the Second World War. The horrors of that conflict were bad enough as they were; the idea of what they might have been like without penicillin's ability to counter gangrene hardly bear consideration. There is some poetry, therefore, in the fact that the Nobel Prize for Medicine for the year of peace, 1945, should have been shared between Fleming, Florey and Chain. When Fleming died in Chelsea ten years later, he was a national hero; flags flew at half-mast and his ashes were interred in the crypt of St Paul's Cathedral, close to Horatio Nelson and the Duke of Wellington.

Fleming's story illustrates almost perfectly the way that so many great and useful inventions have been brought to the benefit of Mankind by the English-speaking peoples, in this case primarily by two Englishmen (Wright and Heatley), an Australian, a naturalised German refugee from Nazi Germany and the Scot, Sir Alexander Fleming. Although there was at times a prickly relationship between the Paddington and Oxford groups, and Wright wrote to *The Times* in 1942 complaining that Fleming was being given insufficient recognition for his discovery, it was the combination of the preparatory research, Fleming's Eureka moment and then the Oxford-based follow-up work that together made penicillin the miracle it became, just in time for when it was most urgently needed.

SIX

Capitalism at Bay

1929−31

The St Valentine's Day massacre − Snobbish anti-Americanism − Protestantism and religious toleration − The Wall Street Crash and the Great Depression − Canadian socialism − The Mystery of Capital − The Empire State Building − 'The Prerogative of the Harlot'

'There are few ways in which a man can be more innocently employed than in getting money.' Dr Samuel Johnson, 1775

'The best way to destroy the capitalist system is to debauch the currency.' V.I. Lenin[1]

At 10.30 a.m. on a cold and windy St Valentine's Day, 14 February, in 1929, a mobster hit-man and occasional golfing partner of Al 'Scarface' Capone called Fred 'Killer' Burke entered a garage on North Clark Street in Chicago, then 'a dreary, nondescript stretch of storefronts and small businesses'. Burke had been hired by Capone at $5,000 (plus expenses) to come from St Louis, Missouri, to murder George 'Bugs' Moran, a racketeer who was thought to have been behind several attempts on Capone's life. Dressed as policemen and pretending to conduct a routine Prohibition raid on Moran's bootlegging operation, Burke and his three accomplices, driving a stolen black-and-white police car, found seven men there, comprising three bootleggers, a safecracker, a racketeer, a bank-robber and a saloon-keeper, as well as a German shepherd dog appropriately enough in that company called Highball, who was tied to a pipe.

With two members of Detroit's Purple Gang keeping look-out from a rented apartment across the street, Burke ordered the seven men to raise their hands and line up against the garage wall. They were first relieved of their weapons and then all of them were gunned down in a matter of ten seconds, using two machine-guns, a sawn-off shotgun and a .45 revolver. Meanwhile, 'Bugs' Moran had seen what he had assumed was a police raid in progress and had just kept walking down the street. Highball was later untied by a policeman investigating the scene, and understandably bolted.

The St Valentine's Day massacre came to symbolise America's scourge of gangsterism, racketeering and gun-crime. It is hard, though, to agree with Capone's sympathetic recent biographer, who blames his subject's prominence in American culture entirely on 'a poisonous but intoxicating blend of the shrill journalism of the 1920s, Hollywood sensationalism, and pervasive anti-Italian prejudice'.[2] In fact, of course, ordering the shooting of seven gangsters up against a wall – especially on the day of the year dedicated to romantic love – would guarantee pretty much anyone a certain degree of prominence. Nor was it the worst case of violence; in September 1937 there were some forty Mafia-related murders within a forty-eight-hour period.

The St Valentine's Day massacre also came to be seen in time as a standing indictment of freebooting American capitalism, however absurd that might be. Communists, socialists and anti-Americans certainly made a great deal of propaganda at the time out of the suggestion that *laissez-faire* free enterprise inevitably led to the kind of Mafia-style violence seen so starkly in Chicago that day. In fact, since it arose largely through the gross restraint of trade involved in Prohibition, one might more profitably blame the rise of gangsterism on the nanny state.

Gun-crime has long been a feature of American society, but it is arguable that the widespread distribution of legally owned firearms is a disincentive to certain types of crimes, such as burglary, which are at relatively low levels compared with non-gun-owning societies. The right to bear arms, as protected by the Second Amendment to the Constitution, is also the final bulwark of American citizens' rights as enunciated in the rest of it.

Although organised crime has also long been evident in the United States, the FBI – which was founded largely in order to defeat it – has had very many high-profile successes, and it is thought to be very much on the wane in the early twenty-first century. Like the scourge of drugs, the problems of organised and gun crime stem from the very freedoms that are inherent in American society, but cannot be blamed on *laissez-faire* capitalism so much as the fact that citizens' rights are protected by the Constitution in the United States to a degree not seen in other countries. A hoodlum's right not to incriminate himself by pleading the protection of the Fifth Amendment is fully respected in this most rights-conscious of societies. Liberty and Order have long struggled for supremacy in America, and the fact that neither has triumphed to the detriment of the other is a sign of the strength of her society rather than of weakness.

The soubriquets of some of the other hoodlums of the Prohibition era are worth recording for their sheer colour, and include those of Jack 'Greasy Thumb' Guzik, Ralph 'Bottles' Capone (Al's elder brother), Frank 'The

Enforcer' Nitti, Richard 'Two-Gun' Hart, 'Machine-gun' Jack McGurn (another failed hit of Bugs'), 'Schemer' Ducci, William 'Klondike' O'Donnell and Enoch 'Knuckles' Johnson. Various other associates of these men were nicknamed 'Rusty', 'Artful Eddie', 'Skimmy', 'Glass-eye', 'Nine-toed', 'Peg-leg', 'Wild Bill', 'Big Angelo', 'Dago', 'Diamond Joe', 'Hinky Dink', 'Rough-house', 'Polack Joe', 'Three-fingers', 'Creepy', 'The Camel', 'Hop Toad', 'Bathhouse' and 'Pretty Boy' (although presumably not to his face); there was even one poor hoodlum who somehow wound up being dubbed 'Golf Bag'. In retrospect, therefore, Fred Burke's nickname of 'Killer' seems somewhat over-literal.

In a short biography of Salvador Dalí published in 2002, the Catalan art critic and historian J. Castellar-Gassol sneered at New York as 'the Babel-like city that has replaced the God of the Founding Fathers with the idol of the dollar' and the United States as 'the young homeland of Coca-Cola'.[3] Refined European art critics all too often develop a form of Tourette's Syndrome when they are faced with American culture, yet it rarely prevents them, like Castellar-Gassol himself who writes for *Reader's Digest*, from pocketing the Yankee dollar they affect to despise.

The most virulent criticisms of America and Americans come from Americans themselves, however. Self-hatred, often through guilt over their supposed materialism and obsession with money, is an abiding defect in the English-speaking peoples, and for some reason especially strong in Americans. 'There is but one word to use in regard to them – ', said the American-born but European-domiciled author Henry James of his own countrymen,

> vulgar, vulgar, vulgar. Their ignorance – their stingy, defiant, grudging attitude towards everything European – their perpetual reference of all things to some American standard or precedent which exists only in their own unscrupulous windbags – and then our unhappy poverty of voice, of speech, and of physiognomy – these things glare at you hideously. .... What I have pointed to as our vices are the elements of the modern man with culture quite left out.[4]

It is noticeable how the 'their' in James' rant changes halfway through to 'our' as the author reluctantly recalls his own birth. Even Walt Whitman felt the need to apologise for Americans for being preoccupied with material matters, excusing them on the grounds that they were a new nation.[5]

A superior, cultured contempt for aspiration – for people working hard to better themselves and their families – also led the modern British writer Zadie Smith to describe Britain as 'just a disgusting place', adding, 'It's the way people look at each other on the train; just general stupidity, madness, vulgarity, stupid TV shows, aspirational arseholes, money everywhere.'[6] The English-speaking peoples' worst critics have long come from within their own society. The politics of the pre-emptive cringe is evident throughout the culture of the English-speaking peoples, who in reality ought to be proud of the way that their citizenry can aspire to better themselves, a legitimate hope that does not make them dollar-idolisers or 'vulgar, vulgar, vulgar', let alone 'arseholes'.

'How', asks Luigi Barzini in Henry James' novel *The Europeans*, 'did a peripheral island rise from primitive squalor to world domination?' Much of the answer lies in Protestantism. Although John Cabot set sail in the reign of the Catholic monarch Henry VII, the great period of Henrician, Elizabethan and Jacobean expansion took place after the English Reformation, when Protestant capitalist merchant adventurers financed the expansion of the British diaspora into Canada, Virginia, Ireland, South America and India. The coincidence of the Tudors' break with Rome taking place at almost the same time as the discovery of new oceanic routes round the Cape of Good Hope, then compounded with the Stuarts' encouragement of the Puritan exodus across the Atlantic, meant that England became an outgoing, mercantile and ultimately imperial Great Power, mother to the English-speaking 'Ocean Nations' such as the United States, Canada and Australia.

'Released from the inherent bureaucratic and doctrinal strait-jacket of Roman orthodoxy,' the historian Arthur Bryant wrote in his 1968 book *Protestant Island*, and unburdened by guilt about usury and wealth-creation, the English looked forwards and outwards to lay the foundations of the modern world. Max Weber had written *The Protestant Ethic and the Birth of Capitalism* in 1901, and Bryant applied its lessons to British history with great relish. He even believed it to be no coincidence that in 'the first century as a fully-committed Protestant nation', England produced her greatest poet in William Shakespeare, her greatest scientist in Sir Isaac Newton and her greatest sailor in Sir Francis Drake.

In the present age, this theory of innate religious superiority would be denounced as at best triumphalist and at worst fundamentalist or even 'discriminatory'. Certainly there have been several Roman Catholic countries that have been successful at capitalism, and it is anyhow impossible to predict whether had Mary Tudor had any Catholic descendants, they might not also have presided over the expansion of England to just the same extent as Elizabeth I's Protestant successors. Their Catholicism certainly does not seem to have held back the nautical ambitions of Philip II's Spain, Christopher Columbus or Portugal's Prince Henry the Navigator. Nonetheless, the Protestant concept of exploration for mercantile profit, as opposed to national prestige or grand strategy – let alone for the propagation of Christianity – was the engine that colonised most of the English-speaking world.

Capitalism works best within the political, social and legal framework perfected by the English-speaking peoples, partly because of the strength of Protestantism in those societies. The connection between Protestant individualism and personal responsibility creates a favourable environment for free enterprise, and the domination of the modern US economy in the twentieth century underlines this. There is little place for the philosophy of 'che sera sera' in Protestantism, which is theologically tailor-made for the concepts of individual enterprise and the free market, in a way that Roman Catholic countries' more *dirigiste* economies rarely are.

Throughout this book it will be apparent that southern Ireland often behaves differently from the rest of the English-speaking peoples. This is in part due to the fact that the Reformation never succeeded in the southern part of the island, and the Irish drum has therefore beaten to a different rhythm. Between 1853 and 1913, no fewer than thirteen million people emigrated from the British Isles, many of them to other parts of the English-speaking world. Very often those who left from Ireland were able to take little with them; the heirloom many stowed away was a burning resentment against the British Crown, which they later carefully unpacked in Boston, Sydney, Montreal, Chicago, New York and other great English-speaking metropolises. In some cases the resentment has been passed down the generations.

As the writer Mary Kenny has noticed with regard to agriculture, 'The Scotch-Irish Protestants who farmed in Ulster husbanded the land with near genius: for some unexplained reason, dour, unimaginative Calvinists make the most productive farmers in the world, from Antrim to Wyoming to the Orange Free State of South Africa.'[7] It is important not to be over-didactic about this phenomenon; of course Protestants do not necessarily make better capitalists than Catholics individually, but in the aggregate they have created economies more carefully attuned to the needs of the free market in which capitalism thrives best.

The Tory thinker Michael Gove has written of Protestantism, employing the past tense almost in the sense of a funeral panegyric, that,

> It affirmed the spiritual without the need for ritual. It relished argument, lived in language, and celebrated a faith that had its beginning in the Word. Its spirit was democratic, with the Bible and church office open for all. Its polar opposite is not atheism, but the New Age 'faiths' that celebrate feeling over thought and privilege a cast of gurus over the questioning congregation.[8]

Religious toleration, which has been uniform throughout the English-speaking peoples since 1900 – except for the poor Irish Jews of Limerick in 1904 and the ethnic cleansing of the early 1920s in which the Dublin Government was not involved – has been a consistent source of strength. Efforts wasted by other societies in attempting to impose clerical orthodoxy have instead been channelled into more productive areas by the English-speaking peoples, who have thereby avoided debilitating wars and unrest on the subject. Atheism and agnosticism have had equal rights before the law as faiths, and English-speaking societies have benefited hugely from not trying to 'open windows into men's souls'. Of course, as has been recently pointed out by the American intellectual James Bowman, since 1900, 'The great atheistic faiths of Communism and Nazism killed far more people than religion had managed to do in a comparable period of time', so the equation that John Lennon made in his popular song *Imagine* between there being no heaven or hell and 'No religion too' with 'living life in peace' is an illegitimate one, however seductive.[9]

Religious toleration has been harnessed by the English-speaking peoples as a weapon in their formidable armoury of social advantages over much of the rest of the world. Although it is not true of several non-English-speaking Protestant churches, Anglicanism and its American version Episcopalianism have religious toleration embedded deep in their DNA. 'It hath been the wisdom of the Church of England,' reads the opening sentence of the Preface to its 1662 *Book of Common Prayer*, 'ever since the first compiling of her Publick Liturgy, to keep the mean between the two extremes, of too much stiffness in refusing, and of too much easiness in admitting any variation from it.' (Although this referred to disputes within Anglicanism, it also points to its enthusiastically non-fundamentalist approach.)

Of course the nineteenth century saw much religious intolerance imposed by the Anglican Church – Roman Catholics were only allowed to sit in the British Parliament in 1829, Jews in 1858 and atheists in 1886 – but by the dawn of the twentieth century freedom of worship was secure. In the United States – where the generally Episcopalian Founding Fathers had deliberately excluded mention of religion from the Constitution – freedom of worship had been enshrined since the Revolution. Throughout the English-speaking peoples it took its place beside its sister freedoms of speech, assembly and the press in the great arsenal of their strengths, and gave them a huge advantage over those countries where particular religions were either promoted or banned by the state. By the late-nineteenth century, energies devoted to socially divisive religious persecution in Europe – Germany's anti-Catholic *Kulturkampf*, France's anti-Semitic Dreyfus Affair, Spain's alternating pro- and anti-clericalism, Russia's anti-Semitic pogroms – were already being channelled into more productive areas of state policy in the English-speaking world.

John F. Kennedy's Roman Catholicism was thought to have been an Achilles heel in his campaign for the presidency in 1960, at least until he addressed the Southern Baptist leaders during the campaign and persuaded them that he was not about to become the political pawn of the Pope. By contrast with this lone Catholic, the American presidency in the twentieth century has seen a quite astonishing array of Protestants. There have been four Baptists (Harding, Truman, Carter, Clinton); three Episcopalians (F.D. Roosevelt, Ford, Bush Snr); two Disciples of Christ (Johnson, Reagan); two Quakers (Hoover, Nixon); two Presbyterians (Wilson and Eisenhower, who was originally a Jehovah's Witness); two Methodists (McKinley, Bush Jnr); one Congregationalist (Coolidge); one Unitarian (Taft) and one Dutch Reformed (T. Roosevelt). It's an eclectic array. So too are the affiliations of British prime ministers; all have been Protestants, but not all were Anglicans. The exceptions included Balfour and Campbell-Bannerman, who belonged to the Church of Scotland; Bonar Law and MacDonald, who were Presbyterians; Lloyd George and Callaghan, who were Baptists; Chamberlain, who was a Unitarian; Wilson, who was a Congregationalist, and Thatcher, who was brought up a Methodist but who worships as an Anglican.

Undoubtedly, some *de facto* religious discrimination continued even beyond the 1960s in the English-speaking peoples, especially in such areas as Northern Irish employment practices and in British and American social institutions such as golf clubs, but although painful for the small numbers of victims involved, these abuses were not state-inspired and certainly did not constitute religious intolerance along the lines of the Nazi execution of Dietrich Bonhoeffer, the Soviet persecution of the Russian Orthodox Church, or Red China's demolition of 2,000 Tibetan monasteries. If like is treated with like, the experience of the English-speaking world has generally been a hugely positive one in terms of religious toleration, to its very great social, material, political and military advantage.

This is all the more remarkable considering the high degree of religiosity in the United States. No fewer than 95% of Americans believe in God, compared to 76% of Britons, 62% of Frenchmen and 52% of Swedes. More than three in four Americans belong to a church, 40% go to a church once a week and one in ten goes several times a week. Furthermore, six in ten Americans say that religion is 'very important' in their lives. 'While European churches are trying to hang on to the few parishioners they can muster,' record the authors of *The Right Nation: Why America is Different*, 'American churches seem to be in a state of permanent boom.'[10] Anti-Americans profess themselves fearful of this phenomenon, but can never adequately explain why they or anyone else should be.

The Wall Street Crash did not happen on one day, but across the course of a week from 23 October 1929, when 19,226,400 shares were sold in 24 hours, to 'Black Tuesday', 29 October. These financial shocks proved three things to the world: the first was that what happened in the United States mattered profoundly to the rest of the globe, both because it pulled the rest of the industrialised world down with it and because an apparently irredeemable

flaw in American capitalism gave Mussolini, Stalin and then Hitler an opportunity to persuade many that they had alternative systems that would serve
the masses better. Secondly, it proved that capitalism – or more particularly
the stock market – had serious flaws that urgently required attention; these
included endemic insider trading, weak corporate disclosure and a pricing
system that failed to reflect important externalities. Thirdly, the English-
speaking peoples would be blamed for a recession, and subsequently a depression, even though it hit them just as badly as everyone else.

Everyone has their own explanation for the inordinate length and dreadful
depth of the Great Depression. Monetarists blame it on the misguided policies
of the Federal Reserve Bank; Keynesians blame it on the inadequacy of final
demand; Free-traders blame the Smoot-Hawley Tariff Act of 1930 (which
inaugurated the highest levels of protectionism in US history, prompting
instant retaliation); the Right blame it on FDR's interference in the market
system; the Left blame it on the flaws and contradictions inherent in capitalism.
These explanations are not mutually exclusive, and all but the last of them has
at least elements of truth. The key point was that democratic capitalism had
within it the capacity to cure its own ills, not least when Franklin Roosevelt
improved the quality of capital markets by creating the Securities and
Exchange Commission in 1934.

There had been financial panics on Wall Street before 1929, of course. The
1893 stock market collapse had led to a four-year economic downturn, in
which iron production had fallen by one-quarter and 22,000 miles of railway
lines had been placed under receivership.[11] The 1907 crash was a serious
short-term blow too.

Just as isolationism had seemed attractive to many Americans in the 1920s,
so too did many non-Americans feel that it was possible to isolate themselves
from what was taking place in the United States. This sense of separateness
from – often also implying superiority to – the experience of the Western
hemisphere had been eloquently articulated by Lord Salisbury back in 1862,
when he wrote that American support for Russia in the Crimean War and the
sepoys during the Indian Mutiny

> was expressed with the most demonstrative cordiality, and voiced with all the
> verbal condiments with which they know how to flavour the insipidity of political
> discussion. Yet we cannot remember that their noisy criticism provoked any
> feeling, good, bad, or indifferent, in London. Nobody knew what the Americans
> were saying, or cared to ask. The opinion of New York upon the subject was of
> no more practical importance than the opinion of Rio de Janeiro. And as a
> question of sentiment, it was a matter of profound indifference to us whether our
> neighbours praised us or blamed us.[12]

Those days were now very definitely over, as the Wall Street Crash proved from Sweden to Egypt and from Lancashire to Poland. What people said and did in New York mattered, as it had never mattered before but was increasingly to in the future. By 1929, the United States was contributing 34.4% of the world's gross production by value, as against Britain's 10.4%, Germany's 10.3%, the USSR's 9.9%, France's 5%, Japan's 4%, Italy's 2.5% and Canada's 2.2%.[13] It was therefore inescapable that a collapse in US share prices would affect the rest of the planet.

From the highest peak of the economic cycle just before the Wall Street Crash in 1929 to the deepest point of the Great Depression in March 1933, the amount of money in the US economy – currency and demand deposits combined – fell by 28% and industrial production fell by 50%.[14] The Dow Jones Index did not regain its 1929 heights until the early 1950s, and trading volume did not recover those levels until 1961. Unemployment during the Depression rose to 25%, while international trade declined by no less than 90%.[15] Behind those terrible statistics lies a world of human misery, blighted lives and wasted opportunities. Just as historians have counted 120 explanations for the decline and fall of the Roman Empire, so dozens of reasons have been given for the Great Depression, but as the official historian of the Federal Reserve Bank, Allan H. Meltzer, records, 'It is now generally accepted that the depth of the Depression, its duration, and its spread through the world economy are mainly the result of monetary actions or inactions.'[16]

There were plenty of tragic near-misses. In 1926 and 1927, the Kansas Republican Congressman, James A. Strong, had tried to amend the 1913 Federal Reserve Act in order to make price stability an explicit policy goal of 'the Fed'. He was opposed in this by Benjamin Strong, the Governor of the Federal Reserve Bank of New York, a powerful figure in the organisation, who feared that it would be interpreted as meaning that the Reserve would end up trying to maintain price stability in agricultural products. The bill was defeated. Today, Alan Greenspan, amongst many others, agrees with Meltzer's contention that if Strong's bill had passed, the Federal Reserve 'could not have permitted the Great Depression of 1929–33 or the Great Inflation of 1965–80'.[17]

In 1928, the year after winning his battle against his namesake, Governor Strong died, which led to a serious dearth of leadership and organisational ability that was to prove disastrous the following year. His successor in New York, George L. Harrison, was ill-equipped to deal with the coming crisis and had none of Strong's authority when it came to influencing members of the crucial Open Market Committee, which took – or in this case failed to take – many of the crucial decisions required of the Federal Reserve during the crisis.

The Nobel Prize-winning economist Professor Milton Friedman has distinguished between the recession of 1929, which was part of the ordinary business cycle, and 'the conversion of that recession into a major catastrophe', which happened as a result of failed monetary policy. The blame for this he places squarely on the Federal Reserve, which had been set up in 1913 precisely in order to avoid the kind of situation in which one-third of all US banks were forced to close. 'And yet,' he writes, 'under the Federal Reserve system you had the worst banking crisis in the history of the United States.' There was nothing endemic in capitalism that led to the crisis of 1929, any more than that crisis necessarily had to lead to the Great Depression. There were therefore no systemic and cyclical 'contradictions' within the system, which inevitably led to depression and thus the collapse in living standards for the working man, only a series of bad fiscal decisions, each one exacerbating the last, taken by the leaders of the Fed.

The Federal Reserve pursued policies that led to a decline in the quantity of money by 28% at precisely the time when it should have been ensuring that the system remained liquid. The tragedy is that, as Friedman puts it, while 'millions of people had their savings washed out, that decline was utterly unnecessary. At all times, the Federal Reserve had the power and the knowledge to have stopped that. And there were people at the time who were all the time urging them to do that. So it was ... clearly a mistake of policy that led to the Great Depression.'

Although it is perfectly true that there was an 'unsustainable speculative bubble' that burst on 'Black Tuesday', these had happened before without the entire economic system careering into meltdown. Bull markets and bear markets, greed and fear, boom and near-bust, upturns and downturns are as natural a part of the free market system as buying and selling itself, but they do not imply, let alone guarantee, complete collapse. Yet the stock market in 1933 ended up at about one-sixth of its highest level of 1919.[18] The reason that October 1929 was not just yet another blip was because of the Federal Reserve's ideologically over-tight monetary policy, and not the stock market fall. After all, business activity had reached its peak in August 1929, two months earlier, and the market had already come far off its top 1929 levels by 29 October.

The reason that a short-term stock market fall turned into a long-term depression was therefore not that capitalism is inherently flawed, but rather because, in Friedman's words,

The New York Federal Reserve Bank, almost by conditioned reflex instilled during the Strong era, immediately acted on its own to cushion the shock by purchasing government securities, thereby adding to bank reserves. That enabled commercial banks to cushion the shock by providing additional loans to stock market firms and purchasing securities from them and others affected adversely by the crash. ... Thereafter the System acted very differently than it had during earlier economic recessions in the 1920s. Instead of actively expanding the money supply by more than the usual amount to offset the contraction, the System allowed the quantity of money to decline throughout 1930.[19]

This decline in the quantity of money up to October 1930 was only 2.6% – hardly anything compared to the decline of one-third between October 1930 and early 1933 – yet it was larger than had happened during or before most of the United States' previous recessions, and was easily large enough to worry people. Runs on banks did not occur until the autumn of 1930, when mid-western and southern banks started to fail, but on 11 December 1930 the Bank of United States – the largest commercial bank ever to go bankrupt – failed too. (Although it was not an official bank, its portentous name confused and scared people, especially abroad.)

The Federal Reserve's policy of tightening money supply and not respond-ing vigorously to bank failures meant that even financially sound institutions like the Bank of United States – which ended up paying its depositors 92.5 cents in the dollar – went to the wall, and with them virtually the whole financial system. Yet there was no inherent reason why the Federal Reserve Bank of New York, along with the New York Clearing House Association of Banks, should have behaved in the blinkered way that they did.

With the Bank of United States allowed to go to the wall without rescue, 352 other banks failed in the month of December 1930. A second banking crisis developed in the spring of 1931, in which the Federal Reserve also allowed events to run their own course. Professor Friedman believes that had the Fed not existed, the same thing would have happened in 1929–30 as had happened in the great banking crisis of 1907, which was over by the following year once confidence had been regained through the restriction of payments and a concerted effort by the sounder, bigger banks under the leadership of J.P Morgan Snr. The very existence of the Federal Reserve meant that there was a 'lulling [of] the community as a whole, and the banking system in particular, into the belief that such drastic measures were no longer necessary now that the System was there to take care of such matters'.[20]

Restriction of payments would have prevented the draining of reserves from sound banks; there might have been large-scale open-market purchases of government bonds, which would have provided banks with the cash to meet their depositors' demands. (Even the Federal Reserve of New York pressed for this, but failed to persuade the other Federal Reserve Banks or the Board.) Any number of other free market responses were possible had the Federal Reserve not been so myopic and ultra-conservative almost throughout the Depression. (It took powerful lobbying from Congress to force the Federal Reserve to institute open-market purchases in 1932, which were then halted by it the moment that Congress adjourned.) Of the Bank, President Hoover wrote in his memoirs, with some understatement, 'I concluded it was indeed a weak reed for a nation to lean on in time of trouble.'

The statistics for the banking and monetary aspects of the Great Depression are extraordinary: failure, merger or liquidation meant that no fewer than 10,000 banks in the United States simply disappeared between 1929 and 1933. The total stock of money in deposits and currency in the hands of the American public fell by more than one-third in the same period. The Fed entirely refused to accept that it was in any way at fault, boasting in its Annual Report for 1933 that, 'The ability of the Federal Reserve Banks to meet enormous demands for currency during the crisis demonstrated the effectiveness of the country's currency system under the Federal Reserve Act.'[21]

Unusually in human affairs, people have learnt the lessons of the past. Under the Federal Reserve chairmanship of Alan Greenspan, every crisis that cropped up on his watch was dealt with by the immediate liquification of the system, and therefore never resulted in a full-scale emergency. Instead, the proper working of the FDR reforms has led to a strong, open, liquid capital market, the fountainhead of global capitalism. 'That men do not learn very much from the lessons of history', wrote Aldous Huxley in November 1959, 'is the most important of all the lessons that history has to teach us.' The Wall Street Crash was a fortunate exception to that otherwise unvarying rule.

Without ever having been there, Karl Marx predicted that the USA would see the first socialist revolution. Considering how badly the Wall Street Crash and the Great Depression affected so many Americans, it is worth examining why it never did; indeed, in 2000 a book was published entitled *It Didn't Happen Here: Why Socialism Failed in the United States*.[22] Explanations that have traditionally been given include America's long history of labour shortages, the social teaching of the Catholic Church, the way that race conflict displaced class conflict, the two-party system, the New Deal and the Great Society, hijacking by out-of-touch intellectuals, widespread working-class investment in property and stocks, and the socialists' own mismanagement of their various political opportunities, especially in trade unionism and local government.[23] Probably the true reason is that Americans have never needed socialism since their society already had the best things that it had to offer – community spirit, neighbourliness, civic-mindedness and resource-pooling – while they had a profound and understandable distrust of the detritus that tends to come with socialism, in the form of over-mighty government, confiscatory taxation, over-regulation, austerity programmes and bureaucracy.

The experience of poverty and deprivation during the Slump and after-wards was by no means universal. 'Queer sight in the Depression,' wrote the Liberal MP Robert Bernays after an evening at the Ritz Hotel in London in November 1932, 'foyer crowded and impossible to get a table for cocktails, and the dining room so packed that the tables were extended into the passageway.' Yet that year unemployment stood at 2.8 million in Britain, 5.6 million in Germany and 13.7 million in the United States.[24] When considering why, despite the appalling privations of the Great Depression, the English-speaking peoples did not turn to extreme right- or left-wing politics in the Thirties in order to ease their plight, it is worth examining the experience of Ontario between 1932 and 1945 for the lessons it gives about the wider Canadian and ultimately the even wider English-speaking experience.

The Canadian socialist party, the Canadian Commonwealth Federation (CCF), only became a significant political force in two of the country's nine provinces during the 1930s, and neither of them were the electorally crucial Ontario or Quebec.[25] Founded in western Canada in August 1932, it seemed as though all augured well for the CCF's future success; the Depression was causing widespread hardship, the Prime Minister R.B. Bennett was becoming unpopular, meanwhile the leader of the opposition, William Mackenzie King, seemed unable to capitalise on the situation. Furthermore, as the historian Gerald Caplan has put it, 'The article of faith that industry and thrift brought their due material reward was suddenly and rudely shaken.'[26]

The CCF was introduced into Ontario by Agnes Macphail and William Irvine, who demanded 'a new social order'. It gained huge publicity overnight and was further boosted when the United Farmers of Ontario (UFO) affiliated with the nascent party. Trade unions in Toronto, Hamilton, Windsor and Kitchener soon followed suit. Then a 'Club Section' for people who were neither farmers nor trade unionists was set up which quickly attracted no fewer than 6,000 mostly middle-class members.

The first major problem of socialism in the English-speaking world – fragmentation – soon struck the CCF. The Ontario Labour Party, the Independent Labour Party, the Socialist Party of Canada, the Trades and Labour Congress and the All-Canadian Congress of Labour each had their own agendas and refused to be swallowed up by the CCF. Even within the CCF itself there was widespread mutual misgivings between the UFO, labour and Club Section in the party. In accusing each other of being overbearing, various class resentments quickly came to the surface. The UFO members did not like being referred to as 'comrade' and naturally feared the nationalisation of land; the labour members thought the Club Section elitist and insufficiently ideologically pure, while the Club considered itself the indispensable intellectual core of the movement, looking down on the other two.

This smouldering mutual suspicion and dislike burst into the open in early 1934, when the CCF had to decide whether actively to support the cause of Rev. A.E. Smith, a communist who was being prosecuted for sedition. While the Club and UFO wished to avoid the CCF being identified with communism, the labour section wanted to fight the indictment. When the latter lost the vote, it declared it would not abide by the result but would continue to protest and demonstrate in favour of Smith.

As a result of the labour group's unilateral decision, the UFO seceded from the CCF altogether, on the basis of the labour leaders being 'too close to the Communists'. This left the CCF provincial council for Ontario equally balanced between Club and labour members. Since the two sides were unable to compromise, the national party then suspended the provincial council altogether. The party's (understandably exasperated) national leader, J.S. Woodsworth, told a 'stunned' audience that a new provincial party would be created in Ontario with a single structure in place of the three autonomous groups. 'That was all,' records an historian of these dramatic – if in many ways also petty – events. 'In ten short minutes, the constructive work of eighteen months had been obliterated.'[27]

Since Macphail had been part of the UFO delegation, and another prominent leader Elmore Philpott resigned shortly after the Council meeting, the CCF was suddenly virtually leaderless in Ontario. Few were willing to step forward to try to rebuild the party on the lines that Woodsworth had set out. In the meantime, the Liberal Party leader in Ontario, Michael Hepburn, moved decisively to the left, at least in his rhetoric, taking many rural votes from the socialists, while the Conservatives constantly denounced the CCF as sinister communist fellow-travellers (which they generally weren't).

Such methods by the old centrist Parties were used throughout the English-speaking world and were particularly effective in Britain and the United States at this time as well. The 1934 provincial election saw Hepburn's Liberals triumphant and the CCF winning only one seat – Sam Lawrence of Hamilton East – with their 7% share of the vote. In the federal elections the following year, the party won only 8%, despite the ravages of the Depression still causing misery and blighting the lives of millions of Canadians. Only seven socialists were elected to the Ottawa Parliament in that election, and in the 1940 federal elections the CCF won a derisory 3.8% of the Ontario vote, its worst-ever performance.

It was not until 1942 that the CCF returned as a major force in Canadian politics. During the years of Depression, stagnation and widespread want, red-blooded socialism had utterly failed to make any electoral headway in Canada, as in so much of the rest of the English-speaking world. Disunity, personality clashes, the stigma of Marxism, lack of leadership, but above all the ability of centre-left parties to reposition themselves effectively, all led to socialism missing its best opportunity of the twentieth century to impose a command economy and dismantle capitalism.

*

'No society on Earth has ever had such a privileged existence as the capitalist West,' wrote the British journalist Anthony Browne in his recent book *The Retreat of Reason*, 'even the lives of the poorest sections of society are immeasurably better in almost all ways than under any other form of economic system.' The material success of Western capitalism is undeniable, but why does it seem to be confined to the English-speaking world and Europe, with so few successful outposts elsewhere, except for places like Hong Kong and Singapore which were effectively English-speaking enclaves, or Japan whose polity was established by General MacArthur in 1945? In 1999, according to UN figures, the assets of the world's three richest people were greater than the combined gross national products of the world's poorest countries the inhabitants of which numbered 600 million people.[28]

In his 2000 book *The Mystery of Capital*, subtitled *Why Capitalism Triumphs in the West and Fails Everywhere Else*, the Peruvian economist Hernando de Soto cited legal title as the primary reason why Third World countries fail. As an experiment, de Soto and his team of researchers tried to establish a garment workshop on the outskirts of Lima. It took them 289 days to get through the bureaucracy. When they then tried to secure legal authorisation to build a house on state-owned land, it took them nearly seven years and involved no fewer than fifty-two different Peruvian government departments and offices. Property is thus widely held 'extra-legally', which is acceptable for small-scale enterprises, but is no basis for the kind of large-scale ones needed to raise the Third World out of poverty.

De Soto further found that the poor of developing countries own $9.3 trillion worth of real estate, but since they have no effective legal title it is, in his words, 'like water in a lake high up in the Andes – an untapped stock of potential energy'.[29] Such 'dead' capital cannot be unleashed to generate wealth because the legal systems of the countries concerned are too sclerotic, arbitrary, corrupt or obscurantist to establish proper title. In the West, and especially among the English-speaking peoples where property rights are enshrined in an ancient national culture, property can become collateral, have a value, be part of other transactions and generally play an active part in the economy. Law, not loans, is thus what the developing world most needs.

On Friday, 1 May 1931, the Empire State Building in Manhattan was officially opened by President Herbert Hoover, who pressed a button to switch on its lights even though he was in Washington DC. Construction had begun on 17 March – St Patrick's Day – in 1930 and the 3,400 strong workforce completed the 102-storey building in only fourteen months. At one point storeys were rising at a rate of four-and-a-half per week. It was, and remains, a triumph of English-speaking engineering and technology. Built of Indiana granite and limestone, trimmed with chrome-nickel steel from the sixth floor up, it is 1,250 feet high, 200 feet taller than the Chrysler Building, representing victory for the Empire State building's mastermind, General Motors' founder John Jakob Raskob, over Walter Chrysler of the Chrysler Corporation.

As well as its efficiency, the Empire State Building also represents the invincible optimism of American capitalism. To have begun such a massive project – its base covers two acres and it has 2.1 million square feet of office space – in the depths of the Great Depression was a signal act of faith in the capacity of America to survive the crisis. Of course the Depression also meant that the building could be erected much more cheaply than otherwise; originally expected to cost more than $50 million, it was finally built for $24.7 million, which once land costs were added – it was built on the site of the old Waldorf-Astoria Hotel – made a total of $40.9 million. The Depression also meant that the seven million man hours that went into its construction included work on Sundays and public holidays. It remained the tallest skyscraper in New York for forty years until the construction of the World Trade Center in 1971.

The organisational genius of the general contractor Starret Brothers and Eken, which worked closely with the architect William Lamb and the owners and engineers, meant that eighty-eight floors were built by October 1930, an astonishing achievement. Sixty thousand tons of steel were brought from Pennsylvania 310 miles away, by trains, barges and trucks, and the building's beams, windows and window-frames were made in factories and only put together on site. Had Adolf Hitler – a non-traveller – visited Manhattan in the 1930s and seen the protean energy, time-and-motion efficiency and economic might of the United States as symbolised by the construction of that building, he might have thought twice before making the cardinal error of declaring war on her a decade later.

The strength of the building was severely tested shortly after 9.49 a.m. on Saturday, 28 July 1945, when a 10-ton B-25 bomber crashed into the seventy-ninth storey at the north side at 200 mph, killing thirteen people as well as the three-man crew, injuring twenty-six, and starting a fire that engulfed eleven storeys, but which was swiftly extinguished. The pilot, Lieutenant-Colonel William Smith, had taken off from Bedford, Massachusetts, and was advised to land at La Guardia because of fog, but instead he headed for Newark and took too southerly a route. The plane crashed through seven walls, one elevator plunged down eighty floors, and one of the plane's engines fell 1,000 feet down. Despite much debris crashing down into 34th Street, miraculously no-one was injured on the ground. Within two days the building was open for 'business as usual', another potent symbol of the resurgent power of American business.

Just as the Empire State Building was constructed on the site of the Waldorf-Astoria, so is New York itself erected over the traces of its own past. 'Overturn,

overturn, overturn! is the maxim of New York,' wrote the nineteenth-century diarist Philip Hone. 'The very bones of our ancestors are not permitted to lie a quarter of a century, and one generation of men seem studious to remove all relics of those who preceded them.' It is true that, as the architectural writer Martin Filler has put it, 'No great world city retains so little of its visible past as New York, not even post-apocalypse Berlin. ... If individual buildings or entire portions of New York stayed miraculously intact for decades or even centuries, it was generally a happy accident.'[30] New York's Federal Hall, where George Washington was sworn in as America's first president in 1789, was torn down less than half a century later, and a whole district of Federal Era brick buildings were levelled to make room for the World Trade Center in 1969–71.

Of course in any city it is impossible to build anything without first razing something else. In his book *The Creative Destruction of Manhattan 1900–1940*, Professor Max Page pointed out that the greatest acts of destruction on the island were the beneficial slum clearances of the first half of the twentieth century that pulled down the squalid houses of the Lower East Side. Yet that was a far cry from the gross vandalism represented by the destruction of the neo-classical Pennsylvania Station, which was so solidly built that it took two years to demolish between 1963 and 1965. Built in 1906–10 and designed by architects McKim, Mead and White, the noble portal to New York was replaced by 'a mediocre office tower, a sports arena and a hopelessly seedy underground terminal', which prompted the architectural historian Vincent Scully to remark that travellers used to enter New York like gods but now must scurry in like rats.[31]

'Is it not cruel to let our city die by degrees, stripped of all her proud monuments, until there will be nothing left of all her history and beauty to inspire our children?' asked Jacqueline Kennedy Onassis when plans were put forward to destroy New York's other palatial railway terminus, Grand Central Station. The station was thankfully preserved from the developers, and after its 1990–8 restoration project stands as one of the architectural delights of the City. The Modernist architect Le Corbusier wrote that 'A considerable part of New York is nothing more than a provisional city. A city that will be replaced by another city.' Fortunately, the architectural heritage movement that was given such a fillip by the catastrophe of the destruction of Pennsylvania Station is in a better position to resist Le Corbusier's sinister prophecy.

When Stanley Baldwin stood up to speak in the Queen's Hall in London on 17 March 1931, he knew that his political life was on the line. He had twice served as prime minister and had seen off the threat of the General Strike, but he knew that this was to be his greatest test. For two years since losing the 1929 general election he had faced the unrelenting hostility of the press barons Lords Beaverbrook and Rothermere, who between them controlled vast swathes of the British print media, but principally the *Daily Express* and the *Daily Mail*.

In the era before television and when radio was in its infancy, newspapers dominated the political culture, and those two men loathed Baldwin's policy of giving dominion status to India and opposing imperial preference. The arrogance of the press barons was such that, in the spring of 1930, Rothermere insisted that Baldwin acquaint him 'with the names of at least eight, or ten, of his most prominent colleagues in the next ministry', which Baldwin publicly denounced as a 'preposterous and insolent demand'. In January 1931, Winston Churchill resigned from the Tory shadow cabinet in order to attack Baldwin over his proposals for representative federal government in India, and on 1 March Baldwin had momentarily agreed to resign the Party leadership, a decision he quickly rescinded in favour of coming out fighting against what he called 'press dictation'.[32]

The personal attacks upon Baldwin reached their crescendo when a *Daily Mail* article, signed by the editor, attacked him for the way he had lost the fortune left him by his father and concluded that, 'It is difficult to see how the leader of a party who has lost his own fortune can hope to restore those of anyone else, or his country.'

The press barons were standing an Empire Free Trade candidate, Sir Ernest Petter, against the official Conservative candidate, Alfred Duff Cooper, in the Westminster St George's by-election to be held in only two days' time, and it was feared that they might win. So on 17 March, Baldwin mounted an attack on the press barons that has been described as one of the most memorable of interwar political orations.[33]

Before he did so he is believed to have consulted his cousin, Rudyard Kipling, who gave him the damning phrase for which the speech is still remembered. Although Kipling had been a great friend of Beaverbrook's, godfather to one of his sons and the man who had designed Beaverbrook's coat of arms when he was raised to the peerage in 1916, they had become estranged in 1918 over Beaverbrook's support for Irish Home Rule, and by the time of the Westminster by-election the two men had not spoken for over a decade.

'The papers conducted by Lord Rothermere and Lord Beaverbrook are not newspapers in the ordinary acceptance of the term,' Baldwin told the meeting at the Queen's Hall. 'They are engines of propaganda for the constantly changing policies, desires, personal wishes, personal likes and dislikes of two men. What are their methods? Their methods are direct misrepresentation, half-truths, the alteration of a speaker's meaning by publishing a sentence apart from the context. ... What the proprietorship of these papers is aiming at is power, and power without responsibility – the prerogative of the harlot throughout the ages.' (The reverberating phrase led one Tory MP to complain to another in the tea room of the House of Commons: 'All very well as rhetoric, but it's probably lost us the tart vote.')

As if to underline the accuracy of Baldwin's criticisms, the *Daily Express* led the next morning with the banner headline 'Sir Ernest Petter's Triumph' and much lower down, over only two columns, 'Mr Baldwin Denounces his Enemies'. In the reportage, written by the editor Beverley Baxter, opinion entirely took over from factual reporting and unsurprisingly the phrase about harlotry was left out altogether.[34]

What the newspapers could not spin or censor, however, was the result of the by-election two days later, which Duff Cooper won by 17,242 votes to Petter's 11,532. The Empire 'Crusade' was over, Baldwin was safe, and the only reliquary in modern Britain of the great struggle between the politicians and the press is the symbol of a crusader which survives to this day on the masthead of the *Daily Express*.

SEVEN

The Second Assault: Fascist Aggression

1931 – 9

Josef Stalin's Great Terror – English-speaking apologists and fellow-travellers –
Mussolini – The Labour Party and pacifism – FDR's New Deal and its enemies –
The Hoover Dam – Rudyard Kipling – The remilitarisation of the Rhineland – The
King abdicates – Depression hits the British West Indies – Neville Chamberlain
and the Respectable Tendency – Appeasement and grand strategy – German Jewish
émigrés – The Sudeten crisis – Satirising the Great Dictators – The Prague crisis –
War

'An Englishman's mind works best when it is almost too late.'

Lord D'Abernon in *Geoffrey Madan's Notebooks*

'It is now quite consistent with progressive thought to speak of the bloody
Hun.' Winston Churchill to Robert Bernays MP, 1933

T he Bishop of Durham, Hensley Henson, described Joseph Stalin in
the Thirties as 'One of those dynamic, semi-civilised prodigies, like
Theodoric and Charlemagne, who may be set for the regeneration of society,
but who may be, like others of the type, raised up for its destruction.'[1] In fact
Stalin had no historical precursor; Theodoric the Great, king of the Ostrogoths
(455–526 AD), might have conquered Italy, oppressed the church and perpetrated the judicial murders of the statesman Anicius Boethius and Pope
Coelius Symmachus, but he committed nothing like the destruction unleashed
by Stalin.

That revolutions devour their own children is a truism, but in Russia it was
sometimes literally so. In his recent biography, *Stalin: The Court of the Red
Tsar*, Simon Sebag Montefiore records occasions on which parents ate their
own babies in the artificial famines that the Bolsheviks deliberately engineered
in order to wipe out their class enemies. Nor was that necessarily the worst;
in the Lubyanka prison in Moscow, he tells us, 'Many of the prisoners were
beaten so hard that their eyes literally popped out of their heads. They were

routinely beaten to death, which was registered as a heart attack.'[2] The people who ordered these ghastly things, and even went so far as to pass a Politburo resolution legalising torture, thought of themselves as decent, idealistic, even moral. Anyone who admires Arthur Koestler's masterpiece *Darkness at Noon* will immediately recognise the syndrome.

The minutiae of daily life inside Stalin's Kremlin still exerts a morbid fascination. No detail is too small to be interesting about Stalin and his cronies: the *dacha* holidays, the movie-watching, the food they liked, the chilling gallows humour, the chronic alcoholism, the sly asides, the problems the *Vozhd* (leader) had with his teeth, the sycophancy, the wives and girlfriends, the hypochondria, the books they read (*The Forsyte Saga* and *The Last of the Mohicans* were favourites), and above all the machinations as they all took part in the *danse macabre* around 'Comrade Koba', until one by one the music stopped for each of them.

On the evening of Tuesday, 8 November 1932, at the dinner to celebrate the fifteenth anniversary of the Bolshevik Revolution, Stalin's wife Nadya committed suicide. After her death he would sit alone for hours and hours, spitting at a wall and vowing revenge against the world, against everyone except the person most responsible for the tragedy – himself. 'The greatest delight', he once told one of his victims, 'is to mark one's enemy, prepare everything, avenge oneself thoroughly, and then go to sleep.'[3] He was a murderous tyrant long before 1932, of course, and bloodletting was already a basic part of the Bolshevik state from the very beginning under Lenin. But from 1932 onwards, killing was Stalin's default position, his essential and primary political tool.

Of the 1.5 million people arrested in 1937 alone, nearly half – over 700,000 – were shot. When Stalin moved against the Red Army, he killed three of the five marshals, fifteen of the sixteen Army commanders, sixty of the sixty-seven corps commanders and all seventeen senior commissars. Small wonder, therefore, that Russia was so unprepared for Hitler's invasion four years later. Right across the Russian empire he created abattoirs for humans, supervising everything down to the best foliage to grow over the mass graves.

Stalin took a great interest in learning how his enemies died, as they were taken from under the Lubyanka across the road to be shot in a purpose-built bunker. His lieutenants would act out the pleadings of the Old Bolsheviks as they begged for their lives just prior to receiving a bullet in the back of the neck. A loathsome twist was added by the Jewish secret policeman Karl Pauker as he profaned his race by acting out the pleadings of Grigori Zinoviev, emphasising his Jewish accent and his (possibly invented) cries to the God of Israel to save him.

Yet there was a kind of village life in the Kremlin as the 'magnates' popped into each other's houses, played with each other's children, voted each other ever-longer holidays at their *dachas* and wrote obsessively enquiring after each other's state of health.[4] Only with the gradual destruction of Stalin's inner circle – the suicide of Nadya in 1932, the assassination of Sergei Kirov in 1934, the poisoning of Nestor Lakoba in 1936, the suicide of Sergo Ordzhonikidze and execution of Abel Yenukidze in 1937 – was the curtain of happy families abruptly wrenched aside to reveal the pathologically sadistic reality.

The members of the Red Tsar's court included the sinister bisexual dwarf Yezhov, who co-ordinated the Terror; the prissy plodder Molotov; the vain fool Voroshilov; the quick, exuberant but cautious Kaganovich; the show trials' sly chief procurator Vyshinsky; the sycophantic survivor Mikoyan; and the executioner-in-chief Blokhin, whose leather butcher's apron protected his uniform from the flying blood. The trumped-up charges on which ordinary Russians were imprisoned included sabotage, treason, counter-revolution and the catch-all Article 58/10: 'Anti-Soviet agitation and propaganda.' (People were still being arrested under that one in the 1980s.)

'There is no cause so vile that some human being will not be found to defend it,' wrote the essayist Norman Douglas. The Russian gulag was attended by scenes of dysentery, dementia, monasteries being converted into torture chambers, starved slaves hacking at permafrost, the killing rooms of the Lubyanka, massacres and mindless, sadistic savagery. One way that the Left in the West has attempted to undermine its legacy is to try to argue that there was a 'moral equivalence' between Soviet communism and English-speaking capitalism. Thus in 2004 the University of California Press published a book by Mark Dow entitled *American Gulag* and subtitled *Inside US Immigration Prisons*, and in 2005 a book entitled *Britain's Gulag* was published about British detention camps in Kenya, written by a Harvard historian named Caroline Elkins, whose blood-libels against Britain won her the Pulitzer Prize.

It is perfectly true that the United States has two million people in prison, which the Left regularly describes as 'a gulag'. Yet however vicious, unpleasant and forbidding those prisons might be – or Abu Ghraib in Iraq and Camp X-Ray in Guantanamo Bay, for that matter – to equate them to the Soviet gulag, which housed many millions of complete innocents and killed five million in the last five years of Stalin's life alone, is to indulge in an outrageous misnomer.[5] Everyone imprisoned in the US is there because a judge and jury convicted them after a trial, and only 1,000 people – in a population of 297 million – have been executed in the last thirty years. To draw a moral equivalence between that situation and the Soviet gulag of the Thirties and Forties is fraudulent history, yet it is done regularly by the Left.

'Intellectuals by and large disgraced the twentieth century,' the British historian David Pryce-Jones has written.

With rare exceptions, they whored after false gods, of which the most odious and overwhelming was power. Writers, artists, philosophers, historians, even musicians and architects, enthusiastically committed their talents to the service of one cause or another. This treason of the clerks spread like an epidemic, diminishing the world's hard-won stock of wisdom and morality, and Civilization is still reeling from it.[6]

With Western intellectuals having so dreadful a record in the twentieth century, it is small wonder, therefore, that the conservative American journalist William F. Buckley Jnr has said that he would sooner be governed by the first 200 names in the Boston telephone book than by the faculty of Harvard University.

In 1931, the Irish writer and intellectual George Bernard Shaw visited the Soviet Union for ten days, celebrating his seventy-fifth birthday at a vast reception in the Hall of Columns in Moscow. 'Russia flaunts her roaring and multiplying factories,' he said on his return, 'her efficient rulers, her atmosphere of such hope and security as has never before been seen by a civilized country on earth.'[7] Symbolically throwing his food out of the window of his train as he crossed into Soviet territory, in order to signify his disbelief in the rumours of mass starvation that had reached the Western press, he was later shown around collective farms and was lauded wherever he went.

The Soviet leaders had deliberately created a famine in the Ukraine that was in the process of destroying the peasants' will to continue to resist collectivisation. This artificial, politically-designed famine is now recognised to have claimed the lives of between six and seven million people, yet Shaw declared that the Russian people were 'uncommonly well fed' and that he had 'never met a man more candid, fair and honest' than Stalin, which was the reason for 'his remarkable ascendancy over the country since no one is afraid of him and everyone trusts him'.[8]

Western journalists and others regularly disbelieved the evidence of their own eyes when it came to the Soviet Union, fearing that any criticism of the Soviet workers' utopia would merely play into the hands of the fascists and capitalists. Taking a trip down the Volga in 1932, Iverach MacDonald of the *Yorkshire Post* saw plenty of evidence of widespread starvation and wrote about it, yet on the same trip the New Zealand cartoonist David Low made no reference to any such hardships in his cartoons, but merely ridiculed the idea that there were mass executions taking place in Moscow.

Perhaps the grossest example of the way in which highly intelligent men and women 'chose to serve power rather than speak truth to it as conscience and an honourable tradition of principled opposition dictated' came when Sidney and Beatrice Webb visited Stalin in 1931. The product of the visit, the Webbs' 1,174-page book *Soviet Communism: A New Civilisation?*, started off with a question mark in the title that was dropped in later editions. In the Webbs' messianic view,

> Unlike Mussolini, Hitler and other modern dictators, Stalin is not invested by law with any authority over his fellow-citizens. He has not even the extensive power which ... the American Constitution entrusts for four years to every successive president. ... We do not think that the Party is governed by the will of a single person, or that Stalin is the sort of person to claim or desire such a position. He has himself very explicitly denied any such personal dictatorship in terms which ... certainly accord with our own impression of the facts.[9]

Nor did the passage of time, and the revelations it brought about the true nature of Stalin's power, alter the Webbs' view. Writing from her home at Passfield Corner in Hampshire to the Labour politician and First Lord of the Admiralty A.V. Alexander, on 4 September 1941, Beatrice announced that, 'Stalin is not a dictator and ... the Soviet Union is an industrial as well as a political democracy.'[10]

Walter Duranty, the Moscow correspondent of the *New York Times* between 1921 and 1934, also actively suppressed the truth about the Ukrainian famines. This tremendously influential journalist, who won the 1932 Pulitzer Prize for his reports on Russia and the Five-Year Plan, wrote that 'any report of famine' in the Ukraine was 'exaggerated or malignant propaganda', even though he knew perfectly well of the millions dying there, having seen the horrors for himself. Duranty did enormous damage to the forces of American anti-communism with reports that he not only knew to be inaccurate but actually a grotesque distortion of the truth. (His Prize has nonetheless not been revoked.)

Western journalists became propagandists; men such as Lincoln Steffens and Walter Duranty were arguing – before the horror of Stalin's Terror became generally known in the West – that the USSR had perfected a system for the future. On his return from interviewing Lenin, where he swallowed unquestioningly every lie the Russian leader told him, Steffens told the financier Bernard Baruch, who was sitting for his bust by the sculptor Jo Davidson, 'I have been over into the future, and it works.'[11] He had none of the instinctive aversion to utopian experimentation which has been one of the greatest strengths of the English-speaking peoples since 1900. Fascinatingly, journalists – whose primary duty is supposed to include the checking of sources – have been some of the most naïve of all witnesses to the twentieth century. People who instinctively distrusted democratically elected politicians of their own country, nonetheless fell into raptures over the self-appointed tyrants of Russia.

(In Steffens' 1931 autobiography – 'A human wizard wrote this book,' said Carl Sandburg in the *Chicago Daily News* – the man they called 'America's greatest reporter' enthused breathlessly about Lenin, who, he wrote, 'had shown himself a liberal by instinct. He had defended liberty of speech, assembly, and the Russian press for some five to seven months after the October Revolution which had put him in power. The people had stopped talking; they were ready for action on the program.'[12] According to Steffens, poor Lenin had been holding out against the Bolshevik Left's plans for terror, but then was subjected to an assassination attempt after which the hard-liners took control from the peace-loving leader. A more gullible interpretation of the true course of events is hard to imagine.)

The All Souls' historian Christopher Hill spent ten months in Russia in the mid-1930s, joining the Communist Party on his return. He didn't seem to notice anything amiss in Stalin's Russia either. Nor was he convinced even decades later that the grain harvest of millions of Ukrainians had been requisitioned by the Soviet authorities, reserving his scepticism for the Western journalists like Malcolm Muggeridge and *The Times'* correspondent in Riga, Reginald Urch, rather than the propagandists of *Pravda*. (Urch managed to produce far more accurate reports about Russia from his base in independent Latvia than the Western correspondents in Moscow itself, yet even he underestimated the horrific levels of starvation and murder being carried out by the Bolsheviks.)

Robert Conquest, author of the 1968 exposé of Stalinism *The Great Terror*, put it best when he commented how the Western apologists for Bolshevism constantly contradicted their own firmly held beliefs in order to lavish praise on the Soviet Union, instancing 'Feminists applauding women who were bowed down by hundredweights of coal, Quakers applauding tank parades, and architects looking at buildings with awe and admiration that had just been put up but were already falling down. They got themselves into a very strange mental mood. I think it was the worst thing about the whole century. Something went wrong with people's minds.'

Barbara Betts – who became the Labour Cabinet Minister Barbara Castle – was sent by the left-wing newspaper *Tribune* to Moscow to write on the position of women there. It was the very height of the Great Terror, yet, as she breathlessly reported, 'No-one grumbles on the trains and buses, because everyone who is so perilously strap-hanging is confident Soviet factories will put an end to all these travel difficulties.' In articles headlined 'Russia Goes Gay With Sport, Play and Dancing', 'Security Brings New Life to Soviet Mothers' and 'Children Who Get a Real Chance', Betts lauded the Soviet system.[13]

No fewer than seven million innocent people were arrested in the Great Terror of the late 1930s, six million of whom were executed or died in the gulag, yet Barbara Betts wrote about how 'No-one can visit Soviet Union's institutions for mother and child, crèches, lying-in hospitals and kindergartens without realising that these services are an integral part of the lives of Soviet women.' When asked half a century later about these places, Ms Betts, by then Lady Castle, admitted that they had been 'showcases', but continued to defend 'the philosophy behind it', adding that there had been 'no atmosphere of repression' that she had noticed in Moscow in 1937. Re-reading her own reports in 1999 she commented, 'I don't think they were too bad actually.'[14] In fact, as David Pryce-Jones has written, such behaviour 'may indeed serve only to warn against messianism, the romancing of violence, and the appeal of power'.

'You're not going to unmask me, are you?' Those were the words which Christopher Hill directed to a researcher, Anthony Glees, who had discovered that during the Second World War he had been working as an 'agent of influence' in the interests of Stalin's USSR when he had been in charge of the Russian desk in the Foreign Office. They agreed that he would not be unmasked as a Soviet mole during his lifetime. During the war Hill had proposed, amongst many other pro-Soviet suggestions, that all White Russian émigrés teaching Russian at British universities should be dismissed, to be replaced by staff who had been approved by the Soviet authorities. After the war was over, Hill further advised that Polish exiles should be sacked also. Innocent people were to be removed from dozens, perhaps hundreds, of teaching posts up and down the country solely because of their racial and political backgrounds. It was the antithesis of academic freedom.

After Hill's death and Glees' revelations, instead of a general denunciation of the man who thought up this disgraceful scheme, there was, in the *Guardian*'s own words, 'a fierce defence of Mr Hill by other academics'. The academic who has led the 'outcry' in Hill's defence was the historian Professor Eric Hobsbawm, who better than anyone else personifies the Left's refusal to acknowledge in full the evils of Stalinism. In 1993, a full four decades after Stalin's death, Hobsbawm told a group of Hungarian students that 'for the common citizens of the more backward countries' of Eastern Europe, the Stalinist era 'was probably the best period in their history'.

On British television's *The Late Show* in October 1994, the philosopher Michael Ignatieff asked Hobsbawm: 'In 1934, millions of people were dying in the Soviet experiment. If you had known that, would it have made a difference to you at the time?' Hobsbawm replied: 'I don't actually know that it has any bearing on the history I have written. If I were to give you a retrospective answer which is not the answer of an historian, I would have said "Probably not".' Ignatieff then pressed him further, asking: 'What that comes down to is saying that had the radiant tomorrow actually been created, the loss of fifteen, twenty million people might have been justified?' Hobsbawm immediately answered: 'Yes.' Later Hobsbawm received the much-coveted Companionship of Honour from the Labour Government.

Of course fifteen or twenty million people weren't 'lost' by Stalin: they were shot, frozen, starved and worked to death. And there were probably more than twenty million of them in total. What Hobsbawm and so much of the British, American and Australian Left vociferously argue is that unlike Hitler's racially motivated genocide against the Jews, Stalin was pursuing a different, class-based agenda to try to modernise Russia by freeing it from the reactionary *kulak* (richer peasant) class. The Left argues that it was not racially motivated, like Hitler's Holocaust, and was thus somehow less evil as a result.

Yet the truth is that as well as his campaign against the *kulaks* and his own Bolshevik enemies, Stalin also committed seven major acts of racial genocide, against the Ukrainians in 1930–2, the Poles, Balts, Moldavians and Bessarabians in 1939–41 and 1944–5, the Volga Germans in 1941, the Crimean Tatars 1943, the Chechens and the Inguches in 1944. He was also planning a pogrom against Russian Jews, bound up in what was called 'The Doctors' Plot', when he fortuitously died in March 1953. Indeed, as Professor Alan Bullock so comprehensively proved in his 1991 book *Hitler and Stalin: Parallel Lives*, the Nazis actually learnt most of their repression techniques from the Bolsheviks.

The fact that so many of those racial killings took place during the Second World War amplifies another reason why Western Leftists still resolutely refuse to recognise Stalin as the precise moral equivalent of Hitler, the Tweedledum and Tweedledee of totalitarianism. Because Stalin's USSR fought on the same side as Britain and America, and lost millions killed in the struggle, 'Uncle Joe' is awarded a place on the side of the angels by his apologists. Many on the Left even argue that the 'Cambridge spies', such as Kim Philby and Anthony Blunt, were right to spy for Stalin in the Thirties and Forties because only the Soviet Union offered opposition to Nazism at a time when Britain and France were intent upon appeasing Hitler.

This argument, weak though it inherently is on moral grounds anyhow, underplays the fact that the USSR signed a non-aggression pact with the Nazis on 25 August 1939, thereby allowing Hitler a free hand to start the war, an opportunity he grasped that very same week. Neither Eric Hobsbawm, who joined the Communist Party in 1932, nor Christopher Hill, who joined it four years later, resigned in protest against the Molotov-Ribbentrop pact, as many communists did. Neither did they resign in 1956, when the Soviet Union brutally invaded Hungary, nor in 1968, when it crushed Czechoslovakia's attempt at 'Socialism with a human face'. When asked why he stayed in the Party in 1956, Hobsbawm said, 'Out of loyalty to a great cause and to all those who sacrificed their lives for it.' The fact that a hundred times more people were murdered by the Soviet Communist Party than willingly sacrificed their lives for it seems to have passed him by. Aged eighty,

Hobsbawm stated what so many on the Left also believe, that communism 'can be, in its ideals but even at times in its practice, a powerfully moral force'. This, despite the fact that it has never been practised without the aid of force and bloodshed.

For all their later complaints that their Marxism somehow held back their careers, Hill was Master of Balliol College, Oxford, from 1965 to 1978 and Hobsbawm was elected a fellow of King's College, Cambridge, at the age of thirty-two and was later a reader and then professor at Birkbeck College for twenty-three years. Both men were loaded down with academic honours and posts; if anything, they did well out of their Marxism. The people whose careers were truly blighted in academia were anti-communists like Robert Conquest, whose revelations about Lenin's and Stalin's crimes and the gulag system were systematically derided and denounced by the Left, almost until the fall of the Berlin Wall, when the subsequent opening of the Russian archives proved that if anything Conquest had erred on the side of caution.[15]

Tragically, all five of the Cambridge group of spies escaped justice. Philby, Burgess and Maclean all escaped to Russia, where they died many years later. Blunt finally confessed when cornered, and the worst that befell him was to be stripped of his knighthood in 1979. Cairncross also confessed and left Britain to work for the United Nations, dying in 1995.

There has been virtually no murderous communist dictator who has not had his apologists on the Left, be it Stalin (worshipped by Sidney and Beatrice Webb and George Bernard Shaw), Chairman Mao (adored across Sixties' campuses), or Fidel Castro (who still receives paeans of praise from some fellow-travellers); even Pol Pot was defended by Noam Chomsky until 1980, who wrote of how 'the evacuation of Phnom Penh, widely denounced at the time and since for its undoubted brutality, may actually have saved lives'.[16]

In October 1924, the British writer Lawrence Welch published a sixpenny book entitled *Fascism: Its History and Significance* under the pseudonym 'L.W.' A committed Marxist, he concluded that,

> The support for Fascism which formerly existed has now to a large degree disappeared. And the fascists will find that, as their ideological support has gone, so has their physical. ... The courage and political sense of the Italian workers have often proved enough; if their leadership and organisation can only reach the same high standards in the coming struggle, Fascism can be crushed and a workers' government set up in Italy.

He could hardly have been more mistaken; in fact, it was twenty years before fascism was defeated in Italy, and probably any time between 1924 and 1941

Mussolini would have won a free and fair election in Italy if he had been unfascist-minded enough to have held one.

International veneration for Benito Mussolini rode high in the inter-war years. Pope Pius XI described him as having been 'sent by Providence', Churchill called him 'the greatest law-giver among living men' and 'one of the most wonderful men of our time', and President Roosevelt regarded him as his 'only potential ally in his effort to safeguard world peace'.[17] Cole Porter's song *You're the Tops!*, in a verse that was understandably subsequently excised, had the following couplet in relation to the beautiful society hostess Margaret Sweeny, later the Duchess of Argyll: 'You're the tops, you're Mrs Sweeny,/You're the tops, you're Mussolini!'

It was his defeat of Italian communism that originally raised Mussolini high in the opinion of the West, but thereafter the combination of nationalism and fascism in his domestic policies and the vainglory of his foreign policy ought to have signalled his true nature to many people. He had an opportunity to behave responsibly at the Stresa Conference in April 1935, when the prime ministers of Britain, France and Italy protested against German rearmament and agreed to act jointly against Hitler in what became known as the Stresa Front, but this collapsed after Italy invaded Abyssinia that October.

When the League of Nations, led by Britain, voted by 50 votes to 4 to impose sanctions on Italy on 19 October over her invasion of Abyssinia, an Italian journalist called on his countrymen to desist 'from such pernicious British habits as tea-drinking, snobbery, golf-playing, Puritanism, clean-shaving, pipe-smoking, bridge playing, and inexplicable apathy towards women'.[18] A mere nine months later, on 15 July 1936, the sanctions were raised by the League, leaving the organisation as, in Churchill's phrase, 'a cockpit in a Tower of Babel'. With no army, navy or air force, the League of Nations was impotent. 'And covenants, without the sword, are but words,' wrote Thomas Hobbes in chapter seventeen of *Leviathan*, 'and of no strength to secure a man at all.' It was as true in the Devil's Decade as when Hobbes had written them in the mid-seventeenth century.

To have foregone the dreams of a new Roman empire in Abyssinia, expanded the Stresa Front against Hitler, opposed *Anschluss* in 1938, not invaded Albania in April 1939 and then sat out the war like his fellow southern Mediterranean fascist dictator General Franco would all have been sensible options for Mussolini, but were each eschewed. When Mussolini made the gross strategic error of declaring war against the Allies in June 1940, it was to cost the lives of 300,000 Italian servicemen and 150,000 civilians. Small wonder, therefore, that when he declared war from the balcony of the Palazzo Venezia in Rome, the police reports stated that 'not a single woman applauded'.

Yet could Western appeasement of Mussolini have worked? Might it have been possible to have kept him out of the Axis orbit, to have prevented him joining 'the Pact of Steel' with Hitler in May 1939? There are some who argue that Hitler could have been stopped if only Mussolini had been comprehensively appeased throughout the 1930s, principally after Stresa. 'Mussolini would have been a slippery and treacherous ally,' argued the historian Richard Lamb, 'but in the face of the Nazi menace his goodwill was essential.' Yet in fact Hitler's overall strategy would not have been affected much by anything Mussolini did or failed to do. He knew that Italy posed no serious military threat to Austria, let alone to Germany itself, and with his southern flank safe Hitler would still have invaded Poland once the Nazi-Soviet pact had neutralised Russia. The Führer once said that although Mussolini was a Roman, his people were only Italians. Not a Great Power and barely a military one, Italy simply did not have the influence to affect events significantly.

Moreover, a full-scale alliance with fascist Italy would have split Britain politically even more than she already was. It is also not hard to imagine how the trade unions, Labour Party and the Liberals would have reacted. When *Il Duce* considered visiting London in 1925, the ASLEF railway-union's general secretary threatened a strike on any train he took. Not only would Mussolini's train not run on time, it would not have run at all.

Once France had fallen in May 1940, Mussolini – always a scavenger rather than a bird of prey – who claimed the French territories of Nice, Savoy and Corsica, would have been instantly bought off by Hitler. Franco showed foresight in resisting the temptation to involve Spain in the war, but Mussolini was too vain, bombastic and greedy to follow suit. Nor was Mussolini instinctively pro-British, as some historians have claimed. He saw Malta, Gibraltar, Cyprus and Egypt as Italy's natural pickings once the British Empire had collapsed under the Axis assault. His encouragement of Indian nationalism, torpedoing of British merchantmen in the Spanish Civil War, virulent anti-British radio propaganda and financial backing for Sir Oswald Mosley hardly betokened any fondness for a workable British *entente*.

Mussolini's sheer unpredictability – one moment he privately supported the French occupation of the Ruhr, the next he publicly denounced it – would have made him an impossible ally. In December 1922 he cancelled a press conference because he did not want to get out of the bed he was sharing with a prostitute in Claridge's Hotel. In return for his adherence to the Stresa Front, Britain and France would have had to have abandoned Abyssinia, a member of the League of Nations, to his mercy in 1935. Far from satiating his appetite – as his subsequent attacks in Spain, Albania and Greece showed – concessions would merely have whetted it. As it was, Mussolini's Pact of Steel with Hitler resulted in Italy's peninsula being used as a battlefield, invaded first by the Allies and then by the Germans. All of this further begs the question why he was not seen through far earlier by otherwise perspicacious Western statesmen such as Churchill and Roosevelt?

The story ended in April 1945 with the crowd in the Piazzale Loreto in Milan laughing at and urinating on the corpses of Mussolini, his mistress Clara Petacci, her brother Marcello and fifteen others, before seven bodies were hung upside down from the steel girders of the petrol station there. It was remarked with surprise by the women present, who were joking and dancing around this macabre scene, that Clara Petacci wore no knickers and that her stockings were un-laddered. (She had not been given time to find her knickers before she was taken away and machine-gunned by the partisans, which seems rather un-Italian behaviour towards an attractive if brainless woman.)

For all its later criticisms of the National Government's appeasement of fascism, the British Labour movement vacillated in its own response to the dictators, between idealistic pacifism and hard-headed analysis of what it would really take to defeat the looming threat. Even in November 1933, ten months after Hitler came to power in Germany, the deputy leader of the Labour Party, Clement Attlee, wrote to his brother Tom, a Christian and Fabian who had been imprisoned in Wandsworth Gaol as a conscientious objector during the Great War:

> The movement is not I think quite clear on the question of sanctions i.e. it has not really made up its mind as to whether it wants to take up an extreme disarmament and isolationist attitude or whether it will take the risks of standing up for the enforcement of the decisions of a world organisation against individual aggressor states.[19]

Attlee's own belief in 'a world organisation', in this case the League of Nations, actually went far further than many people at the time realised. In a revealing letter to Tom the following October, he showed that he was, in his fantasy world at least, nothing less than a neo-uniglober. He wanted to write a book entitled *Peace and War*, but

> I have also an idea for a film on the same subject which came to me all of a sudden. The general idea being the last war wherein two Balkanised despotisms simultaneously wipe out each other's capitals to the horror of the civilised world. Extremely realist scenes of destruction to be filmed. War fomented by rival armaments groups who own the press of the two countries. Son of chief armament monger sees wife and children killed most unpleasantly. Repentance of chief armament monger who gives away workings of the ring to the D[aily] H[erald] just in time to turn general election. Follows creation of international World State, abolition of armaments etc with a postscript some years afterwards illustrating

new world conditions by conversations of members of World air communications at HQ aerodrome in Vienna. Love interest etc can be added if necessary.[20]

For something quite so melodramatic and obviously propagandist it is probable that the addition of love interest might well have been necessary, for as Attlee continued,

Incidentally there is the end of Nazism as Hitlerite dictator intent on war is stopped after 48 hours' consideration by threat of international interference ... collapse of Nazism. It is possible that such a theme well worked up might be acceptable to the lords of the film who mostly belong to the Chosen People. It would not do for it to come out under my name. ... It might be quite valuable propaganda if done sufficiently crudely for the popular taste.

Those four sentences, with their combination of absurd wishful thinking, mild anti-Semitism, pusillanimity and snobbery, neatly sum up the contribution of the British Left to the dire, decade-long emergency caused by the rise of the dictators.

If the twentieth was, as Henry Luce dubbed it, 'the American Century', it was largely due to the Roosevelt cousins. Theodore created a framework to ensure that capitalism did not devolve into unregulated monopolistic cartels while simultaneously thrusting the United States on to the world scene and attaining Great Power status for her, but it was under Franklin D. Roosevelt that she became a superpower. An American aristocrat who could nevertheless connect superbly with voters from all backgrounds, FDR's sphinx-like personality dominated the American political scene, and first his defence of democracy domestically and then his promotion of it globally set the political weather for the rest of the twentieth century and into the twenty-first. His insistence on regime-change in Germany, Austria, Italy and Japan by installing democracy through force rendered him the first American neo-conservative.

In the first of his record four Inaugural Addresses, delivered at the Capitol on 4 March 1933, Roosevelt referred to unemployment in an unashamedly partisan way, blaming Wall Street and by implication his Republican opponents for the Crash and its aftermath. 'Only a foolish optimist can deny the dark realities of the moment,' he said. 'Yet our distress comes from no failure of substance. We are stricken by no plague of locusts. Compared to the perils which our forefathers conquered because they believed and were not afraid, we have still much to be thankful for. Plenty is at our doorstep, but a generous use of it languishes in the very sight of the supply.'[21] A capitalist might have been excused for thinking that he had far more to fear from the new Democratic future than fear itself.

Roosevelt eschewed magnanimity as he laid the blame for the Depression squarely on the shoulders of his political opponents, continuing: 'Principally this is because the rulers of the exchange of mankind's goods have failed, through their stubbornness and their own incompetence, have admitted their failure, and abdicated. Practices of the unscrupulous money-changers stand indicted in the court of public opinion, rejected by the hearts and minds of men.' The biblical reference to the money-changers, who Christ cast bodily out of the Temple, could hardly have been a harsher critique of the Wall Street bankers. Yet it was the English-speaking peoples' ability to throw up leaders of the stamp of Roosevelt, who could turn fury and resentment into non-revolutionary avenues, that saved America from being led down avenues of extremism.

Although Roosevelt refused to let anyone know his intentions in the four-month interregnum period between his election and his inauguration, he acted with tremendous despatch as soon as he had taken the oath of office from the Chief Justice of the Supreme Court, the Republican former Secretary of State Charles Evans Hughes. The very next day – 5 March – he summoned a special session of Congress for the 9th, stopped all gold transactions and declared a four-day national banking holiday from 6 to 9 March, which he then extended until he could be reasonably certain that when the banks reopened deposits and gold would flow into rather than out of them. Congress granted the President control over gold and silver bullion and foreign exchange and passed fifteen major bills over the next three months, a period likened to Napoleon's Hundred Days (except there was no Waterloo at the end of it).

'Roosevelt's unprecedented action in taking the dollar off gold to raise prices changed the international financial landscape', not least because he did it deliberately in order to raise domestic prices.[22] The way he propagated his 'New Deal' to the American people, over radio 'fireside chats', the first of which was broadcast on 12 March, played to his great strength of sounding eminently reasonable even when proposing extraordinary measures that his opponents denounced as dictatorial and revolutionary.

Roosevelt was a notoriously difficult politician to read, and still is. The American playwright and four-times Pulitzer Prize-winning journalist Robert Sherwood, who in 1949 wrote about Roosevelt's friendship with Harry Hopkins, described his 'heavily-forested interior', and his War Secretary Henry Stimson said his mind could flit around 'like a vagrant beam of sunshine'. The task is made harder since Roosevelt confided in no-one and died before he could explain his actions in an autobiography.

There are some historians who see Roosevelt as 'trivial and a lightweight; that he was out of his depth; that he was a mere political opportunist with no long-term plans or goals', but they are few.[23] There are more who argue that America was coming out of recession in March 1933 anyhow and all he did was to ride the wave. A recent book has charged that the New Deal actually retarded the recovery because Roosevelt increased excise taxes, personal income taxes, inheritance taxes and corporate income taxes, as well as introducing a special tax on undistributed profits and a new payroll tax.[24] According to this analysis, far from saving capitalism FDR imposed anti-business regulations, compulsory unionism, banking de-mergers (even though big institutions had better chances of survival) and assaulted property rights. It is certainly true that by 1937 serious cracks were starting to appear in the American economy that were hidden by her need to rearm due to Hitler's (curiously geographically anti-clockwise) manufactured crises in the Rhineland, Austria, Sudetenland and Danzig. Yet the New Deal worked.

The Tennessee Valley Authority, set up in 1933, was not the first example of integrated regional planning; Benton McKaye's Appalachian railway was intended to give a spur to the rural economy in the Twenties. However, the TVA managed to develop the entire Tennessee River basin while preventing flooding, improving navigation and producing cheap electricity. It seemed the way ahead. How far away it was from the post-war era of 'urban renewal', which replaced intimate nineteenth-century street patterns with massive soulless developments, and it remains perhaps the best symbol of the New Deal.

Not everything went Roosevelt's way, however. Charles Evans Hughes' Supreme Court struck down several aspects of the New Deal and its accompanying social legislation, and on 27 May 1935 it found against the National Recovery Administration as being in restraint of trade. *Schechter Poultry Corporation v United States*, popularly known as 'the Sick Chicken Case', was only the best known of these reversals for FDR. After Evans 'spoke with great force and animation' in the unanimous judgment, the liberal Justice Louis Brandeis summoned the senior New Deal administrators Thomas Corcoran and Benjamin Cohen to the Supreme Court robing room and advised them 'to tell the President that he would have to redesign his entire legislative programme'.[25] There can hardly have been a more blatant example of judicial activism, what Robert Bork has recently denounced in his book *Coercing Virtue* as the unwarranted power of 'activist, ambitious, and imperialistic judiciaries'.[26]

Once triumphantly re-elected in November 1936, Roosevelt's attempt to pack the Supreme Court in the first half of 1937, by calling for an additional justice to be appointed for each one aged seventy or over – when no fewer than six of them were – led to what one of his recent biographers has called 'the greatest political defeat of his career. ... Roosevelt, the master politician, made mistake after mistake, while the "horse and buggy" Court, with its aged and obscurantist justices, played brilliantly its defensive hand.'[27] Defeated in even the Democrat-dominated Senate Judiciary Committee, Roosevelt had to rely instead on his longevity in office to pack the Court; by 1941, the only pre-Roosevelt appointments who remained were the liberal Harlan F. Stone and Owen Roberts. Roosevelt's attempts aside, the English-speaking peoples have done well to stand by judicial independence, in a way that non-democracies never do, since it strengthens the trust in the rule of law that has been a mainstay of their power.

On Monday, 30 September 1935, President Roosevelt officially opened the Hoover Dam at Boulder City. As well as the 12,000 people at the ceremony on that warm, sunny day, millions listened on the radio when he said of one of the most astonishing civil engineering feats of Mankind: 'I came, I saw, I was conquered. ... Ten years ago, the place where we are gathered was an un-peopled, forbidding desert. The transformation wrought here is a twentieth-century marvel.'[28] The English-speaking peoples had long excelled at creating the wonders of the modern industrialised world: the *Great Eastern*, the Brook-lyn Bridge, the Sydney Bridge (which when opened in 1932 was the largest single-span bridge in the world), the American, Canadian and Australian transcontinental railroads, the Panama Canal among them; the Hoover Dam was part of that noble tradition.

The Colorado River collects water from over 1,000 square miles, 740 billion cubic feet per annum of it from seven states draining the western slopes of the Rocky Mountains and eventually – after flowing down the Grand Canyon and through two deserts – pouring out into the Gulf of California. To harness the terrific energy of this was the task proposed by Arthur Powell Davis, who in 1922 drew up a plan based on a massive dam at Black Canyon. The concrete-arched gravity-dam was to be 726 feet high with a base of 660 feet of solid concrete.[29]

Despite his reputation as a non-interventionist, it was Herbert Hoover who signed the Boulder Canyon Project Act into law on 25 June 1929, representing by far the largest civil project the US Government had ever undertaken. Six construction companies in the West bid $48,890,955 for the contract, which was accepted in March 1931. Their chief engineer, Frank Crowe, was, he said, 'wild to build this dam. I had spent my life in the river bottoms.' A graduate of the University of Maine back in 1905, he was a working civil engineer whose watchword was 'Never my belly to a desk'. It was to be his task to create the world's largest dam, to hold the water of a lake fifteen miles long and 590 feet deep.

'The weight of water in the lake would surely invoke an earthquake that would destroy the dam and deluge the land with waves hundreds of feet high,' the historian Deborah Cadbury has written of the fevered conjectures surrounding the building of the dam. 'Others speculated that the sheer weight of water could even throw the earth out of orbit.'[30] In order to divert the Colorado, four tunnels 56 feet wide and three-quarters of a mile long had to be dug through the hard volcanic rock. The dynamiting and drilling had to be undertaken in temperatures sometimes as high as 120 degrees Fahrenheit, and a further ten degrees hotter in the tunnels. Working conditions were squalid and dangerous, but strikes were few because there were plenty of men who wanted a job that paid $5 a day during the Depression. With harsh penalty clauses for delays, the Six Companies and 'Hurry Up' Crowe drove the workforce at full speed. In all, no fewer than 107 of the 5,000 people who worked on the dam died. 'Death is so permanent' warned large posters around the site advising of the various dangers.

The three-and-a-half-million cubic yards of concrete was poured with superb efficiency; in March 1934, the team poured 10,401 cubic yards of it on a single day. Today the dam's generators can produce 2,000 megawatts of energy, creating electricity for use in three states. Arthur Powell Davis' vision, Frank Crowe's determination, the Six Companies' profit-motive, the Hoover Administration's ambition, and the workforce's professionalism and courage, has made the Western desert bloom.

Yet for all her ability to build the Empire State Building and the Hoover Dam in the Thirties, America was a minor player in world affairs in that decade. In the first eight years of his presidency, Franklin D. Roosevelt travelled outside the United States only once and, by 1935, the United States Army was still only the world's nineteenth largest.[31] Just as he had spotted the danger of America shirking her post-war responsibilities and failing to engage with the rest of the world, so Rudyard Kipling diagnosed earlier than almost anyone else – including Winston Churchill – the mortal threat that Germany would once again pose to Civilisation. His poem *The Storm Cone*, written as early as May 1932, with Hitler on the eve of becoming Chancellor, began:

> This is the midnight – let no star
> Delude us – dawn is very far.
> This is the tempest long foretold –
> Slow to make head but sure to hold.
>
> Stand by! The lull 'twixt blast and blast
> Signals the storm is near, not past;
> And worse than present jeopardy
> May our forlorn to-morrow be.

Kipling ordered that the swastika, previously just a sign of good luck – '*Svasti*' means 'prosperity' in Sanskrit – was to be removed from all future editions of his books.[32] (In those days the swastika even adorned certain Great War memorials, including the one built in 1922 outside the gates of Balmoral Castle, the Highlands home of the Royal Family.)

Only six years after his death in 1936, Kipling was declared to have 'dropped out of modern literature' by the critic Edmund Wilson. Oscar Wilde had years earlier derided his 'superb flashes of vulgarity', and in 1942 George Orwell announced that 'during five literary generations every enlightened person has despised' Kipling as 'morally insensitive and aesthetically disgusting'. Although Kipling's aesthetic appeal, summed up by the literary critic Philip Hensher as 'his formal skill in verse and his visionary strangeness in prose', has been long re-bunked by critics, his politics have continued to outrage the politically correct.

In fact, far from being a jingoistic drum-banger and racist flag-waver, Kipling held far more complex and subtle views about the Empire he loved. A racist would not have glorified the sepoy Gunga Din in the way that Kipling did, while a gung-ho wider-still-yet-wider imperialist would never have chosen *Recessional* as the title of his poem to mark Queen Victoria's ebullient 1897 Diamond Jubilee, in which he warned of the day when 'Far-called, our navies melt away; On dune and headland sinks the fire.'

Kipling was very often proved right in his predictions. He saw that the Boers would establish apartheid in South Africa if they were permitted to; as early as the mid-1890s, he warned that the Kaiser would unleash an aggressive world war; he predicted that communal genocide in the Punjab would accompany any over-hasty transfer of power in India; and he denounced the appeasement of Adolf Hitler. It is a noble, but by no means exhaustive, list.

It is as the finest phrasemaker in English since Shakespeare that Kipling's works will live, and many of the phrases we associate with the First World War and its commemoration were his. Kipling sacrificed his beloved son John for the British imperial ideal; killed at the battle of Loos in 1915 serving with the Irish Guards, his body was never found. It was therefore a grieving father who chose many of the inscriptions for the war memorials, such as the one for corpses so badly disfigured as to be unidentifiable: 'A Soldier of the Great War Known Unto God'.

Kipling died on 18 January 1936, two days before King George V; it was said that 'His Majesty had sent his trumpeter before him.'[33] Those two men in many ways represented all that was best about the English-speaking peoples in the first third of the twentieth century. The King's reign was beautifully précised by the poet A.E. Housman in an address from the Chancellor, Masters and fellows of Cambridge University, delivered on the occasion of his Silver Jubilee in May 1935. The part that referred to the Great War and its aftermath bears repetition at some length, not least for the cadences of the phrases Housman employed:

The events of your reign, for greatness and moment, are such as have been rarely comprised within twenty-five years of human history. It has witnessed unexampled acceleration in the progress of Man's acquaintance with the physical universe, his mastery of the forces of nature, and his skill in their application to the processes of industry and to the arts of life. No less to the contrivance of havoc and destruction has the advancement of knowledge imparted new and prodigious efficacy; and it has been the lot of Your Majesty to confront at the head of your people the most formidable assault which has ever been delivered upon the safety and freedom of these realms. By exertion and sacrifice that danger was victoriously repelled; and Your Majesty's subjects, who have looked abroad upon the fall of states, the dissolution of systems, and a continent parcelled out anew, enjoy beneath your sceptre the retrospect of a period, acquainted indeed with anxieties even within the body politic and perplexed by the emergence of new and difficult problems, but harmoniously combining stability with progress and rich in its contribution of benefits to the health and welfare of the community.[34]

(Since all such formal messages to and from the monarch have to be vetted in advance by the Home Office, this too was sent there. A reply was sent back to Cambridge with the text unaltered, but with a civil servant's comment attached saying, 'This seems to be good English.')

'An insular country, subject to fogs and with a powerful middle class,' remarked Disraeli, 'requires grave statesmen.' Unfortunately, Britain had a surfeit of grave statesmen in the 1930s. The Respectable Tendency who had ousted the Lloyd George Government at the Carlton Club meeting in October 1922 then ruled Britain – with only two short-lived interludes of Labour Government in 1924 and 1929–31 – until the fall of Baldwin's successor Neville Chamberlain in May 1940. The presiding power was Stanley Baldwin, who had helped bring down Lloyd George by describing him as 'a dynamic force ... a terrible thing' at the Carlton Club meeting. He oversaw the General Strike, ensured that even by 1937 the top rate of income tax was only five shillings in the pound and pulled off the Respectable Tendency's great coup of preventing a sexually active American divorcée from ascending the British Throne. What the Respectable Tendency failed to do, however, was ensure that Germany observed the military articles of the Versailles Treaty.

With the West still in disarray over what to do about Italy's invasion of Abyssinia, in the early hours of Sunday, 7 March 1936, German troops suddenly crossed the border over the Rhineland in direct contravention of both the Versailles and Locarno Treaties. They were greeted with joy by the local inhabitants. The reactions of the military attachés attached to the various embassies in Berlin were extremely instructive, as each of them met the departmental chief of the German General Staff, Rabe von Pappenheim, himself a former military attaché, later that day. The Frenchman, General Daston-Ernest Renondeau, told Pappenheim that if he were the French premier, 'I, my dear friend, would declare war on you!' The British attaché, Frederick Hotblack MC, after jokingly thanking Pappenheim for giving him such a 'lovely Sunday', said that everything should be done to avoid taking rash and unconsidered steps that could not afterwards be retracted.[35] The Americans, Major Truman Smith and his assistant Major Crockett, 'expressed their total approval of the action in the Rhineland and congratulated Pappenheim. At the same time they expressed their fears of reprisals by other powers.' They also made it clear that they were echoing the opinion of the US Ambassador to Germany, William E. Dodd. It was not the English-speaking peoples' finest hour.

In 1944, a don at Pembroke College, Oxford, named Ronald McCallum published a book entitled *Public Opinion and the Last Peace*, which was a spirited defence of the Versailles Treaty. Covering the Rhineland crisis of 1936, McCallum accepted that pre-war public opinion in the West was solidly opposed to acting against Hitler and identified the central paradox in the British position:

> We talk much in England of the sanctity of treaties, but our view is somewhat one-sided. We are shocked when a government signs a treaty binding it not to do something and then breaks the treaty by doing it. We are not shocked when we bind ourselves to do something and then refuse to do it. But those who are affected by our refusal to act do not see the force of this distinction.[36]

In McCallum's view, it was perfectly true that British opinion recoiled at preventing Hitler from moving troops into what was after all part of Germany, yet in 1919 'it had not recoiled when we first took on the obligation to stop the Germans invading their own country'. To the accusation that the Baldwin Government lacked public support in stopping Hitler, McCallum makes the straightforward, but curiously largely unheard – both then and now – points that,

> If this was so it might have done its duty under the constitution and resigned. It might on the other hand have prepared the public mind for the event. It was no secret that Hitler was likely to take advantage of the Abyssinian question. A lead from the Government informing us of our duties under the Treaty of Locarno, a warning that the consequences of failing in these duties would be ... the final success of German rearmament, would have had its effect and at least separated those who were willing from those who were not willing to oppose Fascism.[37]

Baldwin did none of these things.

The Rhineland Crisis was the first of the new reign. King Edward VIII was only forty-one when he ascended the Throne; at the time he seemed the very model of a modern monarch. His remark that 'Something must be done' for the unemployed, made on a visit to South Wales in November 1936, was held to prove that he had a social conscience denied to Stanley Baldwin's Tory-dominated National Government. Yet far from being the spokesman of the dispossessed and marginalised, there was always something disgraceful about the King, who had already made up his mind to abdicate when he made that nebulous, open-ended comment, saying something upon which he knew he would not be present to encourage action.

Edward VIII was deluged with letters of sympathy and support during the Abdication crisis, and the latest historian of the crisis believes that the majority of ordinary people in Britain and the Empire were quite ready for Wallis Simpson to become queen.[38] Most simply wanted the King to be happy. Fortunately, however, such things were not decided by the people's sentiment but by altogether harder-headed analysis by the governing class.

The Abdication saw the apotheosis of the Respectable Tendency, who recognised that the happy, normal domestic life of an upper-class British Royal Family was a tangible asset to the prestige of the British Empire, and that although the Duke and Duchess of York (the future George VI and Queen Elizabeth) provided that, Edward VIII and Mrs Simpson never could. Although it is not difficult to detect a distinct whiff of anti-Americanism to the Respectable Tendency's attitude during the crisis, it is nonetheless hard to imagine that had the King wanted to marry a British divorcée commoner, men like Stanley Baldwin, Neville Chamberlain and the ultra-High Church Lord Privy Seal, Lord Halifax, would have felt very much differently.

The situation was summed up clearly by Chamberlain – as Chancellor of the Exchequer a prime mover in the negotiations – who wrote to his sisters Ida and Hilda on 8 December 1936, two days before the Abdication, explaining that for the King,

> There are only 3 alternatives before him. (1) Marriage with Mrs S[impson] as Queen (2) Abdication and marriage (3) Renunciation of this marriage altogether. Now (1) is already barred because apart from feeling in this country the Dominions have already plainly said they won't have it. The choice is therefore between (2) and (3). The general public will prefer (3) but if the K[ing] is not prepared for (3) there remains nothing but (2).[39]

There was of course also brief discussion of a fourth option, the King marrying morganatically and Mrs Simpson not becoming queen, which the Government dismissed as contrary to British law and custom. It is an interesting comment on the morality of the day that there was no possibility of even the consideration of a fifth outcome, that the King simply carried on keeping Mrs Simpson as his mistress. She would hardly have been the first monarch's mistress in British history, or even the twentieth century, but the King wanted to marry Mrs Simpson, and anyhow public opinion would have been outraged.

Two days after the Abdication, the Irish Dáil (parliament) took full advantage of the crisis to pass its Constitution (Amendment No. 27) Act that removed the British monarch and governor-general from the Constitution regarding Ireland's internal matters, while confirming their right to represent her in external affairs. The age-old saw that England's danger was Ireland's opportunity also applied the following year, in 1937, when the Irish Free State became Eire, with a new Constitution. Although this claimed the entire island of Ireland as its national territory, the laws of the Irish parliament in Dublin were acknowledged as only applying to Eire, which in 1948 changed its name yet again, when the Republic of Ireland Act removed the last constitutional links with the United Kingdom.[40]

While southern Ireland had every right to independence from the British Crown, it had no right to claim in Articles 2 and 3 of its Constitution sovereignty over Protestant-majority areas of the north, which had no wish to be incorporated into it, yet disgracefully enough the formal claim stayed in the Irish Constitution until after a referendum held on 22 May 1998 agreed to drop it. For over sixty years, therefore, extreme Irish nationalists seeking to terrorise the Protestant northerners into a unitary state could (and indeed did) argue that they were merely carrying out the provisions of the Irish Constitution.

The Great Depression hit the West Indies harder than almost any other part of the English-speaking world, with the possible exception of Australia. From Belize to Barbados and from the Bahamas and St Kitts to Trinidad and Guyana there were, in the words of a recent historian of those deeply troubled times, 'strikes, demonstrations, and even insurrections that led directly to near-revolutionary change', not least in the increase in national consciousness that was to lead to demands for independence after the Second World War.[41]

These came to a head in what is today called the Barbadian 'insurgency' of July 1937. While this was primarily a rebellion by black workers, Barbadian historians acknowledge that, in the words of one of them, 'poor white workers were often paid just as badly and worked under conditions just as harsh as blacks'.[42] Poverty, not race, was the prime motivator behind the disturbances, and left-wing agitators ruthlessly exploited class tensions between working-class blacks and Barbados' middle-class mixed-race coloureds for their own political ends.

Before 1937 there were no comprehensive labour unions or mass political parties in Barbados, and few in the Caribbean as a whole. When on 26 March 1937 the Trinidadian left-wing activist Clement Payne arrived in Bridgetown to demand the immediate radicalisation of Barbadian politics, much of what he said fell on willing ears. Payne was a prominent member of Trinidad's civic rights organisation, the Negro Welfare Cultural and Social Association, and was a talented orator. In no fewer than seventeen political meetings, he attacked the coloureds for being 'lackeys of the employers' and described the police as 'dogs of the capitalists'. It was estimated that by the time he was arrested on 22 July 1937, for having made a false declaration of his place of birth on an official document, Payne had garnered around 5,000 supporters.

Even though Payne was released on payment of a £10 fine, protest rallies were held in Golden Square and Lower Green in Bridgetown, which led to a march on Government House to petition the Governor, Sir Mark Young. Payne was soon re-arrested, but won his appeal largely due to his defence by Grantley Adams, an Oxford-educated barrister and Assembly member since 1934. Despite this, Payne was ordered to be deported on 26 July. That night, the (untrue) rumour got about that the police had killed Payne's girlfriend's child, and two days of serious rioting broke out in Bridgetown. 'Shop windows were smashed,' records the historian W. Marvin Will, 'albeit somewhat selectively, with those of merchants known to be specially favourable to the working class being spared.[43] Cars, both on the streets and in showrooms, were overturned or pushed into the sea.

The rioting then spread over much of the island until 30 July, accompanied by mass looting, widespread destruction of property and violence. In all, there were fourteen deaths, fifty woundings and 500 arrests. Payne's lieutenants – including Fitzgerald 'Menzies' Chase, Mortimer 'Mortie' Skeete, Darnley 'Brain' Alleyne and Ulric Grant – were detained in Glendairy Gaol, where they were reportedly subjected to severe police brutality. As in the Mau Mau insurrection of the 1950s, this was largely carried out by black police loyal to the British, rather than by Britons themselves.

Barbados was until 1937 usually considered fairly conservative because of her larger-than-average white-to-black ratios, and because she had strong tourism and commercial and manufacturing sectors of her economy, as opposed to simply sugar and agriculture. Yet that summer the island saw almost one-third of the total riot-related fatalities suffered by the British West Indies. Low wages, a bad drought, collapsing sugar prices – twenty-six shillings per hundredweight in 1923 had fallen down to a five shillings average by 1934–7 – led to extreme distress.

Although Barbados had a twenty-four-member House of Assembly, it was dominated by the white plantation and commercial elites, since the franchise was restricted to taxpaying male property owners who lived in houses valued at £50 or rented one for £15. Non-property holders needed to earn £50 per annum, which few labouring blacks did, so that registered voters made up a mere 3.3% of the Barbadian population in January 1938. In eight Barbadian parishes accounting for sixteen members of the House, there were fewer than 300 voters per parish, out of a total electorate of 6,359.

It was partly in order to raise wages above the income qualification that the

Barbados Progressive League (BPL) was founded in a series of meetings over the fortnight following 31 March 1938, the first being held in the Upper Bay Street home of J.A. Martineau, a prominent labour leader. The title 'party' was deliberately not chosen in the hope that it 'might sound less threatening in conservative Barbados'.[44] C.A. Braithwaite was elected president and Grantley Adams vice-president, and a movement was born which within a year had won between 20,000 and 30,000 adherents. Much of this can be put down to the recruitment skills of Herbert Seale, who served as General Secretary, and to the support of the *Barbados Observer*, published by W.A. Crawford.

Just as the Canadian socialists had found in Ontario after 1932, severe political and personality clashes emerged within a very short period of time. As W. Marvin Will records,

> Major leadership fissures occurred during the League's first year, divisions that involved the organisation's stalwarts: Adams, Seale, Braithwaite, Martineau, and Crawford. The genesis of these conflicts involved strategy and tactics, ideology, and significant personality differences. There is an old West Indian adage that three rodents can't live in a hole. This seems to have applied to these five political animals as well.[45]

These internecine troubles were only resolved with the victory of Adams, one of the relatively few indisputably great leaders to emerge from West Indian politics in this period. Adams, who became President General of the BPL in 1939, was a gradualist who had a natural feel for politics, financial rectitude, a powerful rhetorical technique and genuine charisma.

Adams pressed for the legalisation of trade unions, which came about in 1940 and led to the founding of the Barbados Workers' Union (BWU) in 1941. Although the BWU was also led by Adams, it had established a separate identity from the BPL by 1949. This important differentiation between polit-ical party, trade union and island government was achieved by Adams in a way it never really was on Antigua by Vere Bird or on Grenada by Eric Gairy. By the time Adams left Barbados to assume the premiership of the West Indies Federation in 1958, he had proved that even in the most financially deprived area of the English-speaking world, gradual political and social reforms ach-ieved through constitutional means were in every way superior to the kind of violent revolution preached by the likes of Clement Payne.

Whereas Britons only owned 3,000 domestic refrigerators in 1937, there were over two million such appliances in the United States, a country that was also producing 1.27 billion barrels of oil per annum, eight times more than its nearest rival, the USSR. Of that incredible figure, 517 million barrels were consumed as motor fuel in its nineteen million private cars, more than Britain's 1.7 million, the Third Reich's 1.3 million, Italy's 1.1 million and France's 800,000 all put together and then trebled.[46] Coming out of the Great Depression, the United States had shown that capitalism has an inherent power of resurgence and renewal that should never have been underestimated by its opponents.

Of course no-one knew the inter-war years were that at the time. For all that the Thirties were later characterised as 'the Devil's Decade' and 'the Locust Years', in which the various Western governments sleep-walked towards war, that did not seem the inevitable outcome. So glowering and portentous is the date of 3 September 1939 to us today that it is almost impossible to remember that for people living in August 1939 it meant little more than a hopefully sunny weekend.

The vast majority of people chose not to listen to Winston Churchill's jeremiads against Nazism and his Cassandra-like warnings of the dangers of Hitler, but trusted instead to the Respectable Tendency of British politics to see them through future crises. A glance at Churchill's pre-1939 career explains why the English-speaking peoples seemed justified in supposing that his warnings about Germany sounded suspiciously like the boy who cried 'Wolf'. For this was the man whose judgment had been questioned when he had visited the scene of the Sidney Street siege, who devised the Gallipoli landings, promoted the ill-fated intervention in the Russian Civil War, was thought to have overreacted during the General Strike, who rejoined the gold standard at the wrong time and rate, opposed Indian self-government and supported the King over the Abdication crisis. Perhaps, therefore, Churchill's warnings about German rearmament were simply further evidence of his lack of judgment, allied to an unpleasant stench of warmongering?

(At the time of Churchill's campaign against Indian self-government, Mrs Ogden Reid of the *New York Herald Tribune*, who was placed next to Churchill at a White House dinner, asked him: 'What do you intend to do about the poor Indians?' According to a story told years later by Lord Mountbatten, who admittedly was not an altogether trustworthy source in his anecdotage, Churchill replied: 'Madam, to which Indians do you refer? Do you refer to the brown Indians of the Asian subcontinent, who under benign and beneficent British influence have multiplied alarmingly? Or do you refer to the red Indians of this continent, who under the current Administration are almost extinct?')[47]

In May 1937, the crown prince of the Respectable Tendency, Neville Chamberlain, took over the premiership from Stanley Baldwin. Chamberlain was a man of culture and honour; he founded the Birmingham Symphony Orchestra, read English literature widely, was devoted to his children and knew the names of several hundred birds and plants. He was also a forceful politician and had been a highly effective Lord Mayor of Birmingham, Minister of Health and Chancellor of the Exchequer. However, as the historian Robert

Blake caustically put it, 'When national security is at stake, one does not judge a statesman by his successes in slum clearance.'

Yet the fact remains that to those who had lost family members in the Great War – which comprised a vast cross-section of the English-speaking world – the prospect of a European conflict breaking out all over again was simply an unthinkable obscenity, and Chamberlain expressed that better than any other politician of the day. By the time Chamberlain moved from Number 11 Downing Street next door to Number 10, the pass had already been sold fourteen months before over the Rhineland. After Hitler's bold, bloodless coup there, the West had to face three full years of bluster, bullying and brilliant brinksmanship before Chamberlain laid down the tripwire of Britain's guarantee to Poland in April 1939.

Part of the problem in dispassionately appraising the policy of appease-ment – the word originally had Christian connotations – lies in the character of Neville Chamberlain himself. For all his honourable intentions, he was a vain man, the evidence for which comes straight from his own pen, and he was nothing if not a diligent correspondent. Every week for over thirty years he wrote to his sisters Ida and Hilda letters that were in effect a diary of all he was doing politically. A sign of how remote an individual he was, or perhaps of how formal was the age in which he lived, is the way that he concluded these letters with the words: 'Your affectionate brother, Neville Chamberlain.'

Through the forest of nicknames – Curzon was 'The All-Highest', Lloyd George was 'The Goat', and so on – it is possible to perceive a personality that by the 1930s had more than the normal politician's surfeit of self-regard. Chamberlain was ambitious and keen to replace Stanley Baldwin as premier long before May 1937, by which time he was sixty-eight years old; he was comfortable with power ('As chancellor of the exchequer I could hardly move a pebble, now I only have to raise a finger and the whole face of Europe is changed!'), and boundlessly self-confident; he even referred to 'the Cham-berlain touch' in a conscious reference to the Nelson Touch. He could occa-sionally be caustic, calling the Liberal MP Clement Davies 'that treacherous Welshman' and Wallis Simpson 'a thoroughly selfish and heartless adven-turess'.

Although Chamberlain started off with a credulous attitude towards Hitler when he flew to Hitler's Bavarian mountain retreat of Berchtesgaden to negotiate over the Sudeten-German areas of Czechoslovakia in September 1938 – writing, 'I got the impression that here was a man who could be relied upon when he had given his word' – he always disliked the Führer personally, warning the Cabinet after his second meeting at the spa town of Bad Godes-berg that Hitler 'had a narrow mind and was violently prejudiced on certain subjects'.[48]

Chamberlain was casually anti-American in the manner of a large number of Britons of his background at the time; he said of Roosevelt's special envoy Sumner Welles in March 1940 that he was 'the best type of American I have met in a long time', which was meant as a great compliment. Yet that did at least allow him to maintain the position in 1937 that it was 'always safest to count on nothing from the Americans except words', tragically the only sensible policy for Britain to follow for the next three years.[49] Of course he also distrusted Russia, writing in March 1939 that, 'I have no belief whatever in her ability to maintain an effective offensive even if she wanted to,' and to a lesser extent France also, making any kind of effective encirclement policy against Hitler next to impossible.

Yet if he had become prime minister in the 1930s, or even foreign secretary with a suitably supportive prime minister, might Winston Churchill have been able to prevent the Second World War from breaking out? Many Churchillians have argued that he would have created a 'Grand Alliance' that could have deterred Nazi aggression by guaranteeing a war against Germany on two fronts.[50] They believe he would have persuaded France, Poland, Romania, Yugoslavia, Czechoslovakia and Russia – or if not all at least a significant number of them – to act with Britain in the interests of collective security to build a coalition which would not only have left Hitler impotent but also highly liable to being overthrown by his own generals if he attempted to disturb the European status quo.

Sadly this beguiling thesis falls down in a number of crucial areas. The first involves the suspicion that Stalin felt about Churchill's motives. This lasted throughout the Thirties and indeed right up until the moment that Hitler's Operation Barbarossa offensive against Russia in June 1941 forced Britain and Russia into the same camp. Churchill's attempt to 'strangle Bolshevism in its cradle' in 1918 and 1919, and his subsequent virulent anti-communism, meant that Stalin was very unlikely indeed to trust him.

Furthermore, Russia could not threaten Germany except by marching through Poland, a fiercely independent nation state which hated and feared the Red Army just as much as it did the Wehrmacht. The Poles made it very clear that they would never allow Russian troops – who had a habit of overstaying their welcome – on to their sovereign territory. They had fought the Russians to a standstill in 1920 with huge losses and, as the Katyn massacre of 1940 was soon to prove, they were right to suspect Stalin's motives. Poland also coveted the Teschen region of Czechoslovakia, which she picked up during the dismemberment of that unfortunate state after the 1938 Munich agreement. Romania similarly thought in the short-term, and also wished to profit territorially from the Czech tragedy rather than involve herself in an (necessarily very risky) anti-German coalition.

Just as the 'Grand Alliance' idea was never truly a possibility through its internal contradictions, nor was it conceivable that Churchill would ever hold high office in any ministry except as a result of a crisis such as that of September 1939. He had resigned from the Tory shadow cabinet in January 1931 over the issue of Indian self-government and had then fought a bitter four-year struggle against his own Party's leadership, during which he had effectively accused it of lying over evidence to an official commission on the issue. This had led to him being demonised by the Parliamentary Conservative Party, to the extent that during the Abdication crisis he was even shouted down in the House of Commons.

Although Churchill had held every major office of state except foreign secretary and prime minister by 1931, during the rest of the Thirties there was no real prospect of his holding any important political position in peacetime, especially after the National Government won a landslide victory in the 1935 general election. It took a world war to bring Churchill out of the wilderness, and by then the decade only had another four months to run. His 'Grand Alliance' idea therefore had insurmountable political, geographical and military problems that doomed it, at least until Hitler himself brought it into being by declaring war against Russia and America in 1941.

It is sometimes forgotten that there were deep familial as well as ideological and personality differences between Churchill and Chamberlain, ones that went back to when both men had been teenagers. Both Churchill's father, Lord Randolph, and Chamberlain's father, Joseph, had occupied the centre-left ground of British politics back in the 1880s, and either – or even both – could easily have become prime minister. The fact that neither did was largely up to the other. They worked in tandem as personal friends until Christmas 1886, when Lord Randolph sensationally resigned from Lord Salisbury's Unionist ministry. Had Chamberlain supported Churchill, the latter believed, they could have created a centre-left government between them, but instead Chamberlain stayed loyal to Salisbury, and the scene was set for a half-century struggle between the clans.

Winston Churchill's revenge on behalf of his father came in 1904 just as Chamberlain had almost grasped control of the Unionist Party in favour of Imperial Protection, but Churchill crossed the floor of the House of Commons in support of Free Trade, leading in part to the Unionists' disastrous defeat in the 1906 general election and Joseph Chamberlain's political eclipse, soon after which he suffered a stroke that removed him from politics. Both Neville Chamberlain and Winston Churchill believed his own father had been robbed of a premiership that was rightfully his, and they were both intensely filial.

By the time that Churchill was in the Wilderness during the Thirties, there were myriad reasons why Chamberlain did not want his potential rival for the Tory leadership to re-enter the Cabinet. 'Chamberlain is the shaded background against which Churchill dazzles,' as the historian of their relationship has written, and both men knew it.[51] They had taken radically different stances over Indian self-government, the Abdication crisis, compulsory military conscription, a Russian alliance and Czechoslovakia. Even in the summer of 1939, when Chamberlain could have sent an unmistakable message of defiance to Hitler by bringing Churchill back into the Cabinet as Minister of Defence Co-ordination, he avoided doing so, seeing it as tantamount to an admission of failure.

Only when the appeasement policy did finally collapse, and war was declared on 3 September 1939, was Churchill appointed to the post of First Lord of the Admiralty. Then a fascinating and hugely uplifting event occurred; for, as everyone watched for the old feud to flare up in Cabinet, the two men publicly and privately worked together for the benefit of their embattled nation. Churchill served Chamberlain with commendable loyalty until he replaced him on Friday, 10 May 1940, and thereafter Chamberlain served him with equal steadfastness until his death from inoperable cancer that November. In the national emergency the aristocratic ancestral voices of Blenheim no longer preached dynastic vengeance against the respectable screw-makers of Birmingham, or vice-versa.

Chamberlain was not the naïf that some of his detractors have painted him. Writing a public letter to John Busby, the National Government's candidate in the West Fulham by-election, on 28 March 1938, he made it clear that 'the preservation of the peace of the world must be largely dependent on the strength of our own country. If we are to avert the perils of war the defence programme on which the national Government embarked three years ago must be accelerated even though it may involve sacrifices. But let us always remember that the sacrifices of peace are far less terrible than those of war.' He accepted that the League of Nations 'may some day be the salvation of the world. But we should not be promoting the cause of peace by pretending that the League can in its present weakened condition guarantee collective security.'[52] This was not the stance of the weak appeaser of the kind that many have sought to characterise. His attitude towards the League was refreshingly frank; in the month that war broke out in Europe, the League of Nations in Geneva was busily discussing the standardisation of European railway gauges.

For socialists, there was little hope that the British Labour Party would take the necessary practical steps to stop Hitler. Although Labour was not enamoured of communism, neither was it of much use in stiffening the nation's resolve for the coming clash with the dictators. Having replaced the pacifist George Lansbury as Party leader in 1935, Clement Attlee was writing to his brother Tom in late April 1938 still criticising the Chamberlain Government for being too bellicose, complaining,

The Government is leading the country into war. ... The conversations with France that are now on are directed to the same end. We are really back in 1914. The Government will, I think, continue to allow all the smaller democratic states to be swallowed up by Germany not from a pacifist aversion to war, but because they want time to develop armaments. There is really no peace policy at all. Chamberlain is just an imperialist of the old school but without much knowledge of foreign affairs or appreciation of the forces at work. It is a pretty gloomy outlook.[53]

He underlined this myopic assessment the following February, when he wrote about how 'Neville annoys me by mouthing the arguments of complete pacifism while piling up armaments.'[54] In fact, the policy of buying 'time to develop armaments' was one of the primary reasons that Britain had enough Hurricanes and Spitfires to win the battle of Britain in 1940. That victory should be ascribed to Chamberlain quite as much as to Air Chief Marshal Sir Hugh Dowding, and far more than to Churchill who only became prime minister long after the vast majority of planes had already been produced. The title of Tory MP Quintin Hogg's 1945 book about the Labour Party's foreign policy in the Thirties, *The Left Was Never Right*, can hardly be bettered as an accurate analysis of the effects of its opposition to rearmament, conscription and all the other measures that might have made the fascists take note.

Simply to enumerate the Jewish émigrés who left Germany, Austria and Poland in the 1930s, and who settled in the English-speaking world, is to read 'a more or less endless list of eminence'. They include Geoffrey Elton, Nikolaus Pevsner, George Weidenfeld, Ernst Gombrich, Carl Ebert, Rudolf Bing, Georg Solti, André Deutsch, Claus Moser, Joseph Horowitz, Albert Einstein, Arthur Koestler, Arnold Schönberg, Stefan Zweig, Leo Strauss, Otto Klemperer, Henry Kissinger, Thomas Mann, Isaac Deutscher, Sigmund Freud, Karl Popper, Sigmund Warburg, Max Reinhardt, Emeric Pressburger, Lucian Freud, Elias Canetti, Egon Wellesz, Moses Finlay and any number of others.[55]

That list, shorn as it is of the various peerages, knighthoods, professorships, Orders of Merit and Companionships of Honour for ease of reading, prompts two thoughts: what would post-war British and American cultural and scientific life have been like without these paladins, and what did Central Europe lose because they fled the Continent? Furthermore, if such a tiny exodus was able to contribute so much, what might have been achieved for Civilisation by those millions who instead perished in the Holocaust?

The great talent of the scientists who fled Germany, Austria and Hungary led to 'the enfeeblement of European science and the enormous advantage of Britain and the United States', especially when so many chose not to return to Europe after the war was won.[56] As has been pointed out, their 'international impact was multiplied because so many of the émigrés were teachers who in their turn inspired several generations of younger colleagues and students'. Physicists, biologists, chemists and medical researchers, brilliant scientists such as Hans Bethe, Max Perutz and Ernst Chain, came over and stayed.

These Jewish scientists were present in numbers far out of proportion to the percentage of Jews in the general population. They were 'a prosperous, self-confident rigorously educated elite, sharing similar intellectual and cultural values, who were simultaneously representatives, connoisseurs and patrons of all that was most admirable in German culture'.[57] They were able to take full advantage of the open, democratic academic world they found in the English-speaking world, and often advanced more rapidly than they could have within their own more rigid hierarchies, to everyone's mutual benefit. The warm welcome they found amongst the intellectual elite in the West, and principally in British and American academia, served their new hosts well, especially in the field of atomic research. Had Hitler not persecuted the Jews, forcing German-Jewish nuclear scientists to flee Germany, the Nazis might have produced an atomic bomb. (But then, he would not have been Hitler.)

English-speaking society was chronically infected with the bacillus of anti-Semitism on both sides of the Atlantic, albeit not of the exterminationist kind found in Germany, on the extreme Right in France and in much of the rest of continental Europe. In 1935, the American society hostess Emerald Cunard boasted to the Liberal MP Robert Bernays how the previous night at her dinner party for the Prince of Wales, she had 'suddenly exclaimed that she hated all Jews and immediately the party came to life. "Not that I really hate them," she added, "but I wanted my party to be a success".'[58] Even Lady Violet Bonham Carter, a lifelong Liberal, reacted to her best friend announcing her conversion to Judaism with the retort that to 'become a *Jew* seems to me the most impossibly squalid, cynical antic'. When the American Lady Astor invited Chaim Weizmann to a dinner during the Great War, she embarrassed her other guests by introducing him as 'the only decent Jew I have ever met'.

For all the influx of individual geniuses into the United States, the Thirties were overall a hard time for American Jewry. As Jeffrey Gurock of Yeshiva University has pointed out, 'the doors of America were closed to new arrivals after a century of uninterrupted Jewish migration, and native-born Jews were staying away from synagogues in droves'.[59] Their response was to concentrate on areas of Jewish life, such as day-school education, camping, temple youth work and Zionism, that helped lead to a recrudescence of American Judaism after the Second World War.[60] The doors were not slammed wholly shut, however. Immigration to the US from Europe between 1933 and 1944 totalled 365,955. Before, during and after the war, 132,000 people arrived from Austria and Germany. It is fair to say, however, that the President's 'own attitude to the Jewish catastrophe was at the least negligent or thick-skinned'.[61]

The benefits gained from the influx of Jews into the English-speaking world before the Second World War can be illustrated by the statistics for Nobel Prize winners in the half-century between 1951 to 2000. These show that 32% of Nobel Prizes for Medicine went to Jews, 32% of the Prizes for Physics, 39% of the Prizes for Economics and 29% of all Science Prizes to people from a race that makes up fewer than half of 1% of the world's population.[62] For Central Europe to have lost so many people from that race, and the English-speaking peoples to have gained them, was due to the superiority of democracy over dictatorship, and of religious toleration over persecution.

In a sense, although the governments of the English-speaking peoples produced the finances and facilities for building the atomic bomb – especially at Los Alamos in New Mexico – it was the combined genius of the Jewish Albert Einstein, J. Robert Oppenheimer and Edward Teller, along with the New Zealander Sir Ernest Rutherford, that brought into being the technology that finally ended the Second World War. Hundreds of thousands of Allied soldiers in the Pacific theatre in 1945 owed their lives to the Jewish scientists who had fled Nazism in the Thirties.

In August 1938, President Roosevelt used the occasion of the opening of the new Thousand Islands International Bridge, linking the United States to Canada, to deliver a warning to Hitler. The five bridges and their connecting highways traversing four of the 1,000 islands, bringing no fewer than 200 of them into view, provided a perfect opportunity to dilate upon US-Canadian amity. 'The people of the United States will not stand idly by if domination of Canadian soil is threatened by any other empire,' he said, clearly of the Reich that was seeking the domination of Europe.

The Sudeten crisis came to a head in September 1938, when Hitler supported the right of about three million ethnic Germans in the Sudetenland region of Czechoslovakia to secede from that state and join his Reich. Although Britain had no treaty obligations to protect Czechoslovakian integrity, as she did over Belgium in 1914, Germany's behaviour was rightly seen as dangerously provocative and liable to disturb world peace. In one area the British Government was severely hamstrung at the time of the crisis; had it forced the issue to the extent of going to war with Germany, some of the rest of the Dominions might not have gone along, particularly Canada and South Africa.

In August 1941, when the Canadian Premier William Mackenzie King visited Churchill in Downing Street, they discussed 'the origins of the war and how it might have been prevented'. Mackenzie King had earlier in the day attended a War Cabinet meeting and had been shown around Number 10's concrete-reinforced basement; and after lunch the two men discussed

Munich and Chamberlain's legacy, at a time when Chamberlain's reputation was perhaps at its lowest-ever ebb. Nonetheless, Mackenzie King pointed out to Churchill how he thought

> full justice was not being done to Chamberlain. . . . If he had not gone to Munich, the situation might have been much worse. That certainly was so as far as Canada was concerned; we would never have been able to go to war as a united country. I said I had gone through my Gethsemane knowing that the country ought to go to war and intending to make that my policy, but that I would have lost a good proportion of the Cabinet which would have been divided.

Churchill then asked whether Mackenzie King 'would have lost outright', to which the Canadian Premier replied that he would have, 'but that, of course, I could have made up the majority from the opposite side, but it would have been a divided Canada. That, as a result of Chamberlain's visit and the deferring of war for a year, the nation had got a chance to really see the issue and to become convinced that aggression was the aim.'[63]

Although Canada was not prepared to go to war to prevent this happening, throughout the crisis New Zealand was stalwart. As the National Party leader, Adam Hamilton, told the House of Representatives in Wellington as Chamberlain flew to Berchtesgaden to meet Hitler,

> In the event of Britain being involved, we as an Opposition offer our full co-operation to the Government in seeing that New Zealand does its full duty, and takes its full share of responsibility as a unit of the Empire. We all hope that wiser counsel prevails, and that the war clouds will pass away; but if there should be trouble the Government may rely on loyal support from the Opposition.[64]

On Chamberlain's return from Berchtesgaden, there were two Cabinet meetings held in Downing Street at 11 a.m. and at 3 p.m. on 17 September 1938. At the first, Chamberlain told his colleagues of his meeting with Hitler the previous day. He was satisfied with his agreement to self-determination for the Sudeten German minority, and Hitler's promise not to use force when the majority-German areas seceded from Czechoslovakia and joined the Reich. Yet Chamberlain seemed under no illusions about the personality of the Führer. As the First Lord of the Admiralty, Alfred Duff Cooper, recorded afterwards in his diary, the Prime Minister told the Cabinet 'that at first sight Hitler struck him as "the commonest looking little dog" he had ever seen without one sign of distinction'.[65] For all that Chamberlain was politically taken in by Hitler, he was under few illusions about his opponent's defective personality.

At the second meeting that day, the Lord Chancellor Lord Maugham, who was Duff Cooper's greatest *bête noire* in the Cabinet, argued that, 'according to the principles of Canning and Disraeli, Great Britain should never intervene unless her own interests are directly affected, and unless she could do so with overwhelming force'. This demonstrated a deplorable grasp of understanding of the foreign policy of Canning, who liked nothing better than to intervene in the republics of South America, where British interests were entirely unaffected and where she could not use overwhelming force either.

In answer to Maugham, Duff Cooper argued 'that the main interest of this country had always been to prevent any one power from gaining undue predominance in Europe. That we were now faced with probably the most formidable power that had ever dominated Europe and resistance to that power was quite obviously a British interest.' As the author of a distinguished biography of Talleyrand, Napoleon's foreign minister, which had by 1938 been translated into eight languages, Duff Cooper's views should have been listened to with respect. His argument was that 'if we held to the Lord Chancellor's doctrine of defeatism it meant that we could never intervene again. ... The next act of aggression might be one that would be far harder for us to resist. Supposing it were an attack on one of our colonies. We shouldn't have a friend in Europe to assist us, nor even the sympathy of the United States which was ours today.'[66]

While it might seem surprising, and even rather impressive, that in the midst of a great international crisis the British Cabinet had time to consider the foreign policy of Canning and Disraeli and to analyse the balance-of-power theory since the Armada, several of the themes Maugham and Duff Cooper mentioned on 17 September 1938 have reverberated through the history of the English-speaking peoples since then, and right up to the invasion of Iraq in 2003. They can be separated into the distinct yet overlapping foreign policy strands that stress the importance of isolationism, prestige, the thin-end-of-the-wedge argument, the domino theory and the importance of coalitions.

The problem was not that Chamberlain went off to Munich quoting *Henry IV Part I* – 'out of this nettle, danger, we pluck this flower, safety' – but that he was quite so triumphant when he returned. He and the French Premier, Édouard Daladier, had essentially agreed to Hitler's demand that the Sudetenland should revert almost immediately to the German Reich; the Czechs were not even present at the conference. On his return, Chamberlain quoted from the meaningless document that he and Hitler had signed about 'the desire of our two peoples never to go to war with one another again' and then waved it in the light breeze at Heston aerodrome, before predicting 'peace for our time' from the window of 10 Downing Street. He even went up on the balcony of Buckingham Palace to acknowledge the cheering crowds.

Those whom the gods have marked for destruction they first make hubristic, yet the King and Queen also deserve reproof for inviting the Prime Minister onto the balcony considering that their constitutional duty was to be strictly politically impartial. The Munich agreement was still subject to a vote in Parliament, and Labour and the Liberals voted against it.[67] Newsreel footage

of the crowds outside Buckingham Palace illustrates how popular the agreement was with ordinary Britons, but part of the duty of the monarchy is to
rise above the passions of the hour and observe the proprieties of the overall
situation. (Broadcasters felt no similar obligation: 'Mr Chamberlain stands as
the saviour of peace!' said British Movietone News, recording how the crowds
at the Palace were 'reminiscent of a jubilee or coronation'.)

Churchill was routinely denigrated as a 'fire-eater and militarist', a 'rogue
elephant', or – by the *Daily Express* in October 1938 – as 'a man whose mind
is soaked in the conquests of [the 1st Duke of] Marlborough'. When Stalin
asked Lady Astor about Churchill's future in the Thirties, she confidently
pronounced, 'Oh, he's finished.' Yet Churchill's words in the debate on Munich
should have brought them all to their senses. 'This is only the beginning of
the reckoning,' he told the Commons. 'This is only the first sip, the first
foretaste of a bitter cup which will be proffered to us year by year unless, by a
supreme recovery of moral health and martial vigour, we rise again and take
our stand for freedom as in olden times.'

Disgracefully, the Lord Chancellor Lord Maugham added insult to Czecho-
slovakia's injury on 3 October by arguing that, 'In plain words ... we and
France have been engaged in saving from destruction a State which ought
never to have been created at all.' The same day in the Commons, the Home
Secretary Sir Samuel Hoare had also said, 'I agree that the President of the
Czechoslovak Government was placed in a very difficult position, but I cannot
help saying that, if he had acted more quickly ...', before being cut off by Mr
Wedgwood Benn crying, 'Monstrous.' Chamberlain defended Munich, saying
that the Great Powers 'have averted a catastrophe which would have ended
civilisation as we have known it', but that, 'We must feel profound sympathy
for a small and gallant nation in the hour of their national grief and loss.'

It is clear from the extensive files of deciphered cablegrams from the
Dominions Secretary in London to Robert Menzies, the Prime Minister of
Australia, that Canberra was being kept in close and constant touch with the
Munich crisis as it developed hour by hour.[68] However, this was largely because
Chamberlain knew that the Dominions supported his policy wholeheartedly.
In his report of the Munich debate, the Australian external liaison officer in
London, Alfred Stirling, told Menzies' office that when Chamberlain had said
during his speech 'We no longer think of war as it was in the time of Marl-
borough or the days of Napoleon, or even in the days of 1914', it had been a
conscious 'reference to Mr Churchill, and not inappropriate'.[69]

In a letter of 12 October, Stirling wrote, 'Mr Churchill misfired, but much
more will be heard of him, for his pen is as vigorous and able as his tongue
...' Stirling found Churchill's policy 'hopelessly defeatist'. Friendship with
Russia he thought 'at least as difficult as rapprochement with Germany', and
the state of the Red Army, 'in view of the wholesale liquidation of senior

officers, is also problematical'. Stirling concluded his report by saying that 'Mr Churchill is, if one may use the phrase, ancestor-ridden, and too addicted to history in general.'

The best satirising of the dictators of the Thirties was done by the English-speaking peoples. The New Zealander cartoonist David Low, drawing for London's *Evening Standard* from 1927 onwards, and the London-born Charlie Chaplin, who wrote, directed and starred in *The Great Dictator* in 1940, were sublime, but so too was the 1939 Hollywood classic *The Wizard of Oz*, in which the sound, fire and fury of the great Wizard turns out to have been mere pulling of levers by a small man behind the scenes.

In *The Code of the Woosters* (1938), P.G. Wodehouse produced a superb skit ridiculing the blackshirt leader Sir Oswald Mosley. In it, the hero Bertie Wooster tells Roderick Spode, the leader of 'Saviours of Britain, a Fascist organisation better known as the Black Shorts':

> It is about time that some public-spirited person came along and told you where to get off. The trouble with you, Spode, is that just because you have succeeded in inducing a handful of half-wits to disfigure the London scene by going about in black shorts, you think you're someone. You hear them shouting 'Heil, Spode!' and you imagine it is the Voice of the People. That is where you make your bloomer. What the Voice of the People is saying is: 'Look at that frightful ass Spode swanking about in footer bags! Did you ever in your puff see such a frightful perisher?'[70]

(Spode is finally unmasked as a designer of women's underclothing and the 'proprietor of the Bond Street emporium known as Eulalie Soeurs'.)

Wodehouse was a political naïf, as his wartime broadcasts from Germany amply demonstrated, but in Bertie's denunciation of Spode he displayed acute insight. The wearing of political uniform was considered un-British and, as a recent biography of Mosley attests, 'There was a widening contradiction between the British Union of Fascists' proudly proclaimed patriotism and its mimicking of foreign forms. Like the British Communist Party, it was widely perceived to be an alien implant in the British body politic.'[71] The fighting that broke out at the Olympia stadium in Kensington in June 1934, where hundreds of communists disrupted a Mosley rally of 2,000 Blackshirts among a 12,000-strong audience, 'profoundly shocked the British public. Forgetting that political violence had been commonplace in the eighteenth and nineteenth centuries, many felt that the scenes in Olympia had been "un-British".'[72]

As well as street violence, Mosley was opposed by reasoned argument and facts. In April 1935, Nathan Laski, speaking for the Jews of Manchester, devastated Mosley's central argument about how international Jewry based in London was destroying the Lancashire clothing trade. Laski pointed to the contribution made by Jews to the struggle to keep Lancashire mills working before the Great War and during the Depression against Japanese competition. He spoke of the way that Lancashire Jews sought out new markets in South America, while developing those in the East where they had connections in the markets of Egypt, Turkey, Baghdad, Beirut, Aleppo and Eastern Europe. 'Not a single Jew is on the directorate of the Bank of England and there are only three Jews among all the directors of the [Big Five] banks in London and only six among the directors of all the insurance companies,' he reported, and thus since Treasury permission needed to be sought before any substantial loans could be floated in London, international Jewry hardly dominated British finance.[73]

Membership of the BUF declined dramatically after the Olympia riots, and the numbers on its payroll collapsed from 350 in 1936 to fewer than fifty by 1939. Part of the contempt Bertie Wooster was made to voice about the un-British wearing of political uniforms stemmed from an underlying sense of disdain felt by many British people for the uniform-wearing Italian fascists in the Twenties and Thirties. Italy was despised because she declared herself neutral in August 1914 and then opportunistically declared war against Austria-Hungary, but not Germany, in 1915. During the Second World War, Churchill was greeted with gales of laughter in the House of Commons when he remarked of the famously unimpressive Italian Navy, 'There is a general curiosity in the British Fleet to find out whether the Italians are up to the level they were in the last war, or whether they have fallen off at all?'

The largest membership that the British Communist Party ever enjoyed between the wars was a mere 17,000, although its Moscow funding and high subscription rates meant that it was well off. Its major appeal, like its American counterpart, was not so much anti-capitalism as anti-fascism, but, unlike the situation in the US, British Jews faced a dangerous indigenous enemy in Mosley's violent British Union of Fascists.[74]

While communism made some very limited political headway in Britain – there were five communist MPs elected to the House of Commons between 1924 and 1945 – fascism made none. No-one was elected to the US House of Representatives on an overt communist ticket, although Victor Berger from Milwaukee (1911–13, 1919, 1923–9), Meyer London from Manhattan's Lower East Side (1915–19, 1921–3), Vito Marcantonio from East Harlem (1939–51) and Leo Isacson from the South Bronx (1948–9) were Marxists to all intents and purposes.[75] Occasionally parts of the fascist social programme would be adopted by politicians such as Huey Long, the Governor of Louisiana elected in 1928, whose public works programme was similar to Mussolini's, but overall fascism excited few Americans beyond the Ku Klux Klan.

Communism, which admitted no borders and required the unanimity of the globe before its end of a classless society could be achieved, similarly found little echo among the British working class. In some geographical areas – South Wales, the North-East, parts of London's East End – and in some trade unions it found short-term favour, but the British Labour Party successfully absorbed, adopted and adapted anti-capitalist feeling in a constitutionalist, non-revolutionary manner. Of the 571 candidates that the Communist Party of Great Britain put up in parliamentary elections between 1922 and 1992, 534 lost their deposits.[76] Across the English-speaking peoples, both fascism and communism were considered essentially undemocratic, profoundly obnoxious and foreign creeds.

The seemingly illogical, indeed craven way that the British and American Communist Parties supported the Nazi-Soviet pact of 24 August 1939, when the Soviet Foreign Minister signed a non-aggression pact with the German Foreign Minister Joachim von Ribbentrop, further tended to undermine the communists' credibility as a principled counterweight to fascism, and simultaneously underlined the critique of the Soviets as simply another form of fascists, only this time of the Left. Slavishly to adopt the line from Moscow, as the Communist Party secretary Harry Pollitt and most other British communist leaders did, made them seem quite as un-British as the BUF. As Lawrence Welch correctly observed in his 1924 book, 'The English Fascist movement is commonly regarded as an object of laughter, and as far as its pretensions to the scope of the Italian movement are concerned, correctly so.'[77] This wasn't to change once Oswald Mosley set up his New Party in February 1931.

In the 1931 election, the New Party contested twenty-four constituencies, but lost in all of them; indeed, apart from Mosley himself, their candidates all lost their deposits. They chose not to fight the 1935 general election, but put up candidates in a number of by-elections in 1939 and 1940, which they also lost, before the organisation was proscribed on 30 May 1940. The only MP who had to be interned under the emergency legislation Regulation 18(B) was the anti-Semitic Conservative MP Archibald Maule Ramsay. Of course there were no general elections held between 1935 and 1945, so it is impossible to say how many Britons would have voted for the BUF, but it is safe to say that the failure of both communism and fascism at the polls was a tribute to the common sense of the English-speaking peoples during that hugely fraught period.

The vast majority of the English-speaking peoples welcomed the Munich agreement wholeheartedly. The Dominion premiers cabled their congratulations. In Geneva, Eamon de Valera's response to Chamberlain's decision to visit Berchtesgaden had been: 'This is the greatest thing that has ever been done.'[78] Similarly, President Roosevelt cabled Chamberlain with the (perhaps somewhat patronising-sounding) two words: 'Good man.' Nevertheless, for all the rejoicing, there were many who recognised, in the words of *The Press* newspaper of Christchurch, New Zealand, on 1 October, that despite the talk of 'national self-determination' current at the time,

> When the representatives of four Great Powers, by no title other than that conferred by their strength, redraw the boundaries of a sovereign state and dictate its foreign policy without allowing its government to be heard, the phrase becomes the hollowest of mockeries. What has been done was necessary; but if the lessons of the crisis are to be of any value, it is imperative that the democratic peoples should face the fact that, in order to avert catastrophe, a nation and a principle have been sacrificed. The political methods which on this occasion have saved the peace of Europe are not the political methods by which peace can be made just or permanent.[79]

Appeasement was not simply a political phenomenon. The Church of England supported it on spiritual grounds, ex-servicemen's organisations supported it as a way to avoid war, and the management of corporate Britain embraced it as the best way to avoid damaging Britain's economic strength. Under the auspices of the CBI's predecessor, the Federation of British Industries, Britain's major companies believed that they could play a key role in humanising Hitler's regime through closer trade contacts.

In December 1938, even after the Munich crisis and the *Kristallnacht* pogrom of German Jews, the Federation was happily organising joint conferences in Düsseldorf with the Reich Federation of Industry, where the Nazi economist Herr Ripp told their members that 'Great Britain was part of Europe and that the goal which must be aimed at was the creation of a strong economic unit comprising Germany, Great Britain, France and Italy'. Certainly, the London stock market reacted very positively to the Munich agreement, with the industrial sector leaping 13.3% in the week between Chamberlain leaving for Berchtesgaden and his return from Munich.

In January 1939, Ripp was welcomed to London for the National Chambers of Commerce annual dinner, where the British guest speaker eulogised 'the great achievement if in the near future some ten agreements could be concluded between British and German industry'. Sir William Larke, Director of the Iron and Steel Federation, was to have a joint Reich-British Federation meeting with Herr von Poensgen, the Chairman of the I.G. Farben company – later the manufacturer of Zyklon B gas – but it had to be cancelled when Hitler invaded Prague in March. Undeterred, Sir William was still writing to von Poensgen in June 1939, inviting the entire German delegation over for the Wimbledon fortnight, in order to continue their discussions 'in a purely social manner'.

★

After German forces marched into Prague in mid-March 1939 and occupied the rump of Czechoslovakia, Chamberlain, in a speech in his home city of Birmingham, sought to blame the old scapegoat – Versailles – rather than the actual aggressor, Hitler. 'I have never denied that the terms I was able to secure at Munich were not those I myself would have desired,' he told his listeners, 'but as I explained then, I had to deal with no new problem. This was something that had existed ever since the Treaty of Versailles; a problem that ought to have been solved long ago if only the statesmen of the last twenty years had taken a broader and more enlightened view of their duty.' Chamberlain despised Lloyd George, who had sacked him from his post of Director-General of National Service in 1917, but to have blamed him explicitly, and also Bonar Law, Ramsay MacDonald and Baldwin (not to mention the French statesmen), for being unenlightened about Versailles was a self-exculpation too far.

In fact, Western leaders had taken a far too broad and enlightened view of their duty to Versailles, which was to maintain the Treaty's provisions vis-à-vis a revanchist Germany. As Étienne Mantoux diagnosed shortly before his death, 'For good or ill, the whole structure of the Treaty of Versailles had to rest upon the active and continuous support of all those who had designed it.'[80] Ronald McCallum agreed. Although of course America was no longer in those ranks after 1919, France and Britain certainly were.

Accused for most of his life of lacking both principle and judgment, Churchill had to be radically reassessed by the English-speaking peoples virtually overnight once Hitler had occupied Czechoslovakia, and his many vocal warnings about Nazism suddenly looked principled, prescient and of remarkably sound judgment. Yet most Tory MPs were not about to abandon Chamberlain simply because he had been made to look foolish over Czechoslovakia. On 28 March 1939, the 1922 Committee of backbench Conservative MPs gave a dinner at the St Stephen's Club in his honour. According to the hitherto-unpublished diary of one of their number, the twenty-nine-year-old MP Christopher York, who had only been elected for the Ripon division of Yorkshire the previous month, the Prime Minister

> said in effect 'I'm not such a B[loody] F[ool] as some people think' about Munich. He didn't really believe Hitler would keep his word but had to say so to give Hitler a chance to do what he promised and to give us time to prepare. He declared against Compulsory Service mainly on the grounds that the Trades Unions were doing the job now and that to introduce Compulsory Service would antagonize them and perhaps stop the drive the Trade Unions were putting into re-arming. It was a great speech and very heartening.[81]

Two weeks later, on 13 April, York reported how 'Chamberlain told us that we had agreed to join Greece and Romania into our peace front. He did not deal with Russia because there was no agreement as yet, nor with America because as Quintin Hogg aptly said if we told the USA that she ought to come in, it is certain that she wouldn't, whereas if we said very little she might line up.' It is not good enough to persist in seeing Neville Chamberlain as a naïf, or indeed a B.F.

It is perhaps a necessary attribute in anyone who wishes to be prime minister to believe unquestioningly in one's own brilliance, but how many of them would write – even to their own sisters – 'It really seems as if Providence designed my speeches to be timed at the right moment to create the effect I want at that point', as did Chamberlain on April Fool's Day 1939? In the speech to which York referred, Chamberlain had reassured the trade unions that he would not introduce conscription, only to do exactly that on 26 April. Yet only three days later he was writing with equally invincible self-satisfaction: 'More and more I am convinced that much of the art of statesmanship lies in accurate timing, as the fisherman knows when he is trying to get a long cast out.' (He'd caught a $16\frac{1}{2}$ lb salmon in Hampshire the week before.)

Time ran out for Chamberlain and Britain in the early hours of 24 August 1939. The non-aggression pact signed by Molotov and Ribbentrop in Moscow was a masterpiece of cynicism, opening the way for Hitler to unleash *Blitzkrieg* ('Lightning war') on Poland and subsequently on the West. It also allowed Stalin to grab the eastern half of Poland, and to stand aside when Hitler attacked in the West and hopefully fought a war that would exhaust both of his enemies without Russia needing to involve herself. The differences between the fascist and communist ideologies disappeared before the perceived require- ments of both countries' *realpolitik*. A diplomat in the British Foreign Office put the situation succinctly when he said of the pact, 'All of a sudden, all our isms became wasms.'

Much has been made – rightly – of the tremendous sacrifices of the Russian people in defeating Hitler at the cost of over twenty million dead in 1941–5. Yet against that massive and undeniable contribution to the destruction of Nazism must be placed the fact that Stalin had allowed Hitler to secure his eastern flank in 1939. Fear of a war on two fronts, such as wrecked Germany in the Great War, had been the only thing holding Hitler back in 1939. The Pact removed that fear and thus made war as inevitable as anything ever is in human affairs. The Soviets reaped in June 1941 only what they had sown in August 1939.

The news of the Nazi-Soviet pact came as a thudding blow to the English- speaking peoples. Two days later, on 26 August, Chamberlain wrote to one of his former private secretaries, the backbench Tory MP Sir William Brass: 'I still hope we may avoid the worst, but if it comes we are thank God prepared for it.' Fortunately the year between Munich and the outbreak of war had solidified Commonwealth feeling behind Britain and allowed for major advances to be made in the crucial areas of air rearmament and radar (another vital invention of the English-speaking peoples that could not have come at a better time).

The overwhelming response to the outbreak of war in September 1939 was of sad resignation that Hitler should have behaved in the way he had. The English-speaking peoples entered their second great test of the century without any sense of euphoria, but firm in the knowledge that, as in 1914, the conflict was not of their making. The British parliamentary lobby correspondent Guy Eden recalled how in September 1939, 'The war was coming slowly and coldly upon us. There was no excitement, no war-fever, to divert attention from the meaning of it all. Just the cold, calm, methodical preparation for the storm that was to come, with no illusions about the ordeal that was to be Britain's and the world's.' A few weeks before the outbreak, Eden had met an alderman from one of the London boroughs, who looked 'thoroughly shaken and ill'. On being asked what was wrong, he had 'explained that he had just been to a committee meeting at which plans had been made for the storage of scores of thousands of cardboard coffins, to meet the demands that might arise when intensive air-raids on London began'.[82]

'War is a beastly thing now,' Winston Churchill once ruefully told Robert Bernays in the House of Commons tea room in the Thirties, 'all the glamour has gone out of it. Just a question of clerks pushing buttons.' He was right about the lack of glamour, but wrong about war being confined to clerks pushing buttons. The coming struggle was to see the united English-speaking peoples (except Ireland) fighting Nazism all over the world, facing every kind of adversity. Even so, it was indeed finally ended by two American airmen – not clerks – pushing buttons.

EIGHT

Divided and Faltering

1939–41

Australia declares war – Eire declares neutrality – The scuttling of the Graf Spee *–
London Pride – The IRA's war – The Channel Islands under occupation –
Churchill replaces Chamberlain – The Dunkirk evacuation – Gandhi's suggestions –
The Caribbean contribution – Churchillian oratory – The sinking of the French
Fleet – The Destroyers-for-Bases deal – Air power – The sinking of* The City of
Benares *– FDR re-elected – America as the arsenal of democracy – The London
Blitz – New Zealand heroism – The Murmansk convoys – Pearl Harbor*

'We can never forget how in the hour of trial in 1939 the call to save Civilisation
met with an instant response from the Dominions. ... Hitler, like the Kaiser
before him, learnt that there are the bonds of the spirit much stronger and
more enduring than any material ties, and that freedom unites more surely
than domination.'   Clement Attlee, House of Commons, 30 October 1945[1]

Hugh Trevor-Roper: 'I never met anybody who wasn't quite confident that
we would win the war. People who knew much more about our military
strength or weakness at the top of government might think differently but the
ordinary people, the ordinary officers and people whom I dealt with, both
socially and in my job, never had any doubt at all that we would win.'
Frank Johnson: 'Was that because there was a feeling in Britain that hubris
would strike down the Germans?'
Hugh Trevor-Roper: 'It's difficult to say. But we always do win our wars.'

B ack in 1938, Robert Menzies, then Attorney-General of Australia and
deputy leader of the United Australia Party, earned the unenviable nickname of 'Pig Iron Bob' when he was instrumental in the Government's ill-
fated decision to sell that commodity to Imperial Japan, which was then
viciously oppressing China. Resigning from the Cabinet in 1939, he was

nonetheless elected leader a month later, on the death of the United Australia leader Joseph Lyons, and served as premier for the crucial two years that saw the outbreak of war, the retreat from Dunkirk, the battle of Britain and the decision to reinforce the Middle Eastern theatre.

Menzies was born in Japarit, Victoria, the son of a farmer-politician. He graduated in 1916 with a first-class honours degree in law from Melbourne University and was elected to Victoria's upper house in 1928, before entering federal politics as a member of the coalition United Australia Party. A master of political manoeuvre, he managed to follow generally conservative policies domestically while being progressive in certain areas, such as his support for Australian universities. He never lost an election as prime minister and his contribution to the cohesion of the English-speaking peoples was second to none in the twentieth century.

The Australian declaration of war on Germany took place, to Menzies' great personal satisfaction and pride, only seventy-five minutes after the Cabinet heard that Chamberlain had stated that Britain was at war. In his memoirs he recorded how there was only 'a brief discussion, in which there was complete unanimity', before he broadcast over a network which included every national and commercial broadcasting station in Australia, saying,

> It is my melancholy duty to inform you officially that in consequence of a persistence by Germany in her invasion of Poland, Great Britain has declared war upon her, and that, as a result, Australia is at war. No harder task can fall to the lot of a democratic leader than to make such an announcement. ... I know that, in spite of the emotions we are all feeling, you will show that Australia is ready to see it through. May God in His mercy and compassion grant that the world may soon be delivered from this agony.[2]

When the Federal Parliament met three days later, on 6 September 1939, there was 'no audible dissent' to the Government's declaration of war. (The phrase 'melancholy duty' came from Gibbon's *Decline and Fall of the Roman Empire*, which ascribed it to the historian, who 'must discover the inevitable mixture of error and corruption, which [Religion] contracted in a long residence upon earth, among a weak and degenerate race of beings'.)

Yet two years later, the phrase 'in consequence' from Menzies' broadcast was brought up to be used against him. The form of words he had used, it was said, implied that Australia had had no constitutional say in the matter of whether or not the country went to war in defence of Britain. The sentence does indeed bear out this literal interpretation, but Menzies vigorously defended himself, arguing that the announcement 'expressed the overwhelming sentiments of the Australian people, and they would have been shocked to be confronted by formalities and delay'. It would have been 'an intolerable thought' for Menzies to have left Britain – 'the country in the immediate firing line' – in doubt even for two or three days as to 'whether they were standing alone'. Furthermore, treading slightly trickier constitutional grounds, he argued, 'How could the King be at war and at peace at the same time, in relation to Germany?'[3]

New Zealand acted equally instinctively in supporting Britain. On the day that war broke out, Adam Hamilton of the National Party laid party politics aside, saying, 'This fateful hour demands the clearest possible statement concerning the unity of the Dominion. New Zealand gives unqualified support to the Motherland in her decision to stand with France and Poland against German aggression.' Meanwhile, the Minister for Public Works Robert 'Fighting Bob' Semple, in the course of opening a boot factory, said, 'When the historian comes to write the history of the world, Britain can stand at the bar of international justice and say that it did all that man can do to prevent this threatened calamity.'[4] The Labour Premier, Joseph Savage, broadcasting from his sick-bed, said,

> I am satisfied that nowhere will the issue be more clearly understood than in New Zealand – where for almost a century, behind the sure shield of Britain, we have enjoyed and cherished freedom and self-government. With gratitude for the past and with confidence in the future we range ourselves without fear beside Britain. Where she goes, we go. Where she stands, we stand.[5]

As soon as war was declared, Menzies pursued an active policy of maximum military support for the Old Country, while keeping back forces deemed necessary to protect Australia from a possible attack from Japan. As he recalled in his memoirs, *Afternoon Light*,

> We then called for volunteers for a Second Australian Imperial Force, and got them in great numbers. We dispatched to the Middle East under General [later Field-Marshal Sir Thomas] Blamey the famous 6th Division, which was to fight with great success at Bardia, Tobruk, and Benghazi. We raised and sent the 7th Division, under [Lieutenant-General Sir John] Laverack, which earned fame in Syria, and later the 9th Division which, under [General Leslie] Morshead, was to play a great part in the defence of Tobruk and later in the crucial battle of El Alamein. We dispatched the cruiser *Sydney*, under [Admiral Sir John] Collins, to the Mediterranean, and sent the famous 'scrap-iron flotilla' of destroyers, which for years had been laid up in Sydney Harbour, first to Singapore and then to the Mediterranean.

Trouble was to come between the British and Australian Governments, especially after the attack on Pearl Harbor and the fall of Singapore, when Churchill wanted to employ the Australian 7th and 8th Divisions in the Western theatre or Burma rather than allow them to protect their threatened homeland, but in the early period of the conflict, before Menzies was replaced as prime minister in August 1941 by Arthur Fadden, leader of the Country Party, and five weeks after that by the Labour leader John Curtin, Anglo-Australian relations were good.

During the Second World War, no fewer than five million Commonwealth citizens went to fight for the Allies; 170,000 of them perished. (The 260,000 death toll of Britons, out of the six million who bore arms, was proportionately higher than that of the Commonwealth, however.) The 2.5 million who joined the Indian Army comprised the largest all-volunteer military force in human history. The Commonwealth contribution in theatres as diverse as the Western Desert and Burma, Normandy and the Pacific, the skies of Britain and the Murmansk convoys was crucial in wearing down and eventually destroying the Axis. Indeed, had the British Expeditionary Force been captured at Dunkirk and Britain invaded, there were only two divisions protecting London, both of which were Canadian.

Patriotism was a primary motive for those millions who willingly joined the colours in 1939 – two years before Japan entered the war – even though Nazi Germany alone could not possibly have posed a direct threat to the African and Asian countries, older Dominions and smaller Caribbean and Pacific islands from which so many came. For millions, too, the war provided the best and most regular pay they had ever enjoyed, whilst giving them the chance of taking part in a noble global endeavour.

There was also a genuine sense of gratitude; on 6 February 1940, the centenary of the Treaty of Waitangi, which in 1840 had ended the Maori Wars and formed the basis for a lasting Anglo-Maori peace settlement after much bloodletting, the leading Maori statesman and scholar Sir Apirana Ngata acknowledged that it was unlikely that any native race had ever been treated as well as the Maori had been by the British.

By contrast, Eire declared her neutrality in September 1939 and stuck to the policy right to the end of the war. Eire's actions cannot even be explained, like Sweden and Switzerland's, by a close physical proximity to – and thus a well-justified fear of occupation by – Germany. Neither was it a case of malingering, for even after D-Day, when there was no chance of a German invasion, Eamon de Valera still never once publicly denounced either Hitler or the Nazi regime. When he criticised the invasion of Belgium and Holland, he did not even specify who had been responsible. 'Today, these small nations are fighting for their lives, and I think it would be unworthy of this small nation if, on an occasion like this, I did not utter a protest against the cruel wrong that has been done them,' he said.[6] Quite who had done this cruel wrong was left to the listener to deduce, but from de Valera's language it might almost have been an Act of God. A recent scholarly study of European neutrality and non-belligerency by the Irish historian Eunan O'Halpin has concluded that of all the neutrals, Ireland was 'the most scrupulous', even though they were protected from invasion by geography (and Britain) in a way that countries like Denmark, Norway, Belgium and the Netherlands were not.[7]

Neutrality meant very different things for the seven European states, which, unlike the neutral Benelux countries, managed to avoid German occupation. Where Turkey, Portugal and the Vatican were pro-Allied all along, and Spain instinctively pro-Axis, the roles of Sweden, Switzerland and Eire have been far harder to assess. But all three discovered that in a struggle to save Civilisation from what Churchill rightly called 'a new Dark Age', true neutrality was simply not an honourable option.

In order to preserve strict neutrality, the Irish press was censored, to an almost ludicrous degree. In May 1940, during the battle for Belgium, Holland and France, Irish newsreels instead covered the New York World Fair, US polo stars playing for charity, the Pope canonizing a saint in Rome and an Australian boat race.[8] In the Irish press, the word 'Nazi' was banned from publication; Hitler was always given his title 'Herr'; the battle of Britain was referred to as 'the air battle over Southern England and the Channel'; references to German bombers destroying Stalingrad were altered to unidentified 'planes'; the anti-Nazi movies *The Great Dictator* and *Mrs Miniver* were banned and a scene in one newsreel in which old ladies were shown playing bowls in England was even cut because they were carrying gas-masks, which might have evoked sympathy for Britain's plight.[9]

(It was in May 1938 that Churchill had identified one of the stumbling blocks of Anglo-American amity as 'the powerful and highly organized Irish-American community. They have taken with them across the ocean a burning and deep-rooted hatred of the English name. They are irreconcilable enemies of the British Empire.' Three months earlier, de Valera – who had originally called for sanctions against Italy – had urged the recognition of the conquest of Abyssinia, drawing down Churchill's rebuke that 'Mr de Valera, oblivious to the claims of conquered peoples, has also given his croak in this sense. No sooner had he clambered from the arena into the Imperial box than he hastened to turn his thumb down upon the first prostrate gladiator he saw.')[10]

Yet there were plenty of ways in which the Irish people, as opposed to their Government, showed whose side they supported. Between September 1941 and the end of the war, 18,600 southerners passed through Ulster recruiting offices. Southerners won 780 decorations, including seven Victoria Crosses, serving in British units in the war. (An eighth was won by James Magennis, a Belfast Catholic.) Thousands of southerners worked in British munitions factories. Firemen from the south crossed the border to help during the blitz on Belfast, and meteorological reports also found their way to the north. Allied aircrew who crash-landed in the south were allowed back across the border, especially towards the end of the war.[11]

None of this begins to compare with Ulster's own contribution to the war effort, of course, for as General Eisenhower was to say, 'Without Northern Ireland I do not see how the American forces could have been concentrated to begin the invasion of Europe.' Nonetheless, although overall they accepted their Government's stance on neutrality, very many southern Irishmen took their place in the ranks of Civilisation's line of battle against Nazism.

By contrast, the United States sought constantly to interpret her neutrality in as pro-Allied a manner as possible. 'When peace has been broken anywhere,' warned President Roosevelt in his 'fireside chat' radio broadcast of 3 September 1939, 'the peace of all countries everywhere is in danger.' The Neutrality Act of 1935, as amended in 1937, embargoed shipments of munitions from the United States to warring countries. Once the European war became imminent in 1939, US Secretary of State Cordell Hull called for a major revision, arguing that 'no matter how much we may wish or may try to disassociate ourselves from world events, we cannot achieve this disassociation'.[12] After the revised bill passed the Senate by 63 to 30 and the House by 243 to 172, Roosevelt signed it into law on 4 November 1939. The same day Hull issued a statement that read: 'I desire to repeat with emphasis what I have consistently said heretofore, to the effect that our first and most sacred task is to keep our country secure and at peace, and that it is my firm belief that we shall succeed in this endeavour. I am satisfied that the new Act will greatly assist in this undertaking.'[13] It did not, but it did greatly assist the British Army when munitions stocks worth $43 million were transferred to Britain after the evacuation from Dunkirk in June the following year. Despite the misgivings of the US Chief of Staff, Roosevelt sent over 500,000 rifles, 900 70-mm artillery pieces and 50,000 machine-guns.

As in the Great War, the war at sea was going to be crucial to Britain's survival: her merchant fleet totalled 17.8 million tons in 1939, compared to the United States' 11.4, Japan's 5.6, Germany's 4.4, Italy's 3.4 and France's 2.9 million respectively. Her vulnerability was Germany's opportunity; of the twenty-nine million tons of shipping passing through the Suez Canal in 1939, for example, just over half were British, and of the twenty-seven million tons passing through the Panama Canal, 35% was American and 26% British. As in the Great War, shipping losses could hold the key to Germany's throttling of the British Isles, and between September and December 1939 nearly half-a-million tons were lost by Britain, a further 90,000 by other Allies and 347,000 by neutrals. One of the most dangerous threats in the opening stage of the war was the 10,000-ton German pocket-battleship the *Admiral Graf Spee*, which was one of the most feared raiders of Hitler's fleet.

The *Graf Spee* had been launched in 1936, bristling with armaments and sheathed in heavy armour-plating. Churchill wrote that, along with the other two warships *Deutschland* and *Admiral Scheer*, the ships 'had been designed with profound thought as commerce-destroyers. Their six eleven-inch guns, their 26-knot speed, and the armour they carried had been compressed with masterly skill into the limits of a ten-thousand-ton displacement. No single British cruiser could match them.'

When the Royal Navy's HMS *Exeter* came under fire from the 11-inch shells of the *Graf Spee* at the battle of the River Plate at 6.17 a.m. on Wednesday, 13 December 1939, Royal Marine Wilfred Russell had his left arm blown off and his right arm badly broken in two places, both above and below his elbow. The bombardment had knocked out 'B' turret, either killing or wounding all of the fifteen men stationed there. Further shelling destroyed all the communications on the bridge, killing many of the officers and putting the ship temporarily out of control. One of the survivors, able seaman Jack Napier, years later recalled the horror of the carnage, in particular how a severed head had rolled down a ladder from the bridge to just where he was stationed. He recognised the face.

Despite his horrific wounds, Russell dragged and pushed the wounded survivors from 'B' turret to safety below-decks. He was a strong thirty-two-year-old Devonshire-born man with what his colleagues remembered as 'a zest for life', and when he spotted a midshipman on the shrapnel- and splinter-strewn deck, he asked: 'I wonder if you could tourniquet my arm, sir? Don't bother about the other one – it's gone.'

With the bridge destroyed and the communications system therefore wrecked, *Exeter*'s captain, Frederick 'Hooky' Bell, desperately needed orders to be taken around the ship if she was to be able to fight on. Russell acted as a messenger for the rest of the engagement, during which *Exeter* somehow managed to fire over 180 rounds at the *Graf Spee*. Although only a small percentage of these actually hit home, *Exeter* bought vital time for the rest of Commodore Henry Harwood's 'hunting group H' – the cruisers HMS *Ajax* and the New Zealander ship HMNZS *Achilles* – to close with the enemy.

As the 670 lb shells smashed down from the *Graf Spee*, each of *Exeter*'s 8-inch forward guns were put out of action. Soon she was burning amidships and also listing heavily; survivors later estimated the angle at about 10 degrees. Captain Bell nonetheless decided to keep his ship in action and firing at the *Graf Spee* from the only gun turret still in operation, despite his ship having been hit over 100 times. One man continued to operate machinery despite both his legs having been blown off; before he died, he told an officer that he was 'not doing badly under somewhat adverse circumstances'.

Holes below *Exeter*'s water-line were plugged with blankets wrapped around chair-legs, and the choking smoke from the fires and deafening noise of the guns were constant throughout the rest of the battle. Finally, Bell was forced to disengage the action, but only after failure of air pressure had put his sole remaining gun out of service. *Exeter* was a relatively young ship, built in 1931, but with hundreds of gallons of water pouring into her hold it was clear that she could not keep up the unequal combat for any longer than she already had.

At 7.40 a.m. the *Exeter*, having borne the brunt of most of the punishment that the *Graf Spee* had been doling out for over eighty minutes, turned away to effect repairs. By then the other ships of the hunting group had managed to inflict enough damage on the German raider – totalling fifty hits – that she was forced to sail to Montevideo harbour to refit. Sixty-four of *Exeter's* complement lay dead and many more were seriously wounded, including Wilfred Russell, who died several weeks later on 20 January 1940 in a hospital on the Falkland Islands. He was awarded a posthumous Conspicuous Gallantry Medal, which King George VI presented to his widow. The King, who had himself served in a gun turret during the battle of Jutland, understood better than most the remarkable nature of Russell's gallantry.

Graf Spee's commanding officer, Hans Langsdorff, was also a veteran of Jutland. He was a gentleman warrior, who believed in the Hague Convention and was proud that not a single sailor in the nine Allied ships he had captured and sunk since the start of the war had lost their lives; indeed, he had sixty-two British merchant seamen on board as prisoners of war. He was also cunning; he transmitted false radio messages to throw pursuing ships off the scent and rigged dummy funnels and turrets to make his ship look like the British battleship HMS *Repulse* in silhouette.

So serious had been *Graf Spee's* sinking of British shipping by the autumn of 1939 that Churchill, then First Lord of the Admiralty, took the grave decision to send no fewer than twenty-three Allied warships, including five aircraft carriers, to scour the South Atlantic in search of her. In his war memoirs, he admitted that Langsdorff – who he called 'a high-class person' – had initially out-foxed the Admiralty. 'It was by no means clear whether one raider was on the prowl or two,' he recalled, 'and exertions were made both in the Atlantic and Indian Oceans.'

On 2 December, Thomas 'Digger' Foley was on look-out watch in the crow's nest of the freighter *Doric Star*, which was carrying meat, butter, canned goods and wool from Australia round the Cape of Good Hope to England. At 1.20 p.m. he heard an explosion and half a minute later saw a shell splash about 100 yards to the port side. Captain Stubbs had just ordered the radio-operator, 'Sparks' Comber, to send out a report when the *Graf Spee's* boarding party pulled up with a sign saying 'Stop wireless or we fire'. Stubbs countermanded his order and the *Doric Star* was scuttled by the Germans as soon as the crew had been safely embarked onto the *Graf Spee*. No-one except Comber knew that he had ignored the skipper's countermanded order and had sent the message nonetheless, and that therefore the Admiralty at last knew the whereabouts of the German raider.

Commodore Harwood then had to use his seaman's intuition to predict where the *Graf Spee* would be heading. He narrowed it down to Rio de Janiero, the Falkland Islands, or the River Plate – three targets each separated by 1,000 miles. He ordered his three cruisers *Ajax*, *Achilles* and *Exeter* to converge off the Plate, where they arrived on 12 December and practised concentrating their fire, sending signals and operating the all-important flanking manoeuvres they would need when they met the enemy. Sure enough, at four minutes past dawn the very next morning, out of the darkness on the horizon loomed the *Graf Spee*.

The combatants were eleven miles from one another, and the artillery shells ascended three-and-a-half miles into the air at velocities of up to 2,000 miles per hour, each staying aloft for over a minute before crashing down. At the end of the battle, the *Graf Spee*, with thirty-seven men dead and forty-seven wounded, headed for Montevideo harbour, harassed all the way by *Ajax* and *Achilles*, finally docking near midnight on 13 December. In a further indication of his inherent decency as an officer, Langsdorff freed his British prisoners as soon as he had moored.

That night the world's media descended on the Uruguayan capital to cover the spectacle. After a diplomatic tussle between the British and German Governments, the Montevideo Government decided to order the *Graf Spee* to quit its waters within seventy-two hours. Brilliantly, the British Ambassador had been pressing for only twenty-four hours' grace, thereby making Captain Langsdorff believe that Harwood had been reinforced by the aircraft carrier *Ark Royal* and the battleship *Renown*, which were in fact still several hundred miles away.

In the fine act of deception needed to keep the *Graf Spee* in port as long as possible, the BBC patriotically and marvellously unethically broadcast an entirely untrue report stating that *Ark Royal* and *Renown* had been spotted just beyond the horizon, taking up position for when the *Graf Spee* broke cover. Meanwhile, Langsdorff tried to charter a civilian plane for aerial reconnaissance, but was unable to find one. Low on ammunition and with his ship damaged, the captain considered that Hitler's order to fight to the last was inhumane to the men who served under him.

As the deadline ended on the afternoon of Sunday, 16 December, the sun shone brightly on the crowds of three-quarters of a million people who had gathered along the shoreline to watch the drama. Along with tens of millions of people throughout the world, President Roosevelt listened to the radio commentators broadcasting the whole story from the harbour. Meanwhile, Churchill stayed at his post in the Admiralty throughout the four-day crisis.

The *Graf Spee* slipped her moorings as the evening light fell and steamed towards the estuary mouth. Covered by the British warships, and miles away from the city of Montevideo, Langsdorff nosed his ship onto a mud-bank, where he set charges and transferred his skeleton crew to tugboats. At 8.54 p.m., at the moment of sunset, with the *Graf Spee* silhouetted against the horizon, the charges went off and the huge warship sank.

Hitler was incensed; although Goebbels released a bulletin saying that the Führer had ordered the scuttling in order to avoid an ignominious surrender, there could be no doubt that it was a propaganda disaster for Germany. Three days later, Langsdorff bid farewell to his crew, wrote letters to his wife and parents, retired to his room in the Buenos Aires naval arsenal, covered himself in the Imperial German Navy ensign, and shot himself in the head with his service revolver. It was remarked upon that he had not chosen 'to honour with his death the swastika flag'.

The scuttling of the *Graf Spee* marked the end of that period of the battle of the Atlantic where German surface vessels, rather than U-boats, raided the sea-lanes. 'We feel ourselves more confident day by day of our ability to police the seas and oceans,' Churchill broadcast a few days later, 'and to keep open and active the salt-water highways by which we live, and along which we shall draw the means of victory.' In a speech to the neutral countries on 20 January 1940, Churchill pointed to the wreck of the *Graf Spee* 'as a grisly monument and as a measure of the fate in store for any Nazi warship which dabbles in piracy in the broad waters'.

'Percy Bates bets Roderick Jones one bottle of port (vintage) that Germany invades Denmark before the end of March.' Mr Bates lost his bet at Brooks's Club, but only by a little over a week, as it was on 9 April 1940 that Hitler unleashed his invasions of Denmark and Norway. A month later, after vanquishing Denmark and Norway, Hitler attacked in the West, at dawn on Friday, 10 May 1940.

The bombing of the largely unprotected city of Rotterdam by eighty-four Heinkel He-111 planes on 14 May 1940 led the Dutch to surrender, once the Luftwaffe threatened to bomb other prominent cities including Utrecht. This form of terror-warfare was widely decried at the time as new, but in fact the Germans had shelled Paris itself in December 1870 during the Franco-Prussian War and again during the Great War. (Before one decries this Teutonic desecration of the City of Light, it must be acknowledged that four months later the French Government also bombarded the Communard-held city during the subsequent siege, resulting in no fewer than twenty-seven shell hits on the Arc de Triomphe.)[14] Since then, of course, Zeppelin raids over Britain in the Great War and more latterly the bombing of Guernica in April 1937 during the Spanish Civil War had led to civilian deaths.

In November 1941, Clement Attlee wrote to his brother Tom about the effect that the bombing of cities had upon Churchill personally. In particular, he wrote of the Prime Minister's 'extreme sensitiveness to suffering. I remember some years ago his eyes filling up with tears when he talked of the sufferings of the Jews in Germany while I recall the tones in which looking at Blitzed houses he said poor poor little homes. It is a side of his character not always appreciated.'[15] Indeed it is not, but then Churchill was an extremely lachrymose warlord, who was moved to tears on at least a score of occasions during the Second World War.

'Paris is a beautiful woman,' wrote Disraeli in 1857, 'and London an ugly man – still, the masculine quality counts for something.' If that was true in the mid-nineteenth century, how much truer it was after the Second World War, yet the British capital city's lack of beauty vis-à-vis the French is in part due to the fact that London was pulverized by the Luftwaffe in 1940 and 1941, whereas Paris surrendered. The 'masculine quality' of London certainly counted for much during the Blitz. The architecture of London – with its winding streets, small alleyways, higgledy-piggledy courtyards and street names going back six centuries – represents the legacy of liberty. By contrast, the wide boulevards of Paris were designed by Baron Haussmann so that government artillery could destroy barricades and command great sweeps of the centre of the city. Where huge swathes of Paris could be demolished on Second Empire *diktat* for rebuilding, the legal implications in 1850s' and 1860s' London would have made similarly grandiose projects almost out of the question.

It was this ancient, largely unplanned London that took such a pounding during the Blitz; many are the churches and public buildings with commemorative plaques that record only two eras of destruction – the 1666 Great Fire and the London Blitz. For all its obvious sentimentality, the words of Noël Coward's 1941 song *London Pride*, based on an old English melody that had been appropriated by the Germans, still retain the power to bring a lump to the throat of many Londoners:

> In our city darkened now, street and square and crescent,
> We can feel our living past in our shadowed present,
> Ghosts beside our starlit Thames who lived and loved and died
> Keep throughout the ages London Pride.[16]

The song, which features the line: 'Every Blitz, your resistance toughening', came to Coward after he had been waiting at a London railway station after a particularly heavy bombing in July 1941. 'Most of the glass in the station roof had been blown out and there was dust in the air and a smell of burning,' he later recalled. 'I sat on a platform seat and watched the Londoners scurrying about in the thin sunshine. They all seemed to be to be gay and determined and wholly admirable and for a moment or two I was overwhelmed with a wave of sentimental pride. A song started in my head then and there and was finished in a couple of days.'[17] For some it is his finest.

In May 1940, Seán Russell, the IRA's chief of staff, visited Joachim von Ribbentrop in Berlin and concluded that, 'Our ideas have much in common.'[18] He was quite right; the IRA was essentially a fascist revolutionary organisation in aim, method and organisation, and remained so. There had been close contacts between German Intelligence and the IRA before the war, including a clandestine meeting in Dublin in February 1939 between Abwehr agent Oscar K. Pfaus and several members of the IRA high command, including Russell. Shortly after Pfaus' departure back to Germany, the IRA sent its director of munitions and chemicals, the German-speaker Jim O'Donovan, to Berlin as the IRA's envoy to the Abwehr, making three trips in 1939.[19]

It was Russell's visit to Germany in the summer of 1940 that represented the high-water mark of IRA-Nazi relations. Russell's German minder Dr Edmund Veesenmayer found Russell 'straightforward, strait-laced; a traditionalist who only wanted what was good for Ireland'. He was taught the use of forty-day delay detonators and trained in a laboratory that specialised in designing ordinary objects that contained powerful explosives.[20] A red flower pot in the window of the German Legation in Dublin was to be the signal to Russell for when to start the IRA's sabotage campaign against British targets across Ireland. By early August, Russell was considered a technically highly competent explosives expert.

On 8 August, Russell and another IRA man called Frank Ryan, who had been freed from a Spanish gaol on German request, left the naval station at Wilhelmshaven on board the submarine *U-65* to launch the grotesquely misnamed Operation *Taube* ('Dove'). The U-boat commander, Korvettenkapitän Hans-Gerrit von Stockhausen, had orders to deliver his passengers – codenamed Richard 1 and Richard 2 – to a point near Ballyferriter, County Kerry, south of Ballydavid Head in the bay near Smerwick Harbour, on the Feast of Assumption (15 August), which it was hoped would provide them, their radio transmitter and explosives with the necessary cover as pilgrims to get them first to Tralee and then to Dublin.

Yet shortly after leaving port Russell started to complain of stomach pains, and the submarine had only a former medical student on board who hadn't trained as a combat medic. Russell died on 14 August only 100 miles from Galway, and Ryan decided not to continue with the mission alone. Although the corpse was disposed of at sea, the Germans concluded that 'Richard 1' had died from a burst gastric ulcer. Needless to say, after the war any number of conspiracy theories abounded that he had been murdered, by Admiral Canaris of the Abwehr, by British Intelligence, by Frank Ryan, or others. All that can be known for certain is that for all that Russell 'only wanted what was good for Ireland', in fact there is nothing to suggest that after a successful invasion of the British Isles, the Wehrmacht would have dutifully stopped at the border of neutral Eire. Like so many Irish nationalists before and after him, Russell put his pathological hatred of Britain before his patriotic love of Ireland.

Although a burst gastric ulcer on a submerged submarine could hardly be foreseen, German espionage in southern Ireland during the Second World War was generally characterised by hilarious incompetence. Several spies who were sent over spoke little or no English, and even the ones who did spoke it with a heavy accent that soon gave them away. Dr Hermann Görtz did not know the currency of Ireland and asked the way to Laragh at a local police station while still wearing part of his Luftwaffe uniform. Walter Simon was unable to distinguish between active and long-inactive railway lines, and then asked an undercover policeman if he knew anyone in the IRA. Wilhelm Preetz used the money intended for his undercover work to buy a brand new five-seater Chrysler Saloon, into which he was climbing when he was arrested. Henry Obéd was a Muslim born in Lucknow, who thus looked very different from the other inhabitants of Skibbereen, County Cork, and who led his three-man espionage team straight into a Garda trap. Although the Abwehr forged an Irish passport for Günther Schütz, they forgot to include a visa. Joseph Lenihan – who was at least Irish – failed even to conceal, let alone to memorise and destroy, his radio instructions. This was the Keystone Cops school of espionage.

Although the Germans sent ten agents to Ireland, none supplied them with any useful information. More potentially damaging to the cause of the English-speaking peoples were those Irish officials of the Dublin Government who sought to hedge Ireland's bets in the event of a German victory in the West. The activities of three in particular were particularly dangerous, namely Joseph Walshe, Leopold Kerney and Charles Bewley.

Joe Walshe was Eire's Secretary of External Affairs (i.e. foreign minister), who on 17 June 1940, during the fall of France, held a meeting with Dr Eduard Hempel, the German Minister to Dublin. According to Hempel's report back to Berlin, published after the war in the eighth volume of the official *Documents on German Foreign Policy*,

> The conversation, in which Walshe expressed great admiration for the German achievements, went off in a friendly way ... [Walshe] remarked that he hoped that the statement of the Führer in his interview with Weygand respecting the absence of intention to destroy the British Empire, did not mean the abandonment of Ireland.[21]

Meanwhile, the Abwehr agent Dr Hermann Görtz, for all his incompetence, did manage to secure meetings with such senior members of the Dáil as the

Minister of Agriculture Dr Jim Ryan, the Minister for Posts and Telegraphs P.J. Little and possibly even the Minister for Co-ordination of Defensive Measures, Frank Aiken.

Even before Walshe, the activities of Charles Bewley, the Irish Minister in Berlin from 1933 until 1939, were dangerous for the Allies and useful for the Axis. Bewley had been replaced as Irish consul in 1922 due to his foaming anti-Semitism, but was appointed by de Valera as Minister in Berlin in 1933 and stayed there for the next six crucial years. As well as giving newspaper interviews about his admiration and support for Hitler's regime, Bewley wrote reports to Dublin emphasising how responsible the Jews were for their own tribulations at the hands of the Nazis. He attended the Nuremberg rallies and sent literature about Jewish ritual murders to the Irish Government. Since Bewley also decided who received visas to leave Germany for Ireland, in the opinion of many historians 'he was certainly responsible for the deaths of many Jewish people'.[22]

After resigning in August 1939 in protest over what he believed to be the pro-British stance of the Department of External Affairs, Bewley returned to Berlin as a private citizen the following month. From Berlin and then Rome – where he had previously been Minister to the Vatican – Bewley wrote reports for the German Foreign Ministry on the military value of the IRA and also worked for Goebbels' Propaganda Ministry.[23] (When he fell into American hands in June 1945, he was handed over to the British, who imprisoned him for six months. He was only released without trial after Walshe asked the British to treat him leniently.)

One of Bewley's best services to the Reich he loved – German documents describe him as 'a convinced friend of National Socialist Germany and a fanatical Irish freedom fighter' – was to put Dr Veesenmayer in touch in 1941 with another Irish diplomat, Leopold Kerney. Kerney was the Irish Minister to Spain who helped secure the release of Frank Ryan from Burgos gaol. In May 1940, he established contact with an Abwehr agent, Mary Pauline Mains, and later in the year his personal interventions secured her passage to Galway. Carrying $10,000 to Görtz, Mains – codenamed 'Agent Margarethe' – was also aided by Kerney in her return to Spain, on which she carried a situation report from her Abwehr contact in Ireland.

To have an ex-diplomat like Bewley working for the Nazis was bad enough, but Kerney was Ireland's serving Minister in Madrid, one of the most sensitive posts of the war. In a secret report in early 1941, Bewley described Kerney as the only 'real Irish nationalist' in the Irish diplomatic corps, and soon after-wards a series of five meetings were held between Kerney and 'the group' (German intelligence agents including Veesenmayer himself) between November 1941 and July 1943. In the course of these meetings, Kerney 'agreed that partition was inevitable unless British power collapsed through a

German victory', and that in the event of a German victory in Russia and Hitler once again turning westwards, de Valera – whom Kerney knew well – would then 'announce his claim to the six Northern counties'.[24] This was a very strange interpretation of Irish neutrality.

Whether Kerney spoke with any authority is impossible to prove or disprove, but his consistent message was that 'de Valera was not fanatical about neutrality and would enter the war against the Allies as soon as any chance of liberating Northern Ireland presented itself. If Germany chose to aid this move, she would need to publicly deny any interest in Ireland and German troops would only remain to complete the war against England.' After the war, Veesenmayer himself said that 'the tenor of the conversations could be put this way: to such an extent that if Germany's chances of winning the war improved, so would "official" and "unofficial" Ireland be prepared indirectly to give Germany useful help within the framework of her efforts to gain independence.' Since Eire already had independence, what Veesenmayer was really referring to was the six counties in the north that were still British. (Veesenmayer himself was later responsible for the mass murder of Hungarian Jews towards the end of the war.)

Even though Eire's counter-intelligence service 'G2' had been on to Kerney ever since the Agent Margarethe incident in 1940, and had interviewed him without getting believable answers, the Minister was nonetheless allowed to remain *en poste* until long after it was clear that Hitler was not going to win the war. After the war, but before certain German documents were available that revealed the truth, Kerney successfully sued for libel over allegations that he had behaved treacherously, winning £500, an apology and costs in an out-of-court settlement against the historian Professor Desmond Williams. Only many years later did it become obvious that Williams rather than Kerney had been telling the truth.

At the very least, at a time between 1941 and 1943 when the entire English-speaking peoples were fighting for their existence, the Irish Government was keeping its options resolutely open, while putting the issue of partition above the question of the survival of Civilisation itself. The idea that a Nazi victory in the West would lead to an extension of Irish liberty, sovereignty and independence might be laughable today, but a (fortunately small) section of the Irish governing class was so blinded by Anglophobia that they were willing to take the risk.

The only part of the English-speaking peoples' sovereign territory to be subjected to German occupation during the war were the Channel Islands; but does its experience of collaboration with the Nazi authorities give us any indication about how other parts of the English-speaking world might have responded to a successful invasion? Certainly, when the secret Ministry of

Defence files relating to the Channel Islands' occupation were released in November 1996, commentators were not primarily interested in the Channel Islanders' experience per se, so much as what it might tell us about the way mainland Britons might have behaved under similar circumstances. The sorry tales of Alderney, Jersey and Guernsey women sleeping with Wehrmacht soldiers, for example, carry an extra *frisson* precisely because of the fear that such *collaboration horizontale* might soon have been emulated by women in England, Wales, Ireland and Scotland, had the battle of Britain turned out differently.

It is fashionable to argue that the response of the British people to invasion and occupation would have been no different from that of the French, Belgians or Luxemburgers. According to this view, it was only the geographical fact of the English Channel, and nothing to do with an indomitable national spirit, which saved Britain in 1940. In her 1995 book *The Model Occupation*, the *Guardian* journalist Madeleine Bunting stated that because 'the Islanders compromised, collaborated and fraternised just as people did throughout occupied Europe', it therefore followed that their experience 'directly challenges the belief that the Second World War proved that [Britons] were inherently different from the rest of Europe'.

Reviewing her book, the novelist John Mortimer described the Channel Islands as an 'ideal testing ground for the British character and British virtues under stress', and concluded that, 'The British were put to the test and behaved no better or much worse than many people in Europe.' Other authors have imagined an occupied Britain in which 'slowly a relationship of sorts began to develop between the British people and members of the German armed forces', and a historian has even claimed that 'great numbers of ordinary decent Britons would have begun to co-operate with the Germans in putting down the Resistance just to bring about a sort of peace'.

All this goes to the very heart of the self-perception of the English-speaking peoples. The subliminal question is why, if English-speaking institutions would have been no better than the rest of Europe's in withstanding Hitlerism, should we be so protective of them today? Does the Channel Islands' experience of German occupation between 1940 and 1945 – in which they undoubtedly collaborated and established a *modus vivendi* with the enemy – really mean that the English-speaking peoples are exactly the same as everyone else when it comes to withstanding tyranny?

Fortunately, the Channel Islands' experience in 1940 in fact tells us precisely nothing about the way the rest of Britain would have behaved towards a Nazi invasion, let alone the United States, Canada, Australia, New Zealand or the West Indies. The Islands had been specifically ordered by the War Office not to resist, as their strategic importance was minimal. St Helier and St Peter Port could hardly, as Churchill said of London, have swallowed an entire

German army. One-third of their population, including all their able-bodied men of military age (10,000 of whom served in the war) had already been evacuated. The 60,000 who were left were guarded by no fewer than 37,000 Germans – a ratio which, if translated to mainland Britain, would have required them to station thirty million troops there.

Flat, isolated, rural, lightly populated and with a far higher proportion of Germans per square mile than in Germany itself, with no political parties, trade unions or obvious centres for resistance, the wartime Channel Islands cannot provide any indication whatever for how the East End of London, the mining valleys of South Wales, the Black Country or the Glasgow slums would have reacted towards a foreign invader. If the Germans had landed in Britain, though they might well have won the set-piece military engagements through sheer superiority of weaponry and their revolutionary *Blitzkrieg* battlefield tactics, they would have then been faced with the implacable, visceral enmity of a nation under – albeit somewhat makeshift – arms.

The enthusiasm to fight was unmistakable. On 14 May, Anthony Eden broadcast a call for 'large numbers of men ... between the ages of seventeen and sixty-five to come forward now and offer their services' as Local Defence Volunteers (LDV). Even before he had actually finished speaking, police stations across the land were inundated with calls. The next morning queues formed outside them and within twenty-four hours a quarter of a million Britons had volunteered. By the end of May, the War Office – which had only anticipated 150,000 coming forward in all – had to deal with 400,000 volunteers, and by the end of June no fewer than 1.46 million Britons had applied to join the LDV, nearly ten times the number expected. The Channel Islands, stripped of their able-bodied men who were already in uniform, provide no template whatever for the reaction of the rest of Britain.

Often without waiting for detailed instructions, these Home Guard units immediately began training and patrols, armed only with farm implements, private shotguns and occasionally even home-made weapons. Although only one in six volunteers had a rifle, to counterbalance that around one in three were veterans of the Great War, who had thus seen the terrible realities of conflict back in 1914–18. The experiences of the Spanish Civil War and the Warsaw and Budapest uprisings show how effective even unconventionally armed populations could be. A Molotov cocktail requires little more than a bottle, some petrol, a rag and a match.

In contrast to the War Office's pacific orders to the Channel Islands, by June 1940 Ministry of Information posters in Britain were proclaiming that, 'The people of these islands will offer a united opposition to an invader and every citizen will regard it as his duty to hinder and frustrate the enemy and help our own forces by every means that ingenuity can devise and common sense suggest.' The *Stand Fast* leaflet also distributed at that time even had to discourage over-zealous volunteers, advising that, 'Civilians should not set out to make independent attacks on military formations.'

The experience of the Blitz, though it can only approximate that of a military invasion and attempted occupation, also gives grounds for optimism in terms of British morale. In his biography of wartime London, Philip Ziegler described how 'the population endured the Blitz with dignity, courage, resolution, and astonishingly good humour'. Something of the latter might have evaporated as the nation was terrorised and reprisals were exacted against civilians, but the first three would most probably have held true for longer than might be expected. Tom Harrison of the Mass-Observation movement, who almost made a career out of exploding wartime myths, noted of the urban populations of Britain how, 'They did not let the soldiers or leaders down.' Churchill himself, who believed that 'the massacre on both sides would have been grim and great', intended to broadcast the slogan 'You can always take one with you.' Whatever had happened on Jersey, he would have been heeded on the mainland, and by the rest of the English-speaking peoples.

On 6 May 1940, Chamberlain wrote to Max Beaverbrook thanking him for a 'splendid' article that the newspaper proprietor had written in that morning's *Daily Express* about the campaign in Norway, a Commons' debate on which was scheduled for the following day. A phrase in it shows that Chamberlain wildly underestimated the seriousness of the Norwegian débâcle: 'When so many are sounding the defeatist note over a minor setback, it is a relief to read such a courageous and inspiriting summons to a saner view.' The British people didn't think Norway a 'minor setback', even though there was plenty of precedent for a British expeditionary force being chased off the European mainland. Neither, as soon became clear on 7 May 1940, did the House of Commons.

On 10 September 1939 – a week after the outbreak of war – Chamberlain had written to his sisters that 'What I hope for is not a military victory – I very much doubt the possibility of that – but a collapse of the German home front.' For that sentence alone, it is clear that he should not have been Britain's wartime leader, and by the end of 10 May 1940 he was no longer.

Thirty-four years earlier at Brooks's Club, Mr Mowbray Morris had bet the former Liberal MP David Guthrie 'that Mr Winston Churchill will never be prime minister of England'. On 10 May 1940, he lost his bet. As one of the most interesting political figures right from the start of his career, Churchill appeared often in the betting book. In January 1912, during a crisis in Northern Ireland, R. Morris bet J.K. Fowler two shillings and sixpence 'that if Winston Churchill holds his meeting in the Ulster Hall there will be no bloodshed (bloodshed means more deaths than three)'. He won.

Anecdotes abounded about Churchill's precocious ambition; his friend

Lady Violet Bonham Carter told one about the time that Churchill escaped from the Boer prisoner-of-war camp in Pretoria in 1900 and finally reached freedom after hiding in the bottom of a railway coal-wagon. The British consul who received him gave him a bath and had his filthy clothes burned. 'What a pity,' said Young Winston when he was told afterwards. 'I wanted them for Madame Tussaud's.'[25]

Horace Walpole had once lamented that, 'No great country was ever saved by good men, because good men will not go to the length that may be necessary.' By May 1940, Britain was less interested in whether her saviours were good than whether they were tough and single-minded. Plenty of people had criticised Lloyd George's moral character before he got to Number 10 in the crisis of December 1916, but cheerfully accepted him once he was there, only getting around to attacking him again once victory was safely won. Similarly Churchill's ruthlessness, once thought an incubus by 'good men' of the Respectable Tendency in British politics, was now considered a benefit.

It might be doubtful whether good men such as Baldwin and Chamberlain would have been ready to consider going to such lengths as laying down gas across Britain's south coast in the event of a German invasion, or invading southern Ireland, or even dropping a nuclear bomb on Japanese civilians, but Churchill was willing at least to contemplate all that, and much more. He appointed men of similar ruthlessness; the historian G.M. Young once described Lord Beaverbrook as 'looking like a doctor struck off the roll for performing an illegal operation'.

If the same strict standards regarding financial disclosures had pertained in the 1930s as do in the British Parliament today, it is uncertain that Winston Churchill would ever have made it to Downing Street. 'Sleaze' allegations would have continually dogged him if he had been forced to admit the truth about his financial affairs in any Register of Members' Interests form, which in fact only came into operation in 1975, ten years after his death. He would, for example, have had to admit that in March 1938, a week before his £18,000 debts to his stockbrokers Vickers da Costa were about to force him to sell his country house Chartwell for £20,000, an Anglo-South African businessman named Sir Henry Strakosch, the Chairman of Union Corporation Ltd, had taken on his debts and guaranteed all his investments against further losses for the next three years.

The Moravian-born Strakosch's cheque to Vickers of £18,162/1/10 – around £450,000 in 2005 money – would have taken some explaining to the Committee on Standards and Privileges, along with his covering letter which read: 'My dear Winston, As agreed between us I shall carry this position for three years, you giving me full discretion to sell or vary the holdings at any time, but on the understanding that you incur no further liability.' On receipt of the letter and accompanying cheque, Churchill promptly took Chartwell off the market, even though *The Times* had already announced that it was up for sale.

Although today only former historians of the extremist ilk like David Irving claim that because Strakosch was Jewish, Churchill was the 'hired help' of the anti-Nazi lobby, the press of the late-1930s would have looked very much askance at the Strakosch deal. Fifteen years earlier, on 14 August 1923, Churchill – then out of office but working hard for certain oil interests – attempted to sound out the Prime Minister Stanley Baldwin with an agenda that one of his biographers, Roy Jenkins, describes as 'half-hidden'. It is one that would certainly today be emblazoned across the front pages of the newspapers with a 'Cash for Access'-style banner headline.

For a fee of £5,000 – worth £125,000 today and equivalent to the annual salary of a Cabinet Minister – Churchill was hired by Royal Dutch Shell and Burmah Oil to sound out the Prime Minister about a merger between them and the Anglo-Persian Oil Company, of which the British Government owned a majority of the voting shares. It had actually been Churchill himself who, ten years earlier as First Lord of the Admiralty, had suggested that the Government go into the Persian oil business in the first place, as a way of protecting naval oil supplies during the great changeover from coal-fired ships to oil.

That Churchill knew he was doing something at least slightly outré was confirmed by the fact that, as he told his wife Clementine afterwards, 'I entered Downing Street by the Treasury entrance to avoid comment.' He had already asked a senior civil servant, Sir James Masterton-Smith, about the propriety of what was proposed, who replied that he ought to fight 'very shy of it on large political grounds'. But Churchill was short of money and went ahead, later reporting to his wife: 'My interview with the PM was most agreeable. I found him thoroughly in favour of the Oil Settlement on the lines proposed. Indeed he might have been [the Shell managing director, Sir Robert] Waley Cohen from the way he talked. I am sure it will come off. The only thing I am puzzled about is my own affair.'

Baldwin helpfully put Churchill on to both the First Lord of the Admiralty and the President of the Board of Trade to discuss the merger further. The implications of two oil companies hiring the former decision-making minister to sound out the prime minister and, through him, the other two senior ministerial decision-makers over a huge merger deal involving taxpayers' assets were obvious even to Twenties' sensibilities, let alone to those of a later more 'sleaze'-obsessed age. They obviously did not deter Baldwin, however, who the following year appointed Churchill Chancellor of the Exchequer.

If either the Strakosch or Persian Oil stories had been picked over in detail in the way that present-day politicians' business affairs are, Churchill's career might have ended before the Second World War, with unforeseeable but surely appalling implications for the history of the English-speaking peoples. Nor would his other dubious financial arrangements with men like Sir Ernest Cassel, Bernard Baruch and Charles Schwab probably have borne the kind of microscopic scrutiny that the modern press and parliamentary watchdogs have imposed.

When the British Expeditionary Force was obliged to evacuate from Dunkirk between 28 May and 3 June 1940, it had to leave behind no fewer than 475 tanks, 38,000 motor vehicles, 400 anti-tank guns, 1,000 heavy guns, 8,000 Bren guns, 90,000 rifles and 7,000 tons of ammunition. Twenty-two ships were sunk during the operation, including six destroyers. After the extent of the disaster became known in New Zealand, she immediately shipped no less than half her entire store of rifle ammunition to Britain.[26]

According to the diary of the Tory MP Christopher York, 'During Dunkirk the Home Affairs committee of the Cabinet spent 35 minutes discussing whether the Divorce Laws of Scotland should be applied to India!'[27] Robert Menzies used the time more profitably, making three direct private appeals to President Roosevelt at the time of the débâcle. On 26 May, he instructed the Australian Minister in Washington, R.G. Casey, to stress to FDR the imminent danger to 'the power of Great Britain to defend liberty and free institutions' and thus the danger to 'your English-speaking neighbours on the Pacific Basin'. By sending American air-force planes to Britain immediately, the United States could 'make a decisive contribution without actually itself participating'.[28]

The next day Peter Fraser, the Prime Minister of New Zealand, sent a similar message, Mackenzie King having already said that he had been 'in direct and personal touch' with Roosevelt emphasising the need for urgent aerial reinforcement. Menzies sent a further message on 14 June, telling Roosevelt how if America were to 'make available to the Allies the whole of her financial and material resources ... the whole of the English-speaking people of the world would, by one stroke, be welded into a brotherhood of world salvation,' and Germany would be defeated. A similar appeal was again made on 28 June. Roosevelt's reply was the only one he could constitutionally make at the time, that the United States could not make war directly without the support of Congress, but that 'so long as the British Commonwealth of Nations continue in the defence of their liberty, so long may they be assured that material and supplies will be sent to them from the United States in ever increasing quantities and kinds'. He was as good as his word.

'Hitler, Mussolini, Stalin, the Japanese, the opportunists as well as the Jew-haters, the Anglophobes of the lower middle classes, oily Spanish functionaries as well as the dark peasant masses of Russia – they all had their mean little enjoyments in witnessing the humiliations of Britannia,' so the American historian John Lukacs has written of Britain's defeats in 1940.[29] Yet at least Britannia ignored the advice given to her by Mahatma Gandhi, who during the London Blitz suggested: 'Invite Hitler and Mussolini to take what they want of the countries you call your possessions. Let them take possession of your beautiful island with its many beautiful buildings. You will give all this, but neither your minds nor your souls.'[30] If Britons felt disinclined to go along with the Mahatma's proposal, it was at least consistent with his earlier suggestions to Ethiopians to 'allow themselves to be slaughtered' by the Italians since, 'after all, Mussolini didn't want a desert', and his equally helpful proposition that German Jews ought to make 'a calm and determined stand offered by unarmed men and women possessing the strength of suffering given to them by Jehovah', because he believed that would convert the Nazis 'to an appreciation of human dignity'.[31]

On 16 June 1940 – the same day that Churchill was told that France intended to sue for peace – Churchill dictated a minute to his secretary Kathleen Hill as they drove in the pouring rain to Downing Street from Chequers, requesting that a West Indies regiment be formed 'to be available for Imperial Service; to give an outlet for the loyalty of Negroes, and bring money into these poor Islands'.[32] His wife had visited the West Indies the previous year and written to him about the harsh living conditions there.

A regiment was raised and sent to Italy and the Middle East in 1944, but the war ended before it was required in combat and it was disbanded in 1946. Nevertheless, thousands of British West Indian subjects joined other units in Britain and Canada, and no fewer than 5,000 served in the RAF during the war.[33] Thousands more worked in British factories and agriculture.

Although British ships were torpedoed and mines were sown in Castries harbour, the capital of St Lucia, the war did not affect the West Indies much militarily; yet the vital tourist industry of Bermuda and the Bahamas collapsed and several basic industries had to be kept alive through British subsidy. Furthermore, important raw materials such as bauxite from British Guiana, aviation spirit from Jamaica and some rubber from British Honduras, Trinidad and British Guiana were exported to Britain. (Jamaica provided a wartime home for the civilian population of Gibraltar, which had to be evacuated en masse.)

When he was brought back from 'the Wilderness' by Chamberlain at the outbreak of war in September 1939, Churchill had, as one historian has put it, 'been able to dream, while [the appeasers] had become accustomed to the sober chores of tailoring hope to reality'.[34] As a result, Churchill could enthuse the nation during the Blitz using vocabulary which simply would not have occurred to the workaday Respectable Tendency politicians on the Government Front Bench who had spent the previous decade trying to avoid giving any verbal hostages to fortune.

In his speech of 18 June 1940, in which Churchill coined the phrase 'Their finest hour', the Prime Minister had to try to advance some arguments to persuade the British people that, as he put it, 'there are good and reasonable hopes of final victory', despite the French armistice. The very first one he mentioned was the support of the Dominions, telling his listeners:

> We have fully informed and consulted all the self-governing Dominions ... and I have received from their prime ministers ... who all have Governments behind them elected on wide franchises, who are all there because they represent the will of their people, messages couched in the most moving terms in which they endorse our decision to fight on, and declare themselves ready to share our fortunes and fight on to the end. That is what we are going to do.

After that the list of reasons why Britons should be optimistic rather tailed off; Churchill mentioned 'increasing support in supplies and munitions of all kinds from the United States', a cold winter, the possibility of a sudden German collapse and French resistance, none of which realistically offered much genuine hope of victory. It was in fact to be another year and three days before Hitler's invasion of the USSR gave any logical grounds for optimism.

Not everyone appreciated Churchill's speeches of 1940 and 1941; the travel writer Robert Byron complained to the aesthete Harold Acton of their 'fustian' nature, and said of Churchill's use of the almost-obsolete word 'foe': 'What on earth has it got to do with the enemy?'[35] Such etymological and somewhat precious criticism aside, however, Churchill's language was generally recognised as sublime.

In the peroration of his great 'Finest Hour' speech, Churchill conjured up the vision of a nightmare world in which a Nazi victory produced 'a new Dark Age made more sinister, and perhaps more protracted, by the lights of perverted science'. The Nazis certainly perverted science for military and ideological ends, in a way that even under wartime strictures the English-speaking peoples baulked at. General Sir Ian Jacob, the assistant military secretary to Churchill's War Cabinet, once quipped to me that the Allies won the war 'because our German scientists were better than their German scientists'. While it is true that Werner Heisenberg's atomic programme in Germany lagged far behind the Allies' nuclear 'Manhattan Project' at Los Alamos, nonetheless Hitler's scientists did come up with an impressive array of non-atomic scientific discoveries during the war, including proximity fuses, synthetic fuels, ballistic missiles, hydrogen-peroxide-assisted submarines and ersatz rubber.

In the first two decades of the twentieth century, German scientists won over half of all the Nobel Prizes awarded in every discipline of the natural sciences and medicine, yet once Hitler came to power in 1933 – and especially after his pogroms against the Jews forced many of the most brilliant German scientists into exile – the country could no longer call upon science's best intellects. Instead, by August 1939 Albert Einstein was writing to President Roosevelt to inform him of the incredible potential of uranium. 'This requires action' was FDR's fortuitous response.

François Rabelais wrote that 'Science without conscience is the ruin of the world', and all too often Hitler's scientists ignored the suffering that their work created, including (in Wernher von Braun's case) tens of thousands of people working under slave-labour conditions to build the installations for his weaponry. (After the war von Braun headed President Kennedy's space programme, his rocketry career saved by the fact that he had once briefly been arrested by the SS when Himmler had wanted to take over one of his projects.)

On the same day that Churchill delivered his 'Finest Hour' oration – 18 June 1940 – Otto Abetz, the German Ambassador to Paris and effective ruler of Occupied France after its Fall, reported to Berlin that he had received fifty French politicians, town councillors, *préfets* and magistrates. 'Forty-nine have asked for special permissions of one sort or another, or for petrol coupons – and the fiftieth spoke of France.'[36]

The situation was no better in the Vichy-governed part of France than the German-occupied part for which Abetz had responsibility. Britain's most pressing fear after the Fall of France was that the powerful French Fleet might fall into the hands of her enemies. Two weeks after the armistice, the situation was still unresolved, so at 17.54 hours on 3 July 1940, three battleships of Vice-Admiral Sir James Somerville's Gibraltar-based Force H – HMS *Hood*, HMS *Valiant* and HMS *Resolution* – opened fire on the Vichy Fleet anchored at Mers-el-Kébir, off Algeria.

In thirty-six 15-inch salvoes at maximum visibility range of 17,500 yards, aided by aircraft-spotting bi-planes from the aircraft-carrier HMS *Ark Royal*, the French Fleet was effectively put out of action. Although the firing ceased at 18.04 to allow the French to abandon their ships with the minimum loss of life, 1,299 French sailors died and 350 were wounded. (Force H suffered one officer and one rating injured.) A single French battle-cruiser, the *Strasbourg*, escaped to Toulon.

Although of course the primary reason why this ruthless action by the Royal Navy needed to be undertaken was to, as Admiral Somerville cabled the French Admiral Marcel Gensoul, 'prevent your ships from falling into German or Italian hands', it also had another beneficial result. From that moment onwards, American public opinion was in absolutely no doubt that Britain meant to fight on, come what may.

Although Somerville's actions have long been held as a British war crime by French nationalists, Gensoul had been offered the choices of sailing with the Royal Navy, or turning his ships over to the British, or taking them to French bases in the West Indies, or of scuttling them himself. That he chose none of these options at a time when politicians in Vichy might well have used the French Navy as a bargaining chip with the Germans meant that Churchill – a lifelong Francophile – was forced to take what he called 'a hateful decision, the most unnatural and painful decision in which I have ever been concerned'. Somerville concurred, saying that his officers were revolted at the necessity of disabling the French Fleet, although he also reported how, 'It did not seem to worry the sailors at all.'[37]

If – thanks to the Vichy Fleet operating with the German Navy – Britain had been successfully invaded and the Royal Family, Government, Royal Navy and the Bank of England's gold reserves were forced to move to Canada, as was planned, Ottawa would have been a fine place from which to run the rump of the British Empire. Canada's magnificent Parliament was built in 1867, a fine example of gothic architecture at its grandest. The tremendous dignity of its House of Commons' dark wood panelling, heraldic devices, high ceilings, black-robed officials, stained-glass windows, large public galleries, heavy chandeliers and green leather seats would have reminded British MPs – at least those who had managed to escape – of the Palace of Westminster, albeit with specifically Canadian additions such as the Chamber's bronze representations of the moose, beaver, buffalo and squirrel.

Writing to his Oxford contemporary Mary Fisher from the Shoreham Hotel in New York on 21 August 1940, the British diplomat and fellow of All Souls College, Oxford, Isaiah Berlin, said that, 'The Americans are by now enormously frightened, and if they believed in us sufficiently would certainly hurl themselves to our help with no thought of either yesterday or tomorrow: or rather with vague thought about the day after tomorrow.'[38] The time had come for an historic deal whereby such American trust in eventual British victory could be made tangible.

In May 1940, soon after becoming prime minister, Churchill asked Roosevelt for 'the loan of forty or fifty of your older destroyers to bridge the gap between what we have now and the large new construction we have put in hand at the beginning of the war'. Unfortunately, the President was forbidden by law to dispose of any military matériel unless it was considered useless for America's own defence, so it had to be linked instead to a simultaneous ninety-nine-year leasing to the United States of naval and air bases on various British possessions in the Atlantic and West Indies.[39]

It was by this means that American bases were set up in Newfoundland, the Bahamas, Bermuda, Jamaica, Antigua, St Lucia, Trinidad and British Guiana. This in turn allowed Roosevelt, less than eight weeks before the presidential elections, to sell the deal to the American people. Some reactionary Tory MPs protested at American military bases being extended over British sovereign territory, but by 1940 it was taken for granted that British and American interests coincided, at least in the Western hemisphere where the United States enjoyed unquestioned hegemony. When MPs criticised the deal to Churchill, arguing that in 2039 no American politician would vacate the bases, Churchill answered: 'I would sooner that they have them than a lot of Wops. ... We cannot expect to hold everything.'[40] (It is possible that the MP misheard and Churchill said 'wogs', which would have made more sense in the context of the West Indies than the equally derogatory terms for Italians and southern Mediterranean peoples, who were hardly poised to inherit the Caribbean. Churchill used such phrases freely, however, as was customary for people of his age and background.)

The suggestion to help Britain by donating fifty destroyers was first made publicly in America by General Pershing in a radio broadcast of 4 August. He had obviously been put up to it by the Administration and supported by the influential Kansas editor William Allen White, founding Chairman of the Committee to Defend America by Aiding the Allies, the aim of which was to give 'all aid short of war'. Writing to Supreme Court Justice Felix Frankfurter's daughter Marion the next day, Isaiah Berlin had said that the legal advisor to the British Embassy, another All Souls' fellow John Foster,

> keeps assuring us that the destroyer situation is really acute: that these fifty American boats might well make all the difference between survival and defeat: that if the German invasion succeeds and the American people is then told that but for these few ships it might have been averted they will, justifiably enough, say that we didn't realize that it was as critical as that. ... The Embassy does its best but it is as unimaginative and giftless as a propaganda agency, etc.[41]

Just as isolationists were complaining about the limitless guile and sinister brilliance of British propaganda in America, so those closest to it were constantly criticising it as clod-hopping.

The truth about the fifty American destroyers that were handed over to Britain on 3 September 1940 – the first anniversary of the outbreak of war – was neither so bad as critics of the deal suggested, nor so good as supporters of it like John Foster implied. Although it is true that several were of pre-Great War vintage, and some were nearing obsolescence, nonetheless possession of them did allow the Royal Navy to free up more modern ships for active duty, leaving the older, 'Deal' destroyers for routine patrols. Even so, by February 1941 only nine of them were found fit for service. Nonetheless, in September 1940 they were priceless as a propaganda tool, and it was brave of Roosevelt to take such a bellicose step less than two months before an election in which he was forced to promise – in Boston on 30 October 1940 – 'I have said this before, but I shall say it again and again and again: Your boys are not going to be sent into any foreign wars.' The best that can be said is that he believed it at the time.

The bases that the United States established helped her to protect the eastern approaches to the Panama Canal. Since this was one of the primary routes for war matériel to be transported from the west coast of America, it was of direct benefit to Great Britain. The base on the Avalon peninsula of Newfoundland near St John's also helped protect the north-western approaches to Canada. Bermuda lay within a few hours' flying time of the continental United States and was used by the American air force for patrol bomber operations. The huge base there was made even larger by reclamation from the sea by the use of dredging.

Although some people lost their homes in Jamaica to make way for the bases, work was provided for thousands of inhabitants of the Islands. Ships based in British Guiana also patrolled the whole north coast of South America and were well within range of the shipping routes linking Europe and Africa with South America. The deal was thus of great and mutual benefit to the English-speaking peoples in their struggle against German raiders in the Atlantic and ought not to be treated with the cynicism that nationalist Tories reserved for it.

A sign of the (entirely unwarranted) optimism that was felt about Britain's chances can be found in the entry in the Beefsteak Club's betting book the day after the Destroyers-for-Bases deal was announced, where the Conservative MP for Birmingham, Commander Oliver Locker-Lampson DSO, 'bets H.M. Howgrave-Graham 4 to 1 (in five-shillingses) that Adolf Hitler will be put to death by sentence of an International Tribunal within two years after the cessation of hostilities between Britain and Germany'.

That same month – September 1940 – a man named Roy Hardy of Lincoln's Inn asked prominent Britons to sign a petition of support for Churchill. The Archbishop of Canterbury Cosmo Gordon Lang, the Cardinal Archbishop of Westminster Arthur Hinsley, the Archbishop of York and the Bishop of London all refused, unless he excised a paragraph that read: 'In this, the supreme trust of our national existence, we salute you as the incomparable captain of our destiny. We trust you to the uttermost. We want you to know and to feel that we are standing steadfastly behind you to the victorious end.'[42] The divines would not even allow Hardy to edit it down to 'We salute you. We trust you,' considering it too extravagant praise for Churchill. The Respectable Tendency fought a longer rearguard action against Churchill than is often imagined.

★

The importance of air power has been epicentral to the survival and the triumph of the English-speaking peoples in the twentieth century, and this was never more startlingly displayed than in the battle of Britain between June and September 1940. Had the Luftwaffe managed to establish mastery of the skies over southern Britain for any extended length of period, the history of the twentieth century might have had a very different outcome.

The story has oft been told, both in books and on celluloid, because it is an inspiring one. It is not generally understood, however, quite how close the battle came to being lost before it was ever fought. This was starkly illustrated in an exchange of letters in March 1961 between a writer called Robert Wright and Air Chief Marshal Lord Dowding, the retired former chief of Fighter Command. In one of these (hitherto unpublished) letters, Wright wrote to Dowding quoting from page 38 of Churchill's second volume of war memoirs, *Finest Hour*, in which Churchill had written, 'Air Chief Marshal Dowding, at the head of our metropolitan Fighter Command, had declared to me that with twenty-five squadrons of fighters he could defend the Island against the whole might of the German Air Force, but that with less he would be overpowered.'[43] This figure of twenty-five squadrons had then been used by Air Chief Marshal Sir Arthur Longmore in *The New Cambridge Modern History* and in television documentaries, one of which had been aired the night before Wright wrote his letter.

Wright wanted to know from Dowding whether it was true that he had told Churchill that twenty-five squadrons would suffice, because on 16 May 1940 Churchill had requested additional squadrons to be pledged to the battle of France, Dowding had written to the Air Ministry saying that fifty-two squadrons were 'the force necessary to defend this country', of which he then only had thirty-six. Dowding answered Wright, saying,

> You must remember that the foundation of all Churchill's writings is 'I was never wrong.' Your quotation is one of his most flagrant terminological inexactitudes. I most certainly never made such a ridiculous statement to him. Just take 25 squadrons, and fit them into sector maps from Wick to Bristol, and see what it looks like! ... so far as my memory serves me, I said that 52 squadrons was *the Air Staff's own estimate* of requirements, (and not necessarily mine). ... Of course you couldn't expect Churchill to admit that he was within a hair's breadth of throwing away our last chance of victory by wrecking Fighter Command before the Battle had ever started and somebody has edited the minutes of the vital Cabinet meeting to cover it up nicely. This is Confidential, because nobody is supposed to see Cabinet minutes and I cannot disclose the source of my information.[44]

★

On the night of Tuesday, 17 September 1940, the *City of Benares*, a ship taking Britons to Canada, including ninety child evacuees, was torpedoed with the loss of 255 passengers, among them eighty-three children.[45] 'I am full of horror and indignation that any German submarine captain could be found to torpedo a ship over six hundred miles from land in a tempestuous sea,' said Geoffrey Shakespeare, Under-Secretary at the Dominions Office. 'The conditions were such that there was little chance for passengers, whether adult or children, to survive. This deed will shock the world.'

He was right, at least as far as the English-speaking peoples were concerned: the American Secretary of State Cordell Hull described it as 'a most dastardly act'; a US congressman called Hitler 'the mad butcher'; Robert Menzies anticipated that 'this latest exhibition of savagery by the Nazis' would steel the British to 'defeat the dark spirit for which the Nazi regime stands', and in Canada the minister responsible for receiving the evacuees described it as 'just another demonstration of Nazi frightfulness'. The Australian Consolidated Press predicted that 'This brutal sinking will shock and horrify the whole civilised world.' In America, a Gallup poll showed how Americans were far more ready to risk their neutrality as a result. There was a general assumption made that Kapitänleutnant Heinrich Bleichrodt of *U-38* knew about the cargoes, whereas all he could have seen through his periscope was a lightly armed, sizeable modern passenger liner, leading a formation of nineteen merchant ships. Nonetheless, it was a propaganda victory for the Allies, albeit bought at an horrific cost.

On Tuesday, 5 November 1940, Franklin D. Roosevelt was re-elected president for an unprecedented third term with 449 electoral votes to the Republican Wendell L. Willkie's 82. Furthermore, the Democrats retained a 66 to 28 majority over the Republicans in the Senate and a 268 to 162 lead in the House. It was fortunate for the fate of the English-speaking peoples that FDR was re-elected. An opponent of isolationism, whose secret policy was to try to get the United States into the war against Hitler, Roosevelt's feline touch for politics allowed him to push each incident as far as possible towards interventionism, but never too far to provoke an isolationist backlash. His Republican opponent, Wendell Willkie, though a fine man and a true American patriot, would probably not have been able to massage, embroider and subtly tug the USA towards war in the way that the arch-politician FDR did.

The national census returns for 1940 showed that the population of the United States numbered 131,409,881, a growth of 7% since 1930. Although this made America the fourth largest country in the world in terms of population – after China, India and the USSR – the rate of growth was considerably lower than the 16.1% achieved between 1920 and 1930, due to a lowered birth rate and far lower immigration.[46] (The US Census Bureau put the 2006 population total at 297,888,255, a rise of over 100% since 1940.)

Within the country, huge population shifts were evident in the census. Never before in American history had more than three states shown an overall loss in population between censuses, yet in 1940 this was the case in the drought-stricken Great Plains states of North and South Dakota, Nebraska, Kansas and Oklahoma. A great exodus from the Dust Bowl, reaching all the way from Canada down to Texas, was evident. In all, 587 counties in Montana, Wyoming, the Dakotas, Nebraska, Colorado, Kansas, Iowa, Missouri, New Mexico, Oklahoma and Texas lost a total of 835,978 people, which corresponded to 229 people moving out every day for a decade.

The destination of many of them, and many more, was the Golden State. California's population increased by nearly 1.2 million – 21.1% – during the Thirties, allowing it to take Texas' place as the fifth-biggest state, rivalling Ohio (but far behind New York, Pennsylvania and Illinois). Florida also grew by 27.9%, New Mexico by 24.9% and the District of Columbia by 36.2%. The Depression also meant that fewer Southerners left the South to find work, although the population growth was lower there than in the Pacific Coast and Rocky Mountain regions, which saw the biggest population gains.

For the first time in American history, and for reasons also connected with the Depression, the United States' big cities did not outstrip the rest of the country in population growth, only showing an average gain of 5%. Of the biggest three cities in the country, New York gained 6.5%, Chicago 0.2% and Philadelphia actually lost people. With the automobile coming into almost general use, suburbs grew: small cities of 10,000 to 25,000 inhabitants saw an average 9% gain. Much of the present-day geographical make-up of the United States can be seen as stemming directly from these large post-Depression population shifts. They showed, apart from anything else, that Americans were willing to cross their continent and re-locate themselves and their families in order to look for a better life. It implied a vigour and flexibility to the free market and the American way of life that the dictators would have done well to ponder before declaring war in December 1941.

One indication of the power and confidence of American capitalism was the explosion in commercial aviation in the United States before her entry into the Second World War. The free market had been able to create in America that which even state-subsidised corporations were unable to elsewhere: a viable, profitable non-military aircraft industry. 'Not only are modern planes flying in regions once forbidden to man,' pronounced the *National Geographic* magazine in December 1940,

> but aviation in general today has 'grown up'. Gone are the days of the stunting barnstormer, when flying was a 'thrill', and pilots too often were reckless glamor boys. Air liners now run as soberly and regularly as trains. Private plane owners

fly on fishing trips, to golf games, or to work as matter-of-factly as they might
drive an automobile. .... You can buy an airplane for your own use as cheaply as
a medium-priced car, with elementary flying lessons thrown in for free.[47]

Every day in the United States alone, scheduled airlines such as Trans-
continental and Western Air covered a quarter-of-a-million miles by the end
of 1940. Cruising at four miles a minute more than three-and-a-half miles
above the earth, its passengers breathing compressed air to a density equivalent
to that usually found at 8,000 feet, four-engined Boeings flew from New York
to Los Angeles in fifteen hours, taking thirty-three passengers each by day or
twenty-five by night. The return journey, pushed by prevailing tailwinds, could
be done in thirteen-and-a-half hours. (The story was told of a lady passenger
asking the pilot: 'Well, if tailwinds are so much help, why don't you put them
on all your airplanes?')

Trans-pacific Clipper planes meanwhile ploughed the San Francisco-
Hawaii route, a weekly California-New Zealand route was opened up via
Canton Island, and light planes such as the Piper Club were being produced
at the rate of twenty per day. Clippers also operated the world's longest
commercial route across the Pacific from Hawaii out to Pan-American Air-
ways' bases on Midway and Wake Islands. Just before Christmas 1939, one
transatlantic Clipper plane carried 65,000 letters on one trip across the Pacific.
Because of the international dateline, a letter posted in Nouméa, New Cale-
donia, would be delivered at Canton Island the day before the postmark date.

Thousands of people paid ten cents' admission to jam themselves onto the
spectators' runway at LaGuardia Field, New York, on summer weekends,
where fifteen American Airlines planes arrived and departed. The flying
boom was made possible by incredible advances in aeronautics precision
engineering. The linings of many aeroplane cylinders had to be harder than
glass to withstand wear and accurate to within one-tenth of the thickness of a
spider's thread. Once American research and development was capable of
achieving that, the industrial might was ready to exploit it.

Even before Pearl Harbor, America was able to harness this potential to
create monster bombers such as the Douglas B-19, the world's largest military
aircraft, which could fly the Atlantic and back non-stop at over 200 mph with
a crew of ten. Its tyres alone were 8 feet high and weighed more than some
whole aeroplanes. Its wingspread was 212 feet and the rudder was so huge
that it could not be moved as a whole by the pilot; instead, he manipulated a
small tab on the rudder's rear edge, which activated hydraulic controls to
operate the main rudder. With assembly lines in factories such as that of the
North American Aviation Corporation at Inglewood, California, producing
one military training plane every three hours – in peacetime – it was an act of
suicidal hubris for Japan to bomb the Hawaiian naval base.

In one of his 'fireside chat' radio broadcasts on 29 December 1940, President Roosevelt had told the American people: 'We have the men – the skill – the wealth – and above all the will. ... We must be the great arsenal of democracy.' His offer was made concrete by the Lend-Lease Act that was passed on 11 March 1941, which was rightly described by Churchill as 'the most unsordid act in the history of any nation'. For those who value the unity of the English-speaking peoples, Lend-Lease stands as a totem, but it is also important to establish what it was not. It was not a gift of something for nothing, it was not open-ended, it was not conscience-money and it was not directed solely at Britain. The wording of the Act allowed President Roosevelt to 'sell, transfer title to, exchange, lease, lend or otherwise dispose of' munitions to any country, if he felt it was in the interests of the defence of the United States. In return, America was to receive 'payment or repayment in kind or property, or any other direct or indirect benefit' which the President deemed fair.[48]

As soon as he had signed the Act, Roosevelt ordered the Navy Secretary to send twenty-eight motor torpedo-boats and submarine-chasers to the Royal Navy, along with material needed to arm merchant ships. The Greek Army also received artillery and ammunition. That same month Congress appropriated the first $7 billion of Lend-Lease funds, a further $21 billion was later appropriated and in addition almost $26 billion was authorised from the budgets of the War and the Navy Departments. In all, Britain received over $22 billion in aid by the end of the war, a truly vast amount (especially considering that she had reneged on her Great War loans). Other beneficiaries were Russia, France, China and thirty-eight other countries, half of them American republics, which received a further $19 billion between them.

The Office of Lend-Lease Administration was established at the end of October 1941, with Edward R. Stettinius as its administrator. Yet it did not take long for the British Left to complain about the long-term effect Lend-Lease might have on the economy. A meeting of the Fabian Anglo-American Committee at the Society's headquarters in London's Dartmouth Street in January 1942, for example, listed all the reasons why Americans should resent Britons and Britons should resent Americans.[49] It concluded that Lend-Lease would leave Britain a mere outpost or satrapy of the United States.

Yet this was a chimera, however regularly it is still resurrected by anti-Americans whenever Britain supports the United States anywhere in the world, with the two most popular images being of Britain as America's fifty-first state and of whoever was Britain's prime minister at the time depicted as the US president's poodle. (They are still trotted out today, to criticise Tony Blair's relationship with George W. Bush.) None of these caricatures stand up to scrutiny. The English-speaking nations are each proud of their independence and act according to their own national self-interest; it just so happens that these interests have historically tended to coincide very much more than they have diverged.

Unsurprisingly, considering such strictures from the Fabians and others, great efforts were made in the propaganda sphere to promote warm and close Anglo-American relations. In 1943, to take a typical example, H.L. Gee published *American England*, which was subtitled *An Epitome of a Common Heritage*. This listed hundreds of associations between Britain and America, from the obvious and familiar ones about the Pilgrim Fathers and the Founding Fathers, to the far more obscure – and more interesting – connections such as Eleanor Roosevelt's schooling in Wimbledon, Surrey, Theodore Roosevelt's wedding in St George's, Hanover Square, and how Harvard is named after a Londoner and Yale after a Welshman. 'Let us remember', said the American-born Lady Astor, 'that the American War of Independence was fought by British Americans against a German king and a reactionary prime minister for British ideals.'

On the night of Friday, 27 December 1940, London saw the outbreak of no fewer than 532 fires as a result of the Luftwaffe's bombing, including five serious and thirty-four medium ones. Yet such was the sang-froid of the London Fire Brigade headquarters' official responsible for reporting to the Clerk to the London County Council that he deleted the word 'very' from the phrase 'very heavy night'. The next twenty-four hours were 'perfectly quiet', but then came the raid of Sunday, 29 December, which was to devastate huge areas of the City. (It was on that night that the iconic photograph of St Paul's Cathedral was taken, showing the dome rising above the surrounding smoke. Goebbels reproduced it in the German press with the caption: '*Die City von London brennt!*')

This raid was on the City of London, the 'Square Mile' that housed the financial hub of the Empire. Since ton for ton, fire was four times more destructive than explosives, incendiary bombs – which were also easier to drop than high explosives – were largely used. The Thames was low – 'a minor stream in the centre of an expanse of mud' – and water correspondingly hard to pump.[50] Brewery wells and even children's paddling pools were used during the war by the Fire Service's water officers.

Between the sirens sounding at 6.08 p.m. and the All Clear at 11.45 p.m., the City was subjected to a massive onslaught. 'About 700 fires have been reported,' County Hall was informed, 'of which ten were major fires, 28 serious and 101 medium.' The medieval Guildhall, eight Wren churches, five railway termini and sixteen underground stations were completely or partially destroyed, as well as huge tracts of offices and shops.[51] The fire was visible sixty miles away, and although fire crews raced there from all over south-east England, very many of the beautiful ancient buildings and Livery Halls were razed. Because so few people lived in the City, only 163 died and 509 were seriously injured, relatively few compared to the 1,430 and 1,800 respectively in the last air raid of the Blitz.

It is astonishing how much of London life remained essentially unchanged during these terrible times. The ordinary pursuits of life continued to a degree unimaginable to modern ears. Harvey Klimmer, an American attaché of the US Embassy in London, travelled extensively in Britain but was in London during the air assault of 7 September 1940. 'The East End of London was one vast inferno,' he recalled seven months later.

> Surely, we thought, human beings cannot stand such punishment. Another night went by. Five hundred more were killed. The people stood fast. A week passed … two weeks. By that time, in my opinion, the crisis was over. The people adjusted themselves to the nightly attacks from the skies. They resisted the impulse to flee. They obeyed the Government's injunction to 'stay put'. It may well be that the fortitude of the ordinary people of London in the terrible nights of September will mean the difference between defeat and victory for the British Empire.[52]

What Klimmer witnessed between September 1940 and January 1941, instead of mass panic, was described as 'Business as usual'. Millions of people slept underground in gardens, cellars and underground stations, but when they awoke, traffic moved, street-sweepers swept, taxis operated, porters shone brass, railway stations functioned as in peacetime, bridges were open, buses and trams drove, and although 150,000 people crowded nightly into the underground stations to sleep, services were maintained on all lines. There were 20,000 men who worked every morning to clear the streets of debris after every raid.

Although there were some localised problems for short periods when sub-stations received direct hits, in general gas, water and electricity supplies were kept going throughout the Blitz. The telephone service was badly affected at first, but was back to normal within a few weeks. Telegrams were unaffected. As Klimmer recalled, 'You can still walk into a London store, order practically anything you wish, and tell the clerk to put it on your account. You will be asked to carry your purchase if it is a small one; the clerk will also forego the use of wrapping paper if feasible. Outside of that, shopping is not greatly different from what it was before the war.'

When shop windows were shattered, or shops were partly destroyed, proprietors vied with each other to write perkily defiant remarks, such as, 'If you think this is bad, you should see my branch in Berlin!' A barber shop had the sign: 'Never mind the blasted windows, walk right in. Close shave sir?' Butter, sugar, meat and tea were rationed, and more foodstuffs as the war progressed, but many were not. Nor, crucially, was alcohol. Restaurants and nightclubs stayed open, although theatres and cinemas generally closed at 7 p.m. Milk delivery continued. Weddings were conducted in bombed-out churches.

Racing was curtailed because of the need for horses, but most other sports continued, even though greyhound tracks and rugby and football matches had been machine-gunned.[53] (At Highgate golf course there was a sign warning: 'Time bombs on holes 3, 10 and 18.') Klimmer found it 'almost macabre during a bombing to have an announcer go on the air and proceed calmly with a reading of the cricket scores'. As for the British press, he recalled how insubstantial much of the comment was, with 'Hours given over to discussion of such trivial subjects as how much money Gracie Fields took out of the country, what Noël Coward is doing in Australia, and whether or not the authorities should substitute a bugle call for the air raid siren now in use.' Trivia in the British press; it was business as usual.

There were no major epidemics, as had been predicted. Bronchitis and influenza raged around the bigger shelters – where as many as 8,000 people slept together in crowded conditions – but these were not life-threatening. The London Transport Board Lost Property Office was jammed with as many as 200 gas masks left by passengers on trains, trams and buses every day, but they were not a huge percentage of the four million daily travellers, or the nine million gas masks belonging to Londoners. 'Thus far, at least,' wrote Klimmer in April 1941, 'the tenacity of ordinary life has proved greater than the menace of bombs in Britain.' It was true of the Britain threatened by Zeppelin raids in the Great War, remained true of Britain during the Blitz, stayed true of her under the shadow of the hydrogen bomb during the Cold War, and the thirty-year IRA bombing campaign, and continues to be true under the threat of suicide bombers today.

On the night of 10 May 1941, no fewer than 1,486 Londoners were killed, 1,800 were injured and 11,000 houses were destroyed. It was a 400-bomber raid, taking advantage of bright moonlight. 'London was like a pot of boiling tomato soup,' recalled one Luftwaffe pilot. British fighter aircraft taking off from West Malling airfield were guided by the terrible orange glow on the horizon. Westminster Abbey was hit, the Houses of Parliament were rendered unusable, Waterloo Station was destroyed, the British Museum lost 150,000 books and the St Mary-le-Bow Church was razed to the ground. Two thousand fires were started, many of which burned through to the next day, and on that one night alone thirty-six firemen were killed and 289 injured. The historian of that terrible night has nonetheless concluded that London emerged 'the symbol of the free world, bruised and battered, but unbeaten and as bloody-minded as ever'.[54]

From the moment when Flight-Lieutenant F.L. Litchfield, of New Plymouth, North Island, took off eight hours after war was declared to bomb German warships off Heligoland, New Zealanders were in the thick of the fighting, despite their country being under no imaginable threat from Hitler. The ties that bound the Dominions to the place that they still unselfconsciously referred to as 'the Mother Land' were strong enough to banish any selfish view of antipodean *realpolitik*. Long before Russia entered the war in June 1941, the New Zealand Bomber Squadron had attacked targets from Norway to Italy, particularly over Germany and during the battle of France. Squadron-Leader L.W. Coleman of Havelock North, North Island, won the DFC for bombing Munich during a Nazi Party rally there; Air Vice-Marshal Keith Park from Thames, North Island, flew the last aeroplane out of Dunkirk back to Britain. New Zealanders such as Flight-Lieutenant G.G. Stead of Hastings, North Island, based in Iceland, flew Sunderlands protecting convoys. Others such as Flying Officer Ian Patterson from Auckland ferried American Flying Fortresses over to Britain. Charles Upham won the only VC and Bar of the War.

The first Allied 'ace' of the war was Flying Officer E.J. 'Cobber' Kain, from Wellington, while Flight-Lieutenant Alan Deere, DFC and bar, from Wanganui, North Island, shot down seventeen German planes despite having himself been shot down over Dunkirk. Wing-Commander S.C. Elworthy from Timaru was awarded the AFC in January 1941, the DFC in March and the DSO in April. New Zealand had her own fighter squadron which flew Spitfires that were bought by ordinary New Zealanders through public subscription. They put their money where their hearts were.

On the high seas before the entry of Russia into the war, New Zealanders had distinguished themselves in the *Achilles* against the *Graf Spee*; they had swept the sea of mines in front of fifty Atlantic convoys and had served in raids on the Lofoten Islands, Channel Islands and other enemy-occupied territory. After only five weeks studying to take his naval commission at HMS *King Alfred*, Lieutenant G.M. Hobday from Remuera, near Auckland, boarded an enemy trawler with an armed party of five in the Mediterranean, which he took first to Gibraltar and then back to England despite a limited knowledge of navigation. 'War has taken a heavy toll of New Zealand shipping and of New Zealanders in the merchant navy,' her Premier, Peter Fraser, was told in a report of 12 June 1941. 'Many of them have been killed while bringing the great food ships home to England.'[55]

As well as the New Zealand regiments fighting in Greece and Crete in mid-1941, New Zealanders were to be found in Britain's civil defence corps, women's services, hospitals and the Women's Land Army. A Rhodes Scholar and his wife from Auckland were ARP wardens in Hampstead, a Dunedin woman was a senior officer in the WAAF, and others served in the ATS and WRNS. Britain did not 'stand alone' in the year before Russia entered the war; she stood with Greece and with her superbly loyal Dominions and overseas territories.

Fraser visited Britain between June and August 1941, touring New Zealand bases and 'cities and towns that have been badly blitzed', including Coventry, Birmingham, Liverpool, Bootle and Aberdeen. When he visited Liverpool, he was cheered by the dockers there who were busily unloading New Zealand lamb. (There was a curious dichotomy between the way the Nazis tried to keep their visiting allies and clients away from bomb-damaged areas, thinking it bad for their morale, whereas the British positively encouraged visitors to go to such places.)

At Manchester, Fraser was taken to see the damaged Royal Exchange, Assize Courts, Cathedral and Royal Infirmary, and 'In one badly blitzed, poorer area he stopped and talked with women and children and was cheered for his encouraging words. "We aren't downhearted," he was told.'[56] The Scottish-born Fraser was then taken to Clydeside to see the effects of the bombing there and on to Glasgow, where no fewer than 3,000 people watched him receive the freedom of the city, after which Sir Harry Lauder led the singing of the National Anthem.

Adolf Hitler predicted that when he unleashed his *Blitzkrieg* invasion of Russia, looking for *Lebensraum* for his Aryan people, 'The world will hold its breath.' It certainly was breathtaking in its scale – involving 3.6 million German troops attacking across a 2,000-mile front – but in a sense the British were able to exhale, since they and their allies were, after over a year of standing alone, no longer the sole components of the anti-Hitler struggle.

Although in retrospect it is clear that Hitler's fate was sealed the moment he unleashed Operation Barbarossa on the night of Saturday, 21 June 1941, few thought so at the time. We can today see how an invasion campaign across thousands of miles of steppes and tundra through winter after winter against a vast and implacably hostile population made success almost impossible, but in the summer of 1941 well-informed observers were making estimations that were not flattering to Stalin about the number of weeks that the Soviets could hold out. The US Joint Chiefs of Staff believed that 'Germany will be thoroughly occupied in beating Russia for a minimum of one month and a possible maximum of three months'.[57] In the event, Russia took nearly four years to beat Germany.

There is no more ironclad commandment in human affairs than the Law of Unintended Consequences, and Stalin had not expected the Molotov-Ribbentrop pact of August 1939 to result in Hitler having a free hand to attack the Soviet Union by June 1941. Under the terms of the Nazi-Soviet pact, to which Stalin stuck rigidly to the letter, the USSR had to deliver large amounts of grain, oil and other raw materials to Germany. So unaware was Stalin of what Hitler had planned for the early hours of Sunday, 22 June 1941, that there were trainloads of such supplies actually being taken westwards just as Operation Barbarossa was getting under way. Stalin had ignored British warnings of what was afoot and had naïvely trusted the Nazi leader. As late as 14 June, the Soviet press agency Tass had released a statement saying that 'rumours of Germany's intention to violate the pact and attack the USSR are groundless'. The result was that the 187 million inhabitants of the Soviet Union were utterly unprepared for what happened next.

Certainly, the scale of the assault was astonishing, involving 162 divisions of ground troops – numbering approximately three million men – attacking over a 2,000-mile long front, supported by preliminary air bombardments and featuring spectacularly deep thrusts into Soviet territory, especially by General Fedor von Bock's Army Group Centre. By mid-July von Bock's Panzer tank pincer movements had snapped shut around Minsk and 290,000 Soviet soldiers had been taken prisoner, with 2,500 tanks and 1,400 guns captured. By the end of that month, Smolensk had also yielded up 100,000 prisoners, 2,000 tanks and 1,900 artillery pieces.

It took a fortnight before Stalin, who seems to have suffered some form of mental breakdown when brought the news of his ally's betrayal, was capable of operating effectively. Before the German advance began to run out of momentum in the sub-zero temperatures of November and December 1941, they had captured much of European Russia. No fewer than 665,000 Red Army troops had surrendered at the fall of Kiev alone. Leningrad was subjected to a 900-day siege so dreadful that cannibalism was resorted to, and the Wehrmacht even reached the suburban railway stations of Moscow itself.

The struggle over the next four years saw some of the most bitter and hard-fought military engagements in the history of Mankind, where the Red Army fought with outstanding courage and the Russian people endured incredible hardships for their Motherland. In this merciless Manichean clash, the rules of warfare were discarded, as the two vast armies fought – often house-to-house and street-to-street – across half a continent. Historians estimate that perhaps as many as twenty-seven million people perished in the war in the East, the overwhelming majority – perhaps as many as 90% – Soviet citizens. Of course the Russians were used to brutal treatment: Stalin had sentenced between seven and eight million of his own citizens to execution or a living death in the gulag concentration camps, and the struggle of the Great Patriotic War was an extension of that horror.

Nor did Stalin's war against his own people end when Russia was invaded. In order to instil discipline in the Red Army, Soviet commissars ordered the liquidation of thousands of Russian soldiers. Researching for his bestselling book *Berlin: The Downfall* in the Russian Ministry of Defence archive in Podolsk, the historian Antony Beevor found the files of no fewer than 13,500 soldiers – more than an entire division – who had been shot by their own side for cowardice, desertion, drunkenness, 'anti-Soviet agitation' or treachery.[58]

'Treachery' might simply mean that they had surrendered to the Germans after further resistance was futile.

Once the Germans had been turned back – losing half-a-million men in the battle of Stalingrad in the autumn and winter of 1942 – and forced on to the defensive, Stalin set about ensuring that every country through which the Red Army passed had a pro-Soviet government installed. No genuine independence was allowed to any territory 'liberated' by the Russians: free expression was crushed, opposition politicians were arrested, democracy was stifled. With nine million men mobilised in the Red Army, Stalin could do whatever he wished.

Nonetheless, it was vital for the English-speaking peoples to aid the Soviets to the best of their ability after Hitler's invasion of Russia. On 12 August 1941, even while Churchill and Roosevelt were still meeting at Placentia Bay in Newfoundland discussing how to help Russia, two squadrons of British fighters comprising forty aircraft left Britain on board HMS *Argus* bound for Murmansk. Under the command of a New Zealander, Wing-Commander Ramsbottom-Isherwood, they reached the Soviet naval base at Polyarnoe, near the sea-port which was to become a huge receiving depot for Allied supplies over the next four years. Although the RAF needed every aircraft it could get for home defences in the summer of 1941, even so it transported planes to help the USSR in her desperate moment of trial.

They called it 'the hell run', and no wonder. The seventy-five Arctic convoys of the Second World War faced the horrific combination of sustained German attacks and atrocious weather conditions over four years as they attempted to re-supply Russia between 1941 and 1945. In all, over 3,000 Britons lost their lives in the vital campaign that was aimed at giving the Soviets the necessary tools with which to defeat Hitler on the Eastern Front.

Between August 1941 and the end of the war, no fewer than 5,000 tanks and 7,000 aircraft were delivered to the Red Army by the English-speaking peoples, as well as vast amounts of other vital war matériel, such as fifty-one million pairs of boots. They did much to keep Russia going in her most desperate hour. The first outbound convoys, which all had the codename PQ followed by a consecutive number, started out from Iceland to Murmansk and Archangel via Bear Island. On 28 September, PQ 1 set out loaded with military supplies and large quantities of the vital raw materials that Stalin had requested, including rubber, copper and aluminium. Soon afterwards, Churchill announced that Britain's entire tank production for the month of September would be despatched to Russia. They were badly needed, for on 2 October the Nazis launched their major attack – codenamed Operation Typhoon – on Moscow.

The icy winter of 1941/2, which did so much to destroy Hitler's dreams for turning European Russia into an Aryan colony, also fell heavily on the Arctic convoys. Veterans later recalled the way that ice froze on their breath, and how they had to venture onto the decks regularly in freezing temperatures in order to hack off ice to prevent ships capsizing. The route taken was the hazardous one that comprised seventeen days' sailing around the North Cape above Norway and Finland, through Arctic storms and potentially lethal ice floes, through German air strikes and U-boat attacks.

Yet the schedule of deliveries was not reduced, but rather increased as time went on. On 12 October 1941, twenty heavy tanks and 193 fighter aircraft arrived and only a week later 100 fighters, 140 heavy tanks, 20 Bren guns, 200 anti-tank rifles and 50 heavy guns were unloaded at Murmansk. Three days later, 200 fighters and 20 heavy tanks were docked. In the bitter fighting across the Eastern Front, Allied ammunition and equipment were used by the Red Army in ever-increasing amounts.

On 14 May 1942, the cruiser *Trinidad* was sunk by German torpedo-bombers west of Bear Island, as it escorted Allied merchant ships to Murmansk. Eighty sailors lost their lives, twenty of whom had been injured after HMS *Edinburgh* was sunk on the same route two weeks previously. Yet that month more than 100 merchant ships reached their destination to deposit their precious cargoes of arms, just as the Germans launched their major offensive towards Stalingrad.

What historians have described as 'one of the most serious setbacks of the war' occurred on 4 July 1942, three days after PQ 17 had been spotted by German submarines and aircraft. It was a large convoy, comprising twenty-two American, eight British, two Soviet, two Panamanian and one Dutch merchant ship, protected by six destroyers and fifteen other armed vessels. On the morning of 4 July, four merchant vessels were sunk by Heinkel torpedo-bombers, and fearing that four powerful German warships – including the battleship *Tirpitz* – were on their way towards the convoy, the First Sea Lord, Admiral Sir Dudley Pound, ordered PQ 17 to scatter.

The German warships had been ordered to intercept the convoy, but they were then instructed to turn back. Instead, the scattered convoy was mercilessly picked off from the air and by U-boats. No fewer than nineteen Allied ships were sunk and only eleven reached Archangel. Of the 156,500 tons loaded on board the convoy in Iceland back on 27 June, 99,300 tons had been sunk, including no fewer than 430 of the 594 tanks. It was astonishing that not more than 153 sailors were drowned. Further tragedy was to follow three days later, when the returning convoy, QP 13, ran into a British minefield off Iceland through bad navigation and a further five merchant ships were sunk. There were other such serious setbacks during the war, including the sinking of thirteen of the forty ships of Convoy PQ 18, although it did at least manage to take a severe toll of its attackers, destroying four German submarines and forty-one aircraft.

It was only in late 1943 that the Allies began to win the Arctic campaign; in November and December, three eastbound and two westbound Arctic convoys reached their destinations without loss. Grand Admiral Karl Dönitz despatched the German heavy battle-cruiser, *Scharnhorst*, to try to tip the balance back in the Reich's favour. Instead, after a brilliant naval action commanded by the C-in-C Home Fleet Admiral Sir Bruce Fraser and Rear-Admiral Burnett in HMS *Belfast*, *Scharnhorst* was sunk on Boxing Day, dealing the Germans a devastating blow, just as they were also reeling from a major defeat inflicted by the Red Army on the Eastern Front.

Meeting at Placentia Bay in Newfoundland on 12 August 1941, Roosevelt and Churchill signed a declaration that the *Daily Herald* newspaper two days later dubbed the Atlantic Charter. They stated eight principles 'on which they base their hopes for a better future for the world'. Neither country sought territorial aggrandisement or border changes 'that do not accord with the freely expressed wishes of the peoples concerned', but both respected 'the right of all peoples to choose the form of government under which they will live'. They would open up equal access to trade, improve economic advancement for victors and vanquished alike, and, 'after the final destruction of Nazi tyranny', they would establish a peace in which 'all men in all the lands may live out their lives in freedom from fear and want'. If any such document were signed by a president and premier today, it would be denounced as neo-conservative and utopian. As it was, it was politically brave of Roosevelt to look forward to the 'final destruction' of a 'tyranny' with which the United States maintained a fast-disappearing neutrality, since the forces of isolationism were still strong in the United States.

Roosevelt's isolationist opponent, the aviator hero Captain Charles A. Lindbergh, spoke for many Americans who still clung to the principles enunciated by Washington in his Farewell Address. Lindbergh's $33\frac{1}{2}$-hour, 3,800-mile solo flight from New York to Paris on 20–21 May 1927, aged only twenty-five, was greeted by headlines such as, 'New York Millions Hail His Triumph. Throngs Crowding Streets Cheer as "Greatest Sporting Event in History" Ends Brilliantly. Showers of Paper Flood Streets as Goal is Won. Women, Gripped in Dramatic Tension, Weep.'[59] He was by far the greatest celebrity in the world at the time. 'Thirty million turned out to see his triumphal tour of the USA, a quarter of the entire population, kings and presidents fawned, ladies swooned. Modern celebrities are nothing in comparison: "Lucky Lindy" was half a god, half a figure in a Norman Rockwell painting, simple, direct, confident.'[60] And correspondingly hard for Roosevelt to beat. Lindbergh's popularity increased even further – especially amongst women – after the horror of the abduction of his twenty-month, blond, curly-haired son in 1932. Newsreels of the baby in his cot prior to the kidnapping made harrowing viewing.

For all that Lindbergh claimed that he spoke only as a private citizen, such was his charisma and fame – even fourteen years after 1927 – that his joining the board of the isolationist, anti-war America First Committee in April 1941 was immensely damaging for Roosevelt's brand of internationalism. Lindbergh's confident prediction that Britain would lose the war was similarly influential. Speaking at an America First rally at Des Moines, Iowa, on 11 September 1941, Lindbergh identified the 'powerful elements' attempting to 'entangle' the United States in European affairs as 'the British, the Jewish and the Roosevelt Administration', including 'a number of capitalists, Anglophiles and intellectuals'.[61] Instead of 'agitating for the war', Lindbergh argued, 'The Jewish groups in this country should be opposing it in every possible way, for they will be among the first to feel its consequences. . . . Their greatest danger to this country lies in their large ownership and influence in our motion pictures, our press, our radio, and our government.' Yet despite this kind of mob oratory, after the war Lindbergh won the Congressional Medal of Honor in 1949 and the Pulitzer Prize in 1954.

The same day as Lindbergh's Des Moines speech saw Roosevelt announce that he had signed a shoot-on-sight order, authorising American ships to fire upon any vessel they deemed to be acting in a hostile manner. A statement was published by distinguished public men such as Philip C. Jessup, Edwin S. Corwin, Ray Lyman Wilbur, Charles A. Beard and Igor Sikorsky describing it as 'a grave threat to the constitutional powers of Congress and to the democratic principle of majority rule', yet the order stood. A sign of how knife-edged the debate was can be seen in Roosevelt's demand in October 1941 to amend the Neutrality Act in such a way that US merchant vessels should be armed and that warships should be allowed to enter combat zones. The President got his way, but a shift of only ten votes in the House would have scuppered him.

'The USA won't believe in the brotherhood of man till bombing planes can cross the Atlantic,' quipped the British wit Geoffrey Madan. In fact, it was the Pacific rather than the Atlantic that saw Japanese bomber planes attack America, and the result was not so much a new-found belief in the brotherhood of man as an instinctive tightening of the brotherhood of the English-speaking peoples. After the news arrived at Chequers – over the radio rather than through official channels – of the surprise attack on Pearl Harbor, Churchill was as good as his promise to declare war on Japan within the hour.

The Japanese lost only twenty-nine of the 343 aircraft they sent to bomb the American naval base, as well as one submarine and two midget submarines. American losses were much more severe, but because the two aircraft carriers had been on patrol at sea they survived unscathed to form the nucleus of

American naval vengeance. No fewer than 188 USAAF planes were destroyed and 159 damaged, and 2,403 US servicemen lost their lives. Of the battleships in harbour, *Arizona* was sunk, the capsized *Oklahoma* was a total loss, *West Virginia* and *California* had sunk upright but were re-conditionable after many months' work, *Nevada* was beached, but *Maryland* and *Tennessee* – though damaged – were soon back in action. Moreover, the Japanese had missed the enormous aviation fuel tanks, which had they exploded would have caused terrible devastation to the port.[62] Though undoubtedly severe, the attack was thus not the knockout blow for which the Japanese had hoped.

The following day, President Roosevelt addressed the US Congress, to say, 'Yesterday, December seventh, nineteen forty-one – a date that will live in infamy – the United States was suddenly and deliberately attacked by naval and air forces of the Empire of Japan.' One can make too much of the 'infamy' involved in Japan attacking Pearl Harbor without a prior declaration of war; after all, Churchill had authorised the bombardment of the outer Dardanelles' forts on 3 November 1914, two days before Britain and France declared war on Turkey. In war, at least in the twentieth century, gentlemanly conduct has tended to come second to the advantage of surprise.

Equally, too much is often made of the Americans' entering the Second World War 'late', since it was simply not their conflict until Japan and Germany made it so. Britain and France had only entered the Crimean War a full year after that had started, yet few accused those powers of untoward tardiness. A nation that picks fights unnecessarily is more at fault than one that carefully prepares for those it knows it cannot avoid. The fact that President Roosevelt recognised that one day it would necessarily become the United States' conflict, and prepared America for that as best he could within the letter if not the spirit of the law, is a tribute to his great foresight, political courage and leadership.

The English-speaking peoples can, however, be indicted for allowing their assumption of racial superiority over the Japanese to lull them into the assumption that they had little to fear from Nippon's army and navy, and especially its air force. After their victory at Tsushima in 1905, it did not do to underestimate the Japanese. Yet it was, absurdly enough, popularly believed that they were genetically too short-sighted to make successful fighter and bomber pilots. In this attitude the English-speaking peoples resembled those members of the British War Council of 1915, who had been willing to 'gamble upon the . . . inferior fighting qualities of the Turkish Armies' at Gallipoli.

As well as being – as it turned out, literally – a suicidal decision, Hitler's declaration of war on the United States on Thursday, 11 December 1941, was an unnecessary one. Germany had no treaty obligation to Japan to support her, and indeed had not been forewarned and thus had no moral obligations either (insofar as such matters ever weighed upon the Führer's mind). Unlike

Poland, France or Norway, and possibly Britain, America was a completely un-invadeable country; there were simply not enough stormtroopers to occupy the American continent, so the best the Führer could ever have hoped for was a perpetual armed stalemate. Had the United States not entered the war against Germany, it would have been impossible for Britain to invade mainland Europe in 1943 and 1944, and her strategic bombing campaign would have been far less effective without the USAAF. Furthermore, Hitler would have been able to maintain minimal forces on his western front, deploying virtually the entire Wehrmacht against the Red Army, which would moreover not have been provided with the extensive support in war matériel it received from America.

It is hard to know what was going through Hitler's mind when he made the declaration of war, which played so perfectly into Roosevelt's hands. (During his speech to Congress, absolutely no mention had been made by the President either of Hitler or Germany.) Clues can be gleaned from Hitler's unpublished sequel to *Mein Kampf*, however, much of which was about the United States and the duty of the Nazi Party to prepare Europe for the coming clash with that country, which he viewed as inevitable.

When historians debate why Hitler declared war in December 1941 on a power that he must have appreciated could not be invaded by the Axis – and was thus unconquerable – they should turn to chapter nine of this unremittingly polemical work. It affords a fine glimpse, along with *Mein Kampf*, his Berghof 'table talk' and the verbatim reports of his military conferences, into the thinking about America that was going on inside the diseased mind of the Führer.

The chapter titles in themselves give a hint to the remorseless, unrelenting nature of Hitler's thinking with regard to foreign policy. Chapter two is entitled 'Fighting, not Industry, Secures Life'; chapter three: 'Race, Conflict and Power'; chapter six: 'From the Unification of the Reich to a Policy of Space'; chapter ten: 'No Neutrality'; and so on. Only an utterly unappeasable mind could also have penned a work whose ninth chapter is entitled: 'Neither Border Policies nor Economic Policies nor Pan-Europe'.[63] Utter victory on the dictator's own terms was all that could be accepted. Small wonder that the closer Hitler got to power, the less he wanted his second book to be published.

Although there were a good deal of the usual Hitlerian epithets in this book – 'Anyone who does not wish to be the hammer in history will be the anvil' – there were also some unusual aspects of the Führer. Here, for example, is Adolf Hitler the pacifist: 'Wars that are fought for objectives that by their very nature cannot ensure the replacement of lost blood are an offence against the people and a sin against the future of the people.' Or Hitler the Marxist academic: 'England needed markets and sources of raw materials for its goods. That is the point of the English colonial policy.' Or Hitler the platitudinous modern politician: 'Politics is history in the making. A people, collectively, is only a large number of more or less equal individual beings. Its strength lies in the quality of the individuals who form it.' Or Hitler commending the spirit of 1776: 'The farm boy who emigrated to America 150 years ago was the most determined and boldest in his village.' There was even something of the business-management 'how to' book about the Führer's philosophy: for example, when he wrote that, 'In everyday life a person with a clear-cut life goal, which he strives to reach in all circumstances, will always be superior to others who are aimless.'

Hitler's own 'life goal' was very obvious. 'Any coalition of Powers that turns against Germany can from the outset depend on France,' he states, and thus the destruction of France has to be the first duty of any future chancellor of Germany. As has already been seen, Hitler was under no illusions about the capabilities of American capitalism. He understood how the Ford Motor Company could undercut competitors and mass-produce automobiles cheaply and efficiently. So why did he not apply the logic of American domination of the global motor industry – which he correctly diagnosed as 'a matter of immeasurable future significance' – to the geopolitical realities of declaring war against such a behemoth? If the Americans could produce vast numbers of motor cars, were they not also going to be capable of doing the same thing for guns, tanks and aeroplanes?

Fourteen years after 1928, once he had declared war against the United States, Hitler began ritually and regularly to denigrate its racial make-up, as on the evening of 7 January 1942 when he said,

> I don't see much future for the Americans. In my view, it's a decayed country. And they have their racial problem, and the problem of social inequalities. ... Everything about the behaviour of American society reveals that it's half Judaised, and the other half negrified. How can one expect a State like that to hold together ... a country where everything is built on the dollar?[64]

There was plenty more of this kind of generalised talk about America, combined with some staggeringly naïve factual inaccuracies about the country; he seemed to think that 80% of US revenue was 'drained away by the public purse', whereas federal taxes as a percentage of GDP only took 7.6 cents in the dollar in 1941. Never having visited America, Hitler was also perhaps hoping to encourage his listeners and himself to believe in ultimate victory, despite his ludicrous hubris in having declared war.

The discussion at Hitler's lunch table on 11 December 1941 revolved, unsurprisingly, around America's fighting abilities. General Fritz Halder 'was scornful, drawing from his experiences in the First World War. American officers could stand no comparison to Prussians – they were businessmen in uniform who shivered for their lives. In the art of war they had a long way to go.'[65] A few days later, Hitler, dropping in to the room of the head of his personal household, Sturmbannführer Heinz Linge, in order to listen to popular music on his radio, saw his military attaché General Lieutenant Rudolf Schmundt having a glass of schnapps with Linge. 'Read that, Schmundt,' Hitler said, handing him a report on U-boat sinking of American shipping. 'Do you see how good open war against America is for us? Now we can really strike.' The conversation got round to Hitler's 'contempt for the Americans. He pointed out that an American car had never won an international tournament; that American aircraft looked fine, but their motors were worthless. That was proof for him that the much lauded industries of America were terribly overestimated. They didn't really have to perform, only in an average way, and benefited from lots of discounting.'[66] In fact, it was the Führer who was badly discounting American industrial capacity, leading him utterly to underestimate his foe. This was all the more astonishing considering America's contribution to victory over Germany in 1918, when Hitler was fighting on the Western Front.

NINE

United and Conquering

1942–4

General George C. Marshall and Field Marshal Lord Alanbrooke – Australia looks to the USA – The State of the Union 1942 – The Wannsee Conference – Auschwitz – Singapore surrenders – The fall of Tobruk – The battle of Midway – Disaster at Dieppe – Victory at El Alamein – Maoris at Monte Cassino – The Beveridge Report – Squabbling generals – Churchill's Harvard speech – MacArthur's code-breakers

'We have awakened a sleeping giant and filled it with a terrible resolve that will soon be turned upon us.'   Admiral Isoroku Yamamoto, December 1941

'We Nazis never said we were nice democrats. The problem is that the British seem like sheep or bishops, but when the moment comes they are shown to be hypocrites, and they become a terribly tough people.'
Reinhard Spitzy, Joachim von Ribbentrop's private secretary[1]

The decision of Roosevelt and the US Chiefs of Staff to concentrate on defeating Germany before directing the full might of the United States against Japan represents one of the greatest acts of American statesmanship of the twentieth century. Much militated against it; Japan had attacked Pearl Harbor, after all, whereas Germany had so far deliberately not attacked US interests. Japan was much closer to the United States, had captured the US possessions of Guam and Wake Island the day after Pearl Harbor, and was in the process of attacking the Philippines. (Manila fell in January 1942.) American newspapers and public opinion were clamouring for an immediate punishment of Japan's 'infamy'. Yet Roosevelt and General George C. Marshall held firm and decided to fight a war of containment in the Pacific until Germany had been defeated, and only thereafter snuff out the Empire of the Rising Sun.

The Commander-in-Chief and the Chairman of the Joint Chiefs of Staff had correctly identified Germany as being the stronger and thus the more dangerous threat, and so decided to deal with her first. So, however illogical it might seem at first glance, it was in North Africa that the United States struck the totalitarian Axis powers first, in response to an attack thousands of miles away in the Pacific Ocean. The 'Germany First' policy was the one that Churchill and Stalin desperately needed America to pursue, of course, but it was in her own best interests that she did as well. Although she was fighting a war on two fronts – every strategist's nightmare – she soon proved that she had the necessary financial, industrial and manpower resources, combined with geographical security, to bring both to successful conclusions.

Churchill and Roosevelt spent 120 days – over four months – in each other's company on nine separate occasions during the four years in which they were allies during the Second World War.[2] Overall, considering how much more the United States was contributing to the war effort than Britain in terms of men, money and matériel by the autumn of 1942, it is impressive how often Churchill managed to get his way in the great strategic issues that faced the Western allies. Crucially, in the summer and autumn of 1942, he persuaded the Americans not to undertake a risky cross-Channel invasion in 1943.[3] Although he was very conscious of being the junior partner from 1943 onwards – and was regularly infuriated by it – Churchill defended his corner with tenacity and, except at Yalta, where there was virtually no room for manoeuvre between Stalin and Roosevelt, he generally got his way.

One of the reasons that the English-speaking peoples pursued such an overall sound strategy for winning the Second World War was because of the effective working relationship that the US Chief of Staff, General George C. Marshall, established very early on with the British Chief of Staff, General (from 1944 Field Marshal) Sir Alan Brooke (later Lord Alanbrooke). Born at Uniontown, Pennsylvania, in 1880, Marshall was educated at the Virginia Military Institute rather than West Point and was commissioned in 1901, serving thereafter in the Philippines. He showed his exceptional leadership and organisational talents early on. Before the Great War his superior officer Lieutenant-Colonel Johnson Hagood was asked in a proficiency report whether he would like Marshall to serve under his command. 'Yes,' he replied, 'but I would prefer to serve under his command.'[4]

During the Great War, Marshall was chief of operations for the American 1st Army, advancing to colonel. Between 1919 and 1924 he served as an aide to General Pershing, with whom he developed a mutual admiration and deep friendship. Although Marshall ranked behind twenty-one major-generals and eleven brigadiers, he was appointed US Chief of Staff on 1 September 1939, the day that Germany invaded Poland. He inherited an army of only 174,000 men and 1,064 aircraft, but by 1945 this had grown under his stewardship to one of 8.3 million men and 64,000 planes, the fastest and greatest military mobilisation of any society in human history.

Alanbrooke, an Anglo-Irishman who was born at Bagnères-de-Bigorre in France and educated at the Royal Military Academy, Woolwich, joined the Royal Field Artillery in 1902 and served as a General Staff officer in the Great War. He commanded the 2nd Corps of the British Expeditionary Force in the retreat to Dunkirk, covering the flank left open by the Belgian surrender. Sir James Grigg, later Secretary for War, said that, 'by almost universal testimony, it was due largely to his skill and resolution that, not only his own corps, but the whole BEF escaped destruction on the retreat'. After a period as Commander-in-Chief Home Forces, building up Britain's defences against the expected invasion, he became Chief of the Imperial General Staff (CIGS) in December 1941.

Marshall and Alanbrooke had plenty of hard-fought disagreements during the war, but the two men were able to communicate as soldiers who both felt that they – rather than their political masters – should be directing overall strategy, and they recognised that they could do this far more effectively if they tried to work with one another. Fortuitously, although they differed profoundly over the timing of campaigns, Marshall was willing to accept Alanbrooke's overall strategy for how the war should be won. This was laid out in broad terms by the latter in a post-war note attached to his diary entry for 17 July 1942:

> It was evident that if Russia cracked up, the Germans could concentrate the bulk of their forces in France and make an invasion quite impossible. Under those circumstances our only hope would be to operate in Africa. But in any case from the moment I took over the job of CIGS I was convinced that the sequence of events should be:
>
> a) liberate North Africa
> b) open up Mediterranean and score a million tons of shipping
> c) threaten Southern Europe by eliminating Italy
> d) then, and only then, if Russia is still holding, liberate France and invade Germany.[5]

The key, therefore, to winning the war was continued Russian resistance in the East. Given that, the sequence for the English-speaking peoples was to be North Africa – Italy – France – Germany. This was agreed in broad terms at the 'Symbol' Conference in Casablanca between 14 and 23 January 1943, and although the Americans very much wanted to invade France much earlier than Britain did, it formed the basis of strategy for the rest of the war in the West. Churchill hoped to alter it after the North Italian stage to include an attack in the Balkans and subsequently Austria, but this was rightly stymied by Roosevelt and Marshall.

Sadly for his later reputation, but entertainingly for historians and the general reader, Alanbrooke was an exceptionally caustic diarist. The fact that he and Marshall never fell out permanently over strategy did not mean that he was any kinder in his private estimation of Marshall than he was about virtually anyone else. The highest reaches of the grand strategy of the English-speaking peoples might have run remarkably smoothly during the war, all things considered, but that does not imply that there weren't furies and resentments constantly swirling under the surface, especially under the brass hat of the seemingly imperturbable Ulsterman who was Britain's CIGS.

It is often said that History goes to the victors; in fact it often tends to go to the diarists. Yet Alanbrooke's diary was simply too scathing to aid his repute. A typical entry for 15 April 1942 reads:

> After lunch I had Marshall for nearly two hours in my office explaining to him our dispositions. He is, I should think, a good general at raising armies and providing the necessary links between the military and political worlds. But his strategical ability does not impress me at all!! In fact in many respects he is a very dangerous man whilst being a very charming one!

Yet before Alanbrooke's strictures on Marshall – one of the towering figures of the war rightly described by Churchill as 'the Architect of Victory' – are taken at face value, it ought to be borne in mind that his diary was largely a way for the CIGS to let off steam in a manner that would not result in disaster.

Alanbrooke was cutting about almost everybody, except General Smuts and Joseph Stalin, whom he seems to have admired unreservedly. For the rest, he characterised General Alexander as 'a very, very, small man and cannot see big', George Patton as 'at a loss in any operation requiring skill and judgment', Anthony Eden as 'like a peevish child', Charles de Gaulle as 'a pest', Max Beaverbrook as 'an evil genius who exercised the very worst of influence on Winston', General Weygand as a careerist, Chiang Kai-shek 'evidently [had] no grasp of war', Oliver Lyttelton had 'few ideas in his head', Joseph Stilwell was 'nothing more than a hopeless crank with no vision', General Chennault was 'a very gallant airman with a limited brain', Klimenti Voroshilov had 'nothing in the shape of strategic vision', and Herbert Morrison 'appears to be a real white-livered specimen!' Finally, Lord Mountbatten 'lacked balance' for the job of Supreme Allied Commander in South-East Asia (although admittedly Alanbrooke was right about the last).[6]

Equally, it should not be thought that the Americans were merely the abused victims of British ire. Their undisguised disdain for the British Empire, especially once it began to collapse in the Far East, was perhaps understandable from a people who wrested their own birthright from that same entity, but it was undisguised nonetheless. Lord Mountbatten's SEAC, which stood for South-East Asia Command, was popularly nicknamed 'Save England's Asian Colonies'. Roosevelt joked about Churchill's drinking, and the distinguished historian Christopher Thorne's view of Anglo-American relations during the war was summed up in the title of his history of those years, *Allies of a Kind*.

Much has been made of Alanbrooke's fury with and occasionally disdain for Churchill, as expressed almost constantly in his wartime diaries, but a passage from them ought also to be quoted, where the CIGS wrote after the war:

> Throughout all these troublesome times I always retained the same unbounded admiration, and gratitude for what he had done in the early years of the war. One could not help also being filled with the deepest admiration for such a genius and super man. And mixed with it all there were always feelings of real affection for the better side of him. In reading these diaries it must be remembered that I had a long and trying time with him and that the writing of this diary presented the only safety valve that I had to pent up feelings of irritation which I could share with no-one else.[7]

The 'danger' that Alanbrooke thought Marshall posed in April 1942 was that he was 'going 100% all out' on a plan to invade France, since Admiral Ernest King was 'proving more and more of a drain on his military resources, continually calling for land forces to capture and hold bases', while General Douglas MacArthur, who was in Australia having been flung off the Philippines, was asking for forces to enable him to carry out his vow made to the *New York Times* of 20 March after he ordered the evacuation to Australia: 'I came through and I shall return.'

Alanbrooke accepted that Marshall's insistence on a cross-Channel re-entry onto the European continent was 'a clever move which fits in well with present political opinion and the desire to help Russia. It is also popular with all military men who are fretting for an offensive policy.' But, and here Alanbrooke's sarcasm broke through yet again, 'his plan does not go beyond just landing on the far coast!! Whether we are to play baccarat or chemin de fer at Le Touquet, or possibly bathe at Paris Plage is not stipulated! I asked him this afternoon – do we go east, south or west after landing? He had not begun to think of it!!'[8] (It is more likely that Alanbrooke did not ask whether they should go west, since that would have sent them back into the English Channel, but the point was made.)

For all the undeniable complexity of the shifting relationships between Churchill, Roosevelt, Marshall and Alanbrooke, a delicate balance between soldiers and politicians, Americans and Britons, early and late cross-Channel attack, was successfully struck. Few historians believe that the war could have been significantly shortened by fighting it in a different way; books such as John Grigg's *1943: The Victory That Never Was* are as rare as they are controversial. Despite the enthusiasm of opponents of the Special Relationship for highlighting differences of strategic opinion between the Atlantic allies during the war, overall the grand overarching concepts that led to victory were agreed upon and then followed through. When one considers the momentous issues and numbers of lives at stake, the agreement over the high strategy pursued in the Second World War rates as one of the finest achievements of the English-speaking peoples.

It was all the more remarkable considering that Churchill had to grasp the mortifying fact that his beloved British Empire was ceding primacy to another power, however friendly. After the war, Alanbrooke recorded how in November 1943 there were 'new feelings of spitefulness which had been apparent lately with Winston since the strength of the American forces were now building up fast and exceeding ours. He hated having to give up the position of the dominant partner which we had held at the start.'[9] Of course this attempt to delve into Churchill's mind might have been wildly unfair, but if Churchill had felt resentment at the prospect of the Empire he loved taking a lowlier place in the coming world order, it would only have been natural. Nor was he a man to keep his feelings under control. Equally, although the preponderance of American production of war matériel was undeniable, until the summer of 1944 Britain and the Commonwealth had more divisions in fighting contact with the enemy than had the United States.[10]

On 27 December 1941, while Roosevelt and Churchill were in Washington and only two days after the fall of Hong Kong, the Australian Prime Minister John Curtin wrote a New Year message to the Australian people in the form of an article in the *Melbourne Herald* in which he stated that, 'Without inhibitions of any kind, I make it quite clear that Australia looks to America, free of any pangs as to our traditional links or kinship with the United Kingdom. ... We shall exert all our energies to shaping a defence plan with the United States as its keystone.' This went far beyond the appeals that Menzies had made to Roosevelt in May and June 1940, and understandably, in the words of the historian of Australian foreign policy, 'angered Churchill, aroused misgivings in London and caused controversy in Australia itself'.[11]

Churchill's anger can be gauged from the diary note of his doctor, Charles Moran, which, although it is a flawed source as parts of it were written up much later, certainly seems to ring true for the entry of 9 January 1942: 'The PM is in a belligerent mood. He told us that he had sent a stiff telegram to Curtin. ... The situation in Malaya was making Australia jumpy about invasion. ... London had not made a fuss when it was bombed. Why should Australia?' In fact, the cable was greatly de-fanged before it had been sent, for as Moran commented on Churchill's attitude towards Australians in general, 'He liked them as men and respected them as fighting soldiers.' It was not just in Australia that people predicted an invasion; on 27 January, 1942, Michael

Stewart bet John Lawrence £5 at Brooks's 'that the Continent of Australia is invaded by the Japanese within six months from today'.

The *Sydney Morning Herald* denounced Curtin's remarks as 'deplorable', whereas the *Canberra Times* merely described them as 'realistic' and 'unequivocal'.[12] In fact they were all three and represented the start of a fundamental reorientation of the Australian global outlook, which had already seen the way that power had shifted and had pragmatically altered the emphasis of her protection from the waning English-speaking power to the waxing one. It did not result in greater influence in either Washington or London, however; indeed, it was to be another six months before Curtin even discovered that the Arcadia Conference had decided upon a 'Germany First' policy. It would have been better for Curtin if he had not publicly voiced what everyone knew to be the case anyhow.

Australia's losses in the Second World War amounted to 39,798 killed, bringing to over 100,000 the number of people from that lightly populated country – seven million people in 1939 – who perished in the two great wars of the twentieth century. The name of each of them, and of those who fell in every other conflict since the Sudan in 1885, is recorded on the walls of the majestic Australian War Memorial in Canberra, where families place poppies alongside their loved-one's names. It also houses fifty-eight of the no fewer than ninety-six Victoria Crosses that Australians have won since the decoration was inaugurated in 1856. It is a deeply affecting shrine.

In his State of the Union Address on Tuesday, 6 January 1942, President Roosevelt set out the war production targets of the United States, provoking, in his biographer's estimation, 'perhaps the greatest applause he ever received in his dozens of addresses to the Congress'.[13] Well might they applaud; the figures that he proposed were astonishing and included: 60,000 aircraft in 1942, rising to 125,000 in 1943; 45,000 tanks in 1942, rising to 75,000 in 1943; six million tons of merchant shipping in 1942, rising to ten million the following year. Of the $59 billion budget he submitted on that occasion, no less than $57 billion was to be devoted to military expenditure. 'These figures', said Roosevelt to a huge ovation, 'will give the Japanese and the Nazis a little idea of just what they accomplished in the attack on Pearl Harbor.'

In the event, and from virtually a standing start, these targets were not simply reached but were comprehensively beaten. In all during the war, the United States mobilised 14.9 million people, more than the 12.5 million of Germany and twice the 7.4 million of Japan. She also spent a total of $350 billion, more than Germany's $300 billion, Russia's $200 billion, the UK's $150 billion or Japan's $100 billion.[14] She turned over her huge industrial capability to war production and harnessed the limitless energies of her people to creating the tools with which the Allies finished the Axis. This was not simply done out of fury at Japan's perfidy, but because Roosevelt understood that history stood at a fulcrum moment. In 1928, he had declared in an article in *Foreign Affairs* that only by international collaboration could the United States 'regain the world's trust and friendship' after her 1919 decision to return to isolationism. By becoming the production power-house of the struggle against fascism, he hoped to turn America back towards engaged internationalism.

In February 1945 at Yalta, at an exceptionally festive dinner attended by Churchill, Roosevelt and Stalin, the 'Big Three' – there were forty-five toasts in total – Stalin acknowledged that back in 1941, President Roosevelt's country 'was not seriously threatened with invasion, but he had a broader conception of his national interest and even though his country was not directly imperilled, he has been the chief forger of the implements that have led to the mobilization of the world against Hitler'.[15] It was a valediction as wise as it was generous.

In the process of reaching his ambitious production targets, Roosevelt accreted to himself enormous powers over the American economy and society in the year 1942. In January, the US Office of Production Management (OPM) banned the retail sale of new cars and passenger trucks in order to shift the focus of the automotive industry to the production of military vehicles. (There were 38.8 million private cars in the US at the time, against 2.2 million in Britain.) On 10 February, the last automobile was produced in the USA until 1945; manufacturers instead geared up to produce tanks, aircraft and manufacturing equipment. In April 1942, the OPM was given the power to fix all prices except farm produce and to stabilise the rents in over 300 communities, affecting no fewer than eighty-six million Americans. An Office of Economic Stabilisation was established to monitor wages and salaries, and an Office of Price Administration to fix prices, and on 3 October they were empowered by Congress to freeze prices, wages and rents.

Tyres and gasoline were rationed to conserve rubber stocks; sugar and then coffee were also rationed. Clocks were put forward one hour on 9 February 1942 for daylight saving, where they stayed for the duration. In June, the Office of War Information was created to monitor US news and to broadcast propaganda. The next month Congress approved the creation of Women Accepted for Voluntary Emergency Services (WAVES) in order to bolster US naval reserves, and later in the year a female reserve of the Coast Guard called Semper Paratus Always Ready Service (SPARS).

One exhibition of executive authority that remains controversial was Roosevelt's signing of Executive Order 9066, sanctioning the internment not only of Japanese nationals but also of Japanese-Americans living along the west coast of the United States. (Canada separated families in enforcing a similar policy.)[16] Although there was a dearth of evidence collected by the FBI that these people constituted a potential fifth column, it was a precautionary measure that is entirely understandable given the exigencies of the situation. Certainly, Britain had done the same thing to her German and Italian communities in 1940, without any stigma attaching to the people interned. (In 1988, the Reagan Administration officially apologised and paid $1.6 billion in compensation to the internees or their heirs.)

The unleashing of American enthusiasm, energy and expertise for the war effort achieved some astonishing results. US Army engineers built the Alaska Highway, a 1,523-mile road connecting Dawson Creek in British Columbia to Fairbanks in Alaska, in less than nine months. The Boeing B-29 'Superfortress' made its inaugural flight, the first of 2,180, to fly with the US Army Air Force. In October 1942, the Bell XP-59A, the first US jet, made a successful test flight at Muroc Lake, California, and the giant American Federation of Labour (AFL) and Congress of Industrial Organisations (CIO) made a patriotic announcement that they would not indulge in strike action for the duration. Back in 1928, Hitler had recognised that the industrial power of the United States was likely to make her the most powerful nation in the world. Now, as Roosevelt promised, he was about to discover quite how crushing that power could be when brought to bear against his Reich.

The effect on the American economy of all this centralised action was wide-ranging. Government purchases of goods and services increased fivefold between 1939 and 1945. As the economist N. Gregory Mankiw records, 'This huge expansion in aggregate demand almost doubled the US economy's production of goods and services and led to a 20% increase in the price level (although widespread government price controls limited the rise). Unemployment fell from 10% in November 1940 to about 1% in 1944 – the lowest level in US history.'[17] Back in 1933, it had been 33%. From a near-disaster in 1937, Roosevelt had been able, through the threat and afterwards the reality of war, to spend America out of her difficulties.

A mere twenty days after the State of the Union speech, on Monday, 26 January 1942, the first GI, Private Melburn Henke from Hutchinson, Minnesota, set foot in Britain. As he stepped off his troopship at Belfast, he was welcomed by the Duke of Abercorn, the Governor of Northern Ireland, and the Air Minister Sir Archibald Sinclair, as the band of the Royal Ulster Rifles played *The Star-Spangled Banner*. Over the next two years no fewer than two million American servicemen were to follow him. (Henke, a twenty-three-year-old ex-waiter, was specifically chosen because of the symbolism of his German background. It was widely reported how his father had told him, 'Give them hell.')[18]

Although there were undoubtedly some severe social (and sexual) problems that were thrown up by the presence of large numbers of American servicemen stationed in Britain during the war – 'Of course the American soldiers are encouraged by these young sluts,' complained Admiral Edward Evans of

London's Civil Defence headquarters to the head of Scotland Yard, Sir Philip Game, about the nightly 'vicious debauchery' in London's Leicester Square – there was a good deal of true love too. No fewer than 60,000 British girls became 'GI brides', marrying American soldiers and going to live in the United States after the war.

Meanwhile, on the morning of Tuesday, 20 January 1942, a group of fifteen senior civil servants met in the dining room of a comfortable villa by the side of Lake Wannsee in south-west Berlin. Working hard there they managed to tidy up and centralise, with Teutonic thoroughness, a government operation that had hitherto been characterised by largely ad hoc actions whose efficiency had widely diverged according to local conditions, personnel and commitment. Now everything would be done with suitably modern, time-and-motion-style, industrialised methods, overseen by the competent authorities.[19]

It was a snowy day and after their business was over they congratulated each other on their successful ninety-minute meeting, while they were served with cognac and cigars. At his trial in Jerusalem in 1961, Adolf Eichmann, who took the minutes, recalled that it was 'conducted quietly and with much courtesy, with much friendliness. There was not much speaking and it did not last a long time.'[20] In an hour and a half the process of the genocide of European Jewry had been co-ordinated by the leading Reich agencies involved. They had streamlined the Holocaust.

Before the Wannsee Conference, writes its historian, Mark Roseman, 'the slippage from murderously neglectful and brutal occupation policies to geno-cidal methods took place initially without a comprehensive set of commands from the centre'. After it, however, there was a very clear set of guidelines as enunciated by its chairman, SS-Obergruppenführer Reinhard Heydrich, head of the Reich Security Main Office, who did almost all the talking. The minutes cover such thorny questions as the distinctions between half-Jews, Jews who had won the Iron Cross in the Great War and Jews who were married to German gentiles.

That any record of such a meeting survived at all is a near-miracle. Of the thirty copies originally made of it, all but one were destroyed by the Nazis before *Götterdämmerung* overcame them in 1945. Yet somehow, copy number 16 of the Wannsee Protocol remained in the Foreign Ministry archives, where it was found in March 1947 – helpfully labelled 'Secret Reich Matter' – by American officials searching for just such documentary evidence of high-level complicity in what Heydrich had referred to at the meeting as 'the final solution of the Jewish question'.

The fifteen state secretaries – ten of whom had university degrees – represented a majority of the twenty-seven government departments that were involved in undertaking the Holocaust. There were men from the Interior

Ministry, the Four-Year Plan Office, the occupational government of Poland, the Justice Ministry, the Reich and Party Chancelleries, the 'Race and Settlement Main Office', the Gestapo, the SS and, of course, the forgetful official from the Foreign Ministry.

Apart from Heydrich and his deputy SS-Obersturmbannführer Adolf Eichmann, whose names will resonate through History for as long as Mankind is capable of differentiating between Evil and Good, the individuals present at the conference are not well known. Experts in the period might recognise the names of Herren Doktors Meyer, Liebbrandt, Stuckart, Bühler, Freisler, Luther, Schöngarth, Lange and so on, but few others will. Yet these were people who finalised the details for the institutionalised and industrialised murder of six million human beings, and who at Wannsee also planned to kill five million more. There can be little doubt that, but for the victory of the Allies, they would have succeeded.

Of course the words 'kill', 'murder' or 'gas' appear nowhere in the minutes, only their euphemisms: 'evacuating the Jews to the east', 'dealt with appropriately', 'eliminate by natural causes', 'transportation and resettlement eastwards', and so on. Yet anyone stupid, perversely literally minded or possibly pro-fascist enough to argue that the meeting was indeed only about what was recorded on paper – evacuation and resettlement – has much explaining to do. Why, for example, was SS-Gruppenführer Hofmann recorded as being 'of the view that extensive use should be made of sterilisation, particularly as the mixed-blood half-Jews, presented with the choice of evacuation, would rather submit to sterilisation'? If evacuation only meant exactly that, why should German Jews prefer to be sterilised than to undergo it? If a Londoner had been asked whether he'd prefer to be evacuated east to Hertfordshire in 1942 or be sterilised, he would have given a very different answer.

After the Jews' journey to the east, Heydrich stated how 'Any final remnant that survives will doubtless consist of the more resistant elements. They will have to be dealt with appropriately, because otherwise, by natural selection, they would form the germ cell of a new Jewish revival. (See the experience of history.)' He was clearly not talking about mere sterilisation either.

The numbers of Jews in each country, totalling over eleven million, also shows the extent of Nazi Germany's territorial ambitions in early 1942, before the tide of war turned against it. Although Estonia was proclaimed *Judenfrei* ('free of Jews') already, as a result of the work of *Einsatzgruppen* death squads that had been working there since Operation Barbarossa was unleashed, the Jewish population of every other European country was minutely listed, right down to the 1,300 in Norway, 8,000 in (neutral) Sweden and even 200 in Albania. The Ukraine, reported Heydrich with typical Teutonic thoroughness, was home to 2,994,684 Jews, and with hubristic arrogance he also included Britain's 330,000 Jews in his total. Ireland's 4,000 Jews were also to be liquidated, demonstrating particularly eloquently just how much respect a victorious Germany was planning to have shown for President Eamon de Valera's protestations of Irish neutrality and independence.

'No one raised objections to the proposals for murder,' Roseman states. 'It was too late for that.' Not only was it too late but also several of the fifteen men around the table were committed Nazis who were enthusiastic about getting on with the job. State Secretary Bühler volunteered the opinion that his department 'would welcome it if the Final Solution of this problem could begin' in his area of responsibility, Poland, not least because of 'the particular danger there of epidemics being brought on by Jews'.

The very organised system for mass slaughter introduced at Wannsee also appealed to the bureaucratic minds of the officials, since it was to be an efficient, ordered, less wasteful programme that was to replace the existing system of relatively unorganised mass shootings. Small wonder that the Proto-col was described at the Nuremberg Tribunal as 'perhaps the most shameful document of modern history'. It showed how the lakeside conference represented 'a decisive transition in German policy, a transition from quasi-genocidal deportations to a clear programme of murder'.

Of Adolf Hitler's direct personal involvement and support for the plan, there is of course – not surprisingly considering its nature – no actual signed documentary proof. Yet the circumstantial nature of the evidence suggesting that Hitler ordered and oversaw the Final Solution is overwhelming. Whenever the head of the SS Heinrich Himmler went to visit Hitler for instructions, the Holocaust would enter a new and yet more vicious phase. In the 'co-operative competition' that existed between the officials who ran the Wannsee policy, it was taken for granted that the Führer always supported whoever took the most extreme measures against the Jews. Certainly nobody's career ever suffered as a result of acting upon that assumption. The surprising thing is therefore not that we do not have any signed orders from Hitler specifically authorising the Holocaust, but that we do have memorandum number 16 of the Wannsee Protocol at all.

For all that the civil servants and SS hierarchy gave their orders in the comfort of the Wannsee villa, it required tens of thousands of ordinary Germans to carry them out. This begs the most important question of the twentieth century: how could a people as civilised as the Germans have perpetrated the most ghastly crime of human history? And would the English-speaking peoples, faced with their choices, have behaved any differently? The work done by the Princeton historian Christopher R. Browning on the wartime activities of Reserve Police Battalion 101 – respectable working- and middle-class citizens of Hamburg who became genocidal killers – implies that peer pressure and a natural propensity for obedience and comradeship, rather than Nazi fervour, turned ordinary Germans into foul murderers.

The recruits of Battalion 101 were not selected in any sense for their Nazi ardour: indeed, many joined up largely to avoid active service abroad. They represented a cross-section of German society, and no-one was coerced into killing Jews or ever punished for refusing to do so. Browning does not even think that their motives in executing thousands of Polish Jewish women and children were even primarily anti-Semitic, nor does he believe that there was anything peculiarly German about the Holocaust, except perhaps in the perpetrators' heightened respect for authority and willingness to obey orders. The vast majority of Germans were simply indifferent about what was happening 'out East' and did not want to know. Yet when called upon specifically to help in the genocide, somewhere between 80% and 90% of Battalion 101 acquiesced without undue complaint. After some initial squeamishness, recounts Browning, they 'became increasingly efficient and calloused executioners'.

Only twelve of the Battalion's 500 members actually refused to shoot 1,800 Jews in the woods outside the Polish village of Jozefow on 13 July 1942. During the remainder of that seventeen-hour day of slaughter – interspersed with cigarette breaks and a short midday meal – perhaps another forty-five members or so absented themselves for various reasons. The remaining 85% simply got on with the job of shooting Jewish women and children at point-blank range, even though they knew that there would have been no retribution exacted had they refused. 'At first we shot freehand,' one recalled. 'When one aimed too high the entire skull exploded. As a consequence, brains and bones flew everywhere. Thus, we were instructed to place the bayonet point on the neck.'

The Holocaust was a gigantic group effort, made by many thousands of people who knew precisely what they were doing. It was unique in the way that it combined primal savagery with the most modern techniques. The SS man who switched on the gas faucet connecting the Zyklon B to the underground chambers was only the last in a very long line of the Reich's extermination-workers.

In the United States and Britain there was a disturbing amount of self-censorship in the early 1940s over the fate of the Jews, even in the *New York Times*, which was owned by a Jewish family. The paper devoted only two inches on 27 June 1942, for example, to the news that: '700,000 Jews were reported slain in Poland.' Reports in December 1942 that 'two million Jews had been killed and five million more faced extermination' in Europe appeared only on page twenty. Even when Europe had begun to be liberated, and information was easier to verify, stories about the Holocaust were allocated tiny coverage, tucked away deep inside the paper. When on 2 July 1944 the *New York Times* accurately reported that 400,000 Hungarian Jews had been deported to their deaths, and that 350,000 more were due to be killed in the next weeks, it received only four column inches on page twelve. (The paper found room on that edition's front page to analyse the problem of New York holiday crowds on the move.) This was at a time when much of the rest of the country's newspapers took the *New York Times'* lead.

The SS guards at the extermination camp at Auschwitz used to taunt the Jews there that even after the war was over, and people discovered what had happened there, the world wouldn't care. Although 1.1 million people were murdered by the Nazis there – more than the entire English-speaking peoples' combat and civilian losses during the Second World War – sixty years after its liberation only 55% of Britons have heard of Auschwitz.

When considering why it was morally right for the English-speaking peoples to fight the Nazi regime, even though they were not really initially threatened by it, it is important to consider the testimony of people like Hans Friedrich, a member of the 1st SS Infantry Brigade, which was sent to Poland to reinforce the *Einsatzgruppen* in July 1941, who recalled without emotion that the Jews he murdered in the Ukraine 'were extremely shocked, utterly frightened and petrified, and you could do what you wanted with them. They had resigned themselves to their fate.' That fate involved marching these innocent civilians out of their village to a 'deep, broad ditch. They had to stand in such a way that when they were shot they would fall into the ditch. If someone wasn't dead and was lying there injured, then he was shot with a pistol.' Friedrich openly admitted that he thought 'nothing' when he took part in these massacres. 'I only thought "Aim carefully so that you hit." That was my thought.' He has never had either a bad dream or a troubled conscience since.

In order to kill the maximum number of Jews in the minimum amount of time, the SS tried out several different methods of extermination. Explosives were not chosen both because they were needed for the war effort and because of the difficulty the SS found in getting body parts out of the upper branches of trees. Finally the use of Zyklon B gas, originally used to disinfect prisoners' clothing, was found to be the most efficient means. Moreover, it was the one that caused the SS soldiers themselves the least distress, a factor that – astonishingly enough – seems to have weighed heavily on the minds of both Himmler and the Auschwitz camp commandant, the calmly fanatical Rudolf Höss.

Just as the word 'gulag' has been cheapened by the Left by being attached to policies of the American and British post-war governments, so the word 'Holocaust' needs to be protected too; in 2001, a book entitled *Late Victorian Holocausts* was published seeking to argue that 'opportunistic self-seeking Western powers, especially Britain, [laid] the foundations in ... famine years for a Third World of poverty and dependency'.[21] However bad the late-Victorians might have been, it is a gross error of judgment to compare anything that they might have inadvertently done to the deliberate Holocaust against European Jewry of the 1940s.

Admiral Jackie Fisher, Britain's First Sea Lord both before and for a short period during the Great War, believed that 'five strategic points lock up the world', namely Singapore, the Cape, Alexandria, Gibraltar and Dover. All in British hands before the attack on Pearl Harbor, one of them was now about to fall to the Axis. The American journalist Frederick Simplich's report in the July 1940 issue of *National Geographic* magazine had been emphatic about Britain's great Seletar naval base at Singapore:

> 'Strongest military base in the Far East!' That's what you hear now of Singapore, British-owned island-city off the south tip of the snake-shaped Malay Peninsula. If you could take a three thousand-pound elephant by the tail and throw it 25 miles, then you might sense the power of the giant guns that now defend Singapore. I stood below when they fired one from a hill above me. Overhead the tropic air screamed from friction pains as the colossal shell went growling over mangrove swamps, went howling over distant ships at anchor, finally to splash far out at sea. Why did the British work fifteen years, and spend carloads of gold, to make this tiny faraway equatorial isle so formidable? Because, along with Gibraltar, Malta, Suez, Aden, Colombo, and Hong Kong, it guards her trade routes to the Far East and Australia.[22]

Seletar certainly was impressive on paper: the 1,000-foot dock had been floated out from Britain in the Twenties, the huge wireless towers that connected it telegraphically with the Admiralty in London, the vast oil and munitions storage facilities, the deep minefields guarding them and, as Simplich reported, 'From rooms in the Raffles Hotel, guests hear pistol practice in nearby barracks, and at night the air may be full of roaring planes simulating enemy attacks, while a score of searchlight beams scour the sky looking for the "enemy".' (The word 'enemy' stayed in inverted commas in the nominally apolitical magazine until December 1941, but it would have been clear to any intelligent reader to which power the author referred.)

Simplich went on to say that the reason why Americans were taking such a deep interest in Singapore was that it dominated South-East Asia, from where the United States got '*nearly all* our much-needed rubber, tin, quinine, and certain other strategic commodities' and that 'should any power try to cut off our rubber, tin, palm oil, quinine, hemp, etc, we might be forced to defend these sea routes to the East to maintain our industrial life'. Fortunately, noted the author, 'Singapore's strength as a naval base lies also in its long distance from any possible foe.'

It did not seem like that after Pearl Harbor. The fall of Singapore on 15 February 1942 was not due to Japanese military superiority so much as, in the words of a recent study, 'a gigantic and wholly successful piece of bluff'.[23] The garrison of over 110,000 troops under Lt-General Arthur Percival surrendered to an assault force of only 35,000 Japanese, less than one-third its size. Those who surrendered included 38,469 Britons, 18,490 Australians, 14,382 local volunteers and 67,340 Indians. It was the greatest humiliation of the English-speaking peoples in the twentieth century.

Singapore comprised 600,000 inhabitants, who used $8\frac{1}{2}$ million tons of fresh water per annum, that had to be pumped forty miles from the mountains of mainland Johore. When these water supplies fell into the hands of the Japanese, no amount of huge naval guns pointing out to sea could prevent the low, tree-grown, oval-shaped 26-mile-long island from falling to them.

As often happens in chaotic military débâcles involving civilians, there were many appalling scenes in which the sang-froid of the British and Australians completely disappeared, to be replaced by inexcusably disgraceful behaviour. These incidents, almost as much as the defeat itself, led to the post-war demise of the British Empire in Asia because of the collapse in prestige that they represented in a region where 'face' was all:

> In Arakan, British troops were said to 'run as no deer has ever run' when encountering the Japanese. In Singapore the defending Australian troops were accused of pushing Indian soldiers in front of them 'at bayonet point'. The racist behaviour which marked the British retreat also damaged the colonial mystique, especially the claim to exercise an imperial responsibility over subject peoples. In retreat from Penang, Asians were turned out of escape boats, and the formal surrender was left to a Eurasian race-horse trainer. One escape boat only took three hundred people, including 'stately memsahibs who refused to share cabins'. In Sumatra, there were ugly scenes as refugee Europeans in rags demanded precedence over Asian nurses at meals and in the bathroom.[24]

Not all of these acts were indefensible. 'All troops run sometimes,' said the Duke of Wellington, and when doing so it makes sense to run as fast as possible. Similarly the Australians were right to use bayonets to encourage forward units not to break. At Penang there was little point in British officers staying for a formal surrender if there was a possibility of escape. Nonetheless, the scenes of panic and selfishness throughout those vast parts of British Asia that the Japanese invaded utterly destroyed the carefully created image of the white man as an ineffably superior being, born to rule the yellow.

A common view was that expressed by Gunner Lawrie Birks, who was serving with the 42nd Battery, 14th New Zealand Light Anti-Aircraft Regiment in the Middle East, and who had visited Singapore with his parents in peacetime. He wrote to them on the day the city fell, saying, 'Seems rather appalling, all that labour and those millions of pounds worth nothing in a few days. Maybe all concerned did their best, but it seems to me that there must have been some rank inefficiency somewhere, after the lesson of the Maginot Line, not to mention Pearl Harbor and the Philippines and Hong Kong.' In particular, he distrusted the 'blandly optimistic statements from the Old School Tie crowd' and feared that an attack on Australia would not be long delayed.[25]

Only four days after the fall of Singapore came two Japanese attacks on Darwin in Australia's Northern Territory. They were to be the first of very many. At 10 a.m. and then again at 11 a.m. on 19 February 1942, fifty-four land-based bombers and 188 carrier-based attack aircraft assaulted the harbour and town, hitting shipping, military and civil aerodromes, the hospital at Berrimah and the Royal Australian Air Force base at Parap, killing 243 people while wounding between 300 and 400.[26] (Fearful of the effect on national morale so soon after the fall of Singapore, the Australian Government put out the news that a total of just seventeen people had died.)

During the Second World War, Darwin was bombed on sixty-four separate occasions. It was certainly not the only North Australian town to be targeted either – others included Townsville, Katherine, Wyndham, Derby, Broome and Port Hedland – but it undoubtedly took the worst battering. In the first raid alone, for example, twenty military aircraft, eight ships at anchor and most of the civil and military facilities were destroyed. Because many people expected that the bombing was a precursor to a full-scale Japanese invasion, half of Darwin's population fled southwards, and even three days after the bombings 278 servicemen from the RAAF base were still recorded as missing. This, too, was rightly kept secret of course, and John Curtin released a leaflet entitled *Darwin has been bombed – but not conquered*, in which he warned that 'We, too, in every Australian city can face these assaults with the gallantry that is traditional in the people of our stock.'[27]

On 30 March, Churchill promised Curtin that the 2nd British Infantry Division and an armoured division soon to be rounding the Cape would be diverted to Australia should that country be invaded by 'say, eight to ten Japanese divisions'. Yet there might have been a reference to Curtin's notorious New Year's Message in Churchill's last sentence, which read: 'I am still by no means sure that the need will arise especially in view of the energetic measures you are taking and the United States help.'[28] Although Japanese midget submarines were to penetrate Sydney harbour, the expected invasion never materialised, and the Japanese southward thrust was blunted with bitter fighting in the jungles of New Guinea.

At the time of his disagreements with Curtin over the allocation of Australian troops, Churchill made sure to emphasise how stalwart New Zealand had been over the same issue. On 2 July 1942 he told the House of Commons:

Although I am not mentioning reinforcements, there is one reinforcement which has come, which has been in direct contact with the enemy and which he knows all about. I mean the New Zealand Division. The Government of New Zealand, themselves under potential menace of invasion, authorised the fullest use being made of their troops whom they have not withdrawn or weakened in any way. They have sent them into battle where, under the command of the heroic Freyberg, they have acquitted themselves in a manner equal to all their former records. They are fighting hard at the moment.[29]

Churchill had a particular affection for the New Zealand commander General Sir Bernard (later Lord) Freyberg, whom he called 'the salamander of the British Empire' after the tenth wound of his career was sustained at Mirqar Qaim in 1942. (He won no fewer than three bars to his DSO and was mentioned in despatches five times.) Born in England, Freyberg's formative years were spent in New Zealand after emigrating there aged two in 1891. A pre-war New Zealand swimming champion, 'Tiny' Freyberg swam ashore to light diversionary flares at Bulair during the initial landings at Gallipoli. He won the Victoria Cross on the Somme and became the youngest general in the British Army. However, controversy still dogs his name for the way he allowed German paratroops to capture the strategically vital Maleme airfield on Crete in May 1941, even though he was in receipt of Enigma decrypts warning him of the operation.

For all his incredibly wide travels around the globe – including some of the most far-flung parts of the British Empire – Churchill never visited the Antipodes. Although travel took far longer in the 1920s and 1930s, it is nonetheless a curious oversight in someone who knew the rest of the English-speaking world so well. Might it have been as a consequence of the Dardanelles débâcle, fearing that his reception would not be so warm in Australia and New Zealand as it was on his sixteen visits to the United States, say, or his nine visits to Canada? Whatever the reason, Churchill had a lacuna when it came to Australia, which was reciprocated by an animus against him. This in turn led to some ill-tempered clashes between the British and Australian high commands and senior decision-makers during the Second World War.

After lunch in the White House on 21 June 1942, Churchill and Alanbrooke were standing beside Roosevelt's desk when an official entered with a pink piece of paper containing the news that Rommel had retaken the Libyan city of Tobruk from the Eighth Army. 'Neither Winston nor I had contemplated such an eventuality and it was a staggering blow,' recalled Alanbrooke. 'I cannot remember what the individual words were that the President used to convey his sympathy, but I remember vividly being impressed by the tact and real heartfelt sympathy that lay behind these words. There was not one word too much or too little.' British forces had to retreat eastwards to positions at Mersa Matruh, a further trudge along the *via dolorosa* that was the year 1942, which thankfully was to end with the great Commonwealth victory of El Alamein.

On 4 June 1942, only a month after the inconclusive battle of the Coral Sea, which had nonetheless prevented Japan from taking Port Moresby and cutting the supply line between Australia and Hawaii, the Japanese Admiral Chuichi Nagumo ordered a massive attack on the American base at Midway Island. The truly vast armada that he deployed to achieve this consisted of almost 200 vessels, including 8 aircraft carriers, 11 battleships, 22 cruisers, 65 destroyers and 21 submarines, supported by no fewer than 600 planes.[30] Apprised of Japanese intentions through Intelligence decrypts of messages between Tokyo and Nagumo, Admiral Chester W. Nimitz sent his much smaller fleet – of three carriers *Enterprise, Hornet* and *Yorktown*, but no battleships due to the Pearl Harbor attack – to a station north of Midway. So strategically important was Midway in an ocean with so few strong-points but such vast distances, that whoever won the battle would be in a prime position to dictate the course of the rest of the war in the Pacific. (The very name of the battle of Midway was augury in itself, though none could know it at the time; it was fought during the thirty-third month of a seventy-one month world war.)

On 4 June, Nagumo ordered 108 carrier-based fighter-bombers to attack Midway's military installations, with a further 100 aloft to protect them. After initial success, he ordered a second attack on the island's aerodromes. Shortly after the planes were refuelled for the second strike, American ships were detected 200 miles to the north. Nagumo ordered the attack broken off and a change of course, which for a time foxed the American dive-bombers. By a stroke of luck one American plane spotted Nagumo's main force, however, and a full-scale attack was ordered.[31] Of the first wave of forty-one American torpedo bombers, no fewer than thirty-five were shot down, but thirty-seven dive-bombers attacked in a surprise second wave from *Enterprise* and *Yorktown*. Nagumo's flagship, the carrier *Akagi* and two other carriers *Kaga* and *Soryu* were destroyed. The fourth Japanese carrier *Hiryu* was able to launch planes that sank the *Yorktown*, before it too was crippled beyond repair. The next day the Japanese fleet limped off westwards, and the most important naval confrontation of the Pacific War had been convincingly won by the United States.

By contrast with the decisive victory at Midway, Operation Jubilee, the large-scale amphibious raid on the small French port of Dieppe in Normandy on the morning of Wednesday, 19 August 1942, was one of the worst Allied fiascos of the Second World War. Poor Intelligence work, bad planning, the refusal to abort it when it was compromised and a lack of any proper military objective led to a severe reversal.

The genesis of the raid was political: the Western Allies wanted to prove to Stalin that they were active at a time when the Red Army was being so grievously hard-pressed deep inside the USSR. As with so many operations with a primarily political rather than a strictly military objective – such as the Dardanelles adventure or the attempt to protect Greece in 1941 – it was a disaster.

Although the first rehearsal for the raid had to be called off because of a lack of proper briefing, and the second went very badly, there was no inquest as to why, and the raid was ordered to go ahead anyway by the Director of Combined Operations Lord Louis Mountbatten, whose brainchild it was. The Intelligence was undertaken by his friend, the Cuban playboy and racing driver the Marquess de Casa Maury, who ignored the advice of a special forces unit which had already raided near Dieppe and had concluded that it was the wrong target to attack.

The cracking of the German naval codes by Bletchley Park allowed Mountbatten to know that there was an escorted enemy convoy in the Channel, which could not fail to compromise the operation by warning the German forces in Dieppe of the raid hours before it started. Yet despite this complete loss of tactical surprise the raid still went ahead.

With totally insufficient naval and air cover, a 6,100-strong force of Canadians, British, Americans and Free French – the bulk of the force being from the Canadian Second Division – went ashore on an eleven-mile stretch of coastline, centred on the port itself. They were massacred. Over seven hours of slaughter, the 4,963 Canadians lost no fewer than 3,369 killed, wounded or captured. Meanwhile, the Germans lost 314 killed, 294 wounded and 37 captured.

Of the twenty-four tank-landing-craft, only ten made it to the beaches and deposited a total of twenty-seven tanks. Of these only eleven managed to get to the esplanade because of the loose shingle and a sea wall, and these were then all destroyed one after the other. All seven Canadian battalion commanders became casualties as the bloodbath continued and communications broke down between the barbed-wire-strewn beach and the task force commander.

Mountbatten later tried to blame several other people for the débâcle, but it was he who was personally responsible for every stage. He then tried to say that the Chiefs of Staff learnt a lot about the nature of amphibious warfare that was to prove invaluable at D-Day two years later. However, Sir Ian Jacob, the Assistant Military Secretary to the War Cabinet, acknowledged that Dieppe did not teach the military planners anything about D-Day that was not common sense anyhow. A lance corporal might have been able to advise

Mountbatten not to attack a well-defended and forewarned town without either proper air cover and a naval bombardment or the element of surprise.

In all, there were some 4,100 Allied casualties in Operation Jubilee, the Nazis had been handed an invaluable propaganda coup and a Canadian division had been sacrificed in vain. It was a case of amateurism at its most culpable, but Lord Mountbatten was afterwards rewarded with promotion, once he had successfully if entirely unfairly shifted the blame on to the exceptionally brave Canadian task force commander, Major-General John Roberts. Courage above and beyond the call of duty was common that day; two Canadians – Lieutenant-Colonel Charles Merritt of the South Saskatchewan Regiment and Honorary Captain John Foote of the Canadian Chaplain Service – won Victoria Crosses. (In all, sixteen Canadians won VCs in the Second World War, and no fewer than ninety-four have won them since the award was inaugurated.)

The Canadians had seen service in Hong Kong in 1941 and Dieppe the following year, both defeats through no fault of their own. In April 1942, Mackenzie King had called a referendum on the issue of national conscription, in order to release him from a promise he had made at the outbreak of war that Canadians would not be conscripted to fight abroad. As in the Great War, the Quebecois showed their opposition to a more vigorous prosecution of the war, with 72.9% of them voting against conscription whereas 80% of the English Canadians voted in favour of it, even though the liberation of France was a principal Allied war aim. In the event, only 12,000 conscripted Canadians served abroad, but Quebec made its view known with anti-conscription riots in Montreal.

During the battle of El Alamein, Churchill called an emergency Chiefs of Staff meeting at which Anthony Eden – who had been a brigade major in the Great War – attempted to lecture Alanbrooke on strategy and tactics, arguing that Montgomery was not doing as well as he ought to have been. Alanbrooke supported his protégé and choice as Eighth Army commander resolutely. As he later recorded, 'I had then told them what I thought Monty must be doing, and I knew Monty well, but there was still just the possibility that I was wrong and that Monty was beat. The loneliness of those moments of anxiety, when there is no-one one can turn to, have to be lived through to realize their intense bitterness.'[32] The fact that none of the senior Allied leaders cracked, or allowed the strains of their tasks to show publicly when it could have had a disastrous effect on morale, was a tribute to them.

The ringing of church bells was banned by government fiat on the outbreak of war. When the Archbishop of Canterbury asked for special dispensation for Christmas Day 1940, it was refused by the War Office as too risky. 'When the church bells of England next ring,' wrote Harvey Klimmer, the US

Embassy Attaché in April 1941, 'instead of summoning people to worship they will be calling the citizenry to man the beaches and the fields of Britain to repel invaders.'[33] In fact, they were next rung to celebrate the victory of El Alamein. It was the only time in the entire war that Churchill congratulated Alanbrooke in Cabinet, possibly prompted by guilt at the memory of the jitters he and Eden had exhibited about Montgomery in the opening stages of the battle.

The Commonwealth contribution to victory at El Alamein is often over-looked, but Lieutenant-General Sir Leslie Morshead's 9th Australian Division played a vital role both in the second battle of El Alamein in October and November but also at its hard-fought predecessor, the first battle of El Alamein, when Rommel's drive to the Nile Delta was halted by General Sir Claude Auchinleck in the first week of July. The military historian Gary Sheffield has described this as 'a moment as decisive as any in the Desert War', adding, 'During both first and second Alamein, the contribution of "Nine Div" was of the utmost importance, and resulted in Eighth Army's casualty list featuring a disproportionately large number of Australians.'[34] It was at El Alamein that the previous failures of communication between infantry and armour were finally put right and the Eighth Army managed 'to harness its fighting power to its brain power'.

Of course it is important not to exaggerate the importance of El Alamein to the overall victory, since it was a mere skirmish beside the vast tank battle of Kursk the following year. At no point could the Western Allies have won the Second World War without the contribution made by the Soviet Union. The siege of Leningrad, which lasted 880 days from August 1941 until January 1944, cost the lives of around 1.4 million of its defenders, including 641,000 by starvation. The death toll at Stalingrad on both sides has been estimated at 1,109,000 people in the months between the summer of 1942 and von Paulus' surrender on 31 January 1943. In the entire war, the ground forces of the English-speaking peoples are credited with killing around 200,000 German troops; the Red Army meanwhile killed over three million.

Yet although El Alamein cannot be rated beside the contemporaneous battles going on in Russia, it was the moment that the tide of the war turned for Britain and the Commonwealth. With Midway, Stalingrad and El Alamein taken together, the second half of 1942 gave the Allies good cause for hope. Whilst it was not quite true to say, as Churchill once did, that before Alamein there were no victories and after it no defeats, it certainly represented the fulcrum moment for the forces of the English-speaking peoples.

At what point did Adolf Hitler realise this? In December 1942, the Führer commanded that every word spoken at his conferences with his military chiefs needed to be preserved for posterity. He accordingly ordered Germany's parliamentary stenographers, who had been idle since the Reichstag was

mothballed that April, to take down in shorthand and then transcribe everything uttered at these crucial meetings. Verbatim, unvarnished and contemporaneous, the transcripts provide the pure raw material of history. Far
from the ranting lunatic of Thirties' newsreels, Hitler emerges from them as
a painstaking, calculating, inquisitive dictator, and even rather a good listener.
At least four-fifths of the transcripts consist of the answers given to his incisive
questioning by such senior figures as Admiral Karl Dönitz, Hermann Göring
and Generals Alfred Jodl, Gerd von Rundstedt, Erwin Rommel, Heinz
Guderian and Wilhelm Keitel.

Because they open on 1 December 1942, with the battle of Stalingrad
effectively lost, and close on 27 April 1945, only three days before Hitler's
suicide, the transcripts cover Germany in retreat and eventual defeat, although
it is hard to tell from the Führer's remarks exactly when it dawned upon him
that he would lose the war, and with it of course his own life. Perhaps it came
after his defeat in the battle of the Bulge at the very end of 1944, for on 10
January 1945 he had the following conversation with Göring about new
weaponry:

> Hitler: 'It is said that if Hannibal, instead of the seven or thirteen elephants he
> had left as he crossed the Alps, had had fifty or 250, it would have been more
> than enough to conquer Italy.'
> Göring: 'But we did finally bring out the jets; we brought them out. And they
> must come in masses, so we keep the advantage . . .'
> Hitler: 'The V-1 can't decide the war, unfortunately.'
> Göring: 'But just as an initially unpromising project can finally succeed, the
> bomber will come, too, if it is also – '
> Hitler: 'But that's still just a fantasy!'
> Göring: 'No!'
> Hitler: 'Göring, the gun is there; the other is still a fantasy![35]

Although there were often up to twenty-five people in the room during
these Führer-conferences, Hitler usually had only two or three interlocutors.
There is no noticeable sycophancy in their answers to his ceaselessly probing
questions. Gun calibres, oilfields, plastic versus metal mines, Panzer driver-
training, encirclement strategies; little escaped his interest. 'Can't we make a
special supply of flame-throwers for the West?' he asked just before D-Day.
'Flame-throwers are the best for defence. It's a terrible weapon.' He then
telephoned personally to order a trebling of the monthly flame-thrower production, perkily ending the conversation: 'Thank you very much. Heil! Happy
holidays.' Much of the talk was simply beyond satire, as when Hitler remarked:
'One always counts on the decency of others. We are so decent.'

The atmosphere was uniformly businesslike, even at the end. There was of
course no mention of the Holocaust in front of the stenographers. Some things

were not transcribed, such as his paeans about his German shepherd dog
Blondi and his constant asking of the time – he never wore a watch – but
otherwise his every word was taken down. He only really started rambling
incoherently towards the very end, as the Red Army advanced on his bunker
and he took refuge in nostalgia, wishful thinking and accusations of betrayal.

The contrast with the military decision-making of the High Command of
the English-speaking peoples could hardly have been starker. Of course it is
impossible to know whether Churchill would also have rambled as the
Wehrmacht surrounded his bunker in the Cabinet War Rooms, but if he had,
Alanbrooke would have brought him back to the point with favoured phrases
such as 'Frankly, I flatly disagree with you, Prime Minister.' Similarly, Marshall
kept tight control of the deliberations of the US Chiefs of Staff committee.
The delineation in powers and responsibilities between politicians and Service
Chiefs worked remarkably well during the Second World War, far better than
they had in the Great War, despite having equally strong personalities at the
top of both politics and the Services. Of course virtually any system would
have been superior to the untrammelled authority of a single man, as existed
in Germany, but the checks and balances inherent in the English-speaking
constitutions ensured that military strategy and general war policy were the
result of toughly argued but logically minded debates. In that sense, democracy
established its strategic superiority over dictatorship.

'Marshall absolutely fails to realise what strategic treasures lie at our feet in
the Mediterranean,' wrote Alanbrooke at the Casablanca Conference in
January 1943, 'and always hankers after cross-Channel operations. He admits
that our object must be to eliminate Italy and yet is always afraid of facing the
consequences. ... He cannot see beyond the tip of his nose and it is mad-
dening.' Heated rows between British and American strategists at Casablanca
led to the muddled compromise that was the Italian campaign, culminating in
the bloody morass of Monte Cassino, where no fewer than 100,000 Allied
troops were killed or wounded. Fighting up a narrow, mountainous peninsula –
near-perfect terrain for defensive operations – could not have been more
difficult. Yet the fact remains that the 602 days of the Italian campaign cost
the Axis 536,000 casualties, against the Allies' 312,000.[36] It also bottled up no
fewer than fifty-five German divisions, more than one-fifth of the Wehrmacht.
Even a fraction of those forces might have proved decisive in Normandy, or
could have held up the Red Army on the Eastern Front.

In the second battle of Cassino, between 15 and 18 February 1944, the
Maori battalion of the 2nd New Zealand Infantry Division attacked the railway
station, but was forced back by German tanks, as were the Indian Division's
attacks on Monastery Hill. Had there been no British Empire, Western civilisation would not have been able to call on heroic fighters like these in the

struggle against Nazism. The Maoris suffered one-sixth of their total strength killed in that assault, figures that were almost reminiscent of Gallipoli (where 2,721 New Zealanders had died, one-quarter of the total who were landed there). In the third battle of Cassino, between 15 and 23 March, the Indians and New Zealanders captured two-thirds of the town. The monastery was not to fall until the morning of 18 May.

There was a controversy at the time of the battle of Monte Cassino – on cultural-historical grounds – that General Freyberg ought not to have ordered the obliteration of the ancient monastery. One answer to the aesthetic complaints was made by the twenty-two-year-old New Zealand machine-gunner D.H. Davis, who wrote to his parents on 28 February that, 'Our platoon had a grandstand view of it and although it went against the grain to destroy it, it was a sheer necessity as Jerry was using it as an O[bservation] P[ost] and gun emplacement. Anyway, after weeks of waiting our bombers went in and cleared out the Jerries, which was pretty good for the old morale.'

As well as Maoris, there were Basuto muleteers at the battle of Monte Cassino, a tribute to the reach of the British Empire. In all, 500,000 Africans and 2.5 million Indians served in British uniform, with imperial forces fighting in places as distant as Abyssinia, Iraq, Iran, Madagascar, Shanghai and Sumatra.[37]

Over the issue of the timing of D-Day, all through 1943 British decision-makers feared a Western Front developing in France as in 1916, didn't believe the Americans had enough trained troops and landing-craft, and carefully considered attacking what Churchill called the 'soft underbelly of Europe' through the Balkans and Austria. At Casablanca, the British got their way, not because they were the stronger power but because they had a *de facto* veto, and Sicily was targeted instead. The hope was for the Gibraltar-Suez route to be made safe for shipping, Italy knocked out of the Axis, German troops taken from the Eastern Front and France, and the eventual return of the Mediterranean to the Allies. The Italian campaign must have encouraged Colonel Arthur Murray at Brooks's, who wagered Archibald Hay £1 'that hostilities will cease before the end of 1943'.

The problem bedevilling the early creation of a second front in France was not what was happening in Italy or Sicily in the first half of 1943, so much as the danger of the U-boat threat which was still preventing large-scale transatlantic movement of troops on the scale necessary. By mid-1943, victory in the battle of the Atlantic – enormously aided by the cracking of German naval codes – had largely cleared the oceanic passages, but it was not possible to assemble the American men and matériel necessary for an autumn 1943 offensive.

That vast ocean-wide battle cost the lives of some 30,000 British merchant seamen, not including those who died ashore of their wounds, whom the Registrar-General of British Shipping for some unaccountable reason did not include in the roll. In the year 1941 alone, over 1,000 merchant ships were sunk and 7,000 seamen killed. The following year, 8,000 died. The writer Christopher Lee, who worked as a deck boy on an old wartime tramp steamer in the early 1960s, recalled colleagues who had fought in that terrible, long, hard-fought struggle:

> They talked of the constant fear engendered by U-boats; of mangled and scream-ing shipmates; of recurring nightmares; of donkeymen and greasers who drank because they dreaded the prospect of a torpedo bursting into the engine room and the certain death it brought; and of why the Navy did not start a continuous convoy system until the summer of 1941.[38]

(It was because it wrongly feared that a single U-boat could cause more damage to a convoy; the official thinking took some time to alter.)

One of the problems facing the Navy was that on the outbreak of war President de Valera had refused to allow the Royal Navy the use of the 'Treaty Ports' bases which had been given up by Britain in 1938. This forced ships to use Northern Irish and western British ports, thus drastically cutting down the area of operations that the U-boats needed to cover. No pleas or protests would move him. (De Valera later complained that the American troops stationed in Ulster amounted to 'an army of occupation', yet he made no protest to Germany when Belfast was bombed. 'In March 1943, as the battle of the Atlantic was at its height,' an historian of Irish neutrality has pointed out, 'De Valera went on Radio Eireann to deliver a St Patrick's Day address in which he described the restoration of the Gaelic language as the most important issue facing the nation.')

On 16, 17 and 18 February 1943, the British Parliament debated the Beveridge Report, the document on social insurance that effectively set up the post-war Welfare State. In the *Hansard* report of the debate, there are no fewer than 1,903 column inches devoted to the speeches given on those three days, or over sixteen feet of print.[39] In all of that, only six column inches were devoted to the question of how the enormous increases in social welfare provision were actually going to be paid for. On the first day the Labour deputy leader Arthur Greenwood stated that, 'It would be foolish to attempt to stem the rising tide of opinion in favour of bold plans by attempts to "crab" them on the grounds that we cannot afford them.' They would be financed, he claimed airily, through 'international economic co-operation and considered plans to avoid financial exploitation and to yield the maximum benefit to Mankind'. Quite why the rest of mankind would wish to co-operate economically in 'considered plans' to bring the maximum benefit to the British working man was not explained by him or anyone else.

Few had the courage to speak the plain truth about the genuine capacity of a near-bankrupted Britain to pay for the extensive social provisions of the Beveridge Report. One such was Sir John Forbes Watson, Director of the Confederation of British Employers, who told Beveridge's Commission the obvious and unvarnished, but unpopular and subsequently ignored, truth that Britain entered the war against Germany to preserve freedom, not to improve social services.

The Report was embraced enthusiastically by the Tory Reform Committee, which assaulted the *laissez-faire* principles of the Conservative Party with gusto. The Committee's Chairman, the Conservative MP for North Devon, Lord Hinchingbrooke, wrote in the *Evening Standard* in February 1943:

> Modern Toryism rejects Individualism as a philosophy in which the citizen has few duties in society, but accepts wholeheartedly the initiative and personal enterprise of the citizen in partnership with his friends and neighbours to a purpose agreeable to the nation as a whole. ... It does not detest restriction ... but welcomes it as part of man's obligation to a nation which has given him life and freedom. ... It is hopeful of Planning which it regards as a grand design to bring the aims of man into a true relation with the aims of the community. It is exhilarated by the Beveridge Report. True Conservative opinion is horrified at the damage done to this country since the last war by 'individualist' businessmen, financiers and speculators ranging freely in a *laissez-faire* economy and creeping unnoticed into the fold of Conservatism. ... True Conservatism has nothing to do with them and their obnoxious policies.[40]

In fact, Bonar Law's, Stanley Baldwin's and Neville Chamberlain's pre-war Conservative Party was a straightforwardly pro-free enterprise capitalist party far removed from Hinchingbrooke's social-democratic ideals, and it was in fact *his* view that was 'creeping unnoticed into the fold' of Conservatism. Hinchingbrooke knew this perfectly well, having been Baldwin's private secretary in the inter-war years.[41] Because Baldwin and Chamberlain had been discredited – although solely because of appeasement and not because of their economic or social policies – and because Churchill had long hailed from the liberal wing of the Party on economic issues, the immediate post-war battle went to the anti-*laissez-faire* Tories, who stayed resolutely in control right up until the election of Margaret Thatcher as leader a full thirty years later.

The assumption that the Civil Service based in Whitehall was a generally acknowledged good thing permeated liberal Tory thinking, and as a result the post-war years saw an astonishing rise in the number of people employed by the state, from a total of 580,891 people in April 1938 to 996,274 by April 1960. (It took eleven years of Thatcherism to get the figure back below the 1938 levels.)[42]

★

The year 1943 finally saw the Allies wresting command of the skies from the Axis. That year the RAF and USAAF dropped a total of 200,000 tons of bombs on Germany, whereas the Luftwaffe only succeeded in dropping 2,000 tons of bombs on Britain. In the bombing of Hamburg, the biggest attacks of 1943, 50,000 civilians were killed in late July and early August alone, almost the same number as died in the whole of the German bombing of Britain in the entire war.[43] Yet Allied air superiority did not mean horrifically high losses were not being sustained. In thirty-five raids on German cities between November 1943 and March 1944, Bomber Command lost 1,047 aircraft shot down and a further 1,682 were damaged, often beyond repair.[44]

When one considers the rivalry between several of the senior commanders serving under Dwight D. Eisenhower on the Western Front in the last eighteen months of the Second World War in Europe – but principally George S. Patton, Bernard Montgomery and Omar Bradley – one has occasionally to be reminded that they were considerable generals. In terms of spite, gangings-up, showing-off, bitchiness, pettiness, competitiveness, whining to their superiors and general prima donna-like behaviour, these great soldiers might just as easily have been squabbling teenage schoolgirls. Small motives of pique, pride, lust for fame and intense competitiveness affected the actions of some of the greatest captains of their age to an incredible degree.

'God deliver us from our friends,' said Patton once. 'We can handle the enemy.' Bradley had 'total disdain' for Monty and contempt for Patton, who in turn was 'sickened' when Monty became a field marshal. Monty meanwhile despised Patton and Bradley. Despite constant and extreme provocations, General Eisenhower somehow held the ring successfully between these three talented soldiers all the way to V-E Day.

George S. Patton was one of the few US Army officers who had always had a vision of how armour might be used in battle, insights that he had gained when fighting with the US Tank Corps in the Great War. For all his foresight, something in his personality also meant that he felt the need to wear riding breeches, riding boots and ivory-handled revolvers, and to drive around 'in flashy motor-cades that always ensured that he was noticed and the centre of attention'.[45]

On two infamous occasions in Sicily in August 1943, Patton became more of a centre of attention than even he desired. During the advance on Messina – which turned into something of a race between him and Montgomery – Patton slapped a GI, whom he called an 'arrant coward'. Then a week later he threatened to pistol whip another soldier, whom he called 'a yellow bastard' and 'a disgrace to the army'. In the second incident, at the 93rd Evacuation Hospital, the senior medical officer had to place himself in-between Patton and his victim, Private Paul G. Bennett, as Patton raged: 'I won't have those cowardly bastards hanging around our hospitals. We'll probably have to shoot them some time anyway, or we'll raise a breed of morons!'[46] It is easy, in these inclusive and politically correct times, to criticise Patton for insensitivity, but in combat conditions what was called 'Lack of Moral Fibre' (LMF) cannot be tolerated since it can lead to comrades being let down in moments of crisis.

Patton was very keen on breeding; he himself came from a distinguished line of soldiers and he was as proud of his military ancestors as any samurai or Junker. His grandfather had commanded a brigade in the Civil War and the history of the Confederacy was a living thing for him, as when he likened Operation Torch in 1942 to the battle of Manassas, or when he wondered aloud what Generals Robert E. Lee or Stonewall Jackson would have done in any given strategic situation.

The obverse side of Patton's justifiable pride in his background was a virulent anti-Semitism; he believed in the Bolshevist-Zionist conspiracy and astonishingly his prejudice seems to have been in no way lessened after the liberation of the Nazi extermination camps. To appreciate quite how weird Patton was, he actually believed that he had often been reincarnated, normally as a soldier.[47] His anti-communism was so extreme that Eisenhower occasionally feared that he might land the Western Allies in a conflict with the Red Army. By the end of his career, the US Army had placed a psychiatrist in his camp, disguised as a staff officer, to keep an eye on him; further, they monitored his phone calls and even bugged his residence.[48]

Although Field Marshal Montgomery is popularly believed to have been Patton's greatest rival and bugbear, it was in fact Omar Bradley who reduced Patton to fury, as when after the Sicilian campaign Bradley was selected to command the US First Army – earmarked for the cross-Channel invasion of Europe – instead of him. When Bradley paid a final courtesy call on Patton on 7 September 1943 at his palace in Palermo, he found him 'in a near-suicidal state. ... This great proud warrior, my former boss, had been brought to his knees.'[49] It's hard to escape the conclusion that Bradley went there on purpose to relish every moment.

Of course Patton and Montgomery also cordially loathed each other – Patton called Monty 'that cocky little limey fart', Monty thought Patton a 'foul-mouthed lover of war' – yet these rivalries need to be placed in their overall geopolitical context. By mid-1943, the United States was overhauling the United Kingdom in every aspect of the war effort, a fact that Churchill acknowledged and which led him subtly to adapt his political posture accordingly. However, Montgomery simply could not bring himself to face the new situation and became progressively more anti-American as the power imbalance became ever more evident.

The moment when Montgomery exhibited this in public came after the battle of the Bulge on 7 January 1945, when Supreme Headquarters Allied Expeditionary Force lifted its near three-week censorship restrictions. He gave an extensive press briefing to a select group of war correspondents at his headquarters at Zonhoven in which he presented the story of the Germans' great Ardennes offensive and the way that it had been turned back as implying that his 21st Army Group had had to save the Americans. 'General Eisenhower placed me in command of the whole northern front,' boasted Monty. 'I employed the whole available power of the British group of armies. You have this picture of British troops fighting on both sides of American forces who had suffered a hard blow. This is a fine Allied picture.'[50] Although he spoke of the average GIs as having been 'jolly brave' in 'an interesting little battle', he claimed that he had entered the engagement 'with a bang' and left the impression on his listeners that he had effectively rescued the Americans from defeat. There were some generous references to the courage of the American fighting man, it was true, but hardly any to any American generals other than Eisenhower. (The battle of the Bulge had resulted in 17,200 Germans dead, 34,439 wounded and 16,000 captured, but also 29,751 Americans killed and missing and 47,129 wounded.)[51]

As a result of the Zonhoven briefing, Bradley, saying that Montgomery was 'all-out, right-down-to-the-toes, mad', told Eisenhower that he could not serve under the Briton but would prefer to transfer back to the United States. Patton immediately made the same declaration. Then Bradley started actively courting the press, rarely 'venturing out of his HQ without at least fifteen newspapermen'. Bradley and Patton subsequently leaked information to the American press that damaged Montgomery, and then proceeded, in the words of the insider Ralph Ingersoll, 'to make and carry out plans without the assistance of the official channels, on a new basis openly discussed only among themselves. In order to do this they had to conceal their plans from the British and almost literally outwit Eisenhower's Supreme Headquarters, half of which was British.' It was all rather pathetic.

The Anglo-American rapport between Patton, Montgomery and Bradley from 1943 to 1945 did indeed constitute a special relationship: it was especially dreadful. The true hero was Eisenhower; how he held the ring between these competing, strutting martinets, using his charm, good humour but occasionally veiled threats too, makes a fascinating study in military diplomacy. The four-star American general of German descent had not commanded troops on an actual battlefield, yet in 1944 Eisenhower was appointed Supreme Allied Commander. His infectious grin and cheery good nature were invaluable morale-boosters. Soldiers loved and trusted him, and it was no coincidence that when he successfully ran for president in 1952 the slogan on his campaign buttons simply read: 'I like Ike.' His simplicity and dislike for vainglory was legendary. Few other commanders, on accepting the surrender of the German armies under the command of General Jodl, would have confined themselves to sending the following report back to Washington: 'The mission of the Allied forces was fulfilled at 02:41 local time, May 7 1945. Signed, Eisenhower.'

Despite having no battlefield experience – he was denied the chance of active service in the Great War and commanded the American landings in North Africa in 1942 from a cave in Gibraltar – Eisenhower made remarkably few major errors as supreme commander. There were some; he must take ultimate responsibility for the reverses at Arnhem and Antwerp in September 1944, for example, and for ignoring Major-General Charles Corlett's advice about ammunition allocations for D-Day, but in the overall context these were not enough to affect his reputation.

History's ultimate expression of English-speaking amity came on Monday, 6 September 1943, when, at Roosevelt's invitation, Churchill travelled to Harvard to accept an honorary degree. His speech, loosely based on an article he had written for the *News of the World* newspaper in May 1938, deserves fairly extensive quotation in a book with this title:

> Twice in my lifetime the long arm of destiny has reached across the oceans and involved the entire life and manhood of the United States in a deadly struggle. There was no use in saying 'We don't want it; we won't have it; our forebears left Europe to avoid these quarrels; we have founded a new world which has no contact with the old.' There was no use in that. The long arm reaches out remorselessly, and everyone's existence, environment, and outlook undergo a swift and irresistible change. ... The price of greatness is responsibility. If the people of the United States had continued in a mediocre station, struggling with the wilderness, absorbed in their own affairs, and a factor of no consequence in the movement of the world, they might have remained forgotten and undisturbed beyond their protecting oceans: but one cannot rise to be in many ways the leading community in the civilized world without being involved in its problems, without being convulsed by its agonies and inspired by its causes. If this has been proved in the past, as it has been, it will become indisputable in the future. The people of the United States cannot escape world responsibility. Although we live in a period so tumultuous that little can be predicted, we may be quite sure that this process will be intensified with every forward step the United States make in wealth and in power. Not only are the responsibilities of this great Republic growing, but the world over which they range is itself contracting in relation to our powers of locomotion at a positively alarming rate.[52]

This was not just high-flown, windy rhetoric; just like Roosevelt, Churchill saw the danger of a return to American isolationism after the war and was determined to warn against it.

Two months later, at a dinner given at the British Embassy in honour of Churchill's sixty-ninth birthday during the Teheran Conference in November 1943, Stalin also recognised the greatness of the United States. In proposing a toast to President Roosevelt and American war production, especially that of 10,000 planes per month, the Soviet leader said that this was more than thrice the USSR's rate of production. (In fact even in his tribute, Stalin was considerably exaggerating the number of planes Russia was producing.)[53] By harnessing America's huge industrial capacity – by 1945 she was responsible for over half of the world's GDP – the Roosevelt Administration had indeed turned the United States into what FDR had dubbed 'the arsenal of democracy'.

In the course of 1943, US government interference into the lives of Americans continued apace. The Roosevelt Administration rationed meat, fat, cheese, gas and canned food, and Americans discovered recycling, with waste rubber, metal, paper, silk (for parachutes), nylon, tin cans and fat all being re-used for the war effort. Civilians were banned from buying more than three pairs of shoes a year; the Marine Corps was authorised to establish a female unit; the US Manpower Commission prohibited twenty-seven million workers in essential services from quitting their jobs; Roosevelt appointed the former Supreme Court Justice James F. Byrnes to preside over the Office of War Mobilisation, which co-ordinated the work of all the Government's many agencies; and in December, he ordained that, in order to prevent a national strike, all railroads were to be seized by the federal government.

Yet in the case of *West Virginia Board of Education v Barnette* on 14 July 1943, the Supreme Court ruled that compulsory saluting of the Stars and Stripes was unconstitutional, reversing an earlier ruling. Even in the grip of a world war, American democracy was virile enough to accept that every citizen had the constitutional right not to be patriotic.

Nineteen forty-three also saw the building of Colossus, the computer and code-breaker, which was designed by the English mathematician Alan Turing, along with the engineer Thomas Flowers and academic professor Max M.H.A. Newman, and was built at Bletchley Park in Buckinghamshire. Comprising 1,500 vacuum tubes, it was the world's first all-electronic calculating machine. The exigencies of war created huge numbers of opportunities for ingenuity and invention, which had long been a major reason for the primacy of the English-speaking peoples. Many of these were carried on and fully exploited in peacetime. Edmund Hillary and Tenzing Norgay's ascent of Mount Everest in 1953 was only possible because of oxygen-breathing apparatus that had been developed for high-altitude flying, for example, a perfect illustration of how wartime inventions can so often aid human development and achievement.

Between 1941 and 1945, the following discoveries and inventions – amongst very many others – were made and exploited by scientists of the English-speaking peoples: the synthetic fibres polyester and terylene (Dacron); 525-line televisions; two-blade windmills; nuclear reactors; all-plastic cars; the detection of radio emissions from the sun; magnetic tape; the long-range navigation system based on synchronised pulses; radar equipment on submarines; the audio oscillator which generated high-quality audio frequencies; antibiotic streptomycin; silicone rubber; the synthesising of quinine; elements 95 and 96; the microwave oven; the Pilot's Universal Sighting System; the Harvard University Mark I (the first programme-controlled computer); rockets; turboprop aeroplanes and jet engines with afterburners.[54]

Of course many of these would have been invented over time anyhow, but most of them were hugely boosted by the commitment to invention that the governments of the United States, Britain, Canada, Australia and New Zealand in particular made during the Second World War. Brainpower as well as physical brawn was enlisted into the ranks of the English-speaking peoples, which was vital since German scientists were also making serious technological advances, such as with the V-2 rocket, 68-ton Tiger II tank, Type XXI U-boat (which could operate submerged for up to four days) and very nearly a manned, rocket-powered, vertically launched interceptor fighter.

An operation that certainly employed and stretched the combined brainpower of the English-speaking peoples' cryptographers in the Far Eastern theatre bore spectacular fruit in January 1944, when Douglas MacArthur's Central Bureau cracked the Japanese Army's mainline code. As the plaque on the wall of a private residence at No. 21 Harley Street, Brisbane, records: 'Central Bureau, an organisation comprising service personnel of Australia, USA, Britain, Canada and New Zealand, both men and women, functioned in this house from 1942 till 1945. From intercepted radio messages the organisation made a decisive contribution to the Allied victory in the Pacific War.'[55]

Being able to crack the enemy's codes and listen in to his communications – from Room 40 in the Great War, via Enigma at Bletchley and the Central Bureau in the Second World War, to the CIA and GCHQ in the Cold War, right the way up to the 2003 Iraq War – has always been a particular success of the English-speaking peoples' various Intelligence agencies. In these, the often-unsung Intelligence agencies of Canada, Australia and New Zealand have played useful – and on various occasions key – roles that have long been valued by the CIA, FBI, MI6 and MI5.[56]

Retreating Japanese soldiers of the 20th Division, fearful of attracting the attention of Allied aircraft if they burned the divisional cipher library, including the Imperial Army Code Book and other cryptographical paraphernalia, instead buried them in a steel trunk at Sio in North-East New Guinea. A young Australian engineer, sweeping the area for mines and booby-traps, heard a shrill noise through the earphones of his metal-detector, and demolition experts moved in to dig up what they assumed was a land mine. Instead they found buried treasure far beyond rubies. An alert Intelligence officer with the advance party sent the trove – still soaking wet from its monsoon burial – to the Central Bureau. It was one of the great coups of the war.

Soon Central Bureau was reading thousands of enemy radio messages, and, as their historian relates, 'By imaginative use of state-of-the-art technology, such as early IBM equipment, and the application of ingenuity and creativity, Allied cryptanalysts were able to keep pace with subsequent Japanese changes to these army codes and read hidden messages with regularity.' There were occasional blackouts when the Japanese introduced new key registers, or even entire code books, but Central Bureau managed to establish mastery and also to crack other ciphers, such as the Japanese Army's air force and its military attachés' diplomatic codes. It was a Far Eastern version of what the cryptanalysts of Bletchley Park had achieved with the Enigma enciphering machine, which produced the vital ULTRA decrypts, and it has with only pardonable hyperbole been described by the historian Edward Drea as being 'as authentic a battlefield victory as Midway'.[57]

The units involved in Central Bureau are indicative of the manner in which the entire English-speaking peoples – except of course Eire – came together for the greater good between 1941 and 1945. They comprised the Australian Militia, the RAF, the US Army, the Canadian Army, the Royal Australian Air Force, the Women's Auxiliary Australian Air Force, the Australian Women's Army Service and the Women's Army Corps, USA.

Even as early as February 1944, decision-makers in Washington were looking towards the possibility of the Soviet Union forcing a Cold War upon the world, although it was some time before it acquired that particular soubriquet. Their foresight was demonstrated to an almost uncanny degree by the strongly anti-communist James Forrestal, the Under-Secretary of the US Navy. A private letter sent on 11 February by Isaiah Berlin to the American syndicated newspaper columnist, Joe Alsop, reported how,

> Jim Forrestal's line that the future is going to be a poker game between the US and USSR, and only very affluent players can be allowed – perhaps baccarat is a better example – a real big thing which it would be kinder not to let anyone except the very very rich take part in (eg not Britain), is probably fairly widely felt in what is called influential circles. A very heavily armed US glares at a very heavily armed USSR, buying and selling merrily and preserving world peace for many years to come, with everyone else adjusting themselves to this new unnecessary kind of twin alliance, is the sort of thing.[58]

Some commentators have mistaken the English-speaking peoples' anticipation of the likelihood of a Cold War for their welcoming of it, or even their instigating of it, but they are wrong. To recognise that something is likely to happen and to plan for it is not the same as to will it. A few months later, on

27 July 1944, Alanbrooke displayed similar foresight when he noted in his diary:

> Germany is no longer the dominating power of Europe, Russia is. Unfortunately Russia is not entirely European. She has however vast resources and cannot fail to become the main threat in fifteen years from now. Therefore foster Germany, gradually build her up, and bring her into a federation of Western Europe. Unfortunately this must all be done under the cloak of a holy alliance between England, Russia and America. Not an easy policy and one requiring a super Foreign Secretary!

Again, the recognition of a future danger, albeit one that was not publicly stated until Churchill's 'Iron Curtain' speech of March 1946, does not imply that the Western Allies encouraged the outbreak of the Cold War, merely that their leaders were not naïve about the likely development of Stalin's plans for the future.

As it was, the first shots of the Cold War were fired as early as December 1944, when the communist-controlled Greek resistance organisation EAM-ELAS – amalgamations of the National Liberation Front (EAM) and the National Popular Liberation Army (ELAS) – attempted to take over Greece after British and Greek forces had liberated the country the previous month. When Churchill visited Athens, staying at the British Embassy over Christmas, shooting was still going on. It was to be forty-five years before European communism was to fire its last bullets, coincidentally also at Christmastime, when Romania's Securitate police, under orders from Nicolae Ceauçescu, fought their last, desperate rearguard action in defence of a creed which even they must have realised had already died.

TEN

Normandy to Nagasaki

1944–5

Operation Overlord – Canada's contribution to victory – Churchill is 'dragooned' –
General de Gaulle – The Percentages Agreement – The battle of Leyte Gulf –
FDR's Last Inaugural – The Yalta agreement – The bombing of Dresden – V-
weapons – Iwo Jima – The death of President Roosevelt – De Valera goes visiting –
V-E Day – Hiroshima

'To us is given the honour of striking a blow for freedom which will live in
history; and in the better days that lie ahead men will speak with pride of our
doings. We have a great and righteous cause.'

General Montgomery's message before Operation Overlord

'Our generation has succeeded in stealing the fire of the Gods, and is doomed
to live with the horror of its achievement.'

Henry Kissinger, *American Foreign Policy*

On Monday, 15 May 1944, the entire Anglo-American top brass met at
St Paul's School in Hammersmith for a briefing on the long-awaited
Allied invasion of Nazi-occupied Europe, the cross-Channel invasion of Nor-
mandy. (The school had meanwhile been billeted with Wellington College in
Berkshire.) King George VI, Winston Churchill, Omar Bradley, George
Patton, Jan Smuts, the Chiefs of Staffs, War Cabinet members and pretty
much everyone of any importance in the British, American and Canadian
armed forces were present to hear General Eisenhower, the Supreme Allied
Commander, and General Montgomery outline exactly what their mission
was and what would take place in Normandy three weeks later. By the end of
his presentation, Eisenhower's confidence in victory had transmitted itself to
everyone there. He closed the meeting with a joke, saying, 'In half an hour
Hitler will have missed his one and only chance of destroying with a single
well-aimed bomb the entire High Command of the Allied forces.'

The Normandy landings on D-Day – Tuesday, 6 June 1944 – represent the largest amphibious operation ever launched and were by far the greatest military enterprise ever undertaken by the English-speaking peoples. The invasion of the European mainland involved no fewer than 6,939 vessels (of which around 1,200 were warships and 4,000 were 10-ton wooden landing-craft capable of an upper speed of eight knots), 11,500 aircraft and two million men. No fewer than 156,000 men were landed by sea and air on D-Day alone.[1] 'I hope to God I know what I'm doing,' General Eisenhower told his staff on the eve of operations. Fortunately, he did. The sheer size of the assault was an important factor in its success, overawing all but the best-trained and most battle-hardened German troops, who woke up to see the apparition of a sea and sky almost completely covered with ships and planes.

The intricate and ingenious deception operations that the Allies had set in place had convinced the Germans that the attack would not come at Normandy at all, but in the Pas de Calais region of France closest to the British coast. Operation Bodyguard involved the controlled leaking to the Germans of hints about the Pas de Calais, including by the turned German agent Dujol Garcia (codename: GARBO). In all, no fewer than twenty-nine German agents were successfully turned by the British Intelligence services during the war, an impressive achievement. Two other major strategic deception schemes were Operation Fortitude North, which successfully tied up no fewer than 372,000 German troops protecting south-eastern Norway, and Operation Fortitude South, where half-a-million German troops were kept occupied at the Pas de Calais until 26 June. Stalin readily aided the deception plans with a feigned amphibious assault on the Black Sea coast of Romania and an equally fictitious threat against northern Norway that historians believed might have diverted as many as twenty German divisions from Normandy.[2]

Meanwhile, Lieutenant-General George S. Patton was given command of the entirely fictitious FUSAG – First US Army Group – that was purportedly stationed across the Channel from the Pas de Calais and was even visited by King George VI as part of the ruse. Just before the invasion, a double posing as Montgomery made a visit to Gibraltar, dummy parachutists were dropped near Boulogne, and diversionary attacks and radio signalling were employed up the coast to suggest attacks were taking place anywhere other than the five designated beaches. Over Pas de Calais itself, Group Captain Leonard Cheshire's 617 Squadron dropped thin metal strips, codenamed 'Window', which led German radar operators to believe that a naval force was making its way across the Channel at a steady nine knots.

As a result of all these elaborate deception operations, the German Fifteenth Army was held back over 100 miles north of where it was most needed. The Germans could not be certain whether the likely invasion point was Normandy or the Pas de Calais – or conceivably both. In order to protect the secrecy of the operations, there was a total ban on holidaying in the South of England, all post was delayed for a month, and all diplomatic communications and couriers were halted from 18 April. Such a blanket ban also permitted – both for reasons of morale and security – the truth to be kept hidden about the tragic deaths of 749 American soldiers and sailors who were practising for D-Day at Slapton Sands on the south coast on 28 April, but who had fallen foul of a surprise attack by seven German torpedo-boats. (Ken Small, in the subtitle of his book *The Forgotten Dead: Why 946 American Servicemen Died off the Coast of Devon in 1944*, claims more died, but this is disputed.)[3]

It was partly in order to maintain secrecy on a strictly 'need-to-know' basis that General de Gaulle was not even informed about Overlord until only the day before the landings were actually just about to take place. Earlier operations in which the Free French had been involved, such as the Dakar raid of September 1940, had been badly compromised in the past, though of course no blame attached to de Gaulle personally. Churchill received the Frenchman in a train carriage, parked near Portsmouth station, which a suspicious de Gaulle remarked was 'an unusual notion'.[4] After outlining the plan for Overlord – which de Gaulle naturally welcomed – the talk turned to politics, and his aide, General Emile Béthouart, later recalled how, 'I had felt that de Gaulle was tense, deeply wounded at having been invited in this way as a spectator and at the last moment, without any previous discussion or understanding on the prime question of the exercise of authority in liberated France.'

When Churchill urged de Gaulle to mend his bad relations with Roosevelt, the Frenchman exploded: 'Why do you seem to think that I am required to put myself up to Roosevelt as a candidate for power in France? The French Government exists. I have nothing to ask of the United States of America, any more than I have of Great Britain.' He then went on to complain that the troops had been furnished with 'so-called French money which is absolutely unrecognized by the government of France', by which he meant himself. De Gaulle ended by denouncing the Anglo-American plans to give Eisenhower executive political authority in France, asking, 'How do you expect us to operate on this basis?'

Churchill was not about to be browbeaten. 'And what about you?' he roared back at de Gaulle.

> How do you expect us, the British, to adopt a position separate from that of the United States? We are going to liberate Europe, but it is because the Americans are with us to do so. For get this quite clear, every time we have to decide between Europe and the open sea, it is always the open sea that we shall choose. Every time I have to decide between you and Roosevelt, I shall always choose Roosevelt.[5]

It was about as definitive a statement of the amity of the English-speaking peoples as was ever expressed from 1900 until the present day, and it rever-

Dubliners loyally greeting their beloved Queen-Empress Victoria on her visit to Ireland, April 1900

A group of Italian immigrants about to join the melting pot
waiting in a hall on Ellis Island, New York, 1905

Richard Seddon, founder of modern New Zealand

The first great assault on the English-speaking peoples since 1900; Prussian Militarism
as personified by Paul von Hindenburg, Kaiser Wilhelm II and Erich von Ludendorff

The opposed landing at Anzac Cove on the Gallipoli Peninsula
on the morning of 25 April 1915

A children's jigsaw puzzle featuring the torpedoing of the unarmed American liner *Lusitania* by a German U-boat on 7 May 1915, in which 1,198 people died

Men of the British West Indies Regiment in camp on the Albert-Amiens Road during the Somme Offensive of September 1916. By mid-1916 the 3rd and 4th battalions had volunteered for service on the Western Front

Allied officers outside the Hall of Mirrors watching the signing of the (on balance, rather reasonable) Treaty of Versailles, 28 June 1919

One of the engineering
wonders of the English-
speaking world: the
Empire State building,
built between 1929
and 1931

Another: the Hoover
Dam, built between 1931
and 1935

Imperial Pride: the British Empire marches forth in a 1939 recruitment poster

Canadian Disaster: within six hours of seven Canadian battalions going ashore at Dieppe on 19 August 1942, three-quarters of her 5,100 men were killed, wounded or captured. The man ultimately responsible, Lord Louis Mountbatten, was later promoted

Imperial Fall: Lieutenant-General Arthur Percival (second from right) surrenders 100,000 troops in Singapore to a force of 30,000 Japanese on 15 February 1942

Canadian Triumph: the Quebec Conference, codenamed Quadrant, in August 1943 at which it was confirmed that the Allied invasion of France would take place by 1 May 1944

berated in de Gaulle's mind for many years, certainly right up until the time that he said '*Non*' to British membership of the European Economic Community in January 1963.

On and off over the next few hours, indeed right up until 4 a.m. on the morning of D-Day itself, Churchill and de Gaulle rowed over whether the French government-in-exile would even send a liaison mission with the Allied units about to land in Normandy, because the Free French leader threatened that he would not associate himself with an 'occupation' of France. At one point de Gaulle even refused to broadcast to the French people in support of the invasion, because it might be seen 'to endorse what [Eisenhower] will have said and of which I disapprove'. De Gaulle's biographer considers that it was only the presence of Anthony Eden and the Free French ambassador to London, Pierre Viénnot, that prevented Churchill and de Gaulle actually from coming to physical blows.[6] In one meeting at which de Gaulle was fortunately not present, Churchill told Viénnot that de Gaulle was guilty of 'treason at the height of battle', and ten times over he said that de Gaulle was responsible for 'a monstrous failure to understand the sacrifice of the young Englishmen and Americans who were about to die for France', and that 'It is blood that has no value for you.'

In his broadcast of 6 June 1944, de Gaulle told the French people:

> The supreme battle has begun. It is the battle in France and it is the battle of France. France is going to fight this battle furiously. She is going to conduct it in due order. The clear, the sacred duty of the sons of France, wherever they are and whoever they are is to fight the enemy with all the means at their disposal. The orders given by the French government and by the French leaders it has named for that purpose [must be] obeyed exactly. The actions we carry out in the enemy's rear [must be] coordinated as closely as possible with those carried out at the same time by the Allied and French armies.[7]

After six references to France and the French, therefore, the presence of the Allies in the invasion force was finally mentioned, but as being separate from the French Army, even though on D-Day a total of nineteen Free French soldiers were killed, against a total of 2,500 Americans, 1,641 British, 359 Canadian, 12 Australians, 2 New Zealanders, as well as 37 Norwegians and 1 Belgian.[8] The English-speaking peoples thus lost 98.4% of those Allied soldiers killed on 6 June 1944. When de Gaulle arrived in France on D+8 (14 June), his image had been so strictly banned in Occupied France that he was not widely recognised.

Although the forces that the English-speaking peoples had at their disposal were vast – more than a million men had gone ashore in France by 1 July 1944 – the result was by no means a foregone conclusion. Previous amphibious assaults such as Gallipoli and Dieppe had been very costly indeed, and unlike those beaches, up to two million slave labourers had been constructing the 'Atlantic Wall' for two years up to 1942. The part of Fortress Europe lying between Le Havre and St Malo alone had had eighteen million tons of concrete and two million tons of steel devoted to its defence.[9]

'If the Germans decided to bring their maximum forces to the beachheads,' says Churchill's biographer Sir Martin Gilbert, 'the Allied armies could have been defeated on the shore.' Eisenhower's recent biographer Carlo D'Este concurs. 'Failure was unthinkable,' he wrote, 'but nevertheless entirely possible.' Amongst the unthinkable consequences of failure were the facts that the Germans were developing 1-ton warheads for their rockets, U-boats capable of refuelling others in mid-ocean and new kinds of magnetic mines, and in the event of victory in Normandy they would have been able to transfer about one-third of the Wehrmacht forces in France over to fight the Red Army on the Eastern Front.

Just before the operation, Eisenhower had even written a letter accepting full responsibility in the case of defeat. It read:

> Our landings in the Cherbourg-Havre area have failed to gain a satisfactory foothold and I have withdrawn the troops. My decision to attack at this time was based upon the best information available. The troops, the air and Navy, did all that bravery and devotion to duty could do. If any blame or fault attaches to the attempt it is mine alone.[10]

He tucked this into his shirt pocket and forgot about it until his naval aide, Harry Butcher, found it the next month and saved it for posterity.

Far from still being the efficient military machine that had crushed France in six weeks in 1940, four years later the Wehrmaht was riven with rivalries and confusing split-commands. From February 1944, Field Marshal Gerd von Rundstedt, the Commander-in-Chief West, had overlapping commands with Field Marshal Erwin Rommel, Commander-in-Chief of Army Group B. Neither man had command over either the German Navy or the Luftwaffe. Although Rommel had three Panzer divisions under his command and Rundstedt six, Hitler's personal orders were necessary for their release.

To make matters worse, on D-Day itself Rommel was in Germany celebrating his wife's birthday; he only learnt of the landings three hours after they had begun. Meanwhile, the commander of the Cherbourg battery was on leave, as was the commander of the crucial 21st Panzer division, who was in Paris with a show-girl having misled his staff as to his whereabouts. Most absurd of all, Hitler was not awoken to receive the news of the attack and attended to an Hungarian state visit before he was even made aware of this knife-blow to his Reich's western flank.

By contrast, British planning for an eventual return to the continent had begun in the very same month that the Allies had been expelled from the continent at Dunkirk – June 1940 – convincing evidence of Churchill's invincible optimism. Direct talks between the Americans and British about the operation began as early as April 1942, and after a number of sharp disagreements – with the Americans pushing for an earlier date and the British for a later one – Roosevelt and Churchill agreed at the Casablanca Conference in January 1943 to re-enter Europe in the early summer of 1944. Walking into his new offices in Norfolk House, St James's Square, in March 1943, from where he was supposed to draw up the plans, Lieutenant-General Sir Frederick Morgan found only a pencil that someone had dropped on the floor. The original forces at his disposal were much smaller than finally undertook the operation, and invasion sites fewer, but as work progressed it was appreciated that only a massive undertaking would work. 'Well, there it is,' said Brooke to Morgan of the original scheme. 'It won't work, but you must bloody well make it.'

By June 1944, however, the logistical and supply sides of the operation were formidable; that month southern Britain had some fifty-seven million square feet of covered storage space filled with supplies, including 450,000 tons of ammunition. The training was similarly intensive; it was estimated that by the time of D-Day the US 29th Infantry Division had marched some 3,000 miles, almost enough to take it back to the USA.

On 7 April 1944, Churchill told the senior commanders: 'Remember, this is an invasion, not the creation of a fortified beachhead.' The invasion was to be spearheaded via five invasion beaches across Normandy, from west to east, codenamed Utah (American), Omaha (American), Gold (British), Juno (Canadian) and Sword (British). There was to have been another beach – codenamed Band – the plans for which were shelved after Rommel completely flooded the area behind it, so Utah was chosen instead.

By dawn on D-Day, 18,000 paratroopers had landed behind Sword and Utah, including the British 6th Airborne and American 82nd and 101st Airborne Divisions. In all, 23,400 Allied troops landed by parachute and glider on D-Day. 'The longest day', as it became known, began at 00.16 hours on 6 June when a Horsa glider landed only forty-seven yards from Pegasus Bridge on the Caen Canal, to be followed a minute later by another and then another minute later by a third. These gliders had flown three miles in pitch darkness, navigated by stopwatches and finger-torches. After 180 men of 'D' company of the 2nd battalion of the Ox and Bucks Light Infantry debouched from their planes, the Bridge was captured in less than ten minutes with the element of total surprise, and German hopes of destroying it rather than let it fall into enemy hands were dashed. When the dashing Lord Lovat, commander of the 1st Special Service Brigade who had landed with No. 4 Commando, arrived to relieve it from Sword beach, he apologised that he was some three minutes late. The Allies were able to reinforce the 6th Airborne that had landed on commanding high ground to the east.

Pegasus Bridge was the only possible route to reinforce the 12,000 men of the 6th Airborne Division, who were about to bear the brunt of concerted attempts by the 21st Panzer Division and several other German units to descend on Sword beach and fling the invaders back into the sea before they had established the bridgehead. There followed seventy-four days of continuous combat, the longest period of action that any unit of the British Army saw in either world war.[11] In total, one-third of the men were killed, wounded or captured, but they nonetheless held the Germans off the high ground commanding the eastern approaches to Sword and Juno beaches.

Some criticism has been made of how little inland the British managed to get on D-Day and immediately thereafter, but this is entirely misplaced since they were the hinge around which the Allies were to swing westwards. This was only possible because the 6th Airborne Division held off sustained German attacks for over two months. The life expectancy of a platoon officer in the Division during this engagement was seven days, one-third of that of a lieutenant in the trenches of the Great War.

One unit of the 6th Airborne, the 9th Parachute Battalion, captured the Merville Battery, whose guns – including 100 mm ones – could reach to Sword and Juno beaches. Although only 150 paratroopers were able to be mustered out of the 750 who were dropped, and there were no wire-cutters to slice through the barbed-wire perimeters of the Battery, Lieutenant-Colonel Terence Otway took the harrowing but correct decision to lead his men through a German minefield in order to take the strategically vital Battery. The position was successfully captured and four large guns were disabled, although only sixty-nine British soldiers were left standing by the end of the engagement. For all the lives lost, however, the 9th had protected the men on the beaches from being subjected to a withering fire from the Battery.

As so often in the history of the English-speaking peoples in the twentieth century, air superiority was vital. The RAF and USAAF had made no fewer than seventy air-reconnaissance sorties per day for three months prior to D-Day, and in earlier bombing raids many German radar and radio-interception stations had been destroyed. On D-Day itself, no fewer than 13,688 sorties were made by Allied aircraft, against only 319 by the Germans. Even so, 127 Allied aircraft were lost on D-Day. In the sixteen weeks before 6 June, 22,000 bombing sorties had dropped 66,000 tons of bombs on military installations and rail servicing and repair facilities around Normandy. In the period between 1 April and 5 June 1944, the Allied air force lost 12,000 officers and men, and 2,000 aircraft, and by the end of the Normandy campaign 28,000 aircrew had been lost.

This scale of preliminary bombing had the tragic but inescapable corollary of killing several thousand French civilians, no fewer than 3,000 in the town of Caen alone. When Churchill wrote to Roosevelt to express the British War Cabinet's anxieties about French civilian casualties, he received a reply that underlines the ruthlessness that leaders of the English-speaking peoples are occasionally required to display. 'However regrettable the attendant loss of life is,' wrote Roosevelt, 'I am not prepared to impose from this distance any restriction on military action by the responsible commanders that in their opinion might militate against the success of Overlord or cause additional loss of life to our Allied forces of invasion.'[12]

France would be liberated, but not before many of her innocent citizens had died in the process. There are huge discrepancies between historians' estimations of the total loss of French civilian life during the whole of the Normandy campaign, varying between 16,000 and 60,000. Beside that, the total of American military personnel killed was over 30,000 and the British and Canadians' total was over 26,000. The German loss of life is hard to quantify, but in one of their cemeteries – at La Cambe – some twenty tons of Wehrmacht body-parts are buried.

The landings on the beaches saw, in some cases, some of the toughest fighting of the English-speaking peoples in the twentieth century. At Utah bad visibility had meant that sixty-seven of the 360 USAAF bombers were unable to bomb their targets, but through a number of fortunate circumstances, including a powerful tidal current that washed the invasion force almost two miles south of their target to a less well-defended part of the coast, the Americans landed 23,000 men ashore with only 200 lives lost, including naval and associated personnel. The 82nd and 101st Airborne had engaged the Germans behind the lines, and were reinforced and re-supplied by the troops who came ashore there.

Partly because of the widespread Allied missing of drop-zones – which was not the fault of pilots who had been asked to do the near-impossible – German Intelligence estimated that 100,000 parachutists had landed, rather than the correct figure of around one-fifth of that.[13] The flooding of large parts of the battle area by Rommel accounted for the drowning of 300 parachutists from the 82nd Airborne and the 101st. Carrying up to 150 lbs of equipment each, with parachute harnesses that required both hands to undo, those who landed in water barely stood a chance.

The situation at the four-and-a-half-mile-long Omaha beach was very different indeed from that on Utah. There were high cliffs and bluffs, and the inwards curvature of the coast helped the Germans' fields of fire. Underwater there were sandbars and ridges, which coxwains of the landing-craft often mistook for the beaches themselves. Much of the heavy equipment needed by the men in the first wave was dropped in deep water and lost. Sheer weight of numbers – 40,000 attackers versus what was thought to be 400 defenders –

was expected to tell. Yet only 2,000 yards inland was the veteran German 352nd Infantry Division of 9,000 men, who moved to the beaches with high morale and deadly purpose. (Bletchley Park had warned about the proximity of this unit to Omaha, but the risk was considered worth taking.)

For reasons still disputed by historians, troops were transferred from ships onto assault landing-craft eleven miles away from Omaha beach. Ten landing-craft sank. With no bow to cut through the waves, but only a landing ramp in front, the four-hour journey was a nightmare of seasickness, made far worse by the anti-mustard gas compound with which the men's uniforms were impregnated, which when mixed with sea-water gave off the stench of rotten eggs. A total of twenty-seven out of thirty-two tanks at the eastern end of Omaha sank, and twenty artillery pieces capsized, meaning that the troops who did stagger ashore – seasick and carrying an average of 70 lbs of equipment on their backs – had no heavy armour to punch their way through the German defences. Moreover, the beach bristled with anti-personnel devices both above and below the water. These included contact mines on poles, concrete triangles called tetrapodes, huge impassable 'Belgian Gates', landmines, barbed wire, 'hedgehogs' and logs on supports that had saw teeth to rip up the landing-craft and which had mines attached.

Eight hundred men died on Omaha beach on D-Day, many of them drowned as they jumped off the landing-craft under the huge weight of their equipment. A further 5,500 were wounded. Some units lost 90% of their officers. No fewer than thirty pairs of brothers are buried at the American military cemetery above the beach. Of the 212 men on the roster of 'A' company, 116th Regiment of 29 Division who landed at Omaha on D-Day, not one of the riflemen was left in it by V-E Day. It was recorded that the flecks on the top of waves coming into that beach were still red with the blood of corpses even by D+13.[14]

The Canadians who landed at Juno beach did the best of all in terms of getting inland, covering over seven miles by the end of the day. The Regina Rifles and the Queen's Own Rifles of Canada had the signal honour of being the furthest beach-dropped units inland by nightfall and were the only units to achieve all their objectives on D-Day. The average ages of the men attacking Omaha in the first wave was 20.6 years, Gold and Sword 24 years but on Juno it was 29 years old, and this greater age and experience told. Juno beach, which was four-and-a-half miles long, was attacked by 14,500 Canadians. It included the mouth of Courseulles harbour, and it was vital to capture it to prevent the gap between Juno and Sword beaches becoming dangerously wide. The Canadians had a tough time before the village of Bernières-sur-Mer, because the tanks did not arrive until after the infantry had landed and a 10-foot-high concrete sea-wall on the highest of the beaches severely impaired movement.

West of Courseulles, 'B' company of the Royal Winnipeg Rifles, for example, suffered 85% casualties. Of the 800 men of the Queen's Own Rifles of Canada, part of the 8th Brigade of the 3rd Canadian Infantry Division, 143 men were killed or wounded on D-Day. Over the rest of the war a further 1,000 battle casualties were suffered, including 462 deaths, so each man in the regiment was effectively replaced twice by May 1945. They more than earned Eisenhower's (necessarily off-the-record) remark that man-for-man the Canadians were the best soldiers in his army. (It was a disgrace that no Canadians were portrayed in the generally historically accurate movie *The Longest Day*.)

Six thousand British troops landed at Gold beach, even though they had to contend with force five gales – the worst of any sector – that sank some amphibious tanks and three landing-craft. At King section of Gold beach, the thirty-two-year-old Yorkshireman, Company Sergeant-Major Stan Hollis of the Green Howards, won the only Victoria Cross of the day after he rushed two German pillboxes and personally took twenty-five Germans prisoner.[15]

It is astonishing that so few Victoria Crosses were awarded, but they had become much rarer in the twentieth century than they had been in the nineteenth. Indeed, one fewer VCs were awarded for acts of gallantry in the Second World War – at 181 – than were won during the Indian Mutiny.[16] The Americans were almost equally parsimonious with their gallantry awards. The bridge at La Fière saw the costliest small unit action in the history of the US Army, with 500 men killed over three days' fighting over the strategically vital but tiny bridge across the swollen Le Merderet River. On 9 June, Private First Class Charles DeGlopper, of 'C' company, 325th Glider Infantry, part of the 82nd Airborne Division, won a posthumous Congressional Medal of Honor protecting his comrades during a withdrawal.

DeGlopper, from Grand Island, New York, was at 6 feet 7 inches and 245 pounds reputedly the biggest man in the Division. He had joined the Army in March 1942 and had fought in Sicily and Italy in 1943 and 1944, before landing near Les Forges on 7 June. Outflanked and outgunned, DeGlopper covered his platoon's retreat by drawing the enemy's fire, continuing to fire his Browning automatic rifle even after his left arm had been shot off, and accounting for a total of twenty-five Germans. When his body was recovered later, it had sixty-six bullet wounds.

Of the four men from 'A' company who defended the bridge from three advancing German tanks, three were seventeen years old and the fourth was eighteen. (One of them, a Swede called Leonold Peterson who had managed to reach America in 1942, where he volunteered to serve in the US Army, won the Distinguished Service Cross at La Fière. After the war he was deported from the United States because his immigration papers were not in order.) Of the 212 men of 'G' company who served in the Hill 30 area overlooking Le Merderet, only six were registered as still fit to fight when the unit got back to the United Kingdom that August.

The seaside town of Arromanches was under Allied control by the late afternoon of 6 June, and a massive operation was immediately got under way to build a gigantic artificial harbour there – codenamed 'Mulberry' – parts of which can still be seen to this day. The idea that a floating harbour could be towed to beaches went back to 1917, and in the 1920s a dock had been towed 10,000 miles from Britain to Singapore. The harbour outside Arromanches, however, was to enclose two square miles, comprise 600,000 tons of concrete and one million yards of steel, take 100,000 men nine months to build and 20,000 men to operate, and would need 150 tugs to pull its constituent parts from all over the United Kingdom to Arromanches. It was an extraordinary endeavour. Overall it had the same capacity as the entire port of Dover itself.

Each of the 120 caissons of the Mulberry harbour was the width of a football pitch, the height of a six-storey building and weighed between 3,000 and 5,000 tons. Once in place the harbour could operate twenty-four hours a day in all weathers, and in total between D-Day and V-E Day, no fewer than two-and-a-half million men, half-a-million vehicles and four million tons of supplies came ashore by that route. Even after the fall of Cherbourg, the Mulberry harbour at Arromanches – the one intended for Omaha beach was swept away in a terrible storm – continued to be the principal place of disembarkation, protected by no fewer than 150 anti-aircraft guns. The harbour was essentially an insurance policy, against bad weather closing Cherbourg and thus cutting off the re-supply of the expeditionary force just as the Germans counter-attacked. The fact that the summer of 1944 saw good weather does not detract from that necessity; it was merely a bonus. The English-speaking peoples had once again constructed one of the twentieth-century's engineering wonders of the world.

Indeed, the sheer inventiveness of the English-speaking peoples that was to play such an important part in protecting their twentieth-century hegemony was on full display on 6 June 1944. Seven new inventions were deployed that were championed by Major-General Sir Percy Hobart (pronounced Hubbard), who had commanded the very first British tank brigade in 1934. His career had seemed to be over in 1940, when he was reduced to serving as a corporal in the Home Guard, but Hobart was suddenly appointed by Churchill to command the 79th Armoured Division in 1943.[17] Floating 'Duplex Drive' tanks known as 'Donald Ducks', flame-throwing tanks known as 'Crocodiles', huge flails that exploded mines before them as they went, 'Bobbin' tanks that laid 110-yard-long coils of coconut matting, 'Petard' armoured vehicles that fired a 'flying dustbin' explosive that could knock out even the best protected German pill-box, all and more were employed on D-

Day. Great ingenuity in military technology has long been a feature of the English-speaking peoples at war.

In order to refuel the armies, a giant petrol pipeline was laid from Britain to Normandy. Seventy miles in length, 'PLUTO' had 172 million gallons flow through it before the war's end. (The building housing its installation was disguised as an ice-cream factory on the Isle of Wight and thus had escaped German bombing.)

Although overall Operation Overlord was undoubtedly a magnificent success, and an incredible achievement for the English-speaking peoples, not everything went completely according to plan, as in any large-scale military enterprise. Paratroopers were dropped over far wider areas than planned, few landing directly in their intended drop-zones. In two cases – involving 100 men of the 1st Canadian Parachute Battalion and 150 men of the 82nd US Airborne Division – they were dropped thirty-five miles east and twenty-five miles south of their drop-zones respectively. The city of Caen did not fall till 9 July (except in the *London Evening News*, where it was declared to have fallen on the first day). There were a good deal of what is today called 'friendly fire' incidents; indeed, when the US General Lesley McNair was hit by USAAF bombs that had fallen short in one attack, his 'body was thrown sixty feet in the air and was unrecognisable except for the three general's stars on his collar'.[18]

The Allies were aided by 21st Panzer Division's decision to split into three groups and only to move eleven hours after the invasion, and then in the wrong directions. Had it counter-attacked Pegasus Bridge at eight o'clock on the morning of the invasion, Lovat's brave but lightly armed commandos could not have held it, and once the Bridge was in German hands the 6th Airborne could not have been reinforced. It only took three hours for the first twelve German tanks of many to get to Lion-sur-Mer, within yards of the sea-front between Sword and Juno beaches, but they then pulled back in order to engage the air-landing brigade of the British 6th Airborne, whose 156 gliders passed overhead at the exact moment the German tanks arrived at the coast. These and many other errors were invaluable to the Allies.

The thick hedgerows of Normandy – many of them planted fifteen centuries earlier by the Vikings – were quite unlike hedgerows in Britain or America. They could be up to fifteen feet high and five feet wide and impenetrably thick for anything less than a tank. They therefore provided perfect defensive cover as the Germans made their fighting retreat.[19]

For all of General de Gaulle's personal petulance, his countrymen played an important role in her own liberation behind the beaches; around 3,000 members of the Resistance cut no fewer than 950 rail and road lines of communication, significantly delaying Panzer reinforcements being sent to Normandy from south-western France. They were helped by the British

Special Operations Executive, which also paid a high price: of the 393 SOE agents who served in occupied France during the Second World War, 119 were executed by the Germans or otherwise killed on duty.

The Allied soldiers buried in the war cemeteries of Normandy were interred regardless of rank, religion or unit, the dates on their gravestones relating to the day their bodies were found rather than the day they died. Superbly curated by the United States American Battle Monuments Commission, the Colleville St Laurent-sur-Mer cemetery above Omaha beach contains the remains of 9,386 soldiers, sailors and airmen, including 307 whose identity is unknown. Taken entirely at random to emphasise the eclectic nature of the geographical backgrounds and units of the men who fought together in the Normandy campaign, here are the inscriptions of the first seven graves on section H, row 13, of that vast cemetery, every headstone of which faces west towards the USA:

'PFC Virgil E Jones 38 Inf 2 Div Oklahoma June 25 1944'

'PFC Francis J Donovan 9 Inf 2 Div Massachusetts June 16 1944'

'PVT Joseph J Planster 116 Inf 29 Div Pennsylvania June 29 1944'

'PFC Charles Knobler 430 AAA AWBN New Jersey 5 August 1944'

'2 Lt William L Myers 24 Cav RCN Pennsylvania July 29 1944'

'Here Rests in Honored Glory a Comrade in Arms Known But to God'

'Sgt Elmo C Farrow 17 Engnr Bn 2 Armoured Div Georgia 15 June 1944'.

The next five men are Robert B. Etheridge Jr from Texas, Robert L. Rose from Kansas, Lloyd E. Fenshe from Oregon, Alvin C. Plocker from California and Ethan Rablitt from West Virginia. Their crosses – in Charles Knobler's case his star of David – stand in perfect order across fields and fields, a mute and hugely dignified testament to the readiness of the English-speaking peoples to rid the world of fascist aggression.

It is sometimes easy, especially when considering great historical events of six decades past, to forget the personal side of sacrifice, the immense scale of human tragedy engendered by the Nazi ambition to conquer and subdue. A short walk through any of the war cemeteries of Normandy quickly reminds one of the fact that every man lying buried was someone's son. At the relatively small Ryes war cemetery at Bazenville in Normandy, beautifully tended by the Commonwealth War Graves Commission, 630 Britons, 21 Canadians, an Australian and a Pole, as well as 326 Germans lie buried. The messages engraved on the tombstones by the families of the deceased make moving reading. They could choose the inscriptions on the graves, and this is what some of them wrote:

Private L. Cohen, Gordon Highlanders 21 July 1944, Age 31, 'A broken link we can never replace. Not a day passes but we think of you'

Sapper C.E. Thomas, Royal Engineers, 28 August 1944, Age 22, 'Greater love hath no man than this'

Sapper G. Robinson, Royal Engineers, 10 June 1944, Age 28, 'Not one day
   but every day I remember, George dear. Mum'
Warrant Officer J.S. Elwood, Royal Australian Air Force, 31 May 1945, Age
   20, 'In Proud and Loving Memory of our John'
L. Sgt W.R. Dowden, Royal Canadian Artillery 9 June 1944, Age 25, 'He gave
   his life that we may live, Fondly remembered by Mum, Dad and family'

The astonishingly varied Canadian units represented in that single cemetery
are also testament to the number of fighting and supply corps that existed in
1944, several of which no longer exist – or have been amalgamated to the
point that their individuality has all but disappeared. The titles of the regiments
still have the power to evoke pride and nostalgia, however. In Ryes cemetery,
there are men who gave their lives serving in the Calgary Highlanders, North
Nova Scotia Highlanders, South Alberta Regiment, Royal Canadian Army
Service Corps, British Columbia Regiment, Canadian Scottish Regiment,
Royal Canadian Infantry Corps, 59th (Newfoundland) Heavy Regiment of
Royal Artillery, Regina Rifles, Le Régiment de la Chaudière, the 1st Hussars
and the Royal Canadian Artillery.

On 7 June, thirty-seven prisoners of war, mostly from the North Nova
Scotia Regiment, were murdered after laying down their weapons by the 12th
SS Panzer Regiment in Authie, just some of approximately 200 Canadian
POWs executed in the week after the landings. The SS commander, Kurt
Mayer, was captured in 1945 and tried in 1946. His death sentence was later
commuted to fourteen years, but he was released in 1954. (A successful beer
salesman after the war, he showed no remorse and even returned to Normandy
in 1957 to visit his former battlefield.)

In his recent book *Armageddon*, about the battle for France and Germany
in 1944–5, the distinguished military historian Sir Max Hastings summed up
the moral difference between the Atlantic allies and the totalitarian powers
thus:

> To an impressive degree, the American and British armies preserved in battle the
> values and decencies, the civilised inhibitions of their societies. . . . The Germans
> and Russians . . . showed themselves better warriors, but worse human beings.
> This is not a cultural conceit, but a moral truth of the utmost importance to
> understanding what took place on the battlefield.[20]

Of course, as Hastings is the first to admit, there is no telling whether Atlantic
civilised values might have survived if Britain and America had ever had to
fight the kind of war-to-the-knife that the Soviets did to defend their Motherland and expel the invader.

Canada's contribution to victory in the Second World War was incredible
considering her population of only eleven million people. In the spring of

1939, there were 10,000 men in her armed forces; by the end of the war, over one million had served in them. In the meantime, they had been, in Professor David Dilks' words, 'the only properly organised, trained and equipped military strength in the southern part of England in the perilous summer and autumn of 1940'; had fought in Hong Kong during Christmas 1941, where their commanding officer had been killed; and had seen action at Dieppe, Sicily, Italy, France and the Low Countries. The Royal Canadian Navy had 500 ships in service by 1943, and at one point was the third largest Navy in the world. No fewer than 125,000 Commonwealth air crew were trained in Canada, and of the RAF's 487 squadrons in 1944, no fewer than 100 came from the Dominions (not taking into account those individual Crown subjects from abroad who had enlisted in British units.).

Financially, the Canadian support of Britain was staggering. At a dinner at Laurier House in Ottawa on 30 December 1941, after he had delivered a speech to the joint session of the Canadian Parliament, Churchill was told by Mackenzie King that Canada would be making an immediate present to Britain of $1 billion. Soon afterwards, debts of a further $700 million were converted into an interest-free loan. Much more came later in many different ways, including $2.8 billion in outright gifts. Professor Dilks has calculated that Canadian contributions equated to a quarter of the total coming to Britain under Lend-Lease, despite Canada having a population less than 9% that of the USA. The burden on individual Canadian taxpayers was nearly four times that of the Americans. Blood was far thicker than the waters of the Atlantic.

Although 6 June 1944 saw a triumph of co-operation between the High Commands of the English-speaking peoples, before the month was out a serious disagreement had arisen about how Overlord was to be supported in southern and western Europe. The US Joint Chiefs of Staff wanted to launch Operation Anvil to capture Toulon and invade the South of France, forcing its way up the Rhône Valley. Meanwhile, the British Chiefs of Staff – minus Alanbrooke, who was ill at the time – wanted General Alexander to press on with his campaign to destroy German forces south of the Rimini-Pisa line in Italy. Whereas the Americans wanted to invade the South of France on 15 August, the British wanted to cross the River Po, advance on Trieste and push into the Balkans in September. Since there was no possibility of doing both simultaneously, a stand-off developed between the two committees, neither of which was willing to back down.

Various memoranda were passed between them, but neither side budged. A memorandum of 24 June declared that the 'proposal for commitment of Mediterranean resources to large scale operations in Northern Italy and into the Balkans is unacceptable to the United States Chiefs of Staff'.[21] Two days later, the British Chiefs of Staff replied that, 'Withdrawal of forces from Italy to achieve an Anvil date of 15 August is unacceptable.' The next day, 27 June, Marshall cabled Eisenhower – who supported Anvil – to say that,

> The British proposal to abandon Anvil and commit everything to Italy is unacceptable. . . . It is deplorable that the British and US disagree when time is pressing. The British statements concerning Italy are not sound or in keeping with the early end of the war. . . . There is no reason for discussing further except to delay a decision which must be made.

The following day, the British Chiefs cabled Washington to say they deeply regretted the US Chiefs' stance, 'but it would be unthinkable for want of patient discussion to risk taking a false step at this critical period of the war'.[22] They then reiterated all the arguments for a Balkan strategy, ending with the uncompromising statement that, 'We feel so strongly on this matter that at present we see no prospect of being able to advise His Majesty's Government in the United Kingdom on military grounds in a sense contrary from that we have set forth.' On receipt of this, Admiral William Leahy wrote a memorandum to Roosevelt, asking him 'to despatch a message to the prime minister', which read: 'It seems to me that nothing can be worse at this time than a dead-lock in the Combined Chiefs as to future course of action. You and I should prevent this and we should support the views of the Supreme Allied Commander.' FDR sent it on to London without altering a word.

Churchill replied immediately, saying, 'We take it hard that this should be demanded of us. . . . I most earnestly beg you to examine this matter in detail for yourself. I think the tone of the US Chiefs of Staff is arbitrary and, certainly, I see no prospect of agreement on the present lines. What is to happen then?' That night Churchill sent FDR a twelve-page cable, setting out the advantages of a Balkan strategy against the long distances that any landing in Toulon would have to go before it could engage significant German forces, ending, 'Let us not wreck one great campaign for the sake of winning another. Both can be won.'

The prospect of Alanbrooke being brought from his sick-bed, from where he had described himself as 'weak as a cat', elicited the request from Marshall that 'on no account should we worry the Field Marshal'. Marshall's own view was that 'there is a big part played by the prime minister in the present affair'. This was immediately and categorically denied by Colonel Hollis at the War Cabinet Office. Roosevelt sent his thirteen-paragraph reply to Churchill the next day, 29 June, saying, 'I again urge that the directive proposed by the US chiefs of staff be issued . . . immediately.' With a general election just over four months off, the President ended with an emotional, personal and electoral plea, saying,

History will never forgive us if we lose precious time and lives in indecision and debate. My dear friend, I beg of you let us go ahead with our plan. Finally, for purely political considerations over here I would never survive even a slight setback in Overlord if it were known that fairly large forces had been diverted to the Balkans.[23]

Roosevelt was out-Churchilling Churchill, and it worked. On 1 July, the Prime Minister telephoned the President to admit defeat. Operation Anvil, in which 77,000 men and 12,000 vehicles arrived by sea and 9,000 landed by parachute along the southern coast of France, took place on 15 August 1944 (the 175th anniversary of Napoleon's birth). It had to change its name to Operation Dragoon, for fear of breached security, and the apocryphal story did the rounds that Churchill had chosen the new name because he felt he had been dragooned into it. Although the truth was evident far earlier, the United States was confirmed as the leading power of the Western alliance. The baton had passed from hand to hand, reluctantly and not without bluster, but neither was it wrenched from Britain's grasp. Churchill was to be the last British leader of the Free World.

The literary critic Desmond MacCarthy once described the primary national characteristics of the French as 'stinginess, and blind vindictive self-assertion', and both were certainly apparent when Charles de Gaulle spoke from the Hôtel de Ville in Paris on 25 August 1944. In his speech he proclaimed that Paris had been 'liberated by her own people, with the help of the armies of France, with the help and support of the whole of France, that is to say of fighting France, that is to say of the true France, the eternal France'. No mention was made of any Allied contribution; the myth-making had begun.

Out of the thirty-one divisions assigned to the campaign in Normandy, just one was French, the Deuxième Division Blindée (2nd Armoured Division) under the command of General Leclerc (the *nom de guerre* of Vicomte Jacques-Philippe de Hautecloque). It fought very bravely in the battle to close the Falaise Gap around the Germans in Normandy, but the battle would undoubtedly still have been won without it.

In Führer-Directive No. 51 of 3 November 1943, Hitler had predicted that although territory could be lost to the Red Army on the Eastern Front, 'a greater danger appears in the West: an Anglo-Saxon landing!' He was correct in his etymology at least, in the sense that the landings that started the process by which France was freed were primarily made up from the English-speaking peoples. In a list of his principal worries drawn up just before D-Day, the Supreme Allied Commander General Eisenhower had placed the Free French Commander General Charles de Gaulle at the head, even above the uncertainties over the weather in the Channel. For the previous four years de Gaulle had been a perpetual irritant to Allied decision-makers, insisting upon being treated as a head of state equal in status to King George VI and President Roosevelt, even though he was very clearly no such thing.

After de Gaulle first set foot in France on 14 June, he made a one-day visit to Bayeux, after which he left for Algiers and did not return to French soil until 20 August. In the meantime, General Patton's Third Army had broken out of Avranches at the end of July and had marched through Brittany. The French Resistance, the *résistants* and *maquisards* – a separate organisation from de Gaulle's Free French forces – was doing brave and vital work in support of the Allied forces, especially in hampering German armoured retaliation, but de Gaulle played little part in any of this from his base in North Africa.

Meanwhile in Paris, unbeknown to his own troops, the German commander General Dietrich von Choltitz took the historic and humane decision not to set fire to the city. 'Paris must be destroyed from top to bottom,' the Führer had demanded of him, 'do not leave a single church or monument standing.' The German High Command then listed seventy bridges, factories and national landmarks – including the Eiffel Tower, the Arc de Triomphe and Nôtre-Dame cathedral – for particular destruction. Hitler later repeatedly asked his Chief of Staff: 'Is Paris burning?'

Yet Choltitz deliberately disobeyed these barbaric instructions, and the Germans did not therefore fight the battle of extirpation that they were even then fighting in Warsaw, at the cost of over 200,000 Polish lives and the utter devastation of the ancient city centre. Von Choltitz surrendered as soon as he decently could once regular Allied forces arrived, telling the Swedish diplomat who negotiated the agreement that he did not wish to be remembered throughout history as 'The man who destroyed Paris.'

In all, General Leclerc lost only seventy-six soldiers killed in the liberation of Paris, although 1,600 inhabitants had died during the uprising, including 600 non-combatants. Today the places where the individual soldiers and *résistants* fell are marked all over the city, and none would ever try to belittle their sacrifice, but the fact remains that the only reason that Leclerc was assigned to liberate the city was because Eisenhower could spare the French Second Division from far greater battles that were taking place right across northern and southern France, fought against crack German units by British, American and Canadian forces. For political and prestige reasons, de Gaulle had begged Eisenhower that it be French troops who would be first into the capital, and the Supreme Commander was as good as his word. Nor did Eisenhower visit the capital himself until 27 August, because he did not wish to detract from de Gaulle's limelight.

The Allies did not see Paris as a prime military objective, as opposed to a political one, and they were right not to. As Ian Ousby wrote in his history of the Occupation, 'Paris's concentration of both people and cultural monuments ruled out aerial bombardment and heavy artillery barrages, so taking the city would soak up time and lives in a campaign already behind schedule and high in casualties. Besides, the capture of Paris was not tactically essential.' In his memoirs, General Bradley dismissed it as a 'a pen and ink job on the map'.

It was Eisenhower – not de Gaulle – who gave the order to General Leclerc on 22 August to advance immediately on the capital. He had other units, including the American 4th Division, that could have done it, but he wanted the French to have the glory. De Gaulle merely added to Leclerc that he must get there before any Americans arrived. The first of Leclerc's (American-made) Sherman tanks rolled up the rue de Rivoli at 9.30 a.m. on the morning of 25 August. In the surrender document signed that same afternoon by Leclerc and Choltitz, there was no mention of either Britain or the United States.

The next morning – 26 August 1944 – de Gaulle led a parade from the Arc de Triomphe down the Champs-Elysées to a thanksgiving service in Nôtre-Dame. When Resistance leaders came up abreast of him in the parade, he hissed at them to get further back behind him; the applause was to be his alone. He was cheered to the echo, but of course wartime crowds are fickle: when Marshal Pétain had visited Paris three months earlier, on 26 April, hundreds of thousands of the same Frenchmen had turned out to cry '*Vive le Maréchal!*' What France desperately needed was a myth of heroic self-deliverance. That is what de Gaulle gave them on 25 August 1944, and which they came to believe.

Churchill, a Francophile to his very marrow, set out the way in which the French had stymied the Allied war effort after their defeat in 1940. In a rousing speech to the Canadian Parliament on 30 December 1941, the Prime Minister recalled how the previous year,

> The French Government had at their own suggestion solemnly bound themselves not to make a separate peace. It was their duty and it was also their interest to go to North Africa, where they would have been at the head of the French Empire. In Africa, with our aid, they would have had overwhelming sea power. They would have had the recognition of the United States and the use of all the gold they had lodged beyond the seas. But the men of Bordeaux, the men of Vichy, lay prostrate at the foot of the conqueror.

Of course de Gaulle himself cannot be accused of any such dereliction of duty. His action in flying to London was one of sublime courage for which his name shall rightly resound through history. But once in London he could do little more than stand on ceremony, insist on his rights and plan for Free French operations, which, like the raid on Dakar in West Africa, were almost uniformly militarily undistinguished. Nor was this his fault; he was profoundly hampered by the fact that many more Frenchmen fought for Vichy than for him.

Gaullism might be defined as the process of turning pride and perversity into a political programme, however starkly adverse the geopolitical realities. When de Gaulle set up the Free French in 1940, his country was ruined, yet only five years later and without having won any serious victories of her own, France was almost a great power again, occupying a zone in Occupied Germany with a seat on the Security Council of the United Nations. Whereas Italy – which had joined the Allies in 1943 – was run by the Allies, France – which was run by Pétain until 1944 – was allowed self-government. It was de Gaulle who achieved this almost single-handedly through his constant deployment of ingratitude, intransigence, 'ferocious sarcasm' and 'volcanic eruptions of contempt'.

De Gaulle admitted to feeling 'an anxious pride' for France, and well might he have been anxious for a country that was wrecked so comprehensively in the two world wars. Yet because he assumed that France needed greatness in order to be what he called 'the eternal France', he simply insisted upon it, whatever the economic, political and military actualities. The British diplomat Gladwyn Jebb pointed out how 'undoubtedly the General's chief failing was to cast his country into a role which was beyond her power', yet it worked. Part of the strategy meant that he had constantly to *épater les Anglo-Saxons*, as happened so volcanically the day prior to Operation Overlord.

De Gaulle's ingratitude towards his hosts that year was legendary. 'You think I am interested in England winning the war,' he told his British liaison officer General Spears. 'I am not. I am only interested in French victory.' When Spears simply made the logical remark, 'They are the same,' de Gaulle replied: 'Not at all; not at all in my view.' He needed to prove that the eternal France still had teeth, so he made the hand that fed him his staple diet.

As well as D-Day, June 1944 saw the introduction into the Pacific theatre of the B-29 Super Fortress bomber, one of the technological wonders of military engineering of the age. Even before America was sucked into the war, General Hap Arnold of the US Army Air Corps had devised an air force far more ambitious than merely a support arm of the ground forces. His plan was to create 'a multi-role air force, capable of strategic bombing, of large-scale air transport, of air defence and tactical air defence'.[24] The country of the Wright brothers had stayed in the forefront of aeronautical engineering, as the Flying Fortress and Liberator heavy-bombers, Lockheed Lightning fighter-bomber and the Mustang and Thunderbolt fighters proved. As happened in the battle of Britain, Korea, Suez, Vietnam, the Falklands, the Gulf War, Afghanistan and Iraq, command of the skies proved invaluable to the English-speaking peoples.

★

The month after D-Day, Polish underground forces in Warsaw bravely attempted to wrest the city from German control, in the hope and expectation that the Red Army, just on the other side of the Vistula River, would help them. Stalin's cynicism was once again starkly demonstrated when in August and September 1944 he ordered his soldiers to wait until the SS had destroyed the resistance, and with it much of the city itself. He also refused the RAF and USAAF permission to land in Soviet-held territory, hampering their ability to drop supplies of food and arms to the Poles. Only after the Uprising had been completely crushed did the SS withdraw from Warsaw, after which the Red Army crossed the river and took over the smoking ruins of the city.

Meanwhile, the efforts of the Americans, British, Canadians and French to liberate France continued. The limits of Military Intelligence were brought home in September 1944, when General Sir Frederick ('Boy') Browning utterly disregarded the aerial reconnaissance photographs brought to him by his Intelligence officer Major Brian Urquhart, which showed German SS Panzer divisions refitting near the town of Arnhem. Believing Urquhart to be 'mentally disturbed by stress and overwork', Browning ordered him to go on convalescent leave. The resulting parachute drops of the British 1st Airborne Division led to one of the greatest disasters to befall Allied arms during the Second World War. 'The soldiers of the British and Polish airborne brigades paid a terrible price for the conceit and arrogance that motivated Browning's refusal to acknowledge the accurate information put before him by his Intelligence staff,' concludes an important recent study of Military Intelligence blunders.[25]

On 3 September 1944, the fifth anniversary of Britain and France's declaration of war against Germany, British troops entered Brussels, the Belgian capital. There were huge celebrations by ordinary Belgians in the streets at the end of dictatorship and the return of their freedom. As in Paris and Brussels, similar celebrations took place across Western Europe as Allied troops liberated city after city, bringing democracy in their wake. The Americans liberated Luxembourg on 10 September, the Canadians liberated Calais on the 28th the British liberated Athens on 14 October, and so on.

Memories fade fast. Many of the countries whose publics demonstrated most vociferously in March 2003 against the United States-led coalition forces invading Iraq to install democracy in place of the dictatorship there – such as Holland, France, Denmark, Greece, Austria and Belgium – were the same countries that had themselves been liberated by the United States-led coalition in 1944–5. Many of their citizens argued in 2003 that because of its special traditions of religion and society, the Middle East was not suited, and certainly not ready, for the destabilising concept of democracy. This was of course precisely what the anti-democratic intellectuals of *Action Française*, such as

Charles Maurras, and any number of other European fascist-sympathisers, such as Vidkun Quisling, were arguing in the Thirties and early Forties about such countries as Holland, France, Denmark, Greece, Austria and Belgium. Fortunately for Europe, the United States-led coalition paid no more attention to those anti-democratic intellectuals in 1944 than they did to the anti-war demonstrators of 2003. (It was no coincidence that the European countries that did send troops to Iraq, such as Spain and Poland, were the ones that had not been liberated by the US-led coalition, but had had to exist without democracy until the 1970s and 1980s respectively. Similarly Italy, whose experience of fascism had lasted twenty years, also sent troops to help liberate Iraq from it.)

At a meeting in Moscow on 9 October 1944, Churchill and Stalin undertook what many have since regarded as a supremely cynical act of *realpolitik*, known to history as 'the percentages agreement'. Saying that he did not want to use the phrase 'dividing into spheres', for the reason that 'the Americans might be shocked', Churchill then neatly divided Eastern Europe into spheres behind their back. As long as he and Stalin 'understood each other', Churchill said, he could sell the idea to the Americans.[26] Those who seek to blame Roosevelt at Yalta for naïveté towards Stalin and lack of sympathy for Churchill need to consider the implications of the agreement.

Churchill produced what he called 'a naughty document', in which he listed five countries and the 'proportional interest' that Russia and Britain could expect to enjoy in each after the war. While it is true that had it been known about it would have induced collective apoplexy in the US State Department, it represented a very good deal for the West.[27] For although Romania was divided 90% to Russia, 10% to Britain, Greece – whose German occupiers had escaped – had those percentages reversed. Whereas it was obvious that Romania would fall within the Soviet sphere for geographical and military reasons, Greece – the world's oldest democracy – was teetering towards communism in an increasingly vicious civil war. Yugoslavia and Hungary were listed as 50%-50% and Bulgaria as 75% Russian and 25% 'to the others'. Stalin took his blue pencil and ticked the top right-hand corner of the paper.

Although Churchill immediately felt guilty about the crudity of the percentages agreement, it proved a very successful negotiating tool, primarily for the way it helped save Greece from the Soviet-backed EAM-ELAS communist insurgency. It is easily forgotten that the communist-dominated resistance movements in France, Italy and Greece could all have been turned against the Allies by Stalin if he had commanded it. 'Might it not be thought rather cynical if it seemed we had disposed of these issues, so fateful to millions of people, in such an off-hand manner?' Churchill asked Stalin. 'Let us burn the paper.'

'No,' said Stalin, who was used to disposing of millions of actual people in an offhand manner, 'you keep it.' Today the document can be viewed in the British National Archives at Kew.

On Friday, 20 October 1944, General MacArthur made good his vow to return to the Philippines, undertaken back in March 1942 when he had been forced to flee the Japanese. The fighting on Leyte, one of the eastern Visayan Islands lying midway between Luzon and Mindanao, was desperate and went on until February, only a month before the fall of Manila to US troops. In all, the Americans lost 3,500 dead and 12,000 wounded, but against that the Japanese losses were over 55,000 dead; only 389 allowed themselves to be taken prisoner.

The simultaneous battle of Leyte Gulf, fought between 22 and 27 October 1944, was the greatest sea and air engagement in the history of warfare. No fewer than 218 Allied warships and 1,280 planes were pitted against 64 Japanese warships and 716 planes. Over the six days of fighting, twenty-six Japanese and six US warships were sunk, including four Japanese aircraft carriers and fourteen destroyers. This was a very significant American victory; the huge weight of her war production was telling.

Although the United States was by far the richest country in the world by the 1940s, there were parts of it that were comparatively backward. Whereas per capita income in 1945 was over $1,300 in California, Connecticut, Delaware, Illinois, Washington, Massachusetts, New Jersey and New York (the highest at $1,595), it was under $800 in Alabama, Arkansas, Georgia, Kentucky, Louisiana, North Carolina, South Carolina and Mississippi (the lowest at $556). Three-quarters of a century after the Civil War the victors remained the same. The value of school property per pupil stood at over $500 in Delaware, Illinois, Massachusetts, New Jersey, Rhode Island and New York (the highest at $670) in 1942, whereas it was under $150 in Arkansas, Georgia, Mississippi, South Carolina, Tennessee and Alabama (the lowest at $103).[28]

Any state that had cotton as one of its chief crops was likely to have a high proportion of the population with no local library services (48% in Alabama in 1941, 47% in South Carolina, 56% in Arkansas, 44% in Louisiana), fewer telephones per thousand of population and fewer automobile registrations. In 1940, no fewer than 78,562 homes in Alabama, 65,886 in Mississippi, and 56,956 in Tennessee had no indoor lavatory.

The statistics for lynchings between 1882 and 1944 reveal 346 blacks being murdered in that way in Alabama, 521 in Georgia, 489 in Texas and 573 in Mississippi.[29] Although Franklin D. Roosevelt carried Alabama, Arkansas, Georgia, Louisiana and Mississippi, North Carolina, South Carolina, Tennessee, Virginia, West Virginia and Texas in all of his four presidential elections, life in those states was still easily the toughest in the Union, for blacks and whites alike, but especially for blacks.

In one area, at least, the South was fortunate. The proportion of American deaths in combat during the Second World War was remarkably consistent state by state, with thirty-five states losing between 0.204% and 0.283% of their population in the conflict. Only two states lost a higher proportion – Montana with 0.334% and North Dakota with 0.308% – but with populations of only around half a million each, these were statistically unremarkable. Because of an initial reluctance to recruit and deploy blacks, however, the states that lost the lowest proportion of their populations were Florida (0.150%), Louisiana (0.156%), Mississippi (0.163%), Georgia (0.177%), South Carolina (0.178%) and Alabama (0.182%).[30]

On Tuesday, 20 January 1945, Roosevelt delivered his fourth and last Inaugural Address. It contained a plea for America to stay involved with the affairs of the world, which was as powerful as Washington's Farewell Address had been a plea for her to stand aloof from them. 'And so today,' said Roosevelt, 'in this year of war, 1945, we have learned lessons – at a fearful cost – and we shall profit by them. We have learned that we cannot live alone, at peace; that our own well-being is dependent on the well-being of other nations, far away. We have learned that we must learn to live as men, and not as ostriches, nor as dogs in the manger. We have learned to be citizens of the world, members of the human community.' It was much the same anti-isolationist line that Woodrow Wilson had tried to pursue, unsuccessfully, after the Great War and that Roosevelt had been proclaiming since his article in *Foreign Affairs* in 1928. Roosevelt's internationalist agenda was structured around a series of conferences that took place in late 1944 and early 1945, principally Dumbarton Oaks (on international co-operation), Bretton Woods (on finance, trade and development), Hot Springs (food and agriculture), Chicago (civil aviation) and Yalta, where he persuaded the Russians to take part in the United Nations. 'The challenge of contriving a smooth transition from isolationism to internationalism shaped Roosevelt's foreign policy,' believes the distinguished American historian Arthur Schlesinger, Jnr.[31] It was a tribute to Roosevelt that, apart from a Congress-led slip back into putative isolationism in the late 1970s, the United States has stayed internationalist ever since his death.

If Churchill or Roosevelt required any reminder of Soviet brutality – which they didn't – they needed only to take the mechanical lift up to the second storey of the Livadia Palace in which the Yalta Conference was held in the Crimea between Sunday, 4, and Sunday, 11 February 1945. Built in the Late Italian Renaissance style in 1911 by Tsar Nicholas II as his summer residence, and situated two miles west of the Black Sea port and health resort, the white limestone palace was redolent of the Romanov family. Nicholas' father Tsar Alexander III had died in a previous palace on the estate in 1894, and the medallions over the white marble columns of the main entrance portico bear the initials of Nicholas, Alexandra, Alexei, Olga, Tatiana, Maria and Anastasia Romanov, the imperial family who were later so horrifically murdered by the Bolsheviks in 1918.

The largest room in the palace, the 218-square-yard White Hall in which the plenary sessions of the Conference took place around a large round table, was used for the sixteenth birthday celebrations of the Tsar's eldest daughter Olga. Roosevelt's study was formerly the Tsar's state reception room, his bedroom was the Tsar's study and the English Billiards Room, where the Yalta agreement was signed on 11 February, has a fireplace ornamented with the last Emperor of Russia's initials. The famous photograph of the Big Three was taken just outside that chestnut-panelled room, in the Italian courtyard. Although ostensibly Stalin chose the palace because it met Roosevelt's needs for a ground-floor bedroom and easy wheelchair-access, a more potent symbol of Bolshevik murder and expropriation could hardly be imagined.

Nothing, however, could be more poignant than the upstairs rooms of the Livadia, which recall the happiness of the gentle, well-meaning family that were so foully butchered in a Siberian town less than a quarter-century earlier. The cosy yew-panelled family dining room, the intimate bedchamber of the Tsar and Tsarina, the Empress's boudoir where Alexandra Fedorovna played duets with her daughters, and the children's modest classroom are all at heart-rending variance from the basement room at the Yusopov House in Ekaterinburg where the imperial family were shot and bayoneted to death in March 1918. (One disgusting Bolshevik brute even sexually abused the Tsarina's corpse prior to it being flung down a mineshaft.) If Stalin had wanted to underline to the English-speaking leaders – and he knew Churchill had been the principal proponent of Allied intervention in the Russian Civil War – the completeness of the victory of the October Revolution, he could hardly have chosen a better place.

The Yalta Conference was the second meeting of the 'Big Three' and it remains hugely controversial even to this day. Its deliberations and conclusions sealed the fate of Eastern Europe for a generation, right up to the fall of the Berlin Wall in 1989, and it has long been regarded as a byword for cynical manipulation and the betrayal of brave peoples. Yet can that judgment truly stand what Churchill in another context called 'the grievous inquest' of History?

After five-and-a-half years of Total War, the Third Reich was about to enter its final death-throes. The situation was thus very different from fourteen months earlier, when the Big Three had first met at Teheran in Persia in November 1943 to map the future course of the war. The principal questions that were answered at Teheran had been military; at Yalta they were political. Codenamed 'Argonaut' by British and American planners, the Conference was originally only expected to set out initial temporary arrangements for Europe, yet in fact it set in stone the way Europe was to look for the next forty-four years.

Stalin had disliked flying to Teheran and insisted that his doctors would not allow him to leave the USSR, so very reluctantly Roosevelt – who was in fact far more ill – and Churchill agreed to the meeting taking place in the Soviet holiday resort about thirty miles south-east of Sevastapol. When they arrived there they found that Stalin was absolutely intransigent about many of the great issues that faced the world; more importantly, he was also in an almost invincible negotiating position.

In a mere seven days, the Yalta Conference had to decide upon the fate of post-war Germany; the future of Poland, Czechoslovakia, Hungary, Bulgaria, Romania, Yugoslavia and Greece; the nature of a United Nations Organisation to take the place of the utterly discredited League of Nations; what to do with Nazi war criminals; and when Russia would join the war against Imperial Japan from which she had so far stayed aloof. No summit meeting had a more serious and controversial agenda to address in the entire course of the twentieth century. In the main, Roosevelt and Churchill got what they came for in every area except Polish and Eastern European democracy, which were frankly unobtainable anyhow with the Red Army only forty-four miles from Berlin by February 1945.

The British, who had gone to war for Polish sovereignty in September 1939, were deeply disturbed that Stalin seemed to want to force a Soviet puppet regime – known as the Lublin Committee – upon the Polish people, but since over two million Red Army troops were stationed right across that country, and were fully engaged fighting the Wehrmacht to the west of it, there was nothing that could effectively be done after this.

The Soviet Union had for nearly four years borne the heaviest brunt of the fighting. In that time Allied propaganda had depicted 'Uncle Joe' Stalin and the Red Army as being Europe's principal liberators, which indeed they were. Their brave soldiers and civilians had suffered dreadful losses in 'the Great Patriotic War' against the Third Reich – perhaps as many as twenty-four million killed by the end – and taking any military action against Russia was thus politically unthinkable for either Western leader. The sacrifices of the Russian people during the war had produced in Britain what one historian has called 'the blooming of a thousand committees and societies to publicize the Soviet cause and encourage friendship between the two nations'. Powerful organisations such as the Anglo-Soviet Trades Union Committee and the left-wing intellectuals' Anglo-Soviet Public Relations Committee could be relied upon to put Stalin's actions in Eastern Europe in the best possible light. The

Polish and Czech democrats whom the Soviets were arresting were meanwhile being labelled – and in most cases libelled – as 'reactionary elements', in an attempt to tar them with the fascist brush.

Once it dawned on Churchill that Russia wanted to swallow up and partition Poland once again – just as she had so often done in previous centuries – it was simply beyond his power to prevent it. Although Stalin promised that there would be 'free' elections after the Nazis were expelled from that country, he completely stonewalled all Anglo-American questions over the timing, supervision and conduct of these. Needless to say, when they finally took place they were entirely rigged by the Soviets. Stalin's promises were just a transparent ploy to cover up his huge and tragically unstoppable grasping of power and territory.

It was agreed at Yalta that Germany was to be split into four zones, to be controlled by Russia, America, Britain and France. Within the Russian zone in the east, Berlin itself was to be divided up on a similar basis. There was further agreement over Russia joining the war against Japan, which Stalin promised to do three months after the surrender of Germany. (By then, however, the atomic bombs were a week away from being dropped.)

Although it was later denounced as highly Machiavellian, and an insult to those brave Poles and Czechs who had fought against one totalitarian regime only for their country to fall into the hands of another, Yalta was the best deal that Roosevelt and Churchill could have negotiated in the circumstances. Although the Russian promises of democracy were clearly worthless, at least it was delineated where the Red Army would halt in its march westwards across Europe. Furthermore, Greece and eventually Austria were saved from falling into the Soviet sphere.

'We had the world at our feet,' Churchill said after meeting Stalin and Roosevelt on the first day of the Conference. 'Twenty-five million men marching at our orders by land and sea. We seemed to be friends.' Yet during the conference he was frequently gloomy as Stalin showed how utterly intransigent he could be. Churchill's daughter Sarah, who travelled out with him, recalled how he once looked out over the sun-sparkling sea and described it as 'The Riviera of Hades'. He had glimpsed the way that Europe was inexorably going and saw the drawing down of what a year later he was to describe as 'an Iron Curtain' that separated free Europe from the Soviet bloc.

Yet no statesman, however omniscient, could have altered the sheer fact of Russia having millions of troops on the ground, all across most of the territories under dispute. Valuable work was done at Yalta on the creation of the United Nations, the de-Nazification of German society, war trials, and many other important areas of post-war policy, but over the issue with which the word 'Yalta' will always be connected in history – the condemnation of so many Eastern European peoples to Soviet communist domination for so long – the sad but unavoidable truth is that the United States and Great Britain simply had no choice but to accede to Stalin's *fait accompli*. Never since 1900 were Western statesmen's decisions more important, more long-lasting, more bitter to swallow and yet more impossible to escape.

Debates continue about whether Churchill or Roosevelt was more at fault at Yalta. It is true that the Americans made few anti-Soviet noises and Roosevelt often criticised Churchill to Stalin, and also that in liberal and progressive circles in the United States, Britain's empire in India was considered more of a concern than Russia's soon-to-be-established empire in Eastern Europe. In a letter from the British Embassy in Washington to his friend Lady Daphne Straight of 9 April 1945, Isaiah Berlin pointed out how suspicion and dislike of the British Empire permeated the Roosevelt Administration. Over the question of post-war colonial trusteeships, he told her:

> We are popularly credited, perhaps rightly, with being against any liberalisation of the colonial and mandates system. . . . The Americans believe that the mandatory system was a step in the right direction, whereas we, despite all our talk about regional commissions, have given the impression that we do not. At the moment there is a violent row occurring between the Service departments, who want to annex various atolls, and the State department, who want to internationalise.[32]

'Coming from America', President Roosevelt told Churchill and Stalin on 6 February, he took 'a distant point of view of the Polish question: the five or six million Poles in the United States were mostly of the second generation.' It was lukewarm at best, but even had Roosevelt thirsted for Polish liberty nothing would have dissuaded Stalin short of deploying the as-yet-untested atomic bomb against the West's heroic ally, which was politically unthinkable. Western public opinion would simply not have understood, let alone accepted, any kind of aggressive stance against Stalin at that stage of the war. In the back of Western minds was also the fear that the Soviets might do another deal with the Germans, as they had in August 1939.

In March 1942, Roosevelt had warned Churchill: 'Stalin hates the guts of all your top people. He thinks he likes me better, and I hope he will continue to.' In fact, of course, Stalin didn't like anyone. Yet a telegram sent by Roosevelt to Stalin on 4 April 1945 shows just how little the President was the 'willing dupe' of the Soviets. 'I have received with astonishment your message of April 3 containing an allegation that arrangements which were made between Field Marshals [Harold] Alexander and [Albert] Kesselring at Berne "permitted the Anglo-American troops to advance to the East and the Anglo-Americans permitted in return to ease for the Germans the peace terms".' Pointing out that no negotiations had taken place and that the policy of unconditional surrender still stood, Roosevelt concluded: 'Frankly I cannot avoid a feeling of bitter resentment towards your informers, whoever they are, for such vile misrepresentations of my actions and those of my trusted subordinates.'[33]

It is wrong to argue that the victory over Germany made no difference, that it merely opened the door for five more decades of totalitarian rule over countries such as Poland. To have rid the world of a monster as baleful and dangerous as Adolf Hitler was undoubtedly worth the enormous sacrifices it took to achieve. A successful *Lebensraum* policy and a completed Final Solution, let alone the possibility of a victorious Hitler getting hold of nuclear weapons in the late 1940s, are such nightmare concepts that they outweigh even the tragedy of post-war Poland. The Final Solution planned at Wannsee might well have extended to all *untermenschen* ('sub-humans'), meaning the probable extermination of most Poles and the enslavement of the rest by the end of the decade. However bad life got in Poland under communism, it was never so ghastly as that.

As well as the division of Germany and Berlin into four Allied zones – somehow the French got one of their own, carved entirely out of the British and American ones with the Soviets making no contribution – Roosevelt and Churchill promised Stalin at Yalta that the vigorous bombing of German cities would continue unabated. While Churchill was on his way to the Crimea, the Deputy Prime Minister Clement Attlee had approved a request to bomb Dresden, since Enigma decrypts had revealed that the Wehrmacht 'were withdrawing substantial numbers of first class divisions, including Panzer divisions, from the western front, from the interior of Germany, from Italy and from Norway in order to launch a counter-attack against the Soviet forces in Silesia', and some of that traffic would be going via Dresden.[34] As well as being one of the largest garrison towns in Germany, the 1944 *Handbook of the German Army Weapons Command* stated that Dresden contained 127 factories manufacturing military equipment, weapons and munitions (which only related to large factories and not the many smaller suppliers and workshops).

The order to RAF Bomber Command's Five Group for its operations for Tuesday, 13 February 1945 – two days after the Conference broke up – could hardly have been starker: 'To burn and destroy an enemy industrial centre.' The target chosen was Germany's seventh-largest city, only a little smaller than Manchester. It was, as one report put it, 'by far the largest un-bombed built-up area in Germany'. Dresden was not merely a city, but a work of art in itself, an architectural jewel whose aesthetic attractions had made it Saxony's pride for nearly half a millennium. That long chapter of its history closed when a 1,000-bomber raid created a firestorm that burned for forty-eight hours, consuming virtually the entire city centre and incinerating somewhere between 25,000 and 40,000 people.[35] The before-and-after photographs taken of the raid emphasise the scale of the destruction.

Yet for all that the bombing of Dresden has been regularly denounced as a war crime, the city was a legitimate military target whose bombing was in fact justified in the context of Total War. The high death toll – which was nothing like the six-figure one claimed by Joseph Goebbels and David Irving – was the result not of deliberate Allied policy so much as a number of accidental factors. 'In practical terms,' argues Frederick Taylor in his definitive account of the raid, 'Dresden was one heavy raid among a whole, deadly sequence of massive raids, but for various unpredictable reasons – wind, weather, lack of defences and above all shocking deficiencies in air raid protection for the general population – it suffered the worst.'[36]

When the Nazi *gauleiter* of Dresden, Martin Mutschmann, fell into Allied hands in 1945, he quickly confessed that 'A shelter-building programme for the entire city was not carried out', since 'I kept hoping that nothing would happen to Dresden.' (He had, however, taken the precaution of having a shelter built for himself, his family and his senior officials.) In all, half the 955,000 tons of bombs that were dropped on Germany during the war fell on populated areas. Air Chief Marshal Harris believed that 'de-housing' German workers would hinder war production significantly, nor is there any evidence on which to gainsay him.[37]

Equally, there was no reason why Mutschmann should have thought that alone of large German cities Dresden should have been immune to Allied bombing, because its railway marshalling yards were huge and it had a con-glomeration of war industries – particularly in the optics, electronics and communications fields – in and near the city. Although it was fervently pro-Nazi – the demonstrations held after the attempt on the Führer's life in July 1944 were huge and spontaneous – once the Lancaster bombers had wrought their terrible destruction, the political outlook of many Dresdeners, and other Germans, changed. The respected German historian Gotz Bergander believes that whereas before Dresden the concept of accepting unconditional surrender was unthinkable to ordinary Germans, 'The shock of Dresden contributed in a fundamental way to a change of heart.'[38]

In the hard-fought and frequently ill-tempered debate over the precise numbers who died that awful night and morning, it is likely that the generally accepted figure of between 25,000 and a maximum of 40,000 is correct.[39] This actually makes the raid less lethal than those on Hamburg, and proportionately less deadly than those on places like Pforzheim and Würzberg. It was certainly not, as is often alleged 'the German Hiroshima'. If anything, the attack on Dresden was 'routine'. Although the Russians had requested the city be bombed, this did not prevent Leonid Brezhnev denouncing it on its fortieth anniversary as a Western war crime.

In 1967, in an hitherto-unpublished letter to his former Chief of Intelligence, Major General Kenneth Strong, Eisenhower wrote about Dresden that, 'I remember well that you advised that it was not a profitable target, but as I recall (and here my memory may be at fault) [the commander of US strategic air forces in Europe, Carl] Spaatz or one of the other airmen maintained that the Russians thought it was a very remunerative target and should be attacked.'[40] The blame for Dresden has fallen very largely on the shoulders of Air Chief Marshal Sir Arthur 'Bomber' Harris, but if blame there is, it certainly ought to be spread around many more figures in the Allied High Command.

No-one has ever denied the horror of what happened to that once-beautiful city on the night of 13 February 1945; the 1,000-degree heat from the firestorm could be felt by the RAF aircrew at more than 10,000 feet above the city. But tribute should also be paid to the bravery of Bomber Command – of men like Mike Tripp and Bruce Wyllie of the RAF and Alden 'Al' Rigby of the USAAF – who night after night heroically undertook missions from which so many of their comrades never returned. Among every 100 men who flew in RAF Bomber Command during the war, more than 50 died and fewer than a quarter completed a tour of thirty operations.[41] Disgracefully, the ground crew of Bomber Command received no campaign medal unless they had served overseas, and Churchill failed to commend the bombers properly in his victory speech of 13 May 1945. Dresden was, in Taylor's words, 'a functioning enemy administrative, industrial and communications centre that by February 1945 lay close to the front line'. They did what they had to do.

On Thursday, 29 March 1945, anti-aircraft gunners in Suffolk shot down the last of the V-1 'flying bombs' launched against Britain during the war. Called the *Vergeltsungswaffe-Ein* by the Germans, meaning 'Reprisal Weapon One', they were nicknamed 'Doodlebugs' or 'Buzz-bombs' by Britons.

The V-1, for which Hitler had announced high hopes on its inception on Christmas Eve 1943, was certainly an horrific weapon. Powered by a pulse jet mechanism using petrol and compressed air, it was 25 feet 4 inches long with a 16-foot wingspan, and it weighed 4,750 lbs. Its warhead was made up of no fewer than 1,874 lbs of Amatol explosive, a fearsome mixture of TNT and ammonium nitrate. Launched up 125-foot concrete ramps stationed right across Occupied France from Watten in the north to Houpeville to the south, they flew at up to 360 mph, which was slow enough to have a proportionately greater surface blast effect for its warhead size than the V-2 rocket bomb.

With a maximum range of 130 miles, London and south-east England were their main targets. Flown by autopilot from a preset compass, the nose propeller operated a log which measured the distance flown. Once it reached the correct range, the elevators in the wings were fully deflected and it dived, cutting out the engine as it did. Much of the terror that V-1s evoked came from the sinister way that the noise of their propulsion suddenly stopped at this preset moment, meaning that they were then going to fall on the people below. To hear the noise continue on meant that the V-1 would carry on flying overhead, but to hear it stop brought the certainty of an imminent, devastating explosion. It is estimated that about 80% landed within an eight-mile radius of their targets.

Between 13 June 1944 – a week after D-Day – and 29 March 1945, no fewer than 13,000 V-1 bombs were launched against Britain. Because their cruising altitude of between 3,500 and 4,000 feet was too low for heavy anti-aircraft guns to be able to hit them very often, yet too high for lighter guns to reach, it was often down to the RAF to try to deal with this grave new threat. Radar-guided fighter aircraft attempted either to shoot them down or to tip them over by lightly tapping their wings. It took outstanding courage to fly so close to over one ton of aerial explosive, yet that was the way it was done. Barrage balloons were also employed.

'I was eleven or twelve when I had my first experience of a doodlebug raid,' recalled Thomas Smith who lived in Russell Gardens, London N20, during the last two years of the war, along with his mother and eight brothers and sisters. 'It was 6.30 a.m. on Friday, 13th October 1944. We were all lying in bed, when we heard the flying bomb come over. We knew it could drop anywhere as we could hear it flying over the house. We were terrified. I was sharing a bed with my four brothers and we all huddled together under the bedclothes.' Mr Smith's father was serving in the British Army abroad, which was at that time attacking and shutting down launch sites in northern France.

'The bomb missed the house,' recalled Smith, 'but it dropped 120 yards away, in Russell Gardens. The force of the bomb caused the roof and ceilings of our house to fall in and the windows were also blown out by the blast. Despite the bomb, my mum still sent me to school.' In all, over 24,000 Britons were casualties of the Führer's vicious 'secret weapon', of whom 5,475 were fatalities.

The attacks came round the clock, allowing for no respite. Whereas the Luftwaffe had long since confined their attacks to night-time, when their bombers could be cloaked in darkness and hidden from RAF fighters, the pilot-less bombs came all through the day and night. At one point during the initial assault in July and August 1944, no fewer than 10,000 homes were damaged every day. By late August, over 1.5 million children had been evacuated from the south-east.

The sheer area that a single V-1 could devastate – a quarter of a square mile, or the size of twenty-seven football pitches – made them a particularly dangerous weapon, although the defenders quickly adapted. Between June and September 1944, for example, no fewer than 3,912 were brought down by anti-aircraft fire, RAF fighters and barrage balloons. It soon became clear that the German High Command, which had hoped that V-1s might destroy British morale and force the Government to sue for peace, had been wrong.

Secret papers released in the British National Archives in February 2005 revealed how flawed Intelligence brought the Government to the verge of panic over the threat of the V-1's equally fiendish sister weapon, the V-2 rocket bomb. An emergency committee set up in the Ministry of Home Security in Whitehall to report on the likely damage came up with the projected figure of 100,000 Londoners killed and a similar number severely injured in the first month of bombardment.

In early 1943, MI6's spies believed that one 10-ton warhead could strike the capital every hour, day and night (i.e. the TNT equivalent of one Hiroshima bomb detonating in the capital every seven weeks).[42] Churchill was warned by Sir Findlater Stewart, the head of the home defence executive, that 'Some 1,200 missiles, assuming that there are no overlap in their effects, would, to all intents and purposes, lay waste the county of London.' By comparison, the worst nights of the Blitz caused 1,750 deaths on the night of 16 April 1941 and 1,450 on 10 May 1941. The emergency services, warned Stewart, would be 'overwhelmed on the second day', and within a week over half-a-million Londoners could be rendered homeless.

In fact, however, the V-2 only carried a 1-ton warhead, and although 1,115 were fired at England in the 202 days between 8 September 1944 and 29 March 1945, only 517 fell on London, which averaged 2.5 per day rather than the predicted rate of nearly ten times that. They nonetheless left 2,754 Britons killed and 6,523 seriously injured, a tragic total but only a fraction of the 167,250 that the committee's predictions would have assumed for the period.[43]

The very fact that V-1s and V-2s were still falling on Britain as late as 29 March 1945 – the V-1s were launched from modified Heinkel He 111 bombers after the fall of the northern France launch sites – shows just how fanatical the Nazis were in the closing stages of the war. Hitler had barely five weeks to live and the Red Army was closing in on Berlin, yet still the Germans were trying to crush Britain's will to fight on. Those who argue that the Allies were too harsh in their bombing campaign against German cities ought to recall quite how long the dogged V-1 and V-2 campaign against London went on for, even after most rational Germans realised that all was lost. Without D-Day and the English-speaking peoples' liberation of Western Europe, there can be no doubt that the V-1 and V-2 bombardment of Britain would have carried on with ever-mounting civilian losses.

If Japanese ambitions were halted at Midway and Guadalcanal, and defeated at Leyte Gulf, they suffered a further reverse on the tiny volcanic island of Iwo Jima between 19 February and 26 March 1945. Japan had built two airstrips on the eight-square-mile volcanic island less than 1,000 miles from its southernmost shores, from which they could intercept American B-29s on their way to the mainland. American war planners recognised the invaluable advantage both of denying Iwo Jima to the Japanese and establishing a US base there for fighter support and as a haven for B-29s that had developed problems on the gruelling 2,800-mile round trip.[44]

Major-General Harry Schmidt's V Amphibious Corps, combining the 4th and 5th Marine Divisions, with the 3rd in reserve, attacked the island at 9.02 a.m. on 19 February 1945. Iwo Jima had been bombed daily between 8 December 1944 and the day of the invasion, representing 'the single most prolonged bombing action of the war'. Yet most of the Japanese installations remained intact, and around 25,000 Japanese fought almost to the last man. Despite the American task force numbering around 100,000 troops, almost all the advantages lay with the well-entrenched defenders.

Iwo Jima lived up to its name, 'Sulphur Island'. As well as sixteen miles of tunnels connecting 1,500 'chambers', near-impregnable gun emplacements covering the island's western shoreline where the Marines landed, the Japanese had a fanatical willingness to die for their Emperor and country. The Marines kept pressing forward for thirty-six days until the last Japanese pillbox was taken. Although most of the island was secured by 26 March, and American bombers were flying regular missions from there by early May, some Japanese continued to hold out in the hills until late May. In all, the Marines had suffered 5,931 killed and 17,372 injured; only a few hundred Japanese were taken alive. Admiral Nimitz said of the battle, 'On Iwo Jima, uncommon valour was a common virtue.' A question began to thrust itself to the forefront of the minds of Allied decision-makers: if a small island could hold out with such fanaticism for so long, what would it take to capture Tokyo?

The death of Franklin Delano Roosevelt at Warm Springs, Georgia, on Thursday, 12 April 1945, removed from the world scene the man who, in the words of his latest, definitive biographer Conrad Black, defeated the Great Depression, suffused the US Government with 'determination and optimism' and then 'triumphed over every foreign and domestic enemy'. As a result, because he 'brought America to the rescue and then the durable protection of the civilised world, FDR belongs in the same pantheon as Washington and Lincoln'. He was also the central political architect of the twentieth century, his internationalist legacy still felt long after his demise.

Roosevelt and Hitler took up office within weeks of one another in early 1933 and died within days of one another in April 1945. Roosevelt deserves great credit for spotting the true nature of Nazism early on and the danger it posed to Civilisation. Although he could not alter the underlying opposition of the American people to entering foreign wars, he did all in his power to provoke the Axis powers to declare war on the United States, knowing that it would be utterly fatal for them to do so. FDR's political courage – against Tammany Hall, Wall Street, America First, the Axis and others – mirrored the physical courage that he displayed in overcoming the disabilities that polio imposed upon him.

There is a lack of hard evidence about whether FDR pursued an adulterous relationship with his private secretary, Marguerite 'Missy' LeHand, but between 1925 and 1928 he spent 116 out of 208 weeks away from home, of which his battleaxe wife Eleanor was present for only four and the attractive, utterly devoted twenty-five-year-old Missy was with him for no fewer than 110, many of them spent cruising on his yacht *Larooco*. In fact 'the truth will never be known, and more than cursory speculation is unseemly', but one certainly hopes that the great man did indeed find some happiness with his lissom secretary, as well as with his mistress Lucy Merier Rutherfurd.

Roosevelt was succeeded by Harry S. Truman – the 'S' was invented to give the Kansas City failed-haberdasher some much-needed gravitas – who astonished his contemporaries by becoming one of the great American presidents. A judge from Jackson County, Missouri's eastern district, from 1922, he was elected to the Senate aged fifty in 1934. Re-elected in 1940, Truman attained national prominence as chairman of the special committee investigating national defence expenditure, where he was said to have saved the US taxpayer $1 billion before Roosevelt put him on his 1944 ticket as his vice-presidential candidate.

As soon as Adolf Hitler had committed suicide in Berlin on Monday, 30 April 1945, Eamon de Valera personally visited the German envoy to Dublin, Eduard Hempel, to express condolences on the death of the Führer, without even waiting for official confirmation of the news. Buchenwald had already been liberated by that time and the genocidal nature of the Nazi regime had been revealed beyond dispute. Moreover, de Valera had not done the same thing at the American Legation after Roosevelt had died eighteen days earlier. The Irish Taoiseach told the Department of External Affairs in Dublin not to react to international criticism of his action, since, 'An explanation would be interpreted as an excuse, and an excuse as a consciousness of having acted wrongly. I acted correctly, and, I feel certain, wisely.'[45] The Allied press reacted with rage, which went largely unreported in Eire's very heavily censored press.

Churchill's attitude towards Ireland's neutrality was forcefully expressed a fortnight later when he recalled how during 1940 the southern Irish 'Treaty' ports and airfield had been closed to the Royal Navy, and thus the approaches to Britain's western ports were threatened by hostile aircraft and U-boats, and 'if it had not been for the loyalty and friendship of Northern Ireland we should have been forced to come to close quarters with Mr De Valera or perish from the earth'. Yet out of self-restraint to which 'history will find few parallels', and despite the fact that they could 'fast and violently have obtained what they wanted', the British Government 'left the de Valera government to frolic with the Germans and later with the Japanese representatives to their heart's content'.[46]

At 3 p.m. on Tuesday, 8 May 1945 – V-E Day – Churchill broadcast an historic radio message to the British people and the wider world. Huge loudspeakers had also been rigged up in central London to carry his words to the crowds of over a million people who thronged the streets. He told them that the previous day the German general, Alfred Jodl, had signed an unconditional surrender and thus hostilities would end officially at one minute after midnight, and that 'The German war is therefore at an end.' The cheering in Trafalgar Square and Parliament Square could be heard in the Cabinet Room at Downing Street, from where Churchill was speaking.

In his brief summing-up of the war, the Prime Minister spoke of how after Russia and America had entered the conflict in 1941, 'Finally almost the whole world was combined against the evil-doers, who are now prostrate before us.' The diarist Harold Nicolson, who was in Parliament Square, noted how 'the crowd gasped' at that phrase. The full enormity of the moment only seemed to be slowly entering the consciousness of a people who had experienced such suffering and sacrifice. (It had only been three months since the last flight of V-bombs had fallen.)

Churchill continued: 'We may allow ourselves a brief period of rejoicing; but let us not forget for a moment the toil and efforts that lie ahead. Japan, with all her treachery and greed, remains unsubdued.' He ended with the peroration: 'Advance, Britannia! Long live the cause of freedom! God save the King!' It was the signal for the British people to celebrate in a way they never had for any coronation or jubilee, or even on receipt of the news of the relief of Mafeking in the Boer War. 'In all our long history,' Churchill told them on V-E Day, 'we have never seen a greater day than this.' Since he was known to be an historian, they rightly took his word for it and danced till dawn.

After speaking to Parliament, attending a service of thanksgiving at St Margaret's Westminster, and signing a boy's autograph album with the words 'That will remind you of a glorious day,' Churchill spoke to a vast crowd in Whitehall from the balcony of the Ministry of Health, employing almost pantomime repartee in reminding them of the twelve months between Dunkirk in June 1940 and Hitler's invasion of Russia in June 1941: 'There we stood, alone. Did anyone want to give in?' The crowd roared back 'No!' 'Were we downhearted?' 'No!' they cried in response.

Joan Bright Astley, who worked for Churchill's military secretary General Hastings 'Pug' Ismay, recalled how on V-E Day the Prime Minister had invited the Chiefs of Staff committee 'to celebrate with him the first news of Germany's defeat and with his own hands had put out a tray of glasses and drink'. At this point an historic opportunity was missed to toast the greatest-ever Englishman at the moment of his ultimate triumph. As Astley records,

> It was a sad example of human imperceptiveness that neither the Chief of the Imperial General Staff, nor the First Sea Lord, nor the Chief of the Air Staff saluted him in a toast. General Ismay, in his modesty, in their presence would never have done so. Mr Churchill drank to them, each one in turn. It is possible that they were shy, it is certain they were British, it is probable that they reacted as a committee, a body without a heart, and that each waited for the other to take the initiative. Whatever the reason it was an opportunity missed that the Grand Old Man, who had been the architect of the victory they were marking, did not receive a tribute from his three closest advisors.[47]

Much worse was soon to befall Churchill at the hands of the electorate.

'We are getting near the end of this general election now,' the Deputy Prime Minister Clement Attlee reported to his brother on 3 July 1945. 'I can't find that my opponent is getting much support. ... Winston keeps slogging away ... but I don't think that he gets the better of these exchanges with me.'[48] He certainly did not; the results were catastrophic for the Tories, who won 180 fewer seats than Labour. Stalin might have joked at Yalta that 'I don't think that Mr Attlee looks like a man who would seize power,' but he did indeed take office on Friday, 27 July 1945.

Back in August 1941, with Churchill at Placentia Bay conferring with Roosevelt, Attlee had written to his brother: 'I had to take the place of the PM last week as the reviewer of the war situation, no easy thing to follow such an artist. I eschewed embroidery and stuck to a plain statement. It is no use trying to stretch the bow of Ulysses.'[49] Yet now that Attlee was himself prime minister and had to carry the bow himself, his first task was to construct a lasting peace settlement. (At the Beefsteak Club, Lord Wimborne morbidly bet John Maude KC fifty guineas 'that on or before 1974 England will again be at war with Germany'.)

'I must confess I find found the event of Thursday rather odd,' wrote Churchill to his former best man Lord Hugh Cecil after the general election debacle; 'there was something pent up in the British people after twenty years which required relief. It is like 1906 all over again.' The results of the election – Labour 393, Conservatives 213, Liberals 12 – seemed so overwhelming that a generation of Tories were lulled into thinking that the British voter had 'gone Bolshie' and, as one analysis has put it, 'For thirty years British politics was played out in terms of the Keynesian consensus'.[50]

There had been ten years since the last election, which the Tories had won in 1935 with a landslide; the appeasement policy was largely blamed on the

Tories but that would only last one election since Chamberlain was dead and the issues had changed. The popular vote broke down 47.8% for Labour, 39.8% for the Conservatives, which was not irretrievable. Yet the Conservatives panicked and went down the collectivist path on the assumption that the electorate would settle for nothing less than a full-blown Keynesian Welfare State, with ever-higher costs financed by ever-higher taxes. It was not until Margaret Thatcher was chosen to lead the Tory Party thirty years later that that assumption was seriously questioned.

At 8.16 p.m. on Monday, 6 August 1945, the world changed suddenly and irrevocably. Out of a cloudless blue sky a USAAF B-29 plane named *Enola Gay* (after the mother of its pilot, Colonel Paul Tibbets) dropped a 4-ton isotope uranium-235 fission bomb on the Japanese city of Hiroshima, at the southern end of Honshu. Forty-three seconds later, 1,900 feet above a bridge over the Ota River, the bomb exploded with a yield equivalent to 20,000 tons of TNT, instantaneously raising the temperature below to over 4,000 degrees centigrade.[51] A gigantic mushroom-shaped cloud rose above the city.

About 60% of Hiroshima was destroyed in the explosion and subsequent firestorm, which burnt out 4.4 square miles, killing over 70,000 Japanese, but also injuring a similar number and inflicting long-term radiation poisoning on more still. Three days later, on 9 August, a plutonium bomb was dropped on Nagasaki that killed between 35,000 and 40,000 people, and injured a like number.[52] Although the raids on Tokyo that March had killed more people than in either Hiroshima or Nagasaki, the deployment of this terrifying new weapon suddenly opened up the possibility of Japan suing for peace. On 14 August, Japan surrendered. 'I did the job,' Colonel Tibbets said later, 'and I was so relieved it was successful you can't understand it.'

As with so much else about the history of the English-speaking peoples since the Wright brothers' invention, air power was central. If the Japanese had been in a position to shoot down the *Enola Gay*, the strategic situation in August 1945 would have been very different. As it was, while there was some anti-aircraft fire, there was simply not the fighter support necessary to protect the skies over one of Japan's most populous cities.

The literary historian Paul Fussell, who had been poised to take part in the projected invasion of Japan, recalled what it was like when he heard the news of the surrender:

> We learned to our astonishment that we would not be obliged in a few months to rush up the beaches near Tokyo assault-firing while being machine-gunned, mortared and shelled, and for all the practiced phlegm of our tough facades we broke down and cried with relief and joy. We were going to live.

The same emotions were felt right across the Allied armies and the decision to employ the atomic bombs was widely supported. Yet it was not long before doubts emerged, and a large number of arguments began to be made by historians and others that together amounted to the accusation that the dropping of the atomic bombs on the Japanese cities of Hiroshima and Nagasaki was the worst war crime ever perpetrated by the English-speaking peoples. The fiftieth anniversary of the use of the bombs, for example, was accompanied by a proposal for the Smithsonian Institution in Washington to mount an exhibition which, in the words of the *Wall Street Journal*, presented a 'besieged Japan yearning for peace', lying prostrate 'at the feet of an implacably violent enemy – the United States'.

Far from deciding to drop the atomic bombs in order to end the war against Japan with as few Allied casualties as possible – which was his only true motive – President Truman's action has been called into question by a series of revisionist historians, anti-American polemicists and journalists. The literature questioning and criticising Truman's decision is now immense, led by Gar Alperovitz's *Atomic Diplomacy*, but including the work of Norman Cousins, P.M.S. Blackett, Carl Marzani, D.F. Fleming, Martin J. Sherwin, Barton J. Berstein and many others. The main thrust of the attacks came in the 1960s, once the memory of the Allied troops' peril had dimmed and America's motives and actions in everything related to the Cold War were automatically held to be suspect in the light of the Vietnam War. Today, writers such as the *Financial Times'* Tokyo bureau chief David Pilling take it for granted that 'Dropping the atomic bomb on Japanese civilians is arguably the vilest single act one set of human beings has ever perpetrated on another.'[53] The historian Joanna Burke has also described the bombings as an 'atrocious aggression'.[54] (Since *The Shorter Oxford Dictionary* defines 'aggression' as 'begin the attack', her attention should, in fact, be directed to what happened at Pearl Harbor, not Hiroshima.)

The eruption of the Cold War only a matter of months after Truman's decision has encouraged writers to argue that, as one of them puts it, 'The bomb was dropped primarily for its effect not on Japan but on the Soviet Union. One, to force Japan to surrender before the USSR came into the Far Eastern war, and two, to show under war conditions the power of the bomb. Only in this way could an intimidation be successful.'[55] Truman has also been accused of reversing Roosevelt's accommodation policy with the Soviets, refusing to entertain Japanese peace feelers and insisting on Japan's unconditional surrender only so that he could test the United States' nuclear programme in real-death conditions. The lack of anything other than circumstantial evidence for these arguments, and the wealth of proof that Truman took the decision solely in order to force Japan to surrender, has in no way dimmed the enthusiasm of their adherents.

On the question of whether Truman's decision was morally right, and whether it was genuinely necessary to secure Japan's defeat, there has also been an avalanche of historical debate. This hinges on the number of American lives that it is estimated would have been lost in an invasion of Japan's home islands. Truman's critics argue that they were not so high as to justify dropping the bombs.[56] In his memoirs, *Year of Decisions*, Truman wrote that he believed an invasion would have cost half-a-million American lives, which was considered an overly conservative estimate by both Secretary of War Henry L. Stimson and Secretary of State James Byrnes, who in their memoirs estimated one million lives and one million casualties respectively. (It was Stimson who saved Kyoto from atomic destruction, probably not for the official reason given – that it was 'a shrine to Japanese art and culture' – but because he had spent his honeymoon there in 1926 and 'had fond memories of the place'.)[57]

The high figures of anticipated casualties have been ridiculed by the revisionists, yet there is ample contemporaneous evidence to support them. The US Joint Chiefs of Staff undertook a study in August 1944 that projected that an invasion of the mainland would 'cost half a million American lives and many more that number in wounded'. A memorandum also exists from Herbert Hoover to Truman in May 1945 which argues that a negotiated peace would 'save five hundred thousand to one million lives'.[58] In mid-June, General Marshall asked General MacArthur for a figure of expected casualties for just the invasion of Kyushu alone (Operation Olympic) and received the answer of 105,500 dead and wounded in the first ninety days, plus 12,500 American non-combatants.

Intercepts showed that the Japanese air force had 10,000 planes to defend the homeland. Furthermore, as well as kamikaze pilots, the Japanese had a large number of other suicide-weapons, including flying bombs, human torpedoes, suicide-attack boats, midget suicide submarines, and navy swimmers to be used as human mines, all of which had been deployed at Okinawa and the Philippines with lethal results against American servicemen.

The closest analogy with the projected invasion of Japan's home island was the battle for Okinawa (Operation Iceberg), which had begun on Easter Sunday, 1 April 1945. The attack on the sixty-mile long and two-to-eighteen-mile wide island by the US Tenth Army had lasted nearly three months. The fanatical resistance of the Japanese – General Ushijima and his Chief of Staff General Cho finally committed *hara-kiri* in their cave bunker on 21 June – cost the American ground forces 7,343 killed, 31,807 wounded and 239 missing, or 35% of the entire force. Furthermore, 36 ships were sunk, 368 were damaged, 763 aircraft were lost, 4,907 US sailors were killed or missing and another 4,874 were wounded.[59] The Japanese meanwhile lost 107,539 killed. With attrition rates that high for Operation Iceberg alone, it is safe to assume that the invasion of the 'sacred' Japanese mainland would have been far worse. Assaults on Luzon, the largest and northernmost island in the Philippine archipelago, and on Iwo Jima in the Bonin Islands had both been highly costly. Being defeated and accepting it were two very different things in the Pacific War.

With 766,700 assault troops assigned for the invasion of the mainland region of Kyushu, an attrition rate of 35% yields more than a quarter-of-a-million dead, always assuming it was victorious. Since Kyushu is only the southernmost island of the main archipelago, not on the mainland that houses Osaka or Tokyo, it would not have seen the end of Allied sacrifice if Japan fought on. As it was, the US Chiefs of Staff's calculations were based on Kyushu being defended by eight Japanese divisions, or fewer than 300,000 men. Instead, by the end of July it had become clear through intercepts that there were in fact 525,000 men ready to defend it, which was to reach 680,000 shortly afterwards. On 31 July, a medical estimate projected American battle and non-battle casualties who would need treatment at just under 400,000, an horrific enough figure in itself but one that entirely excluded those killed. American decision-makers such as Truman, Stimson, Byrnes and Marshall had every reason to suppose that Kyushu might turn into another Okinawa, with simply unacceptable levels of casualties.

Nor is it true that Japan was just about to surrender anyhow, and thus it was unnecessary to have dropped the atomic bombs. The revisionist argument states that since the Japanese were putting out peace feelers, it was only the Allies' unreasonable demand for unconditional surrender that elongated the war. Yet although logic stated that Japan's position was dismal, there was still the will to fight on until a hoped-for tactical victory in the battle for the homeland forced the Allies to negotiate a peace. Some commanders even 'felt that it would be far better to die fighting in battle than to seek an ignominious survival by surrendering the nation and acknowledging defeat'. On 8 June, the Japanese Government had pledged, in the presence of Emperor Hirohito, that 'The nation would fight to the bitter end', and the Prime Minister, Kantaro Suzuki, supported the army's plan to carry this out, as being 'the way of the warrior and the path of the patriot'.

Any peace feelers that Japanese diplomats were trying to put out via the Soviet Union ran up against the granitic fact that the Japanese military, not civilians, had ultimate control, and they had no intention of surrendering. Nor would it have been possible for Truman simply to have dropped the Allied demand for Japan's unconditional surrender, originally made by his predecessor Roosevelt. 'Practically all Germans deny the fact that they surrendered during the last war,' Roosevelt had said, 'but this time they are going to know it. And so are the Japs.' The demand was a popular one in America, and moreover had been agreed by all the Allies. Furthermore, there is no indication that the Japanese High Command made any distinction between surrendering conditionally or unconditionally.

Those who argue that Japan was desperately looking for a way to end the war must explain the astonishing fact that she refused to surrender even *after* Hiroshima was destroyed. The Americans made it clear that the city had been devastated by a new type of weapon, one that would be deployed again, yet still the Japanese military refused to accept defeat. The declaration of war by the USSR and her immediate invasion of Manchuria on 8 August did not alter matters either; instead, the War Minister General Korechika Anami denied that Hiroshima really had been attacked by an atomic bomb.[60] Others in the Government argued that the Americans had no more bombs, and that world opinion would anyhow stop the United States from deploying any more.

Both those arguments were comprehensively rebutted on 9 August, when 1.8 square miles of central Nagasaki were destroyed by the second atomic blast. This persuaded Anami that 'the Americans appear to have a hundred atomic bombs ... they could drop three per day. The next target might well be Tokyo.' Yet even then a meeting of the Imperial Council on the night of 9 August concluded that, in the words of the Chief of the Army General Staff Yoshijiro Umezu, Japan still had the 'ability to deal a smashing blow to the enemy' and therefore 'it would be inexcusable to surrender unconditionally'. The Chief of the Naval Staff similarly stated that, 'We do not believe it possible that we will be defeated.' Since Suzuki and other civilians had been brought round to the inevitability of surrender by the atomic bombs and the Soviet attack, it was decided to ask the Emperor for a decision, which was duly done at 2 a.m. on 10 August.

Considering this level of resistance by the armed forces, it is inconceivable that, as the revisionists claim, a mere blockade of Japan and continued conventional bombing might have forced it to surrender. After all, an attempted *coup d'état* by army officers determined to continue the war only failed when Anami refused to support it; he committed suicide instead. Meanwhile, the commander of Japanese forces in China telegraphed Tokyo on 15 August to say, 'Such a disgrace as the surrender of several million troops without fighting is not paralleled in the world's military history, and it is absolutely impossible to submit to the unconditional surrender of a million picked troops in perfectly healthy shape.' The will to fight on was there.

Emperor Hirohito's Imperial Rescript of 14 August 1945 made it perfectly plain that the dropping of the atomic bombs was absolutely epicentral to Japan's decision to surrender. In explaining the decision, he told his people:

> The enemy has begun to employ a new and most cruel bomb, the power of which to do damage is indeed incalculable, taking the toll of many innocent lives. Should we continue to fight, it would not only result in an ultimate collapse and

obliteration of the Japanese nation, but it would also lead to the total extinction of human civilisation. Such being the case, how are We to save the lives of millions of Our subjects? This is why We have ordered the acceptance of the provisions of the Joint Declaration of the Powers.[61]

While the reference to 'innocent lives' might represent the height of hypocrisy – coming from the head of state of a nation responsible for the Nanjing massacre, the bombing of Shanghai, the Bataan death march, the concentration camp beheadings, Kanchanaburi and other death camps, Changi Gaol, the Burma-Siam railway, Biological Warfare Unit 731's medical experiments on POWs, the 'comfort women' brothels of Korea, and very many more such organised atrocities – this part of the Rescript does at least prove that the revisionists are wrong to belittle the necessity of the atomic bombs.

As the writer Allan Massie has put it, excerpts from the diaries that some of the POWs held by the Japanese managed to keep 'make you realise that the Japanese camps, like the Nazi death camps, were all that we have imagined of Hell translated to the surface of the earth and made reality'.[62] One statistic alone sums up the cruelty of the Japanese; compared to the 4% of Allied POWs who died in German captivity during the war, no less than 27% of Japan's 140,000 British, Australian, American and Dutch military prisoners perished. Although terrible cruelties were visited on some British and Commonwealth POWs in Germany, as recently documented in Sean Longden's book *Hitler's British Slaves*, what took place in Japan was by an order of magnitude more ghastly.

Some critics of Truman have gone on to argue that the bombs ought not to have been deployed against cities, but instead the awesome power of the weaponry ought to have been demonstrated to the Japanese in some uninhabited desert. An immediate problem with such a scheme was that, contrary to General Anami's fears, the Americans only had two atomic bombs, with several months before the next ones could be produced. If one were to be wasted on a demonstration that failed to convince the Japanese to surrender, that would only leave one other, and since the Japanese decided to fight on after Hiroshima a second was needed. The 'display' theory also fails properly to take into account the mind-set of the people who took the important decisions in Imperial Japan in the summer of 1945. No amount of scientists and other observers who had witnessed the destructive power of the bombs during harmless demonstrations would have persuaded the hard men of the Japanese General Staff. As it was, it took two devastated cities to convince the politicians and their Emperor.

Still other critics have sought to argue that there is something so particularly vile about nuclear weapons *per se* that they ought never to have been deployed against Japan, almost as if nuclear fission has a uniqueness that sets it apart from TNT and renders its use in weaponry morally wrong. The horrifically vast numbers killed and maimed on one day at Hiroshima are held to be simply too many. Yet here, too, the facts belie the prejudice. For on the night of 9 March 1945, Major-General Curtis Le May's xxi Bomber Command, consisting of over 300 B-29s, attacked Tokyo at low altitudes with incendiaries and napalm destroying 267,171 homes, a quarter of all the buildings in Japan's capital. More than 83,000 people were killed – more than at Hiroshima – and over 40,000 were injured, with over a million people left homeless. Similar raids were launched against Nagoya, Kobe, Osaka, Yokohama and Kawasaki, destroying an urban area of 257.2 square miles. In all, over 41,592 tons of bombs were dropped by 6,000 B-29 sorties, with the loss of 136 USAAF planes.[63] Atomic bombs need to be seen in the overall context of what Japan had brought upon herself since her 1931 invasion of Manchuria.

Furthermore, as the former Yale history professor Robert Kagan has pointed out, 'The Japanese had plans to kill Allied prisoners of war as the fighting approached the camps where they were being held, so the swift surrender brought on by the bomb saved still more American lives.' With the official figure for expected American casualties for the attack on Kyushu, one writer has stated that 'only an intellectual could assert that 193,500 anticipated casualties were too insignificant to have caused Truman to use atomic bombs'.[64] Fortunately, the English-speaking peoples' wars are fought by professional soldiers under the direction of elected politicians, with intellectuals having very little to do with them until they are safely won, after which they can criticise with hindsight and moral superiority.

It is also often forgotten how the nuclear bombs did not just save Allied lives, but hundreds of thousands of Japanese lives too. As Kagan cogently states,

> The experience of Luzon, Iwo Jima, and Okinawa showed that [Japanese] casualties would have been many times greater than those suffered by Americans – invasion or no invasion. American planes would have dealt with many more Japanese cities as they dealt with Tokyo, and would have repeated their attacks on the capital as well. The American navy would have continued its blockade, and starvation would have taken off thousands of civilians. In sum, the total would have been greater than those exacted by the bombs.[65]

It took a brave former president of the Japanese Medical Association to enunciate it, but it is true that, 'When one considers the possibility that the Japanese military would have sacrificed the entire nation if it were not for the atomic bomb attack, then this bomb might be described as having saved Japan.' The fact that Japan has been a peaceful, indeed almost pacifist, decent, democratic and law-abiding power ever since Hiroshima, with no *revanchist* tendencies, is largely down to the events of 6 and 9 August 1945. Furthermore, the undeniable destructive power unleashed on those two days has meant that no-one during the Cold War and since has been under any illusions about the reality of nuclear warfare. In that sense, far from being the English-speaking peoples' greatest war crime, the attacks on Hiroshima and Nagasaki were its most signal service in bringing about the relatively peaceful world of the past six decades.

It has been argued that President Roosevelt would not have used the bomb if the decision had been left to him, but this is to ignore the fact that he commissioned, at vast expense, its development. The project was considered so vital that, under his direct order, no expense was spared to protect its secrecy. Its progress was one of the two secrets of the war, Enigma being the other, that even the Vice-President and the Supreme Allied Commander in the Pacific were not kept briefed upon. Repeatedly, Roosevelt's approval of the strategic bombing of Germany underscored the level of civilian casualties that he was willing to accept in the pursuit of victory. His answer to Churchill's telegram about the deaths of French civilians at the time of Operation Overlord implies that Roosevelt would have had no more hesitation than had Truman in dropping the bombs on Japan.

Once the Japanese started to surrender, millions of them did so without committing *hara-kiri*. 'Many behaved obsequiously,' as Allan Massie has pointed out, 'they even started bowing to the walking skeletons they had held captive and so abominably maltreated.'[66] The contempt they had shown their prisoners for four long years for not fighting to the death in 1941 was not shown to them in 1945. They were fortunate that the English-speaking peoples generally conformed to different mores of warfare.

On the flight to Nagasaki on the aeroplane *Bock's Car*, the radio operator Sergeant Ralph Curry asked the journalist William Laurence, 'Think this atomic bomb will end the war?' Laurence, who had been an observer at the atomic tests at the 'Trinity' site at the Alamogordo Bombing Range in New Mexico, answered that he thought it probably would, or if not this one, 'then the next one or two surely will. Its power is such that no nation can stand up against it.' Reflecting on the conversation later, Laurence asked himself, 'Does one feel any pity or compassion for the poor devils about to die? Not when one thinks of Pearl Harbor and the death march on Bataan.'[67] It might sound harsh to modern ears, but it was a completely understandable reaction, and undoubtedly the view adopted by an overwhelming majority of the English-speaking peoples at the time. 'I have never lost a night's sleep over my actions,' Paul Tibbets said in an interview shortly before his death, 'and I never will.'

The fact that the nuclear bombs were dropped on Japan in 1945 meant that they were not only the first two ever to be used against humans, but also the last. Had the horror only been observable in the laboratory conditions, there would have been huge pressure on any number of world leaders since 1945 to deploy the bomb in order to end wars. The Sino-Indian War of 1962 certainly might have turned out differently.

The last word should perhaps go to the US Secretary of War, Henry L. Stimson, who had fought as a lieutenant-colonel in the Great War and who had had severe reservations about the carpet-bombing of Germany. On Hiroshima, Stimson wrote in his 1948 memoir, *On Active Service in Peace and War*:

> My chief purpose was to end the war in victory with the least possible cost in the lives of the men in the armies which I had helped to raise. In the light of the alternatives which, on a fair estimate, were open to us I believe that no man, in our position and subject to our responsibilities, holding in his hand a weapon of such possibilities for accomplishing this purpose and saving those lives, could have failed to use it and afterwards looked his countrymen in the face.

ELEVEN

The Third Assault: Soviet Communism

1945‒9

The human and financial cost of Hitlerism – The Nuremberg Trials – George Kennan's 'Long Telegram' – Cold War planning – The United Nations – The Keynes loan – The British Commonwealth builds a bomb – The Marshall Plan – The end of empire in India – The creation of the State of Israel – The Berlin airlift – McCarthyism – The foundation of NATO

'I think I can save the British Empire from anything – except the British.'

Winston Churchill[1]

'Close with a Frenchman, but out-manoeuvre a Russian.'   Horatio Nelson

'It is my earnest hope,' said General Douglas MacArthur on taking the Japanese surrender on board the USS *Missouri* in the Bay of Tokyo in September 1945, 'and indeed the hope of all Mankind, that on this solemn occasion a better world shall emerge out of the blood and carnage of the past – a world dedicated to the dignity of Man and the fulfilment of his most cherished wish for freedom, tolerance and justice.' The blood and carnage of the Second World War had indeed been grievous, though for the English-speaking peoples – who escaped invasion everywhere but in the Channel Islands – nothing like so vast as in the Great War. Great Britain lost 244,723 people killed and 277,090 wounded; the rest of the British Commonwealth suffered 109,929 killed and 197,908 wounded. The United States saw 230,173 of her citizens killed and 613,611 wounded. Germany, meanwhile, had lost three million soldiers and civilians killed with a further million wounded. Her invasion of the Soviet Union had resulted in over twenty million Russian deaths.

Britain led the list in terms of merchant shipping losses, with 11.38 million tons, although Germany had lost 8.32 million, the United States 3.31 million and the rest of the Allies 5.03 million. The naval losses were staggering: the Royal Navy lost 8 aircraft carriers, 5 battleships, 26 cruisers, 77 submarines and no fewer than 128 destroyers. Meanwhile, the US Navy had lost 5 aircraft carriers, 2 battleships, 16 cruisers, 52 submarines and 71 destroyers. Germany lost 7 battleships, 7 cruisers, 25 destroyers and no fewer than 974 U-boats. The biggest naval losses of all were sustained by Japan, including the destruction of 15 aircraft carriers, 4 escort carriers, 12 battleships, 36 cruisers, 125 submarines and 126 destroyers.[2]

It is an accepted truism that the victorious Allies of 1945 had learnt the lesson of Versailles and consequently treated the defeated Axis powers – principally Germany – much more generously, with the result that they turned into the liberal, decent, pacific democracies that they have been throughout the second half of the twentieth century and beyond. This received wisdom further perpetuates the Keynesian myth of the 'abhorrent and detestable' Versailles settlement. Yet is it true?

After 1945, the Allies split Germany into two separate countries and her capital into four sectors, an arrangement that was to stand for over half a century. They hanged German soldiers, diplomats and journalists in 1946, whereas after the Great War there had been very few cases tried by the German supreme court in 1921, most of which had resulted in acquittals. Moreover, the Allies stationed hundreds of thousands of troops throughout Germany for over four decades – the British Army of the Rhine was not repatriated until 1992 – which did not happen after the Great War. Although there were no harsh formal reparations to pay, the Soviet Union transported vast amounts of East Germany's industrial plant to Russia in 1945, in a manner that was not possible in 1919. German rearmament, which was an important factor in the 1919 peace, was not one in 1945. If anything, the 1945 terms were actually tougher on Germany than the 1919 ones about which Keynes had written so passionately, yet they have produced the longest period of European peace that the continent has seen since the Dark Ages. Far from proving Keynes right, the toughness of the Yalta and Potsdam settlements on Germany imply that his strictures against Versailles were hyperbolic. The story is told of the American General Mark Clark, who, on being told by an aide that another 'Carthaginian peace' could not be imposed on Germany, answered: 'Well now, you don't hear too much from those Carthaginians nowadays.'

Back in July 1944, Alanbrooke had described the best way of dealing with Germany was to 'foster her, and gradually build her up, and bring her into a federation of Western Europe'. Yet the fostering of Germany was only going to be attempted if it was clear that the new Germany would be kept physically incapable of threatening the peace of Europe for a third time that century. The successful reintroduction of West Germany, Austria, Italy and Japan into the democratic world was one of the greatest contributions of the English-speaking peoples to twentieth-century civilisation. They were not to know whether the Nazis might stage an insurgency campaign lasting years, or whether the Japanese parliament would baulk at their demands for – amongst other things – the enfranchisement of women; nonetheless they insisted on democratic constitutions after imposing regime change, and these very quickly took root.

As has recently been pointed out, 'Compare the subsequent fate of Hungary or North Korea under Communist rule with that of Japan under Western democracy.'[3] The result of the imposition of constitutions based on the English-speaking peoples' democratic model has been that the Japanese and German economies and societies have flourished in a way that would have seemed inconceivable to anyone walking through the razed cities of either country in 1945 – 'Year Zero'. Had the Soviets been left to their own devices, neither West Germany nor Japan would have risen to such economic power in such a short space of time.

For the whole of the post-war period, both countries have been peaceful, democratic model states, as well as tremendously successful economies. In terms of Gross Domestic Product, Japan and West Germany have for decades occupied the second and third places in the world ranking of economic powers, after the United States and before Great Britain and France. To have helped raise their former deadliest foes to such a place is a tribute to the magnanimity of the English-speaking peoples in not going down the path of mass despoliation that Stalin envisaged for both countries, and which he carried out against much of the industrial plant of East Germany.

The trial and subsequent execution of the leading Nazi war criminals represented an important departure for the English-speaking peoples, establishing several principles of international law that were to serve them well in the coming years. A central paradox of the Second World War was that in order for the pathological murderers of the Third Reich to be defeated, Civilisation had to call in aid Joseph Stalin and a Bolshevik regime that had also massacred innocents on a similarly vast scale since grasping power in 1917. Despite the gaping irony that their Soviet allies had no moral justification to sit in judgment on questions of genocide, the Nuremberg Tribunal sent an unmistakable message to the world about the consequences of waging aggressive war, a message that dictators such as Slobodan Milosevic and Saddam Hussein failed to learn, to their eventual cost.

Between their capture in the summer of 1945 and the time they went before the Tribunal on 20 November that year, the victorious Allies subjected virtually the whole elite of the German Government, Foreign Office, military High Command and concentration camp administration to thousands of interrogations. The defendants exhibited virtually every conceivable human response to their predicament, as recorded in Professor

Richard Overy's book *Interrogation: The Nazi Elite in Allied Hands*. There was defiance (Hermann Göring), 'hysterical amnesia' (Rudolf Hess), qualified apology (Albert Speer), suicide (Robert Ley), chillingly bureaucratic exactitude (Rudolf Höss, commandant of Auschwitz), mental collapse (Joachim von Ribbentrop), regret (Hans Frank, gauleiter of Poland), and terror (the appropriately named Walther Funk), along with much attempted blame-shifting.

At times the interrogations produced truly stomach-churning testimony, as when the Auschwitz guard Otto Moll spoke of the fate of the babies whose mothers had left them hidden in discarded clothing outside the gas chambers: 'The prisoners had to clean up the room after it had been cleared of people, they would then take the babies and throw them into the gas chamber.' Elsewhere he was asked to estimate the length of time in which the Zyklon B gas took effect: 'The gas was poured in through an opening. About one half minute after the gas was poured in, of course I am merely estimating this time as we never had a stop-watch to clock it and we were not interested, at any rate, after one half minute there were no more heavy sounds.' Question: 'What kinds of sounds were heard before that?' Answer: 'The people wept and screeched.'

As well as such face-to-face interviews, the Allies recorded the Germans speaking to one another in private, even in the lavatories, yet going to such lengths was hardly necessary since none of them refused to reply to the interrogators, even those who must have known that only the noose awaited. After Ley's suicide in October 1945, a guard was posted outside each cell to check once a minute that the other prisoners were not also about to take the easier way out.

Class-consciousness seems to have survived the traumas of defeat, national destruction and self-disgust. At Nuremberg, the aristocratic former Foreign Minister Konstantin von Neurath despised the foamingly anti-Semitic rabble-rouser Julius Streicher as much for his low birth as for his generally foul personality. Human nature was such that even after a world war and despite *Götterdämmerung*, genocide and Year Zero, snobbery will always out.

It was the British barrister and politician Hartley Shawcross QC who brilliantly and concisely put Civilisation's legal case against the leaders of the Nazi regime at the International Military Tribunal, and who demanded the death penalty for them at the end of the seven-month trial. Few men could have been better equipped to undertake the immensely complex task of collating the vast amounts of evidence necessary to indict the Nazi leaders, and then to present it in the most compelling way possible. Shawcross was a legal genius, the winner of the Certificate of Honour after he took first place in his Bar Finals examination. He was called to the Bar in 1925 and was successively a senior law lecturer, Recorder and Chairman of the Enemy

Aliens Tribunal by 1940. He was then appointed a regional commissioner, responsible for administering the North-West of England between 1942 and 1945.

In the Labour landslide election victory of July 1945, Shawcross became Labour MP for St Helens, and Clement Attlee acknowledged his legal eminence by immediately appointing him Attorney-General, where he stayed until briefly becoming President of the Board of Trade in 1951. As the senior government law officer in the Commons, Shawcross was responsible for overseeing the incredibly detailed legal aspects involved in nationalisation and the creation of the Welfare State, but he also knew that he would be largely judged by posterity on his performance against Göring and the rest of the captured Nazis.

So conscientious and professional were the British legal team at Nuremberg that they also helped the French and Russians frame their indictments. The Americans, however, with a staff of attorneys that outnumbered the other three Allied teams put together, did not present their case in a way that Shawcross considered wholly competent. Long before the trial began, Shawcross complained to the new British Foreign Secretary Ernest Bevin that the United States' chief prosecutor, Justice Robert H. Jackson, was rarely to be seen in Nuremberg. He feared that since the Americans' primary job was to establish 'that the prisoners conspired together to wage a war of aggression', which was a vital part of the case, this might go by default.

American administrative foul-ups led to Shawcross himself having to drive three doctors, who were due to examine Rudolf Hess's mental health, from Belgium to Nuremberg in rain, sleet and fog in a car which had no windscreen wipers. When, just before the trial began, Jackson tried to swap one member of the Krupp industrialist family for another when it was discovered that the elder one was senile, Shawcross delivered the magisterial rebuke: 'This is a court of justice, not a game in which you can play a substitute if one member of the team falls sick.'

Tall, handsome and distinguished, with a profoundly authoritative speaking voice, Shawcross was impressive when he rose to put the British case for the indictments. As one of the American prosecutors later recalled, 'He cut a fine figure at the lectern, displayed a well-controlled vocal delivery, and his text was well organised and crisply written.' Whereas Jackson had, as Shawcross had feared, made a crusading speech about how international law might ideally develop in the future once the defendants were found guilty, Shawcross concentrated on arguing that the Nazis had broken the law of nations as it already existed, citing a series of international agreements from the Covenant of the League of Nations to the Kellogg-Briand pact of 1928 (to which Germany had been a signatory). As a recent historian of the Trials has commented, 'His speech was beautifully composed; it made a great impression. His style, his manner of delivery, above all his intellectual approach were entirely different from Jackson's. He was a first-class practising barrister, whose primary instinct was to win cases.'[4]

Shawcross opened by stressing the value of having a trial at all, saying,

> There are those who would perhaps say that these wretched men should have been dealt with summarily without trial by 'executive action', but that is not the view of the British Government. Not so would the rule of law be raised and strengthened on the international as well as upon the municipal plane, not so would a world be made aware that the waging of aggressive war is not only a dangerous venture but a criminal one.

By the end of his peroration, the Nazis in the dock were visibly shaken.

Much of the day-to-day cross-examination, including that which broke Hermann Göring, was undertaken by another politician-lawyer David Maxwell-Fyfe, but it was Shawcross who made the British Government's one-and-a-half-day closing speech in July 1946. He went through the defendants one by one arguing that 'each of these defendants is legally guilty' and showing how the Nazis never declared war before invading countries. 'Every one of these men acquiesced in this technique, knowing full well what it must represent in terms of human life. How can any one of them now say he was not a party to common murder in its most ruthless form?'

Of this speech Göring said, 'Compared to Shawcross, Jackson was down-right chivalrous.' Ribbentrop agreed: 'Compared to him, even Jackson was a charming fellow.' Hans Frank loudly cursed 'that damned Englishman'. Only Albert Speer said he 'was delighted' with Shawcross, 'after listening to all the stupid nonsense of the defence attorneys'.

(Shawcross did not always choose his words with such forensic care as at Nuremberg. In a Commons debate on the third reading of the highly contentious Trades Disputes and Trade Unions Bill on 2 April 1946, he uttered the words: 'We are the masters at the moment, and not only at the moment but for a very long time to come.' Once this was truncated and misquoted as 'We are the masters now!', it was held up by the Conservatives as an example of socialist triumphalism and hubris, and was denounced on Tory platforms across the country for many years.)

The Cold War was only cold in the sense that the English-speaking peoples never went to war with the USSR directly. It was certainly not cold for Namibians or Afghans, Koreans or Vietnamese, Mozambiquans or Malaysians, or any of the other numerous peoples who got caught up in the communists' relentless drive to destabilise capitalist societies around the globe for over four decades. Because, in the words of a recent commentator, 'hundreds of thousands, perhaps millions died in it', the Cold War can with reason be described as 'the Third World War', which nonetheless never saw the two main protagonists fight one another except through proxies.[5]

According to a Council of Europe document presented in January 2006, between 1917 and the present day communism has been responsible – at a conservative estimate – for 94.5 million deaths. This was broken down as follows: 65 million killed by Chairman Mao and his successors; 20 million Soviet victims including party purges, mass murder, deportations, Ukrainian starvation policies in the Thirties and wartime reprisals; North Korea 2 million; Cambodia up to 2 million; Africa 1.7 million; Afghanistan 1.5 million; Vietnam 1 million; Eastern Europe 1 million; Latin America 150,000.[6]

There are nonetheless those who still argue – despite the almost 100 million deaths for which Marxism-Leninism was responsible during the twentieth century – that the Cold War was an over-reaction by the West, and specifically the English-speaking peoples, who were paranoiac about the ambitions and strength of the Soviet bloc. If so, it was a paranoia fully shared by Western Europeans, the Japanese, South Africa, Latin America and even on occasion by communist Chinese leaders. 'If the Soviet threat was a hoax,' points out the historian Michael Lind, 'it was a hoax that depended on the collaboration of vast numbers of quite different people on many continents for half a century.' Although it is perfectly true that the Soviet Union could not have physically conquered the world, victory in the Cold War would have allowed it – through imitation and subversion in many countries – to exercise hegemony over it, allowing her views on diplomacy, trade, democracy, human rights, the rules of war, property rights and so on to prevail, perhaps for decades.

Although the general attitude of the Americans towards the British Empire in the immediate post-war world was that it was an impediment to the rationalisation and democratisation of the globe, and that the lion's tail needed to be twisted in order for Britain to disgorge her far-flung territories, this altered with the international situation. Loosening the British grip on her colonies was never as high on the Truman Administration's agenda as containing the ambitions of the Soviet Union. A good example of this came after 9 February 1946, when Stalin made an aggressive speech at the Bolshoi Theatre in Moscow denouncing the West, saying that, 'The development of world capitalism proceeds not in the path of smooth and even progress but through crisis and the catastrophes of war.'

Immediately after this, Isaiah Berlin in Washington detected a softening of the United States' opinion towards the British Empire, especially in the writings and broadcasts of the columnists and radio commentators Walter Lippmann, Raymond Gram Swing and William Shirer, who, he told Frank

Roberts at the Foreign Office, 'before were talking of mediation and twisting our tail for unbridled imperialism' but 'were now complaining that Russia had once more destroyed their hopes and that maybe, after all, an alignment with the British in defence of the Western world was inevitable. . . . They are still most anxious not to "underwrite the British Empire" and simply follow in our wake, but if they can perceive themselves that their own interests are being menaced by what goes on in, say, Turkey and China, they will react in a really useful manner.'[7]

In fact, the most significant development from Stalin's speech was not the altering of American commentators' perception of the usefulness of the British Empire, so much as the sea-change that overcame Administration thinking as a result of Stalin's sabre-rattling. The Secretary of State James Byrnes asked George F. Kennan, the Kremlin specialist who was then American Chargé d'Affaires in Moscow, for a summing-up of Soviet intentions, and on 22 February he received what has become known to history as the 'Long Telegram', an 8,000-word document that argued that there could be 'no permanent peaceful co-existence' with the USSR. It was one of the first significant warnings to the Truman Administration of the sad realities of the Cold War, a conflict that had not been sought by America but which she could not ignore, let alone risk losing.

Writing under the soubriquet 'Mr X' in the influential American intellectual journal *Foreign Affairs* in July 1947, Kennan reiterated his argument and pointed out how the Soviet Union's antagonism towards America was not due to some kind of honest misunderstanding between the countries, but was down to Russia's ingrained hostility towards her rival. Kennan believed that the Kremlin needed hostility with the West to survive, as it provided it with excuses 'for the dictatorship without which they do not know how to rule, for cruelties they do not dare not to inflict, for sacrifices they feel bound to demand'.

'Short of becoming a Communist country, there was nothing the United States could do to gain the Kremlin's trust,' a modern commentator has accurately précised Kennan's argument. 'The Soviets could not be appeased, only contained.'[8] Stalin's conduct became explicable to hitherto-perplexed Americans; it was not their fault that the world was being plunged into a Cold War, but his deliberate policy. The more that we understand Soviet post-war foreign policy as a result of the opening of their archives since *glasnost*, the more this view is shown to have been the correct one.

Soviet naval strategy and shipbuilding programmes between 1935 and Stalin's death in 1953 certainly imply much more than a concern merely for the defence of the USSR. Throughout this period Stalin was obsessed with, in the words of a recent study, 'the drive to construct a sea-going navy equal to that of the Royal Navy or the US Navy, one which had no real rationale either in Russian geography or history'.[9] It did, however, make perfect sense in terms of his ambitions to fight a global Cold War against capitalism, one that he perhaps might even heat up if and when it suited him.

As early as 1935, Stalin had embraced the 'big navy, big ship' party in the Soviet High Command that believed that a world superpower had to have a large navy and powerful battle fleet, rather than a navy that was little more than a defensive 'off-shore extension of the Red Army'. Since Russia had four coastlines – the Baltic, Black Sea, Pacific and Arctic – of which only the Pacific opened onto an ocean (and which was ice-covered during winter), Stalin's drive for a large navy was likely to have been an integral part of his plans for the extension of communism in the post-war period.[10]

It was in response to Soviet expansionism and provocations that Britain took the decision to stockpile nerve-gas after the war, in order to retaliate against any Russian chemical warfare attack. 'Nerve gasses are the outstanding chemical warfare development of modern times,' wrote Brigadier Cedric Price, secretary to the Joint Chiefs of Staff, to the Labour Secretary for War, Emmanuel Shinwell, in September 1950, 'and existing experimental data shows that they have tremendous potentialities against troops and against tanks.'[11] At the time, Britain had 70,000 110 lb bombs filled with the nerve gas Tabun that had been captured from the Germans, half-a-million obsolete mustard gas shells, and 33,000 huge phosgene bombs weighing 500 lb each that had been stored out in the open and were unlikely to last another three years.

When Sarin was tested on servicemen at the Porton Down chemical weapons research centre in 1953, one of them, Ronald Maddison, was killed. Although a nerve-gas plant was built at an RAF base in Nancekuke, Cornwall, it never went into full-scale production, as Britain renounced chemical weapons' production in 1956. Of course such were the Cold War exigencies that production continued in the United States, employing British expertise gleaned from tests and trials. Poor Aircraftman Maddison, whose death was judged unlawful at an inquest in 2004, therefore had not died in vain.

On 13 April 1946, a report was delivered to the Joint Technical Warfare Committee (JTWC) by Dr Henry Hulme, the scientific advisor to the Air Ministry. The JTWC was a specialised, top-secret sub-committee of the Chiefs of Staff that had been set up in November 1943 to 'co-ordinate and direct the technical study of . . . operational projects and problems'.[12] Hulme's twenty-page report, entitled simply 'Preliminary Note', was the first official attempt to project and quantify what would happen in a nuclear war between Great Britain and the USSR. It makes chilling reading.

Although Churchill's Iron Curtain speech, delivered in March 1946, was widely condemned by British and American politicians, especially in the Labour and Democratic Parties, there was a team in Whitehall already working on precisely the scenario that he envisaged and warned against. For all that they hoped for the best, the Government had already started preparing for the worst; it was the only wise course when faced with Soviet strategy.

Hulme produced what he called 'a suitable simple model' that was intended 'to provide a very tentative estimate of the situation ten years hence'. He made the assumptions that both sides possessed high-level manned bombers that could fly up to 500 mph for 2,000 miles with fighter support, that neither side had defensive armaments that radar gave effective warning over 200 miles, and that 180 British and 540 Soviet defensive night-fighters could be deployed. Under those circumstances, he estimated that twenty-six raids would be needed to attack sixty-seven major Soviet centres containing 88% of the urban population. In order to minimise the amount of time the RAF would need to be in Soviet airspace, twelve targets would be attacked from England, thirteen from Cyprus and one from Peshawar.[13]

Hulme estimated that 370 bombs would have to be despatched in order to deliver the 242 bombs required to destroy all the targets. Maps and the list of target cities were appended: Moscow and Leningrad – which together constituted 28% of the Soviet population – were to be obliterated by bombers flying from Britain. There was also a detailed analysis of the destruction that would be caused by a Soviet attack on Britain, where the area to be destroyed was far smaller and the population density far greater than in the USSR.

On the basis of what had happened at Hiroshima and Nagasaki, it was calculated that,

> an atomic weapon falling on a city in the United Kingdom would obliterate or damage irreparably an area of three square miles. If half the population were indoors, one-third in shelters and one-sixth in the open, this would result on average in ten thousand fatalities from all causes and the permanent de-housing of seventy thousand survivors in a city of one hundred thousand inhabitants or more; but these totals would increase by a factor of $2\frac{1}{2}$ if detonation occurred over a central built-up area.[14]

There could only be one upshot from these terrible consequences of being attacked: Britain needed a credible nuclear deterrent as fast as possible.

Before this, the United States provided a nuclear 'umbrella' over Britain. In July 1963, the Archbishop of Canterbury Michael Ramsey received a sharp letter from Harold Macmillan in response to a sermon criticising nuclear weapons. The Prime Minister reminded the Archbishop how until Attlee had commissioned the building of the British bomb in 1947, only the Soviet Union had nuclear capability in Europe, and 'At the time had it not been for America's nuclear superiority there could be little doubt that the Free World would have been severely threatened by the Communist flood.'[15] Instead, the bomb 'gave Britain a front rank in world councils and added political and military strength to the Western Alliance. A deterrent means what it says. It is not a weapon of offence – a weapon used to start a war – it is a weapon of defence.' Furthermore, 'It is not sufficiently well known that we have with the United States an agreement that they would never use nuclear weapons in any part of the world, whether Europe or not, without first consulting the British Government.'

On Wednesday, 10 October 1945, 7,000 people packed into the Royal Albert Hall in Kensington, London, to hear the Prime Minister Clement Attlee, the American Secretary of State Edward R. Stettinius, the Shadow Foreign Secretary Anthony Eden, Lord Cecil of Chelwood, Megan Lloyd George and several other speakers extol the virtues of the United Nations. The octogenarian Cecil had been President of the League of Nations Union since 1923 and had won the Nobel Peace Prize in 1937, and he spoke with passion when he said,

> Don't listen to people who tell you that the atomic bomb can be rendered harmless or even advantageous to peace. Be well assured that it will be used unless we prevent it. . . . And even if some remedy for it could be discovered a fresh infernal machine would be produced. Either we must combine to enforce a stable and lasting peace or we must face irredeemable disaster. We must either combine or perish.[16]

In fact the history of the past sixty years has confounded the predictions of this venerable man; the atomic bomb has not been used since Cecil spoke and the possession of it by the West has been advantageous to peace, in that it successfully deterred any conventional attack by the Soviet Union and her satellites. It has not been necessary to combine all nations together in order to prevent ourselves from perishing; all that was needed was to build a bomb.

The next speaker was Attlee, who proclaimed that it was

> the firm intention of His Majesty's Government to make the success of the United Nations the primary objective of our foreign policy. The Charter is our first line of defence. This country helped to lay its foundations, there's much in it of our thinking and our traditions, as much as those of any other country in the world. It's not perfect, nothing made by human beings ever is, but there's only one way of improving it, and that is to use it, and to use it to the full.

To show cross-Party unity, Anthony Eden added that the UN was 'the world's last chance' and that it was 'indispensable to the peaceful ordering of the modern world', without which the world would 'assuredly destroy itself utterly'.

With such support, how could the United Nations have failed as badly as it has since the end of the Korean War in 1953? It is perfectly true that the United Nations started off very much under the auspices of the English-speaking peoples. The United States determined not to make the same mistake that it had in not joining the League in 1919, and Britain and all the other English-speaking peoples joined in the first tranche of members. Yet as the organisation progressed, and especially after it faced significant threats to its authority, the high ideals began to be tarnished as the inescapable realities of power-politics in a Cold War environment reasserted themselves.

Historically, the English-speaking peoples were tremendously fortunate in their dealings with the organisation since Britain and twenty-eight other nations ratified its charter on 24 October 1945. It is true that during the Cold War, the Eastern bloc-Third World majority occasionally passed insulting resolutions, such as the one in favour of ending 'colonial rule' in Gibraltar in 1968, or tried to embarrass Britain for not doing enough to destabilise the Government of Rhodesia, but when military crises arose, Britain was generally lucky at the UN.

Between 1939 and 1945, the Gross National Product of the USA had increased by two-thirds to $215 billion.[17] She was by far the richest country in the world, yet she needed foreign customers, and most of them were hovering on the brink of national bankruptcy after six years of Total War, former friends as well as well as former foes.

John Maynard Keynes had been dying of heart disease since 1944, yet instead of convalescing he undertook in August 1945 perhaps the most important and strenuous task of any twentieth-century Briton in the field of finance. With no official position besides his acknowledged genius and the (occasionally wavering) support of the Attlee ministry, he went to Washington to try to prevent Britain from going bankrupt as a result of her wartime exertions, by securing a $3.75 billion loan from the Americans.

There had been little in the way of an Anglo-American Special Relationship in economic terms after President Truman abruptly terminated Lend-Lease only two days after the Japanese surrender in August 1945. Seeing an opportunity peacefully to dissolve the British Empire's Imperial Preference and Sterling Area systems, which they regarded – not unreasonably – as a giant organisation for the restraint of international *laissez-faire* economics, the Americans demanded extensive free trade agreements as reciprocation for their loan, on top of the 2% interest. 'Why do you persecute us like this?' Keynes asked the American chief negotiators at one point. 'You cannot treat a great nation as if it were a bankrupt company.' They could and did because it was, but at least it averted the threat of widespread malnutrition in the United Kingdom. (The four months of hard haggling left Keynes fanatically

anti-American, telling Lady Violet Bonham Carter in his snobbish Bloomsbury manner that Americans were 'a rare breed of sub (or super) dagos,
speaking no known language intelligently'.)[18]

There was a definite air of intellectual superiority about the British in
Washington at the time; All Souls' fellows abounded, including the British
Ambassador Lord Halifax. An Embassy ditty from late 1945 went:

'In Washington Lord Halifax whispered to Lord Keynes,
"It's true they have the money-bags, but *we* have all the brains."'

Yet in the loan negotiations there was little doubt which mattered more, and
Britain had to accept a deal that in effect destroyed the Sterling Area and
paved the way for American post-war global commercial domination. It was
absurd to believe that the Americans would or could have allowed a system of
Imperial Preference to survive that put them at a disadvantage in vast global
markets after a war that had largely been won on their massive loans, taxpayers'
expenditure and military production.

As an historian of US wartime aid has pointed out, 'In March 1941 the
British economy had been unable to withstand the armed threat from Germany
and Italy, and she had insufficient dollars to buy what she required from the
US. In September 1945 Britain's economy was unable to meet her civilian
needs, and she had insufficient dollars to buy what she required from the
US.'[19] Under those circumstances, and considering that Churchill had, both
in the Atlantic Charter of August 1941 and in article seven of the Mutual Aid
Agreement of February 1942, agreed to co-operate with the US and pursue
a freer world economy, it was certain that Imperial Preference and the Sterling
Area would be targets for post-war American Administrations of either political stamp.

In return for the Keynes loan, Britain had to ratify the Bretton Woods
Agreement and enter the International Monetary Fund, to agree to further
talks to reduce trade barriers and eliminate Imperial Preference, and to promise
to make sterling convertible after 1947, as well as other concessions. There
was no alternative.

The British decision to build a nuclear deterrent was taken by an ad hoc
committee codenamed 'General 75' of half-a-dozen senior Cabinet ministers.
Neither the rest of the Cabinet nor Parliament were included in the decision-
making process and the relevant files are still classified.[20] As well as Attlee and
Bevin, the members of General 75 were Stafford Cripps, Herbert Morrison,
Arthur Greenwood and Hugh Dalton, with the Minister of Supply John
Wilmot being included later. At the committee's third meeting, on 3 October
1945, Bevin insisted that the Prime Minister must have final authority in all
matters concerning atomic energy, who would 'consult from time to time with
those of his colleagues principally concerned', by which he meant himself and

occasionally Cripps and Morrison. It was thus a tiny group of men who decided that Britain must become a nuclear power.

October 1945 also saw the decision of General 75 to set up an atomic research establishment at Harwell, under the direction of the distinguished nuclear physicist John Cockcroft. At this stage, however, the option was left open as to whether to produce nuclear bombs, although on New Year's Day 1946 the Chiefs of Staff had told Attlee that only the threat of retaliation could be a reliable defence against attack, and therefore it was essential for Britain to have her own stock of A-bombs.[21]

The key moment came on 1 August 1946, when the US Congress passed the MacMahon Act, under which, although 'basic scientific information in specified fields could be freely disseminated, related technical information was to be controlled by a Board of Atomic Information. Dissemination was subject to the proviso that such information was not of value to the national defence.'[22] Cockcroft's hitherto regular visits to the US Atomic Energy establishments therefore had to cease virtually overnight. Bevin's response was unequivocal: 'We have got to have this thing over here whatever it costs. . . . We have got to have the bloody Union Jack on top of it.' Nonetheless, it was not until January 1947 – fifteen months later – that General 75 decided to go ahead with bomb production, again without consulting anyone.

Sir William Penney, who had been involved in developing nuclear bombs in America during the war, returned to Britain with the nuclear formulae largely in his head and began work at the former wartime airfield at Aldermaston in Berkshire on 1 April 1950. In less than three years, the first trial device was successfully tested on the Monte Bello islands off Australia. Soon afterwards, the RAF took its first delivery of the United Kingdom's nuclear bomb. Much of the rest of the English-speaking world was involved in the attempt to build the bomb, particularly Australia but also South Africa, New Zealand and Rhodesia, though not of course Eire.

What was euphemistically called 'The Joint Project' was top-secret and remained so for a long time after it ended. As Carl Bridge of the Sir Robert Menzies Centre recently wrote,

> Visitors to the hydroelectric scheme in the Snowy Mountains of New South Wales are always impressed by the power stations deep underground, which, as guidebooks say, supply cheap electricity to the cities and water to the thirsty inland plains. What the books don't reveal is that they were built principally to generate power for Australia to develop its own atomic bomb.

This great collaborative effort involved uranium mining in central and northern Australia, atomic tests on Australian soil at Monte Bello and Maralinga, and rocket-firing experiments at Woomera in South Australia.[23] Furthermore, the Labour Government of Joseph Chifley and the Liberal one of Robert

Menzies were intimately involved for over a decade, providing vast tracts of the Outback for research, development and ultimately testing.

It was only in the late Fifties, when the Americans feared that the Soviet Union was closing the missile gap, that the Eisenhower Administration agreed to provide Britain with ready-built nuclear bombs, and also promised them to Australia in the event of war. The Joint Project was then dismantled and Australia never became a nuclear power. After further successful tests of British hydrogen bombs in the 1950s, culminating in the 'Grapple' series of trials in 1958, the United States signed an agreement in the July of that year to enable the collaboration of both countries on nuclear issues that has since proved the cornerstone of world security. The agreement gave Britain access to the Nevada Test Site underground facilities for nearly thirty years, until testing ceased in 1991.

Yet Britain had only become a nuclear power in the face of American opposition and discouragement; the passing of the MacMahon Act was indeed perhaps the lowest point for the Special Relationship. It was short-sighted of the Americans to believe that they might remain the only nuclear power; nonetheless, it is entirely understandable that they did not wish to share with anyone – even their wartime partners and co-inventors – the ability to unleash 'the power of the Gods'.

The story of aviation has not been one of unremitting success and technological breakthroughs for the English-speaking peoples; there have been mishaps, dead-ends, mistakes and occasionally tragedy along the way. The price of success has been endless expensive experimentation, but their willingness to commit vast amounts of money and effort to research and develop the biggest, fastest and best has paid off spectacularly since almost the dawn of the twentieth century.

One of the most expensive flops was Howard Hughes Jnr's vast 'Flying Boat', also known as 'H4 Hercules' or 'The Spruce Goose'. The biggest aircraft ever built at the time, and for decades afterwards, it had eight engines, a 320-foot wing span and, although it was constructed out of birch wood, it still weighed 200 tons. With two decks it was capable of accommodating 700 people and was conceived during the war as a means of transporting troops far above the U-boat-infested oceans. The $25 million it cost was split $18 million by the Government and $7 million from Hughes, who told a Congressional hearing: 'I have my reputation rolled up in it, and I've stated several times that if it's a failure, I'll probably leave this country and never come back and I meant it.'

When Hughes himself flew the plane at Long Beach Harbor, California, on 2 November 1947, it reached a top speed of 80 mph, came seventy feet off the water and flew a mile in less than a minute before making a perfect landing.

Nonetheless, the Flying Boat was considered to have failed as a going concern and was towed to a dry-dock on Terminal Island, where it was kept, perfectly maintained as an active aircraft, until 1980. (Today it is on display at the Evergreen Aviation Museum, McMinnville, Oregon.)

Talking about his way of working, and the various errors made while building the Boat, Hughes said, 'I am by nature a perfectionist, and I seem to have trouble allowing anything to go through in a half-perfect condition. So if I made any mistake it was in working too hard and in doing too much of it with my own hands.' Yet it is only in trial and error – and thinking big – that the English-speaking peoples have made the regular and astonishing technological breakthroughs they have enjoyed since the Wright brothers, and which guarantee them so much of their present world status.

'What is Europe now?' said Churchill in 1947, 'a rubble heap, a charnel house, a breeding ground of pestilence and hate.'[24] He was right, yet later that same year all that started to change, when General George C. Marshall, the American Secretary of State who earlier that year had succeeded James F. Byrnes, turned what was intended to be a standard Harvard honorary degree acceptance speech into the most important official announcement of the immediate post-war era. 'Our policy is not directed against any country or doctrine,' he said on 5 June 1947, 'but against hunger, poverty, desperation and chaos.'

Washington journalists, and even the British Embassy there, were slow to recognise the deep significance of Marshall's words, but Leonard Miall, the BBC's first Washington correspondent between 1945 and 1953, had been briefed by Dean Acheson, and he spelt out the implications of what Marshall had said. After listening to Miall's *American Commentary* programme, Bevin responded immediately. Marshall's European Recovery Program, universally known as the Marshall Plan, was designed to invest in the regeneration of Europe; it was hoped this would work against communism, as was made perfectly clear by the US Ambassador to Paris, Jefferson Caffery, who privately stated that if the communists got into government, France could not expect 'a single dollar bill'.

Over the next five years, America subsidised eighteen European countries to the tune of over $13 billion, or nearly $200 billion in 2005 money.[25] At one point over 150 American vessels were arriving in European ports each day, in an operation larger than almost any undertaken in wartime barring D-Day itself. Between 1949 and 1951, four-fifths of Europe's wheat was being imported from dollar-zone countries. In the first year of its operation, Marshall Aid accounted for 14% of the total national income of Austria. Marshall himself won a well-deserved Nobel Peace Prize in 1953.

Isaiah Berlin likened the European demeanour to 'lofty and demanding beggars approaching an apprehensive millionaire', but when the Americans even wiped out France's ten-billion-franc deficit they realised that an extraordinary, even historically unprecedented act was taking place. Denouncing it as 'a programme for the enslavement of Europe', Stalin refused to allow any Soviet bloc countries to accept Marshall Aid, thereby missing an opportunity both to benefit financially and possibly even to stymie the entire scheme. Molotov said 'No.K.' – believing it to be the antonym for O.K. – and walked out of the Paris discussions.

Cynics and anti-Americans have constructed an explanation for the Marshall Plan that has managed completely to exclude any altruism whatever. According to them, America only tried to regenerate the European economies in order to forestall communism, build a consumer base for her exports and enforce her own concept of monopoly capitalism on a recalcitrant continent. Lord Beaverbrook and others believed the Plan was driven by the desire to dump excess American commodities. Similarly, no less than 47% of Frenchmen believed it was primarily intended to extend US export markets. The true reasons were both to forestall communism and to try to make Europe safe from a fascist revival; both of them noble impulses.

By the second half of 1950, Western European output was running at 35% above its pre-war levels. The Marshall Plan achieved in the economic sphere what the military-based Truman Doctrine did in the strategic, a confident, secure Western Europe. The fact that the British largely wasted the money and the opportunity, in a manner chronicled unpityingly in Correlli Barnett's 1995 masterpiece *The Lost Victory*, is largely irrelevant to the story of American decency, generosity and far-sightedness.

When Attlee became prime minister in July 1945, between V-E Day and V-J Day, the greatest long-term threat to his country was that she would squander the opportunities she had won and thereby hamstring future generations of Britons. Over a quarter of her national wealth had been lost in the previous six years of war, so the vast sums of Marshall Aid that were being directed from America desperately needed to be spent rebuilding her industrial and transport infrastructure and making her economy competitive again. Instead of doing that, Attlee effectively wasted it on trying to build the utopian society which socialists in those heady days called 'the New Jerusalem'. Instead of copying Germany and investing Marshall Aid in the crucial tasks of rebuilding infrastructure and modernising industry – and Britain was the largest beneficiary of Marshall Aid in Europe, getting one-third more of it than Germany – Attlee instead spent much of it on the Welfare State.

On his way to watch the Derby at Epsom Downs in 1904, the sixteen-year-old Harrovian schoolboy Jawaharlal Nehru read about the naval battle of

Tsushima and was inspired by the ability of an oriental power to humiliate a proud, hitherto-feared imperial force. Over forty years later, due in part to the Japanese humiliation of the British Empire in South-East Asia in 1941–3, and the subsequent collapse in the prestige of the European empires throughout the region, Nehru became prime minister of a self-governing India. Never in the history of the English-speaking peoples has such an act of voluntary self-abnegation taken place, yet unfortunately the transfer of power was largely botched by the Viceroy of India who had been appointed by Attlee, Lord Mountbatten.

The partition of and transfer of power in India, and the subsequent massacres that took place in the Punjab and the North-West Frontier, cost the lives of somewhere between 750,000 and one million Indians.[26] This represents one of the most shameful moments in the history of the English-speaking peoples in the twentieth century, yet the post-war Labour Government always boasted of the handover as a great achievement. This was partly because of Lord Mountbatten's brilliant public relations success in massaging the figures lower and lower. He himself told a London audience in November 1947 that, 'Only a hundred thousand people had died and only a small part of the country had been affected.' The use of the word 'only' is reminiscent of Stalin's remark that 'One death is a tragedy but a million is a statistic.' Nehru called independence a 'tryst with destiny', but for all too many it was only a tryst with death.

The Punjab Boundary Force that was charged with keeping the peace there as the British withdrew consisted of 15,000 men by 1 August 1947, to police 17,000 villages covering 38,000 square miles. It was moreover, in the opinion of the historian Patrick French, 'hopelessly under-equipped and under-armed for the task that lay before it'. With around one million dead and seventeen million homeless, French states that 'unconscionable horror' took place. In one Lahore refugee camp, the vultures feasted so well on corpses that they could hardly get off the ground.

'I've just been seeing Mountbatten's private secretary who gave me an account of the extraordinary scenes at the handing over,' Attlee reported to his brother on 18 August 1947. 'Mount B was surrounded by a quarter of a million Indians all violently enthusiastic for him. He has certainly captured the Indian imagination. I doubt if things will go awfully easy now as the Indian leaders know little of administration, but at least we have come out with honour instead, as at one time seemed likely, being pushed out ignominiously with the whole country in a state of confusion.'[27] It was at precisely the time that Attlee was writing these words – four days after the independence celebrations – that the Mountbatten Plan was collapsing in inter-communal massacre and bloodshed across northern India.

The blame for what took place of course principally lies with the Sikh,

Hindu and Muslim gangs who actually carried out the atrocities, but also the Congress Party leaders who insisted that the British Army withdraw entirely from India virtually overnight in August 1947, the Labour Government in Britain for going along with Mountbatten's over-hasty plans, and the lawyer Sir Cyril Radcliffe who altered the Indo-Pakistan frontier at the last minute. Of course, Mountbatten was ultimately the person responsible for partitioning the sub-continent on exactly the same day it became independent, without arresting the communal ringleaders or having enough troops to keep peace in the Punjab, or letting anyone know where the new borders were, but many others besides him contributed to the tragedy.

That other legacy of Lord Mountbatten's partition policy – Kashmir – still suppurates sixty years later. The promise that Nehru made to hold a referendum in his ancestral homeland was never honoured, and United Nations' resolutions on the subject have been even more resolutely ignored. Delhi rightly fears that the Muslim majority in Kashmir might not, especially after a long dirty war which has witnessed appalling human rights violations, vote to stay in India.

Britain's role has all too often been to back the stronger, winning side, but Britain has a special responsibility over Kashmir, for as Professor Akbar Ahmed has written in *Jinnah, Pakistan and Islamic Identity*, 'It was a British viceroy who left the Kashmir issue hanging in mid-air between India and Pakistan in 1947 as the British hastily packed to leave.' In *Thy Hand, Great Anarch!*, the great Indian writer Nirad Chaudhuri commented that if generalship were judged by the same rose-tinted criteria as history has so far judged Mountbatten's statesmanship, Napoleon's greatest achievement would be regarded as ordering the retreat from Moscow.

'Come what may, self-knowledge will lead to self-rule', wrote Lord Macaulay in his famous minute on Indian education in 1835, 'and that would be the proudest day in British history.' It was, perhaps, the first governmental mission statement. During the fiftieth anniversary celebrations of 'the midnight hour' when India and Pakistan became independent, much was made of the phrase 'The proudest day'. Nonetheless, to consider the day that Britain lost her Indian empire as its most proud is nonsense. To take and hold a vast empire might be a legitimate cause for pride; to return it for want of personnel, resources, morale, money and willpower – let alone over political ideology – should only be grounds for grief. To attempt to rationalise and perhaps palliate the end of imperial glory by declaring it some kind of *ex post facto* victory is a form of denial, the very opposite of Macaulay's 'self-knowledge'. In fact, the proudest day came on Tuesday, 12 December 1911, at the Delhi Durbar, and to believe ceding power is better than exercising it is an infallible sign of national degeneracy.

When the British quitted India, they left behind monuments to their greatness that far surpassed anything that the French or Germans left in Africa, the Portuguese and Dutch left in Asia, or the Spanish left in Latin America. The town-planners of New Delhi, architects of the genius of Sir Edwin Lutyens and the technical proficiency of Sir Herbert Baker, in their own words, wanted 'the main features of the new city [to be] as interesting, after centuries, as the older buildings in the neighbourhood area'.[28] They succeeded triumphantly.

In place of the formal empires of the past, the future would consist of 'informal empire' whereby trading and security arrangements delineated whether a country was in the Western capitalist or Eastern communist sphere of influence. As the German economist Moritz Julius Bonn wrote in 1947, 'The United States have been a cradle of modern anti-imperialism, and at the same time the founding of a mighty empire.'

Clement Attlee wrote to his brother Tom from Chequers in December 1945, reporting on the Foreign Secretary Ernest Bevin's report of his discussions about Palestine in the United States. 'It appears that Zionism is defined as a Jew who collects money from another Jew to send another Jew to Palestine. The collector, I gather, takes a good percentage of his collections.'[29] His contemptuous tone is indicative of the British Government's attitude towards the Middle Eastern Jews, before its withdrawal from the Palestine Mandate created the *de facto* conditions for the founding of the State of Israel. The primary motivation behind the Government's withdrawal was its desire, for pressing financial reasons, to demobilise as many soldiers as it could in as short a time as possible. That month Attlee was amused to receive a typical card that read: 'A very good Christmas this year but a beastly one next year if we are not demobbed.'

The creation of the State of Israel – which was immediately recognised and welcomed by the Truman Administration – must be seen in the context of the imperial mind-set that still pertained between the Balfour Declaration and the end of the Palestine Mandate thirty years later. 'I think Balfour assumed that the Middle East would be run by some outside power, whether it be the Ottoman or Britain,' said the historian Hugh Trevor-Roper.

> In a way, both the Israeli state and the danger to it are products of the end of Western and Ottoman imperialism in the Middle East. Great empires in history have been the best protection of vulnerable minorities. ... When a great empire is defeated in war, then it can turn on its minorities. But great empires in the period of their prosperity on the whole are the best guarantee, because the great enemy of empires is nationalism and nationalism is the antithesis of a cosmopolitan empire.

Although Armenians, who suffered badly, would take exception to that generalisation being extended to the Ottoman Empire, the Jews were not treated badly by it, except in isolated incidents over short periods and usually without direct sanction from Constantinople.

The process of creating a Jewish homeland in an area where other peoples were already living was always going to be a complicated and delicate business, and one for which Britain as the Mandated power had a profound responsibility, and about which since the Balfour Declaration of 1917 she had made solemn promises. Yet instead of keeping a large number of troops on the ground throughout the birth pangs of the State of Israel, Britain hurriedly withdrew all her forces virtually overnight on 14 May 1948, thus facilitating an Egyptian invasion the very next day. In the end, no fewer than five Arab nations – including one whose armies were actually commanded by a former British Army officer, Glubb Pasha – invaded the nascent state from which the British had scuttled away so ignominiously only hours beforehand. Less than four years earlier, Britain had landed division after victorious division in Normandy; now 'Partition and flee' was the Attlee Government's policy, one whose consequences were still plaguing the world half a century later in Kashmir and the Middle East.

Yet despite that imperial scuttle, Britain also failed to cash in what might be seen as her 'peace dividend'. She was still spending ludicrously large amounts on defence, as much as 8% of her Gross National Product by 1950. In order to try to maintain the illusion of still being a Great Power on the scale of the other victors Russia and America, Attlee invested vast amounts in unnecessary status symbols such as a domestic civil aviation industry. Fourteen days after the Germans surrendered in May 1945, they had the Berlin bus system up and running again; that same day the London buses were on strike. The pusillanimity shown by the Attlee, Churchill, Eden, Macmillan and later Governments towards the trade unions until 1979 ensured that grossly restrictive industrial practices were preserved throughout the 1940s and into the long-term future, all to promote a myth of industrial consensus.

Attlee constantly looked back to the problems of the Thirties – primarily unemployment – rather than trying to look forward to those of the Fifties and Sixties, such as falling productivity, widening trade gaps and declining competitiveness relative to Britain's economic rivals. Because 'full employment' had been such a shibboleth for William Beveridge and the other 'New Jerusalem' social reformers, especially Attlee, it was pursued as a goal regardless of the distortions it wreaked on other parts of the economy. Rigidity in the labour market, wage-induced inflation and tardiness in technological adaptation were the entirely predictable results.

To add to the terrible problems that were loaded on to what Professor Correlli Barnett has described as 'a war-impoverished, obsolescent and second-rate industrial economy', Attlee introduced sweeping measures of nationalisation. Coal mines, railways, gas, electricity, civil aviation, road haulage, steel, cable and radio services, as well as the Bank of England, were taken into public ownership, ensuring that the management in these vital industries became almost completely inured to the danger that they might lose their jobs through inefficiency or incompetence.

An inability to discern new markets was the first noticeable effect of nationalisation, but plenty of even worse ones followed. When nationalised industries turned into lame ducks, as almost all of them did over the following decades, they were subsidised by the taxpayer, often through the sale of long-term bonds. The last of these Attlee bonds was finally paid off by Gordon Brown in June 2002; the twenty-first-century British taxpayer had thus been shouldering half a century later the debts blithely taken on by Attlee in his offer of a New Jerusalem.

Of course as soon as the European economies could afford to, they also instituted comprehensive national health schemes, which have turned out to be in almost every case far superior to Britain's National Health Service. By then, however, they had established clear economic superiority. In 1950 under Attlee, Britain was investing only 9% of her GNP in industry and infrastructure, against Germany's 19%. Small wonder that once Germany had surged ahead, she was able to create a better health system that she could afford. By contrast Attlee had, in Barnett's words, built 'a lavish and expensive Welfare State in the aftermath of a ruinous war, on foreign tick, while paying huge defence costs on the back of an un-modernised industrial system'.

On 25 May 1948, MI5 handed Clement Attlee a top-secret report on the activities of Major Wilfred Vernon, who had been elected Labour MP for the Dulwich division of Camberwell in London in the 1945 general election. Attlee minuted that he was 'shocked by its content', which 'came as a complete surprise'. The file contained information that Vernon – an electrical and aeronautical specialist who had been a technical officer in the Air Ministry between 1925 and 1937 – was a Soviet spy. His position as a Member of Parliament gave him immunity from prosecution, and it was only after he lost his seat in the 1951 general election that the security services could even interview him, 'and then only to augment their historical knowledge of Soviet intelligence operations'.[30] (He died, unprosecuted, in 1975.)

Having unmasked Vernon, MI5's next operation was to hide his activities from the FBI, in order to protect the Special Relationship. Rather than prosecute a Soviet spy, therefore, MI5 concentrated on trying to prevent their Americans allies finding out about him. Captain Guy Liddell, the head of MI5 counter-intelligence, wrote two copies of their report, one for 'internal consumption' and one for the FBI, which omitted the vital fact that Vernon was an MP. 'Please do not tell the FBI that the Wilfred Vernon mentioned in this report is the MP for Dulwich,' wrote another senior MI5 officer on the file. The impression that the FBI, and later and in greater measure the CIA, were to gain that the British security services were either incompetent or insufficiently tough on subversion was not – as the disgraceful story of Major Vernon amply displayed – wholly unfair.

Major Vernon was working for a power that was on the march. In mid-June 1948, the Soviet authorities began to refuse entry to freight trains carrying coal to the two million inhabitants of West Berlin from West Germany, on the grounds of 'defective cars'. Back on 20 March 1948, the Soviet representative on the Four-Power Control Council that governed Germany had walked out of a meeting and not returned; now Stalin's grip was about to be significantly tightened. Every second passenger train was sent back owing to 'crowded stations'. Next, the autobahn that went through 110 miles of Soviet territory was closed for 'urgent repairs'.[31] When on 24 June all overland routes were cut and West Berlin's electricity supply was shut off, it was clear that the resolve of the Western Allies was about to be severely tested. The city had thirty-six days' supply of food, after which it was feared that starving West Berliners would be forced to beg for admittance to the Soviet bloc.

General Lucius Clay, the Military Governor of the US zone, wanted to send an armoured column through to West Berlin, believing that the Red Army would back down. Other ideas put forward at the time were to close the Panama Canal to Russian ships and to blockade Vladivostok. What the English-speaking peoples – Britain and the Commonwealth were at one with the United States throughout the crisis – could not do was to abandon the Berliners to the Soviets. 'We stay in Berlin, period,' Truman told his Cabinet, some of whom, such as the Defense Secretary James Forrestal, seemed to be wavering. Sixty B-29s with nuclear bombing capacity were flown to Britain, while the news was judiciously leaked.

The answer to the problem came to Robert Lovett, an Under Secretary of State, on 30 June. During the war, he had been involved in the airlift of 72,000 tons of military equipment from Burma to China, which had to be flown over the Himalayas.[32] (Berlin was a mere 110 miles.) After preliminary discussions with two other veterans of 'the Burma Hump', the distinguished USAAF commanders Generals Curtis LeMay and Albert Wedemeyer, Lovett conceived the idea of feeding the two million inhabitants through a Berlin Hump. President Truman was swiftly converted to the plan before the military could raise detailed objections, and the Western Allies then proceeded to undertake the massive, and massively expensive, task of relieving West Berlin from the air. 'Can feed by air; cannot furnish coal,' recorded Lovett in his diary. If

Stalin had launched his strangulation policy in the winter, rather than mid-summer, things might have gone differently, but in the event vast amounts of coal were also flown in.

At the height of the airlift, planes landed at Berlin's Tempelhof Airport every three minutes forty-three seconds, delivering 4,000 tons of food and other essentials per day. Twenty-thousand West Berliners built a third airport, virtually 'with their bare hands'. There was severe hardship, of course, but ultimately the West proved that Stalin would not starve West Berlin into surrender. The airlift continued until 30 September, as supplies needed to be stockpiled. The last flight was the 276,926th, flown by Captain Perry Immel. In total, the 321 days of the operation had transported 227,655 people in and out of Berlin, and delivered 2,323,067 tons of supplies (mostly food and coal) at a cost of $345 million to America, £17 million to Britain and 150 million Deutschmarks to the Germans. Seventy-five American and British lives were lost in the operation.[33] As a result of the crisis, and the message it sent about Soviet assumptions and intentions, the United States began to build up her nuclear arsenal massively: in 1947 she had only thirteen bombs, in 1948 fifty, but by 1949 no fewer than 250.

With the Soviet Union behaving in the way she had over West Berlin in the summer of 1948, it was hardly surprising that the United States should have been gripped by fear of communism. Between Tuesday, 3 August 1948, and Wednesday, the 25th, the House Un-American Activities Committee (HUAC), meeting in the Ways and Means Committee room, the largest for public hearings on Capitol Hill, heard a sensational story from a former Bolshevik-spy-turned-*Time*-writer-turned-political-defector named David Whittaker Chambers. He claimed that he had for several years helped a Department of Agriculture civil servant named Harold Ware to run a secret communist network, which reached into several government departments and New Deal agencies. Chambers named Nathan Witt, John Abt, Lee Pressman, Victor Perlo, Alger Hiss and his brother Donald, Charles Kramer and Henry Collins as members of the Ware cell.[34] The group had been connected to Soviet Intelligence through J. Peters (whose real name was Jozsef Peter), an official of the Communist Party of the USA (CPUSA), for whom Chambers claimed to have worked for ten years before breaking with Moscow in 1937.

The communist conspiracy that Chambers claimed to have worked for stretched into the Agricultural Adjustment Administration (a key New Deal agency that promoted government interventionism), the National Recovery Agency, the Office of Price Administration, the Treasury, the National Labor Relations Board, the Office of Strategic Services, the Farm Security Administration and crucially – through Hiss and another employee Julien Wadleigh –

the State Department. Chambers claimed that for four years Hiss, Wadleigh and the Assistant Secretary of the Treasury, Harry Dexter White, had stolen secret government documents and passed them to him and another Soviet agent, John Herrmann.[35]

Fearing assassination, Chambers had got very little sleep the previous night at his 'safe house', with the result that he hardly looked at his best before the media the next morning. 'Chambers took his place at the witness table amid a starburst of flashbulbs and the blaze of klieg lights,' records a recent biographer. 'He had had at most three hours' sleep. The next morning's photographs showed a man who looked newly emerged from the sinister depths of the underground, his suit wrinkled, his expression haunted, his eyes averted from the camera as if in guilty flight.'

By total contrast with Chambers, Alger Hiss was handsome, charming, intelligent, well-connected and talented, and his career had been meteoric. Educated at Johns Hopkins and Harvard Universities, after success in a series of high-profile jobs he had joined the State Department in September 1936, becoming Director of the Office of Political Affairs. He attended the Yalta Conference, was privy to large numbers of important State Department papers and went on to serve for a short time as the first Secretary-General of the United Nations in April 1945. 'At 41, slim with chiselled features,' records the historian Arthur Herman, 'Hiss stood at the brink of becoming America's premier diplomat.'[36] (Had Roosevelt died six months before he did, and his Vice-President Henry Wallace had succeeded him, Hiss might well have become Under-Secretary of State, along with the NKVD agent Laurence Duggan as Secretary of State and Harry Dexter White as Secretary of the Treasury.)

In 1947, Hiss had left the State Department to preside over the hugely prestigious Carnegie Endowment for International Peace, and when Chambers went public with his accusations Hiss dismissed them as the politically motivated untruths of 'a self-confessed liar, spy and traitor'. Harry Dexter White similarly defended himself before HUAC, dying of a heart attack after the session, which created even more sympathy for Hiss. Many leading lights of the American liberal establishment wrote depositions testifying to Hiss's upstanding character, including Felix Frankfurter, Adlai Stevenson, Eleanor Roosevelt, both the majority and minority Senate leaders, the President of Johns Hopkins University Isaiah Bowman, and Columbia University professor Philip Jessup. Even Dwight D. Eisenhower, Dean Acheson and John Foster Dulles offered their support.

Meanwhile, Chambers was subjected to the full weight of the liberal media's ire, with innuendoes being made that he was unbalanced and of course 'a self-confessed liar, spy and traitor'. Other allegations – that he was homosexual and driven by envy of Hiss – probably had a basis in fact, but were hardly germane. The HUAC investigators had one factor on their side denied to the great and the good of the liberal political and media Establishment of the day: the truth. Hiss and the others were indeed traitors who had been spying for the Soviet Union. Hiss might nonetheless have escaped detection had it not been for the terrier-like single-mindedness of the first-term California Republican congressman Richard M. Nixon, the son of a Whittier grocer.

Chambers had hidden undeveloped film of over sixty documents on his Maryland farm, which, because he had concealed them in a hollowed-out pumpkin, became known to history as 'the Pumpkin Papers'. These were secret State and Navy Department papers that Hiss or his wife had made typed copies of in 1938, using a converted Woodstock 230099 typewriter, and given to Chambers to pass on to the USSR. It was largely due to the evidence contained in these that, after two sensational trials in which twenty of the twenty-four jurors believed Chambers, Hiss was convicted of perjury in 1950 and imprisoned for forty-four months. His defence, that he had not owned the typewriter in 1938, was undermined by the man to whom he had given it and as the trial went on it was clear that Chambers was telling the truth. Hiss's claim never to have met Chambers led to some sensational moments during his cross-examination by Nixon. The only reason that Hiss was not convicted for treason as well as perjury was because the United States had a statute of limitations for treachery that had expired, a truly eccentric concept for a great nation.

Even after Hiss's conviction, many on the Left in America continued to believe, as an absolute article of faith, that there were powerfully extenuating circumstances to explain Hiss's actions, which they persisted for years in not seeing as cardinal sins. Because Hiss continued to pronounce his innocence, he was also given the benefit of the doubt by people who were willing to suspend their critical faculties regardless of the evidence. As Hilton Kramer, editor of the *New Criterion*, has written of the seeming paradox:

> Hiss, though sent to jail for criminal acts that publicly identified him as a Soviet espionage agent, was firmly established as a political martyr. By the same token, Chambers, who had risked his hard-won career and indeed his life by exposing a Soviet spy network at the very heart of the Washington bureaucracy, stood condemned as a turncoat, and a villain. In the court of liberal opinion, informing on a fellow conspirator was deemed to be a far greater crime than belonging to a clandestine Communist spy apparatus and stealing government documents for a foreign power. Thus, in the crazy logic of the case, Hiss – convicted of crimes that showed him to be a liar, thief and a traitor – was judged innocent even if guilty, and Chambers – the self-confessed renegade who recanted on his treachery – was judged to be guilty even if he was telling the truth.[37]

Chambers was indeed brave to testify. His friend, the Russian former communist Walter Krivitsky, was due to testify before HUAC when his corpse was found by a maid at the Bellevue Hotel, Washington, on 10 February 1941 with a bullet in his temple and a .38 calibre pistol in his hand. Krivitsky's three suicide notes were suspected to be forgeries, his blood had washed the weapon clean of fingerprints, his wife testified that he had never owned a pistol, he was neither depressed nor in financial straits, and he had told HUAC that he feared he might be murdered.[38]

It was not until after the fall of the Berlin Wall and the subsequent opening up of former communist archives that it was revealed beyond even the ultra-liberals' doubt that Hiss had indeed been working for Stalin. In 1995 and 1996, the 'Venona' traffic of more than 2,000 cables between US-based Soviet agents and Moscow in the 1940s, but intercepted by American counter-intelligence and decoded, were released by the National Security Agency, and they directly implicated Harry Dexter White, Victor Perlo, Laurence Duggan and Alger Hiss.[39]

Unfortunately, the well-founded fear of treachery and communism felt in the United States was exploited one stage too far by the Right. In June 1951, Senator Joseph McCarthy of Wisconsin painted a picture for the American people in which the liberal Establishment, which he always considered was willing to appease communism, would so weaken the United States that 'this nation, this Civilisation, will pass from the face of the earth as surely as did those other great empires of the past which were destroyed because of weak leadership which tolerated corruption, disloyalty and dishonesty'. It was hyperbolic stuff, and as McCarthy and, quite separately, HUAC extended their investigations into every area of American life, they badly overreached themselves. True, there were indeed communist-sympathisers in the State Department who needed investigating and unmasking, but once the US Army and even Hollywood were subjected to accusations of un-American activities, it was clear that McCarthyism was played out. On 9 December 1954, McCarthy was censured by a vote of 66-22 in the Senate and his power was broken.

For all that one might have felt sympathy for those who lost their jobs or felt they had to leave America in order to preserve their artistic freedom during the McCarthyite period, it is worthwhile considering what kind of America would have resulted in the political victory of the CPUSA. Arthur Miller likened McCarthyism to the seventeenth-century Salem witch-trials, but this ignores the crucial difference that there were no such thing as witches, yet there were some Soviet agents at the highest levels of American administrative life. There was an unpleasant class-based undertow to McCarthy's accusations, an attempt to condemn the East Coast Establishment as unpatriotic. Nonetheless, in a public opinion poll taken in the summer of 1954 fewer than

1% of respondents said that they were worried about the erosion of civil liberties in the USA.[40]

The fact that the FBI director J. Edgar Hoover and several other US government agencies expended vast amounts of FBI manpower, money and resources spying on illustrious German anti-fascist émigré writers such as Bertolt Brecht, Anna Seghers, Heinrich Mann and Leon Feuchtwanger has brought the contempt of generations of historians upon Hoover, not least for his use of the terms 'CommuNazis' and 'Red fascists'. It is true that his FBI was largely incompetent in its fight against the Mafia, and that it missed a number of Soviet espionage efforts, but the FBI also amassed vast amounts of material in its investigations of Lachlan Currie, Laurence Duggan, Harry Dexter White, Alger Hiss and the Rosenbergs, all of whom did indeed turn out to be NKVD agents, although none were unmasked by McCarthy, who did not in fact discover a single communist.[41]

In the so-called McCarthyite 'terror', no-one was sent to any gulags or forced to till the permafrosted soil of Alaska, and there were no deportations, tortures, internments or attempts to revoke the US citizenship of people who had moved to America, even of the pro-Stalinist Brecht who was prominent in the Free Germany movement that Hoover, correctly, believed 'has as its aim the establishment of a post-war German government favourable to Soviet Russia'.[42] People who were forbidden to practise their professions and threatened with gaol for ideological crimes did have their careers stymied, it was true, and many went abroad. However, this was usually as a result of private actions taken, for example, by the Hollywood studio cartel, rather than by government action. The McCarthyite era was an unpleasant, disgraceful one in American history, but one that needs to be seen in perspective. Those who claim a moral equivalence between the United States under McCarthyism and what was happening behind the Iron Curtain can only do so by entirely ignoring the contemporaneous and truly murderous assault on human rights conducted by Lavrenti Beria and the KGB.

The propensity of ordinary Americans to stand by their commander-in-chief in difficult moments internationally had been demonstrated in November 1948, when President Truman beat his rival Thomas E. Dewey in the presidential race, making a mockery of Dr Gallup's polls forecasting a Republican victory. In the congressional elections, the Democrats reclaimed majorities in the Senate by 54 to 42 and the House by 262 to 171. In the dangerous shoals of his presidencies – the dropping of the atomic bombs, Bretton Woods, the Marshall Plan, the Truman Doctrine, the Berlin airlift, the Korean War, the sacking of Douglas MacArthur – each found a skilful navigator in Harry Truman. Far from the 'everyday American' that he presented himself as, he was a gifted leader of the English-speaking peoples at an anxious and potentially perilous period in their history.

Perhaps Truman's longest-lasting legacy – aside from a democratic, pacific Japan – was the North Atlantic Treaty Organisation (NATO). On Monday, 4 April 1949, the representatives of the twelve founder nations signed the North Atlantic Treaty at a long mahogany table in the auditorium of the State Department building on Constitution Avenue in Washington DC. The US Marine band played Gershwin tunes, including *Bess, You Is My Woman Now*, in honour of the First Lady, Mrs Truman, who sat in the front row of spectators.

NATO came into being as a response to the vast Soviet military presence in Eastern Europe and the economic blockade of Berlin that had been in operation since June 1948. Of its fourteen articles, the most important was number five, which stated that 'an attack against any signatory in Europe or North America shall be considered an attack against them all'. As well as the English-speaking countries, America, Great Britain and Canada, the other signatories were the Benelux countries, Denmark, Italy, Norway, Iceland, Portugal and France (although France effectively withdrew in 1966).

It is not too much to claim, as the economist and political thinker Sir Rodney Leach has recently, that 'NATO established the USA for fifty years after the Second World War as a European power', and that as a result, 'The American presence saved Europe from succumbing, whether politically or militarily, to the Soviet Union.'[43] It was a noble achievement and honour should go to its founders, men like George C. Marshall and Ernest Bevin. After Greece and Turkey joined in 1952 and West Germany in 1955, NATO became the most successful military alliance in history, deterring the USSR from directly attacking any of its member states for the half-century it took for Soviet communism to collapse and die. Afterwards, in 1999, it took Poland, Hungary and the Czech Republic into its fold.

Eire – true to form – declared on 8 February 1949 that she was unable to participate in NATO while the island remained divided, thereby once again failing to take her place in the ranks of Civilisation against the undoubted threat posed by totalitarianism. Two months later, Eire withdrew from the British Commonwealth. If she had hoped to make an impact with such grandstanding, she largely failed; as one account of the Commonwealth economies put it in 1953, 'The harmonious relationship and volume of trade between member countries of the Commonwealth and the Irish Republic – and the ease with which this trade is conducted – are such that her precise status since April 1949 is apt to be overlooked.'[44]

Of course as ever there were accusations made against the USA for being unilateralist and aggressive, as there have always been against whichever is the greatest world power at any time. In a 1949 conference at the Waldorf-Astoria attended by such prominent literary and artistic figures as the composers

Aaron Copland and Dmitri Shostakovich, the playwrights Arthur Miller, Clifford Odets and Lillian Hellman, and the novelist Norman Mailer, 'US warmongering' against the USSR was condemned, as were 'a small clique of hate-mongers' in Washington who had turned the US into 'a state of holy terror'.[45] Odets said that the Truman Administration comprised the 'enemies of Man' and Copland said its policies 'will lead inevitably into a third world war'. All this came only months after the Berlin airlift. Strident expressions of anti-Americanism are therefore hardly new, are directed against both Democratic and Republican presidents, and are often most virulent when coming from Americans themselves.

TWELVE

Cold War Perils

The 1950s

The Korean War — Human Rights — Encounter — 'The New Elizabethans':
Edmund Hillary and Michael Ventris — Operation Ivy — Conquering polio — The
Warsaw Pact — The Suez Crisis — The Common Market's impact on the British
Commonwealth — America's civil rights struggle in perspective — The Nixon-
Khrushchev 'Kitchen Debate'

'You have to go back to George Washington to find another American who
was first in war, first in peace, and first in the hearts of his countrymen.'
Richard M. Nixon's eulogy on Dwight D. Eisenhower, March 1969[1]

'England is the only great country whose intellectuals are ashamed of their
own nationality.'
George Orwell, *The Lion and the Unicorn*

On 12 March 1947, the President announced a plan, which subsequently
became known as the Truman Doctrine, to give aid to Greece, then still
suffering from the communist insurrection, and to Turkey, which also felt
herself to be under severe pressure from the USSR. The Doctrine was held
to state that democratic states under threat from communism could look to the
United States for support and sustenance, thus pre-dating John F. Kennedy's
promise to 'pay any price, bear any burden' by sixteen years. It was under the
Truman Doctrine that Berlin was protected by the airlift.

At dawn on Sunday, 25 June 1950, several armies of communist North
Korea invaded South Korea from all along the border, the 38th Parallel,
marching southwards as fast as possible. Two days later, true to his Doctrine,
President Truman ordered US forces in Asia to resist this unwarranted
aggression, while sending reinforcements there to help the beleaguered South
Koreans. Nonetheless, the following day the South Korean capital, Seoul, fell
to the invaders.

That month, the temporary boycott of the United Nations by the USSR
meant that the Security Council could pass a resolution condemning the

North Korean invasion of her southern neighbour and promising 'to furnish such assistance as may be necessary to meet the armed attack'. The entire war was thus fought under UN auspices, something which helped Clement Attlee and the Labour Defence Minister Emmanuel Shinwell to sell it to the internationalists in their Party. But what was widely celebrated as a cardinal diplomatic blunder by Stalin might also have set a dangerous precedent for the West. It helped establish the suspicion, which is today an established canon of liberal internationalist thought, that military action without the consent of the UN is somehow illegitimate, and to be demonstrated against. Under this precept, the English-speaking peoples seemed to have lost their right to go to war for their own interests, regardless of the views of the rest of the Security Council of the United Nations.

By 1 July 1950, the first UN forces landed at Pusan in the South Korean peninsula, and a week later General Douglas MacArthur was appointed to command them. This in itself sent a powerful message that the English-speaking peoples were deadly serious about preventing the peninsula from falling to the communists, and that they would once again stand up for the rights of little nations threatened or invaded by their neighbours. Cape Colony in 1899, Belgium in 1914 and Poland in 1939 had already benefited from this readiness of the English-speaking peoples, as later were South Vietnam in 1964, the Falklands and Belize in the 1980s and Kuwait in 1990.

Born in Little Rock, Arkansas, in 1880 and educated at West Point, Douglas MacArthur had been commissioned into the Corps of Engineers in 1903. He had been familiar with the Far East ever since accompanying his father, General Arthur MacArthur, who had been the chief US observer in the Russo-Japanese War, to Tokyo in 1905. (He and his father are the only father and son ever to have won the Congressional Medal of Honor.) During the Great War, Douglas MacArthur was decorated thirteen times and cited a further seven times for bravery, becoming, by November 1918, the youngest divisional commander in France. After becoming the youngest-ever superintendent of West Point the following year, by 1930 he was a general and Chief of Staff of the US Army.

MacArthur's connection with the Far East resumed five years later when he became head of the US military mission to the Philippines and in 1941 the Commander-in-Chief of US forces in the whole theatre of operations. As Supreme Commander in the south-west Pacific, based in Australia, MacArthur developed the 'leap-frogging' strategy by which the Philippine archipelago was recaptured from the Japanese, to be finally liberated in July 1945. It was he who took the surrender of Japan in the name of the Allied powers and then exercised almost unlimited plenipotentiary powers there, giving Japan her new Constitution and reforming the country according to the best practices of the English-speaking peoples. (He erred, however, in his failure to punish Japanese war criminals with anything like the severity they deserved.)

When this man was appointed to command the United Nations' force in Korea, therefore, it was obvious that the prestige of the United States was intimately bound up in the cause. In all, the UN force was made up of seventeen nations and seven non-combatant allies, but it was the English-speaking peoples who spearheaded the outside effort to protect South Korea from what, as the subsequent history of its northern neighbour has proved, would have been a truly horrific national fate.

On 15 September, MacArthur's forces made a surprise amphibious landing at Inchon, west of Seoul, which forced the North Koreans to retreat. Seoul was liberated on 26 September, and five days later UN and South Korean forces crossed the 38th Parallel into North Korea, capturing the capital, Pyongyang, on 20 October. Although on 24 November the UN forces launched an offensive into north-east Korea, the whole situation altered radically two days later when Red China entered the war, forcing the UN to retreat southwards.

A myth has developed that during this period President Truman considered delegating to MacArthur the decision as to whether to employ the atomic bomb against the Chinese invaders, and that it was only Clement Attlee's emergency flight to Washington that dissuaded him. This fabrication was repeated as recently as October 2001 by the British left-wing politician Tony Benn, who argued that it should be used as a precedent for Tony Blair to exercise restraint on President Bush in the aftermath of 9/11.[1] In fact, the Chiefs of Staff had already rejected the idea of using nuclear weapons in October 1950 and were not even going to suggest it to the President. American troops were being hard pressed by the Chinese at the time, and so the President did not specifically rule out the possibility of using the bomb at his press conference on 30 November 1950.

At 4.19 p.m. that day news came through to the British Foreign Office that Truman had answered a reporter's question about the bomb with the words, 'There has always been active consideration of its use', adding that he hoped it would not be necessary, but declining to state that it would not be used against cities. (Of course to have done so on moral grounds would also have called into question the ethics of his own decision to bomb Hiroshima and Nagasaki.) Truman also said that commanders in the field would be in charge of the weapons. According to his biographer, Robert J. Donovan, Truman answered the questions in a way that allowed frightened people fed by excited news stories to believe that the atomic bomb might be used in Korea and at MacArthur's discretion.[2] Yet Truman's circumlocution makes perfectly good strategic sense; why let the enemy know which weapons he was or was not willing to use? (In fact, under the

1946 Atomic Energy Act, only the president could authorise the use of nuclear weapons.)

Attlee's private secretary pointed out that since the questions were unlikely to have been deliberately planted by the Administration, 'The whole tone of the reply reads as though he were more concerned to placate the Republicans, and to reduce the mounting pressure on him to get rid of Acheson than the actual situation in Korea.[3] (Dean Acheson, who had taken over from Marshall as Secretary of State in January 1949, was considered too soft on communism by several American senators.) This explanation seemed to be reinforced by a telegram later that day from the British Ambassador in Washington Sir Oliver Franks, who had been contacted by the White House and who reported their reassurances: 'It should be emphasised that by law only the president can authorise the use of the atomic bomb. ... In brief the replies to the questions at today's press conference do not represent any change in the situation.'[4] Donovan believed that given the intense competition between the three wire services then prevailing at the White House, exaggeration and speculation were intensified without justification.

Nonetheless, the Labour backbenchers were thoroughly rattled by the news reports, and scores of them signed a secret (but soon-leaked) round-robin letter to Attlee calling for the bomb only to be employed under the auspices of the United Nations, in whose name the war was – at least nominally – being fought. Once the letter was leaked to the press by MPs, Attlee decided to visit Washington himself, where he quickly established for himself that Truman was in complete control of the higher direction of the war. He found that he did not even need to bring up the issue of the atomic bomb with the President, which was only mentioned when the conference was over and the communiqué was being drafted. Yet somehow Attlee has been given the credit in some quarters for restraining a gung-ho American warmonger itching to unleash devastation on innocent civilians.

On 27 December 1950, China refused a United Nations appeal for a cease-fire; on New Year's Day 1951, Chinese forces broke through the UN lines at the 38th Parallel and three days later the communists retook Seoul. The United States, Britain, Australia, New Zealand and Canada – but not Eire – were now effectively fighting a proxy war against the vast People's Liberation Army, thousands of miles from home. The United States bore by far the greatest brunt of the fighting – barring the South Koreans themselves – although between 22 and 25 April 1951 the British 29th Brigade, including the 1st battalion of the Gloucestershire Regiment ('the Glorious Glosters'), and the 3rd Battalion of the Royal Australian Regiment, helped to halt a huge communist offensive along the Imjin River in some of the bitterest fighting of the war. In retrospect, it was astonishing that the British death toll in Korea was kept to 1,078 and Australia's to 340. The

United States' was far higher, with 54,246 killed – 36,574 of them in combat – and 103,284 wounded.

Allied prisoners of war faced appalling treatment at the hands of the Chinese and North Koreans, who only signed the Geneva Convention in 1956 and 1957 respectively, after the war was over. Their treatment recalled what POWs had suffered at the hands of the Japanese between 1941 and 1945. 'All the old familiar torments – such as malnutrition, inadequate accommodation, beatings up, and recurring attacks of *beri-beri* were present,' records an historian of the POWs' tribulations, 'plus a newcomer to the list, brainwashing.'[5] Incorrect answers to questions such as 'Give Lenin's five contradictions undermining capitalism' were typically punished by twenty-one days in solitary confinement. Derek Kinne's 1955 book *The Wooden Boxes* detailed how 5' x 3' x 2' boxes were used to incarcerate POWs in the searing heat at the whim of sadistic guards of Camp 1. At the opposite temperature extreme, POWs were marched barefoot to the Yalu River in a minus 20 degree frost, then made to stand on ice while buckets of water were poured over their feet, in order to encourage them to 'reflect on their crimes'.[6] A wounded soldier captured on 24 August 1951 received no medical treatment until 4 June 1953. A number of the Glosters were beaten with a club between 9 a.m. and 3 p.m. one day, forced to stand to attention throughout. Between January and August 1951, an estimated 1,600 UN troops died in captivity.

In all, the communists captured 80,000 South Koreans, 10,000 Americans and 2,500 other UN troops during the war. Of the 150,000 North Koreans and 20,000 Chinese captured by the UN, 23,000 requested not to be sent back and were found homes in South Korea and Taiwan. By contrast only twenty-one Americans and one Royal Marine Commando refused repatriation at the end of the war, all of whom might be said to be peculiarly vulnerable to psychological techniques; fifteen were under twenty-one years old, sixteen were loners, ten had lost their mothers in early childhood, nineteen had problems with their fathers or step-fathers and only two were married.[7]

On 4 March 1952, the Chinese Government accused United States forces of using germ warfare, lies that were disgracefully supported by visitors to Korea such as the Australian journalist Wilfred Burchett and the British biochemist and Cambridge don, Joseph Needham, a Sinologist and linguist who was willing to do Beijing's bidding. Needham headed an 'International Scientific Commission', which conducted no field investigations of its own, but published a 669-page report completely supporting the Chinese claims on the basis of witnesses. They turned out to be completely untrue, yet on being honoured in Beijing on his ninetieth birthday in 1990, Needham – who in the meantime had served as an elected Master of a Cambridge college – repeated the charges.

It was said that US aircraft had dropped flies, fleas and spiders infected with anthrax, cholera, encephalitis, plague and meningitis in 'germ bombs', which later turned out to be typical American propaganda pamphlet carriers. In January 1998, documents in the Russian presidential archive proved that the charges had been entirely fraudulent, and that the North Koreans and Chinese had invented them as a way of blaming America for outbreaks of infectious diseases in their countries.[8] (By then, however, Needham and Burchett were dead.)

Such propaganda was by no means all. Two virulently anti-American journalists, Alan Winnington and Michael Shapiro, who wrote for the British *Daily Worker*, contrasted the supposed comfort in which POWs were held by the communists with 'the misery experienced ... by Communist prisoners held by the UN'. Shapiro even threatened American POWs with instant execution, something the foul Burchett was also reported to have done.

The truth was foreign to these communist fellow-travellers, such as the London solicitor Jack Gaster and Monica Felton, whose ceaseless work for communist front organisations won her the Stalin Peace Prize but probably lost her a job with the Stevenage Development Corporation (in Britain's worst act of McCarthyite terror). This was further indication since the Thirties, and more would be coming during the Vietnam War and later, that there is no cause so rank that it will not find propagandists and apologists to support it from among the ranks of the English-speaking peoples, usually on the anti-American Left.

In March 1953, days after Stalin's death, Powys 'Tojo' Greenwood bet the historian Sir John Wheeler-Bennett £5 at Brooks's 'that a shooting war will exist between America and Russia before the end of July'. The same bet was offered to all members and enthusiastically picked up by five of them for £100. Instead, however, international tensions eased, and on 27 July 1953 delegates from the United Nations, North Korea and China signed an armistice at Panmunjom, after an estimated three million people had perished, and a two-and-a-half-mile-wide demilitarised zone across Korea was accepted by both sides, which has remained in place ever since.

Much more representative of his country than Wilfred Burchett was Australia's contribution of two infantry battalions, an air-fighter squadron, an air-transport squadron, an aircraft carrier, two destroyers and a frigate; similarly New Zealand contributed an artillery group and two frigates, and Canada contributed an infantry brigade, an artillery group and an armoured battalion. Each country therefore continued their tradition of active participation in the armed struggles of the English-speaking peoples since 1900, however far from home these might be.

In January 1953, Canada joined with Britain and the US in the Radioactive Resources Agreement on the sharing of uranium ore produced in Australia, and the following year she announced joint plans with the US to construct a Distant Early Warning system radar line across Arctic Canada. This and other ventures led to the establishment in 1958 of the North American Air Defense Command, whose headquarters were in Colorado Springs. By 1963, US nuclear missiles were installed on Canadian soil under a joint control agreement. This role continued until the Trudeau Government ended it in 1972.

On 8 March 1951, having won one election but with another soon to loom since there was only a four-seat majority, the Attlee Government hurriedly ratified the European Convention on Human Rights. Britain was the first country to sign up, ten months before Norway and twenty-three years before France. It was ratified in the teeth of opposition from British lawyers, especially the Lord Chancellor Lord Jowitt, who, in the words of a recent study, believed it deeply alien to the British legal tradition.[9] Understandably, considering the Convention's wide-ranging stipulations,

> Most [British] lawyers hated it, because they didn't like foreigners interfering, because they thought human rights were perfectly secure in Britain anyway (they had, after all, been invented there), because they felt that such rights shouldn't be defined (British judges knew what they were instinctively), and because they believed that if they were defined they would be exploited by 'Communists, crooks and cranks'.[10]

What was more, Britain still had an empire in 1951, and since anti-imperialists took it for granted that empire was merely an abuse of human rights per se, it seemed to be a recipe for endless court cases to be brought against Britain. Sure enough, one was launched in 1956 by Greece over Cyprus.

Britain was rightly proud of the way that individual liberty was protected in the colonies, as were minority rights. In 1946, the Foreign Office listed personal freedom in her colonies as one of Britain's 'main contributions to Civilization', and, except during States of Emergency, it was protected. The main reason why the Foreign Office rode rough-shod over the Colonial Office and other doubters in its desire to sign up was because of the way it would allow Britain to draw attention to the gross violation of human rights then taking place in the Soviet Union, a perfectly commendable motive. It was also hoped to de-fang the anti-colonialist movement, but this was frankly naïve. However, when colonies did gain their independence from Britain, the Convention was often incorporated into the constitutions of the successor states, although they often didn't last long. Today, the Convention on Human Rights is incorporated into British domestic law, with exactly the results warned of in 1951, with 'Communists, crooks and cranks' suing the Government claiming that their human rights have been breached. (In 2005, the

Prison Service was sued by a convict because he was not allowed pornography in his cell, which he claimed violated his human rights.)

President Truman decided not to stand for a third presidency in November 1952, but instead to retire to Independence, Missouri. Dwight D. Eisenhower accepted the Republican nomination for president and won a landslide victory over Adlai Stevenson in the November 1952 elections. The Republicans regained control of the House by 221 seats to 211 and the Senate by 48 to 47.

In October 1953, a new monthly magazine entitled *Encounter* was published in Britain and America, with the words 'Literature, Arts, Politics' emblazoned below its masthead. Virginia Woolf (posthumously), Albert Camus, Christopher Isherwood, J.K. Galbraith, Edith Sitwell, Irving Kristol and Cecil Day Lewis all wrote for the first issue, and over the next six months they were joined by W.D. Yeats, Arthur Koestler, W.H. Auden, Rose Macaulay, Robert Lowell, Aldous Huxley, Vita Sackville-West, Bertrand Russell, Leonard Woolf, Kenneth Tynan, Dylan Thomas, Raymond Aron, C.V. Wedgwood and Lucian Freud, a virtual *Who's Who* of Fifties English-speaking high culture. Very few of those literary luminaries would have contributed, however, if they had known that the publisher, the Congress for Cultural Freedom, was a CIA-front organisation.

Certainly, *Encounter* made no secret of its profound anti-communism. The opening editorial on the first page of its first issue frankly celebrated Stalin's death and looked forward to 'the destruction of the Marxist-Leninist creed', saying, 'Now, perhaps, words will once again mean what they say, and we shall be spared the odious sophistry by which despotism could pose as a higher form of freedom, murder as a supreme humanism.' One of the political articles, by the young literary critic Leslie A. Fiedler, was entitled 'A Postscript to the Rosenberg Case', and analysed the way the Left had managed to turn the atomic spies Julius and Ethel Rosenberg – who had been executed at Sing Sing for treason in June – into posthumous heroes. 'Rabid animals', was Jean-Paul Sartre's phrase for Americans after the Rosenbergs were electrocuted, demanding that France 'break all ties that bind us to America'. (Few today seriously dispute that the Rosenbergs were guilty.)

A superb intellectual journal had been created, which bears re-reading half a century later, even though the contributors were kept in the dark, by Isaiah Berlin amongst others, about where its ultimate funding came from. The staff, such as the literary editor Stephen Spender and Frank Kermode, regularly inquired, but they believed the editor Melvin Lasky when he untruthfully told them that it was not the CIA.[11]

Overtly political articles were relatively rare in *Encounter*, but a CIA operative in the Congress for Cultural Freedom had the ultimate power to veto contributions which strayed too far leftwards.[12] Spender resigned when the truth was discovered in 1967, despite Berlin's attempts to intervene on behalf of the CIA. With many intellectuals worried about their reputations, Berlin instead tried to minimise the damage. This was only revealed after his death in 1997, and his posthumous reputation suffered on the Left as a result. 'Liberty is liberty,' wrote Berlin in 1958 in *Two Concepts of Liberty*, 'not equality or fairness or justice or human happiness or a quiet conscience.' Berlin was one of the unacknowledged architects of a series of Cold War initiatives by the Americans – of which *Encounter* was one – 'designed to quarantine Western intellectuals against Communism'. Fortunately, he was not alone.

When in July 1996 it was revealed that George Orwell had supplied the covert propaganda unit of the British Foreign Office, the Information Research Department, with a list of communist-supporting journalists and writers, the Left was also outraged. The Labour politician Gerald Kaufman professed himself 'sickened' that the 'untrustworthy' author of *1984* 'was a Big Brother too'. Yet the moral dilemmas which intellectuals such as Berlin and Orwell faced in the Cold War period must be seen in their proper historical context. Berlin, who had written a biography of Karl Marx in 1939, who had family in Russia and who had worked in the British Embassy in Moscow between September 1945 and January 1946, fully understood the threat that the Soviets posed to world freedom. Along with George Kennan, Berlin perceived their true intentions accurately and early. As a fellow of All Souls College, Oxford, since 1932, he also had the intellectual self-confidence to discard all the wishful thinking then prevalent about the USSR, even before Churchill's 'Iron Curtain' speech was delivered at Fulton, Missouri, in March 1946.

Intellectual self-confidence is one thing; moral courage quite another. Rather than confining himself to philosophising about liberty from an ivory tower, something which as a future Chichele Professor of Social and Political Theory at Oxford University he could easily have done, Berlin committed himself to the murky world of the Cold War cultural struggle, even though he knew perfectly well that his side was in part being financed by the CIA. That he had deep concerns and worries about what he was doing should only heighten the regard in which we should hold him, because he had the courage to put the struggle against communism above any career scruples of his own. Here was a man who wrote and lectured about Liberty and the importance of defending it, but who furthermore practised precisely what he preached.

'Few new truths have ever won their way against the resistance of established ideas,' Berlin wrote in *Vico and Herder* in 1976, 'save by being overstated.' In the Manichean, life-or-death struggle between totalitarianism and democracy during the Second World War, black propaganda and psychological warfare were accepted weapons of engagement on both sides. It is naïve to expect that when much the same struggle had to be continued afterwards, against a different totalitarian enemy, all such deception could, would or should cease. If in the process a world-class literary, artistic, intellectual and critical magazine was published that would otherwise not have been, then that was a welcome side-effect. The ends of the cultural and propaganda war against a brutal, genocidal regime were more than fully justified by the means of not letting some talented intellectuals know who was ultimately financing their articles and poems.

Sophisticated and expensive Soviet disinformation campaigns were being carried out in the Western media throughout the Cold War; it would have been gross negligence not to defend against them and occasionally to counter-attack. By promoting what was called the Non-Communist Left, men like Berlin and Kennan were providing a crucial philosophy for democratic resistance to the evil empire which the Soviet Union had already been for thirty-six years. If a certain degree of misinformation was required for that to be successful, it was perfectly morally justified, considering the threat posed by communism in the twentieth century. Set against that level of human calamity, Isaiah Berlin's little deceptions were as dust in the balance.

Put at its most brutal, as one cultural commentator has stated, 'In the century before the Iron Curtain came down, America had managed to produce no one of the calibre of Dostoevsky, Chekhov, Tchaikovsky, Mussorgsky, Stravinsky or Diaghilev.'[13] As the American thinker Charles Murray was similarly forced to acknowledge about his countrymen in his book *Human Accomplishment*, 'Much as we may love Twain, Whitman, Whistler and Copland' – and he might have added Robert Frost – 'they are easily lost in the European *oeuvre*.'[14] Yet it was America, not Russia, that won the important cultural battles of the Cold War, through the simple use of artistic freedom, something that could not be permitted in the Soviet Union for ideological reasons.

McCarthy's own crude attempt to restrict artistic expression fortunately only blighted the lives of those relatively few people it affected for half a decade, whereas in the USSR all forms of artistic expression were subjected to the state for nearly three-quarters of a century. This did not prevent a recent CNN series on the Cold War likening the sufferings of the Hollywood communists under McCarthy to 'torture by the Inquisition'.[15]

As a result of artistic censorship, several of Russia's finest dancers, instrumentalists, writers and cinematographers left for the West, and nothing could be a worse advertisement for Soviet 'freedom' than that. Rudolf Nureyev gave his bodyguards the slip at Le Bourget airport, Paris, in June 1961; Alexander Solzhenitsyn was eventually expelled from the USSR in 1974 after the publication of *The Gulag Archipelago* in France the previous year. To make matters worse, both chose to live for at least part of the time in the hated

United States. 'To Soviet politicians culture was not an end in itself; it was a tool. In using it so crudely they not only blunted it, but effectively wasted it altogether.'[16]

On Wednesday, 6 February 1952, King George VI died in his sleep at Sandringham, and Princess Elizabeth – who was in Kenya at the time – became Queen Elizabeth II. 'George VI's reign will go down in history', wrote the characteristically waspish English novelist Evelyn Waugh to a friend, 'as the most disastrous my country has known since Matilda and Stephen.' As the sixteen-year period had spanned Hitler's rearmament programme, the annexation of Austria, the Munich crisis, the evacuation from Dunkirk, the defeat in Crete, the fall of Singapore, the Austerity programme, the transfer of power in India, the devaluation of sterling, the outbreak of the Cold War and the Korean War, and the beginning of the end of Britain's great power status, Waugh certainly had a point, although of course the blame cannot be laid at the door of the King.

For many people, Queen Elizabeth's Coronation was the first time they had watched television, and they admired the medium almost as much as the event. Although there were only 344,000 households with television licences in 1950, by 1960 this had risen to 10.5 million. There was plenty else that was booming in Britain in the Fifties: real domestic product per head rose by 19% between 1951 and 1957; unemployment never rose above 468,000, and in July 1955 was as low as 185,000; the number of cars produced in those seven years rose by 65% and steel production by 33% as the index of industrial production rose by 33%.[17] All this would be tremendously impressive were it not for the fact that, relatively speaking, they were 'dwarfed by analogous statistics in Britain's European rival states'. Britain was not failing in absolute terms, she was just lagging far behind in comparison with the other countries that were enjoying miraculous post-war recoveries.

When the New Zealander Edmund Hillary and his Nepalese *sherpa*, Tenzing Norgay, reached the 29,028-foot-high summit of Mount Everest, the world's highest mountain, at 11.30 a.m. on 29 May 1953, the achievement was celebrated throughout the world, but especially by the English-speaking peoples and the Nepalese. The two men had climbed to 'the top of the world', mastering a mountain that represented a permanent challenge to the adventurous spirit of Mankind. The final 1,500-feet climb had taken the two men five hours, and they were carrying sixty pounds, over half a hundredweight. As the journalist Bill Deedes recalls,

News of it first reached the crowds lingering around Buckingham Palace late in the evening before the Queen's Coronation in Westminster Abbey. Though *The Times* had bought exclusive rights to the story, in those days it was thought appropriate to share such news, so on Coronation Day newspaper bills were proclaiming: 'All this – and Everest too!' Here surely was a sign that talk of a new Elizabethan Age was not altogether in vain. We had gone through lean years since the end of the Second World War; this achievement was an echo of earlier and ampler times.[18]

Edmund Hillary had attended Auckland Grammar School and then worked in his parents' bee-keeping business before joining the New Zealand air force in 1944. He had very nearly died in the Pacific War and spent months in the severe burns unit of an American military hospital in Guadalcanal. After the war he climbed New Zealand's highest peak, Mount Cook, before reconnoitring the South Col side of Everest in 1951. He was a genuine hero of the English-speaking peoples, who later led a team to the South Pole and jet-boated up the Ganges to its Himalayan source.

Someone else who could genuinely qualify as a 'New Elizabethan' hero was Michael Ventris. In 1953, the thirty-year-old English architect published his revolutionary academic paper *Evidence for Greek Dialect in the Mycenaean Archives*, which announced his successful decipherment the previous summer of Minoan Linear B, Europe's oldest language (in use between 1,500 and 1,200 BC). Ventris worked off clay tablets from the Late Bronze Age found in the palaces of Knossos, Pylos and other Mycenaean sites. His decipherment is generally considered to have represented, 'together with Milman Parry's Oral Poetry Hypothesis, the greatest advance of classical scholarship in the twentieth century'.[19] (Milman Parry was the assistant Harvard professor who before his death aged thirty-three demonstrated that the Homeric style was characterised by the extensive use under the same metrical conditions of fixed expressions adapted for expressing a given idea.)

Cracking Linear B was the intellectual version of Edmund Hillary's conquest of Mount Everest the same year. The famous labyrinth of Knossos, which had housed the Minotaur, was no more maze-like than the intellectual paths down which Ventris had to tread in his attempt to decipher the ancient Minoan language. The hieroglyphics – hitherto completely incomprehensible doodle-like markings – on hundreds of stone shards from the Minoan palace had baffled the world's finest academic and cryptographic minds for decades, and indeed Ventris' own ever since he had visited the Royal Academy as a Stowe schoolboy in 1936 and innocently asked the archaeologist Sir Arthur Evans, 'Did you say the tablets haven't been deciphered, sir?'[20]

The breakthrough came in late May and early June 1952. Ventris' wife Lois was woken by him in their modernist Highgate flat, Highpoint, at 2 a.m. 'with a long story about place names like Amnisos and symbols for chariots and so on, all of course with illustrations'. Early in June, the

Ventrises invited two fellow-intellectuals, Michael and Prudence Smith, to dinner, but Ventris failed to appear. The Smiths got hungry and a little drunk on endless sherries and Lois apologised for his absence again and again. Eventually, Ventris burst into the room, his normally neat hair ruffled, full of apologies but also excitement. 'I know it, I *know* it. I am certain of it,' he said, and he was right.[21]

Ventris exhibited any number of the signs of genius: the capacity to take infinite pain, depression, fluency in four languages from childhood, a mother who committed suicide, lack of interest in human (or even family) relationships, an ill and remote father, a mathematical, logical but above all compartmentalised mind, and no tertiary education. His lack of academic post or classical linguistic education also probably helped give him the supreme confidence to reject Sir Arthur Evans' then generally accepted theory that the Minoan civilisation dominated the Greek world seven centuries before what we think of as the glory days of Ancient Greece, and instead assume that the Minoans spoke a form of Greek.

Seeing the Minoans as the conquered rather than the conquerors allowed Ventris the clue necessary to crack the fantastically complex Linear B 'code'. Utterly useless in commercial terms, it stands as one of the greatest intellectual achievements of the English-speaking peoples since 1900. As the great French scholar Professor Georges Dumézil said of him, *'Devant les siècles son oeuvre est faite.'* ('In the centuries to come his reputation is made.') This genuine 'New Elizabethan' died in a car crash in mysterious circumstances four years later, aged only thirty-four, at a time of great depression and problems with his marriage.

The so-called New Elizabethan Age – so disappointing in many other ways – also recorded huge successes in the spreading of the English language around the world, in accord with the deliberate policy of the British Government to try to ensure that as much of the world spoke it as possible. 'Within a generation from now,' read one Foreign Office report in 1954, 'English could be a world language – that is to say, a universal second language where it is not already the primary tongue. Its expansion should take place under Commonwealth and US auspices.' The British Council, whose object when it was set up in 1934 had been 'to make the life and thought of the British peoples more widely known abroad', developed contacts and held meetings with the representatives of the radio station Voice of America, which had been putting the US Government's case abroad since February 1942, initially with the help of a BBC liaison unit. By the early 1960s, the BBC and Voice of America were meeting regularly to produce joint plans to promote English, rather than Russian, as the global language of the future.

When Singapore became independent, English was chosen as the best neutral language through which her Malay, Chinese and Indian populations could communicate, and this became a pattern with Commonwealth countries after independence, with English seen as a harmonising social force. Rudolf Hess had predicted before the Second World War that English would become a minor dialect of no importance; instead, by the mid-Sixties, 350 million people – 10% of the world's population – spoke it, and it became the primary language of stock exchanges, business, air-traffic control and economic development. The Ford Foundation and other bodies in the United States financed projects to promote English; as Francis X. Sutton, who worked for the Foundation between 1954 and 1983, recalled, 'There was an early appreciation that English was the language of economic progress and international connection and all kinds of things were done in English that were necessary to these countries.'

Soon English had overhauled German as the language of science and French as the language of diplomacy. In 1977, when the Voyager 1 was sent into space, it contained a message from the Secretary-General of the United Nations to any aliens who might be listening, which was recorded in English. Today, a quarter of Mankind is able to communicate in the language. This did not happen by chance; it was part of a deliberate policy by governments of the English-speaking peoples to combat communism, and it is one of their most pervasive legacies, allowing developing nations to modernise faster than they otherwise would have. After the fall of the Berlin Wall, English also became the *de facto* second tongue of countries right across Eastern Europe.

On Monday, 1 March 1954, the United States launched Operation Ivy, the testing of the first hydrogen bomb, on a remote chain of coral atolls in the Pacific Ocean, the Marshall Islands. The H-bomb was 500 times more powerful than that dropped on Hiroshima and was easily the greatest explosion ever witnessed. The island itself was completely destroyed, leaving a crater three-and-a-quarter miles in diameter, big enough to contain twenty St Paul's Cathedrals. The world had entered the thermonuclear era, which – once the Soviet Union had developed her own H-bomb – meant that, owing to the promise of mutually assured destruction, no war broke out between the superpowers.

For the longest period since the Dark Ages, relative peace descended on the European continent that had hitherto been the cockpit for conflict for centuries. The role of the English-speaking peoples in policing this unprecedented peace, based on the ever-present threat of horrific, suicidal destruction, should be a cause for great pride amongst them. By continually developing the very best and latest means of nuclear annihilation, and staying ahead in the arms race, the United States served the cause of relative global peace for over sixty years and deserves commendation for it.

Although the world's first major outbreak of poliomyelitis (polio) broke out in Sweden in 1905, it was in 1916 that an epidemic in the United States killed no fewer than 6,000 people, one-third of them in New York City. In 1927, Franklin D. Roosevelt, the Governor of New York and himself a sufferer, set up the Georgia Warm Springs Foundation for its treatment, but the production of vaccines for such viral diseases as polio was impossible until the American biologist Ernest Goodpasture was able to grow viruses in eggs four years later. A further epidemic killed 1,151 Americans in 1943 and left many more than that crippled for life.[22]

It was not until 1947 that the American physician Jonas E. Salk succeeded in isolating the polio virus, and the following year three American virologists, John F. Enders, Thomas H. Weller and Frederick C. Robbins, grew the mumps virus in test tubes and later used the same technique to propagate polio. Yet their successes did not come soon enough to produce a viable vaccine by 1952, when over 50,000 people were stricken by an epidemic in the United States, which left 3,300 people dead and thousands more permanently disabled. The latest history of the defeat of the disease is entitled *Polio: An American Story*, and as its subtitle implies it was finally subdued by American physicians. (Lack of vaccines make it still prevalent in Africa, however.)

In February 1954, Salk inoculated children in Pittsburgh, Pennsylvania, and that same year Enders, Weller and Robbins were jointly awarded the Nobel Prize for Medicine for the cultivation of polio in various types of tissue. On 12 April 1955, Salk's poliomyelitis vaccine was released for general use in the United States, having been successfully tested in forty-four states.[23] It was one of very many diseases for which doctors and scientists from the English-speaking peoples have found a cure since 1900, far more than the physicians of any other linguistic or political grouping before or since.

After three days of *pro forma* discussion in Warsaw, on 14 May 1955 the Soviet Union signed a 'Treaty of Friendship, Co-operation and Mutual Assistance' with Albania, Bulgaria, Czechoslovakia, East Germany, Hungary, Poland and Romania. The Warsaw Pact had little to do with friendship, co-operation or mutual assistance, but everything to do with the USSR reiterating her military authority over her satellite states. It was under the provisions of the Warsaw Pact, 'negotiated' by the Premier, Nikita Khrushchev, and the Foreign Minister, Nikolai Bulganin, that Russia was to invade Hungary in 1956 and Czechoslovakia in 1968. As well as the premiers of all the satellites, including Ukraine, Belorussia, Latvia, Lithuania and Estonia, the Red Army Marshals Georgi Zhukov and Ivan Koniev were also present, as was an observer from

Red China. The Pact, which gave Moscow the right to station its troops throughout its Eastern European empire for twenty years, was renewed in 1975 and again in 1985.

The Pact provided for a unified command headquartered in Moscow, the standardisation of weaponry and uniforms, the introduction of Red Army military manuals, and for joint training and annual manoeuvres. It also announced that it would be willing to welcome as new members any more countries that might wish to declare 'their readiness to assist the efforts of the peace-loving States for the purpose of safeguarding the peace and security of nations'. None ever did.

From 1955 onwards, and especially after Hungary was invaded the following year for attempting to withdraw from it, the Pact represented the political and military arm of Soviet imperialism.[24] This new Comintern – the original having been abolished in 1943 as a sop to the Allies – was to pose a threat to Western civilisation for more than three decades. Never could NATO's guard drop or the English-speaking peoples' vigilance be weakened, for fear of an overland invasion of Western Germany and beyond by forces that always heavily outnumbered those of NATO. Only once Soviet communism had been defeated in the open market-place of ideas, and near-bankrupted by the Reagan Administration's refusal to lessen defence spending, purposely designed to cause such an economic collapse, was the Warsaw Pact defeated. With telling appositeness, it was finally abolished by its constituent members in July 1991 in the capital of Czechoslovakia, the city that it had ravaged during the Prague Spring uprising twenty-three years earlier.

'The only thing I have in common with Winston', the newly elected British Prime Minister, Anthony Eden, told his press secretary William Clark, when hiring him in 1955, 'is that I like [to hire] people I know.'[25] Otherwise there was little that united the men, and it is extremely unlikely that Colonel Gamal Abdel Nasser would have suddenly renationalised the Suez Canal if Churchill had still been premier. Equally, had Churchill not selfishly clung to office for so long in the 1950s, Eden might have had longer in the job and been able to handle the Suez crisis better. 'Winston has all the virtues a statesman needs except unselfishness,' Herbert Asquith once said. 'He is so wrapped up in himself that he feeds upon his own vitals.' Perhaps the best line about the prima donna-like behaviour of Churchill and Eden when in each other's company came from the Permanent Under-Secretary of the Foreign Office, Sir Alexander Cadogan, at the time of the Yalta Conference. Poking his head around the ship's cabin of Churchill's doctor, Charles Moran, the night before they departed in the Crimea, Cadogan said, 'I never bargained to take [the Italian coloratura soprano, Luisa] Tetrazzini and [the Australian soprano, Dame Nellie] Melba round the world together in one party.'[26]

A popular inter-war description of Egypt was 'the British Empire's Clapham Junction'; it was a vital strategic entrepôt and hub of imperial communications during the seventy-two years of British suzerainty between the invasion and occupation of 1882 and her withdrawal in 1954. Egypt had also provided an important military base in both world wars. Although she was never formally annexed, she was a British protectorate between 1914 and 1922. She was, as Lord Hankey accurately if unimaginatively described it, 'the jugular vein of World and Empire shipping communications'.

Colonel Nasser had come to power in Egypt in November 1954, when he overthrew General Mohammed Neguib, with whom he had carried out a coup against King Farouk two years earlier. After the 1952 coup, all Egyptian political parties had been banned; Nasser was therefore a dictator. On Thursday, 26 July 1956, he suddenly announced the nationalisation of the Suez Canal, through which some 80% of Western Europe's oil passed. Egypt had specifically promised not to do this, as part of the price of British withdrawal from Egypt; now Nasser was not only reneging on the 1954 agreement, but had also chosen the most prominent Western asset in the Middle East on which to stake his adventurous claim. The news came through to Downing Street just as Eden was entertaining King Faisal II and Prime Minister Nuri-es-Said of Iraq to dinner there. Nuri's advice was that the British 'should hit Nasser hard and quickly'.[27] The Labour leader Hugh Gaitskell was also present, and after the Foreign Secretary Selwyn Lloyd suggested sending 'an old-fashioned ultimatum', Gaitskell said 'that I thought they ought to act very quickly, and that as far as Great Britain was concerned, public opinion would almost certainly be behind them'.[28]

That same evening at 11 p.m., Eden invited the American Chargé d'Affaires Andrew Foster – Ambassador Winthrop Aldrich having left London a little earlier in the day for a short vacation – to join a meeting of his advisors to discuss what to do next. The following day Eden wrote to his wartime friend and colleague President Eisenhower ('Dear Friend'), telling him that he had reviewed the situation with the Cabinet and Chiefs of Staff and that, 'We are all agreed that we cannot afford to allow Nasser to seize control of the Canal in this way, in defiance of international agreements. If we take a firm stand over this now, we shall have the support of all the maritime powers. If we do not, our influence and yours throughout the Middle East will, we are convinced, be irretrievably undermined.'[29] As so often, prestige was a vital component in the crisis.

Eden went on to point out to Eisenhower the glaring fact that, 'The immediate threat is to the oil supplies to Western Europe, a great part of which flows through the Canal', and that he was prepared to 'take this opportunity to put its management on a firm and lasting basis as an international trust'. He did not want to 'allow ourselves to become involved in legal quibbles' over whether Nasser could nationalise what was technically an Egyptian company, or in financial discussions about whether they could pay the compensation offered, preferring to stay on 'broader international grounds', such as whether Nasser should be allowed 'to expropriate it and to exploit it'. He warned Eisenhower explicitly that, 'As we see it we are unlikely to attain our objective by economic pressures alone,' and that as a result, 'My colleagues and I are convinced that we must be ready, in the last resort, to use force to bring Nasser to his senses. For our part we are prepared to do so,' adding that he had asked the Chiefs of Staff to draw up a plan to that effect. He then called for a tripartite meeting with the French Foreign Minister Christian Pineau and the US Secretary of State John Foster Dulles either in London or Washington as soon as possible.[30]

Eden could thus not have been firmer or clearer. Although the Americans later complained that they were not kept properly informed of British thinking during the crisis, they could have had no excuse for not appreciating the high likelihood of a military outcome right from the start. In the light of America's relations with the Middle East since 9/11, Eisenhower's reply to Eden – 'While we agree with much that you have to say, we rather think there are one or two additional thoughts that you and we might consider' – was unfocused, shortsighted and pusillanimous. A firm swatting of Egypt's pretensions by the Western powers acting in concert – possibly leading to Nasser's overthrow – would have sent an unambiguous message to the Egyptians about what would be tolerated by the West. Instead, the United States helped hand them a victory that was to inflame Arab nationalism for decades to come.

By 31 July 1956, Eisenhower, who faced an election in early November, was writing to Eden setting out his opposition to Britain and France using force until a conference had been called, one which 'should have a great educational effect on the world'. American public opinion, he warned, would otherwise be 'outraged'. The President was frank in giving Eden 'my personal conviction, as well as that of my associates, as to the unwisdom even of contemplating the use of military force at this moment'.[31] He was less frank in stating that 'employment of United States forces is possible only through positive action on the part of the Congress', which was then adjourned. In fact, of course, the Commander-in-Chief has wide discretionary powers under the Constitution to engage American forces without prior Congressional approval, just as Truman had committed US troops to Korea only six years earlier.

Eden's reply to Eisenhower put Nasser's actions in their wider political and geographical context. He described the sudden nationalisation as 'unpleasantly familiar', thus obliquely referring to Hitler's and Mussolini's incursions of twenty years earlier, without needing to bother to mention them by name. 'His seizure of the Canal was undoubtedly designed to impress opinion not only in Egypt but in the Arab world and in all Africa too,' Eden warned the President. 'By this assertion of his power he seeks to further his ambitions from Morocco to the Persian Gulf.' He then quoted Nasser's hyperbolic speech of four days earlier, in which he had stated at Aboukir: 'We are very strong because we constitute a limitless strength extending from the Atlantic Ocean to the Arab Gulf.' It was just such pan-Arabic militancy that Eisenhower should have discerned as a far worse long-term threat to American interests than the continued influence of the European imperial powers in some parts of the Middle East.

Eisenhower's response had the effect of forcing Eden to wait for an unconscionable amount of time while trying to negotiate with Egypt via the United Nations and other multilateral bodies. All this achieved was to build up non-aligned and Third World opinion against Britain and France, not least because they could see the benefits of a precedent being set whereby they might nationalise international assets with impunity. It also allowed domestic opposition to military action to build in Britain, with anti-war demonstrations taking place in London and other cities, although the majority of Britons supported the Government's tough stance throughout the crisis.

Eden finally agreed to a morally questionable scheme by which Israel attacked Egypt, after which British and French troops landed in the Suez Canal Zone in a 'police action' ostensibly designed to 'separate the combatants'. In fact, this was an elaborate ploy to force Egypt to disgorge what she had forcibly taken. If only the British delegates to the secret meeting at the former French Resistance 'safe' house in Sèvres had refused to initial any pieces of paper afterwards, they might have been able to justify the meeting publicly, as Eden's Foreign Secretary Selwyn Lloyd later tried to do, as merely a non-binding contingency planning operation. There was an unconscious irony in the meeting taking place at Sèvres, since it had been in that Parisian suburb that the treaty had been signed that tragically divided up the Ottoman Empire and created half the problems of the modern Middle East.

'Collusion', as the secret policy was called, was in fact a diplomatic victory for the Eden Government, as it forced the Israelis to 'accept the opprobrium of aggression followed by the indignity of capitulating to an ultimatum'. It also incidentally saved Jordan from a probable Israeli attack, something for which Eden is rarely given any credit. If the clichés then prevalent about a New Elizabethan Age really meant anything, Eden should have been congratulated for a plan that was Drake-like in its cunning and audacity. Instead, he was reviled for not revealing the plan to the House of Commons, and his reputation never survived.

Because the presidential elections took place within days of the landing of British troops in Egypt, Eisenhower and Dulles were forced to adopt a more anti-imperialist and therefore anti-British stance than they might otherwise have, and a perceived threat that the US posed to sterling forced the Eden Government to halt midway through the military operation, when only about half of the Zone was secured. Churchill's advice to ministers on leaving the premiership in April 1955 – 'Never be separated from the Americans' – was ignored by Eden only fifteen months after it was given. Tumultuous scenes in the Commons – in which Labour MPs yelled 'Murderer!' at the Prime Minister, sittings were suspended and ministers were shouted down – were mirrored in the rest of British society with families falling out over what was undoubtedly the most divisive political issue since the Munich Agreement of 1938.

In January 1957 at the Beefsteak Club, the historian Kenneth Rose wagered Colonel Martin Gilliat £2 to £1 'that Anthony Eden will not be prime minister on May 1 1957'. Rose won easily, as Eden resigned later that same month. Suez was all the more astonishing because throughout his political life Eden had been, as he put it in a prime ministerial broadcast in the middle of the crisis on 3 November 1956, 'a man of peace, working for peace, striving for peace, negotiating for peace. I am still the same man, with the same convictions, the same devotion to peace. I could not be other, even if I wished, but I am utterly convinced that the action we have taken is right.'

Eden's credibility was all the stronger because he had resigned as foreign secretary in 1938 over the Chamberlain Government's appeasement of Mussolini, and had been Churchill's wartime foreign secretary from 1940 to 1945 and also during Churchill's peacetime premiership from 1951 to 1955. Indeed, it had been he who had negotiated Britain's withdrawal from the Suez Canal Zone with Nasser in 1954. On first entering the Commons in 1923, the twenty-six-year-old Eden was given a piece of advice by the Tory leader Stanley Baldwin about how to deal with Labour MPs: 'Don't ever make fun of the party opposite; you may have a better education, but they know more about unemployment insurance.'[32] It used to be said of Baldwin that 'He always hits the nail on the head but it never goes in any further.' In this case it did, and it was not until the Suez crisis thirty-three years later that Eden genuinely infuriated the Labour Party.

Suave, intelligent, handsome, an Old Etonian who won a Military Cross in the Great War and who championed the cause of internationalism between the wars, in 1999 Eden was nonetheless voted the worst premier of the twentieth century (something of an achievement considering other contenders included Edward Heath and John Major). Eden's greatest mistake, as he himself later admitted, was to waste too much time over the various London conferences of the Suez Canal Users' Association (SCUA), which was still meeting in late September, even though the Canal had been nationalised a full two months earlier. (Some of that time had been spent trying to find a suitable acronym for the organisation itself after the first suggestion, 'CASU', turned out to be a Portuguese vulgarity. 'Various combinations were tried,' recalled a weary Selwyn Lloyd. 'Almost all of them meant something revolting, usually in Turkish.')

Eden's latest biographer believes that 'Suez brought out the worst in nearly all who were involved in it, either as participants or commentators.'[33] Eden himself was forced to lie to Parliament about collusion; John Foster Dulles gave different messages publicly from privately; the First Sea Lord Lord Mountbatten opposed the operation despite commanding the naval side ; the Defence Secretary Walter Monckton was privy to and did not oppose the decision to use force in Cabinet, but distanced himself from it whenever he could; Eden's own press secretary William Clark constantly briefed against his master; the Labour leader Hugh Gaitskell initially compared Nasser to Hitler, but then opposed the operation, meanwhile, the Lord Privy Seal Rab Butler and Chancellor of the Exchequer Harold Macmillan had their eyes set firmly on Eden's downfall and their own scramble for power after it. Indeed, Macmillan only started to voice doubts over the financial threat to sterling once troops had been despatched, and then exaggerated them to the Cabinet, while conspiring with the American Ambassador, Winthrop Aldrich, behind Eden's back.

According to American records made available under their all-embracing Freedom of Information Act, at 4.27 a.m. on 24 October 1956 – five days before the Israelis attacked Egypt – Washington received a report from Aldrich in London to say that Monckton had informed him about Eden's invasion plans. Four days later, Patrick Dean, Deputy Under-Secretary at the Foreign Office and Chairman of the Joint Intelligence Committee, dropped a broad hint to the CIA liaison officer in London about the 'trouble' ahead, adding significantly, 'and it isn't because of Hungary'. With Cabinet colleagues and senior Intelligence officers leaking like those two, Eden hardly needed the enmity of half the British political class too, although according to Gallup he retained the support of the majority of the British people throughout the crisis.

Australia came out in support of Britain and France's actions. Sir Robert Menzies led an international mission to Cairo to try to persuade Nasser that the Canal should be controlled by an international body established by treaty and associated with the United Nations. These talks collapsed when Eisenhower unnecessarily and unexpectedly declared that the United States could not countenance the use of force against Egypt, seemingly letting Nasser off the hook completely. When the British and French later bombed Egyptian installations, Menzies opposed a UN resolution censuring Britain and France for their actions.[34]

The Left have long held up Suez as 'no end of a lesson', arguing that the adventure proved that Britain could not act without the imprimatur of the

United Nations any longer, and that Eden fell as a result of his unhinged demand for unilateral – or bilateral with France – action against Nasser's perfectly justified demand for an asset that was built on the sweat of the Egyptian peasantry.

There is another version. This one puts the priority of British national interest over high-minded liberal internationalism. Its heroes are not Hugh Gaitskell, Lord Mountbatten and Lady Violet Bonham Carter – who opposed the Anglo-French 'police action' – but Julian Amery, Captain Charles Waterhouse and Fitzroy Maclean of the Suez Group of Tory MPs, who supported it. This revisionist view holds that Eden was right to resist the unilateral and piratical confiscation of Britain's greatest single overseas asset, that had been bought in hard currency by Benjamin Disraeli in 1875.

On the eve of victory, just as General Hugh Stockwell telegraphed Downing Street to say that within forty-eight hours the entire Canal Zone would be in British hands, Eden was stabbed in the back by a cabal of unscrupulous Cabinet colleagues, short-sighted allies and a small and unrepresentative group of Tory liberal internationalists. It is undoubtedly true that Suez tragically proved that Britain was no longer a Great Power, but this was their fault, not Eden's. The cabal – by threats and falsehoods – forced Eden to call a ceasefire only days before Stockwell's objectives of Ismailya and the town of Suez were attained.

If the Suez operation had succeeded, Nasser would probably have fallen, as many discredited anti-Western adventurers have before and since. This would not have preserved Britain's status indefinitely, but it would certainly have slowed the scuttle of the Western colonial powers from Africa and Asia. Over-hasty decolonisation, which brought vicious civil wars and dictatorships to much of Africa over the next three decades, might have been avoided. Had the 'informal empire' system, by which American and British companies shepherded the Arab oil economies towards mutually beneficial co-operation, not been dealt such a blow at Suez, the vicious oil price hikes which did so much to dislocate the Western economies in 1973 might have been blunted or even prevented. In October 1973, a barrel of oil cost $3.02; by December, it was $11.65, because OPEC suddenly quadrupled prices virtually overnight. The result was a huge economic downturn for the West and disastrous ripple effects for the rest of the world.

There was nothing inevitable about Muslim fundamentalist and Arab nationalist victories in places like Iran, Iraq and Libya in the Sixties and Seventies. Britain had regularly put down such revolts, such as those of Arabi and the Khalifa, ever since Gladstone's original invasion of Egypt in 1882. Yet after 1956 she was in a far weaker position to protect Arab rulers from revolution. The coup in Baghdad on 14 July 1958 saw the murders of both King Faisal II and Nuri-es-Said, within two years of their advice to Eden to 'hit Nasser hard and quickly'. The subsequent history of Iraq, and especially her recent history, would have been very different if Nasser had been toppled.

There is even the tantalising possibility that the Eisenhower Administration privately wanted Nasser overthrown and was only criticising Britain and France in public because of electoral considerations and what was happening in Hungary. On 18 November, only days after Eden had called off the military operation, Selwyn Lloyd and the British Ambassador to Washington visited Dulles in hospital, who asked him: 'Selwyn, why did you stop? Why didn't you go through with it and get Nasser down?' Lloyd, with commendable restraint under the circumstances, answered: 'If you'd so much as winked at us we might have gone on.'[35] Yet wink had come there none.

The hypocrisy of both the Americans and the Labour Party over Suez is best illustrated by their reaction to the nationalisation of British oil interests by Dr Mossadeq of Iran in 1951. For all its later moralising about Third World national self-determination, the Labour Government had been perfectly ready to topple Mossadeq, but feared, as the Cabinet minutes of the time put it, 'the attitude of the United States government'. (Over at the Beefsteak Club, Mr Christopher Sykes had bet 'Mr Gore one guinea that the American government intervenes officially in favour of American purchase of Persian Oil before the end of 1953'. Since the coup was very definitely an unofficial intervention, Sykes lost.)

American politicians of the Eisenhower Administration, who had acquiesced in the joint CIA/MI6 coup which had finally removed Mossadeq in 1953, started mouthing banalities about the brotherhood of man when it came to the presidential election of November 1956. The worst part of the liberal internationalist mantra over Suez, however, was their outraged moral sensibilities over collusion with Israel. Without secret diplomacy and alliances, let alone plans of attack, Britain could not have won the Napoleonic Wars, or managed to stave off involvement in every European conflict between 1856 and 1914.

For Labour politicians who hid the existence of the Chevaline nuclear deterrent from their own Cabinet colleagues nevertheless to denounce Eden for misleading Parliament was grotesquely hypocritical. Without collusion, the Israelis would not have destroyed one-third of the (Soviet-built) Egyptian air force, which would otherwise have been directed against British servicemen. (Israel's participation was certainly not without a *casus belli*: she had been subjected to a number of cross-border attacks from Egyptian territory, highly threatening statements from Nasser, and the blockading of the Suez Canal and Gulf of Aqaba to her shipping.)

Liberal internationalists – who routinely draw an outrageous equivalence between Britain's protection of the Suez Canal with the USSR's brutal suppression of Hungary – prefer occupying the moral to the strategic high ground. Far from being, as it is always hailed, the wittiest remark of the crisis, Aneurin Bevan's description of Eden as being 'too stupid to be prime minister' was mere vulgar abuse. Many in the party of nationalisation actually sympathised with Nasser, who, it is often forgotten, went on to nationalise sixty-nine British banks in Egypt, sixty-four insurance companies and even the private property of many British residents.

There is also much in the Left's Suez myth that is misleading about Britain's relationship with Egypt. For nowhere, apart from India, was her colonial record brighter. Since 1882, the British protected the Egyptian people from Ottoman oppression, French financial rapacity, Muslim fundamentalist uprisings, corrupt Khedives, Sudanese invasions, attempted military coups and, most recently, Field-Marshal Erwin Rommel's Afrika Korps. Modern Egypt was largely created by Lord Cromer, and in every constitutional arrangement made between 1882 and 1954, the Suez Canal was treated differently from the rest of the country.

Another abiding myth about Suez was that Eden's comparison of General Nasser to Hitler and Mussolini was hyperbolic and intended to create hatred in a people who had only eleven years earlier escaped the shadow of the real thing. Soon after 9/11, the British left-winger Tony Benn mocked Tony Blair on the BBC current affairs programme *Newsnight* for likening Saddam Hussein to Hitler, saying that Eden had done the same thing. Yet in fact it had been the leader of Tony Benn's own Labour Party, Hugh Gaitskell, who had first made the analogy, telling the House of Commons when Nasser seized the Canal: 'It is all very familiar. It is exactly the same as that we encountered from Mussolini and Hitler in those years before the war.'[36]

Eden made mistakes; he should not have withdrawn all the remaining British troops from the Canal Zone so precipitously in 1954, trusting Nasser to behave honourably. He should also not have believed that the Eisenhower Administration would behave rationally in an election half-year. Above all, he should not have believed Harold Macmillan's figures about the American threat to sterling, which the Yale historian Diane Kunz has shown to be largely products of the Chancellor's imagination and ambition.[37] In the post-war world, just as in the pre-war, Eden fully deserved Churchill's famous 1938 accolade, that he was 'standing up against the long, dismal, drawling tides of drift and surrender, of wrong measurements and feeble impulses'. (Eden's doctors' medical notes also dispel the myth that the Prime Minister was virtually a Benzedrine junkie during the 168-day crisis. Though in poor health, he was simply not knocking back the 'uppers' and 'downers' of Suez mythology.)

The effect of Suez was almost as cathartic on France as on Britain. As Konrad Adenauer said to the French Premier, Guy Mollet, 'Europe will be your revenge.' De Gaulle was disgusted by Britain 'running out' on France before the whole of the Canal Zone was captured, at the behest of the Americans, and punished Harold Macmillan's Government for it in 1961 when it applied to join the Common Market but was rejected by an uncompromising French '*Non*'.

The modern-day analogy of a prime minister called Anthony committing British troops in the Middle Eastern theatre in the face of much domestic opposition is too obvious to be laboured here, although it is noticeable that Blair seems to have learnt from Churchill's dictum, as vouchsafed to Violet Bonham Carter: 'We must never get out of step with the Americans – never.' If only Eden had paid more attention to the sensibilities of the Eisenhower Administration as it faced the 1956 presidential elections, much might have gone differently. Certainly Eisenhower himself years later admitted that not supporting Eden had been his greatest foreign policy mistake.

The embers of Suez took a long time to cool in Britain, especially on the Right, where it resuscitated a strain of Tory anti-Americanism that had not been much in evidence since the Twenties. Even as late as November 2004, after David T. Johnson of the US Embassy in London had said that America had historically been prepared 'to stand by your nation, through thick and thin', a letter appeared in *The Times* consisting of only one word: 'Suez?'[38]

It did not take long after the crushing of the Hungarian Uprising in 1956 by the Soviet Union for the Left to indict Britain and France as being in part morally responsible. Their argument went that because the Suez crisis ran concurrently, the USSR was presented with the perfect opportunity to attack, whilst the world looked elsewhere. In fact, Anthony Eden had given Russia a free hand in Hungary, but it was at the Yalta Conference in February 1945 when that country was consigned to the Soviet sphere of influence, not in 1956, when its struggle for independence would have been smashed, Suez or no Suez. Yalta condemned the Hungarian people to catastrophe, not the fact that the West was 'too busy' in the Middle East to respond. The year 1956 proves how it is possible in international affairs to have pure coincidences, simultaneous occurrences without direct causal connections.

The Eisenhower Administration had, like Western governments so often before and since, irresponsibly led the resisters to believe that they could expect material rather than just moral support. Yet when 200,000 Red Army troops and 4,000 tanks rolled across the Hungarian frontier, Eisenhower, in his biographer's words, 'did nothing, because there was nothing he could do without precipitating a third world war'. Eisenhower told his National Security Council in late October that he was haunted by the memory of how Hitler had known from early February 1945 'that he was licked. Yet he carried on to the very last and pulled down Europe with him in his defeat.' He wondered aloud whether the Soviets might also pursue such a scorched-earth policy and 'precipitate global war' rather than lose their satellite? He thus resolved to treat them with the tenderest of kid gloves. As he put it in his memoirs: 'We could do nothing' for Hungary, which was 'as inaccessible to us as Tibet'.

There would be nothing wrong with this stance if the Voice of America and Radio Free Europe had not simultaneously, ceaselessly and energetically urged the Hungarians to rise. Translations of Dulles' speeches about the 'liberation' of Eastern Europe were beamed to Hungarian radios, and CIA and MI6 agents even trained would-be resisters in forests outside Budapest. Every subliminal message from America led Hungarians to believe that if and when they rose against communist tyranny, the West would in some tangible way come to their aid. In the event, Dulles merely said that America was 'a sincere and dedicated friend who shares their aspirations', but he did nothing more than censure Russia's actions in the United Nations. As another of Eisenhower's biographers put it, 'Liberation was always a sham. Eisenhower had always known it. The Hungarians had yet to learn it.' Standing on a ticket which precluded American boys fighting in foreign wars, Eisenhower was politically and strategically hamstrung, and the Russians knew it. Indeed on election day itself, he forbade U-2 spy-planes flying over Russian air space in case a 'scared and furious' Kremlin started 'a major war'.

As usual there were those in the West who used their freedom of expression to support the Soviet crushing of Hungarian aspirations for freedom. In Britain, the historians Eric Hobsbawm, A.J.P. Taylor and E.H. Carr, publisher Isaac Deutscher, playwright Sean O'Casey and the 'Red Dean' of Canterbury, Hewlett Johnson, all supported the Kremlin's policy, the last describing the Hungarian freedom-fighters as 'troublemakers'. (Among A.J.P. Taylor's other gross errors of judgment were his demand for the British Army to withdraw from Northern Ireland, his denunciation of the Nuremberg Trials as 'nauseating' and his estimation that Rudolf Hess was sentenced 'for the sole crime of being a premature advocate of NATO'. His anti-Americanism was so virulent that he said he would 'sooner die' than lecture there.) In the USA, the playwright Lillian Hellman and singer Paul Robeson also obediently toed the Moscow line.

'We cannot', Eisenhower said in the election campaign, 'subscribe to one law for the weak and another for the strong.' Yet that was precisely what he did when he imposed financial penalties against Britain and France for the Suez adventure whilst effectively ignoring Russia's crushing of the Hungarian rebellion. The moral equivalism drawn between Suez – where a demagogue had illegally grabbed a legitimate Western interest – and Hungary, where a nation was fighting for democracy and independence, was taken one stage further by Jawaharlal Nehru, the Prime Minister of India. He managed to condemn what he called a 'gross case of naked aggression' at Suez whilst simultaneously voting against the UN motion calling for free Hungarian elections and the withdrawal of Soviet troops, because, he said, it was 'a very confusing situation'.

Western powers have long been perfidious towards oppressed peoples in their willingness seemingly to offer but not actually to deliver real help. Lord John Russell encouraged the Poles and Danes to stand up to the Germans and Russians, and then promptly did nothing for them. Kaiser Wilhelm II encouraged the Boers to defy the British Empire and then left them in the lurch. Successive French governments supported the Algerian *pieds noirs*, before de Gaulle dumped them once he had come to power in 1958. The South Vietnamese were left in the lurch by the US Congress' withdrawal of support in 1972–5, the Shah of Iran by the Carter Administration. George Bush Snr personally encouraged the Iraqi Kurds and Shias to rise against Saddam in March 1991, but then the United States permitted him to use confiscated helicopters to crush them; 60,000 were killed. Czechs, Cambodians, Iraqi Marsh Arabs, Kurds and Bosnian Muslims have also learnt painful lessons about the emptiness of Western rhetoric of support.

On 25 November 1947, Salvador Dalí's private newspaper, *The Dalí News*, carried an item on page four entitled 'Tastes and Prophecies for the Next Ten Years', in which the painter predicted such forthcoming occurrences as 'The *Summa Theologiae* of St Thomas shall be revised by the atom cooked three times' and 'An American art critic of Irish blood shall win fame defending the Dalínian theory of painting.' He also predicted that 'Belgium shall know glory in legislation and finance'. Sure enough, ten years later, that country became the focus for a new Great Power in the world, one that was intended to counter the influence of the English-speaking peoples and which succeeded in splitting off Britain from her former Dominions as significant trading partners. If not exactly glory, a new entity ruled from Belgium was to wield enormous power over both legislation and finance. If sometimes the works of the bureaucracy of the European Union have seemed somewhat surreal since 1957, it ought to be recalled that they were first envisaged by Salvador Dalí himself.

On 25 March 1957, treaties creating the European Economic Community (EEC or Common Market) and the European Community of Atomic Energy (Euratom) were signed in Rome by representatives of France, West Germany, Italy, Belgium, the Netherlands and Luxembourg. The text of the Common Market treaty stated that its goal was 'ever-closer union', one that has been adhered to rigidly for nearly half a century. Britain and the United States welcomed what many saw as a laudable opportunity to banish the endemic threat of war – particularly a third post-1870 Franco-German war – from the European continent. A few longer-sighted commentators saw it as perhaps laying the ground for a European super-state that might one day threaten Anglo-American interests, however. 'There is no doubt that many of the architects of the European project intended from the very beginning to create a rival to the United States,' the British author Gerald Frost has written, 'and that anti-Americanism played a major part in their motivation.'[39] At no point did the US State Department appreciate this central feature of the European project, but instead did everything in its power to encourage deeper and closer European integration.

Britain behaved disgracefully towards Australia and New Zealand from the moment the Treaty of Rome was signed and Harold Macmillan applied for Britain to join the EEC. It is impossible to excuse what the distinguished British historian John Ramsden has described as the 'duplicity, bad faith and general obstructionism with which the British behaved in the face of the Australian government's attempts to find out exactly what the British Government intended to do when the EEC was formed in 1957'.[40] What went for Australia also went for Canada, New Zealand and the West Indies, but not Eire, which joined the EEC in January 1973.

At one point during Macmillan's gross obfuscation of what EEC membership would mean to future trading relations, Menzies even considered appealing over the heads of the Macmillan Government to British MPs and the British people, 'with the (perfectly justified) allegation that Macmillan was betraying Australians and not admitting the fact'. It is likely that if he had, Macmillan would have had serious difficulties, because his Government was already being accused of selling out the white Rhodesians at the time, and Menzies was hugely popular in Britain. At heart, though, Menzies was just not that kind of disruptive politician, and after de Gaulle rejected Macmillan's application the moment passed. It was, though, in Ramsden's words, 'a shameful episode in British diplomatic history of which few Britons have ever known'.[41]

Despite being, in his own phrase, 'British to the boot-heels' and regretting the way the Commonwealth was heading, Menzies deftly negotiated the ANZUS Treaty with America in 1951 and a controversial trade treaty with Australia's former enemy Japan in 1957, which were both to Australia's long-term advantage. He also bought American strike aircraft rather than the British TSR2 in 1963, 'hedged against the Sterling Area and diversified Australia's immigration intake'.[42] He kept Australian national interest firmly at the forefront of his foreign policy, developed close diplomatic ties with Indonesia and Malaysia, and generally 'quickly fashioned a new, regional, post-imperial identity' for his country.[43] If Britain could no longer be relied upon in trade, investment or defence, Australia had her own way of making the transition, which she did with aplomb and considerable success. Unlike Curtin in 1942, however, Menzies made the necessary adjustments without unnecessarily wounding British *amour propre*.

Meanwhile, Canada attempted to move in the opposite direction. In July 1957, the incoming Progressive Conservative Government of the Prime Minister John Diefenbaker, who had won the previous month's general election, made a surprise announcement. In an attempt to reverse Canada's slip into the American orbit, and to reassert her ties to Britain and the Commonwealth, Canada would henceforth divert 15% of her imports from the United States to Britain.[44] For all the patriotic intent behind it, great statements of economic policy like this were unrealistic in the modern free market, and there were problems that autumn in negotiating a mutually beneficial Anglo-Canadian free-trade agreement, since Britain's economic future was increasingly being attached by the Macmillan Government to the EEC.

Despite the Diefenbaker Government's laudable increases in aid for Commonwealth countries under the Colombo Plan and its creation of a worthwhile Commonwealth scholarship programme, the importance of the Commonwealth to Canada continued to wane in the Fifties, as the power of the North American market waxed. As so often before and since, Canadians' hearts drew them in the opposite direction to their perceived economic self-interest, and the latter tended to win. Canadians rightly felt themselves increasingly spurned by British governments of both stripes – but particularly the Conservatives – which, despite the pull of shared struggles, cousinage, language, history and monarch, tended to defer to the large European markets rather than the smaller Commonwealth ones.

On Thursday, 1 December 1955, a forty-two-year-old seamstress called Rosa Parks left work for home on the Cleveland Avenue bus in Montgomery, Alabama. She was a long-time civil rights activist who ever since 1943 had been refusing to follow racial segregation rules on city bussing, and as an official of her local branch of the National Association for the Advancement of Colored People (NAACP), she had been fighting the segregated seating rules since the late Forties. Although she has been depicted as an ordinary woman, the truth was that 'Parks was not someone who one day, out of the blue, decided to defy the local custom of blacks sitting at the back of the bus. … She made a deliberate decision to take up the fight. There was nothing spontaneous about this.'[45] Her political activism was cannily timed; the campaign was successful and it triggered similar protests across the South. When she died in 2005, her body lay in state in the rotunda of the Capitol, a signal honour.

Although the ill-treatment of the Black American has long been held to represent an indelible blot on the escutcheon of the English-speaking peoples, the way in which it was ended goes some way towards counter-balancing this.

For in retrospect it is fascinating just how well embedded the civil rights movement was in the established politics of the English-speaking peoples' tradition of protest. Without that tradition, going back as far as John Hampden's protest against Charles I's imposition of Ship Money in the 1630s, Dr Martin Luther King's movement would have been forced down avenues that would have been bloody, counterproductive and ultimately perhaps even futile.

King's movement – as opposed to that of Malcolm X and the Nation of Islam – drew its primary inspiration from Mahatma Gandhi's civil disobedience campaigns against the British Empire in India, which in turn looked to the British experience of Great War conscientious objection and the struggle for female suffrage. Facing almost any other opponent in the 1930s, Gandhi's movement would have suffered far worse privations and oppression than they received from the 'boyish tyranny' of the imperial British. 'Shoot Gandhi,' was Adolf Hitler's advice to the former Viceroy of India, Lord Halifax, in 1937, 'and if that does not suffice to reduce them to submission, shoot a dozen leading members of Congress; and if that does not suffice, shoot two hundred and so on until order is established.'[46] A glance at what Stalin was doing to various Soviet ethnic minorities at the time, or the way the Japanese were behaving in Manchuria and the Italians in Abyssinia, shows how important it was for Gandhi that he was faced by the English-speaking peoples, who were governed by customs of law, decency and fair play.

After the 1955–6 Montgomery bus boycott, Dr King was asked by NAACP to say which books had most influenced his thinking; of the five he chose one was Gandhi's autobiography, another was a biography of Gandhi and a third was the American pacifist Richard Gregg's 1934 work *The Power of Non-Violence*, which was influenced by the author's time with Gandhi in India in the 1920s.[47] In all five works, the doctrine of non-violence only worked so long as the pacifists maintained the moral superiority over their 'oppressors', which itself gave them tangible political power vis-à-vis their enemies. That simply was not the case with totalitarian opponents like the Nazis, who cared nothing for whether their opponents, or the world in general, thought them morally inferior. Non-violence only worked against the governments of the English-speaking peoples, which respected law, such as the British Government in India in the Twenties and Thirties or the Eisenhower, Kennedy and Johnson Administrations of the Fifties and Sixties.

Of course Dr King was by no means the first person to consider attaching the lessons of Gandhi to the problems of American civil rights; as a recent study has pointed out: 'Black intellectuals were fervently debating the meaning of the Indian independence movement as early as the 1920s, and black newspapers regularly covered the Mahatma's activities.'[48] Yet King did take the argument to a new level; he was also (like Gandhi) an astute enough politician to appreciate the likely pitfalls.

Thus when in the mid-to-late 1950s Martin Luther King spearheaded the movement for civil rights for blacks, he did so by putting the two-centuries-old American democratic values first, even before the far more recent doctrine of universal human rights that had only been promulgated a decade earlier by the nascent United Nations. Couching his rhetoric firmly in the language of the Founding Fathers' greatest documents, primarily of course the Declaration of Independence and the Constitution, King forced American whites on to the legal and moral defensive. In his evocation of the universality of these most sacred texts of the English-speaking peoples, he placed himself squarely in the apostolic line of dignified dissent stretching back to Magna Carta and John Hampden's quarrel with King Charles I.

Of course the essential paradox – or hypocrisy – of the Revolutionary Era had been highlighted by Dr Johnson in his 1775 pamphlet *Taxation No Tyranny*: 'How is it that we hear the loudest yelps for liberty among the drivers of Negroes?' Although the United States fought the Second World War in the name of liberty and democracy, she did so with segregated armies. In Georgia, black soldiers – often northerners unaccustomed to southern segregated seating arrangements – sometimes refused to sit at the back of buses, actions that led to their arrest, and in 1944 fifty black Savannah State College students bought tickets for all the seats on a city bus and then refused to give them up for whites, which resulted in the arrest of two of them on a riot incitement charge.[49] During the Second World War, the paradox began to cause concern to more and more Americans.

In 1944, Professor Gunnar Myrdal wrote *An American Dilemma* about the contradictions between many white Americans' assumptions about the paradigm ideals of liberty and justice and the reality of segregation and effective disenfranchisement. Many whites, Myrdal discovered, held views that might be logically contradictory but were nevertheless genuinely held. Governor Eugene Talmadge of Georgia, for example, apparently saw no inherent inconsistency between his statement that 'No religious or social prejudice has a place in a Christian heart' and another remark of his, 'I like the nigger, but I like him in his place, and his place is at the back door with a hat in his hand.'[50]

It was only because the United States had such high national ideals that the discrepancies became so obvious. As Lincoln had made clear in the opening sentence of his Gettysburg Address, the Civil War was fought to uphold the high ideals of the Declaration of Independence. No fewer than 670,000 Americans had died in that conflict, proving the United States' commitment to its relevance and meaning. Thus it was only because America had such a commitment to the literal meaning of its founding documents, and democracy as an overriding concept, that her black population achieved the vote.

Racial minorities in Eastern Europe, Africa, Asia and the Middle East suffered horrific privations in the 1940s, whereas by 1946 the NAACP had forty branches in Georgia alone and a total membership of more than 13,000. It was successfully sponsoring court cases establishing blacks' right to vote in Democratic primaries and equal pay for black schoolteachers. Outside the countries of the English-speaking peoples that same decade, racial minorities such as the Chechens, Ingushi, Crimean Tatars, Karachai, Balkars, Kakmyks and Volga Germans – totalling 1,332,000 in all – were being deported en masse from the regions they had in some cases inhabited for centuries.[51]

In his speech to the Twentieth Congress of the Soviet Communist Party on 25 February 1956, Nikita Khrushchev admitted that, 'This deportation action was not dictated by any military consideration. No man of common sense can grasp how it is possible to make whole nations responsible for inimical activity, including women, children, old people ... to use mass repression against them, and to expose them to misery and suffering.' The plight of black Americans, real though it undoubtedly was, needs to be seen in context of the much worse contemporaneous suffering of persecuted minorities outside the English-speaking peoples' framework of protection under the law.

Reading Martin Luther King's Record for the Degree of Doctor of Philosophy that was signed off on 5 June 1955 at Boston University Graduate School, one thing stands out. Although he scored a straight 'A' in Personalism and History of Christian Doctrine II, an 'A Minus' in Philosophy of Religion, Directed Study in Systematic Theology and History of Recent Philosophy, and a 'B' in Religious Teachings of the New Testament, King only managed a 'C' in Formal Logic.[52] Yet it was the rock-solid intellectual logic of his political campaigning stance that brought him victory, by pointing out to whites the inherently illogical nature of trying to deny blacks equal rights in a country whose Declaration of Independence states immediately after its preamble not only that all men are created equal but that this truth was 'self-evident'.

'Integration is best because the whole idea of America and democracy is expressed in the statement contained in the Bill of Rights which declared that all men are created equal,' King wrote to a schoolgirl, Marjorie Huelle, on 21 December 1960 in reply to her question asking him for 'very good reasons why you think Integration is best and Segregation isn't'.[53] He might have got the documents of the Founding Fathers mixed up – the Bill of Rights makes no mention of equality – but there was nothing mixed up about his policy of rooting his movement in the liberties they guaranteed.

On 15 June 1960, four months before the presidential election, Bayard

Rustin, the special assistant to the Southern Christian Leadership Conference, wrote to King, then its president, with 'the main points I think we should demand from the platform committees of both parties'. There followed seven demands, but foremost among them was that 'both political parties repudiate the segregationists within their ranks, and make a forthright declaration that racial segregation and discrimination in any form is un-Constitutional, un-American and immoral'. The term 'un-American' had powerful political overtones at the time, having recently been used by Senator Joe McCarthy as a synonym for communist treason, and although Martin Luther King was denounced as a communist, there was no truth in it. When in April 1960 former President Truman had publicly stated that civil rights sit-ins were communist-inspired, King wrote him a sharp letter, stating:

> I have worked very closely with the students in this struggle and the one thing that I am convinced of is that no outside agency (Communist or otherwise) initiated this movement, and to my knowledge no Communist force has come in since it started, or will dominate it in the future. The fact that this is a spiritual movement rooted in the deepest traditions of non-violence is enough to refute the argument that this movement was inspired by Communism which has a materialistic and anti-spiritualist world view.[54]

Nothing would be more likely to alienate potential white support than communist connections, which was why King avoided them and why FBI smear tactics regularly alleged them.

Without much support from the Kennedy Administration, but with enormous support from the liberal press – King thrice graced the front cover of *Time* magazine, the first time as early as February 1957 – the civil rights movement could be ruthless on occasion. The schools' boycott and 'children's crusade' placed children in the front line, and sometimes in gaol itself, radicalising a generation. Moronically heavy-handed police tactics, especially from the Birmingham, Alabama, police chief Eugene 'Bull' O'Connor, also played into the movement's hands. Nonetheless, both tactically and strategically the campaign was masterful, and its key feature – staying within the non-violent tradition of dissent of the English-speaking peoples – ensured its ultimate success.

At the foot of Martin Luther King's Boston University report is typed the laconic summation, 'Deceased 4.4.68', and of course other black campaigners besides King paid with their lives for civil rights. However, it was almost incredible that such a fundamental transformation was made across such a huge swathe of the United States in such a short period of time with relatively so few people killed. During the same period in Algeria, French forces killed thousands, and on one day alone – 21 March 1960 – no fewer than sixty-seven Africans were shot dead at Sharpeville by South African police. For all their violence and intimidation, the white police of the American South behaved in a far more restrained manner than non-English-speaking security forces such as the French or Afrikaans.

Amongst South African whites, from the inception of the policy of apartheid in 1948, it was the English-speaking community that tended to oppose it and that generally treated blacks in a far better way than the Afrikaans-speaking community. Anglo-South African business interests took a fundamentally different approach to the issue of separate development to the smaller-scale Boer businessmen and farmers. If South African politics, the army and the police force had been dominated by the Anglo-South Africans rather than the Boers – or if the Anglophile Jan Christian Smuts' United and Labour Party coalition had defeated the Nationalists in the May 1948 elections – it is likely that arrangements would have been put into place that extended the franchise to blacks on a gradualist model that could both have reassured the majority population and ensured social stability.

Four major factors have had the effect of hugely increasing the proportion of Americans eligible to vote in presidential elections since 1900, with enormous implications for politics. At the beginning of the twentieth century, fewer than half of adult Americans were eligible to vote, but the situation changed dramatically because of the 24th Amendment of 1964 that decoupled voting from tax-paying, the 1965 Voting Rights Act and subsequent laws for easier voter registration, and the lowering of the voting age to eighteen in 1971.[55] Yet for all the quantity of democracy that these changes have brought about, it is debatable whether the quality of governance has improved much. And voter turn-out in American elections is still amongst the lowest in the democratic world, even when the result was as finely balanced as in 2000.

Vice-President Richard Nixon arrived in Moscow at 2.50 p.m. on Thursday, 23 July 1959, on one of the USAF's new 707 Superjets, breaking the air-speed record between the two capitals. (It had been Howard Hughes' idea that he travel like that, in order to impress the Soviets.) Nixon had been invited to the first-ever United States trade exhibition to be held in Moscow, but rather than allow it to become, in the words of his biographer, 'a superficial chore of vice-presidential flag-waving', the scene was set for the most scintillating set-piece capitalism-versus-communism verbal clashes of the post-war era, known to history as 'the Kitchen Debate'.

It came at a time of rising tension; the year 1959 saw no fewer than forty-three Communist Parties around the world receive $8 million in secret donations from the USSR and four years later, after the Cuban missile crisis, eighty-three were handed $15 million, including the Parties of tiny San Marino and Réunion. The various ways that assistance was given to supporters and agents of influence were ingenious. Foreign communist leaders, such as

Britain's Harry Pollitt, were awarded large royalties on the Soviet editions of books that did not sell in Britain, and the art-collector wife of the pro-Stalin American Ambassador to Moscow, Joseph E. Davies, was allowed to buy paintings from the Tretyakov Gallery at artificially reduced prices.[56] (In his 1942 book *Mission to Moscow*, Davies recalled how, 'I was startled to see the door . . . open and Mr Stalin come into the room alone. . . . His demeanour is kindly, his manner almost depreciatingly simple. . . . He greeted me cordially with a smile and with great simplicity, but also with a real dignity. . . . His brown eye is exceedingly kindly and gentle. A child would like to sit in his lap and a dog would sidle up to him.')[57]

Because only days before Nixon landed in Moscow the US Congress had passed its annual 'Captive Nations resolution', strongly criticising the way that 'Communist Imperialism' oppressed twenty-four Eastern European peoples, the Soviet leader Nikita Khrushchev was already in a towering rage even before they visited the exhibition together. 'The resolution stinks,' the General-Secretary of the Communist Party of the USSR yelled at Nixon. 'It stinks like fresh horse shit, and nothing smells worse than that!' Superbly unperturbed by such astounding rudeness, Nixon countered that in his Californian countryside experience, pig shit actually smelt worse, which Khrushchev was willing to admit.[58] It was nonetheless not an ideal start.

The situation was no calmer the next morning when the two men toured the exhibition together, each hoping to use the exhibits celebrating American consumerism to score points off the political system represented by the other. Cameras captured the high-point of the debate – in reality a festival of aggressive, hard-nosed, ideological point-scoring – when the two men arrived at a full-scale replica of the home of the average American worker. This was so full of Western 'luxuries', such as central heating, fitted carpets, en-suite bathrooms, and – in the kitchen – a washing-machine, tumble-drier and refrigerator, that *Pravda* had ridiculed it as 'the Taj Mahal', implying that it was no more truly representative of a normal American worker's home than the great monument to love was of the average habitation in India.

Khrushchev was just as contemptuous of the model home as the Soviet press, pointing to an electric lemon squeezer and calling it a 'silly gadget'.[59] 'Anything that makes women work less hard must be useful,' countered Nixon. 'We don't think of women in terms of workers – like you do in the capitalist system,' replied Khrushchev. Since the kitchen was stuffed full of labour-saving devices which were indeed common to the average American worker's home, Nixon had an advantage. 'To us, diversity, the right to choose, the fact that we have one thousand builders building one thousand different houses, is the important thing,' said Nixon. 'We don't have one decision made at the top by one government official. This is the difference.'[60] The debate went back and forth, with Nixon aggressively putting the case for free enterprise and the free market. His equally articulate opponent was, he said afterwards, 'a bare-knuckled slugger who had gouged, kneed and kicked'. To many in the TV audience, however, Nixon had won on points.

That evening, invited to broadcast over Russian radio untrammelled by censorship, Nixon told his listeners that America's 46 million families owned 56 million cars, 50 million TV sets and 143 million radio sets. (These at least could not be described as silly gadgets.) Thirty-one million American families owned their own homes, and no fewer than twenty-five million of them lived in houses or apartments actually larger than the model one in the exhibition. All this proved, said Nixon, that 'The United States, the world's largest capitalist country, has from the standpoint of distribution of wealth come closest to the ideal of prosperity for all in a classless society.'

Khrushchev, who was in the audience, called out '*Nyet! Nyet! Nyet!*', but Nixon simply spoke over him, stating that in America, 'We are free to criticize our government and our President. ... We live and travel where we please without travel permits, internal passports or police regulations. We also travel freely abroad.' According to some accounts, this was the point at which Khrushchev said, under his breath, '*Ëb' tvoyu babushky.*' ('Go fuck your grandmother.')[61]

The 'Kitchen Debate' – in which Khrushchev had also argued that whereas the USSR had thousands of peasants who could afford the model home they preferred the state to spend the money on rockets, which were superior to American rockets – neatly encapsulated the capitalism/communism divide. Even when the supporters of communism were forced to admit that their secular religion was unable to deliver the material comforts that capitalism could, they argued that it was morally superior. Nixon was accused of being a bourgeois triumphalist, only concerned with providing silly gadgets, whereas socialism could provide solidarity, fraternity, human decency and scientific advances. Yuri Gagarin's flight into space the following year seemed to under-line the last of these.

The accusation that the English-speaking peoples are merely unromantic, consumerist and heartless bourgeois crops up throughout the twentieth century, often levelled by those whose own economic systems cannot compete. In this analysis, the English-speaking peoples are presented as so obsessed by wealth-creation that they lack soul, character, courage and humanity. Wilhelmine Germany called Britain 'the land without music', Nazi Germany condemned Anglo-American capitalism as 'only considering private interests', Soviet communism described it as morally despicable and Al-Queda as deserv-ing of eternal hell-fire.[62] Yet far from being spiritually devoid, utterly materi-alistic consumers, a far higher proportion of Americans attend divine service than do Germans, Russians or Frenchmen.

It was primarily dislike of Anglo-Saxon liberal economics that persuaded the French people to vote 'no' to the proposed European constitution in their May 2005 referendum. For all this century-long demonising of consumerism and free-market capitalism, ordinary people have clamoured to belong to the bourgeoisie and there have been scores of would-be immigrants to English-speaking countries for every emigrant. Yet for all Nixon's victory in the 'Kitchen Debate', it still took three more decades for Soviet communism to collapse. Being right was not enough; pressure needed to be actively exerted against the Soviet system.

THIRTEEN

Civis Americanus Sum
The 1960s

Eisenhower's Farewell Address − The Cuban missile crisis − Rudolf Nureyev defects −
Britain finds a role − Menzies and Macmillan correspond − American political
assassination − South African apartheid − Fighting escalates in Vietnam − The
funeral of Sir Winston Churchill − France withdraws from NATO − The IRA's
war against statuary − Chairman Mao − The danger of Deconstructionism − The
media and the Vietnam War − Australian 'black-green' nationalism −
The fast-food revolution − The Stars and Stripes on the moon

'If we look into history, we shall find some nations rising from contemptible beginnings and spreading their influence until the whole globe is subjected to their ways. ... Soon after the Reformation a few people came over into the New World for conscience sake. Perhaps this (apparently) trivial incident may transfer the great seat of empire into America. It looks likely to me.'

John Adams to Nathan Webb, October 1755

'A bipolar world loses perspective for nuance; a gain for one side seems like an absolute loss for the other. Every issue seems to involve a question of survival.'   Henry Kissinger, *Central Issues of American Foreign Policy*[1]

On 17 January 1961, President Eisenhower delivered his Farewell Address to the American people, and in so doing he coined a phrase that was to become staggeringly unhelpful for American policy-makers ever since, handing a propaganda coup to conspiracy theorists in a way that he of all the commanders-in-chief should have recognised to be unconscionable. In the part of the Address in which he spoke of future security threats to the United States he stated,

[The] conjunction of an immense military establishment and a large arms indus-try is new in the American experience. The total influence – economic, political, even spiritual – is felt in every city, every statehouse, every office of the federal government. We recognize the imperative need for this development. Yet we must not fail to comprehend its grave implications. ... In the councils of government, we must guard against the acquisition of unwarranted influence, whether sought or unsought, by the military-industrial complex.

The speech was a gift to all conspiracy theorists, and the phrase 'military-industrial complex' entered the lexicon, just as the Sixties got under way. Yet Eisenhower was doing a disservice to those Americans who were toiling away to keep the United States at the forefront of military technology, upon which the primacy of the English-speaking peoples depended. Far from endangering American liberties or her democratic processes, the US military and its industrial suppliers have guaranteed both, with remarkable efficacy.

If chemicals, electrical equipment and automobiles were the key markets in the 'third wave' of America's economic expansion between 1890 and the Great Depression, the leaders of her 'fourth wave' were electronics, communications and aerospace. Once more the English-speaking peoples stayed at the forefront of the vital aerospace inventions and developments. By the 1950s, six Western cities – Seattle, Los Angeles, San Diego, Fort Worth, Dallas and Wichita – accounted for almost all of the United States' aircraft production.[2] Huge contracts from the Department of Defense sustained Lockheed, General Dynamics, McDonnell-Douglas, Northrop and several more. The power of the Pentagon to fuel research and development in areas vital for the continued hegemony of the English-speaking peoples far beyond simply aerospace can hardly be over-estimated. As *The Oxford History of the American West* has pointed out,

> Federal contracts have been the basic support for the development and utilization of new electronics and information technologies in the newly high-tech cities of the West. Stanford Industrial Park in 1951 was the first planned effort to link the science and engineering faculties of major universities to the design and production of new products. ... Federal contracts, especially from the Department of Defense, have been a mainstay of Silicon Valley. A broader definition of high-tech, based on a high ratio of research and development to net sales, includes such industries as aircraft, guided missiles and space vehicles, computing machines, communication equipment, electronic components, and drugs. The federal government has been a primary customer for all but the last.[3]

When one considers all the extraordinarily high-tech equipment needed in a modern jet-fighter, the possible non-military spin-offs are correspondingly enormous. Rather than criticising 'the military-industrial complex', Eisenhower ought to have lauded its enormous contribution to American life and commerce.

It did not take long for misplaced fears about the supposed 'military-industrial complex' to appear. During the Cuban missile crisis of October

1962, the US Chiefs of Staff, chaired by General Maxwell Taylor, looked into every possible alternative scenario for winning the dangerous stand-off that had developed, but they always appreciated that in the end the decision would be one to be taken by politicians rather than they. Yet today they are regularly presented – especially the Chief of Air Staff Curtis LeMay – as having been thirsting for war against the USSR, on behalf of the 'military-industrial complex'. (At one point in a recent movie about the crisis, Kevin Kostner's *Thirteen Days*, the word 'coup' is even mentioned.) Since both LeMay and Taylor were Kennedy appointees, about whom he was extremely complimentary after the crisis was over, this analysis owes far more to paranoia than to historical fact.

Despite the fact that John F. Kennedy was only president for as long as the completely obscure Millard Fillmore or Warren Harding, he still rides high in any popularity contest between the former presidents.[4] Much can be put down to his inspirational oratory, promoting ideas and aspirations that were never empirically tested due to his sensational death. In his Inaugural Address, President Kennedy famously promised that America would 'pay any price' that was needed, and to 'support any friend, oppose any foe ... in order to assure the survival and success of liberty'. This rousing cry – which would be vigorously denounced for its neo-conservatism if uttered by a modern Republican – was not turned into action by Kennedy in Cuba or Laos because he was, as a recent study of 'Kennedy's wars' states, 'more tentative and non-committal than his activist rhetoric implied'.[5]

Inaugural Addresses are intended to be heard more as poetry than prose, but Kennedy's oratory did encourage Third World leaders to believe that the United States would actively support nationalist movements against the colonial powers. Yet as one historian has noted, 'The Nehru-Kennedy meeting was a disappointment, and [Kwame] Nkrumah [of Ghana] and other African leaders (including more stable ones) were dismayed when Kennedy failed to prevent the murder of the Congolese Premier Patrice Lumumba in 1961 and seemed to abandon nationalists in Angola in their struggle against the Portuguese.'[6] Kennedy's foreign policy was just as much dictated by the exigencies of *realpolitik* as that of any earlier president, and it was somewhat naïve to believe that because of a superb series of high-sounding phrases in his Inaugural Address – today known as 'sound-bites' – very much would change.

What did change, or at least what did start to change under Kennedy, was the doctrine of 'massive retaliation' that had underpinned Cold War doctrine during the Truman and Eisenhower Administrations. Under the prompting of Maxwell Taylor and others, the US developed an alternative strategy of 'flexible response', which was necessary once the Soviet Union had demonstrated her capacity to deploy long-range nuclear missiles. Under 'flexible response', the President would have a range of options for how to deal with Soviet provocations and insurgency short of full-scale nuclear war. Ultimately it was merely a common-sense response to the new threat posed by the Soviet missiles.

After the Bay of Pigs débâcle of April 1961, Kennedy tended to proceed cautiously in Cold War matters. His first meeting with Khrushchev, in Vienna in June 1961, inaugurated a policy of keeping lines of communication open at all times. Despite this, two months later on the night of 17 August 1961, East German building workers began constructing a near-impregnable wall designed to seal off West Berlin and thus preventing Eastern Europeans from escaping their workers' paradise. Despite East German claims that it was a defensive measure to protect them from NATO incursions, it was the most obvious manifestation imaginable of the superior quality of life in the West, and remained so for twenty-eight years until it was physically torn down by ordinary Germans on both sides. On hearing of the Wall being built, Kennedy sent his Vice-President, Lyndon B. Johnson, to West Berlin the very next day to reassure the inhabitants that the United States guaranteed their liberty.

If anything, Kennedy's under-reaction to the building of the Berlin Wall stood him in good stead the following year over Cuba. The phrase 'at the height of the Cold War' is one of the most overworked in the history-writing profession and has been variously applied to the Berlin airlift, the Korean War, the building of the Berlin Wall, Vietnam, the U-2 spy-plane incident involving Gary Powers, the invasion of Afghanistan, the shooting down of Korean Airlines Flight 007 and any number of similar incidents. In fact, it should only ever be attached to the Cuban missile crisis of October 1962, which was the moment that Russia and America came closest to fighting one another directly, rather than through proxies. Even then, however, the odds never approached evens that they actually would, largely due to Kennedy's televised statement during the crisis that left the USSR under no illusions about the seriousness of the situation: 'It shall be the policy of this nation to regard any nuclear missile launched from Cuba against any nation in the Western hemisphere as an attack by the Soviet Union on the United Stares, requiring a full retaliatory response upon the Soviet Union.'

Cuba was emblematic of Kennedy's presidency; it demonstrated his fine skills as a crisis-manager, communicator, team leader and public-relations genius, but it was not a genuine victory for the United States. His 1960 campaign had stressed the failure of the Eisenhower Administration to prevent the rise of Castro, but when he came to power in January 1961 Kennedy inherited Eisenhower's secret invasion plans using militant Cuban exiles with American support, as well as a plot to assassinate Castro himself. 'The minute I land one Marine, we're in this thing up to our necks,' Kennedy said. 'I can't get the United States into a war and then lose it, no matter what it takes. I'm not going to risk an American Hungary.'

Despite the promise to 'pay any price, bear any burden, meet any hardship, support any friend, oppose any foe' in his Inaugural Address, Kennedy minimised the political risk to himself by refusing to commit American troops to Cuba. He also denied exiled Cuban pilots the right to use American airfields. Partly as a result, the whole operation was a disaster. When on 17 April 1961 1,500 Cuban exiles, who had been trained by US military instructors and supported by the CIA, landed at the Bay of Pigs and the expected Cuban uprising against Castro failed to materialise, after three days the small force were all either killed or captured.

In his press conference on 21 April, Kennedy took full responsibility, saying, 'There's an old saying – victory has a hundred fathers and defeat is an orphan. . . . I am the responsible officer of the Government.' He could hardly have acted otherwise, but nonetheless his popularity surged so much afterwards that he joked: 'It's just like Eisenhower. The worse I do the more popular I get.' With the Bay of Pigs fiasco hanging over him, however, Kennedy put a programme to Congress for an extra $3.4 billion in extra defence spending to meet the Soviet worldwide threat, which was to increase the US Army by no fewer than 250,000 troops. He also gave the order to his CIA operatives to 'do something about Castro's regime and Castro'.

On 16 October 1962, photographs from US reconnaissance spy-planes from the previous day clearly showed that the Soviets had installed twelve SS-5 and three SS-4 launch-pad sites on Cuba. Some of these medium- and intermediate-range nuclear ballistic missiles were estimated to become operational as early as December. Kennedy's first reaction was to set up a special Executive Committee (ExComm) to offer him advice. This consisted, besides the two Kennedy brothers, of Vice-President Lyndon B. Johnson, Secretary of State Dean Rusk, Secretary of Defense Robert McNamara, CIA Director John McCone, Treasury Secretary Douglas Dillon, NSC Special Assistant McGeorge Bundy, Chairman of the Joint Chiefs of Staff General Maxwell D. Taylor, presidential counsellor Theodore C. Sorenson, Under-Secretary of State George W. Ball, Deputy Under-Secretary of State U. Alexis Johnson, Assistant Secretary for Latin America Edward Martin, Deputy Secretary for Defense Roswell Gilpatric, Assistant Secretary for Defense Paul H. Nitze, White House Press Secretary Pierre Salinger, and Special Advisor for Soviet Affairs, the former Ambassador to Moscow Llewellyn Thomas.

Since only JFK and possibly his brother Robert knew the meetings were being tape-recorded, the President was advised with candour all through them, the transcripts of which emerged twenty-three years later under the Freedom of Information Act. 'We're going to have this knife stuck right in our guts,' he told the Joint Chiefs of Staff Committee on 19 October. At first he and the majority of ExComm favoured a conventional air strike to destroy the missiles, until General Walter Sweeney, commander-in-chief of Tactical Air Command, told them on 21 October that such attacks could not guarantee destruction of all the sites.

Instead, in a televised broadcast at 7 p.m. on Monday, 22 October, Kennedy announced a limited quarantine on shipments of hardware to Cuba to prevent further missiles getting there, with the unanimous support of the Organisation of American States. Since these were of foreign ships in international waters, it could be considered a declaration of war. One hundred thousand US troops and 500 aircraft were rushed to Florida. All US forces were placed on DefCon (Defence Condition) Two. Since DefCon Five represents normal and DefCon One denotes maximum readiness – or wartime footing – this was very serious. He then calmly and rationally pursued all lines of communication with the Russians.

The United Nations Secretary-General U Thant suggested on 24 October that the USSR voluntarily suspend shipments for two weeks in exchange for a US suspension of the quarantine. Since this played into the Soviets' hands by not including a suspension of construction work on the sites, Khrushchev readily accepted the proposal, which Kennedy was embarrassingly forced to reject.

A public exchange of letters between Kennedy, Khrushchev and the British philosopher Bertrand Russell helped even less, with Khrushchev's contribution appearing in the *New York Times* on 25 October and stating, 'If the way to the aggressive policy of the American Government is not blocked, the people of the United States and other nations will have to pay with millions of lives for this policy.'

On Friday, 26 October, Kennedy received a long and rambling private letter from Khrushchev suggesting that if Kennedy declared he would not invade Cuba, 'Then the necessity of the presence of our military specialists in Cuba will be obviated.' The same message was transmitted privately to the ABC News correspondent John Scali by Alexander Fomin of the Soviet Embassy. This seemed to break the impasse; however, before Kennedy could respond positively, a second letter arrived on 27 October demanding that the US 'will evacuate its analogous weapons in Turkey'. (The Turks owned fifteen Jupiter intermediate ballistic missiles under NATO control to which the US had custody of the warheads.) The Russians had significantly raised the stakes.

To make matters yet more tense, a U-2 spy-plane piloted by Major Rudolf Anderson Jnr was shot down over Cuba on 27 October by Soviet forces under General G.A. Voronkov, who was acting without Khrushchev's approval. (Although Kennedy did not know that Khrushchev had not ordered it.) Although Robert Kennedy later claimed in his memoir of the crisis, *Thirteen Days*, that the shooting down of the U-2 produced, in a splendid mixed metaphor, 'the feeling that the noose was tightening on all of us, on Americans, on Mankind and that the bridges to escape were crumbling', in fact the tapes reveal that ExComm took the attack very much in its stride, with McNamara even saying, 'I think we can forget the U-2 for the moment.'

Nonetheless, there was an obvious danger posed to further spy-plane over-flights. ExComm briefly discussed air strikes against Cuban surface-to-air missile sites to protect them, but the President swiftly changed the subject back towards a Turkish deal. In retrospect Saturday, 27 October was undoubt-edly the key day of the crisis. The sense of urgency at ExComm was palpable; at the United Nations, the US Ambassador Adlai Stevenson argued that unless the second Khrushchev letter was dealt with immediately the initiative would be forfeited; it was feared that the Turks – who disliked being equated with Cubans and treated like pawns – might unilaterally derail any deal by refusing to allow the Jupiters to be withdrawn anyhow; work was continuing round the clock in Cuba to make the nuclear sites operational; difficulties were multi-plying in maintaining the quarantine around Cuba without a serious naval incident; the FBI reported that morning that Soviet personnel in New York were preparing to destroy sensitive documents; the Russian leadership was felt to be unpredictable, and political support for the USA worldwide seemed to be eroding. Yet despite these pressures, and throughout the crisis, at least on ExComm, there was an atmosphere of calm efficiency.

The release of the ExComm tapes in the 1980s exploded a few (largely Kennedy-created) myths about the crisis, which by then had become embed-ded in the public mind. The first of these was that there was a split between hawks and doves, with a gung-ho military insisting on military action against a pacific White House. Ball later claimed that Nitze, Dillon and Taylor showed 'demonstrably increased ferocity' and Sorenson recalled the hawks as being 'rancorous' and 'vigorous'. Yet the tapes and transcripts reveal no high temper or intransigence in the ExComm meetings, no raised voices, no hawk-dove grouping, nor even people fighting corners either for themselves or for the organisations they represented. Even though sometimes ExComm was in virtually permanent session – on the twelfth day of the crisis it sat for twelve hours continuously – the members simply gave their best advice possible to the one man who had to take the ultimate decisions. 'I heard no voice raised in anger or rancorous exchange,' said McGeorge Bundy when he transcribed the tapes years later. Courtesies were observed; Adlai Stevenson was referred to as 'Governor Stevenson', for example. There were even jokes and laughter, as when one voice says at the end of a meeting, 'Suppose we make Bobby mayor of Havana?' In the crisis that brought the English-speaking peoples the closest they ever got to a nuclear war, their leader was advised with dignity, calm, reason and some commendable foresight.

No-one at any stage expressed any enthusiasm for attacking Cuba. 'Rather,' as an authoritative analysis of the tapes by the *International Security Bulletin* in 1987 has reported, 'each participant seemed to be working mightily, and often quite creatively, to find a diplomatic solution to the Crisis.' The furthest that General Taylor – the only military man out of the seventeen on ExComm – ever went was to say, 'My personal view is that we . . . [must be] ready to invade but make no advance decision on that.' This was no more than common-sense if the pressure was to be kept up on the Soviets. For all the later attempts to find a villain in 'the military-industrial complex' impelling the USA towards nuclear confrontation, there simply wasn't one.

For many years, Kennedy's aides claimed that the crisis was resolved because the President took the brilliant step of responding positively to Khrushchev's first letter (of 26 October) and simply ignoring the second one, of 27 October. This was simply not true. Instead, unknown to ExComm, at 7.45 p.m. on Saturday, 27 October, a private one-to-one conversation took place in the Department of Justice between the Attorney-General, Robert Kennedy, and the Soviet Ambassador, Anatoly Dobrynin, in which Kennedy threatened to invade Cuba, but also offered to remove the Jupiters within six months, i.e. full acceptance of the terms of the second letter. Simultaneously, at 8.05 p.m. Kennedy accepted the offer contained in Khruschev's first letter of 26 October. The Kennedys were therefore publicly accepting the first offer and privately accepting the second, a brilliant manoeuvre, but far removed from what they later claimed to have done.

According to Khrushchev's memoirs edited by Strobe Talbott, *Khrushchev Remembers*, Dobrynin reported that Kennedy looked exhausted, complained about not having slept properly for six nights and was worried that the two countries might slip into war through 'an irreversible chain of events [which] could occur against his [brother's] will'. The President himself had been looking for a peaceful way out of the crisis since the third day; he himself had even brought up the idea of a Turkish solution in ExComm before the Russians alighted upon it. He repeatedly stated the view to ExComm that the Jupiters were obsolete, which they were not, and claimed that NATO would blame the United States if there were a war.

With the U-2 shoot-down, Soviet ship movements and the second letter from Khrushchev, Kennedy repeatedly expressed the fear to ExComm that war or a deal were the only two likely outcomes. With no fewer that 250,000 US ground troops and Marines, 1,000 planes and 250 naval vessels in place to attack Cuba by Sunday, 28 October, the opening stage of a war would probably have gone America's way. Whether the Soviets would really have plunged the world into a global nuclear winter for the sake of Fidel Castro, who Khrushchev deeply distrusted by this stage, is doubtful.

By 9 a.m. on Sunday, 28 October, Washington was informed that the Kremlin had accepted Robert Kennedy's terms and within three hours, after a Moscow Radio broadcast, workmen in Cuba began dismantling the missile sites. The Americans then ended the quarantine and ratcheted down from DefCon Two. One of the prime stipulations Robert Kennedy made to Dobrynin was that the Cuba-Turkey connection had to be kept secret, since otherwise it would have smelt to the American people too much of appeasement. If the Russians publicised it, he warned, Washington would entirely deny the existence of any agreement and keep the Jupiters in place. The truth did not seep out for twenty years. In the absence of that connection, the crisis could be presented to the world as a Kennedy victory, which is duly what happened. In October 1964, Khrushchev fell from power, replaced by Leonid Brezhnev as First Secretary and Aleksei Kosygin as Prime Minister.

For all his later manifestation as a dove, McNamara was one of the toughest members of ExComm, but he and his colleagues were over-ruled by the President, who hankered after a deal that could be presented as a victory. Ever since the Bay of Pigs invasion, Moscow and Havana had been fearful that the next time there would be direct US military interference, which would succeed in overthrowing the Castro regime. When Khrushchev asked his Defence Minister, Rodion Malinovsky, how long Castro could survive against a full-scale American invasion, the answer came: two days. They therefore wanted to try to terrify the Administration with the prospect of a nuclear exchange, to such an extent that they could extort a promise from the Kennedy brothers to allow the political status quo in Cuba to continue indefinitely.

'We have to weigh the gains against the losses,' said Khrushchev long after the crisis. 'Our aim was to preserve Cuba. Today Cuba exists. So who won? It cost us nothing more than the round-trip expenses for transporting the rockets to Cuba and back.' At no stage did the Russians ever so much as begin to contemplate unleashing nuclear war. Any attack on America, Khrushchev later admitted, would have resulted 'in a counter-attack equal to, or even greater than, ours'. When on 26 October Castro seemed to suggest a first strike against the United States to the Politburo, Khrushchev recalled how 'I, and all the others, looked at each other, and it became clear that Fidel had totally failed to understand our purpose'.

It was Russian policy to raise the stakes over Cuba, which the crisis certainly did. An unexpected but very welcome by-product for Moscow was the deal over Turkey, a promise that was shortly afterwards honoured in full. In short, the Russians got more than they could possibly have hoped, considering their aims from the start were far more limited than were recognised at the time. Meanwhile, the Americans merely returned to the *status quo ante*, except that they had given up their Turkish-based missiles and given an undertaking that communism was safe only ninety miles from their shores. In the event, the Castro regime long outlived the Soviet Union itself. The Cuban missile crisis was thus a Soviet victory, which the Kennedy White House – by keeping the peace terms secret – managed to spin into an American victory instead. Yet if anyone 'blinked first', it had been JFK.

Although he did not intervene in Laos in the spring of 1961, Kennedy sent 500 'advisors', including Green Berets, to South Vietnam to help with the counter-insurgency operations there against the North Vietnam-supported Vietcong guerrillas. This was 'flexible response' in action. Walt W. Rostow, Deputy Special Assistant to the President for National Security Affairs, wanted many more than that number to be committed, joking, 'We are not saving them for the junior prom.'[7] In all, there were 16,000 American military 'advisors' in South Vietnam by the time that Kennedy was assassinated in mysterious circumstances in November 1963. By then, over seventy US servicemen had already died trying to preserve the part of Vietnam to the south of the 17th Parallel from falling to the communists. For all his commitment to keeping America's foreign policy options open, there is no reason to believe – as many Americans still do – that Kennedy would not have been drawn into Vietnam in much the same way that his successor Lyndon Johnson was.

It was not originally intended to allow the Kirov Ballet's twenty-two-year-old star Rudolf Nureyev to travel to Paris to dance in *Sleeping Beauty* at the Palais Garnier on 16 May 1961. The Ministry of Culture distrusted him because he was 'anarchic, hyper-individualistic, fascinated by the West'.[8] Yet after the Paris impresario Georges Soria had telegraphed the Kirov management saying that young blood was necessary if the production was to be a success, he was allowed to go. Even so, and despite a superb performance at the dress rehearsal, the director granted the opening performance to another dancer. Nureyev responded with typical flamboyance, preferring to listen to Yehudi Menuhin playing elsewhere than attending the Kirov's opening night. Convinced of his own artistic genius, and moreover being regularly compared to Vaslav Nijinsky himself, Nureyev was about to register a far greater protest and a declaration of independence.

Because of the way that Nureyev was spending his time in Paris, being shown around by Frenchmen and Britons and staying out late, the KGB station head Captain Strizhevski suggested to the Leningrad Deputy Minister of Culture that he be recalled. Yet no fewer than three demands for Nureyev to be sent home – 'taking all the necessary precautions' – were blocked by senior officials of the Kirov, who thought they had detected an improvement in Nureyev's behaviour, and who appreciated his having been awarded the coveted Nijinsky Prize.[9] It was to be the greatest mistake of their lives.

On Friday 16 June the company was due to move to London, when finally it was decided that Nureyev would indeed be sent back to Moscow. On the bus taking them to Le Bourget airport outside Paris, the Kirov's director

Sergei Korkin told Nureyev that he would be flying back to the USSR because his mother was ill, and would take part in 'important concerts' there. As the Moscow flight took off two hours later than the London one, Korkin left Nureyev in the hands of Strizhevski. At the airport, Nureyev excused himself to make a phone call. Through an intermediary he contacted Clara Saint, who had been the fiancée of the French culture minister's son. She then warned the French police what was about to happen.

In a development as dramatic as any he had performed professionally, Nureyev ran over to two French plain-clothed policemen who were ordering coffee at the airport bar, crying, 'I want to stay!' One of his KGB minders then attempted to grab the dancer, but was intercepted by one of the policemen, who said, '*On est en France ici*.' When Soviet Embassy officials arrived soon afterwards to demand Nureyev's return, the dancer kept repeating, '*Nyet! Nyet!*' This was the first high-level cultural defection of the Cold War and was quickly recognised as a disaster for Soviet prestige, since it drew attention to the lack of artistic freedom behind the Iron Curtain.

Soviet propaganda put out that Nureyev 'consorted with homosexuals' and that Clara Saint was a CIA agent. Then French communists attempted to ruin his performance in *Sleeping Beauty* six days later with catcalls, tomatoes and pepper bombs. Nureyev had struck a powerful political blow against communism, even though he himself later stated that, 'My leaving Russia was purely artistic and not political.'[10] The cultural elite of the English-speaking peoples drew Nureyev to their heart; in February 1962, he danced with Margot Fonteyn in *Giselle* for the Royal Ballet in London, where the tickets were oversubscribed by 70,000 applications. Their partnership continued for another fourteen years, despite her being twenty years his senior. He also danced in America, to the fury of the Soviet Establishment. For decades behind the Iron Curtain, art had been impossible to distinguish from politics.

In 1961, the playwright John Osborne, the first of the 'Angry Young Men' school, wrote a letter to the left-wing newspaper *Tribune*, entitled 'Damn you, England'. For all its focused fury, it caused not a particle of the indignation a year later when, on 5 December 1962, Dean Acheson, the former US Secretary of State, delivered a speech at the West Point military academy in which he said,

> Great Britain has lost an Empire and has not yet found a role. The attempt to play a separate power role – that is, a role apart from Europe, a role based on a 'special relationship' with the United States, a role based on being the head of a 'Commonwealth' which has no political structure, or unity, or strength … this role is about to be played out.[11]

Winston Churchill and Harold Macmillan in particular were incensed by these remarks, coming as they did from a perceived Anglophile. To make matters worse, Macmillan was due to meet President Kennedy in Nassau that December, to patch up a serious rift in the Special Relationship when the US pulled out of an agreement to supply Britain with the Skybolt nuclear weapons system, as had been previously agreed. Sceptics of the Special Relationship meanwhile hailed Acheson's remark as a brilliantly incisive *aperçu* which ought to send Britain straight to the psychiatrist's sofa in search of a new, more meaningful identity.

Yet Acheson's prediction has been proved to be wildly off the mark. In fact, Britain has managed to balance her commitments to the Special Relationship, the European Union and the Commonwealth – as well as to NATO, G7, Gatt and the Security Council of the United Nations – with remarkable assiduity. Far from being 'played out', over four decades after Acheson's speech she was in as strong an international position as she has enjoyed in any period since the Suez crisis, without having had to ditch her commitments. In the European Union, but not in the euro currency or federal constitution; 'shoulder-to-shoulder' with the United States in the War against Terror; a nuclear power with an assured seat on the UN Security Council; a leading member of the (expanding and largely democratic) Commonwealth; and the world's fifth-largest economy despite having only 1.3% of global population, Britain protected her status well between 1945 and 2005. Acheson, meanwhile, had joined the long list of statesmen whose own obituary appeared before the one that they had written of the Special Relationship.

The Kennedy White House moved to distance itself from Acheson's remarks. With the President's full authority, his National Security Adviser McGeorge Bundy instructed the State Department to make it clear to the press that 'US-UK relations are not based only on a power calculus, but also on deep community of purpose and long practice of co-operation. Examples are legion. . . . "Special relationship" might not be a perfect phrase, but sneers at Anglo-American reality would be equally foolish.'[12]

Only slightly less foolish, however, was Macmillan's own characterisation of Britain's relationship with America correlating to Ancient Greece's relationship with Ancient Rome, by which he meant that Britain would soften and civilise her stronger colleague, planing off her rough edges with her superior diplomatic expertise. Macmillan's beloved mother was American, but as well as patronising the United States (while paying due obeisance to her power), the historical conceit – surprisingly from a Balliol classicist – ignored the fact that the Greeks were usually slaves in Roman households. As it was, the Nassau meeting went well, and the United States agreed to furnish Britain with the Polaris nuclear weapons system, which has served her superbly well – complete with occasional necessary updates – ever since.

★

In early 1962, there took place a fascinating exchange of letters between Harold Macmillan and Sir Robert Menzies, then still Prime Minister of Australia. For the disarmingly honest way in which the letters tackled the issues of immigration, the Cold War and the Commonwealth they bear fairly extensive reproduction. 'My dear Harold,' wrote Menzies on 15 January 1962,

> Australian immigration policy is aimed at avoiding internal racial problems by the expedient of keeping coloured immigrants out ... the new Commonwealth has nothing like the appeal for us that the old one had. ... I know that we have prided ourselves on having a genius for compromise and for pursuing pragmatic politics. But we can of course follow these lines too far. ... When I ask myself what benefit we of the Crown Commonwealth derive from having a somewhat tenuous association with a cluster of republics some of which like Ghana are more spiritually akin to Moscow than to London, I begin to despair.[13]

Britain was even then preparing to apply to join the European Coal and Steel Community that March and, although General de Gaulle replied '*Non*' the following January to her application to join the Common Market itself, Menzies was understandably concerned at the way events seemed to be going; as he told Macmillan, 'Great Britain's entry into the European Community will bring about a drastic change in the Commonwealth relationship. I am sure that an immigration debate will produce deep and perhaps permanent changes in the Australian attitude.'

Macmillan replied three weeks later in his typically chatty and avuncular manner. 'Here we are, my dear Bob,' he wrote on 7 February, 'two old gentlemen, prime ministers of our respective countries, sixty years later, rubbing our eyes and wondering what has happened ... [to the] Empire of free, independent, advanced, civilised, Christian people that you now correctly call the Crown Commonwealth.' (In fact, 'Crown' Commonwealth really meant 'Old' or 'White' Commonwealth in this context, since several of the black-majority countries Menzies was deprecating also had the Queen as their head of state, though not Ghana, which had become a republic in July 1960.) Macmillan, in that pessimistic way that actuated so much of his politics and world outlook, blamed the way that the two world wars of the twentieth century had 'destroyed the prestige of the white people', adding, 'What we have really seen since the war is the revolt of the yellows and blacks from the automatic leadership and control of the whites.'

It was this defeatist analysis, rather than any inherent belief in the likelihood of true Westminster-style democracies blooming in arid African political soil, that had led to Macmillan's sensational 'Wind of Change' speech in Cape Town in February 1960. Even the language of the most famous sentence of that speech had contained a sub-clause that let his true feelings be glimpsed, when he said of the rise of African consciousness: 'The wind of change is blowing through this continent, and, whether we like it or not, this growth of national consciousness is a political fact.'

In his reply to Menzies, Macmillan was happy to admit that the Commonwealth used to be 'like a small and pleasant house party. Now it is becoming a sort of miniature United Nations.' Yet it was impossible to, in his words, 'chuck it', not least because 'our Canadian friends would not agree'. A glimpse into why the British Government was not keen to adopt the same stringent immigration policies as Australia can be found in further remarks that Macmillan made about the Cold War in that letter, in which he said that 'both the Communists and the Free World must try to attract the unaligned nations to their side by any means'. More than a million Britons emigrated to Australia under the assisted passage programme that was in place between 1946 and 1972; they were nicknamed 'Ten Pound Poms' as that was the nominal amount that the ticket cost, the rest of the fare being subsidized by the British and Australian governments.

In 1962, Macmillan feared that strict immigration policies based on skin colour would hardly endear African and Asian countries to the anti-communist Western cause, as he wrote to Menzies: 'This ideological struggle dominates everything ... [and] puts a tremendous blackmailing weapon into the hands of quite unimportant countries in the Afro-Asian camp who, if it were not for the tremendous rivalry between the East and the Free World, would not be able to sell their favours so dear.' Macmillan readily accepted that 'Ghana is very dictatorial and almost crazy today', but he nonetheless concluded by saying that the Commonwealth was, in his opinion, 'certainly worth doing while the Communist/Free World division really holds the front of the stage. Indeed in this situation we are forced to try.' It was hardly a ringing endorsement, let alone the kind of lofty reference to 'the wider vision of the Commonwealth' that Macmillan had used at London Airport when he was trying to shrug off the resignation of his entire Treasury team as a 'little local difficulty' back in January 1958.

The exchange of letters with Menzies was sent to the Queen and also to the Foreign Secretary, Lord Home. The latter commented that, 'It may just be possible to hold the modern Commonwealth together but our European children are more sensitive than the mother country and they will deeply resent any interference by the coloured brethren with their affairs.' The Canadians, Australians and New Zealanders were therefore to be handled cautiously by Britain over the question of black immigration – which was after all entirely a domestic matter for each of them – while Britain herself continued to pursue an almost 'open door' policy till the last possible moment, partly out of fear of angering relatively unimportant black dictatorships in case they responded by veering towards the communist bloc in a time of very high Cold War tensions. The Cuban missile crisis broke out later in the same year as the Menzies-Macmillan exchange of letters.

So it was not simply a fit of absence of mind that explains the strange case of imperial implosion that took place between 1948 and 1971, but also a dread of antagonising New Commonwealth countries at a time that it was feared that they might leave the Western camp for the Soviet one. Britain became a multiracial society partly because of the Conservatives' initial lack of energy and focus on the question, and later partly because of their (justifiable) fear of communism.

The assassinations of John F. Kennedy in November 1963 – the fourth American president to die that way – and his brother Robert in June 1968, and two months after that of Martin Luther King, created a doleful trio of senior American political figures to be gunned down in the space of only half a decade. While there have not been any similarly high-profile successful political assassinations in recent years, serious attempts were made against Presidents Truman in 1950, Nixon in 1974, Ford twice within three weeks in September 1975, Reagan in 1981 and George Bush Snr in Kuwait in 1993, as well as against the lives of one president-elect (FDR in 1933), three presidential candidates (Theodore Roosevelt in 1912, Robert Kennedy in 1968 and George Wallace in 1972), eight governors, seven senators, nine congressmen, seventeen state legislators, eleven mayors and eleven judges.[14]

Other than a close escape for Margaret Thatcher on 10 October 1985, when the IRA destroyed the hotel in which she was staying during the Conservative Party Conference in Brighton, the rest of the English-speaking peoples have been spared this unhappy political phenomenon in the twentieth century, at least at a high level. The murders of Field-Marshal Sir Henry Wilson MP in 1922, Lord Mountbatten and Airey Neave in 1979, and Ian Gow in 1990 – all at the hands of Irish republicans – were important and terrible events, but none was a senior politician.

The accusation is made, especially by anti-Americans, that the United States might indeed be a democracy, but it is also an inherently violent society with many unstable people living in it who have easy access to firearms and who have been able to kill many of the brightest and most talented public servants. It is pointed out that no other nation of over fifty million people has had the same level of murder of politicians in the twentieth century, unless of course one counts those countries who have deliberately executed theirs (in which case the United States is very low on the league-tables indeed).

Nor is it simply a twentieth-century phenomenon. Although assassination was unknown during the colonial period, Andrew Jackson only narrowly survived a point-blank pistol attack in the Capitol's rotunda in January 1835, and during Reconstruction between 1865 and 1877 – admittedly hardly a normal period historically – no fewer than thirty-six political officials were attacked, twenty-four of them fatally.

The stated motives of those who have attempted to assassinate twentieth-century political figures admit of no uniformity. They have ranged from anarchism via opposition to Israel to an obsession with the actress Jodie Foster. Of the eleven presidential assailants between those of McKinley and Reagan, only two had regular employment, only one was married with children, few planned their attacks carefully and all but two fired pistols, which were only effective at close range. With 297 million citizens in a rich country which has the constitutional right to bear arms, the United States was inevitably likely to have occasional assassinations. (On the basis that nothing was beyond the bounds of Brooks's betting book, Lord Sherwood bet Alistair Buchan £10 each way that '[the murderer of Lee Harvey Oswald] Mr Jack Ruby will be alive legally on 11 March 1967'. With Ruby dying in prison on 3 January 1967, Sherwood narrowly lost.)

An accusation regularly made against the English-speaking peoples is that they helped prop up the white-minority government of South Africa for decades before the release of Nelson Mandela from his prison on Robben Island in 1994 and the subsequent all-race elections that inaugurated black-majority rule. As with American support for some right-wing dictatorships in parts of Latin America, Asia and elsewhere in Africa in the post-war period, this stemmed solely from the overwhelming strategic necessity of countering the Cold War advances of Civilisation's ultimate enemy: Soviet communism. Stability mattered more than democratic principles, for the simple reason that if the Soviets won, there would be no future hope for the extension of democracy. Only once that threat had imploded in 1989–91 could democracy safely be exported to places like South Africa, when it was indeed actively championed by governments of the English-speaking peoples.

In December 1951, Nelson Mandela addressed the annual conference of the African National Congress Youth League, telling it that,

> In Africa the colonial powers Great Britain, Portugal, France, Italy, Spain and their servitors in South Africa – are attempting with the help of the notorious American ruling class to maintain colonial rule and oppression. So-called geological and archaeological expeditions are ... in reality the advance guard of American penetration. ... There is also noticeable a growing affinity among the English, Jewish and Afrikaner financial and industrial interests. ... The possibility of a liberal capitalist democracy in South Africa is extremely nil.[15]

It is therefore hardly surprising that, given this crude Marxist analysis of Southern Africa and its problems, it was not considered in Western interests to encourage black-majority rule there at the time. It was a very different

Mandela indeed who finally became President of South Africa a third of a century later and promoted precisely the capitalist democracy that he had earlier condemned as a tool of Western oppression. In his first State of the Nation address in the Houses of Parliament in Cape Town on 24 May 1994, Mandela also declared that the South African Cabinet had applied to join the British Commonwealth, and 'This important community of nations is waiting to receive us with open arms.'

In May 1964, Michael Ramsey, the hundredth Archbishop of Canterbury, wrote to the British Prime Minister Sir Alec Douglas-Home protesting against British arms sales to South Africa, which he alleged were being used for internal repression. The reply from Downing Street neatly summed up Britain's Cold War dilemmas, but above all the central fact that,

> We have strategic interests in South Africa which make it difficult for us to cut off the supply of arms altogether. We attach importance to the communications link with the Middle and Far East which we have in South Africa as a result of the facilities which we are enjoying at the Simonstown naval base and our over-flying and staging arrangements for military aircraft, especially when Nasser is so hostile. ... The danger areas are the Aden Federation, India and Malaysia. If we had to send a lot of troops the facilities in South Africa would be necessary.[16]

South Africa's vital role in the defence of the sea routes around the Cape in any future conflict meant that warships, high-performance Canberra and Buccaneer aircraft, helicopters that could carry torpedoes, and so on, had to be sold to her, despite the moral objections of the Church of England.

Yet Britain's strategic role east of Suez that necessitated taking such a comprehensive world-view was about to be ditched once Douglas-Home narrowly lost the October 1964 general election to the Labour Party led by Harold Wilson. Britain did not retreat from east of Suez because of the sterling devaluation of 1967, as some historians have implied; it was far more deep-seated than that. The Defence Review Studies Report that concluded that Britain should quit her east of Suez bases by the mid-Seventies was published in July 1967, months before the devaluation. Shortly after coming to power in 1964, the Wilson Government had placed a £2 billion ceiling on defence expenditure for the rest of the decade and further financial stringency led to the scrapping of several defence contracts with east of Suez implications, such as tactical-strike and reconnaissance aircraft and a new aircraft carrier.[17]

In the 1960s, Britain spent 6.8% of her Gross Domestic Product on defence, twice West Germany's figure and four times Japan's. Her economic growth meanwhile lagged badly behind those of her competitors. She had 3.2% growth in terms of GDP per person per annum between 1950 and 1973, compared with Japan's 7.6%, Germany's 6%, Italy's 5.5% and France's 5%.[18]

The withdrawal from east of Suez further distanced Britain from Australia, which had troops stationed in Singapore until 1988, and New Zealand, which had them in Malaysia until 1989. The very phrase contained a reminder both of Britain's humiliation of 1956 and – at least subliminally – the superpower that had ultimately been so antagonistic. When President Johnson was to ask Harold Wilson for help over Vietnam, memories of the Suez crisis were fresh in the minds of British policy-makers, who said no.

On Sunday, 2 August 1964, the destroyer USS *Maddox* was attacked in the Gulf of Tonkin ten miles off the North Vietnamese coast by three North Vietnamese torpedo-boats. Despite three torpedoes being launched and machine-guns being fired at the destroyer, the *Maddox* sustained no casualties. In reprisal, one of the boats was sunk by US Navy fighters from the aircraft-carrier *Ticonderoga*. President Johnson sent Hanoi a warning of 'grave consequences' that would flow from any further attacks. The following day *Maddox*, along with a second destroyer USS *C. Turner Joy*, continued to operate within eight miles of the coast. Owing to crew members misreading electronic instruments whose accuracy was affected by heavy thunderstorms, both ships believed that they had again come under attack, opening fire on numerous apparent targets but without any confirmed sightings of any assailants.

Although it is today thought very unlikely that the *Maddox* had indeed come under a second attack on 3 August, the key point was that the ship itself, and thus the White House and US Congress, genuinely believed that she had. Ancestral voices echoed in the minds of American policy-makers; attacks such as those on the *Lusitania*, Pearl Harbor and across the 38th Parallel could not be allowed to go unavenged. The Joint Chiefs of Staff recommended that sixty-four US Navy fighter-bombers attack oil facilities and naval targets in North Vietnam, in which a pilot – Lieutenant Everett Alvarez of San Jose, California – was shot down and taken as the first American prisoner of war. On 5 August, an opinion poll showed that 85% of Americans supported Johnson's decision to bomb North Vietnam, and two days later Congress authorised him – by 98 votes to 2 in the Senate and unanimously in the House – to 'undertake all necessary measures to repel any armed attack against forces of the United States and to prevent further aggression'.

In fact, although the Tonkin Gulf incident has been made much of, the United States was fairly heavily engaged in Vietnam long before it took place. In January 1961, Khrushchev had pledged his support for 'wars of national liberation' throughout the world, and especially for Ho Chi Minh's escalation of the armed struggle to unify Vietnam under communist control then being spearheaded by Vietcong guerrillas. In May, the Kennedy Administration sent

400 Green Beret 'special advisors' to train South Vietnamese troops in counter-insurgency techniques. After some 26,000 Vietcong launched several attacks across the border that autumn, President Diem – who Vice-President Johnson had earlier hailed as 'the Winston Churchill of Asia' – asked for more military aid from the United States. In October, General Maxwell Taylor and Walt Rostow visited Vietnam and reported that, 'If Vietnam goes, it will be exceedingly difficult to hold South-East Asia.' They recommended sending 8,000 troops.

By the end of 1961, the United States was spending over $1 million per day training and supporting South Vietnam's 200,000-strong army. In his State of the Union speech on 11 January 1962, Kennedy avowed that, 'Few generations in all of history have been granted the role of being the great defender of freedom in its maximum hour of danger. This is our good fortune.' Yet four days later, when asked in a press conference whether any Americans were engaged in fighting in Vietnam, Kennedy answered, 'No.' Far from the supposed falsehoods and tergiversations of the Johnson and Nixon Administrations, the true mendacity began with Kennedy's desire to fight Vietnam as an unofficial war, rather akin to some of the United States' interventions in Latin America, rather than as the anti-communist crusade it genuinely was. By the end of Kennedy's presidency, there were over 16,000 American 'military advisors' in Vietnam, many of them taking part in serious military operations on a regular basis.

As was seen during the Cuban missile crisis, one of the strongest hawks in the Kennedy Administration was Robert McNamara, a signatory – along with Secretary of State Dean Rusk and National Security Advisor McGeorge Bundy – of a memorandum dated 24 May 1964 urging Johnson to 'use selected and carefully graduated military force against North Vietnam' for as long as it supported the Vietcong insurgency. Three weeks later, on 16 June, Johnson was tape-recorded saying that there were those who would like the US to withdraw from Vietnam, at which McNamara said, 'I just don't believe we *can* be pushed out of there, Mr President. We just can't *allow* it to be done. You wouldn't want to go down in history as having ...', at which Johnson interrupted, saying, 'Not at all.' The historian who transcribed the taped conversation for publication, Michael Beschloss, described McNamara's tone in the recording as 'pressing Johnson very hard'.[19]

A Chinese proverb of 1100 BC stated that, 'It is foolish not to lag behind when the elephant approaches a new bridge.' Britain certainly lagged behind in the mid-to-late 1960s, when the United States began to get ever more closely involved in the affairs of Indo-China, and it did not prove to be foolish. That region had long been considered in the British Foreign Office as lying in the French zone of influence; Malaysia was free of communism since the British-fought Malayan Emergency campaign in the Fifties – for which American forces had not been requested – and since Thailand was one of the only countries in that part of the world never to have been occupied by any European Great Power, she was thought to be capable of defending herself.

Although Harold Wilson refused President Johnson's requests for British troops to fight in Vietnam – 1968 was the only post-war year that the British Army lost no soldiers on active service – Vietnam of course counts as a major war of the English-speaking peoples. The US troop levels there were as follows: 1965: 154,000; 1966: 169,000; 1968: 563,000; 1969: 484,000; 1970: 335,000; 1971: 158,000; and 1972: 24,000.

As a member of the South-East Asia Treaty Organisation (SEATO), Australia contributed forces ranging from a battalion to a brigade group and their supporting elements (about 7,500 troops in all) between 3 June 1965 and 18 December 1972. A total of 520 Australians died in the Vietnam War, the first conflict they had engaged in without Britain, and as such it was an important part of their historical development as an independent nation. New Zealand also contributed to the effort to turn back the communist incursions into South Vietnam. The United States remembers the contributions of both countries with gratitude. 'The Australians have been with us in all our wars since their Federation; we honor that,' former US secretary of state Colin Powell told the author in January 2006.

The introduction of conscription in March 1966, for the first time in Australia's peacetime history, provoked anti-government demonstrations and the election in December 1972 of Gough Whitlam's Labor Party, which withdrew from Vietnam and ended the draft in the same month as its election. The next month, however, Australia reaffirmed her ties with SEATO and the USA. New Zealand also contributed an artillery battery to the Vietnam war effort, which was later increased to a battalion of 550 men. This too withdrew only when the main body of American troops left in 1972. The next time that Australia pledged support to a foreign military endeavour would be in February 1998, when her Prime Minister John Howard said that she would take part in a US-led attack on Iraq should Saddam Hussein continue to hinder UN arms inspections.

As James C. Bennett has pointed out in his book *The Anglosphere Challenge*,

Few have sufficiently appreciated the extent to which Australian pressure was responsible for Lyndon Johnson's decision to commit the United States to the Vietnam War. The US-Australian-New Zealand alliance in that war, combined with British reluctance openly to support that effort, weaned Australia away from Britain even more. The fact that many Britons fought in Vietnam in Australian and New Zealand uniforms served to emphasize that alienation rather than counter it, as that aspect of inter-Anglosphere cooperation remained invisible.[20]

★

The 1964 presidential elections, which Lyndon B. Johnson won by a landslide within a year of Kennedy's assassination, and in which the Democrats retained their majorities in the House by 295 to 140 and the Senate by 68 to 32, were remarkable for the level of personal abuse the Democrats and their supporters in the media heaped upon the Republican candidate, Senator Barry Goldwater of Arizona. Governor Pat Brown of California spoke of 'the stench of fascism. . . . All we needed to hear was "Heil Hitler"'; CBS linked Goldwater's invitation to Bavaria to Adolf Hitler; Martin Luther King said, 'We see dangerous signs of Hitlerism in the Goldwater campaign'; San Francisco mayor John Shelley said the Republicans 'had *Mein Kampf* as their political bible'; the *Chicago Defender* ran the headline, 'GOP Convention 1964 Recalls Germany 1933'; the NAACP leader Roy Wilkins said, 'Goldwater's election would bring a police state'; and the black baseball player, Jackie Robinson, said, 'I now believe I know what it felt like to be a Jew in Hitler's Germany.'[21] Utterly undeserved and disgraceful though such accusations were, they tended to link Goldwater in the public mind with extremism, especially once his opponents had comprehensively negatively 'spun' his own speech accepting the Republican nomination, in which Goldwater had said: 'I would remind you that extremism in the defence of liberty is no vice. And let me remind you also that moderation in the pursuit of justice is no virtue.'

The funeral of Sir Winston Churchill on Saturday, 30 January 1965, marked the end of a distinctive epoch in British history, one that had been as glorious as it was long. When General de Gaulle was brought the news of Churchill's death, he remarked, with little evident regret, 'Now Britain is no longer a Great Power.' Much of the British press agreed with his analysis; with little remaining of the colonial empire by 1965, the uninspiring Harold Wilson in Downing Street, the winding-down of commitments east of Suez and recurrent problems with sterling which were in 1967 to force a humiliating devaluation, it was hard not to fit Churchill's passing into an overall picture of post-war national decline, even malaise. As the historian Sir Arthur Bryant wrote in the *Illustrated London News*, 'The day of giants is gone forever.' Churchill's detective agreed, saying, 'If the king dies you can say "Long live the King", but now Sir Winston's gone, who is there? There's no one of his stature left.'

Such sentiments were also echoed by A.L. Rowse, a fellow of All Souls College, Oxford, who was profoundly moved by the sight of the train carrying Churchill's coffin passing through the railway station there. 'The sun is going down on the British empire,' he wrote. The novelist V.S. Pritchett, writing in the *New Statesman*, thought: 'There was, I suppose, an undertone of self-pity. We were looking at the last flash of Victorian aplomb; we were looking at a past utterly irrecoverable.' The *Daily Mail*'s editorial on the day of the funeral concentrated on the end of greatness as well; its last sentence read: 'And now it is over.'

The funeral arrangements had to be constantly updated over twelve years due to Churchill's longevity. 'The problem was', Lord Mountbatten joked, 'that Churchill kept living and the pall-bearers kept dying.' Churchill himself played relatively little part in planning the event, although he promised Harold Macmillan that 'There will be lively hymns', and he said to his last private secretary, Anthony (later Sir Anthony) Montague Browne, 'Remember, I want lots of military bands.' In the event, he got no fewer than nine.

Across the English-speaking world, flags flew at half-mast, newspapers printed lengthy obituaries and black armbands were worn. In Britain football matches were rescheduled, shops closed and the National Association of Schoolmasters even cancelled a strike. Another break with tradition was the Queen's decision to attend personally, a rare mark of royal favour. There were six sovereigns, six presidents and sixteen prime ministers present at St Paul's Cathedral that day.

In all, some 350 million people saw the funeral on television, including a larger American audience than watched President Kennedy's funeral just over a year earlier. No fewer than 112 countries were represented at St Paul's; only Red China refused to send a representative and only Eire failed to broadcast the occasion live. Laurence Olivier provided some of the ITV coverage, but it was Richard Dimbleby's commentary on the BBC that won the most plaudits. After the ceremony, President Eisenhower and Sir Robert Menzies gave impressive broadcasts to the American people and the Commonwealth respectively.

Just as Churchill had promised Macmillan, there were indeed some 'lively' hymns. Hymns were not sung at Wellington's funeral because they were considered unsuitable for solemn occasions, but 113 years later they were a central feature in Churchill's. His half-American parentage and his belief in the potency of the English-speaking peoples were reflected in the choice of *The Battle Hymn of the Republic*. The pageantry was solemn, superb, sublime. As the *Sunday Times* commented, the funeral was 'an act of history in itself'.

After the official ceremony, a private burial took place in Bladon in Oxford-shire, near to Blenheim Palace where Churchill had been born over ninety years before. (Lady Churchill had gently talked him out of his original inten-tion, which was to be buried on the croquet lawn at his country house, Chartwell in Kent.) Invited to mourn with the Spencer-Churchill family was Anthony Montague Browne. The occasion brought on in him 'black melancholy thoughts of the decline and decay of so much of what Churchill had stood for. Well might the nation mourn him.'[22] As if to underline this moral decay, once Montague Browne got back to London from attending the private family burial at Bladon, he found that his flat had been burgled.

On 7 March 1966, the newly appointed French Ambassador to Washington, Charles Lucet, was understandably nervous as he brought a personal letter from President de Gaulle to President Johnson. It contained the news that France was going to request the removal of NATO bases from French territory in three days' time, which she duly did. He was received politely and entirely without undue emotion, for a reason that would have infuriated him had he known why. The Americans had not only been apprised of French intentions by a high-ranking mole in the French foreign office, the Quai d'Orsay, but they had even been forewarned of the text of the General's letter.[23] A high-ranking diplomat had for three years been leaking de Gaulle's plans because he was alarmed about the President's desire for *rapprochement* with the Soviet Union. Johnson's reply was a model of its kind; he directed Secretary of State Dean Rusk to ask de Gaulle: 'Does your order include the bodies of American soldiers in France's cemeteries?' (No fewer than 30,922 bodies of Americans killed in the Great War, and 93,245 of those killed in the Second World War, are buried in Europe, the majority in eleven huge cemeteries in France.)

On 1 July 1966, France withdrew her forces from the NATO command structure altogether. This was not the disaster it seemed, since their nuclear plans included a *pre-stratégique* strike, in effect a 'demonstrative' strike, which once outside NATO made France a further complication for Russian planning. Throughout the Cold War, NATO, in the words of the historian David Miller, 'regarded battlefield nuclear weapons either as a reasonable response to Soviet first use or as a last resort in the face of imminent conventional defeat. In addition, the West had plans to use a very small number of nuclear weapons in a "demonstrative" capacity.'[24]

The English-speaking peoples who led the NATO alliance were always willing to countenance the massacre of millions of Russians in order to preserve their independence, just as they had indeed used the nuclear bomb on Japan to save several hundred thousands of American lives in 1945. This mental and moral toughness disgusted some – 'The white race is the cancer of human history,' wrote Susan Sontag in *Partisan Review* in the winter of 1967 – but it ensured that the Cold War stayed at that temperature, to the overall benefit of humanity.

If the Russians had launched the attack on Western Europe that they planned for, trained for, raised vast armies and navies for, bought advanced weaponry for and finally bankrupted themselves over, they might have succeeded through sheer weight of numbers. Yet Russia had fared very badly militarily over the previous century, having been defeated in the Crimean War, the Russo-Japanese War and the Great War, fought to a standstill in the Russo-Polish and Russo-Finnish Wars, and only victorious in the Great Patriotic War at the cost of twenty-four million killed. What can be certain is that a third world war would have killed very many more, and it is a tribute to the United States and British willingness to station large armies on the Rhine for nearly four decades after 1945 – backed up by the credible threat of nuclear war – that such a conflict never broke out.

When Julius Cæsar was assassinated on the Ides of March 44 BC, he fell at the foot of the statue of Gnaeus Pompeius Magnus (Pompey), his great rival and enemy. Despite his triumph over Pompey at the battle of Pharsalia four years earlier, such was his confidence in his own greatness that Cæsar had not seen fit to have the statue removed or destroyed in the meantime.

At 1.30 a.m. on Tuesday 8 March 1966, the 134-foot-high Nelson Pillar that had dominated the Dublin skyline since 1809 was blown up by Irish republicans to mark the half-centenary of the Easter Rising. The bomb, placed two-thirds up the internal spiral staircase, shattered the column. Pieces of granite were blown 100 yards away, but no-one was hurt. The following Monday, Eire's armed forces demolished the rest of the Pillar with an explosion that blew out windows in O'Connell Street. Nelson's head was promptly stolen from the Dublin Corporation's depot.[25] Thereafter, republicans sung paeans about how the 'wicked eye' of 'England's Admiral' (who had after all saved Ireland from Napoleon's rule as well as the rest of the British Isles) no longer stared out over Dublin.

In June 1948, the Irish had already removed their statue of Queen Victoria from the courtyard of Leinster House in Dublin, the seat of their parliament, shortly before Prime Minister John Costello's proclamation of the Irish Republic. (After spending nearly forty years in the Royal Hospital, Kilmainham, it was donated to Sydney in 1987.) She was lucky to survive at all: William III was blown up in College Green in 1929, George II in St Stephen's Green was destroyed to mark George VI's coronation in 1937, and the great Anglo-Irish soldier Lord Gough was blown up in Phoenix Park in 1957.

Some nations have no psychological problems about monuments of the past; Britain, France, the United States, Spain and many other countries have any number of symbols of earlier unpopular – even tyrannical – regimes and rulers that they did not see fit to destroy. The fate of Britain's imperial statuary provides an interesting commentary on the political maturity of the successor governments. A magnificent statue of George V by Charles Jagger was removed from the heart of New Delhi to Old Delhi, in the Coronation Gardens close by the Durbar Grounds. Many miles north of Lutyens' great city, it is neglected and unvisited. In the back yard of a museum a few miles outside Lucknow, there is a veritable elephant's graveyard of imperial statues, all clustered together, collected from their imposing sites around the Punjab. Queen Victoria, Edward VII, George V, Lord Curzon and various others are all there, communing together as in a very important but anachronistic privy council meeting. By contrast, the splendid Queen Victoria Memorial in Calcutta has recently been renovated and is thronged with Indian and foreign tourists. 'The passage of time makes these relics exciting, not threatening or domineering,' opined one correspondent to *The Times*. 'The once-Communist state of Bengal showed wisdom in retaining them and in now exhibiting them as tourist attractions.'[26]

Because Nelson was, in the words of a recent biographer, 'the very epitome of the greatness of Britain, a founder of its security and worldwide influence', aggression towards his statues has become a focal point for Britain's detractors and enemies. In 1940, Hitler discussed shipping his column in London's Trafalgar Square back to Berlin as an unmistakable symbol of Britain's defeat, and both Bridgetown in Barbados and Montreal in Canada have discussed dismantling their Nelson statues and moving them to more obscure locations, and thus 'less capable of offending nationalist sentiment'.[27]

In April 2004, provincial councillors in the Eastern Cape in South Africa proposed that the statue of Queen Victoria in Port Elizabeth be removed, since they claimed it was a symbol of colonialist oppression. 'I wonder if the new rulers of South Africa will be removing any of the other leftover relics of British colonialism,' asked a letter in *The Times*, 'such as democracy, freedom of speech and the rule of law, not to mention the transport infrastructure, hospitals and the English language – and the very notion of a South African state?'[28]

Destruction of statuary has long been an intensely political act. On 25 June 1940, the day that Adolf Hitler visited Paris after the Fall of France, he ordered that the statue of the French Great War general Charles Mangin, one of the heroes of Verdun, be removed from the Place du Président Mithouard, behind the St Francis Xavier church. He also demanded the destruction of the statue of Nurse Edith Cavell in the Tuileries Gardens, who had been executed by the Germans for helping Allied soldiers to escape from German-occupied Belgium. (Both have since been replaced.)[29]

The English-speaking peoples rarely resort to such historical vandalism. Admittedly, after George Washington captured Fort Duquesne during the French and Indian War, he renamed it Fort Pitt in honour of Pitt the Elder (now Pittsburgh), and in 1664 New Amsterdam did become New York – in honour, not of the English city, but of Charles II's younger brother the Duke of York – but such alterations are rare. A surprisingly large number of British names were retained after the American Revolution too. The main street of Williamsburg, Virginia, is still Duke of Gloucester Street. The states of Georgia, North Carolina, South Carolina, Virginia and Maryland were all named after British sovereigns or their consorts, and states like New Jersey and New Hampshire are all obviously derivative of British places and were retained. King's College in New York changed its name to Columbia, of course, but that seems to have been something of an exception.

Nor did a triumphalist English-speaking America change the names of cities and towns founded by non-English-speaking people. The biggest cities in California are Los Angeles, San Diego, San Francisco and San Jose, for example. Although there were never significant numbers of Spanish-speaking people in Nevada, Colorado, or Montana before the twentieth century, all three states retained names they had been first given. French names fill out much of the middle of the country, dating from eighteenth-century trading posts: Detroit, Beloit, St Louis, Des Moines, Dubuque and Terre Haute among them. What had been the Kaiser Wilhelm Street in San Antonio, Texas – because it had been heavily settled by Germans in the mid-nineteenth century – was renamed King William Street during the First World War, however.

Eschewing any such cultural revolutions, and showing reverence for the past whoever it belonged to, has been a feature of the English-speaking peoples and sets them apart from several of their rivals. On Monday, 1 August 1966, Chairman Mao Tse-tung wrote to the student groups that were beginning to style themselves 'Red Guards' to tell them of his 'fiery support' for their promises to 'be brutal' in 'trampling' his enemies. He simultaneously circulated copies of his letter to the Central Committee of the Chinese Communist Party, ordering them to aid the Red Guards in the 'Cultural Revolution' that he was about to unleash. What was to be known as 'Red August' had begun.

The first known death by torture of the Cultural Revolution took place four days later, on 5 August, at a Beijing girls' school. In the words of Mao's latest biographers Jung Chang and Jon Halliday,

> The headmistress, a 54-year-old mother of four, was kicked and trampled by the girls, and boiling water was poured over her. She was ordered to carry heavy bricks back and forth; as she stumbled past, she was thrashed with leather army belts with brass buckles, and with wooden sticks studded with nails. She soon collapsed and died.[30]

Starting in Beijing, the state-sponsored violence against teachers soon spread to every school and university in China. Then it extended to the custodians of Chinese culture, such as writers, artists and singers, and finally against anyone held to represent 'the old culture'. As Chang and Halliday record,

> Many of those raided were tortured to death in their own homes. Some were carted off to makeshift torture chambers in what had been cinemas, theatres

and sports stadiums. Red Guards tramping down the street, the bonfires of destruction, and the screams of people being set upon – these were the sights and sounds of the summer nights of 1966.

Of the 6,843 public monuments still standing in Beijing in 1958, no fewer than 4,922 were destroyed by Red Guards.[31]

The Chinese Communist Party also declared war against the private sphere of life, in particular the institution of the family. Millions were forced to watch the public executions of thousands, and an acceptance of cruelty was inculcated into the everyday existence of generations of Chinese. Despite the seventy million who died as a result of Chairman Mao – three million in the Cultural Revolution alone – his mausoleum and picture still overshadows all else in Tiananmen Square in Beijing. Mao had a uniform plan for the whole of Earth, one which was fundamentally opposed in every way to the philosophy of the English-speaking peoples that values family, friendship, private property and the private sphere far higher than the state, and which venerates no human being as a living deity.

Yet as so often before and since, many amongst the English-speaking peoples, though guiltless themselves of the atrocities, looked the other way, or worse. The true nature of the Maoist regime in China was for many years deliberately downplayed by journalists and academics. Mirroring what Walter Duranty had perpetrated in Russia during the Thirties, reporters in China such as Theodore White of *Time* magazine, Brooks Atkinson of the *New York Times* and Arch Steele of the *Herald Tribune* sought to portray Mao and his supporters in the best possible light, concentrating on their 'agrarian democratic' credentials, rather then their murderous Marxist-Leninist beliefs. 'We were reluctant to paint them as real Communists', admitted Steele years later, 'because we knew that would go against the American grain.'[32]

The newsreels were no better: the breathlessly impressed British Movietone News broadcast of 13 October 1966 was entirely typical in its report of how nearly a million teenagers had completely filled the huge square of the Gates of Heavenly Peace, and in completely uncritical terms stated how, 'Their adulation for Mao and his teachings far exceeds hero-worship; singing songs in his praise, reciting poems from his verses, they appear to regard him something very like a god.' Self-censorship amongst those whose duty it was to tell the unvarnished truth meant that the English-speaking peoples were generally well disposed towards an emerging superpower that itself saw democracy and capitalism as bitter ideological foes.

Among the very worst Western apologists for the genocidal Chinese Communist Party were: the American journalist Edgar Snow, whose 1938 book *Red Star Over China* was hugely influential in the West, where people did not know that Snow had submitted his manuscript to Mao to be rewritten; the

Master of Caius College, Cambridge, Joseph Needham, who falsely accused the Americans of using biological weapons in Korea; the future Canadian premier Pierre Trudeau, who wrote a paean to Mao appositely entitled *Two Innocents in Red China*; François Mitterrand, who visited the country in 1961 and credulously repeated Mao's claim that, 'There is no famine in China'; the Cambridge economist Joan Robinson, whose 1969 book *The Cultural Revolution in China* argued that Mao's policies were the solution to Third World poverty, even though she admitted that China hadn't published any economic statistics since 1960; Felix Greene, who fawningly interviewed Chou en-lai for the BBC and believed every word he was told in a manner he never would have of a Western democratic politician; Simone de Beauvoir, who declared that Mao was 'no more dictatorial than, for example, Roosevelt was' since 'New China's constitution renders impossible the concentration of authority in one man's hands'; and her sometime consort Jean-Paul Sartre, who described Mao's 'revolutionary violence' as 'profoundly moral'.[33] Finally, in his 1973 book *A China Passage*, John Kenneth Galbraith stated that the moribund China had 'a highly effective economic system' whose claim of over 10% annual growth 'does not seem to me implausible'. China has indeed seen growth rates of that order, but only thirty years later when it effectively ditched Maoist-Leninism for *laissez-faire* capitalism.

The manner in which Western intellectuals were able to convince themselves that there was 'a greater truth' than the simple truth, one that was deserving of their utter fidelity in spite of the power of any contrary evidence, was hugely aided in the late 1960s by the philosophical theory of Postmodernism. In 1967, Jacques Derrida, an Algerian-born French intellectual, wrote two publications, *Writing and Difference* and *Of Grammatology*, in which he laid out the main themes behind this idea. The father of Deconstructionism and post-structuralism, Derrida pioneered a branch of critical analysis in the late 1960s that was to have a profoundly baleful effect on Western philosophy, anthropology, literature, linguistics, law and even architecture.

By arguing that language had multiple layers, and thus multiple possible interpretations, Derrida advanced the notion that speech was therefore not a direct form of communication and that the author of a text was not necessarily the author of its meaning.[34] By 'liberating' the written word from the structures of language, thereby opening up the possibility of endless textual inter-pretations of any piece of writing, Deconstructionism struck at the very heart of Western morality and ethics. It was not long before Derrida and his disciples attempted to apply Deconstructionism to political and moral values, arguing in effect that accepted mores were a gigantic fraud. The whole body of Western learning was thus denounced as 'nothing but the ideology of dead, white males', and as a result, as the Australian historian Keith Windschuttle has put it, 'Multiculturalists want the curriculum of higher education to be rewritten from a "gender specific" or "Afrocentric" perspective.'[35]

Derrida taught philosophy at the Sorbonne University in Paris from 1960 to 1964, but from the early 1970s he spent much of his time at American universities such as Yale, Johns Hopkins and the University of California at Irvine. There is something indefinable about the great English-speaking universities that welcome those who are in effect their intellectual enemies. The complexity and seemingly deliberate obscurantism of his written style led many philosophers of the English-speaking peoples, 'many of them reared in the tradition of plain-speaking Anglo-Saxon thought', to consider Derrida a fraud, and in 1992 twenty philosophers, including the world-renowned logician W.V. Quine, wrote a letter to Cambridge University protesting at the decision to award Derrida an honorary doctorate there.[36] A vote was then taken, which Derrida won. Nonetheless, one of the 40% who dissented, Howard Erskine-Hill, described the decision as 'symbolic suicide for a university'.

Derrida and Deconstructionism on their own would hardly matter were their disciples not in a position by the late 1960s to lead a long march through Western educational institutions, first the university humanities departments and then later the secondary schools. Because intellectuals saw Western society as fundamentally irrational, they became highly dismissive of it. This led them to assume that anywhere that the English-speaking peoples went to war – in Vietnam, for example, or latterly in Iraq – they were necessarily in the wrong. When a physical as opposed to a philosophical attack was made at the heart of Western civilisation, in Manhattan on 11 September 2001, Derrida refused to describe it as a terrorist act, arguing that 'an act of international terrorism is anything but a rigorous concept that would help us grasp the singularity of what we are trying to discuss', despite the fact that the attacks certainly seemed a rigorous enough concept for the people of New York to grasp at the time.[37]

Postmodernism, which the historian Richard Pipes has described as 'the latest poison to come out of France', further underlined the concept that nothing is objective, nothing 'is' but only 'seems to be'. Although this self-immolating culture flirts on the boundaries of nihilism, and tends to glorify destruction, Derrida always denied that he himself was a nihilist. In the 1960s, universities across the English-speaking world were to see department after department captured by the radical Left, whose grip on appointments and tenured posts was then near-impossible to loosen, lasting until even after the collapse of communism across Europe in 1989.

From the late 1920s until his death in 1937, the Italian communist intellectual Antonio Gramsci had preached the importance of attaining 'cultural hegemony' within Western institutions in order to promote Marxism. He believed that the bourgeois powers, by which he primarily meant Britain, America and France, could best be undermined through capturing the high ground of intellectual and elite leadership, which he thought as important economic factors in the class struggle. Thus capitalism could be destroyed from within, even without the need for recurring financial and economic crises. Gramsci was perhaps the most important communist thinker in the West since Marx himself, whose views he modernised and adapted for the twentieth century, and nowhere were his ideas followed more effectively than in academia.

In many faculties and in several fields, such as sociology, English literature and philosophy, the Left has dominated since the 1960s, teaching Western culture in terms of a series of crimes against humanity. They have thence taught the teachers, who have perpetuated these myths in the schools. Those opposed to Deconstructionism and Postmodernism, who argue that empirical evidence is vital and one can go from facts to truths, have all too often been ridiculed and sidelined in the Left's pursuit of Gramscian cultural hegemony.

In the field of history, Postmodernism can be particularly corrosive. Followers of the French philosopher Michel Foucault assert that historical accounts are merely narratives imprisoned by a language that Derrida had already proven was incapable of providing meaning, thus it is impossible to know the past, which can only be created. As the writer Patrick West recently pointed out in the *Times Literary Supplement*, 'According to these Poststructuralist relativists, we cannot even be sure that the Holocaust took place.' Some of the more extreme postmodernists even call for the abolition of history as an intellectual discipline altogether. This has led to the widespread conviction amongst educated people that postmodernists are all too often, as West calls them, 'merely disillusioned ex-Marxists who, despairing at the failure of the socialist experiment, have sought refuge in apathetic solipsism'.

In a number of respects, the Korean and Vietnam Wars were similar. An aggressive Marxist-Leninist dictatorship in the north of a peninsula attacked without warning a weaker neighbour to its south, with United States-led forces intervening to try to stem the advance of communism and defend if not democracy, then at least a relatively benign capitalist society. The great difference between them, which largely explains why the former war ended in victory whereas the second one ended in defeat, was that the Korean War began before the debilitating flood of counter-culture defeatism sapped America's will to win in the Sixties and Seventies.

Yet if anything, the Korean War, which today has none of the perceived illegitimacy of the Vietnam War because of the involvement of the United

Nations, was the more brutal. As Michael Lind points out in his revisionist work, *Vietnam: The Necessary War,*

> US bombing in Korea was less discriminating. . . . By the end of the Korean War, almost every city, town, and village in the Korean peninsula had been damaged or destroyed. . . . An estimated three million people died in the Korean peninsula in only three years. . . . By contrast, an estimated two million died in the Vietnam War during the space of a decade and a half. Seventy percent of those killed in the Korean War were civilians, compared to 45% in the Vietnam War. . . . The most concentrated, indiscriminate, hellish fighting took place, not in Indochina between 1959 and 1973, but in the Korean peninsula between 1950 and 1953.[38]

Because of the supposed moral inviolability of the United Nations, the first US-led protection of democracy is considered a righteous crusade, whereas the second equally honourable one is generally thought of even today as little better than a prolonged 'dirty war'.

Contrary to the received view of the Vietnam War, the United States was never defeated in the field of battle. General Vo Nguyen Giap, the Commander-in-Chief of the North Vietnamese Army (NVA), does not give interviews, but Bui Tin, who served on his General Staff, recalled in 1995 how he had acknowledged that the Tet offensive of 1968 had been a defeat for his forces.[39] General Colin Powell agrees: 'Judged in cold military terms, the Tet offensive was a massive defeat for the Viet Cong and North Vietnam. Their troops were driven out of every town they attacked, with horrific losses, estimated at 45,000 of the 84,000 men committed.' Yet the bitterness of the fighting at Hué and the fact that the NVA had managed to infiltrate Saigon itself, helped turn the all-important US media against the war. Absurdly optimistic bulletins put out by the US War Department between 1965 and 1968 – the so-called 'Five o'clock follies' – which reporters could see on the ground and hear from soldiers were largely untrue, also led directly to the media's disillusionment.

After suffering serious reverses at Tet and Khe Sanh, the Hanoi Politburo decided to change their strategy in May 1968 and prepare for 'regular war in which main forces would fight in a concentrated manner'. To achieve this, however, the Vietcong would have to be left to their own devices while North Vietnam built up her strength for when the Americans left. Negotiations were begun to that end with the United States, which culminated in a ceasefire agreement between the United States, North and South Vietnam and the Vietcong, signed in Paris on 27 January 1973, after which the last American troops 'left Vietnam, flags flying and bands playing' on 29 March. American POWs were subsequently also released.

It was not until over two years later – on 30 April 1975 – that Saigon and the South Vietnamese Government fell; in the North Vietnamese offensive that precipitated it, no significant main-force engagements involved American military units. This was a disaster since South Vietnamese forces were fighting at their best by 1973, beating NVA forces in key battles when they had the support of US air power. In fighting against Soviet and Chinese personnel who flew planes and manned missile sites, the South Vietnamese desperately needed US war planes, which were withdrawn in 1973. America had prevented the Vietcong insurgency from overthrowing the Saigon Government before then, but the US Congress would not prevent a North Vietnamese conventional invasion from achieving the same goal. The United States had forced her enemy radically to alter his strategy on the military front, but she lost her will to continue the fight because of protest and opposition on the domestic front.

Although the Great War had seen tremendous carnage and loss of life, the newspapers carried no images of the dead. Even in the Second World War, the first photograph of a dead American soldier – lying in the surf of a Pacific beach – did not appear in *Life* magazine until the US War Department had held it up for a full nine months. It was in Vietnam that graphic photo-journalism of American corpses was used to try to turn domestic support of the war into opposition.

The Left's campaign against the Vietnam War had been tireless and had regularly stooped to outrageous depths of mendacity. In 1968, the left-wing intellectual Susan Sontag published *Trip to Hanoi*, in which she wrote that the North Vietnamese 'aren't good enough haters' in their struggle against the United States, because,

> They genuinely care about the welfare of the hundreds of captured American pilots and give them bigger rations than the Vietnamese population gets, 'because they're bigger than we are,' as a Vietnamese army officer told me, 'and they're used to more meat than we are.' People in North Vietnam really do believe in the goodness of man ... and in the perennial possibility of rehabilitating the morally fallen.[40]

Of course anti-Americanism was by no means confined to America during Vietnam; the German historian Steven Ozment recalls watching student protesters at Tübingen University in 1968 daubing the buildings with 'USA', with the 'S' written as a swastika. Nonetheless, it was in America that it mattered most, because it was there – rather than in the jungles, paddy fields and deltas of the Indo-Chinese peninsula – that the war was truly lost.

The English-speaking media had long criticised the armed forces in times of war. Even though the British Commander-in-Chief Lord Raglan had entirely ignored him on campaign, *The Times* correspondent William Howard Russell had been a significant irritatant to the High Command during the Crimean War, writing detailed reports of the inadequacies of the commissariat and medical arrangements. Although *The Times* had banged the drum for the war before it started, it was soon denouncing what it called the 'incompetence, lethargy, aristocratic hauteur, official indifference, perverseness and stupidity' of the High Command.[41] Yet Russell, who in modern terms was 'embedded' with the British Army, was genuinely campaigning for a more efficient pursuit of the conflict, not for peace at any price. Much of the American news media during the Vietnam War had a very different agenda.

While the news coverage of the First and Second World Wars was generally undertaken highly responsibly, with journalists loath to report anything that might be of use to the enemy, as the Vietnam War progressed it became clear that some of the media was indeed a prime enemy of the conflict itself. The Left lost a sense of proportion which it failed thereafter to regain, as exemplified in 1989 when Ted Turner of CNN likened the tragic deaths of four students who were shot dead by national guardsmen at Kent State University while protesting against the bombing of Cambodia in May 1969 to the killing of over 2,000 pro-democracy demonstrators in Tiananmen Square in Beijing.[42]

It would be wrong to assume that the Vietnam War was opposed by a majority of Americans at any stage, except perhaps after it was lost. When Chicago's Mayor Richard Daley's policemen broke up an anti-Vietnam student demonstration outside the 1968 Democratic Convention there, and a poll asked Americans' views, 66% replied that they approved and only 20% disapproved.[43]

Today the Vietnam War is seen almost entirely through the eyes of the Left, as an unmitigated, ignoble disaster for the United States. This has been exacerbated by Hollywood's treatment of the conflict, which relentlessly and excessively portrayed every negative aspect to the virtual total exclusion of any positive ones. A large number of movies such as *Apocalypse Now, Full Metal Jacket* and *Platoon* made far better anti-Vietnam propaganda than anything the Vietcong could have produced, and the John Wayne movie *The Green Berets* provided virtually the sole response. In his 1988 book *The Hollywood History of the World*, George Macdonald Fraser, who had himself seen plenty of jungle action in the Far East in the British Army in the Second World War, gave this account of the unit portrayed in *Platoon*:

> They are brutal, degraded, nasty, hysterical, drug-sodden slobs, without decency or discipline, apparently hating each other, despising their leaders, and generally disgracing the profession of arms. Their evil genius is bad sergeant Barnes, who murders good sergeant Elias, a witness to Barnes' atrocities against Vietnamese villagers. Barnes in turn is murdered by the platoon innocent, an incident that gives point to the film's closing monologue: 'The enemy was in us; we fought ourselves.'[44]

Similarly in *Full Metal Jacket*, 'the platoon butt, after being beaten in cowardly fashion by his loyal comrades, goes mad, shoots the sergeant, and then commits suicide'. Anyone considering *Platoon*, *Full Metal Jacket*, or most of the other movies in the canon to be a realistic portrayal of the US Army in Vietnam has fallen under the delusion that Hollywood set out to create. Except for a very limited amount of 'fragging' – the murder of zealous officers by their own men by fragmentation grenades – the US Army was not its own enemy, nor did it fight itself. It had a perfectly obvious enemy, which it fought victoriously in every set-piece battle of the war. Certain units experienced demoralisation over certain periods in a long war, it is perfectly true, but the Vietnam War created conditions in which the sadistic fantasies of a large number of entirely unrepresentative film directors could be given full rein.

In fact, as McGeorge Bundy has pointed out, the Vietnam War gave South-East Asia an invaluable twelve-year breathing space in which to develop their societies peacefully; Singapore's long-standing leader Lee Kuan Yew was certain that it saved the region from falling under the communist maw. Maxwell Taylor's fear that 'If Vietnam goes, it will be exceedingly difficult to hold Southeast Asia' has been sneeringly referred to on the Left as 'the domino theory', yet because the United States held up communist insurgency in the Vietnamese peninsula for over a decade, the rest of South-East Asia did not 'go'. The incredible economic success of South Korea once it became democratic might be an indication of what could have happened in South Vietnam if it had not fallen to the communists in 1975.

Any American president would have risen to the challenge of protecting South Vietnam, as Kennedy was doing when he was shot, and Johnson did thereafter. Writing to LBJ on 17 February 1965, Eisenhower said that, 'The US has put its prestige onto the proposition of keeping SE Asia free. ... We cannot let the Indo-Chinese peninsula go.' The former Supreme Commander added that he 'hoped it would not be necessary to use the six to eight divisions mentioned, but if it should be necessary then so be it'. If the United States was prepared to defend South Korea and Taiwan in 2005, then of course, in the words of an American historian of Vietnam, 'it makes no sense to argue that it was irrational for the United States to defend its Indochinese protectorate at the height of the Third World War'.[45]

That protection cost the United States dearly; of her 205,023 combat casualties, she suffered 46,226 killed in action or dying of wounds, 10,326 non-battle deaths, 153,311 wounded and 5,486 missing, out of a total of 3.3 million Americans who served.[46] The maximum deployed strength was 625,866, on 27 March 1969. The rest of the Free World – principally Australia, South Korea, New Zealand, the Philippines and Thailand – suffered over 17,000 casualties. The South Vietnamese lost nearly 200,000 killed in action, and the North Vietnamese and Vietcong an estimated 2.5 million combat casualties of whom 900,000 were killed in action. These are horrifying figures, but if the Great War and Second World War and Korean War were honourable conflicts which the English-speaking peoples could not allow to be lost, so too was the Cold War, where there was just as much at stake.

Vietnam was in the tragic position of being in the front-line of the Cold War struggle, just as Korea had been, but to have simply abandoned her to a communist fate without attempting to protect her would have signalled to other countries in the region, indeed throughout the world, that the United States would not support her allies and could therefore not be relied upon when pressure came to be placed on them by Soviet imperialism. Prestige and credibility was as important to Washington in the Cold War as ever it was to London back in the days of the British Empire.

America's massive loss of prestige when her proxy was overrun on 30 April 1975, coming exactly two years to the day after Nixon's acceptance of responsibility for Watergate, represented the most humiliating period in the history of the English-speaking peoples since the fall of Singapore and Manila in early 1942. Yet the war was still justified, even if there were doubtless better ways to have fought it. 'America's involvement stretched over seven presidencies and was a unique succession of misjudgements,' is Paul Johnson's verdict, 'all made with the best intentions.'[47] It was not, however, an Unjust War, and there were plenty of fine American generals who were engaged in fighting it, including William Westmoreland, George Forsythe, Bernard Rogers, Creighton Abrams, Walter 'Dutch' Kerwin, William E. DePuy and Bruce Palmer.

There were alternatives to the way that the Vietnam story went, by which the United States would probably have had to help defend the border for a very long period after a long-term, low-intensity conflict. (She has defended South Korea in this manner for over half a century, one of her great oft-unspoken services paid to a free people.) Since an invasion of North Vietnam could have resulted in a Sino-American War, President Johnson was probably right to have avoided that option, but the 'massive, high-tech war of attrition' was probably in retrospect a mistake. What was certainly impossible, however, was simply to have allowed what one historian has called 'a panicked bug-out from Indochina in 1965'.[48]

The retreat from Vietnam had profound implications across Asia and the Pacific, not least in Australia where there had been an active anti-war movement. 'In the late 1960s and the 1970s came a remarkable surge in two strands of Australian nationalism,' records the distinguished Australian historian Geoffrey Blainey, 'the black and the green. Both groups pointed back to a golden age in which the Aborigines reportedly lived in harmony with the environment. Both groups regretted the arrival of a materialist European society which destroyed that harmony.'[49] Environmentalism and the Aboriginal rights movement grew side by side, were mutually supportive and had the same enemy.

The greatest victory of the 'black-green' coalition came in the still-controversial Mabo decision by the Australian High Court in 1992, which hugely extended Aboriginal land rights into geographical areas where they have only the barest historical (and allegedly religious) interests. The nursing of long-term historical grievances has tended to be a disastrous way for minorities to look to their future in Western societies, tying them to the past and producing a victim culture that is only of limited use in doing much more than giving them a sense of moral superiority over the dominant English-speaking one.

The Maoris, Roman Catholic Irish, French Quebecois, American Blacks and Native Americans have all dealt with this issue in their varying ways in the countries of the English-speaking peoples, and the most successful of them have tended to be the ones who simply got over their (often perfectly genuine) historical grievances and instead looked resolutely to the future, embracing competition rather than endlessly remonstrating and effectively self-ghettoising. In Australia, the Aboriginals and British had a 'long-standing difficulty during the first generations of contact ... the two peoples were so far apart in attitudes to land, kinship, work, the accumulation of possessions and many other facets of life'.[50] But since the English-speaking entrepreneurial capitalist culture was firmly established in Australia long before the 1960s, and certainly impossible to dislodge by then, the Aboriginals' best interests were to try to adapt to it.

It is astonishing how often anti-Americans will refer to American eating habits in their list of the supposed failings of their *bête noire* nation; a sign, perhaps, of how intolerant they often are. To despise a nation because a section of its population does not eat sufficiently nutritiously is a sure sign of a prescriptive, fundamentally illiberal outlook. The 'Big Mac' was the brainchild of Jim Delligatti, one of the earliest franchisees of Ray Kroc, the founder of McDonald's, who by the late 1960s operated a dozen stores in Pittsburgh. It proved successful and in 1968 the burger was introduced across the company and revolutionised fast food for ever. 'Yet fast food was not invented by Americans,' as the anthropologist Professor Sidney Mintz has noted. 'People have been eating it for centuries, and probably millennia. Anyone who watches cooks at work in the canoes on Lake Xochimilco, or on the kerbsides of busy New Delhi streets, or in the market squares or subway stations of Hong Kong's New Territories, knows that. But something special happened in the American case. Americans linked fast food to an entirely mobile clientele.'[51]

Fast food liberated ordinary working people from the time-consuming prepared meal, should they wish to use their free time differently. The concept of a slice of meat between two pieces of bread, that had originated with the Earl of Sandwich in the eighteenth century to help him gamble at Brooks's Club, was perfected by the Americans in 1968 (albeit with the addition of one extra piece of bread and meat and two of cheese).

As the twentieth century dawned, Asa Candler was selling 200,000 gallons of 'Cola' syrup, each gallon capable of making 400 Colas, or a total of eighty million servings per annum. Cola had been invented by John Pemberton in Atlanta in 1869, and in 1888 the recipe was sold to Candler. After two further entrepreneurs, Benjamin F. Thomas and Joseph B. Whitehead, bought a contract to bottle it, the system was in place that was to make Coca-Cola an international corporation operating in over 200 countries. It was at the St Louis World Fair in Missouri that the ice-cream cone was invented in 1904, setting in place all the necessary ingredients for America's great fast-food revolution, six decades before it actually took place.

The November 1968 presidential election, won by Richard Nixon and Spiro T. Agnew, was testament to the rich quality of American political life and debate. As well as Nixon himself, the Republican contenders were Governors Nelson Rockefeller of New York, George Romney of Michigan and Ronald Reagan of California, while the Democrats fielded Vice-President Hubert Humphrey and Senators Robert Kennedy, Eugene McCarthy and George McGovern. Almost any of them could have made a fine peacetime president, but the American people needed a wartime one and chose Richard Nixon. Disastrously for him as it turned out, the Democrats retained large majorities in the House (243 to 192) and the Senate (58 to 42).

On Sunday, 20 July 1969, Neil Armstrong, commander of the *Apollo* 11 space module, walked on the surface of the Moon, the first human being to do so. Rather sadly he fluffed his words, missing out the indefinite article by saying, 'That's one small step for man, one giant leap for Mankind', rather than 'a man', but since the pun he was articulating was pretty banal anyhow, given the august circumstances, it hardly mattered. He was soon afterwards joined by Edwin 'Buzz' Aldrin, and together they installed and operated the first lunar seismograph at Tranquility Base, spending a total of twenty-one hours and thirty-seven minutes on the Moon's surface, while their colleague Michael Collins remained orbiting the surface in the command module. The lunar landings rate as the single greatest technological triumph of the English-speaking peoples since 1900.

The policy to put a man on the Moon had raced up the US Government's agenda on 12 April 1961, when the twenty-seven-year-old Soviet cosmonaut Yuri Gagarin became the first man in space. He spent 108 minutes there in

Vostok 1. Rather more poetically than Armstrong, Gagarin reported that, 'The Earth looked a delicate blue floating in a black sky.' Professor Sir Bernard Lovell, chief of the Jodrell Bank radio-astronomy observatory in Cheshire, England, warned that the Russians, who had 'broken the barrier that binds Man to Earth', would probably land someone on the Moon within seven years, which was 'a lesson that the peoples of the West disregard at their peril'. Lovell added that Gagarin's achievement represented one of the greatest events in the history of Mankind, opening the way to the exploration of the solar system.

A central figure in putting America's first satellite – *Explorer* 1 – into orbit on 31 January 1958 had been the New Zealander Sir William Pickering, who headed California's Jet Propulsion Laboratory between 1954 and 1976. Pickering was involved in many space missions, including *Mariner* 11 which explored Venus, and was one of the few non-politicians to appear on the cover of *Time* magazine twice.[52] With men like Pickering and Lovell urging that a lunar landing was well within America's capabilities, and rightly mindful of the propaganda advantages of American science and technology being shown to be superior to the Soviets', the very next month President Kennedy committed $2 billion to the project of 'landing a man on the Moon and returning him safely to earth before the decade is out'. It was achieved with only five months to spare, but capitalist American technology had beaten communist Russian technology. As the historian Anthony Pagden put it in his book *Peoples and Empires*, 'What mattered most was national prestige. The disinterested pursuit of science was transformed into a new kind of ideology, and the scientist became a new kind of hero.'[53]

The deaths of all seven astronauts aboard the *Challenger* space shuttle on 28 January 1986 underlined the heroism still involved in this hazardous method of travel. Indeed, as one of the men who walked on the Moon has pointed out, today there is more technology under the dashboard of a rental car than there was guiding man to a safe landing on another planet, and the technology that had crews escaping earth's gravity in the Sixties and Seventies wouldn't work a modern laptop computer.[54] Spacecraft can now reach the Moon in nine hours, whereas in 1969 it took three days.

Despite it being the Stars and Stripes that were planted on the Moon, as opposed to the UN insignia, *Apollo* 11's achievement had the effect of promoting the concept of the planet having universal interests. 'Other, earlier civilisations and empires established a dominant culture within their boundaries,' the historian Hywel Williams has recently pointed out. 'Those boundaries seemed to them the limits of culture itself. But the capacity to see a single world with one pair of eyes – a moment that perhaps first arrived when humans first went into space in the 1960s – did have a deep emotional impact on such parochialism.'[55]

It is true that space travel has come nowhere near directly repaying financially the huge investment made by the US taxpayer, although there have been considerably more advantages gleaned from it than merely non-stick frying pans. Without satellites in space, for example, our telephone communications would be far less advanced, our weather forecasts even more unreliable and the Global Positioning System has proved invaluable for soldiers, explorers and even taxi-cabs. Also, Osama bin Laden was deeply irritated that 'the Infidels survived the blasphemy of walking on the moon', so it wasn't all money wasted.[56] It is nonetheless very hard to see how the US taxpayer can possibly start to justify – let alone recoup – the trillion dollars that the manned Mars mission announced by President Bush in January 2004 is expected to cost.

Despite the Sixties ending on such a bright, hopeful note for the English-speaking peoples, with their language being the only one spoken on another body in the cosmos, the next decade was to witness a series of debilitating retreats and defeats, the worst of the century so far.

FOURTEEN

The Long, Dismal, Drawling Tides

The 1970s

The Dawn of Modern Terrorism – Edward Heath's U-turn – The Nixon Administration – Henry Kissinger – The US State Department and Europe – President Allende overthrown in Chile – The Rock of Gibraltar – The Yom Kippur War – Spiro Agnew and 'Sleaze' – Watergate – Hollywood Paranoia – The dismissal of the Whitlam Government – The Fall of Saigon – Détente – The Wilson Premierships – The Carter Administration – US hostages in Teheran

'The ancient insanity of governments: the mania of wishing to govern too much.'
                                                            Maximilien Robespierre

'Much of the world today, including the United States, is still living in the social, cultural, and political aftermath of Britain's cultural achievements, its industrial revolution, its government of checks and balances, and its conquests around the world.'   Thomas Sowell, *Conquests and Cultures*

The only British prime minister since the Second World War to doubt the value of the Special Relationship was Edward Heath, whose keenness to establish his name in history as the man who took Britain into the European Union left him antagonistic towards the United States. On 8 September 1970, Heath wrote to his Foreign Secretary, Sir Alec Douglas-Home, saying that the Nixon Administration had asked him 'on a number of occasions to agree to facilities of one kind or another ... in connection with our remaining island dependencies', such as Ascension Island, Diego Garcia and Mauritius. Heath wanted to know 'what we get in return for all this from the Americans. I may be wrong but so far this seems to me to be very much a one-way movement.'

Ten days later, the Foreign Office furnished Heath with an eight-page document marked 'Secret' that ought to have stilled the criticisms of any but the most fervently anti-American of politicians. The help the United States furnished Britain in the area of defence alone, the memorandum stated, was 'ranging over the fields of operations, training, communications, intelligence, and technical information of a wide variety'. Douglas-Home pointed out that 'the complexity and breadth of these exchanges is impressive' and Britain's possession of island dependencies around the world 'enables us to make a considerable contribution to an alliance which is important to both of us but in which otherwise our respective contributions might be very ill-balanced'. He went on to mention 'the massive American military, technological and intelligence machine to which we can ourselves contribute usefully but from which we in return derive benefit which no-one else could give us'.

Specifically, Douglas-Home listed the fact that the US had allowed British nuclear weapons to be transported from the Far East across the continental USA; the intelligence provided by the US Coastguard to the Bahamas Patrol; the use of landing fields on American territory for RAF aircraft during the Antiguan emergency; help over the evacuation of British citizens from Jordan in 1969; shared facilities in the Philippines; intelligence data contributions from the National Security Agency that were described as 'irreplaceable'; non-NATO defence research and development collaboration in which 'we have gained more than we have given'; 'Since 1958 we have had a deep and intimate exchange of information with the Americans on nuclear weapons technology, involving frequent exchanges of information and equipment . . . and occasional use of each others' facilities for specific purposes', and the joint monitoring of French and Chinese nuclear testing. There were also bilateral Anglo-American projects covering fuel cell research, advanced lift engines, the use of beryllium in aero-engines, the ballistic early-warning system, the Mallard tactical trunk communications system and much more besides. On top of that, 'the importance to Britain and the Western nations afforded by the US nuclear capability and the deterrent needs no emphasis'.

Finally, there was the Technical Co-operation Programme, which was founded in 1957 and which by 1969 encompassed most major English-speaking countries – the US, UK, Canada, Australia and New Zealand. This covered widespread joint work and information exchange on 'guided missiles, chemical-biological defence, defence against ballistic missiles, undersea warfare, aircraft and aero-engines, electronic devices, infra-red, radar techniques, military space research, nuclear weapons effects, ordnance, materials, electronic warfare, counter-insurgency research, communications techniques, ground mobility, and biological, bio-chemical and social sciences'. The Programme is still very much active today, with 6,000 scientists attached to it worldwide in 2006, and represents further proof that the United States sets great store by the special relationship that she continues to enjoy with the rest of the English-speaking peoples. The sheer comprehensiveness of the Foreign Office's response seems even to have blunted Heath's scepticism about the United States for a short while. Whatever else it was, it was not a 'one-way movement'.

★

On Sunday, 6 September 1970, the English-speaking peoples entered a terrifying new phase of their existence: the era of international terrorism. Leila
Khaled, a member of the Popular Front for the Liberation of Palestine
(PFLP), hijacked an El Al flight from Tel Aviv to London. At the same time
three other planes were taken over in the air, one of which was flown to Cairo
and the other two to Dawson's Airfield, near Amman, Jordan. Fortunately,
the passengers overwhelmed Khaled and she was arrested and held at Ealing
police station in London, while the other terrorists demanded her release,
using the passengers on the other three planes as hostages and threatening
bloody reprisals against them.

The Conservative Government in Britain under Edward Heath, which had
been elected that June, decided to appease the PFLP and release Khaled,
which in the long term was quite the worst decision it could have taken,
unleashing copy-cat hijackings and much worse over more than three decades.
The British politician Enoch Powell warned at the time that Heath's action
was 'not only wrong in itself but fraught with grave consequences for the
future', and he was soon proved right. Would-be terrorists around the world
were led to believe that Western governments – the Germans and Swiss also
acceded to the PLFP's demands – would capitulate if enough hostages were
taken. As a result, the Seventies saw a terrifying upsurge in such acts. For all
his trumpeting of his support for Winston Churchill in the Thirties as a
teenager, Heath had clearly not learnt the central message from that terrible
decade: that if you 'feed the crocodile', you might get eaten last, but your
ultimate fate remains the same.

Heath knew the risks; even British Movietone News reported the Khaled
story with the comment: 'Whether you approve a policy of appeasement or
not, once the principle is established, once the sky-jackers are allowed to hold
the world to ransom, we will have paid a price to lawlessness that can never
be redeemed.'[1] Within two years the Black September group of Palestinian
terrorists murdered eleven Israeli athletes at the 1972 Munich Olympics and
terrorism became an established part of global political life. An absolute refusal
to deal with terrorists by Heath and all other Western leaders in 1970 might
have led to the deaths of some air passengers in the initial stages of the
campaign, but sooner or later the message would have got through to the
terrorists that such outrages did not pay. If so, the atrocities of 9/11 might
even have been avoided.

There was no doubt that the foundation of the State of Israel in 1948 left
many Palestinians feeling dispossessed, but instead of following the prescribed
democratic, constitutional and internationalist routes for the relief of their
grievances, their political leadership embraced a strategy of terrorism, entering
a political cul-de-sac from which they had not emerged even over half a
century later. They failed in 1948 to accept the offer of a two-state solution,

proposed by the UN, even though Israel (reluctantly) accepted it. When the opportunity arose again following the 1967 Six-Day War, when Israel proposed withdrawal from the newly occupied West Bank and Gaza Strip in return for Arab recognition of Israel's right to exist, the Arab summit answered with the infamous 'Three No's', namely no peace with Israel, no recognition of Israel and no negotiations with the Jewish state. Three decades later, they rejected another opportunity of a land in exchange for peace deal, presented at Camp David and brokered by Bill Clinton in the latter days of his presidency. As at Oslo such rejections by the Palestinian leadership condemned their people to waging a war they simply could not win, against a Jewish people who would sooner fight to the end than return to their almost 2,000-year-long nightmare of statelessness and recurring persecution.

Heath's surrender to PLFP blackmail set a leitmotif for the rest of the 70s in Britain. During that decade, there was a tangible sense among the governing classes that the UK was in irreversible decline and that there was nothing much she could do about it. Indeed, it was feared by the front benches of both political parties that even attempting to take steps to reverse that decline would merely exacerbate the situation. Politicians, civil servants, businessmen, academics, writers, opinion-formers and diplomats tended to assume that Britain's time as a major Power had passed, and the best that they could do was to manage her inevitable decay in as civilised a way as possible. Although many senior political decision-makers in those days – including Heath himself – had had 'a good war' between 1939 and 1945, they often evinced a moral cowardice when it came to fighting the problems that beset the country in peacetime. The worst of these by far, throughout the 1970s, was the issue of industrial relations.

After a national dock strike in July 1970, the Heath Government declared a national State of Emergency (the first of three), but then it paid off the dockers. Despite the Trade and Industry Minister John Davies assuring the Conservative Party Conference that, 'I will not bolster or bail out companies where I can see no end to the process of propping them up' in October 1970, the very next month, £48 million of taxpayers' money was given to Rolls-Royce Ltd in order to offset losses. In December, in the depths of mid-winter, the power workers started a work-to-rule for a 25% pay increase. Christmas lights were blacked out and there was even a run on candles. Later Heath's Government nationalised the aero-engine and marine divisions of Rolls-Royce, the first such action since Clement Attlee twenty-two years previously; meanwhile, top income-tax rates were at 75 pence in the pound and British Rail went on work-to-rule for a 16% pay claim.

The worst act of appeasement – barring the Khaled surrender – came in 1972 when the Heath Government performed a spectacular and comprehensive U-turn on all its major industrial policies, ditching every commitment it had made only two years earlier in its election manifesto. Yet this *volte face* was swallowed by a Conservative Party that could not bear confrontation. All its promises, of tax cuts, free-market reforms, immigration controls, law-and-order measures and legislation to control the trade unions, were abandoned overnight in an act of mass collective funk.

On 6 November 1972, the Government announced a ninety-day freeze on all prices, wages, rents and dividends. Every customer was invited to be a nark, reporting shopkeepers who sneaked so much as a penny on to a can of beans. As under socialism, a huge bureaucracy geared up to administer every aspect of the economy. A Pay Board and a Prices Commission set incomes and prices, rather than the free market; ministers were now ultimately responsible for what everything cost in the shops. Politicians meeting in the Treasury decided the levels for plumbers' rates, taxi fares and the rents on furnished versus unfurnished flats. Even in wartime the Government had never interfered in the minutiae of economic decision-making to such an extent. Meanwhile, subsidies of £175 million were paid to the National Coal Board in December 1972, as well as a debt write-off of £475 million.

The Conservative Party went along with this complete role-reversal and rejection of manifesto commitments and election pledges, although the private views of many of them were articulated in public in the House of Commons on 7 November 1972, when Enoch Powell asked the Prime Minister, 'Does my right honourable friend not know that it is fatal for any Government or party or person to seek to govern in direct opposition to the principles on which they were entrusted with the right to govern? In introducing a compulsory control of wages and prices, in contravention of the deepest commitments of this party, has my right honourable friend taken leave of his senses?'

The U-turn, despite its comprehensive nature, did not save Heath's Government. Gas supplies were simply cut off for a whole month in February 1973; the mortgage rate reached 11%; petrol-rationing coupons were distributed and street lights were extinguished early to save energy. On 13 December 1973, a three-day working week was instituted, industrial suicide for any nation that survived on trade. The following February, the miners began an all-out strike in support of a pay rise of over 30%, which led to regular nationwide power-cuts. Heath then called a general election on the question 'Who Governs Britain?', to which the electorate replied: 'Anyone but you'. The 1970–4 Conservative Government holds the record for being the sole ministry since the war which once elected for the first time has failed to be re-elected.

One person at least had learnt the bitter lesson of the disastrous Heath U-turn of 1972. Eight years later, at the 1980 Party Conference, when unemployment was far higher than the one million that it had been under Heath and the Conservatives were trailing badly in the opinion polls, Margaret Thatcher nevertheless defiantly said, 'For those waiting with bated breath for that favourite media catchphrase "the U-turn", I have only one thing to say. You turn if you want to. The lady's not for turning.'

Perino's restaurant in Los Angeles on the evening of Thursday, 15 July 1971, was the venue where Richard Nixon, Henry Kissinger, Bob Haldeman, John Ehrlichman and White House Press Secretary Ron Ziegler flew by helicopter for a dinner to celebrate the announcement that the President would be making a state visit to China. After twenty-two years of more or less outright hostility, relations between China and the United States were to be 'normalised'. Nixon and Kissinger discussed what wine would be appropriate to celebrate this astonishing diplomatic and political coup, deciding on a magnum of 1961 Lafite-Rothschild, which cost $600, but which they nonetheless got for $300.

The dinner was a success and toasts were drunk to the Chinese venture as it was agreed that there would be no further discussion with the press about the details of what 'normalisation' might involve in practice. After the meal, Nixon spoke affably to other diners and tourists, who were surprised to see the President there. Meanwhile, the press were told that the wine that his party had drunk cost $40. It was a small, unnecessary lie and it was not found out at the time, but it demonstrates the paradox at the core of the Nixon White House. A fine, dramatic political initiative – bold, brave and counter-intuitive – was immediately accompanied by a grubby, demeaning, unnecessary untruth.

The historian and biographer of FDR, Conrad Black, has made the case for the Nixon Administration succinctly, pointing out that when it came into office in January 1969,

> the United States had 545,000 conscripts in an undeclared war at the ends of the earth, no arms control negotiations with the USSR, no worthwhile contacts with China, and almost no relations with the Arab powers apart from Jordan, Tunisia and Morocco. The Soviet Union and its allies were generally gaining ground in their conflict with the West and almost casually suppressed efforts at liberalisation in the Eastern bloc, as in Prague in 1968. After five years, the United States had negotiated its departure from Vietnam with as much dignity as the circumstances allowed. It had negotiated an arms limitation agreement with the USSR, the beginning of a civilised relationship with China, and the US-brokered beginning of the normalisation of relations between Israel and some of its Arab neighbours, with the consequent elimination of almost all Soviet influence in the Middle East.[2]

It was indeed a formidable achievement, one that has been partly overlooked by history because of the Watergate scandal.

During his five-and-a-half-year presidency, Richard Nixon watched no fewer than 535 movies, mostly screened at the White House, Camp David,

Key Biscayne and his Californian ranch at San Clemente. He watched his first one, *In the Shoes of the Fisherman,* on his third night in the White House in January 1969 and his last, *Around the World in Eighty Days* (for the third time), in July 1974, just before his final departure. He never spoke, ate or slept during them, and always watched them through to the end, whatever he thought of them.[3] If Franklin D. Roosevelt was a difficult president to read, Richard Nixon was near-impossible; Nixonologists write of his 'inscrutable ungainliness'.

Yet Nixon had been known to the US people as vice-president for eight years. He certainly didn't accept that the modern vice-presidency had to be simply, in the words of the old gag, a question of 'waking up every morning and inquiring after the health of the president'. By the end of his time as Eisenhower's deputy, Nixon was an instantly recognisable figure both nationally and internationally. He was helped in this by the fact that Eisenhower was seventy by the time he left the White House and perfectly willing to leave much of the more arduous travelling to him. He was also helped by the violent demonstrations against him that took place in Peru and Venezuela in May 1958, which understandably angered Americans.

'In all the modern history of man's inhumanity to man,' the Secretary of the US Navy John H. Chafee told the US Naval Institute in July 1971, 'there is no example of crueller or more inhuman treatment than that being dealt to our prisoners of war and their families by the North Vietnamese.' He was right; although North Vietnam had signed the 1949 Geneva Convention, she argued that since the United States had not formally declared war, the Americans she captured were 'criminals' or 'air pirates' but not prisoners of war, who were thus afforded no legal protection.

There were virtually no tortures too loathsome for the North Vietnamese or Vietcong to visit upon American servicemen. They were hung from the ceilings of cells, dragged along the ground with broken legs, deprived of food and sleep, subjected to months and on occasion years of solitary confinement (the maximum allowed under the Geneva Convention was thirty days), beaten, trampled and kicked, interrogated under extreme duress, incarcerated in foul and tiny cells averaging 8 feet by 8 feet, and denied proper medical treatment.[4] Some lost their minds, many died, and no fewer than 700 signed statements condemning the capitalist warmongers of the White House and Wall Street's 'aggression against the peace-loving people of Vietnam'.

In his 1976 autobiography *When Hell Was In Session,* Rear-Admiral J.A. Denton described how he wrote, 'Dear Ho Chi Minh, I am sorry I bombard your country. Please forgive me', something for which he never forgave himself, but which, considering the physical and psychological pressures on him in the French-built Hoa Lo Prison – ruefully nicknamed the Hanoi

Hilton – was perfectly understandable. (Furthermore the ungrammatical present tense betrayed the fact that it was dictated to him for signing.) Other inmates, like Lieutenant-Commander John M. McGrath, who recorded his experiences in *Prisoner of War: Six Years in Hanoi* in 1975, refused to condemn 'the imperialist warmonger United States' and was forced to kneel with a small rock under each knee for thirty hours.[5]

The man chosen by Nixon as his National Security Advisor, Dr Henry Kissinger, remains one of the most controversial public servants of the English-speaking peoples since 1900. Coming to office at the beginning of a time of collapsing US power, yet trying to conserve it globally for eight tumultuous years in any way he could, Kissinger has been regularly accused of almost every possible moral, political and war crime imaginable, both by the Left and – because of his policy of US-Soviet engagement known as '*Détente*' – sometimes also by the Right. Both sides virulently accuse him of being insufficiently concerned about the global abuse of human rights.

Yet Kissinger needed no lectures on human rights. On a visit to Bavaria in 1975, the German Government issued a press release about his meeting the family he had in the land of his birth. 'What the hell are they putting out?' he asked. 'My relatives are all soap.' He himself left Germany for America in 1938, but no fewer than thirteen family members perished in the Holocaust. Kissinger's experiences in the Second World War, his 1954 Harvard PhD thesis on the Congress of Vienna (later published as *A World Restored*), his seminal *Foreign Affairs* magazine article of 1955 on limited nuclear war, his work on the Committee on Foreign Relations and his reading of the Cuban missile crisis left Kissinger a convinced supporter of the concept of *realpolitik* long before he was appointed National Security Advisor by Nixon in January 1969 and put it into active practice. As such, there could have been no better person to advise Presidents Nixon and Ford during so much of the Seventies, a period when – through no fault of his own – American prestige was to reach its post-1900 nadir.

'Vietnam may be one of those tragic issues that destroys everyone who touches it,' wrote Kissinger, who had opposed Kennedy's initial despatch of 1,600 'advisors' there, although he supported LBJ sending in combat troops in 1965. Kissinger spent two weeks there that year staying with Ambassador Henry Cabot Lodge and meeting Nguyen Van Thieu, returning twice the following year. He quickly developed the stance that withdrawal would be disastrous for American prestige but also that negotiations were inevitable, a view that he essentially stuck to right up until he signed the peace treaty eight years later. In July 1967, Kissinger secretly opened up a connection between Washington and Hanoi via a French friend of Ho Chi Minh. During these short-lived and inconclusive talks, he developed several diplomatic traits that were to stay with him for ever. The first was a taste for secret diplomacy that cut out the State Department, the media and public opinion. The second was for face-to-face 'shuttle' diplomacy in which one-to-one deals were made based on personal understanding. Both fitted in perfectly with his Metternichian theories of relations between powers, and they were ultimately to stand his country in good stead.

'We could not simply walk away from an enterprise involving two Administrations, five allied countries, and thirty-one thousand dead as if we were switching a TV channel,' Kissinger has said of Vietnam. The military, political, diplomatic and moral issues at stake there could hardly have been higher. When Nixon took office, America had 545,000 troops stationed in Vietnam, who were dying at the rate of 200 per week. However disgraceful it was for Daniel Ellsberg to have leaked 'The Pentagon Papers', they did show that the strategic thinking in the JFK and LBJ Administrations was remarkably similar to Nixon's: the war was only winnable through massive engagement of forces over a wide front for a long time.

The 'secret' bombing of Cambodia that began in February 1969 involved scores of B-52s flying 1,045 missions and dropping 108,823 tons of bombs.[6] It was perfectly justifiable strategically, since the enemy had taken up positions on the flanks of the central battle, within neighbouring Cambodia, and modern wars cannot be kept localised on one side only. However, constitutional issues were raised about the extent to which Congress needed to be kept informed about the bombing of a (strictly speaking neutral) country's territory, even if the parts of Cambodia that were bombed were effectively occupied by the Vietcong.

'I can't believe that a fourth-rate power like Vietnam doesn't have a breaking point,' said Kissinger, but it appeared that the totalitarian Marxist-Leninist regime was capable of absorbing however much punishment the US-led coalition forces meted out for its southward incursions. As Colin Powell, who served in Vietnam with distinction, later wrote of the Tet offensive, 'It did not matter how many of the enemy we killed. The Viet Cong and North Vietnam had all the bodies needed to fling into this conflict and the will to do so. The North simply started sending in its regular army units to counter the losses.'

The Nixon Doctrine, enunciated in Guam in July 1969, provided for a progressive withdrawal from Vietnam, but this created the classic diplomat's problem for Kissinger: how to negotiate from a position of increasing weakness? Kissinger himself said of his opposite number in the negotiations, North Vietnam's Le Duc Tho: 'He had not suffered in prison for ten years, and fought wars for twenty years, to be seduced now by what a capitalist fancied to be his charm.' The negotiations were tough but eventually bore fruit.

Nixon's 1972 election triumph was soon followed by the 'Christmas bombing' of Hanoi and Haiphong in 1972, in which ninety-three US airmen and perhaps 1,600 North Vietnamese died. Nixon and Kissinger had reduced the number of GIs in Vietnam from 545,000 in 1969 to 27,000 by 1972, who were facing 140,000 North Vietnamese and a similar number of Vietcong troops in the south. The peace deal was signed on 27 January 1973, after the 'Christmas bombing' proved that the Administration still had the ability and will to strike against North Vietnam if necessary. On 29 March 1973, the last American troops left Vietnam. All the peace itself could do, however, was to provide 'a decent interval' – in the event, two years – between the US Army's withdrawal and South Vietnam's defeat. Afterwards a new bout of American insularity – the most dangerous since 1919 – beckoned.

In his 1968 essay, *Central Issues of American Foreign Policy*, Kissinger wrote a paragraph that might be taken as his diplomatic *raison d'être* over the following hard-fought decade:

> The greatest need of the contemporary international system is an agreed concept of order. In its absence, the awesome available power is unrestrained by any consensus as to legitimacy; ideology and nationalism, in their different ways, deepen international schisms. ... A new concept of international order is essential; without it stability will prove elusive.'[7]

As National Security Advisor between 1969 and 1975, and Secretary of State from 1973 to 1977, Kissinger's period as a formulator of US foreign policy coincided with many of his country's most perilous post-war predicaments. The threats to American prestige posed by the loss of will in Vietnam, the Yom Kippur War and its aftermath, nuclear proliferation, the overnight quadrupling of the price of oil, Watergate, Soviet expansionism in Africa, Asia and Latin America, and even, at one point, Cambodian high seas piracy, all landed in his in-tray. Any one of those crises might have broken a lesser diplomat.

At the time of the 1973 oil price hike, Kissinger and the Defense Secretary James Schlesinger looked into the possibility of seizing Arabian oilfields in retaliation – 'securing the oil by our own means on a continuing basis' – but contented themselves with a deal with King Faisal of Saudi Arabia by which the US Navy was still supplied for the anti-communist struggle in Vietnam. Kissinger proved himself a good wartime and peacetime *consigliere* because of his appreciation of *realpolitik*; however, when the time finally came to ditch *Détente*, to go on the offensive and put ideology before stability, in the Eighties, different personnel were needed.

Kissinger presents his policy as having been a judicious balance between the ethically driven foreign policy of the East Coast Establishment Left and the crusading anti-communism of the neo-conservative Right, for both extremes of which he displayed a caustic contempt. American national self-interest was always his watchword, although he was unduly hyperbolic in the last volume of his autobiography, *Years of Renewal*, when he accused the neo-conservatives, who, under Ronald Reagan, did after all preside over the collapse of the USSR, as advocating the taking of 'unwarranted risks in the face of the consequences of nuclear war'.

From 1969 to the nadir of America's humiliation in April 1975, Kissinger set out to ensure that wherever the tectonic plates of Russia's and America's informal empires rubbed against one another, America would not come off second-best. Yet neither did he significantly challenge the Brezhnev Doctrine, by which Russia never gave up what it had once taken. By making him out to be little less than a modern Mephistopheles, admittedly with Woody Allen's gift for one-liners, Kissinger's critics level charges so vitriolic as to damage their own case against him. In fact he did what any diligent and responsible American Secretary of State would have done while the Free World was in overall retreat before the forces of communism. The atrocities undoubtedly committed by anti-communist governments during his watch – such as in Indonesia and Honduras – were after all not the fault of the United States or Kissinger, but of the regimes that actually carried them out, and need to be seen against the backdrop of the enemy that they were facing.

Overall, the Seventies were a disastrous period for the Free World, but it was partly down to Nixon and Kissinger that they were not worse. They could be highly manipulative, secretive and occasionally ruthless, but America desperately needed modern-day Machiavellis in the uniquely dangerous cir-cumstances of the time, when the US Congress point-blank refused to stand up for American interests abroad. No-one today decries Talleyrand for being 'manipulative', since it is occasionally part of a foreign minister's job. In trying to appreciate the fantastically difficult line Kissinger had to tread, it is worthwhile considering what he was up against domestically.

Just as Wilfred Burchett and many others had supported the North Koreans during the Korean War, so some Americans went so far as effectively to support the North Vietnamese twenty years later. The actress Jane Fonda, for example, on a visit to North Vietnam in 1972, even sat in the firing-seat of an NVA anti-aircraft gun and posed for photographs looking admiringly at the battery's crew. On the evening of 22 August 1972, the thirty-four-year-old Fonda made a long radio broadcast from the Hotel Especen in Hanoi, during which she said,

> I cherish the memory of the blushing militia girls on the roof of their factory, encouraging one of their sisters as she sang a song praising the blue sky of Vietnam – these women, who are so gentle and poetic, whose voices are so beautiful, but who, when American planes are bombing their city, become such good fighters.[8]

She then went on to describe Nixon as 'a true killer' and 'neo-colonialist', who committed war crimes, and recommended he read Vietnamese poetry, 'particularly the poetry written by Ho Chi Minh'. The fact that the United States never wished to colonise, or 'neo-colonise' North Vietnam, but only wished to preserve South Vietnam from her neighbour, was simply ignored.

Although Jane Fonda's work with and financial support for organisations such as Vietnam Veterans Against the War – which numbered 7,000 members at its highest, out of the two-and-a-half-million Americans who served there – and Fuck the Army (which encouraged soldiers to desert), were within her constitutional rights, her Hanoi visit resulted in her informing the news media on her return to the United States that American prisoners of war were being well treated. When the following year returning POWs comprehensively disproved her, she called them 'hypocrites and liars'.

Nor did it end with the war. After the fall of Saigon in 1975, Fonda returned to Hanoi for the victory celebrations. During them, she christened her new-born son Troy, after a Vietcong fighter Nguyen Van Troi who in 1963 had attempted to assassinate US Secretary of Defense Robert McNamara on a visit to South Vietnam. Bui Tin, who served on the NVA General Staff and who received South Vietnam's unconditional surrender on 30 April 1975, told the *Wall Street Journal* twenty years later that the American domestic anti-war movement had been 'essential to our strategy' and recalled how the political and military leadership of North Vietnam used to listen to the US evening news bulletins, in order 'to follow the growth of the American anti-war movement'.[9]

In 2005, in the pre-publication publicity for her autobiography, Fonda said that she regretted posing for the photographs, but stated that, 'The majority of Americans opposed the war. It was a desperate time.'[10] This was not true: in poll after poll during 1972, and indeed right up until 1975, more Americans supported the war than opposed it. If it was 'a desperate time', it was helped to be made so by the vociferous domestic opposition to the war that could not help but deflate the morale of the men fighting in the field to preserve South Vietnam's independence. A recent letter to the *Times Literary Supplement* from the KGB defector Oleg Gordievsky, someone who knows a great deal about totalitarianism, put the Vietnam situation in its proper historical context. Denying that the United States had been militarily defeated, he wrote:

> North Vietnam was a totalitarian country, while South Vietnam was a liberal and pluralistic society, needing protection. If the USA had really wanted to win that war, it could have smashed North Vietnam to ashes, as it had done to Germany in 1944–45. Eventually, in 1975, North Vietnam, in breach of the Paris Accord, seized the whole South, killing democrats, liberals, Buddhists and Catholics, and installed a totalitarian regime. And it is still there, twenty-nine years later, which is a collective shame for the rest of the world.

Gordievsky is substantially correct, if one assumes that he meant that South Vietnam was as liberal and pluralistic as any wartime country whose capital was being regularly subjected to terrorist attacks ever reasonably can be. (There was no general election held in Britain between 1935 and 1945.)

In a speech at the Albert Hall in London on 14 May 1947, Winston Churchill said that in the future,

> United Europe would form one major regional entity. There is the United States with all its dependencies; there is the Soviet Union; there is the British Empire and Commonwealth; and there is Europe, with which Great Britain is profoundly blended. Here are the four main pillars of the world Temple of Peace. Let us make sure they will all bear the weight that will be imposed and reposed upon them.

From this and from everything else he said on the subject, it is clear that Churchill never intended Britain to be an integral part of the 'United Europe', despite what its propagandists have since claimed.

In the early debates surrounding Britain's applications to join the European Common Market, Tory politicians went out of their way to allay British and Commonwealth fears about what it would mean for the future of trading relations. In a radio interview broadcast in September 1962, Harold Macmillan, the then Prime Minister, said, 'My point of view is a very simple one. Commonwealth life depends on Britain remaining a powerful country. Economically, if Britain enters the Common Market, I think everybody admits – all the great industrialists and economists as a whole – that Britain will strengthen her power. ... It's much better to be even part of a growing market than a large proportion of a shrinking market.'[11] Churchill, too, wrote to his Woodford constituency chairman, saying that he had long supported European unity, 'But we have another role which we cannot abdicate, that of leader of the British Commonwealth. In my conception of a unified Europe I could never contemplate the diminution of the Commonwealth.'

Yet of course Britain's entry into the EEC on New Year's Day 1973 in fact hugely diminished her connections with the Commonwealth. 'The failure to defend Singapore,' wrote one commentator in 2001, 'and the dismantling of earlier systems of imperial preference caused by Britain's entry into the EU, are the most obvious examples of the way in which both countries have felt a loosening of former ties to the "mother country".'[12] Until 1967, Britain was Australia's largest single export market; by 1998, it was only her eighth-largest accounting for a mere 3% of her total exports. It has been a cataclysmic collapse, and not one for which Australia or New Zealand were in any way responsible. Britain has not been an actively abusive parent, so much as a coldly neglectful one.

As part of his 1972 campaign for entry, Edward Heath had categorically stated that, 'There is no question of eroding any national sovereignty. There are some in this country who fear that in going into Europe, we shall in some way sacrifice independence and sovereignty. These fears I need hardly say are completely unjustified.' We now know that Heath was in receipt of a letter from the Lord Chancellor, Lord Kilmuir, who had been specifically asked by him to report on 'the constitutional implications of our becoming party to the Treaty of Rome'. Kilmuir's letter concluded that 'in my view the surrenders of sovereignty involved are serious ones'.[13] Yet still Heath made his soothing statement. Heath also promised that Britain would not enter the Common Market without 'the full-hearted consent' of the British people, yet the second reading of the European Communities Bill was only passed by 309 to 301 votes. Whatever that might have represented, it was hardly full-hearted consent.

Britain's entry delighted the US State Department, whose policy towards European integration has been incisively summed up by Henry Kissinger as having 'urged European unity while recoiling before its probable consequences'.[14] As early as 1968, Kissinger was able to see that, 'In the recent past, the United States has often defeated its purposes by committing itself to one form of European unity – that of federalism. It has also complicated British membership in the Common market by making it a direct objective of American policy.'[15] Yet that was to remain the American view until long after victory in the Cold War. Throughout the period up to 1973, the USA had been pressurising Britain to abandon preferential Commonwealth tariffs and instead fully to absorb herself into the Common Market. For Britain's part, Edward Heath was the only post-war British prime minister not to place any special value on the transatlantic alliance, yet it was his essentially anti-American geopolitical agenda that the United States continually praised and advanced.

The reasons that America welcomed closer European integration were several: it was believed to provide a bulwark against communism, it was hoped to become a richer outlet for American exports and it was thought to guarantee internal peace, with no more European wars into which American soldiers might be drawn.[16] These were undeniably powerful incentives for American policy-makers, but after the fall of the Berlin Wall in 1989, the relative decline in European GDP in the 1990s, the outbreak of war in the former Yugoslavia, and the refusal of France and Germany to bring more than mild diplomatic pressure to bear on Saddam Hussein in 2002/3, the situation had materially changed. Yet US State Department policy did not; in February 2005, President George W. Bush was only persuaded at the last moment to excise from his speech to the European Union's parliament a passage that welcomed the proposed European Constitution, equating it with the American Constitution. It was an absurd analogy; along with over two centuries of amendments the entire (readable and easily intelligible) US Constitution can be printed out onto twelve pages of A4-sized paper; the (unreadable and impenetrably complicated) proposed European Constitution ran to 265.

A few intellectuals in the United States have spotted the long-term danger posed to the English-speaking peoples by the European Union, such as the Indian-born economist Professor Deepak Lal, who has recently written that, 'It must be in the interest of the United States to see that a politically united Europe does not emerge ... [but] ... to see Europe remain a congeries of independent states, happy, as in the past, to be free riders in the world order maintained by the US imperium.'[17]

When Britain effectively turned her back on the rest of the Commonwealth on joining the EEC, there were unpleasant emotional consequences. As James C. Bennett has put it,

> The severing of Britain's economic ties with its Commonwealth partners as a price of European entry further strained ... relationships. Today, Germans arriving at London's Heathrow airport breeze through the domestic arrivals line, while Australians who fought against the Germans at El Alamein for Britain's sake wait in the foreigners' line with the Japanese. As many Australians noted ... 'There were no bloody queues at Gallipoli; no bloody queues at Alamein.'[18]

In November 2004, it was announced that a thirty-four-year-old New Zealander submariner, Leading Seaman David Kayes, who had served with the Royal Navy for seventeen years, was forced to leave the service because the Ministry of Defence had ordered him to take British nationality. There were 8,000 servicemen and women from the fifty-three Commonwealth countries in the British armed services, and those in 'sensitive' jobs were instructed to take British citizenship or lose them. 'I've been decorated three times,' said Mr Kayes. 'If I were a genuine spy or a terrorist and I had been asked to become British, I would have done it. So it doesn't make sense.'[19]

On Tuesday, 31 October 1978, the Spanish parliament, the Cortes Generales, agreed the wording to the Spanish Constitution, which was subsequently ratified by the people in a referendum on 7 December and signed into law by King Juan Carlos. 'The political form of the Spanish State is that of a Parliamentary Monarchy', it stated, and also provided for human rights, 'democratic co-existence', a 'State of Law' and a 'democratic and advanced society', thus finally ending the chaos, civil war, anarcho-syndicalism, fascism and limbo that had followed one another since the 1930s.

Although they are ancient states, many of the constitutions of European countries are very young indeed, far younger than those of Britain's constitutional monarchy (1688–9), America's democracy (1776), Canada's responsible government (1848) or even Australia's Federation (1900). By stark contrast, the French Constitution establishing its Fifth Republic was only promulgated in 1958, Germany's Basic Law was passed in 1949 (and amended in August 1990), Italy's was adopted in 1949 (and amended thirteen times since) and Portugal's became law in 1976 (to be revised in 1997). In Eastern Europe, the shelf-lives are shorter still; most Baltic states adopted their Constitutions in or around 1992.

It is small wonder therefore, with these Constitutions being so young, that they do not have the same purchase on the imaginations of their populations as do the English-speaking peoples' constitutions, which – with the obvious exception of Eire's *Bunreacht na hÉireann* of 1937 – reach long beyond the memory of anyone alive. The concept of an over-arching European Constitution is therefore much easier for European (and Irish) minds to embrace than for British. Sometimes the US State Department ignores this key difference when it questions why Britain cannot integrate more tidily into the European Union.

On Tuesday, 11 September 1973, a military *coup d'état* in Chile overthrew the democratically elected Marxist Government of Salvador Allende. That is pretty much all that can be stated with factual certainty before political claim and counter-claim between Right and Left takes over, in one of the most controversial episodes of the post-war period. The Chilean Supreme Court and Congress had declared that the Government was acting outside the Constitution, and a junta initially led by the Navy – but which was subsequently run by the Army Chief of Staff General Augusto Pinochet – used this as a justification for its coup.

During the 1970 elections, Allende had misled the electorate about quite how left-wing his Unidad Popular (Popular Unity) Government would truly be, and he won by 36.2%, against the opposition candidate Alessandri's 34.9%, with a third candidate receiving 27.8%. Under the Chilean Constitution as then pertaining, if no candidate had an absolute majority the decision had to be made by Congress. Support for Allende was withheld until he signed a 'Statute of Constitutional Guarantees' before assuming office, affirming that his socialist reforms would not undermine any element of the Chilean Constitution. Only after he signed did parliament vote for him. It was not long before he broke its provisions, however.

As economic problems heightened after his nationalisation programme, excess profits taxes, moratorium on foreign debt repayments, defaulting on international loans and price-freezes, Allende tried to rule by decree, using what he called *resquicios legales* (legal loopholes), which ignored Chile's Christian Democrat and National Parties that dominated Congress. He also angered the Judiciary by refusing to permit the police to carry out judicial sentences that he felt ran contrary to 'the revolutionary process'.

There were massive shortages of basic foodstuffs, industrial production collapsed, foreign currency reserves dried up, two huge strikes brought the country to a near halt and hyper-inflation was met with strong price-control measures, which nonetheless failed. Chaos descended on Chile as *cordones industriales* (workers' committees) failed to run the nationalised industries successfully and political violence started to become an almost daily occurrence.

The Allende Government also attempted to suppress free speech. Radio stations were nationalised, newspaper offices and university departments were occupied by violent pro-government demonstrators, and state advertising budgets were offered and withheld in order to influence editorial comment. Throughout 1972, the Chilean Supreme Court, the country's highest legal authority, as well as her Comptroller-General (whose duty it was to protect the Constitution), declared that the Government was acting in defiance of the Constitution over the way it condoned the illegal and often violent seizures of private land by peasant communes. (In all, there were about 2,000 such incidents.) The peso, meanwhile, reached an official (i.e. artificially low) inflation rate of 600%, triggering fresh crises.

On 8 July 1973 the presidents of both houses of the Chilean legislature – the Senate and Chamber of Deputies – issued a statement which said that 'neither the laws nor institutions are respected' by the Government. In a formal resolution of the Chilean Chamber of Deputies on 22 August 1973, the Allende Government was accused of systematic human rights abuses and was declared to be illegitimate. Among other things, this resolution called upon the 'Secretaries of State, members of the Armed Forces and the Carabineros Corps' – i.e. the service chiefs of the armed forces, who were then members of the Cabinet – to 're-establish the rule of law'. It was signed by Patricio Aylwin, the President of the Senate.

Mounting evidence that arms were being accumulated by the *cordones industriales*, and of the creation of militias heavily infiltrated by foreign military 'advisors', persuaded the Chilean armed forces to act. In August 1973, there were attempts to provoke mutinies in the Navy, and the Chilean Admiralty accused the Left of conspiring with sailors against their officers. Law and order was breaking down in Chile, and the danger of a full-scale Marxist-Leninist revolution was ever-present. There were many violent incidents in Santiago, some sparked by the far-Left Movimiento de Izquierda Revolucionaria (MIR), which idolised Che Guevara and openly advocated a Cuban-style revolution.

With armed foreigners entering the country, many of them Cubans sent by Fidel Castro, there were fears that the Government would suspend the Chamber, which might provoke civil war. The spectre of what happened in Spain between 1935 and 1939 hung over Chile in 1973, with many of the same issues and events being replayed. That war had cost the lives of over one million people, albeit in a far larger population. The choice in Chile by September 1973 was between a coup or a civil war. At one point, thousands of impoverished housewives held a demonstration in which they threw chicken-feed at the soldiers guarding the Army headquarters in Santiago, because of the military's continued reluctance to intervene and restore order.

As the Christian Democrat politician, Eduardo Frei, a man of impeccable socialist and democratic credentials, told the interviewer Luis Calvo of the Spanish newspaper *ABC* in October 1973,

> The Marxists, with the knowledge and approval of Salvador Allende, had brought into Chile innumerable arsenals of weapons which they kept in private houses, offices, factories, warehouses. . . . The military were called in, and they justified a legal obligation because the executive and the judicial power, the Congress and the Supreme Court, had publicly denounced the president and his government for their infractions of the Constitution. I tell you this, when a government refuses to fulfil the social laws, ignores the warnings of the Bar Association, insults and disobeys the Supreme Court, scorns the great majority of Congress, provokes economic chaos, arrests and kills workers who go on strike, crushes individual and political liberties, depletes the market so as to direct food and other goods to the Marxist monopolists in the black market; when a government behaves in this way . . . then the right to rebel becomes a duty.

The Chilean Army had a long record of not intervening in domestic politics, unlike those of many other states in Latin America. On 11 September 1973, however, it acted swiftly and ruthlessly in overthrowing the Government. As his doctor later testified, President Allende committed suicide in the presidential palace, La Moneda, appropriately enough using a machine-gun inscribed with the words: 'To my good friend Salvador Allende from Fidel Castro.'

The fact that Allende had been democratically elected meant that for the Left his overthrow was automatically illegitimate. Under the Chilean Constitution, however, there was one force superior to that of the will of the people: the rule of the nation's law. Because Allende flouted that, as both the Congress and the Supreme Court attested, his Government's overthrow was ultimately justified. Of course that does not mean that he should have paid with his life, but since he and his supporters naturally opposed the coup with force, there was little viable alternative to bloodshed. It was the first time in history that a Marxist-Leninist government had lost power, proving that there was nothing historically inevitable about the victory of the dictatorship of the proletariat after all.

The accusation levelled by the Left against the United States, and in particular Richard Nixon and Henry Kissinger, was that it brought about the necessary conditions for the overthrow of the Allende Government (or 'regime', since it was illegal when it fell). After Pinochet assumed power, Kissinger, referring to the coup, told Nixon that the US 'didn't do it', but that it had 'created the conditions as great as possible'. In fact, though, the US never imposed economic sanctions on Chile, nor did it manage to persuade the Paris group of lenders to make re-financing of Chilean debt conditional on compensation for nationalised foreign assets. The US never even vetoed any proposed loans to Chile (which were admittedly few anyhow because of the economic policies of the Allende Government).

In the early 1970s, nearly every South American country had violent left-wing revolutionary movements, and the legend of Che Guevara had a strong hold. Furthermore, Castro's Cuba was very active in the region, supported by massive economic aid by the USSR. The continent was thus, despite the Monroe Doctrine, a front line in the Cold War. Had Chile, with its long coastline and Pacific port of Valparaiso, gone communist in 1973, it would have altered the entire strategic balance of the area in the Soviet Union's favour, to the severe detriment of the English-speaking peoples and the rest of the Free World.

The CIA therefore undoubtedly supported opposition to Allende; the National Security Council approved $7 million for covert action in Chile, which largely went to the Christian Democrat and National Parties, trade unions and the truckers' strikes in 1972 and 1973. Yet as the Church Committee's investigation showed, the CIA was not responsible for the coup itself, and certainly the activities pursued by the private militias financed by Cuba, Czechoslovakia and even North Korea were acting much further outside the law than anything done by the Americans.

In October 1988, General Pinochet voluntarily held a plebiscite to decide whether he should remain in power for another eight years. He lost this; although 44% of Chileans voted against the restoration of democracy, many of those who had supported the coup in 1973 felt that after fifteen years it was time for elected politicians to return. Although there were precedents for right-wing dictators voluntarily giving up power to democracy, there were none for communist governments. It is worth noting, therefore, that if Chile had fallen to communism in 1973, it is most unlikely that she would have returned to democracy, as happened under Pinochet after the December 1989 election was won by the Christian Democrat Patricio Aylwin. Pinochet then handed over the presidency of what was by then the most stable and prosperous nation on the continent.

William Makepeace Thackeray described the Rock of Gibraltar as 'The very image of an enormous lion, crouched between the Atlantic and the Mediterranean.' At its narrowest, fewer than nine miles separate Europe from

Spanish Morocco. The British naval base there has been one of the most important nodal points in the maritime power of the British Empire, whose strategic importance declined significantly after the Wilson Government abandoned its commitments east of Suez in the mid-Sixties. Spain had claimed the peninsular fortress connected to the Spanish mainland by a thin causeway, yet rather like other British territories claimed by the adjacent country – such as the Falklands and Northern Ireland – the majority of people living there preferred to stay British, as the best way of preserving their way of life. All too often, powerful voices in British governments of both political complexions have argued that the best way to rationalise Britain's modern place in the world was to 'tidy away' these post-imperial problem areas, regardless of the democratic wishes of the overwhelming majority of their inhabitants.

For over half a century, Spain has agitated for the return of Gibraltar. Spanish newspapers denounced the royal tour of Gibraltar undertaken by the Queen and Prince Philip in 1954, when the Spanish Foreign Ministry even suggested that it should not go ahead because Spaniards might attempt to disrupt it.[20] In the 1960s, Spain managed to raise the position of Gibraltar in the United Nations' Decolonisation Committee, and the colony's leading elected representatives, Joshua Hassan and Peter Isola, had to go to New York to state that, 'Nothing could be further than the truth than to suggest that the people of Gibraltar are subjugated or exploited by a foreign power.' To prove this, a referendum was held in 1967 in which 96% of all electors participated, with over 12,000 votes cast against sharing sovereignty with Spain and only forty-four votes cast in favour. Thirty-five years later, on Thursday, 7 November 2002, another referendum was held in which 17,900 people (i.e. nearly 99% of the electorate) voted against co-sovereignty and 187 people (just over 1%) voted in favour.[21] Despite the best efforts of the British Foreign Office over thirty years, the tercentenary celebrations of British Gibraltar in 2004 were held on a staunchly British rock.

In 1973 – four years into Spain's unlawful and aggressive eleven-year total blockade of the Rock – the Heath Government tried to hand over sovereignty on a complicated 'leaseback' deal, because the Foreign Office thought of it as merely 'a historical and geographical anomaly' that they would like to clear up. For all his much-vaunted anti-fascism of the 1930s, Edward Heath was willing to consign 26,000 Crown subjects over to the rule of the Spanish caudillo, General Franco. 'We cannot go on defending this historical and geographical anomaly,' wrote the Ambassador to Madrid, Sir John Russell, in 1973. 'Colonial anachronisms have been cleared up all over the world. Gibraltar is the only one left in Europe.' Every sentence of that seemingly measured statement contained a plain factual inaccuracy. Of course Britain could go on defending Gibraltar, as she did for the next three decades. Nor was the Rock an 'historical and geographical anomaly', since there are plenty of other places where outposts have been left under the control of one country despite being adjacent to or contiguous with another. Nor is Gibraltar the only one left in Europe.

Sweden accepts Finnish sovereignty over the Swedish-populated Aland Isands; Denmark accepts German sovereignty over the Danish-speaking parts of Schleswig-Holstein; Spain regards the enclaves of Ceuta and Melilla in North Africa as Spanish; the Channel Islands are almost within French territorial waters, yet the French do not claim them; yet the Spanish do Gibraltar even after three centuries of British control. Nonetheless, in notes written for the minister concerned, Alan Goodison, head of the Foreign Office's Southern European Department, stated: 'We hope that within ten years the European Community will become a political and defence union. When that time comes Gibraltar will be neither British nor Spanish. It will be European.'[22] Thirty years later Gibraltar is indeed European, but she remains British first.

On Saturday, 6 October 1973, Egypt and Syria launched a full-scale surprise attack on Israel while Jews were observing the holiest day of their year, Yom Kippur (the Day of Atonement), and much of the population was fasting or attending synagogue. Taken by surprise, Israeli armoured units suffered heavy losses and over the next two days had to pull back into central Sinai. However, Syrian forces thrusting across the Golan Heights towards the Jordan Valley were forced back in the Wadi Harridan, and by 11 October Israeli forces had entered Syria.

With the USSR starting to airlift military supplies to Arab states on 10 October, and Iraq, Saudi Arabia and Jordan joining the war against Israel by the 13th, American policy-makers had to decide whether and how they would support the tiny Jewish state which was only a quarter of a century old. On 14 October, there took place in Sinai a tank battle that was second only in size to the Russo-German battle of Kursk thirty years earlier; Israel's victory drove the Egyptians back to the bridgeheads over the Suez Canal, which two days later they crossed. Within eight days of being subjected to a massive surprise pan-Arabic assault, Israel had troops occupying territory inside two of the five Arab nations that had attacked her.

There is still contention about who in the Nixon Administration wanted to support Israel, when, and by how much. What is certain is that on 13 October American transport planes began flying the first of 550 sorties to deliver massive amounts of equipment and ammunition to the embattled Jewish state. Fifty new Phantom jets were sent, and in two weeks a greater tonnage of supplies arrived in Israel than even during the Berlin airlift.[23] On 19 October, the President asked Congress for $2 billion in military aid for Israel, and two days later Henry Kissinger met Leonid Brezhnev in Moscow for talks to end the conflict. By 24 October, both Egypt and Syria had accepted a cease-fire.[24] The aftermath of the Yom Kippur War tested Kissinger's 'shuttle' diplomacy to the full; he undertook eleven visits to the Middle East and in one month – March 1974 – he flew 24,230 miles in thirty-four days, visiting Jerusalem and Damascus no fewer than fifteen times each.

A nation of only six million people surrounded by countries of over 200 million who want it obliterated has often needed the support of the United States to survive. It is a sign of America's altruism that she has never abandoned the small democracy that she helped to found. America's relations with most Middle Eastern countries would have been far better if she had repudiated Israel at some point since 1948, so it is a mark of her selflessness and commitment to democracy that she never did. Nor was this simply because of the politically powerful Jewish lobby in the United States, as has often been made out by Israel's enemies. The great majority of American Jews have voted Democrat in most general elections between 1944 and 2000, yet Republican administrations have been just as solidly supportive of a secure home for the Jewish people in the Holy Land.

That the threat to Israel has if anything grown in the early twenty-first century can be seen from the recent remarks of the President of Iran, a country that by 2006 was on the way towards possessing a nuclear bomb. Mahmoud Ahmadinejad, speaking on 'Jerusalem Day', 26 October 2005 – a moment set aside in the Iranian calendar by Ayatollah Khomeini to be dedicated to excoriating the Jews – said that Israel 'must be wiped off the map' and that destroying Israel 'is the prelude of the battle of Islam with the world of arrogance'.[25] For 'the world of arrogance', it is legitimate to read 'the English-speaking peoples'. Less than a month later, he described the Holocaust as 'a myth' (which is curiously illogical of him since, with his views, he ought to be glorying in it rather than denying it ever took place).

The resignation on 10 October 1973 – four days into the Yom Kippur War – of Vice-President Spiro Agnew over charges of tax evasion was a glaring exception to the rule that politicians in the English-speaking world are generally not corrupt. They do not enter politics for the money. Few places in the world – and certainly not in Africa, Latin America, many parts of Asia, and much of Eastern Europe and Russia – can boast such high standards of honesty both in politics and public administration. The Victorian ideal of public service, by which public servants were repaid in honours and social standing for what they missed out on in financial remuneration, still generally holds good across the English-speaking world. This is partly due to the vigilance of the unfettered media and its relish for 'sleaze' stories, and to strict parliamentary rules for the policing of wrongdoing, but also because public service has never been seen as a lucrative source of income.

Although there have been, and always will be, occasional high-profile cases such as that of Agnew, the very fact that we know about them implies that the system works. Of the 107 ministerial resignations from British governments between September 1903 and May 1994, only five concerned financial impropriety, of which only that of the Conservative Home Secretary, Reginald Maudling, in July 1972 involved a politician of the front rank.[26] With kleptocracies being the rule rather than the exception in so many countries of the world, it is certainly an attribute of the English-speaking peoples that such scandals are generally rare and on an absurdly small scale. One British minister had to resign in June 1993, for example, for not declaring the gift of a wristwatch from a crooked businessman.

As the historian Paul Johnson has noted,

> For more than one hundred years one overriding principle has governed British public life: the fastidious separation of public and private interests. Those who have worked for the state – whether in the armed forces, the Civil Service, as MPs, or in some other way – have never used their office for private gain or any other selfish purpose. ... There have of course been many individual lapses from this high ideal; but the system itself has been extremely robust, surviving throughout the twentieth century.[27]

(It is perhaps worth noting that neither of the two most corrupt parts of the English-speaking world on the American continent – Louisiana and Quebec – were originally colonised by the English-speaking peoples.)

Richard Nixon never lacked personal courage; in April 1958, during a visit to Caracas when vice-president, his limousine was attacked by 'a very hostile mob' wielding iron bars and baseball bats, which smashed in the supposedly bullet-proof windows. His interpreter, Colonel Vernon Walters, who was cut on the lip by flying glass, later recalled how Nixon remained perfectly cool, and 'When his secret service man became excited and drew his gun, Mr Nixon told him to put it away. The Vice-President then asked me if I was hurt and I replied that I did not think so. He said, "Well you are bleeding at the lip. Spit out that glass. I have some more things to say to these Venezuelans today."'[28]

Nixon needed all his sang-froid and more in the second half of 1972 when he discovered that – entirely without his prior knowledge – key White House aides had indulged in a very serious act of political espionage. The first he knew of what became known as the Watergate Affair was when he read a small item in the *Miami Herald* while on holiday in Key Biscayne, Florida. Both the Kennedy and Johnson Administrations had engaged in dirty tricks: JFK most probably stole the 1960 general election in Illinois; it was during Robert Kennedy's period as Attorney-General that the largest number of wiretaps without warrants were ordered pre-9/11, including of the telephones of Martin Luther King and of a reporter who was writing a book about Marilyn Monroe; LBJ's lieutenants bugged the Republican Party headquarters during Barry Goldwater's campaign; Democrats routinely used the Internal Revenue Service to institute wide-ranging field audits of prominent Republicans' tax returns, and so on. Nonetheless, they had not been caught, and the act of placing electronic listening devices inside the Democratic Party headquarters in the Watergate apartment complex in Washington was impressive even by international standards of political skulduggery.[29] (The fact that one of the bugs didn't work and the other only picked up secretaries' gossip about dates and boyfriends is rather immaterial; it was the thought that counted.)

The thirty-one-month-long tragedy of Watergate was drawn out between the early hours of Saturday, 17 June 1972 – when three District of Columbia policemen arrested five men who were rearranging the bugs that had originally been planted the previous month – and 1 January 1975, when former White house aides H.R. Haldeman, John D. Ehrlichman and John N. Mitchell were found guilty of their Watergate offences. The actual genesis of the affair started considerably earlier, however, at the time of the disgraceful leaking of the deservedly top-secret documents relating to Vietnam, known as the Pentagon Papers, to the *New York Times* by a former staffer of Robert McNamara's called Daniel Ellsberg. A unit was created that was intended to prevent such leaks in future. This small group, nicknamed 'the Plumbers' and loosely connected to the Committee to Re-Elect the President – a fantastically mistitled organisation, owing to its acronym – then went on the rampage, but without the President's prior knowledge.

If Nixon had entirely repudiated the Plumbers and their White House protectors from the very start – as he certainly should have done – he might have survived, but he put loyalty to his aides before a strict interpretation of the law and briefly considered attempting a cover-up. Once it was admitted on Friday, 13 July 1973, that all his conversations were routinely taped, his chances looked bleak, especially after 30 April 1974 when he released a 1,308-page edited version to the relevant Congressional committee.[30] Occasional taping of Oval Office conversations had began with FDR, and both Kennedy and Johnson indulged in it routinely. The Democratic-controlled House of Representatives forced Nixon to disgorge the tapes; thereby, in Paul Johnson's words, 'making a frontal assault on the "Imperial Presidency"'.[31]

The tapes revealed that on 21 March 1973 Nixon had briefly considered paying one of the Plumbers, Howard Hunt, and the other Watergate defendants to stay silent, before he specifically ruled out any such payments, which were not therefore in the end made. Although on 23 June 1972 Nixon suggested that the CIA intervene in the Watergate investigation, in order to stymie the FBI's own probe, nothing came of that either. Nor is it true that the

White House deliberately erased eighteen-and-a-half minutes of supposedly incriminating taped conversation. As Nixon himself later wrote, 'It begs credulity to believe that I or my staff would erase this one segment of tape and yet leave untouched dozens of hours of other frank and earthy conversations that I clearly would have preferred not to see made public.'[32]

For his patriotic decision not to contest the result of the 1960 election, and his equally patriotic decision not to drag the country through a presidential impeachment process – which several commentators believe would have led to his eventual acquittal by the Senate – Nixon fully deserved his successor Gerald Ford's presidential pardon under Article II section 2 of the Constitution, 'for all offences against the United States which he, Richard Nixon, has committed or may have committed or taken part in during the period from January 20, 1969 through August 9, 1974'.

The reason given by Ford was that any court hearings would have had to have been held over a year in the future, and 'The prospects of such trial will cause prolonged and divisive debate over the propriety of exposing to further punishment and degradation a man who has already paid the unprecedented penalty of relinquishing the highest elective office of the United States.' It wasn't a good enough reason for everybody, and it certainly damaged Ford's hopes of re-election in 1976, but at least his presidency started off with an atmosphere of clemency. A line seemed to have been drawn under the Watergate débâcle, but of course its wider implications did – and still do – infect the American body politic.

The first and most profound upshot of Watergate – coinciding as it did with Spiro Agnew's resignation – was a sense of distrust in politicians and the political process that extended far beyond the borders of the United States, to affect the whole of the English-speaking world. Watergate pushed healthy scepticism about the political process over the edge into chronic cynicism about politicians' motives, where it has tragically stayed ever since. Howard Baker's questions, 'What did the President know, and when did he know it?', were henceforth repeated at several future instances, primarily of Ronald Reagan during the Iran-Contra affair, of President Clinton during the Whitewater scandal and of President George W. Bush over the question of Iraqi weapons of mass destruction. A generation grew to voting age who admired not the statesmen who led the country so much as the journalists who led the investigations into them. The suffix '-gate' once attached to virtually any noun immediately implied sleaze, cover-ups, governmental corruption and wrongdoing in high places.

For Hollywood, Watergate opened up a vast opportunity to blame dark forces in government for every disaster to overcome ordinary people. The whispered phrase, 'This stretches to the top, to the very top,' became a cliché of cinema-going, as film after film was made with the underlying promise that the United States Government was essentially the covert enemy of the average American. This has spawned a genre of 'paranoia' movies, in which treachery and corruption are virtually the only characteristics attributed to politicians, policemen, generals and others previously considered as American role models.

Typical examples of such paranoia movies, taken entirely at random, include: *The Forgotten* (2005), in which the National Security Agency co-operated with aliens who erased Julianne Moore's memory of her nine-year-old son. 'We just try to minimise the damage,' says the NSA agent. In *Rambo: First Blood Part II*, Sylvester Stallone was not picked up from the extraction site after his mission behind enemy lines because the CIA did not want the United States to learn that thousands of POW-MIAs were still being held in Vietnamese captivity. In *The Day After Tomorrow* (2004), the destruction of the eastern seaboard of the United States by a gigantic tidal wave is the fault of the environmental policies of a (very thinly disguised) Dick Cheney, who puts commercial interests before the lives of millions of Americans. In *Jaws* (1975), the Mayor of Amity Island, Larry Vaughn, insists on keeping the beaches open on the Fourth of July weekend for commercial purposes, despite being acquainted of the danger by the shark-expert Richard Dreyfus and the principled police chief Roy Scheider. Another Steven Spielberg movie, *ET: The Extra Terrestrial* (1982), has spooks from the Pentagon trying to capture the lovable alien for experimentation.

In the 2005 remake of the John Carpenter-directed 1976 classic movie *Assault on Precinct Thirteen*, the attack on the soon-to-be-decommissioned police station is undertaken not by the hoodlums of the 'Street Thunder' gang (as in the original movie), but instead by no fewer than thirty-three corrupt cops from the Detroit Police Department led by Gabriel Byrne. Children's movies are similarly infused with the assumption that Americans in uniform are invariably villains; the highly decorated US naval officer in Vin Diesel's 2005 film *The Pacifier* turns out to be a traitor in the pay of North Korea, for example. In *State of the Union* (2005), the ultra-right-wing secretary of defense played by Willem Dafoe attempts a military coup against a JFK-lookalike president, and is only prevented by a gang of black hoodlums and car-jackers led by Ice Cube and Samuel L. Jackson. 'The fate of the free world is in the hands of a bunch of hustlers and thieves,' says one, only to be answered, 'Why should tonight be any different?' The concept that politicians and those in authority are generally 'hustlers and thieves' underlies much of Hollywood's output.

In *Danger Island* (1992), the CIA's scientists – 'sick people working for sicker people' – have a 'dirty little secret': they have deliberately created e-coli bacteria and a serum that turns people into murderous scaly sea-monsters. 'Why aren't I surprised?' asks one of the characters when it is discovered that

US Intelligence is behind the plan to experiment on the peace-loving native inhabitants of a Pacific island in order to test biological weapons and mind-control drugs.

In *Mission Impossible III* (2006), Ethan Hunt, played by Tom Cruise, is betrayed by a senior US secret serviceman working on behalf of the military-industrial complex that wants to invade a Middle Eastern country to win lucrative contracts. And so it goes endlessly on.

Individually, of course, most movie plots mean next to nothing, but taken together they can create, over decades, a baleful collective groupthink about the essential untrustworthiness and corruption of the political Establishment that represents a powerful social and cultural phenomenon damaging to American democracy. The days when the villains in a Hollywood movie were fascist or communist spies, Mafia chieftains or mere hoodlums are long gone; today they are more likely to be US government agencies, politicians or policemen. It is only under such a culture of paranoia that so much of the Western public could seriously believe that George Bush and Tony Blair would have taken the US and Britain to war against Iraq in 2003 whilst knowing that Saddam Hussein did not possess weapons of mass destruction.

So powerful is the post-Watergate fondness for conspiracy theories in America that a significant proportion of the US population today believes that President Roosevelt knew that Pearl Harbor was about to be attacked; that President Lyndon Johnson was somehow involved in the assassination of his predecessor; that the FBI and CIA are covering up the crash-landing of a UFO at Rockwell, New Mexico, in 1947; that the Israeli intelligence agency the Mossad was behind 9/11, etc, etc. Fuelled by popular TV programmes such as *The X-Files*, the American public has become far more gullible and paranoiac than ever before. The Trilateral Commission, the Davos Economic Forum, the Bilderberg Conference, the Warren Commission, Bohemian Grove, Le Circle and others have been invested with sinister motives and powers that they simply did not and do not possess. The effect of these baleful and generally absurd theories upon trust in the elected representatives of the English-speaking peoples has been enormous and has not lessened in more than three decades since Nixon's resignation.

The disaster that Watergate wrought on America's self-image and her standing abroad could largely have been avoided if the United States had had a con-stitution like Britain's that did not require an impeachment to remove an elected political leader. Similarly, a monarchical system would have allowed for the replacement of President Allende's Government in Chile without the need for General Pinochet's coup and subsequent military dictatorship. In Britain, all that would have happened was the dismissal of Richard Nixon or Salvador Allende by the Queen and his replacement with someone else who could have commanded the confidence of the legislature. (It is a mistake to think that the Queen's constitutional role is to choose the *best* person for the role; anyone capable of commanding the confidence of the House of Commons is enough, which is why her choice of Alec Douglas-Home over R.A. Butler in 1963 was constitutionally correct.)

The months-long, painful, damaging process of forcing Nixon to resign through the threat of impeachment would have been over in the course of an afternoon in Britain and any of the Commonwealth countries where the Queen retains her prerogative. The advantage of having an ultimate constitutional arbiter entirely above politics or the merest suspicion of partisanship – which the Supreme Court cannot be owing to its method of recruitment – is inestimable.

In Australia the year after Watergate, on 11 November 1975, the Governor-General, Sir John Kerr, acting in the name of the Queen, dismissed the democratically elected Labor Prime Minister of Australia, Gough Whitlam, for his chronic mismanagement of the Australian economy and other misdemeanours. Under section 64 of the Australian Constitution, ministers 'hold office during the pleasure of the Governor-General', and by then his pleasure had most certainly run out. Born in Balmain, Sydney, the son of a boiler-maker, Kerr had been Chief Justice of the New South Wales Supreme Court before Whitlam chose him as 'my Governor-General', only offering one name to the Queen to choose for the appointment in July 1974. Kerr has been described as 'a bear of a man: tall, broad, strong-featured . . . on that famous silver mop a homburg looks out of its class and a topper simply ridiculous'.[33]

During a constitutional crisis that followed a series of political disasters for the Labor ministry, Kerr was not someone to be swayed from his decision to exercise the Queen's prerogative power as Governor-General to remove Whitlam. To finance extraordinary government borrowing of US$4 billion, Whitlam had negotiated with a Pakistani wheeler-dealer called Tirath Khemlani to raise a loan that would not need to be signed off by the Australian Loans Council. Then the Deputy Prime Minister Jim Cairns appointed a woman named Juni Morosi, whose company had been dissolved with $40,000 liabilities, to be the Government Treasurer's private secretary, which finally led to Whitlam sacking Cairns. After a series of other scandals and mishaps, the Liberal-Country opposition, which had a majority in the Senate, decided to refuse Supply, an unprecedented act.

As Whitlam tried to explore ways to govern without finance, and then offered deals to the opposition which were refused, Kerr decided to act. At a meeting at 1 p.m. on 11 November he demanded of Whitlam: 'Are you prepared to recommend a general election?' After Whitlam answered that he was not, Kerr said: 'In that case, I have no alternative but to dismiss you.' Immediately after Whitlam left, Kerr swore in the opposition leader Malcolm

Fraser as prime minister of a caretaker government whose first action was to call for a general election as soon as the necessary Appropriation Bills had passed the Senate, which took less than an hour.[34]

Whitlam was incandescent at what he saw as a constitutional coup: 'Well might we say "God save the Queen",' he fulminated later, 'because nothing can save her Governor-General.' Yet in the subsequent general election on 13 December 1975, the Australian people showed their approbation of Kerr's actions by leaving Whitlam's Labor Party with only 36 of the 127 seats in the House of Representatives.[35] In the general election two years later in December 1977, once the voters had had time to ponder the constitutional implications of what Kerr had done, Fraser was re-elected with the loss of only two seats. The system of constitutional monarchy that would have saved Chile from bloodshed in 1973, and America from protracted humiliation in 1974, thus proved its inestimable worth in Australia in 1975.

The escape of the last US personnel from Saigon (thenceforth Ho Chi Minh City) by helicopter from a roof in the American Embassy compound on Tuesday, 29 April 1975, marked the lowest point in the history of the United States in the twentieth century, almost equivalent to the fall of Singapore for the British Empire. The South Vietnamese Government collapsed the following day, to be followed by massacres of its members and supporters and millions of South Vietnamese being sent to re-education camps.[36] (The exact numbers can now never be known.)

As he sat alone in his White House office after the last Americans – but not 400 Vietnamese – were helicoptered off, Kissinger admits being 'torn' with wracking doubts about whether he could have managed America's extrication differently over the previous six years. 'Enveloped by the eerie solitude that sometimes attends momentous events,' he later recalled, his prodigious mind worked its way back through the crises and negotiations, leaving him worried and unsure about his historical role.[37] He then emerged from his office to take on the US press corps and, of course, reverted back to his confident, gladiatorial self.

By then the West's agenda had moved on. On 1 August 1975, the Helsinki Conference on Security and Co-operation in Europe issued its 'Final Act', signed by thirty states, including the USSR. It was the high point of the policy of *Détente* by which the West attempted to contain the Soviet Union. It acknowledged each state's equality and individuality, abjured the use of force in settling disputes, but most crucially it demanded respect for certain inalienable human rights. Russia's signature provided the West with a perfect propaganda tool with which to hold the Soviets' and their allies' human rights abuses up to sustained international criticism.

There are those on the Right who argue that since the Soviets' promise to respect human rights was meaningless, and Helsinki ratified Soviet post-war annexations, it was a retrograde step. With West Germany subsidising East Germany for two decades and Western banks keeping Poland and Hungary afloat throughout the Seventies, they argue that communism could have been ended ten years earlier than 1989. In fact, it was necessary for the West to get through the disastrous Seventies in sufficiently good order before true leadership could bring down communism in the Eighties, and *Détente* helped to buy that time. George Kennan had prescribed Western policy towards Russia in the late 1940s as, 'Stand up to them, but not aggressively, and give the hand of time a chance to work.' It was excellent advice.

Finland herself, in whose capital the Final Act was signed, acted rather like the Egyptian plover bird throughout the Cold War, an animal that picks the teeth of the crocodile but does not get eaten for providing the service. 'Finlandization' became the recognised term for a country that retains its nominal independence, but whose essential self-determination in terms of defence and foreign policy was ultimately controlled from Moscow. This was not through cowardice, however; the small country that had resisted Stalin so bravely and at such high cost in 1939 and 1940 could not survive another clash.

For all that it must have seemed for countries like Finland that the brutal power of their eastern neighbour would go on for ever in the mid-Seventies, there were a small number of people who did not subscribe to such a pessimistic view. One of the many astonishing facts about the fall of the Soviet system was that so few Western commentators, analysts, Intelligence officers, intellectuals or politicians predicted it until it was actually under way. Yet in 1976 a French social scientist named Emmanuel Todd published a book entitled *La Chute Finale* ('The Final Struggle'), which predicted the collapse of Russian communism between ten and thirty years from then. Analysing the USSR's infant mortality figures since 1920, Todd detected a rise that by 1974 had so embarrassed the authorities that the statistics were no longer made available. This, along with other economic and social indicators, led him to assume that the days of Soviet communism must be numbered. His book was badly received by the Left, as might be expected, but neither was it much welcomed by the Right, which also considered the Soviet Union a permanent feature on the international scene, hence Ford and Kissinger's policy of *Détente*.

The more that is known about the British Governments led by the Labour Prime Minister Harold Wilson in 1964–70 and 1974–6, the more squalid they seem. In November 1998, Joe Haines, the former Downing Street Press Secretary, revealed that Wilson (via two intermediaries) actually blackmailed George Wigg, one of his own former ministers, into altering his autobiography, using secretly obtained information about Wigg's sex life. Wilson had had Wigg followed to collect the necessary information about Wigg's use of prostitutes, which was then used to persuade him to excise passages of his book critical of Wilson. These appalling facts even seemed to have surprised Haines, who nonetheless excused it as merely 'a case of Harold behaving badly'.[38] (Wigg himself had been the Labour MP who had harried the courteous and gentlemanly John Profumo during a similar scandal a few years earlier.)

Blackmail might almost serve as the paradigm for that dreadful decade, the Seventies. David Astor, former editor of the *Observer*, writing in the magazine *Index on Censorship*, also revealed how the print unions used to exercise *de facto* editorial control over what went into some newspapers, simply through the threat of selective industrial action. For trade unions to undermine the freedom of the press is no less a disgrace than for government to do so, yet it went on in the Seventies.

In 1971, the British journalist Bernard Levin published *The Pendulum Years*, his history of the 60s which contained a powerful indictment of Harold Wilson for taking lack of principle 'to lengths undreamed of by almost any other politician alive'. He quoted Wilson's self-revelatory remark that, 'A lot of politics is presentation, and what isn't presentation is timing.' Yet even Levin could not have known that Wilson would stoop to hiring private detectives to snoop on his political opponents' private lives.

If anything, the Wilson premiership that came after *The Pendulum Years* was published was worse than the one before it, culminating in the notorious 'lavender' resignation honours list of May 1976; in this, with a few honourable exceptions, people were recommended for peerages and knighthoods that insulted Parliament and the Crown and brought the entire honours system into disrepute. Totally unsuitable people, some of whom later wound up in gaol or committing suicide, were given high honours for the dubious services they had rendered the Prime Minister over the years.

We now also know that Wilson used MI5 to bug his political enemies in the seamen's union during their perfectly legal industrial action in 1966. He denounced them as 'a tightly-knit group of politically-motivated men' (as though any trade union executive should ever be anything else), and assumed that this justified treating them as potential traitors or terrorists. In 1995, the former *Times* editor William Rees-Mogg revealed how Wilson had tried to get *The Times'* political correspondent David Wood fired from the newspaper as the price of not referring Lord Thomson's bid for *The Times* and the *Sunday Times* to the Monopolies Commission. (They faced him down.) Small wonder that the honourable Labour MP John Freeman believed Wilson was 'immoral'.

Wilson's 1964 cabinet was, at least on the (admittedly less-than-infallible) criteria of Oxbridge firsts, the cleverest of the twentieth century. After the

1966 general election, it also enjoyed a ninety-nine-seat majority in the House of Commons. Wilson himself had a fine academic brain and won four out of his five general elections. Yet for all his cleverness, after enjoying as many premierships as William Gladstone, he claimed as his greatest achievement the setting up the Open University.

Here was someone who in 1948 had claimed in a speech in Birmingham that 'more than half the children in my class had no boots or shoes to their feet', a downright falsehood; someone who resigned with Aneurin Bevan and John Freeman over the twin principles of National Health Service (NHS) charges and rearmament, yet who in office increased NHS charges and bought the Chevaline nuclear deterrent without informing the Cabinet; someone who opposed the EEC when it was politically convenient, and then embraced it as soon as the wind had changed; someone who wrote anti-American pamphlets and then shamelessly fawned over President Johnson (who called him 'that little creep'). 'He isn't honest and he isn't a man of principle,' was Bevan's estimation, 'but a slimy, resolute careerist, out for himself alone.'

As a result of Wilson's preference for manoeuvre over statesmanship, Britain was saddled with George Brown's disastrous National Plan, a 'pound in your pocket' which was devalued at the wrong time and the wrong rate, inflation running at 26.9% by August 1975, surtax and super-tax, the so-called Social Contract with the trade unions that went tragically unobserved, and a sense of national malaise which was as moral as it was economic. 'This'll be good for forty seconds on the TV tonight,' Wilson murmured to Freeman during a photo-opportunity outside Number 10. 'Ted Heath's speaking at Gravesend and I haven't got an engagement.' As the diaries of the Labour Cabinet Minister Richard Crossman show, Wilson would even engineer visits to Balmoral Castle solely in order to draw attention from his Conservative opponent.

He was not the only senior Government member to behave disgracefully. When the Foreign Secretary George Brown visited British embassies abroad, his Principal Private Secretary Murray MacLehose would draw the Ambassador aside to warn:

> If you don't know it already, this man is an alcoholic. In the course of the next forty-eight hours he is bound to insult you, your wife, and probably everyone on your embassy staff. There is no point in creating a fuss or resigning. It will achieve nothing. Just grin and bear it. He will be gone before the weekend and you can relax and pretend his visit never happened.[39]

On those few occasions that Wilson did concern himself with strategy rather than simply tactics, as in 1968 when he supported Barbara Castle's initiative to curb the unions, as set out in her ground-breaking policy document *In Place of Strife*, he was badly let down by leading figures on the Party's centre-right, such as James Callaghan and Roy Jenkins. Otherwise it was a case of tough talking against Rhodesia whilst turning a blind eye to sanctions-busting by British firms, or making noises about 'the white heat of the technological revolution' and then appointing the Transport and General Workers' Union leader Frank Cousins to be Minister of Technology, thus guaranteeing that the trade unions would dictate its terms of reference.

When Wilson died in 1995, Tony Blair – perhaps somewhat self-referentially – said that he had 'personified a new era, not stuffy and hidebound, but classless, forward-looking, modern'. The more that is revealed about Wilson, the more we realise that he did indeed personify the Seventies, an era that was deceitful, defeatist and distinctly grubby. For all his cleverness in debate and skill in political opposition, Harold Wilson brought British politics down to a new low, just at the time when his country needed true leadership.

Wilson's surprise resignation in March 1976 was the sole piece of positive news during the bleakest two years in the post-war history of the English-speaking peoples. In the year to August 1976, wages had increased by 13.8%, triggering spiralling wage-induced inflation, which had peaked at nearly 26%, which in turn led to sterling falling from well over $2 to $1.58. The depth of the crisis was plumbed during the Labour Party Conference that year. A sudden plunge in the pound caused the Chancellor of the Exchequer, Denis Healey, to be summoned back from Heathrow Airport en route for an International Monetary Fund (IMF) meeting to try to stabilise the situation. In the manner of a Third World banana republic, Britain had to apply for a £2.3 billion loan from the IMF without which, Healey claimed, Britain would have to have 'economic policies so savage that they would lead to rioting in the streets'. (In an interview in January 2006, Healey nonetheless claimed: 'When it comes to post-war chancellors, I place myself near the top.')

Across the globe during 1976 the West was on the defensive, suffering some of its worst defeats of the Cold War at the hands of communism. Most of Angola fell to the Marxist MPLA guerrillas; the Khmer Rouge took over Cambodia and started their campaign of genocide that killed over a quarter of its citizens; Euro-communists formed part of the new government of Italy; absurd yet foul dictators like Idi Amin (who declared himself Uganda's president for life) and Jean Bokassa (who declared himself emperor of the Central African Republic) were able to play the capitalist West and the communist East off against each other.

James ('Jimmy') Carter – easily the least effective American president of the twentieth century – was elected in November 1976. A political outsider who escaped labels, gained support from across the Democratic Party and was free of links to its leaders, he was, however, in the view of the historian Stephen Graubard, 'self-righteous, concealing his arrogance by pretending to be concerned only with others. A consummate actor, he offered himself as an ordinary American representing millions of others.' During the campaign Carter announced, 'I have committed adultery in my heart many times', perhaps the first public example of the stomach-churning new American penchant for Oprah-style 'over-sharing' that has since dominated 'celebrity culture'. Yet such was the humiliation that Americans felt over Watergate that Richard Nixon's party was very unlikely to win the subsequent election, however personally clean the Republican candidate. Carter, the Governor of Georgia and a successful farmer there, won 297 electoral college votes to Gerald Ford's 241, and the Democrats retained their majorities in the House by 292 to 143 and the Senate by 63 to 31.

November 1976 was also to see someone run for the US presidency who categorically opposed *Détente*, because he thought that the policy had tended to build up the Soviet Union when it was more in American interests to humble it. Ronald Reagan, the Governor of California, was an easy person for intellectuals to ridicule. He had been an actor, in movies with names such as *Bedtime for Bonzo*, and his grasp of international affairs was assumed to be shaky at best, mere anti-communist caricature at worst. Yet Reagan possessed something that those who scoffed at him did not: an instinctive belief in America's capability to win the Cold War, because of the desire of those trapped behind the Iron Curtain to live in liberty.

This allowed Reagan to frame the issue of anti-communism in stark, black-and-white moral terms, entirely eschewing the nuanced chiarascuro of *Détente*. Reagan did not win the Republican nomination, which predictably enough went to the incumbent President Ford, but in the campaign he advanced a view of Soviet communism as something that could be faced down and ultimately defeated as a matter of immediate US policy, a near-revolutionary concept in post-war American politics.

By the mid-1960s, the United States Air Force was in serious need of a new generation of bomber, or Advanced Manned Strategic Aircraft (AMSA), to update the B-52. After a huge amount of work was done on the aircraft specifications – the AMSA was dubbed by wags 'America's Most Studied Aircraft' – in 1971 the USAF finally placed an order for four B-1As, a four-engined, swing-wing aircraft capable of flying at high altitudes on Mach 2, the first prototype of which flew successfully in 1974.[40] When the Carter Administration took office on 1 January 1977, it commissioned a negative report on the project, which was then cancelled in its entirety that June, with nothing being asked from the Soviets in return. Only when Ronald Reagan defeated Carter in 1980 was the project resuscitated, and 100 B-1Bs entered service in 1985, with payloads stretching from air-launched cruise missiles and short-range attack missiles to nuclear gravity bombs.

The cancellation was only the first in a series of weak messages that the

Carter Administration was to give to Cold War opponents in the Kremlin, which taken together implied that the United States was wearying of the task of protecting the Free World and was no longer inclined 'to pay any price, bear any burden'. In a letter to Senator John C. Stennis on 11 July 1977, during the public debate over enhanced-radiation weaponry, Carter wrote that, 'A decision to cross the nuclear threshold would be the most agonizing decision to be made by any President. I can assure you that these weapons, that is to say, low-yield, enhanced radiation weapons, would not make that decision any easier.' Carter's public agonising about the decision to use nuclear weapons might have eased his conscience, but it looked pusillanimous to the Soviets, who understandably suspected that he might put his Christianity before the policy of massive retaliation. When Carter initiated a review of strategic nuclear-war plans in 1977–9, it was expected to result in major modernisations and alterations. 'In the event, however,' records the historian of the Cold War, David Miller, 'it led only to a refinement of the previous plan, together with the element of rather more political sophistication.'[41]

The Vietnam War had left the United States with a neo-isolationist consensus between Congress, the media and the American intelligentsia that encouraged the Soviet Union and her Cuban and Vietnamese proxies 'to engage in empire-building in the Third World without fear of American reprisal'.[42] Henry Kissinger had done what he could to hold this back, but the Carter Administration allowed the perception of a waxing Soviet Union and a waning America to seep into the global consciousness. Two prominent manifestations of this were Western European appeasement of Moscow and the willingness of Third World states to support the Soviets in the UN General Assembly.

It was the Democratic-dominated Congress that had prevented the Ford Administration from stopping Angola, Mozambique, Somalia and Ethiopia from falling under pro-Soviet rule in 1975, and over the subsequent three decades the fate of each country was to jostle the others in the stakes for which was to be the poorest in Africa in terms of per capita income. In Cambodia, the Khmer Rouge embarked on an amazingly comprehensive campaign of extermination between 1975 and the fall of Pol Pot in 1979, which resulted in the murder or deliberate starvation of somewhere between one-and-a-half and two million people, a huge proportion of the 1975 estimated population of seven million.[43]

Meanwhile, the late Seventies were also to see the Soviet Navy setting up a base at Cam Ranh Bay in Vietnam and in South Yemen; East German and Cuban soldiers on the Red Sea; a pro-Cuban guerrilla government in Nicaragua; a full-scale Soviet-backed insurgency war in El Salvador; and the deployment across Eastern Europe of the new SS-20 missile. Yet in 1977 Carter told a Nôtre-Dame University Commencement Address, 'We are now free of that inordinate fear of Communism which once led us to embrace any dictator who joined us in our fear.' As one acute commentator has written in a book about the various Soviet incursions during the late Seventies, 'The Ford Administration, like a man in a nightmare, had tried to react, but could not move through the congressional goo. The Carter Administration disdained even to try.'[44] The unprovoked Soviet invasion of Afghanistan in December 1979 and the eight-year war that resulted was to lead to perhaps as many as half of the population of that country being killed or fleeing to Pakistan.[45]

Carter himself had an astonishingly naïve view of the true nature of Soviet ideology and diplomacy, cancelling the neutron bomb programme in June 1977 without asking for reciprocity from the Soviets, and later complaining of the Kremlin's 'unfriendly rhetoric' towards him. He said he supposed that the 'Soviets perhaps have some political reasons for spelling out or exaggerating the disagreements', but that nonetheless he believed that 'Calm and persistent and fair negotiations with the Soviet Union will ultimately lead to increased relationships with them.'[46] Such language flew in the face of everything that the English-speaking peoples ought to have learnt about the Bolsheviks and their successors since 1917, not to mention Chamberlain's efforts with Hitler.

Over in Brooks's in March 1978, Carter's position looked so insecure that Mr David Karmel 'wagers Mr Brian Nicholson one bottle of 1969 Dom Perignon that President Carter will not be the Democratic candidate at the next presidential election (death by assassination or otherwise to void the bet)'. When Carter spoke of Western 'malaise' at a town meeting in Bardstown, Kentucky, on 31 July 1979, he little recognised how perfectly he himself personified it. The longest-lasting result of the impotence of his Administration came with the fall of the pro-American Shah of Iran in January 1979. It was a comprehensive disaster for Western interests in the region, yet the following month Carter nonetheless warned against 'the temptation to see all changes as inevitably against the interests of the United States, as a kind of loss for "us" or a victory for "them". .... We need to see what is happening not in terms of simplistic colors of black and white, but in more subtle shades.'[47] By August, the constituent assembly in Iran was under the control of the Shi'ite ultra-fundamentalist cleric the Ayatollah Ruhollah Khomeini, who turned out not to be an aficionado of subtle shades.

In obedience to the Ayatollah, on 4 November 1979 a crowd of about 500 Iranian students seized the US Embassy in Teheran, taking sixty-three Americans and forty others hostage, of whom fifty-two remained in captivity for 444 days. The United States was about to embark on a *via dolorosa* as humiliating as the two-year period after April 1973. In a speech delivered soon after the Embassy fell, Carter's Secretary of State said, 'Most Americans now recognize that we alone cannot dictate events. This recognition is not a sign of America's decline. It is a sign of growing maturity in a complex world.'

Fortunately for the future of the English-speaking peoples, that kind of defeatism struck no chord with the vast majority of Americans. By 1980, no fewer than 60% of Americans responded positively to polls asking whether the United States was spending too little on defence, when only five years earlier the figure had been 18%. In September 1980, while seeking re-election, Carter was reported in the *Dayton Daily News* telling fifty Chicagoans that his Republican opponent Ronald Reagan's calls for 'a strong military [was to] just show the macho of the United States'. That kind of language, when fifty-two US diplomats were languishing in captivity in Teheran, showed a tin ear for legitimate American sensibilities, and that autumn Americans voted for the candidate they thought would best prosecute the Cold War, defend the United States and bring the hostages safely back from Iran.

On Tuesday, 4 November 1980, the sixty-nine-year-old Reagan won 489 electoral college votes against Jimmy Carter's 49, along with control of the Senate and an extra thirty-three seats in the House. The humiliating retreats of the dismal decade were finally over. This was underlined when the hostages were released in time for Reagan's inauguration on 20 January 1981. Dealing with the United States after that date was to be a very different experience for enemies of the English-speaking peoples, who had made the Seventies their own.

FIFTEEN

Attritional Victory

The 1980s

*Ronald Reagan becomes president – The Falklands War – Australian Anglophobia –
Margaret Thatcher becomes prime minister – The 'evil empire' – Withdrawal from
Beirut – Cruise and Pershing Missiles Deployed – The invasion of Grenada –
African Marxist-Leninism – Reagan Re-elected – Working with Mr Gorbachev –
David Lange threatens ANZUS – The Reykjavik Summit – Colonel Gadaffi tamed –
The Iran-Contra affair – The World Wide Web – The fall of the Berlin Wall
and defeat of Western Communism*

'To win one hundred victories in one hundred battles is not the acme of skill.
To subdue the enemy without fighting is the acme of skill.'

Sun Tsu, 5th century BC

'Empires collapse because they have been beaten in wars. The Soviet Empire
was unique in simply falling to pieces – with an undefeated army of four
million, plus a nuclear arsenal, and six hundred thousand in its KGB and
security services.'

Robert Skidelsky[1]

Ronald Reagan's assumption of office suddenly opened up the dazzling
prospect of the West not simply fighting for a continued stalemate in the
Cold War, but instead actively attempting to win it. 'Politics', wrote Algernon
Cecil in his 1927 *oeuvre* on British foreign secretaries, 'is one long second-
best.' This is usually true, but it was not like that for the English-speaking
peoples in the 1980s.

For those who still doubt the efficacy of Reaganomics, these key statistics
from the Federal Reserve Bank ought to be instructive: between January 1981
when Ronald Reagan took office and January 1989 when he left it, US inflation
dropped from 12% to 4.5%, the Standard & Poor 500 Index rose from 130
points to 285, unemployment inverted itself from 7.5% to 5.7%, the mortgage
rate fell from 13.1% to 9.3%, while the top rate of personal tax plummeted
from 70% to 33%.[2] In Reagan's second term alone, eighteen million new jobs

were created and the prime interest rate fell nearly six points to 9.32%. Fostering private enterprise, reducing the size of government and cutting taxes produced a virtuous circle for the American economy. This prosperity allowed the Reagan Administration to spend enough on military rearmament to leave the Soviets little alternative but to sue for peace in the Cold War.

On coming to power, Reagan ordered a similarly comprehensive strategic review to the one that Carter had asked for in 1977, but one that came up with very different conclusions. Correctly, the Reagan Administration believed that the Soviet Union respected readiness and strength far more than political sophistication. The 1981 review led to an entirely new update of the Single Integrated Operation Plan (SIOP). It listed 40,000 potential enemy targets, including Soviet nuclear bunkers, conventional military forces, military and political command posts, communications systems, and economic and industrial targets, 'providing the National Command Authority (the President and his immediate advisors) with an almost limitless range of options', including those of the pre-emptive, launch-on-warning and launch-on-attack type of strikes. Beneath the public denunciations of Reagan's supposed 'warmongering' in *Pravda, Tass, Izvestia* and the rest of the Russian media – enthusiastically endorsed by the left-wing media in the West – a new tone of respect was discernible in private contacts with the Kremlin leadership. It was fairly soon accepted that the days of profitable Communist provocations of the English-speaking peoples were over.

Ronald Reagan and Margaret Thatcher proved the most successful duo in leading the English-speaking peoples to victory over totalitarianism since Churchill and Roosevelt four decades earlier. Yet such was the visceral hostility against Reagan on the Left internationally that when he died a headline in the *Guardian* newspaper on the very day of his funeral in June 2004 read: 'He Lied and Cheated in the Name of Anti-Communism.'[3] The Left regularly underestimated Reagan, which he never minded since it tended to play into his hands politically. Continually written off by them as breezy, confident but near-moronically stupid – the veteran Democratic White House counsellor Clark Clifford called him 'an amiable dunce' – Reagan was perfectly happy to lull his opponents. He laughed off the endless malapropisms and verbal gaffes he made, understanding, as two-term Eisenhower had before him and two-term George W. Bush was to after him, that the American people do not always esteem cold intellect in their president more than warm affability.

Yet as the recently published collection of his handwritten speeches and radio addresses between the end of his term as Governor of California in 1975 and the start of his race for the presidency in 1979 have shown, Reagan was far from the intellectual lightweight that his opponents constantly portrayed him.[4] As one reviewer of *Reagan In His Own Hand* put it, 'The periodic newsflash that Reagan was no dummy is more a commentary on the gullibility of those who thought him a dummy in the first place than it is a genuine discovery.'[5] Like Margaret Thatcher and most other successful politicians, Reagan was not so much an original thinker as an inspired interpreter and populariser of the ideas of others. Both Reagan and Thatcher had the gift of being able to make the ideas of Nobel Prize-winning economists – often the same ones – easily comprehensible to people with little grasp of higher economics. Thus Reagan's 1981 economic address shortly after entering the White House read:

> Some say shift the tax burden to business and industry, but business doesn't pay taxes. Oh, don't get the wrong idea, business is being taxed, so much so that we are being priced out of the world market. But business must pass it's [sic] costs of operation, and that includes taxes, onto the customer in the price of the product. Only people pay taxes – all the taxes. Government just uses business in a kind of sneaky way to help collect the taxes.

So what if Reagan couldn't differentiate between 'it's' and 'its', so long as he could make such an important idea intelligible to millions of ordinary electors?

Although Reagan's image as a Westerner might have been obnoxious to many aesthetes in Europe who despised the genre of the Western movie, and wished to portray him as a slouching gun-slinging cowboy, most Americans had a far less prejudiced approach. *The Oxford History of the American West* points out how, 'With a straight-talking, straight-shooting reputation won from his career as an actor, Ronald Reagan parlayed his western persona into political capital to become governor of California in 1966. As president from 1981 to 1989 he continued to draw on his western image as a source of popular appeal.'[6] For all his supposed trigger-happy gun-slinging, the cowboy is also recognised as a hard-working, independent-minded American, his image thus an enduringly popular one in marketing and advertising.

For the American public the Westerner had long had generally positive connotations, not least because the movies in which John Wayne and Ronald Reagan acted tended to include strong moral messages that projected values of individualism, patriotism, the family and, of course, law and order. It was only much later, with the advent of Hollywood's counter-cultural film-making in the mid-Seventies, that the sheriff's motives were held up to ridicule against those of the avenging lone stranger. Reagan hailed from an earlier, better age, which was generally recognised by Americans but never properly understood by many Western Europeans. (Neither was the West the reactionary backwater that many Europeans supposed; Western states bettered the national average in voting for the 1972 Equal Rights Amendment that was intended to win political equality for women, but which failed to secure ratification by enough states over ten years to ensure inclusion in the US Constitution.)

★

The Falkland Islands, at the eastern entrance to the Magellan Straits in the South Atlantic east of the Argentine province of Patagonia, was a strange place for a major test of the English-speaking peoples' will and sense of unity in 1982. In December 1981, a military junta under General Leopoldo Galtieri had seized power in Argentina and revived a long-standing claim to the islands that they called 'Las Malvinas', which had been under British control for nearly a century and a half since 1833. (When in December 1981 the Foreign Office had suggested that Mrs Thatcher congratulate the Argentine junta on taking office, she replied that British premiers 'Do not send messages on the occasion of military takeovers.') The 1,813-strong population of the Islands of 4,618 square miles were 97% British, but since the Islands had been taken and held principally by *force majeure*, there was technically a legal question over who legitimately owned them. Although Britain could claim the first recorded landing on the Islands in January 1690, they had also belonged to both France and Spain in the meantime. Even though they were 250 miles away from the Argentine coastline, they were fully 8,000 miles from Britain.

As a result there had been long, tenuous, drawn-out negotiations going back several decades between Britain and various Argentine regimes over the Islands' future.[7] By the early Eighties, these were given relatively low priority by a Foreign Office that had other important decolonisation issues, such as Rhodesia, to concentrate upon. As the then Foreign Secretary Lord Carrington was to state in his 1988 autobiography, 'The gulf between what is theoretically desirable and what is practically attainable is so wide that it is sensible to concentrate almost exclusively on the latter.' (Admittedly, that might have been written by pretty much any foreign secretary at any period since the Great War.)

What no-one could have foreseen was that the Galtieri Government would suddenly and unilaterally decide to take matters into Argentina's hands and invade the Islands virtually overnight. There had been discussions about shared sovereignty, an Anglo-Argentine condominium, a ninety-nine-year leaseback arrangement and various other schemes, all of which faced the intractable and almost unanimous opposition of the Islanders, who wished to remain British in perpetuity. The endemic political strife in Argentina, which had long oscillated between periods of dictatorship and uneasy democracy, merely confirmed them in this desire. Yet once the Islands had been invaded, none of these plans had any further relevance because subjects of the British Crown had been attacked, and it was the clear duty of Her Majesty's Government to come to their aid, regardless of the vast distance between them and metropolitan Britain.

After the Falklands conflict, it was assumed that there must have been a grievous breakdown in the competence of British Intelligence that the invasion was not predicted, yet in fact there was no such failure, because Galtieri's three-man junta did not itself decide to attack the Islands until 30 March 1982, and the British were apprised of it the very next day. This happened because British Intelligence, as so often in the century since the days of Ewing and Room 40, was eavesdropping on foreign signals traffic. The superiority of the signals interception and deciphering departments of several of the English-speaking peoples' Intelligence agencies has time and again proved invaluable in maintaining their hegemony, but as late as 11 a.m. on Wednesday, 31 March, the Joint Intelligence Committee was of the opinion that 'The Argentine Government does not wish to be the first to adopt forcible measures.' By 6 p.m. that same day, it was clear that in fact an invasion plan was actually under way.

As the then Secretary for Defence John Nott later recalled in his memoirs, 'A series of intercepted signals and other intelligence ... left little doubt that an invasion was planned for the morning of Friday 2 April.'[8] An Argentine submarine had been deployed to the Falklands capital, Port Stanley, the Argentine Navy was assembling for invasion, an army commander had been earmarked as the commander of an amphibious force and the Argentine Embassy in London had been ordered to destroy its documents prior to hostilities breaking out. Nott immediately set up a meeting with the Prime Minister in her room in the House of Commons.

Although much was later made of Nott's decision back in June 1981 to withdraw the Royal Navy's Arctic survey vessel HMS *Endurance* from service, as having sent a signal of weakness to the Argentine leadership, in fact the Argentinians decided to attack not because of anything that Britain had done or not done, but rather from the classic Bonapartist tendencies of an authoritarian regime keen to divert attention from its domestic – in this case largely economic – failures.

The sudden nature of the surprise attack can be illustrated by the fact that on Wednesday, 31 March, Lord Carrington was in Israel, the Chief of Defence Staff Admiral Sir Terence Lewin was in New Zealand, the Chief of the General Staff General Sir Edwin Bramall was in Ulster and the Fleet Commander-in-Chief Admiral Sir John Fieldhouse was in Gibraltar. One senior figure who fortunately was in London that evening, returning from visiting a naval establishment in Portsmouth, was the First Sea Lord, Admiral Sir Henry Leach, who arrived at the Commons' meeting in full naval uniform. 'The sight of a man in uniform always pleases the ladies', wrote Nott rather patronisingly much later, 'and Margaret, very much an impressionable lady, was always impressed by men in uniform.'

Whether Thatcher was impressed by what Leach looked like or not, she could not help but be impressed by what he had to say. The Admiral argued forcibly for sending every element of the fleet of any possible value. ... This required a powerful force, not just a small squadron, with an amphibious capacity and a full commando brigade. It should also include two aircraft carriers HMS *Invincible* and HMS *Hermes*, as well as the appropriate number of destroyers and frigates as escorts. Enough, in short, for a war rather than just a 'police action'.[9]

Leach added that the fleet could set sail after the weekend, which since the invasion was expected to take place that coming Friday naturally impressed the ministers present. Thatcher then asked Leach the key question: could we recapture the Islands if they were invaded? In the finest traditions of 'the Senior Service', stretching back to Nelson and far beyond, Leach replied: 'We could and in my judgement (though it is not my business to say so) we should. Because if we do not, or if we pussyfoot in our actions and do not achieve complete success, in another few months we shall be living in a different country whose word counts for little.' Nothing could have had greater effect on Margaret Thatcher, as Leach had doubtless calculated.

Nott, who had had a long series of disagreements with Leach and considered him 'not exactly a cerebral man', was far more timid about the consequences of sending a fleet 8,000 miles without air cover from land-based aircraft, even before he was briefed on the capabilities of the Argentine air force. Nott later wrote that he had in mind the disastrous First Afghan War of 1838–42 in which his ancestor, Major General Sir William Nott, had served, and also the latest Defence briefing which had indicated considerable uncertainty about Britain's ability to recapture the Falklands, and lastly 'the impact which the disaster of Suez had upon me whilst I was at Cambridge'.[10]

In fact, the Falklands were, after more than a quarter of a century, finally to put Suez behind Britain. It was briefly to reappear as a bogey in the left-wing press at the time of the Iraq War in 2003, but by then the successes of the Falklands, the Gulf War and Kosovo had anyhow eclipsed Suez as an effective shibboleth of liberal internationalism. It was not the least of Margaret Thatcher's revolutions that she allowed British policy-makers to think offensively again.

In an emergency debate in the House of Commons on Saturday, 3 April, the leader of the Labour opposition, Michael Foot, fastened on to the United Nations as the central justification for the use of force by Britain. 'We are supposed to act under the authority of the UN,' Foot said. 'Indeed it is the only authority under which we are supposed to act.' Enoch Powell demurred, pointing out that Britain's right to protect the Queen's subjects was 'inherent in us' and was anyhow 'one which existed before the United Nations was dreamt of'.

Once again, the diplomatic cards fell in a fortunate way. With an Anglophile American president, an unpopular right-wing military junta which communist

Russia and China could not possibly condone, a *prima facie* case of unprovoked invasion, and tough diplomatic bargaining by some of the finest professionals in the Foreign Office, such as Sir Anthony Parsons at the UN and Sir Nicholas Henderson in Washington, Resolution 502 was passed, demanding 'an immediate and unconditional withdrawal' by Argentina.

The memory of Suez, the worries of the Ministry of Defence, let alone the lessons of the First Afghan War, had far less effect on Margaret Thatcher than on John Nott. Instead, a remark made by Powell in the debate of 3 April made a huge impression on her. Referring to the soubriquet 'Iron Lady' that the Russians had bestowed on her, Powell said that in the coming weeks Thatcher herself and the rest of the world 'would learn of what mettle she is made'. General Galtieri had entirely underestimated the calibre of Mrs Thatcher, who showed outstanding resolution and sterling leadership throughout the crisis.

Despite Nott's hesitations and misgivings about Leach's judgment, the Government gave the Admiral authority to prepare what became known as the 'Falklands Task Force', prior to its being ordered to set sail south on the very uninspiring-sounding codename Operation Corporate. (Whereas nowadays public-relations reasons dictate that military endeavours must boast heroic, uplifting titles such as Operation Desert Storm or Operation Restore Hope, this has not always been the case. During the Second World War, they were often given more homely nouns, such as Operations Matchbox, Dartboard, Periwig, Husky and Market Garden. But 'Corporate' was lacklustre even by those standards.)

'Britain may have dashed to the South Atlantic with nuclear-powered-submarines and Sea Harriers with advanced Sidewinder air-to-air missiles, and its efforts may have been almost undone by the sea-skimming Exocet air-to-ship missile,' wrote the official historian of the campaign, Professor Sir Lawrence Freedman, recently, 'but the final struggle for the Falklands relied on old-fashioned soldiering to a remarkable degree.'[11] Much about the Falklands campaign seems old-fashioned: with underlying causes that stretched back to the seventeenth century, it was about sovereignty and 'what we have we hold' rather than -isms or ideas; the actual territory involved was far less important than the issues of prestige, honour and whether Britain would be 'a different country' if she had not at least tried to free Crown subjects, and it was in large part a Royal Navy operation, with all the atavistic flavour that implied. Finally, twenty-six years after Suez, the Empire seemed about to strike back.

Because a number of the vessels that were needed in the Task Force had to come straight from an exercise near Gibraltar, and speed was of the essence to wrest back the diplomatic and strategic initiative, nuclear depth-charges were taken down to the South Atlantic on HMS *Brilliant* and HMS *Broadsword*, although for safety's sake they were transferred from those frigates onto the two aircraft carriers, which both also carried their own. There was never any intention to use nuclear weapons though, and it is a tribute to the Royal Navy's commitment to secrecy that it was only announced that they were even there twenty-three years afterwards when Freedman's two-volume *Official History of the Falklands Campaign* was published in 2005.

The stalwart help provided to the Task Force by General Pinochet of Chile and US Defense Secretary Caspar Weinberger at the Pentagon were both significant, but deliberately underplayed at the time. Chile had bad relations with Argentina, but did not want to advertise the extent of the support she provided against her neighbour, while some prominent Americans, such as the US Ambassador to the UN, Jeanne Kirkpatrick, and the powerful Senator Jesse Helms of North Carolina, feared that the Falklands dispute might derail the United States' anti-communist mission in Latin America. Meanwhile, the US Secretary of State General Alexander Haig tried to undertake 'shuttle' diplomacy so as to engineer a solution. This produced one of the most splendidly convoluted statements of American policy towards Britain ever to delight aficionados of diplomatese, when he said, 'The United States had not acceded to requests that would go beyond the scope of customary patterns of cooperation based on bilateral agreements.'[12] (It wasn't true, either.)

While personally very much inclined towards the British, and towards his friend Margaret Thatcher rather than General Galtieri, who had refused to take his call trying to prevent the invasion, Ronald Reagan did all he reasonably could do not permanently to alienate Argentina, a useful regional ally. For all that he had to be seen to be relatively neutral, his Administration nonetheless showed that the Special Relationship had teeth. As is clear from Freedman's official history, the United States quietly provided Britain with invaluable logistical, weaponry, intelligence and satellite support, which because of its sensitive nature was not at the time trumpeted for what it truly represented – a reaffirmation of the unity of spirit of the English-speaking peoples.[13] In particular, Caspar Weinberger

> felt that if the British were going to mount a counter attack and try to retake the islands, we should, without any question, help them to the utmost of our ability. … I therefore passed the word to the Department [of Defense] that all existing requests from the United Kingdom for military equipment were to be honoured at once; and that if the British made any further requests for any other equipment or other types of support, short of our actual participation in their military action, those requests should also be granted, and honoured immediately. I knew how vital speed would be for the extraordinarily difficult operation they were about to undertake.[14]

The United States provided extra fuel to support the British air supply effort at the Wideawake base on Ascension Island, despite the press pointing out how it technically broke the Americans' official line of impartiality in the conflict. By 19 April, presidential authority had been gained to release six Stinger surface-to-air missile launchers equipped with twelve missiles to British forces, plus night goggles. On 2 May, the strongly Anglophile Weinberger told the Foreign Office that, 'We would supply them with everything they needed that we could spare, and that we were able to do it very quickly.' Two days later, the US Secretary for the Navy, John F. Lehman, visited the British Minister of State for the Armed Forces, Peter Blaker, and made it clear that he would speed up the supply of any equipment that was not coming through fast enough.[15] He even indicated that he would be prepared to move one of the carrier battle groups then in the Caribbean to the South Atlantic, and the possible transfer of a US aircraft carrier was also discussed, before it was found to be impractical on re-training grounds.

'The British did not want to imply that Corporate could not be sustained without US assistance,' Freedman has concluded, 'but that it would help Britain conduct operations with greater despatch and effectiveness, and bring about the earliest possible resolution of the conflict.' Although Nott told the press that 'no assistance was needed at this time', in fact he was also writing to Weinberger asking for the sale-or-return of large amounts of US equipment should the conflict prove a long one. In the short-term, he asked for two Vulcan/Phalanx gun systems and 300 AIM 9L Sidewinder missiles. Within nine days, the first 100 missiles were delivered to Ascension Island and the gun systems were on board HMS *Illustrious*.[16]

'During May, Britain procured some $120 million of US material made available at very short notice (often 24 hours) and frequently from stocks normally earmarked for US operational requirements,' records Freedman. This included some 4,700 tons of airstrip matting for Port Stanley, Shrike missiles for use by the Vulcans, helicopter engines, submarine detection devices for use on Sea King helicopters, Stinger ground-to-air missiles and a large amount of ammunition. 'From the start of the conflict,' concludes the official historian, 'Caspar Weinberger had supported the British and paid scant attention to the delicate line of impartiality along which the secretary of state Alexander Haig trod.'[17] Anyone responsible for hold-ups or delays immediately felt the blow-torch ire of one of the most formidable defense secretaries the Pentagon had seen since the Second World War.

Early on in the conflict the British had announced the creation of a twelve-mile Total Exclusion Zone around the Islands, inside which she demanded that Argentine vessels not sail. The reason that the sinking of the Argentinian cruiser the *General Belgrano*, with the loss of 321 Argentine lives, on 2 May 1982 caused such an outcry in political circles in Britain was because of the confused way that the news was communicated. If it had been stated clearly at the time that the British submarine HMS *Conqueror* had sunk the enemy vessel when she was outside the Total Exclusion Zone rather than within it, and while she was steaming away from the Falklands rather than towards them, but that Admiral Sandy Woodward still considered she could pose a threat to British lives, there would not have been such a controversy during a war that was overwhelmingly popular with the British people. (Freedman makes it clear that the ship was sunk because the Navy considered her a threat, and not in order to derail various Peruvian and American peace proposals, as anti-Thatcher conspiracy theorists have constantly alleged.)

The Navy was concerned that the *Belgrano* and her supporting destroyers represented the southern part of a pincer movement, the danger of which could only be removed by sinking the cruiser. As the *Belgrano*'s captain stated in 2004, his orders were to attack any British ships he enountered. 'What it successfully achieved was to persuade the Argentinians to withdraw their carrier fleet from action (for fear of being torpedoed),' summed up one British commentator of the *Belgrano*'s sinking, 'thus sparing our own indispensable carriers from the likelihood of being sunk.'[18]

By 11.45 Greenwich Mean Time (GMT) on 2 May, when the War Cabinet took the unanimous decision to sink the *Belgrano*, the ship had changed course and was steaming towards the shallow water of the Burdwood Bank straddling the Exclusion Zone south of the Islands, where *Conqueror* might well have lost contact with her. They therefore changed the Rules of Engagement to allow the sinking, in a decision that the Deputy Prime Minister Willie Whitelaw described as among the easiest he had ever had to take in politics. 'Particularly compelling', writes Freedman, 'was the question of what the politicians would say if they had refused the military request when the *Belgrano* could have been sunk, and the cruiser then went on to sink a British carrier with hundreds of casualties.'

Conqueror only received the new Rules of Engagement at 17.10 GMT as *Belgrano* changed course once more, by which time the decision to sink her was Woodward's alone. Since she had already made no fewer than three major changes in direction over the previous nineteen hours, there was no telling whether she might not make a fourth back towards the Burdwood Bank, to which she was still perilously close when she was sunk at 18.57 GMT. As events turned out, Woodward's decision proved the correct one, and doubtless saved hundreds of lives on both sides, by persuading the Argentine fleet to stay in port for the rest of the conflict and not risk a major battle against the Royal Navy in the open sea. It was thus like a miniature Jutland.

The fighting on land and sea during the Falklands War saw exactly the same kind of tremendous bravery and sacrifice that readers will by now appreciate has been the almost universally standard practice of the troops of the English-speaking peoples in wartime. The Argentine Skyhawk pilots who attacked the disembarking British units at San Carlos Water, Fitzroy and Bluff

Cove were also highly courageous. In all, 253 men from the Crown forces and 649 Argentines lost their lives in the conflict. Yet although the Argentine writer Jorge Luis Borges quipped that the war reminded him of 'two bald men fighting over a comb', in fact Britain fought for a principle, rather than merely a few South Atlantic rocky outcrops. She fought so as not to become 'a different country whose word counts for little'. Britain's victory in the conflict soon led to the fall of the Galtieri regime in Argentina, a fortunate by-product of the conflict.

Had British foreign and defence policy been decided by Europe back in 1982, rather than by the Thatcher Cabinet, the Falkland Islanders would probably not be Crown subjects today. Even if the pro-Argentine objections of the Spanish and others had been overcome in the European Community, the delays, rows over financing and inevitable demands for a peaceful solution would have wrecked any hope of swiftly liberating the Islands. If there had been a single currency, Britain would also have had to try to persuade the other countries to countenance the huge expenditure of the operation, which would have had a bad effect on inflation and thus the euro's exchange rate.

Between 1948 and the end of the Cold War, Finland was allowed to stay nominally sovereign, but her Treaty of Friendship and Mutual Assistance with the USSR effectively prevented her acting in a pro-Western way. The proposals contained in the 2004 European Constitution would – had they been in operation in 1982 – have Finlandised Britain with regard to her defence and foreign relations, a state of affairs which would be as humiliating to her pride as it would be damaging to her interests, and to those of the rest of the English-speaking peoples, since she would not have been able to come to the aid of her countrymen and allies in wartime. 'In my lifetime all our problems have come from mainland Europe,' said Margaret Thatcher in 1989, 'and all the solutions have come from the English-speaking nations who have kept law-abiding liberty for the future.'

Australian Anglophobia continued its lone march through the institutions of that country during the 1980s, spearheaded by indentured intellectuals in the universities and schools. Charles Manning Hope Clark, whose six-volume *A History of Australia* published between 1962 and 1987 was hugely influential, did much to vilify the ideas and institutions that Australia derived from the Mother Country, Britain. Even Clark's own publisher, Peter Ryan, admits that the six volumes are 'almost unbelievably prolix ... a vast cauldron of very thin verbal soup ... gaseous verbal excess ... it has given long-windedness and self-pity a bad name.'[19] Yet the work came 'to *be* Australian history, defining Australia as much as the Old Testament defines Judaism'.[20]

As with so many such feasters on the hand that feeds them, C.M.H. Clark had had a first-class education at the British taxpayers' expense. The son of an Anglican clergyman, he attended some of the best schools in Victoria and Melbourne University, before being awarded a scholarship to Balliol College, Oxford, in 1938. In his autobiography *The Quest for Grace*, Clark scoffed at his own belief in the 1930s that 'British institutions and the Protestant religion were the essential conditions for a high standard of material well-being, liberty, the rule of law, tolerance, fair play and decency'.[21] England was instead now a 'dead tree', and everything about its inhabitants seemed to infuriate him.

Travelling on free first-class return Orient Line tickets to Europe given to him and his fiancée by Melbourne University, Manning loathed his English fellow-passengers, who called Gibraltar 'Gib' as though 'they owned the bloody place'. (It seems otiose to point out that Britain did in fact own 'Gib' and had done since 1713.) Manning found *The Times* to be 'unctuous and self-righteous', the *New Statesman* to be written by 'over-civilised men and women who wrote of themselves as paragons of the civilisation a war would be fought to preserve'. (That war would not be fought by Clark himself, a mild epileptic who was nonetheless a successful sportsman.) The 'supercilious', 'bloodless' and 'arrogant' English furthermore had 'high-pitched voices' that he hated and which made them sing 'like eunuchs'. Any Australian who showed friendliness, let alone respect, to any Briton was accused of 'toadying' and 'grovelling'; indeed, those two words appeared so often in Clark's work that the distinguished historian Professor Claudio Véliz, reviewing one volume, wrote that 'powerful proclivities must be at work here, for Professor Clark has declined to make use of any other twenty-eight serviceable alternatives offered in Mr Roget's useful compilation'.[22]

Clark agreed with Germans who 'were bitter about "perfidious Albion"' and opined that, in September 1939, 'Hitler at least was aware of the solemnity of the moment', whereas Chamberlain 'spoke like Arnold of Rugby School'. Churchill was even worse, speaking 'for an England that was a museum piece in the age of the masses' and, 'like Chamberlain, he made no serious analysis of the reasons for the crisis in Western society in 1940'.[23] Clark considered Stalin, on the other hand, to be admirable at analysis and furthermore, 'The Communist party was for many the conscience of Australia.' Even by 1990 Clark was still referring to the Great Terror as 'Yezhov's Terror' rather than that of Stalin.

On his return to Australia in July 1940, Clark taught the boys at the elite Geelong Grammar School that 'maybe a victorious Russia, a Russia that had rediscovered the humanism of Marx, would light a cleansing fire in Australia', which would in turn lead to the end of the 'apologists for Englishmanism in Australia'. Despite his alcoholism, communism and contempt for what he sneeringly described as 'objectivity, impartiality, detachment, cool reason', Clark became hugely influential in Australian academia. 'He was the idol of a generation of Australian history students, and a great and influential patron in the appointment of academic historians.'[24] Nonetheless, the chance of this virulently anti-British racist writing objectively about the contribution of the Mother Country to the development of his homeland was absolutely nil.

In 2001, an enjoyably bitter debate took place in the pages of the *Times Literary Supplement* as to whether or not Clark had ever been awarded the Order of Lenin by the Soviet Union, as the *Brisbane Courier-Mail* had alleged five years earlier.[25] Tempers frayed, high horses were clambered upon. It turned out that he might well have been awarded the Order, but he was certainly given the less-prestigious Lenin Jubilee Medal in 1970. In his acceptance speech, Clark described the genocidal Russian leader being commemorated as a 'teacher of humanity' and said that only when communism conquered the world could all men be brothers. Furthermore, 'We are lucky to live in a time when this tenet is being verified by life.' In his 1960 book *Meeting Soviet Man*, Clark described Lenin as 'Christ-like, at least in his compassion' and 'as lovable as a little child'.[26] Whether or not Clark was actually awarded the Order of Lenin, he most certainly deserved to be.

In the estimation of the author Geoffrey Partington, 'By the 1970s many history departments in Australian universities were dominated by Clark's creed. ... By 1980 he ranked among the leading conductors of a large scholastic chorus that daily poured out hatred on much of the past and present of both Britain and Australia.' Clark dined with Australian cabinet ministers, was awarded the Order of Australia and even became Australian of the Year; when he died, the Federal Parliament adjourned their ordinary business in order for the Prime Minister to praise him and for a special condolence motion to be passed. The intellectual fight-back was not to take place for another two decades, by which time many anti-British, anti-capitalist, anti-monarchist and anti-bourgeois assumptions had been pumped through Australia's academic system.

It was paradoxical that the Labour Government that was closest to the trade unions, that of James Callaghan, who succeeded Harold Wilson in April 1976, was brought down by them in what was called 'the Winter of Discontent' of 1978/9. A hospital supervisors' strike was the first of a large number of mainly public-sector disputes. Tanker drivers, teachers, sewage workers, janitors, water and electricity workers, ancillary health-service staff and even gravediggers came out on strike, in pursuit of wage increases as high as 25%. That winter turned out to be the coldest for sixteen years and along with blizzards there were, in the words of one of Callaghan's obituaries in January 2005, 'blocked roads, undelivered fuel, closed public buildings, a paralyzed health service, bin bags piled high in the London squares and even the dead unburied'. Union picketing became extremely violent, and when Callaghan returned from a G7 summit in the West Indies and said, 'I don't think that other people in the world would share the view that there is mounting chaos,' *The Sun* newspaper paraphrased him with the headline: 'Crisis? What Crisis?'

The strikes, unrest and economic disruption of that winter were followed on 28 March 1979 by the Government losing a confidence motion in the House of Commons by 311 votes to 310, and subsequently the election of a Conservative Government under Margaret Thatcher with an overall majority of 43 that May. When years later Mrs Thatcher was asked what she had changed in politics, she answered, 'Everything.' In another politician that might have sounded absurdly egotistical; with her it was mere historical accuracy. She was the first prime minister of the twentieth century to have an '-ism' attached to her name.

For a long time the 1980s were decried as 'the Decade of Greed' and 'the Me Decade'; only now can we see that it was in fact a splendid period in history. Just as the Regency period followed the French Revolution, the Roaring Twenties ensued from the Great War and the Swinging Sixties came after the respectable Fifties, so the exuberance of the Eighties was a reaction against the dour, drab, defeatist Seventies. One of the criticisms levelled – especially on the Left – against the Eighties was that *laissez-faire* capitalism ran riot during that period. It was the time of the New York takeover arbitrageurs Ivan Boesky and Michael Milken, of the highly paid Wall Street 'masters of the universe' (taken from Tom Wolfe's *Bonfire of the Vanities*), of the Oliver Stone movie *Wall Street*, in which Michael Douglas' character Gordon Gekko says, 'I create nothing; I own' and 'Greed is good', and of *Pretty Woman*, where the asset-stripper played by Richard Gere is equated with a prostitute played by Julia Roberts. The Left has written off the decade as one of heartless, self-indulgent, arrogant materialism, and those among them who did not own shares celebrated when on 'Black Monday', 19 October 1987, nearly one-quarter of the value of the US and UK stock markets was wiped out overnight.

Yet in fact the Eighties were one of the most innovative and exciting decades in the history of the Free World since 1900. Even on 'Black Monday', with a rising sense of panic in some quarters that there might be another Wall Street Crash of 1929, complete with another Great Depression, it was reasoned that the underlying profits made by soundly based companies meant that capitalism was simply not in some kind of terminal crisis. Within a few months the crisis was over and by June 1989 the stock market had risen to levels higher than the ones from which they had fallen on that day.

The financial excesses of the Eighties – and of course both Boesky and Milken wound up in gaol for breaking the law, so the regulations did work – were merely the froth and spume on the top of the great waves of wealth creation that were unleashed by Ronald Reagan and Margaret Thatcher during that astonishing decade. The sense of well-being that those two statesmen engendered in consumers, through cutting taxes and expressing confidence in the future, unlocked a virtuous economic circle which in turn led to further tax cuts. In 1979, the top rate of income tax in Britain was a confiscatory 83p in the pound – 98p for unearned income – but by the end of the Eighties that had been brought down to 40p. This in turn unlocked the energy, innovation and enterprise of the British people, just as exactly the same phenomenon was being seen in the USA.

Massive and painful alterations in British industry – essentially from an uncompetitive manufacturing base towards a successful services one – were effected during the Thatcher counter-revolution. By far the most serious attempt to defeat her reforms was made by the Leninist Arthur Scargill's unconstitutionally named National Union of Mineworkers' strike of 1984–5, which was first outmanoeuvred, then physically faced down and finally defeated after nearly twelve months of bitter struggle. In all, Mrs Thatcher won three successive general elections and stayed in power for eleven-and-a-half consecutive years, a record for any twentieth-century British prime minister.

The sight of Western prosperity during the Eighties provided an image of capitalism that proved irresistible for the peoples still imprisoned behind the Iron Curtain. Technological innovations such as satellite dishes permitted Czechs, Poles and other Eastern Europeans to watch German, British, French and Italian television for the first time, with the result that millions of people in communist countries recognised how far behind their system really was in terms of delivering material benefits.

Yet the physical separation between East and West was still manifest. Nearly forty years before the fall of the Berlin Wall, Margaret Thatcher told the electors of Dartford, 'We believe in the democratic way of life. If we serve the ideal faithfully, with tenacity of purpose, we have nothing to fear from Russian Communism.' The electors rejected her at that election, but she continued adumbrating her uncompromising message until the Soviet press dubbed her 'the Iron Lady' for her militant anti-communism. On Friday, 29 October 1982, she saw the Berlin Wall for the first time, and what she called 'the grey, bleak and devastated land beyond it in which dogs prowled under the gaze of armed Russian guards'. That afternoon she made a prediction that few other Western politicians were prepared to at that time, when she declared,

You may chain a man – but you cannot chain his mind. You may enslave him – but you will not conquer his spirit. In every decade since the war the Soviet leaders have been reminded that their pitiless ideology only survives because it is maintained by force. But the day comes when the anger and frustration of the people is so great that force cannot contain it. Then the edifice cracks: the mortar crumbles. . . . One day liberty will dawn on the other side of the Wall.[27]

Although Mrs Thatcher could see the day dawning, there were very many who could not. The noted Harvard economist Professor John Kenneth Galbraith visited Moscow to receive an honorary doctorate along with the writer Graham Greene in 1984, only five years before the collapse of the Soviet economic system. Galbraith reported: 'That the Soviet economy has made great material progress in recent years – and certainly in the near-decade since my previous visit – is evident both from the statistics (even if they are below expectations) and from the general urban scene, as many have reported. One sees it in the appearance of solid well-being of the people on the streets, the close-to-murderous traffic.' He went on to argue that in terms of the Cold War, 'The Russians could well be even more frightened than we are.' Galbraith once said that, 'In economics, the majority is always wrong.' He was certainly part of the overwhelming majority that failed to spot the cracks in the communist system until just before they sundered the Soviet Union into pieces.

The London School of Economics don Philip Windsor remarked when the Berlin Wall came down that it meant the end of two ideologies: communism and political science. Certainly, for all the vast well-funded political science departments in universities across the West, none of them predicted the sudden end of European communism. It took Margaret Thatcher, who many political science professors so despised, to do that.

Those same political science professors descended into raptures of fury when on 8 March 1983 Ronald Reagan, addressing the annual convention of the National Association of Evangelicals in Orlando, Florida, said that in discussion of the nuclear freeze then being proposed by the Soviet Union,

> I urge you to beware the temptation of pride – the temptation of blithely declaring yourselves above it all and label both sides equally at fault, to ignore the facts of history and the aggressive impulses of an evil empire, to simply call the arms race a giant misunderstanding and thereby remove yourself from the struggle between right and wrong and good and evil.[28]

This was held to represent Reagan's dangerously unsophisticated Manichean world-view, and the adjective 'evil' when applied to the equally pejorative (to American and Left-liberal ears) word 'empire' was denounced for bringing moral absolutes into a complex world. Yet as Reagan went on to affirm,

> While America's military strength is important, let me add here that I've always maintained that the struggle now going on for the world will never be decided by bombs or rockets, by armies or military might. The real crisis we face today is a spiritual one; at root it is a test of moral will and faith. . . . I believe we shall rise to the challenge. I believe that Communism is another sad, bizarre chapter in human history whose last – last – pages are being written.

American Democratic and European commentators professed themselves fearful at the 'strident' and seemingly apocalyptic nature of Reagan's sentiments, claiming that he had heightened Cold War tensions and handed the Soviets a propaganda victory, yet by the end of the decade he had been conclusively proved right and they wrong. As the American journalist George F. Will was to joke in 2005, 'Today there are more Marxists on the Harvard Faculty than there are in Eastern Europe.'

As has already been seen with the attacks on the *Lusitania*, at Pearl Harbor and the perceived attacks on USS *Maine* and USS *Maddox*, and was later to be demonstrated again after 9/11, America avenges herself fully on sudden unprovoked assaults made against her. Very sensibly the Russians avoided any such incidents during the Cold War, although the shooting down of Korean Airlines Flight 007 on 1 September 1983 almost qualified. It certainly provoked absolute and justifiable outrage in the United States, since the Boeing 747 had left Anchorage in Alaska on the way to Seoul and there were Americans on board, including Republican Congressman, Lawrence Patton McDonald.

The plane's south-westerly course should have taken it over Japanese airspace, but due to unwitting pilot error it in fact flew into Soviet airspace. The Russians tracked it for two-and-a-half hours as it flew a straight-line course at between 30,000 and 35,000 feet, in a manner in which only civilian airlines fly. At one point the Korean pilot gave his position to Japanese air-traffic control as being east of Hokkaido, Japan, entirely unaware that he was more than 100 miles off course. Instead of acting in accordance with civilised standards of behaviour, Soviet jet-fighters were scrambled from their base on Sakhalin Island, and one of the pilots, called Osipovich, shot two rockets at KAL 007, bringing it down and killing all 269 people on board, including sixty-one Americans.

The Russians' paranoiac and brutal reaction proved to the world once again that Soviet communism was indeed the 'evil empire' that Reagan had denounced. In the emergency Politburo meeting immediately after the incident, the leadership was told that Osipovich 'claimed he had been unable to distinguish a passenger aircraft from a military spy-plane', an absurd lie.[29] None of those present at the meeting showed any remorse nor, in the words of the Russian historian Dmitri Volkogonov, 'even the expression of remorse'. They decided to try to stick to the official line, that the USSR had protected herself from what she believed to be an attack by hostile forces.

In the Politburo discussion, their youngest member, Mikhail Gorbachev, said, 'We have to show precisely in our statements that this was a crude violation of international conventions. We mustn't remain silent at this moment; we must take up an offensive position. We must support the existing version, and develop it further.'[30] The Politburo therefore put out the statement that: 'The measures taken in connection with the violation of Soviet airspace by the South Korean aircraft on 31 August are approved. It proceeds from this that the violation was a deliberate provocation by imperialist forces ... capable of distracting from the USSR's peaceful initiatives.'

In reply, President Reagan broadcast to the American people the facts of what happened, adding the comment:

> Make no mistake about it, this attack was not just against ourselves or the Republic of Korea. This was the Soviet Union against the world and the moral precepts which guide human relations among people everywhere. It was an act of barbarism, born of a society which wantonly disregards individual rights and the value of human life and seeks constantly to expand and dominate other nations. They deny the deed, but in their conflicting and misleading protestations, the Soviets reveal that, yes, shooting down a plane – even one with hundreds of innocent men, women, children, and babies – is a part of their normal procedure if that plane is in what they claim as their airspace. ... Memories come back of Czechoslovakia, Hungary, Poland, the gassing of villages in Afghanistan. If the massacre and their subsequent conduct is intended to intimidate, they have failed in their purpose. From every corner of the globe the word is defiance in the face of this unspeakable act and defiance of the system which excuses it and tries to cover it up.

At 6.22 a.m. on Sunday 23 October 1983, a smiling Shia Muslim with a bushy moustache drove an 18-ton Mercedes truck loaded with explosives equivalent to 18,000 tons of TNT over a barbed- and concertina-wire obstacle into the lobby of the headquarters of the US Marines' Battalion Landing Corps 2/8 in Beirut. It was the nerve-centre of the Multi-National Force (MNF) that had been stationed in the Lebanon since August 1982, in the aftermath of the Israeli invasion of that country.[31] American, French, Italian and British soldiers made up the MNF, and on that day 242 Americans and thirty-eight Frenchmen lost their lives. It represented the largest number of Americans to lose their lives to enemy action on a single day between the end of the Second World War and the attacks of 9/11.

The previous month both houses of the US Congress had held hearings on the War Powers Act and passed a law saying that the Marines could only stay for another eighteen months. The Corps' commander, General P.X. Kelley, had asked Congress not to set a public schedule for the withdrawal of US forces, saying, 'If the time is too short, our enemies will wait us out; if it is too long, they will drive us out.' As he reiterated it twenty-two years later over Iraq, 'Never tell your enemies your plans. Ambiguity in war is essential.'[32]

The American evacuation from the Lebanon in February 1984 was accompanied by a large number of Western commentators opining that the United

States did not have the political will to take heavy losses, and that unlike Russia or China even a relatively small number of body-bags returning home would always influence American opinion against active engagement beyond her borders. Yet this was always based on a fallacy, one that later opponents such as the Taliban and Saddam Hussein were to believe at their peril. In fact, the American people have historically tolerated tremendously high casualty levels in war – the American Civil War, the Second World War, Korea and Vietnam being cases in point – but only so long as victory was in sight or their leaders had developed a clearly articulated strategy to get there. Once, as happened in Vietnam in 1968, the leaders of America could not quite define what victory looked like; then and only then were casualty rates considered uppermost.

With nothing more than nebulous peace-keeping operations in Lebanon, victory seemed an indefinable goal, and the Americans left. As a result, the message was received in the Middle East that spectacular acts of terrorism against prominent US targets paid off, thus ensuring that there would be more. Osama bin Laden was later to cite President Clinton's evacuation from Somalia in March 1994 as proof that the Americans could be terrorised into a general Middle Eastern withdrawal.

(Sometimes the English-speaking peoples are very slow to learn from their errors. Lord Mountbatten's publicly announced final date for the transfer of power in India led directly to widespread massacres; Congress' demands for a publicly announced timetable for US withdrawal from South Vietnam invigorated the Vietcong. Yet still in 2005 there were calls in Congress for publicly-stated dates for US forces to quit Iraq.)

Even whilst President Reagan was withdrawing from Lebanon, 1983 also saw him announcing the start of the Strategic Defense Initiative (SDI), intended to build a defensive shield behind which the United States could defend herself from incoming Soviet nuclear missiles. 'The Soviet Union became extremely agitated about SDI,' records an historian of the Cold War, David Miller, 'since it threatened to negate the value of its vast stocks of intercontinental and submarine-launched ballistic missiles.'[33] Reagan himself, using a characteristic analogy in his memoirs, likened the strategic doctrine of Mutually Assured Destruction to 'two westerners standing in a saloon aiming their guns at each other's heads. Permanently.'[34] SDI seemed to offer a way towards a far better situation, in which satellites could forewarn the US of incoming Soviet ballistic missiles, which could be destroyed in the air.

As Caspar Weinberger sarcastically recorded in his autobiography, international outrage greeted the news of SDI, since 'the idea that any country might try to defend itself against the nuclear missiles of another country was not only revolutionary, it was sacrilegious'.[35] For all the fury of liberal internationalists against the SDI project, which they alternately (and

contradictorily) denounced as expensively unworkable and strategically destabilising, the effect on the Soviet Union was to demoralise them even further.
Reagan later described SDI as 'the single most important reason, on the
United States' side, for the historic breakthroughs that were to occur' in the
period between 1983 and the fall of the Berlin Wall.

After failing to persuade the USSR to stop deploying SS-20 missiles in
1983, NATO began its deployment of Pershing II missiles in West Germany
and ground-launched cruise missiles both there and in Britain. For all the vast
demonstrations organised by anti-nuclear groups in the West, it was this show
of determination by NATO that was finally to bring the Soviets back to the
negotiating table, particularly at Reykjavik in October 1986, thus opening up
the prospect for the Intermediate-range Nuclear Forces (INF) Treaty, which
was signed in December 1987.

In his *Annual Report to Congress for the Fiscal Year 1987*, Weinberger spelt
out the four factors necessary for effective nuclear deterrence, namely:

> Survivability: our forces must be able to survive a pre-emptive attack with
> sufficient strength to threaten losses that outweigh gains;
>
> Credibility: our threatened response to an attack must be credible; that is, of
> a form that the potential aggressor believes we can and would carry out;
>
> Clarity: the action to be deterred must be sufficiently clear to our adversary
> that the potential aggressor knows what is prohibited; and
>
> Safety: the risk of failure through accident, unauthorized use, or miscalculation
> must be minimized.

What the Reagan Administration delivered in general, and its tough Secretary
of Defense provided in particular, was credibility and clarity, after a Carter
Administration that seemed to wring its hands in the face of Soviet and Soviet-
backed incursions and provocations. Carter had once graphically described
the destruction of the civilised world as capable of taking place during 'one
long, cold, final afternoon'; that it never did was largely down to the resolution
of the leadership of the English-speaking peoples during the 1980s.

Two days after the Beirut bomb-blast, Americans went into action in a
completely different part of the world. 'Unexpectedly,' wrote Margaret
Thatcher in her memoirs, 'the autumn of 1983 turned out to be a testing time
for Anglo-US relations. This was because we adopted different attitudes
towards crises in the Lebanon and in Grenada.'[36] That was a huge understatement. When the US Marines Corps landed on the 133-square-mile island
of Grenada in the eastern Caribbean in the early morning of Tuesday, 25
October 1983, to overthrow the Marxist Government of General Hudson
Austin, Margaret Thatcher was incandescent and told Ronald Reagan so in
unmistakable terms. A Commonwealth country of which the Queen was head

of state had been invaded by the US with the very minimum of warning.

The Americans were not acting unilaterally, but in accordance with the unanimous and publicly expressed wishes of the Organisation of Eastern Caribbean States (OECS). Furthermore, although the thugs who had taken over Grenada on 19 October were only replacing another pro-Castro Marxist government under Maurice Bishop, there were signs that the Bishop regime was moving towards pragmatism. Bishop himself had even visited the US earlier that year, but had been executed in the Austin coup. A new airfield capable of landing large aircraft was being built by well-armed Cuban construction workers and was due for completion in January 1984, after which the island would have fallen deeper into Castro's maw. As even Mrs Thatcher accepted, for the Cubans the new airport 'would be a way of managing more easily the traffic of their thousands of troops in Angola and Ethiopia back and forth to Cuba. It would also be useful if the Cubans wished to intervene closer to home.'[37]

For these perfectly good reasons, the Reagan Administration decided to intervene in Grenada, regardless of the British Government's view. The whole trend of geopolitics in the Western Hemisphere ever since the Destroyers-for-Bases deal of 1940 made it clear that Grenada now lay well within America's rather than Britain's sphere of influence, whatever the older ties of the Commonwealth – which of course meant nothing to the Grenadan Marxists – might have suggested. *Realpolitik*, as so often and so immutably in the twentieth century, counted for far more than sentiment. Nor would Margaret Thatcher's initial fury at her friend's actions skew her overall strategic judgment, for as she wisely put it in her memoirs, 'I had wider objectives as well. I needed to ensure that whatever short-term difficulties we had with the United States, the long-term relationship between our countries, on which I know Britain's security and the free West's interests depended, would not be damaged.'[38] After the help that the United States had provided during the Falklands crisis only the previous year, it would have been illogical for Mrs Thatcher to have acted in any other way.

Neither was it true, as anti-American elements in Britain have regularly alleged since 1983, that President Reagan launched the attack on Grenada without warning Downing Street that it was going to happen. On 22 October Margaret Thatcher received a report of the conclusion of the US National Security Council meeting about Grenada, telling her that the USS *Independence* carrier group had been diverted to the Caribbean, along with a different, amphibious force of 1,900 Marines. At 7.15 p.m. (London time) on 24 October, Reagan asked Mrs Thatcher for her 'thoughts and advice' on the situation, saying that he was giving serious consideration to the OECS request for military intervention. She only had a draft reply ready before he sent a second message at 11.30 p.m. saying that the attack would be going ahead. It was cursory consultation at best, but it was not undertaken entirely without warning the British Government.

The attack began at 5.36 a.m. (local time) on 25 October, when 400 Marines from Guam landed by helicopter at Pearls Airport on the western shore of Grenada, and it was wholly successful. In all, 5,000 US soldiers took part in the liberation of the island. The 800 Cuban so-called 'construction workers' employed heavy anti-aircraft fire and AK-47s, but were overcome after two days' fighting. Their arms caches were discovered to include enough automatic rifles, machine-guns, rocket launchers, howitzers, artillery, armoured vehicles and coastal patrol-boats to arm a force of 10,000 men, further indication of Castro's plans for the island and beyond.[39] The United States lost nineteen killed and 115 wounded. The defenders sustained fifty-nine killed and twenty-five wounded. Forty-five Grenadians were killed and 337 wounded during the fighting. All of Bishop's murderers were arrested and order was swiftly restored. The invasion proved America's readiness to swat communist threats to her hemispheric hegemony, as Reagan made clear in a televised address two days later.

The KAL 007 tragedy, Beirut disaster and Grenadan invasion were on the face of things not much linked, but they came within two months of each other. Part of Ronald Reagan's political genius was to be able to draw such seemingly disparate events into what one of his biographers has called 'a single message of patriotism and anti-communism'. Thus on 27 October he told the American people:

> The events in Lebanon and Grenada, though oceans apart, are closely related. Not only has Moscow assisted and encouraged the violence in both countries, but it provides direct support through a network of surrogates and terrorists. ... You know, there was a time when our national security was based on a standing army here within our own borders and shore batteries of artillery along our coasts and, of course, a navy to keep the sea-lanes open for the shipping of things necessary to our well-being. The world has changed. Today, our national security can be threatened in faraway places.[40]

Here was yet another repudiation of Washington's Farewell Address, of the Senate's 1919 isolationism and of President Carter's post-Vietnam insularity. Reagan was acknowledging that for a superpower like America, locked in struggle with an enemy devoted to the global unanimity of communism, there was almost no such thing as a 'foreign war'. His speech was in the direct apostolic line from that of the Roosevelt cousins, and particularly FDR's internationalism. The Polisario guerrillas in the western Sahara, the MPLA government in Mozambique, the Shining Path terrorists of Peru, the muja-hadeen in Afghanistan, the lorry-bomber in Beirut, the fighter pilot from Sakhalin Island, the Cuban soldiers in Grenada, and literally dozens of other such local fighters – all of them needed to be seen in the wider context of the struggle between Totalitarianism and the Free World. Reagan did the latter a signal service in emphasising the geographical seamlessness of the struggle. No longer could Western leaders speak, as Chamberlain had at the time of Munich, of 'a quarrel in a far away country between people of whom we know nothing'.

Perhaps nothing better illustrated the utter ideological poverty of Marxism-Leninism as applied to Africa than the famine in Ethiopia that killed over one million people in 1984. In common with Stalin and Mao, the communist dictator Mengistu Haile Mariam saw starvation and the fear of it as political weapons with which he could reduce the numbers of his enemies and terrorise the rest of his population. Mengistu's greatest fear in 1984 was that the truth about the famine might reach the outside world and that therefore relief might arrive in the secessionist province of Tigray before it was starved into submission.

Much the same methods have been used more recently by his fellow Marxist, Robert Mugabe, in Zimbabwe, one of whose officers explained his food policies to the people of Matabeleland thus: 'First you will eat your chickens, then your goats, then your cattle, then your donkeys. Then you will eat your children and finally you will eat the dissidents.'[41] (However bad the human rights abuses got in the Soviet Union, at least they never treated their dissidents as a food source.)

In his 2005 work *The State of Africa: A History of Fifty Years of Independence*, the historian Martin Meredith showed quite how many African countries had undergone appalling (and largely unnecessary) privations since the British and other colonial rulers had departed with such optimism and fanfares in the 1950s and 1960s. While it might perhaps have been credible to blame the white colonialists for these countries' problems for a decade or two after independence, it defies credulity that African, Asian and occasionally Latin American countries still hold them primarily accountable for multifarious catastrophes over half a century afterwards.

It was often the over-hasty nature of decolonisation, before there was a large enough domestic middle class, that wrecked democracy in many of these places, but the foreshortened timetables were forced upon the colonial powers by the agitators, who usually hailed from those same classes. In the late 1950s, black Africa's 200 million population included only 8,000 who had attended secondary school. Too often local leaders moved straight from prison cell to presidential palace without the intervening stage of administrator's office.

In a review of Meredith's book, the British historian Piers Brendon took his readers on a brief *tour d'horizon* of African dictatorships, including those of

'Redeemer' Kwame Nkrumah [of Ghana] inheriting one of the richest countries of Africa, with a competent administration and an established parliament; he reduced it by 1965 to a corrupt and bankrupt dictatorship. ... Kenya's Daniel arap Moi and Nigeria's Sani Abacha looted on a huge scale, starving hospitals, schools and other amenities of funds. Mobutu, the 'Zairean Caligula', ran the grossest kleptocracy of them all, chartering Concorde for his personal use and seizing a third of the state's revenue. ... Uganda's Idi Amin dumped 'truckloads of corpses' in the Nile and claimed to be the 'true heir to the throne of Scotland'. Francisco Macias Nguema ... turned Equatorial Guinea into the 'Dachau of Africa'. In the Central African Republic Jean-Bédel Bokassa denied charges of cannibalism but parts of a mathematics teacher were found in his fridge. In Liberia, Samuel Doe practised *juju* rituals which included drinking the blood and eating the foetuses of pregnant girls.[42]

Whatever criticisms might be directed at the administrators of the British Empire in Africa – including the legitimate one that they occasionally drew straight lines on maps which sometimes cut through tribal groupings – they rarely ate mathematics teachers.

European powers scored their highest economic growth rates in the decades after they had shed their empires, despite the fact that, in Geoffrey Wheat-croft's phrase, there was 'a widespread belief ... that Europe enjoyed its comforts thanks to the efforts of distant coolies'. Simultaneously, after independence, 'the former colonies suffered terrible economic decay, relatively in many cases, absolutely in some, suggesting that empire had been burden more than blessing for the imperial powers'.[43]

On 19 October 1984 a thirty-seven-year-old pro-Solidarity Roman Catholic priest, Jerzy Popieluszko, was abducted in Warsaw. After eight days the Polish authorities admitted that he had been murdered by their internal security service. His corpse was found in a reservoir on 30 October, and the following day three secret policemen were charged with his killing. The murder of Father Popieluszko serves to remind us that communism in Poland, Czechoslovakia and elsewhere did not necessarily have to end in 'a Gorbachevian whimper', but that in fact 'brutal repression, in the style of Deng Xiaoping or Nicolae Ceauşescu, remained a plausible alternative until quite late'.[44]

The policy followed by Ronald Reagan and Margaret Thatcher of encouraging the Solidarity trade union in Poland might easily have been met by an horrific response, as in Budapest in 1956 or Prague in 1968, and it took statesmanship of the highest level to ensure that it was not. For the Solidarity underground movement, much of the leadership of which spent the winter of 1984/5 in prison, the Popieluszko murder taught them that new ways were needed to mobilise the disheartened Polish people, in order to force the regime to negotiate. Otherwise – as one influential article in the main underground weekly press put it at the time – their movement would just consist of 'a chronicle of martyrs'. The following May Day saw demonstrations in which 10,000 Solidarity members clashed with police in Gdansk.

On 6 November 1984, the Soviet leadership's fears were realised when Ronald Reagan won a landslide electoral victory over his Democratic challenger Walter Mondale, with 525 electoral votes to 13. A month after Reagan's re-election, on 15 December 1984, the youngest Soviet Politburo member, Mikhail Gorbachev, visited London as part of a Soviet parliamentary delegation. Relations between Russia and America were as bad as they had been at any time since the Cuban missile crisis; there was, in the words of one of Gorbachev's arms-control advisors, 'a macabre dark environment' to international affairs at the time. When the opportunity arose to invite Gorbachev – a Kremlin technocrat whom the Foreign Office had correctly identified as a possible reformer – to Britain, Margaret Thatcher recognised a chance to become a conduit between the Americans and the Russians in such a way that might reduce Cold War tensions without reducing pressure on the Soviets.

'Mrs Thatcher thought out Mr Gorbachev's programme very carefully,' recalled her foreign policy advisor, Charles (later Lord) Powell. She invited Gorbachev and his wife Raisa to Chequers, the prime ministerial country residence in Buckinghamshire, where she and her husband Denis made 'a warm welcoming impression'. Malcolm Rifkind, a junior Foreign Office minister, remembered how the Thatchers and Gorbachevs 'chatted, joked, were relaxed', although he himself was embarrassed when Raisa asked him which modern Russian novelists he enjoyed, and he could only think of the (then still banned) Alexander Solzhenitsyn.

On the car journey down from London together, Gorbachev commented on the farms, fields and hedgerows he saw en route, which led to a discussion of Russian agricultural organisation, during which 'Mrs Thatcher left him in no doubt about her view of collective farming.' The discussion through lunch about the relative merits of the capitalist versus communist systems was, in the words of one present, 'continuous, vigorous and good-humoured'. In the drawing room afterwards over coffee, Thatcher spoke of the West's sincerity in the search for arms control, during which, in the view of Powell – the only official present – Gorbachev 'revealed himself to be a different character from his Soviet predecessors'. He showed a willingness to engage in open-ended discussion, only occasionally consulting some handwritten notes, whereas men like Leonid Brezhnev, Yuri Andropov and Konstantin Chernenko had tended to speak from pre-written scripts. 'I like Mr Gorbachev; we can do business together,' Thatcher told the BBC soon afterwards.

Within a week, Mrs Thatcher had seen Ronald Reagan to report on her meeting. Britain was of course never going to be the major player in the great task of prising Russia away from communism, but she did nonetheless play an important role in persuading Reagan to re-engage, and to interpret Russia to America and vice versa. Western triumphalism at the USSR's discomfiture, or a refusal to negotiate over the endgame of communism, could have set back the process by which first Eastern Europe and then Russia herself threw off communism in the late 1980s and early 1990s. That this did not happen is largely down to the foresight of Margaret Thatcher and her foreign policy advisors, and the willingness of Ronald Reagan to exploit the historic opportunity to the full.

Of course, as President Kennedy said after the Bay of Pigs débâcle (quoting Mussolini's foreign secretary and son-in-law Count Galeazzo Ciano, understandably without attribution), success finds a hundred fathers while failure is an orphan. As soon as the Berlin Wall fell, the Left rushed to try to claim credit, or at least to deny it to the Right. In her Keith Joseph Memorial Lecture delivered to the British think tank, the Centre for Policy Studies, on 11 January 1996, Margaret Thatcher would have none of it. 'During most of my life, freedom in this country was under a direct challenge from fellow-travelling Socialists and an aggressive Soviet Union,' she recalled.

> These challenges were overcome because the Conservative Party in Britain and other right-of-centre parties – under the international leadership of Ronald Reagan – proved too much for them. The fashionable expression is that Communism and indeed Socialism 'imploded'. If that means that their system was always unviable, so be it, though many of the people who say this scarcely seemed to believe it was true before the 'implosion' occurred. But, anyway, let's not forget that the system collapsed because it was squeezed by the pressure that we on the Right – I repeat on the Right – of politics applied. And the Left should not be allowed to get away with pretending otherwise.[45]

Just as the English-speaking peoples were beginning to thaw Soviet communism, one of its constituent parts broke ranks. In February 1985, the New Zealand Labor Government of David Lange plunged the ANZUS alliance into turmoil when it banned American warships from its ports, on the basis that they might be nuclear-powered or armed with nuclear weapons. Although President Reagan confined himself to telling the New Zealand Ambassador to Washington, Sir Wallace Rowing, that 'withdrawing from shared responsibilities would not help in achieving the common objective of nuclear disarmament', the American Ambassador to New Zealand, Monroe Browne, was much more forthright, complaining to an audience of Hawera Presbyterians on 5 March 1985 that, 'The very ships which defend New Zealand in time of war may not enter New Zealand's ports in time of peace.'[46]

Meanwhile, at the Geneva disarmament conference, David Lange said,

'New Zealand is a small country and remote, but if there should be a nuclear war then New Zealand will join the company of those who have destroyed themselves.' Declaring New Zealand 'nuclear-free' was to employ one of the favourite weasel words of liberal internationalism, since in any conflict it is up to the enemy, not oneself, to decide who remains free from nuclear attack. There is no indication that the Soviet Union altered her nuclear strategy in order to take Mr Lange's new policy into account.

The tense situation was massively antagonised by Mr Lange on Saturday, 2 March 1984, when he took part in a televised debate at the Oxford Union, proposing the motion 'That this House believes nuclear weapons are morally indefensible'. In the course of it, Lange argued that, 'It makes no sense for New Zealand to ask allies to deter enemies which do not exist with the threat of nuclear weapons', and 'The means of defence terrorises as much as the threat of attack.'[47] Most unnecessarily unfair to the United States, Lange even said, 'To compel an ally to accept nuclear weapons against the wishes of that ally is to take the moral position of totalitarianism, which allows for no self-determination.' The United States was not requesting New Zealand to accept nuclear weapons on her soil, merely to allow American sailors on nuclear-powered warships and submarines time for rest and recuperation in her ports.

On his return, Lange assured his countrymen of the Americans: 'If the balloon goes up, they will be there.' His Government, it seemed, could have its cake and eat it in a morally uncompromised nuclear-free manner. Part of New Zealand's motivation was the desire, as her newspapers put it, to make Australia 'realise that we are something more than a putative seventh state', since Australia was considered to be far more pro-American than New Zealand.[48]

The Americans retaliated by cancelling future ANZUS operations and by cutting New Zealand off from future Intelligence material. The Australians responded by calling off that year's ANZUS council meeting in Canberra, but Lange refused to accede to the demand of the leader of the opposition, Jim McLay, for an emergency ANZUS summit meeting, saying that the nuclear-free policy was 'an act of national self-determination', and 'It is inconceivable that the United States will allow us to be overrun.'[49]

Despite Margaret Thatcher telling Mr Lange that while she disapproved of his actions it would have no effect on UK-New Zealand relations, there were cartoons in New Zealand newspapers of Uncle Sam and John Bull ripping pages out of Soviet books entitled 'How to Bully an Ally'. Lange even claimed to a press conference that the US was seeking to destabilise his government. Self-importance merged with low-level paranoia to create a serious rift in the English-speaking world in the South Pacific.

By claiming that New Zealand's enemies 'do not exist', Lange effectively cut his country off from the concerns of the rest of the English-speaking peoples, whose genuine opponents and rivals in the shape of the Soviet Union and her allies and China did exist. He underlined this by saying, 'I feel safer in Wellington that I ever could do in London or New York,' and 'The collisions and confrontations that take place in Europe are very far from us,' both of which statements were undeniably true, but which would never have occurred to former New Zealand premiers such as Richard Seddon or Joseph Savage or Peter Fraser. Physical distance was anyhow becoming less of a factor in nuclear strategy, considering the reach of the long-range intercontinental ballistic missile.

It was perhaps always possible after eighty-five years that New Zealand would one day finally separate herself from the rest of the English-speaking peoples' global defence posture, but it came at a time when the United States was in deep negotiations with the Soviet Union over intermediate nuclear weapons and so could hardly have been worse timed. Lange was right when he said that his country's new departure 'will not be seen as an act of provocation'; instead, it was rightly viewed as a meaningless but self-indulgent piece of gesture politics that had more to do with national identity and self-esteem than any genuine regional defence concerns. This was all the more absurd since New Zealand had long been an honoured component of the English-speaking peoples; any inferiority complex that her left-wing politicians and newspaper editors might have had was all in the mind.

Mikhail Gorbachev met Ronald Reagan for his first face-to-face encounter in Geneva in November 1985. The General Secretary naïvely believed that he had got the better of Reagan in these talks, which was reflected in the telegram he afterwards sent to Fidel Castro, Kim Il-sung, Li Xiannian of China and other communist leaders, which read: 'The talk with Reagan was a real skirmish. [Donald] Regan – Reagan's closest aide – later said that no-one had ever talked so frankly and with such force to the President before. ... In Geneva we had no intention of letting Reagan get away with just a photo session, which he loves so much.'[50] The Politburo resolution concluded that Gorbachev's diplomacy had 'placed the present American Administration on the defensive and landed a serious blow on the ideology and policy of their "crusade"'. It was so gross a miscalculation as to be risible; however it did no harm for Gorbachev to think himself the victor.

On his return from meeting Reagan in Iceland the next year, Gorbachev delivered another upbeat assessment of his own success, telling the Politburo on 14 October 1986:

At Reykjavik we have scored more points in our favour than we did after Geneva. But the new situation demands new approaches in our military doctrine, in the security of our armed forces, their deployment and so on, and in the defence industry. ... The meeting at Reykjavik showed that, in the representatives of the American administration we are dealing with people who have no conscience, no morality. ... In Reagan at Reykjavik we were fighting not only with the class enemy, but one who is extremely primitive, has the looks of a troglodyte and displays mental incapacity.[51]

For the balding, portly General Secretary to criticise the looks of the former film star, who – along with JFK – was perhaps the most handsome of all the US presidents, was perhaps indicative of the poverty of options left open to him by five years of resolute American foreign policy, and the prospect of at least another three to come. The conclusion Gorbachev drew from Geneva, that the Soviet Union needed to continue her massive arms spending, was of course precisely the one that was to drive the 'evil empire' into near-bankruptcy and eventually to break the will of her leadership to continue to refuse her people freedom and democratic rights.

On 10 April 1986, the La Belle discotheque in West Berlin, much frequented by US servicemen, was bombed, killing two people and injuring scores more. The outrage was authoritatively traced back to Libyan involvement, only three months after the United States had already imposed sanctions on Libya for other links with international terrorism. So, five days later, planes from US warships and bases in Britain – France having refused them permission to over-fly her airspace – bombed various targets in Libya, killing over 100 people, including Colonel Muammar Gadaffi's infant daughter.

There are many who argue that US bombing of Arab targets can never achieve positive results and only ever worsens matters. Yet the Libyan experience disproves that. Before the attack, Gadaffi broke off relations with Saudi Arabia because of the 'US occupation' there, he bought huge amounts of arms from the USSR and his MIG jets 'played chicken' with American planes near the Libyan coast. When oil was discovered between Libya and Malta, he declared that the territorial waters of Libya included everything up to twelve miles from the Maltese shore. He invaded Chad and sent death squads into Nigeria. As one recent account recalled, he 'also had an affection for resistance fighters and revolutionaries everywhere, no matter whose side they were on. To give you some idea, he funnelled money and arms to *Scottish* revolutionaries.'[52]

The list of 'national liberation' movements that Gadaffi supported financially or with arms represented a virtual *Who's Who* of extremist groups, and included the Moros in the Philippines, the Palestine Liberation Organisation, radical Native American groups, the New Jewel movement in Grenada, the IRA, the Basques, the Kurds, Louis Farrakhan's Nation of Islam, and the Black Panthers, as well as guerrilla movements in Chad, Eritrea, Lebanon, the Canary Islands, Egypt, Sudan, Corsica and Sardinia, and even Wales.[53]

Yet immediately after the US bombings, Gadaffi cut back massively on almost all his support for international terrorism and was the first Muslim leader to condemn Al-Queda after 9/11, declaring, 'Irrespective of the conflict with America, it is a human duty to show sympathy with the American people and be with them at these horrifying and awesome events which are bound to awaken human conscience.' Although those tears were doubtless somewhat crocodilian, especially coming from the man whose agents were responsible for the deaths of 271 mainly Americans and Britons in the sky above and on the ground at Lockerbie in Scotland on 21 December 1988, they certainly represented a major shift in emphasis.

At the time of the US response to 9/11 in Afghanistan and Iraq, Gadaffi went far further and announced a complete cessation of his support for terrorism, his weapons procurement projects, and he invited the West to inspect the decommissioning of his weapons of mass destruction, including a nascent nuclear programme. (Today those weapons can be seen in a museum in Oakridge, Tennessee.) This was a hugely important and positive by-product of George W. Bush's vigorous prosecution of the War against Terror, and a sign that rogue states can respond positively to firm treatment. That welcome process had started fifteen years earlier, under Ronald Reagan.

Far less happy for President Reagan was the revelation of the Iran-Contra affair, which burst onto a completely unsuspecting American public on 25 November 1986. On that day the President and his Attorney-General Edwin Meese went public with the news that two seemingly separate international events were in fact closely connected. The previous month, a C-123K cargo plane had been shot down by a surface-to-air missile over Nicaragua, then a Marxist-Leninist state controlled by the pro-Castro Sandinista Government. The only survivor of the crash, Eugene Hasenfus of Marinette, Wisconsin, went on television in Managua to state that the CIA was supplying arms to the right-wing Nicaraguan Democratic Resistances Forces (i.e. the 'Contra' guerrillas), who were attempting to overthrow the Sandinista Government.[54] This activity was illegal, since in October 1984 a Massachusetts Democratic congressman named Edward Boland had passed a resolution specifically banning 'the CIA, the Department of Defense, or any other agency or entity of the US' from helping the Contras.

Just as Hasenfus appeared on television screens across the world, a hemisphere away the National Security Council (NSC) officer in charge of the support operation, Lieutenant-Colonel Oliver North, was involved in top-secret negotiations with representatives of the Iranian Government, by which the US agreed to sell her former enemy arms in return for the release of a number of hostages held by the terrorist organisation Hezbollah. North was a former Marine Corps lieutenant-colonel and Vietnam veteran who was driven by ideological anti-communism. (Amongst his codenames during the operation were 'Blood and Guts' and 'Steelhammer'.) North's activities of course made a nonsense of Reagan's vow to the American people that 'America will never make concessions to terrorists.' Iran, which was then at war with Iraq, was similarly able to swallow her publicly expressed distaste for America ('the Great Satan'), just as she had privately swallowed her distaste for Israel when she bought arms from her in 1980 and 1981.

The 'smoking gun' in the Iran-Contra affair was a memorandum written by North on 4 April 1986, later known as the 'diversion document'. Entitled 'Release of American Hostages in Beirut' and unsurprisingly labelled 'Top Secret – Sensitive', it explained that the previous September, Rev. Benjamin Weir had been released in Beirut only forty-eight hours after Israel had sold 508 Tube-launched, optically-tracked, wire-guided (TOW) missiles to Iran with the endorsement of the US Government, arms that the US Government would then replace for Israel.[55] As part of a further deal, of more arms for more hostages, North wrote that of 'the residual funds from this transaction', $12 million 'will be used to purchase critically needed supplies for the Nicaraguan Democratic Resistance Forces'.

Once the 'diversion document' came to light in November 1986, the Administration called an immediate press conference in which Reagan and Meese revealed that there had been a scheme 'to skim millions of dollars from arms sales to Iran to finance the Contras', in violation of the Boland Amendment, and that as a result North had been fired and the National Security Advisor, Admiral John Poindexter, had resigned. The story did not stop there, however. As a result of Congressional hearings that continued for many months, it became clear that several other Administration officials had known about the operation, and that many foreign governments – including those of Israel, Saudi Arabia, South Africa, China, Taiwan, Panama, Costa Rica, Guatemala, El Salvador and Honduras – had been involved to some extent or another.[56]

Although any amount of important constitutional issues were raised as a result of the Iran-Contra affair, to do with the Executive's ability to run an 'off-the-books' foreign policy without Congress knowing, secrecy versus open government, and so on, none of the in-depth investigations managed to ascertain that Reagan himself had authorised anything. Unlike Nixon during Watergate, Reagan made no attempt to cover up what his errant staffers had done, or to protect them from prosecution. In December 1981, he had ordered the CIA to spend $19.95 million to finance a 500-man paramilitary 'action team' of Nicaraguan exiles to fight the Sandinistas, as part of the fight-back against communism for which he was elected, but that had been years before the Boland Amendment was passed.

Yet just as in 1973 Congress would not go along with further anti-communist activity in South-East Asia, so resolutions such as the Boland

Amendment hamstrung the Administration's struggle against international communism in the Eighties. Yet for true Cold War warriors such as North, Poindexter and the CIA director William Casey, the opportunity of simultaneously getting hostages released, keeping America's potential enemies Iran and Iraq fighting each other, and – best of all – financing anti-communist guerrillas in Nicaragua, was too good to miss. It is rare enough in politics to be able to kill two birds with one stone; this policy killed three. As Poindexter told the Iran-Contra hearings in 1987, 'My object all along was to withhold from the Congress exactly what the NSC staff was doing.' North was similarly pugnacious, saying, 'We all had to weigh in the balance the difference between lives and lies.' North was sentenced to 1,200 hours of community service and a $150,000 fine.

Between 1987 and 1988, Saddam Hussein, the dictator of Iraq, launched attacks on no fewer than forty Kurdish villages in the north of that country, using them as testing grounds for new mixtures of mustard gas and various nerve agents that his scientists were developing, such as Sarin, Tabun and VX. (Ten milligrams of VX on the skin will kill most people; a single raindrop weighs eighty milligrams.) According to Dr Christine Gosden of Liverpool University, who developed treatments and research programmes for survivors,

> Iraqi government troops would be surrounding the attack site and they would have chem-bio suits on ... included would be doctors and interested observers. ... They would go in and find out how many people were dead ... and how many survived. What ages ... did men, women or children or the elderly suffer more? From there they would shoot the survivors and burn the bodies.

The worst attack came on Wednesday, 16 March 1988, when the Kurdish town of Halabja was attacked. The Iraqi troops methodically divided it into grids, in order to determine the number and location of the dead and the extent of injury, thereby enabling them scientifically to gauge the ability of various different types of chemical agents to kill, maim and terrorise population centres. One of the first war correspondents to enter the town afterwards, Richard Beeston of *The Times*, reported that, 'Like figures unearthed in Pompeii, the victims of Halabja were killed so quickly that their corpses remained in suspended animation. There was a plump baby whose face, frozen in a scream, stuck out from under the protective arm of a man, away from the open door of a house that he never reached.'[57] They were the lucky ones.

Between 4,000 and 5,000 civilians, many of them women, children and the elderly, died within hours at Halabja, through asphyxiation, skin burns and progressive respiratory shutdown. However, according to Dr Gosden, a further 10,000 were 'blinded, maimed, disfigured, or otherwise severely and irreversibly debilitated', who afterwards were subjected to neurological disorders, convulsions, comas and digestive shutdown. In the years to come, thousands more were to suffer from 'horrific complications, debilitating diseases, and birth defects' such as lymphoma, leukaemia, colon, breast, skin and other cancers, miscarriages, infertility and congenital malformations, leading to many more deaths.[58]

It was to prevent the violent overthrow of Saddam Hussein by the English-speaking peoples and their allies that millions of anti-war protesters marched in massive demonstrations across the US and Europe in late 2002 and early 2003.

On Thursday, 11 June 1987, Margaret Thatcher led the Conservatives to a third consecutive general election victory, winning an overall Commons' majority of 101. Back in February 1986, over at the Beefsteak Club, the Earl of Onslow had bet Earl De la Warr a bottle of vintage Dom Perignon 'that Mrs Thatcher – if still leader of the Tories – will win the next General Election with an overall majority'. Victory tasted good to the Conservative Party that night, as their Labour opponents were forced to consider ditching red-blooded socialism as the price of having any hope of returning to office. Seven years later, after having lost a fourth general election fought entirely on Thatcher's legacy but against a different Conservative leader, Labour did just that.

In 1989, Timothy Berners-Lee, a thirty-four-year-old London-born mathematician and physicist working at CERN, the European particle physics laboratory in Geneva, invented what he called the World Wide Web, which was formally launched two years later. By allowing people to share information in a web of 'hypertext' documents, Berners-Lee's invention allowed the internet to branch out from 600,000 users in 1991 to over forty million users only five years later, and an estimated one billion by 2005. 'There was never a feeling of "Heh, heh, heh, we can change the world",' recalls Berners-Lee. 'It was, "This is exciting, it would be nice if this happened", combined with a constant fear that it would not work out.'[59]

Berners-Lee was the son of computer mathematician parents; he liked to build computers out of cardboard as a child and as an undergraduate at Queen's College, Oxford, he was banned from using the nuclear physics laboratory computer after hacking into it for a 'Rag Week' prank, so he built his own with an old television set and miscellaneous spare parts. When he invented the World Wide Web, utterly revolutionising the way that the human race communicates with itself, he refused the billions that could have been rightfully his through patenting, insisting that everyone should be able to access it for free. 'He designed it,' as *Time* magazine put it when listing him as one of the twentieth-century's 100 most influential people. 'He loosed it on the world. And he more than anyone else has fought to keep it open, non-proprietary and free.'[60]

Of all the great inventors of the English-speaking world, apart from Lord Rutherford, Sir Alexander Fleming and Logie Baird, few can have had a more profound and immediate effect on global life than Berners-Lee. 'We have to recognise that every powerful tool can be used for good and evil,' he acknowledged in respect of the internet's attraction for fascists, paedophiles and fraudsters, but its efficacy in 'breaking down barriers' would, he hoped, 'support a fair and just planet, but remember it's not the technology's role to make the rules or enforce them. ... We can't blame the technology when we make mistakes.'

Yet it was not just Berners-Lee's genius alone that had helped create the internet: in the 1960s, US military analysts 'saw the potential for a fault-tolerant command-and-control network in the event of all-out nuclear war'.[61] The Pentagon's collaboration with major universities such as University College, London, funded MILNET, which a decade later turned into the internet. Cash as well as brains were necessary, and the American taxpayer provided much of both.

On 12 June 1987, Ronald Reagan stood at the Brandenburg Gate in Berlin and demanded: 'General Secretary Gorbachev, if you seek peace, if you seek prosperity for the Soviet Union and Eastern Europe, if you seek liberalisation: Come here to this gate! Mr Gorbachev, open this gate! Mr Gorbachev, tear down this wall!' Two-and-a-half years later, on Thursday, 9 November 1989, the Berlin Wall was finally torn down, after twenty-eight years of standing as a potent symbol of communist oppression of the peoples of Eastern Europe. Suddenly millions of people who since the Second World War had lived under totalitarianism and dictatorship were allowed to enjoy the benefits of representative institutions, property rights and freedom of speech, worship and association. It was the greatest single moment of liberation in the history of Mankind since V-J Day forty-four years earlier.

There are two persuasive but mutually antagonistic explanations for the collapse of communism, ones that the British historian David Pryce-Jones has characterised as the 'High Road' and the 'Low Road' rationalisations:

> The High Road argument is that the implosion of Communism was all Mikhail Gorbachev's doing, and that he should receive praise for his nobility or blame for his stupidity, depending on one's outlook. He happened to believe in the perfectibility of Communism, and that he was the man for the task. In the nature of things, this mindset could bring only contradictions fatal to the system. The alternative Low Road argument is that through the fraught years of the Cold War, the United States established the superiority of its institutions and values,

obliging the Soviet Union to accept that it could not compete in the long run. Through NATO, the United States built and maintained a coalition of democratic allies. More than anything else, the costs of military technology in general, and of meeting the challenge of [the US's Strategic Defense Initiative] Star Wars in particular, exposed the Soviet Union's centralised economy as an inefficient sham.[62]

So which is the correct analysis? Should Gorbachev be given the credit, or were the English-speaking peoples and their pro-democracy allies the prime movers, led by such people as Ronald Reagan, Margaret Thatcher, Pope John Paul II, Alexander Solzhenitsyn, Arthur Koestler, Andrei Sakharov, Lech Walesa, Vaclav Havel, Irving Kristol, Roger Scruton and Robert Conquest?

Détente, the policy of accommodating communism that earlier Western governments had adopted, was finally ditched by the English-speaking peoples in the Eighties, with spectacular results. It had served its purpose in the early Seventies in keeping communism at bay while the West was in retreat on so many fronts; but by the late Seventies it had outlived its usefulness and was merely extending the life of a demonstrably 'evil empire'. *Détente* had anyhow meant very different things in the East and the West. The West saw it as a way of lowering tensions, 'in the hope that it might disengage from the dreadful and even apocalyptic tests of strength it was inflicting on the rest of the world'.[63] By contrast, in 1976 Leonid Brezhnev stated, '*Détente* does not in any way rescind, nor can it rescind or alter, the laws of class struggle. We do not conceal the fact that we see in *détente* a path towards the creation of more favourable conditions for the peaceful construction of socialism and communism.'

The way that Ronald Reagan and Margaret Thatcher sought to defeat European communism was by simultaneously proving to the peoples of the Soviet Empire that capitalism was simply superior in delivering material benefits, and simultaneously deploying advanced weapons systems such as Cruise missiles and Pershing II. Moscow foreign policy analysts agree that if the huge Soviet effort to prevent these deployments had been successful, 'The Kremlin leadership, already almost convinced of their ability to disarm much of Europe psychologically, would have adopted a particularly dangerous and aggressive stance.'[64] Fortunately, NATO's deployment of Pershing II and Cruise missiles was not prevented by the Campaign for Nuclear Disarmament, German peaceniks and neutralists, Soviet 'agents of influence' and other – in Lenin's phrase – 'useful idiots' in the West.

Instead, on 23 March 1983, President Reagan proposed a Strategic Defense Initiative system for the United States, using satellite technology to detect and destroy incoming nuclear missiles, which was promptly nicknamed 'Star Wars' after the 1977 George Lucas movie. If successful, SDI would nullify at a stroke the Soviet Union's entire nuclear strategy against the West. On 4

February 1985, the Pentagon's defence budget included provision for the trebling of expenditure on the SDI research programme, an unmistakable sign to the Soviets that the Americans believed it might well work. At the Reykjavik mini-summit in October 1986, Gorbachev insisted that the project be abandoned, which Reagan refused to do. The following February, Gorbachev dropped his demand for the curtailment of the programme, proposing instead a separate agreement to abolish intermediate-range missiles in Europe, which Reagan enthusiastically took up, and a treaty to eliminate them was duly signed on 1 June 1988.

In 1989, the US Delta Star satellite was launched, which successfully detected and tracked test missiles shortly after they were launched. Suddenly, the theoretical prospect opened up of the United States having a laser defence system that could destroy Soviet ICBMs (intercontinental ballistic missiles) in midair, before they reached America, while the US was still capable of devastating the USSR. Although SDI was in fact very many years from fruition – and debate still rages about whether it was ever technologically achievable anyhow – Gorbachev did not know that at the time. (Those tests that were carried out seem to have been artificially arranged as demonstrations for the benefit of the KGB, rather than genuine tests of a workable prototype.) Whatever the capabilities of SDI really were, the Soviets knew that they could compete neither in terms of technological know-how nor the huge cost. The former Soviet dissident Natan (formerly Anatoly) Sharansky has likened Reagan's confrontation with the USSR over the arms race in space to challenging 'a Soviet pensioner on his deathbed . . . to run a marathon'. Years later, records Sharansky, 'close advisors of Gorbachev admitted that the realisation that the USSR could never compete with Star Wars made them finally accept demands for internal reform'.[65]

The fact that the US abandoned the Initiative in May 1993 implies that the Clinton Administration assumed that it was leading nowhere expensively, but by then the Russians' bluff had been called. Gorbachev had blinked first and SDI had served its invaluable political purpose of persuading the Soviets that they could not compete technologically with the Americans in this strategically vital sphere. (Today there is another US missile defence programme, based in Alaska, which the Canadian Government turned down the chance to join in February 2005, the first time in decades that it had refused to take part in a strategic project designed to protect the North American continent. The Canadian Premier Paul Martin's decision was a sadly retrograde step for the amity of the English-speaking peoples, and in the future an American president will not now be obliged to protect Canada from incoming missiles if he chooses not to, which also cannot be in Canada's national interest.)[66]

Once Gorbachev accepted that his country's archaic command economy had been woefully overhauled by the capitalist market economies of the West, the pressure for change grew until it became irresistible. The moral case against the Soviet Union had been made much earlier, but was still being made by Russian and East European dissidents who were persecuted relentlessly for speaking truth to power. These brave men and women numbered thousands, but special mention should be made of the courage of Cardinal Frantisek Tomasek of Prague, the imprisoned and tortured Cardinal Jozsef Mindszenty of Hungary, Jiri Muller in Brno, the intellectual Gaspar Tamas of Hungary, Gabor Demsky (later Mayor of Budapest), the Catholic youth movement leaders Frantisek Miklosko and Jan Carnogursky of Slovakia, and the composers Henryk Gorecki and Arvo Paart. On 16 January 1969, the Czech student Jan Palach had set himself on fire in Wenceslaus Square in Prague to protest against the Soviet occupation of his country.

Although Gorbachev initially hoped to strengthen communism by reforming it, in breaking the grip of the Party he unleashed forces such as nationalism that spelt doom for the very system he was trying to save. He and his allies hugely underestimated the power of the forces they were uncorking, and he was quickly overwhelmed by events. Although the West became love-struck with him, making him *Time* magazine's Man of the Decade and awarding him the Nobel Peace Prize amongst any number of other accolades, the Russian people knew perfectly well that he had in fact been out-manoeuvred by his own reforms. The iron law of unintended consequences operated once again. There was no capitulation, no conspiracy, simply the realisation that the game was finally up. As Janis Jurkans, the Latvian Foreign Minister, said of Gorbachev's Politburo, 'They created the instrument that destroyed them.' Vaino Valyas, the Party leader in Estonia, agreed, saying, 'He wanted a more efficient Soviet Union but finished with no Soviet Union at all.'[67]

Of course it did take Gorbachev to cut away support from the hard-line communist puppet rulers in the Eastern European satellites and then to allow communism to fall in Russia without blood being shed, and he deserves credit for that, but as with his decision to allow East Germany to join NATO in 1990, it is hard to see any alternative policies that would have worked in the long run, except for one that turned him into another Nicolae Ceauşescu, the Romanian dictator whose Securitate killed over 1,000 people during the Christmas revolution as he fell from power.

The various 'popular fronts' and cadres of Party officials that Gorbachev set up to create a reformist base turned into nationalist movements that wanted freedom from Moscow even more than the reform of communism. That would never have been necessary if the leadership of the English-speaking peoples had not taken up a steely post-*Détente* stance against Soviet communism. This forced the Soviet Union to finance, in the words of the British intellectual Noel Malcolm, 'a military-industrial complex which was gobbling the country's wealth like a greedy cuckoo chick in a hedge-sparrow's nest'.[68]

Ronald Reagan and Margaret Thatcher dedicated themselves to undermining and eventually defeating European communism, and without their efforts it could have been further decades before it collapsed under the weight of its own internal contradictions – to borrow a phrase Lenin used about capitalism – and the days of *glasnost* (openness) and *perestroika* (reconstruction) dawned.

The shattering end came – with a timing that was truly poetic – just days before the close of the decade that President Reagan and Mrs Thatcher had made their own, the 1980s. In that sense, October 1989 was just as important a revolutionary moment as July 1789 or October 1917. As one post-communist Bulgarian put it, 'Lenin said that the system that guaranteed higher productivity would prevail, but that turned out to be Capitalism.' Russia had taken the Low Road to Freedom.

SIXTEEN

The Wasted Breathing Space

1990 – 11 September 2001

The New World Order – Iraq invades Kuwait – The Fall of Margaret Thatcher –
The Gulf War – Gorbachev totters – The World Trade Center attacked – Black
Hawks down – Whitewatergate and other scandals – The French language stages a
counter-attack – Hollywood Anglophobia – Bosnia – Tony Blair – Hong Kong –
NATO expands – The Death of Diana, Princess of Wales – Bin Laden escapes yet
again – Sinn Fein-IRA (effectively) surrenders – Kosovo – USS Cole – George
W. Bush becomes president – Australians reject a republic

'*Annuit coeptis, novus ordo seclorum.*'
('He gave his approval to these beginnings, a new world order.')
The Great Seal of the United States of America

'American superiority in all matters of science, economics, industry, politics, business, medicine, social life, social justice, and of course the military was total and
indisputable. Even Europeans suffering the pangs of wounded chauvinism looked
on with awe at the brilliant example the United States had set for the world as the
third millennium began.'   Tom Wolfe, *Hooking Up*

The West's victory in the Cold War brought a general lowering of tensions
across the globe, except in one particular region, where it led to them
being heightened. According to the 2005 *Human Security Report*, which was
sponsored by five governments and drawn up by a Canadian team under the
supervision of the Australian National University professor and UN security
advisor Andrew Mack, 'Since the early 1990s, there has been an 80% decrease
in the number of battle deaths per conflict per year. Where back in the Fifties,
the average number of battle deaths per conflict per year were between thirty
and forty thousand, by the early 2000s this number was down to around six
hundred.' The former Foreign Minister of Australia, Gareth Evans, com-
menting on the *Report*, pointed out that, 'More civil wars have been ended by
negotiation in the past fifteen years than in the previous two centuries.'[1]

The disappearance of the power of the Soviet Union and her allies to intervene in and destabilise pro-Western governments led directly to this 'New World Order' (although the liberal internationalist Evans absurdly put it down to, *inter alia*, the success of the United Nations). Yet in the Middle East, where the USSR had generally exerted power against Islamic fundamentalism due to her own internal security concerns with her own large Muslim populations, Russia had overall been a force for stability. The Soviet defeat in Afghanistan had led the victorious Islamicist mujahadeen to believe that the world's other superpower, the United States, could also be forced to quit the region.

The first post-Cold War threat to the influence of the English-speaking peoples in the Middle East came early, within a year of the fall of the Berlin Wall. The invasion of Kuwait by Saddam Hussein's Iraq at 2 a.m. on 2 August 1990 forced her emir, Sheikh Jaber al-Sabah, to flee to Saudi Arabia. With $70 billion of foreign debt, more than half of which was owed to Kuwait and Saudi Arabia, Saddam had discovered a way of wiping out Iraq's financial problems at a stroke. Two days later, Iraqi troops and tanks massed on the border of Saudi Arabia. On 7 August, President George Bush Snr sent US military forces to Saudi Arabia to try to prevent an Iraqi invasion, and two days after that Iraq announced her formal annexation of Kuwait. The spectre suddenly appeared before Western policy-makers of some 40% of the world's output of oil – the Iraqi, Kuwaiti and Saudi fields combined – being controlled by a single Middle Eastern dictator.

Both the Gulf and the Iraq Wars have been described as being merely 'all about oil', but it was perfectly legitimate for powers such as the English-speaking peoples, whose economies largely ran on oil, to ensure that their supply was not disrupted or cornered by a single unpredictable dictator. The prosperity and employment of millions of industrial workers in the West depended on Saddam Hussein not being able to dominate the world's oil resources, and any lesser response by the two Bush Administrations would have been an abdication of their responsibility to protect the livelihoods and wellbeing of their citizens.

Once again, the Intelligence services had failed to predict Saddam's likely action. The British Foreign Secretary, Douglas Hurd, discovered what had happened not from the CIA or MI6 but from the radio news.[2] As with the Falklands invasion, there is only a certain amount any Intelligence agency can accurately surmise about the actions of an unpredictable dictatorship.

Its enemies like to portray the Anglo-American Special Relationship as that of a poodle trotting obediently at the heels of its master. It was certainly the demeaning image much favoured by the opponents of the Iraq War in 2003, regularly displayed on the posters carried by demonstrators in Britain and abroad. Despite the lie being given to that characterisation on any number of occasions during the twentieth century, notably at Suez, Vietnam and the Falklands, it is still pushed by the opponents of English-speaking amity today. Yet in August 1990, Margaret Thatcher told George Bush Snr, in relation to the Kuwait crisis, 'This is no time to go wobbly, George.' Far from resembling a poodle, the British Prime Minister used her influence and considerable powers of personal persuasion to ensure that there was no backsliding once Saddam had made clear his intention to take and hold Kuwait and, further, to menace Saudi Arabia.

The month after Iraq's invasion, Bernard Lewis, the most knowledgeable Western thinker on the Middle East, wrote an essay entitled *The Roots of Muslim Rage* that pointed out why the United States, as the heir to the European imperial powers and the greatest exporter of the new world political culture, had become the primary focus of the frustrations and rage of the Islamic world. In it he wrote,

> The Muslim has suffered successive stages of defeat. The first was his loss of domination in the world, to the advancing power of Russia and the West. The second was the undermining of his authority in his own country, through an invasion of foreign ideas and laws and ways of life and sometimes even foreign rulers and settlers, and the enfranchisement of non-Muslim elements. The third – the last straw – was the challenge to the mastery in his own house, from emancipated women and rebellious children.[3]

The military retreat of Islam had been continuing since the defeat of the second Ottoman siege of Vienna in 1683, so it was not surprising that when Saddam Hussein presented himself as the first Muslim leader since Colonel Nasser to stand up to and possibly even humiliate the West, he should have found tremendous popularity in the Arab streets, especially those of the Gaza Strip and the West Bank. He was nonetheless going to need more than that to deter the readiness of the English-speaking peoples-led coalition to stand up once more for the rights of the small nation – as with Belgium in 1914, Poland in 1939, South Korea in 1950, South Vietnam in 1964, the Falklands in 1982 and now Kuwait.

George Bush appreciated this; in a speech he gave outside the River Entrance of the Pentagon on 14 August 1990, with both General Colin Powell and General Norman Schwarzkopf in the audience, the President said that the United States would make a stand 'not simply to protect resources or real estate, but to protect the freedom of nations'. He compared Saddam Hussein to Hitler, which, as Schwarzkopf noted in his memoirs, 'did not sound like a leader bent on compromise', adding, 'There is no substitute for American leadership, and American leadership cannot be effective in the absence of American strength.'[4] Yet Bush did indeed compromise, by letting Saddam survive the Gulf War in power and thereby passing the problem on to the next generation. Literally so, in the shape of his son.

Just before the blow against Saddam fell, Margaret Thatcher was ousted from power by her own Conservative Party, on 22 November 1990. Even though she won more votes than her challenger Michael Heseltine, the arcane procedure used by the Tories to choose a successor meant that she could not carry on as really leader, and thus prime minister. Here is the view of the (Left-leaning) novelist Robert Harris about those dramatic events, written in May 2005:

> I wonder how many of those 168 Tory MPs who either voted for Heseltine or abstained would have rushed to depose their most successful twentieth century leader if they had known what the future held? For although the short-term effects were beneficial, the long-term consequences have been catastrophic. Of those fifteen years since Thatcher fell, the Tories have been behind in the polls for nearly 13. For eight there has been a Labour Government. ... It is a thoroughly bad idea for a minority party cabal to bring down an elected prime minister. The Liberals did it to Asquith in 1916 and never gained power again. The Tories did it to Thatcher in 1990 and have since suffered three successive election defeats – a calamity previously unknown to them for ninety-five years.[5]

That calamity was very largely brought about by a minority of Conservative MPs whose commitment to what they called 'the European Project' outweighed their loyalty to their Party and their gratitude to the woman upon whose coat-tails they had thrice been elected. 'When you see the way she was done down,' wrote the former Labour leader Neil Kinnock soon afterwards, 'you are bound to think that the people who organised the coup must have had a conscience by-pass.' Their success, and the Party's subsequent choice of a personally insignificant replacement for her in the shape of John Major, spelt doom for Toryism, even if the Conservative Party benefited electorally in the short-term by narrowly winning the 1992 general election.

In November 1990, those nebulous concepts 'the world community' and 'international opinion' came together at the United Nations to authorise the use of force to expel Iraq from Kuwait. Even the Russians supported, but did not participate in, the UN action. Security Council Resolution 678 mandated its members 'to use all means necessary' to force Iraq's complete withdrawal from Kuwait. To complaints that Resolution 678 sounded somewhat euphemistic, Colin Powell replied: 'It did not matter. A bullet fired through a euphemism is still a bullet.'

President Bush had already built up an enormous anti-Saddam coalition. Under the overall command of the US General H. ('Stormin'') Norman

Schwarzkopf were military contingents from the United States, Britain, Canada, France, Italy, Kuwait, Saudi Arabia, Egypt, Syria, Qatar, Bahrain, the United Arab Emirates, Oman, Morocco, Czechoslovakia, Pakistan, Bangladesh, Senegal and Niger. There were also Belgian and German contingents based in Turkey. Unfortunately, the very breadth, width and depth of the coalition was to prove fatal to its proper purpose.

As Lord Salisbury had written to Lord Curzon, the Viceroy of India, in 1901, 'In the last generation we did much as we liked in the East by force or threats, by squadrons and tall talk. But we now have "allies" – French, German, Russian: and the day of free, individual, coercive action is almost passed by.' There is a moment when having too large a coalition begins to work against the best interests of the purpose for which it was originally created, and such a thing took place in early 1991.

The English-speaking peoples have long been expert at building and maintaining coalitions. The British brought no fewer than seven into being against Napoleon; the Crimean War was fought alongside France, Austria and Sardinia; the Great War and Second World War were fought together with France and Russia; NATO started life with twelve nations; Korea was fought under the auspices of the UN, and Vietnam saw contributions from Australia, New Zealand and several others. Yet in each of those conflicts the participants had a common purpose, which turned out to be not quite the case in 1991.

Facing the coalition forces were over 550,000 Iraqi troops in Kuwait and south-western Iraq, organised into forty-two divisions, with 4,200 tanks, 250 helicopters and 550 combat-ready aircraft.[6] They seemed a formidable foe, at least on paper. Iraq having failed to meet the United Nations' deadline to withdraw from Kuwait by midnight on 15 January 1991, the US-led coalition forces immediately commenced the air-offensive part of its Operation Desert Storm. The coalition's air superiority proved virtually unchallenged and by 22 February over 35,000 sorties had been flown over enemy territory. This air supremacy both denied the Iraqis any aerial intelligence and also allowed the coalition to strike ground targets at will, which in turn provoked mass desertions and a general collapse in enemy morale.

Saddam responded by launching Soviet-made surface-to-surface Scud missiles against Saudi Arabia and Israel. Although Israel was not a belligerent, he hoped to peel off Arab members from the coalition if she retaliated, but in the event she exercised statesmanlike constraint in not doing so. US Patriot surface-to-air missiles also managed to destroy many Scuds before they landed, resulting in surprisingly few Israeli casualties.

After a feint attack by several Iraqi battalions into Saudi Arabia between 29 and 31 January, which was beaten off by Saudi, Qatari and US Marine units, the ground war proper began at 4 a.m. local time on 23 February 1991. With three major simultaneous assaults – an outflanking movement through southern Iraq, a thrust northwards along the coast towards Kuwait City and an attack north-eastwards through south-west Kuwait – the coalition defeated the Iraqi army and crushed all organised resistance over the next seventy-two hours. The result was in sharp contrast to those predictions from journalists such as Robert Fisk, who had foreseen another Vietnam, with huge casualties and a collapse in American national morale. In Britain, the former politicians Denis Healey and Edward Heath also predicted a disaster in which tens of thousands of coalition troops would die.

Despite Iraq having no fewer than 2,800 tanks in Kuwait and southern Iraq – twice as many as the coalition – her Soviet-built T-72 was vastly outperformed by the main US battle tank, the Abrams M1A1. A few days later, on 26 February, the Iraqi army's convoy at Mitla Ridge on the Jahra-Basra Road was dive-bombed to destruction by coalition air power. As one writer, Justin Wintle, has judiciously put it, in comparison with earlier conflicts such as Vietnam, 'the war against Saddam was rapid and in terms of lives lost, virtually cost-free. Above all the carnage of Mitla Ridge, in which tens of thousands of poorly equipped retreating Iraqi infantry were exterminated by American air power, was the triumph of a honed technology.'[7] Total coalition casualties numbered ninety-five killed, 358 wounded (many of them through 'friendly' fire, rather than Iraqi action) and twenty missing. Iraqi losses were estimated at between 30,000 and 50,000 killed and a similar number wounded, with 60,000 captured. Not since General Prendergast had captured Burma for the loss of twenty-two men in 1885 had there been such an uneven victory for the English-speaking peoples.

On 27 February, President Bush made a statement on television announcing the cessation of hostilities. 'Kuwait is liberated,' he said. 'Iraq's army is defeated. Our military objectives are met.' It was true, but they were the wrong objectives. For the road was now wide open for the coalition – or at least as much of it as was willing to continue – to move straight on to Baghdad, to depose Saddam Hussein and the Ba'athist regime, and to install a provisional Iraqi government. Throughout March there were unco-ordinated anti-government uprisings taking place across Iraq, especially by Iraqi Shi'ites in the south and the Kurds in the north of the country. It was then that the English-speaking peoples showed a disastrous lack of will and flexibility.

Surprised by the speed and completeness of their victory, President Bush Snr and John Major failed to take advantage of the unique opportunity now offered of overthrowing Saddam Hussein. Because the terms of Resolution 678 had been fulfilled, and because the pan-Arabian breadth of the coalition meant that it did not want a long-term English-speaking military presence in Iraq, Bush and Major passed up the chance of ridding the Middle East of that destabilising tyrant in 1991. There is even the suggestion that they feared that the massacre on the Basra Road would provoke sympathy for the enemy.

When the English-speaking peoples fight for specified, attainable objectives – such as capturing Pretoria, Manila, Berlin, Pyongyang, Seoul (twice), Port Stanley, Grenada or Baghdad (in 2003) – they achieve them, and in so doing they usually win the political ends desired. When, however, they deliberately hamstring themselves for moral scruples or other reasons, and refuse to march on an objective despite being militarily capable of it – the southern end of the Suez Canal Zone, Hanoi and Baghdad (in 1991) are cases in point – they wind up failing in the longer run. In Iraq's case, it was to take another twelve years before the English-speaking peoples and their allies finally put an end to Saddam's monstrous rule. In the meantime, however, brigades from the 5th Motorised Division of the Iraqi Republican Guard brutally surprised the Shi'ite uprisings in Basra and An-Nasiriyah and then joined with units from Baghdad to crush the Kurdish revolt in the north of the country. Saddam was in control and in no mood to forgive those who had risen against him.

On 19 August 1991, communist hardliners, led by Gennady Yanayev, staged a *coup d'état* against Mikhail Gorbachev, who was placed under arrest in his Crimean dacha. Radio and television stations were suspended from broadcasting and military rule was imposed in many cities. One of the reasons the coup failed was a very Russian one, born of the inefficiencies of the communist era: the plotters were unable to cut the phone lines properly because the Soviet-era telecommunications technology in Moscow was so primitive. The English-speaking peoples did what they could – denouncing the coup – but they could only watch as the Russians decided for themselves and in their own way what their future would be. After only two days the coup collapsed following widespread popular demonstrations, courageously led by Federation president, Boris Yeltsin, and on 22 August Gorbachev returned to the Kremlin.

An indication of how the collapse of Soviet communism emancipated its neighbours could be seen in Mongolia, which adopted a democratic constitution in 1992. 'Without a bullet being fired, without tanks in the streets,' her prime minister, Elbegdorj Tsakhia, later enthused, 'we laid the groundwork for building a new society based on democracy, the rule of law, and free-market economic reforms. ... Thanks to support from the United States, as well as other countries and international financial institutions, we were able to make the transition to a free-market economy.'[8] From being a command-economy dictatorship, by 2005 more than 80% of Mongolia's GDP was derived from the private sector. Once synonymous with inaccessible remoteness, even Mongolia has been touched by the English-speaking peoples' counter-revolution.

With a population of only 2.5 million, Mongolia would never have been capable of wresting herself from under the Soviet heel unless there had been a sudden power vacuum in Moscow. 'To give our students an advantage in international business,' Tsakhia recently announced, 'we have made English our official second language.' As a further part of the peace dividend for the English-speaking peoples, in 2003 Mongolia sent troops to, in her Prime Minister's words, 'create free societies and fight terrorism in Iraq and Afghanistan'.[9]

At noon on Friday, 26 February 1993, a bomb in a rental truck exploded in the basement of the North Tower of the World Trade Center (WTC) in New York, killing six people, injuring 1,000 but only slightly damaging the structure. The target was chosen because it was an iconic building at the very heart of global – and in particular American – capitalism. The attack was carried out by Islamic fundamentalists, one of whom said they had 'hoped to kill 250,000 people'.[10] This twenty-strong cell was connected to the Egyptian cleric Sheikh Omar Abdel Rahman. In the course of a series of court cases between 1993 and 1997, other targets were identified including the FBI headquarters in Manhattan, the UN skyscraper there and two tunnels under the Hudson River. Yet still the English-speaking peoples, led by President Clinton, failed to wake up to the fact that a war was being made upon them that needed to be prosecuted without delay and with maximum force deployed for their protection.

In 1990, Sheikh Omar, a Sunni who had taken refuge in Sudan, had obtained an entry visa to the United States, where he had preached hatred and called for a *fatwa* whilst living in Jersey City. Another member of the New York City cell, a Pakistani named Ramzi Ahmed Yousef, was involved in a thwarted conspiracy in 1995 to destroy eleven hijacked American airliners over the Pacific Ocean. Hijacking was clearly no longer a precursor to the plane being used as a bargaining tool.[11] The man who mixed the chemicals for the 1993 WTC attack, Abdul Rahman Yassin, was unaccountably awarded bail, which he of course skipped and fled to the protection of Iraq.

Eight months after the World Trade Center bombing, on Monday, 3 October 1993, in Mogadishu, in southern Somalia, two US Black Hawk helicopters on UN peacekeeping duty were shot down by the insurgent forces of the warlord General Muhammed Faraf Aydid. A total of eighteen American soldiers were killed then and in the subsequent fighting that day, some of their corpses dragged through the streets of Mogadishu by triumphant mobs. President Clinton afterwards gave the order to withdraw US forces and, as newspaper columnist Mark Steyn laconically remarked a decade later, 'We know what conclusion Osama bin Laden drew.'[12]

Operation Restore Hope consisted of a US force of 30,000 troops that was sent to Somalia by President Bush Snr shortly before he left the White House. It succeeded in its short-term objective of preventing famine by securing supply routes through Somalia against the chief warlord General Aydid, but then in March 1993 the US force was withdrawn, to be replaced by a United Nations force, albeit under US leadership, called UNOSOM II, which itself withdrew two years later. Far from a 'New World Order', Somalia became the first indication that the collapse of the bi-polar world had merely ushered in a new form of global disorder.[13]

The heroism displayed by the US Rangers and Delta Force in Somalia, not least in the immediate crisis following the downing of the two helicopters, was exemplary, and in marked contrast to the pusillanimity shown by the Clinton Administration in the aftermath of the tragedy. The Rangers had hoped to arrest a number of Aydid's senior lieutenants in central Mogadishu, which they succeeded in doing, but the two Black Hawks were shot down during the operation.

For the following sixteen hours the Rangers fought to live up to their proud motto that 'No man gets left behind', as a huge gun battle took place in the centre of Mogadishu that claimed the lives of hundreds of Somalis. Nonetheless, Aydid was greatly strengthened by the deaths of so many US Rangers. A UN military spokesman after the fighting summed up the campaign as, 'We came, we fed them, they kicked our asses.'[14] After UNOSOM II left in March 1995, its $160 million headquarters was so comprehensively ransacked that even the concrete foundations were looted. (As is often the case with UN débâcles, the report of the official inquiry into the disaster was suppressed, as was a report about how a safe containing the wages of the local staff was stolen from the heart of the UN fortress.)

It is hard to quantify how badly the various financial and sexual scandals that were to engulf President Clinton's presidency actually affected the performance of his duties. On 24 March 1994, allegations were made in the US Congress that Bill Clinton and his wife Hillary had used their part ownership of the Whitewater Development Corporation in Arkansas for improper purposes in connection with the failed Madison Guaranty Savings Bank. The couple testified under oath about the affair in June and a Congressional committee began hearings on it in July. A special prosecutor, Kenneth Starr, interviewed the couple in April 1995, and again in July.

That month, a Senate panel began hearings on the affair, and in January 1996 Hillary Clinton testified before a Grand Jury. In spite of President Clinton testifying for the defence of his former business partners, Jim and Susan McDougal, and former Arkansas Governor, Jim Tucker, in April 1995, the following month all three were found guilty of fraud and conspiracy in relation to the failure of the Whitewater property company, and in June the

First Lady was strongly criticised in a Senate report over her role. At the very least the Clinton Presidency, which had done relatively well in areas of the economy despite its lacklustre performance abroad, was badly tarnished in the eyes of many Americans as a result of the constant speculation about the President's sexual and financial misdemeanours.

In January 1998, an investigation was launched into whether President Clinton had urged a twenty-four-year-old White House intern, Monica Lewinsky, to lie under oath and deny she had had an inappropriate sexual dalliance with him. The next two months saw an avalanche of accusations and denials that most Americans found excruciatingly embarrassing for their country, leaving a general feeling that Clinton had brought the venerable office of the Presidency into disrepute. Presidents and prime ministers of the English-speaking peoples have committed adultery while in office – including Woodrow Wilson, Asquith, David Lloyd George, FDR, JFK, John Major (but probably not Eisenhower, as has been alleged), and no fewer than five Australian premiers – yet never before had the lights of the international media been shone into a leader's private life with the ferocious glare that the modern electronic media could then concentrate. Nor, it might be added, are previous presidents thought to have indulged in sex acts in the immediate environs of the Oval Office itself, as in Clinton's case.

The results were particularly unedifying and included testimony from a former campaign volunteer Kathleen Willey on the CBS programme *60 Minutes* about being kissed and fondled in 1993, a four-year lawsuit from Paula Jones who alleged Clinton had sexually harassed her in 1991, accusations about his sexual use of cigars and even a semen-stained dress. Although Richard Nixon avoided impeachment hearings over Watergate, Bill Clinton did not over Miss Lewinsky; indeed, he was successfully impeached by the House of Representatives before going on to be acquitted by the Senate. The scandal and innuendoes dragged on with increasing rancour until 1 April, when US federal judge Susan Wright dismissed Paula Jones' lawsuit. Clinton's Presidency was severely damaged, but then his career outside the economic field had largely been, in the historian Paul Johnson's sage words, 'an extraordinary example of how far a meretricious personal charm will get you in a media age'.[15]

On 29 July 1994, the French Assemblée Nationale passed the Loi Toubon, which was designed to protect the French language from imprecations from the English tongue. It was signed into law by President Mitterrand six days later. Named after the culture minister who framed it, Jacques Toubon, the law's twenty-four articles provided that French would be mandatory across the fields of 'instruction, work, trade and exchanges and of the public services', also for 'the designation, offer, presentation, instructions for use, and description of the scope and conditions of a warranty of goods, products and services as well as bills and receipts. The same provisions apply to any written, spoken, radio and television advertisement.' Furthermore, 'Any inscription or announcement posted or made on a public highway, in a place open to the public or in a public transport system and designed to inform the public must be expressed in French.' All contracts 'may neither contain expressions nor terms in a foreign language where a French term or expression with the same meaning exists'.[16]

Since 1994, the Loi Toubon has been used against several American and British companies, such as the Disney Store on the Champs-Elysées and the Body Shop retail store, which had labels in English. The French Government's attempt to outlaw 'le weekend', 'les drinks', 'l'aftershave' and 'le babysitter' on pain of hefty fines collapsed in ridicule; nonetheless, in October 1996 the principle was extended into cyberspace when the French Government asked the Organization for Economic Co-operation and Development to take up the issue of regulating language content on the internet. Two months later, a lawsuit was brought against Georgia Institute of Technology on the grounds that the website of its campus in Lorraine was in English, even though all the teachers there came from Atlanta and all the students had to be fluent in English to enrol for the courses, which were all taught in English.[17]

No less a writer than Voltaire himself regularly frenchified English words. Writing in 1756 about his lawns, he refers to a 'boulingrin', which came from 'bowling green' and which had been an officially recognised French word since 1663. His word for pony was 'haquenée', a transliteration of the English word 'hackney', which had similarly been part of the French tongue since 1360.[18] If the French Academician and author of *Candide* employed English expressions such as these, sanctified by centuries of usage, why did M. Toubon consider himself superior to them? One is tempted to quote Churchill speaking in another context in 1906: 'The recognition of their language is precious to a small people', yet whatever else they might be, the French are certainly not a small people.

Rather than allow their beautiful and ancient tongue to compete in the open market-place of global languages, France set up a further scheme for linguistic protectionism in January 1997, when another law required pop music radio stations to play French-language songs for at least 40% of the time. (Although it was the Ancient Persians who invented the guitar, since the 1950s the vast industry of pop music, and especially of rock'n'roll, has been almost completely dominated by the English-speaking peoples.) By contrast, the United Kingdom attempts to protect the various non-English tongues spoken within her borders. Although only one-fifth of the residents of Wales speak Welsh, the language is given equal status and authority throughout the principality, which is entirely bilingual in road-signs, public institutions and in its Assembly. Manx, Cornish and Gaelic are also nurtured.

According to a European Commission report in 2001, English was being spoken by more than one in three of the 350 million citizens of the European Union, whereas fewer than one in ten spoke French outside France itself. Four years later, France's Higher Audiovisual Council ordered television channels to translate the titles of popular programmes and cartoons into French. *Popstars* was instructed to become *Vedettes de Variétés*, *La Star Academy* became *L'Écoles des Vedettes*, *Funky Cops* was transformed into *Des Flics dans la Vent* and, most unwieldy of all, *Totally Spies* became *Des Espions à Part Entière*.[19] The Council also sought to halt the advance of franglais terms that had crept into French TV culture, such as 'le prime de samedi soir'.[20] Despite the obvious absurdity of such campaigns, they demonstrate the deep unease with which the French Establishment views the future of its tongue whenever it comes into competitive contact with English. In a couple of hundred years, the French language might well have to be protected as a linguistic curio, like Manx or Cornish.

Nothing has advanced what Professor Niall Ferguson has dubbed 'Anglo-balization' faster than the adoption of English as the second tongue of many countries around the world. Standard English was brought to the British Isles in the fifth century AD by Germanic warriors, and evolved via the Anglo-Saxons, Chaucer, Shakespeare and Dr Johnson. Since then 'non-standard English' has evolved into what the Cambridge lecturer Freya Johnston has called 'the busy, flexible, everyday language, including regional dialects and international idioms, slang, e-mail, internet-speak and text-messaging'. Today the English language comprises over half-a-million words, more than thrice the number of any other tongue.

The philologist Robert Claiborne has discerned a political explanation for the way that

> our language and literature and our basic philosophy of government developed in parallel: if the English-speaking people have been writing well for over four centuries, the reason is not simply that they wrote in English but that they have had a lot to write about – and could write it, generally speaking, with relatively little interference from government or anyone else.[21]

The phrases 'liberty of conscience' (1580), 'civil liberty' (Milton in 1644) and 'liberty of the press' (1769) were all first expressed in English. 'The tongue and the philosophy are not unrelated', argued Claiborne, since

> both reflect the ingrained Anglo-American distrust of unlimited authority, whether in language or in life. As long as the English and the Americans continue to distrust unchecked power, public or private, and retain the courage and

determination to move against those who have or seek it, Anglo-American civilisation will ... remain worth loving, whatever mistakes or even crimes its leaders
may commit.[22]

Even as early as the sixteenth century the poet Edmund Spenser was
complaining, in a letter to his friend the Cambridge rhetorician Gabriel
Harvey, 'So now they have made our English language a gallimaufry or
hodgepodge of all other speeches.' It was true; today there are three non-
native English speakers for every native one. 'This hungry creature, English,
demanded more and more subjects,' until today, with 1.5 billion speakers
worldwide, it is poised for global hegemony.[23] The historian David Crystal
points out that English is 'the most etymologically multilingual language on
earth', with an omnivorous appetite for digesting foreign words. In the very
heterogeneous nature of our linguistic inheritance lies its strength.

Nearly strangled first by the Danes and then by the Normans, the special
genius of English has been its ability to morph its enemies into itself, like some
monstrous sci-fi extra-terrestrial growing ever stronger by gobbling up its
opponents. As in any great adventure story there have been nail-biting
moments, such as when King Alfred saved the language by beating the
Norsemen, or the battle for survival that Old English had to fight after the
Norman Conquest, struggles that could easily have resulted in English ending
up as a fringe language like Gaelic. Yet because 'English's most subtle and
ruthless characteristic of all is its capacity to absorb others', the tongue simply
soaked up 10,000 French words into its vocabulary and survived for the three-
and-a-half centuries it took before British monarchs spoke English again. Now
English is wreaking its revenge upon its eleventh-century tormentor; by 2005,
Europeans who speak English now outnumber those who speak French by
three to one. Once the Chinese embrace English – and already 750 million
people speak it as their second language – it will be history's first true world-
tongue.

The long history of English has plenty of heroes: men such as William
Caxton and his bestselling author Geoffrey Chaucer, and John Wycliffe, whose
samizdat version of the Bible in English made him so unpopular with the
Latin-dominated clergy that his corpse was dug up from consecrated ground.
A modern hero of the English tongue is Bill Gates, whose Microsoft Cor-
poration is doing for the dissemination of the language what the monks of
Lindisfarne did in the seventh century and William Tyndale's King James
Bible did in the seventeenth.

Although English is a living and constantly growing organism, of the 100
most commonly used words in English worldwide almost all come from the
Old English of 1,000 years ago. 'We can have intelligent conversations in Old
English,' the British intellectual Melvyn Bragg points out, 'and only rarely

need to swerve away from it.' Of course the fact that the United States speaks English was central to the language's twentieth-century burst of second-wind, but the Anglo-English kept the flame alive through the perilous Dark Ages.

Despite Britain having just 1.3% of the world's population – and taking up less than 0.2% of the world's land area – English is today both the language of wealth and, just as importantly, of aspiration to wealth. It is not enough that many hundreds of millions should speak English as their first or second language, but the people who do so have on average higher per capita incomes than those who speak the other great world languages. Although there are many more Mandarin-speakers than English-speakers, they are only worth £448 billion in total. Against that Russian-speakers are 'worth' £801 billion, German-speakers £1,090 billion, Japanese-speakers £1,277 billion, but English-speakers are worth a staggering £4,271 billion – more than all the rest put together. The statistics allow of no other interpretation than that the English tongue is poised for world domination; soon, Bragg predicts, there will be 'the possibility of a world conversation, in English'.[24]

As Bragg has pointed out in his book *The Adventure of English,*

> English is the first language among equals at the United Nations, at NATO, the World Bank, the International Monetary Fund. It is the only official language of the Organisation of Petroleum Exporting Countries, the only working language of the European Free Trade Association, the Association of Baltic Marine Biologists, the Asian Amateur Athletics Association, the African Hockey Federation ... while it is the second language of bodies as diverse as the Andean Commission of Jurists and the Arab Air Carriers Association.

Before we feel revolted by this linguistic version of Jabba the Hut, Bragg explains how English is of course an infinitely beautiful thing, capable of constructing from its hundreds of thousands of words – there were only 25,000 in the Old English vocabulary – the very highest of mankind's cultural achievements. Even by the early seventh century there was a twenty-four-letter alphabet (but no letters J, Q, V, X or Z), a construct that was, he writes, 'like discovering intellectual fire'.

There are dangers inherent in this success. 'The more English spreads,' Bragg tells us, 'the more it diversifies, the more it could tend towards fragmentation.' Just as we reach the tantalising possibility of that single global *lingua franca*, therefore, local dialects all over the world might so pidginise English that the opportunity slips away from us. At present, English is the official language of the following countries and territories: Australia, Bahamas, Belize, Botswana (where the national language is Setswana), Canada (federally, with French), Fiji (with Bau Fijian and Hindustani), The Gambia, Hong Kong (with Chinese), Guyana, India (with Hindi and fourteen other languages), Kenya (with Kiswahili), Kiribati, Namibia, Nigeria, Pakistan,

Papua New Guinea (with Tok Pisin and Motu), Republic of Ireland (with Irish), South Africa (with Afrikaans, Ndebele, Northern Sotho, Sotho, Swazi, Tsonga, Tswana, Venda, Xhosa and Zulu), New Zealand (an official language by custom; the other by law is Māori), Singapore (with Malay, Tamil and Chinese), the Philippines (where the national language is Filipino), Trinidad and Tobago, the United Kingdom, the USA and Zambia. It is a formidable springboard from which to launch a bid for global linguistic domination.

Under a 1990 law all Spanish children are taught English from the age of eight, and in some regions from the age of six. In Madrid alone, there are twenty-six bilingual schools and colleges, where courses are – with the sole exceptions of Spanish literature and mathematics – taught in English.[25] Similarly, English-teaching has exploded across the former Soviet Union. Lord Macaulay's wisdom in encouraging the spread of English in nineteenth-century India has also borne fruit. When Muhammad Ali Jinnah made his demand for Pakistan at Lahore in 1940, he did so in English, despite calls from the audience demanding it to be done in Urdu. His answer was that since the world's press was covering the occasion, he needed to speak in a world language.

In India, English-language daily newspapers have a circulation of 3.1 million copies and each is often read by several people. In Indian academia, English continues to be the premier language. Careers in business and commerce, high government positions, and science and technology continue to require fluency in English. It is also almost mandatory for those students who wish to study overseas. All large cities in India and many smaller ones have private, English-language middle schools and high schools. Even government schools run for the benefit of senior civil servants use English because only that language is an acceptable medium of communication throughout the nation. As India develops economically, this trend is set to continue. Should either China or India one day dislodge the English-speaking peoples as the foremost world power, it will have to do so using the English language as a primary tool. Indeed, according to the British Chancellor of the Exchequer Gordon Brown, 'in two decades, China's English speakers will outnumber native English speakers in the rest of the world'.

The modern Irish poet Michael Hartnett, explaining the almost terminal decline of the Irish tongue in the nineteenth century, explained that English was the better language 'in which to sell the pig'.[26] Similarly, the decline of Welsh, Scots, Gaelic and Breton, and the perhaps terminal illnesses of Manx and Cornish, are largely down to economic factors. Today there are 6,000 languages spoken in the world, but 52% are spoken by fewer than 10,000 people and 28% by fewer than 1,000.

That does not mean that the English-speaking peoples can be complacent over their language's omnivorous rise. As the linguistics historian Nicholas

Ostler has written in his history of the world's tongues, *Empires of the Word*, Akkadian and Egyptian, Sanskrit and Persian, Greek, Latin and French all seemed irresistible in their own day.[27] Perhaps the fate of Ozymandias hovers over English, but it is, as the historian Tim Blanning points out, 'the first language to achieve domination in an age of global communication'.[28] That, if nothing else, will make it hard to destroy as, for example, Alexander the Great was able to end the four-centuries-old domination of Aramaic.

According to UNESCO's recent list of the world's top ten most translated authors, each with over 1,500 translations, writers from the English-speaking peoples make up half, despite accounting for a mere 7.5% of global population. The number one position is taken by Walt Disney Productions, which since the collapse of the Soviet Union has sold its stories throughout Eastern Europe, allowing it to nudge ahead of its closest rivals, Agatha Christie and the Bible. V.I. Lenin comes next, but since he is no longer being translated anywhere he is fast descending the tables. The next, in descending order, are Jules Verne, Barbara Cartland, Enid Blyton, William Shakespeare, Hans Christian Andersen and the Brothers Grimm.[29]

In the half-century between 1890 and 1940, the proportion of lawyers to the general population in the United States remained at the steady (and healthy) level of one lawyer for every 730 people, or thirteen per 10,000 Americans. Yet by 1990, American law schools were producing 35,000 new lawyers per annum. That year there were no fewer than 281 lawyers per 100,000 Americans, an over 100% increase in only half a century. This contrasts with 111 German, 82 British and 11 Japanese lawyers per 100,000. Total cases filed in all US federal courts rose from 68,000 in 1940 to over 300,000 by the mid-1980s, and in 2000 the United States was home to one million lawyers for the first time in her history.

The effect on politics has been astounding; where there were 500 registered lobbyists in Washington during the Second World War, today there are over 25,000. The number of interest groups listed by *The Encyclopaedia of Associations* exactly doubled in twenty years from 10,300 in 1968 to 20,600 in 1988.[30] The hybrid of the lawyer/lobbyist problem – the Washington lawyer – used to number fewer than 10,000 in 1970; today there are over 45,000 of them, in which time the number of political action committees has grown from fewer than 100 to over 4,000.

Similarly, in 1947 there were fewer than 2,500 staff members working in the United States Congress; by the year 2000 this had grown to almost 18,000. Congress meets all the year round whereas once it took long holidays; it has dozens of full committees and over 200 sub-committees, whereas once it had hardly any; seniority is no longer the all-important touchstone it once was; almost all Congress business is done in public whereas in 1947 it was in private, and all the sessions of both Houses are now televised. The alteration of almost every aspect of American political life since the Second World War has been described as, 'A revolution without revolutionaries, without a revolutionary ideology or revolutionary manifesto or call to arms. It has in fact been one of the most peculiar revolutions in the history of the world. . . . It has been a revolution by accumulation, by inadvertence, by miscalculation, by demographic destiny – a revolution of good intentions run amok.'[31]

On Friday, 6 May 1994, the Channel Tunnel was officially opened and for the first time in scores of millennia Britain was once again connected to Continental Europe. Isambard Kingdom Brunel had begun construction of a tunnel in Victorian times, although nothing came of it then. In April 1916, Sir James Fowler of the Beefsteak Club had bet Mr Reginald Morris '£1 that one or other of them crosses to France through a tunnel', but it was not until September 1963 that an Anglo-French report favoured a twentieth-century attempt. An agreement was signed the following February for a rail link, and definite plans were agreed in 1966, but the project was abandoned in 1975 due to escalating costs.

In 1980, true to Thatcherite principles, the British Government announced that a private consortium might build the tunnel, so long as it did not expect public money. Construction began in November 1987, and the French and English sections were connected in December 1990. It was an astonishing undertaking, comprising twin tunnels over thirty miles long and twenty-five feet in diameter located 130 feet below the seabed. The final cost was £12 billion, and the Anglo-French company that built it, Eurotunnel plc, did not announce a net profit until March 1999.

On 16 March 1995, President Clinton met Gerry Adams, the leader of Sinn Fein, at the White House and permitted him to raise funds in the United States. This was before the IRA-Sinn Fein had renounced the armed struggle against the British Government in Northern Ireland. The 1990s were the period when it became clear that terrorism paid, and the more violent the terrorism, the more it paid. Of the ninety-eight conflicts that took place worldwide between 1990 and 1996, only seven were between recognised states, despite their collectively causing over five-and-a-half-million deaths (of whom over three-quarters were civilians). Of course, this was relatively peaceful compared to the 1945–90 period, which included such spectacular bloodbaths as the Chinese Civil War and the Korean and Vietnam Wars, but the Nineties were the decade when the world enjoyed a 'peace dividend' through the reduction of Cold War expenditures.

'Where force is necessary, there it must be applied boldly, decisively and completely,' wrote Trotsky in 1932, 'but one must know when to blend force with a manoeuvre, a blow with an agreement.' 17 November 1997 saw sixty-eight people killed when Islamic fundamentalist terrorists blew up two tourist buses in Luxor, Egypt. Whereas the IRA-Sinn Fein was embarked on the same route originally mapped out by Trotsky and later perfected by Mao Tse-tung, the new strain of Islamic terror was not. Under the old Trotsky-Mao strategy, terror is used first to raise the political consciousness of the population; then to force society to choose sides; then to create a backlash which unmasks the power of the state; then to isolate, demoralise and destroy 'collaborationists'; and finally, by 'blending force with manoeuvre', blows are exchanged for an agreement. What the Clinton Administration failed to spot was that the new strain of Islamic fundamentalism was essentially fascistic and nihilistic in nature, and did not want an agreement, so much as to kill as many of the infidel – and preferably the English-speaking peoples – as possible, without any logically achievable goal in sight.

The Trotsky-Mao system had served as a template for the movements which brought Archbishop Makarios to power in Cyprus, Yassir Arafat to the presidency of the Palestinian Authority, Jomo Kenyatta to lead Kenya and even Nelson Mandela to form *Umkonto we Sizwe* (Spear of the Nation) as the guerrilla wing of the African National Congress in March 1960, for which he was charged with conspiracy and sabotage in October 1963. Without a readiness to shed the blood of innocents, Zimbabwe would doubtless still be Rhodesia and South Vietnam would be an independent state. 'Freedom-fighters' were simply terrorists who had won.

At 9.02 a.m. on 19 April 1995, a bomb ripped through the Alfred P. Murrah Federal Building in Oklahoma City. No fewer than 168 people were killed in what was the worst terrorist attack on American soil before 9/11. Ninety minutes later, the twenty-seven-year-old Timothy McVeigh was arrested at a routine traffic stop in Billings, Oklahoma. McVeigh stated that he had been inspired to murder those innocents – many of them infants since there was a day-care centre in the building – by a book entitled *The Turner Diaries*, written by a West Virginian neo-Nazi called William Pierce. At his execution in June 2001, McVeigh quoted W.E. Henley's poem *Invictus* as his last words, 'putting the final touch', as the *Times Literary Supplement* put it with heavy sarcasm, 'on what was quite a literary affair'.[32] McVeigh's belief that the American federal government in all its guises was evil – even, in this case, its day-care centres – stemmed directly from the paranoia culture that had infected American society since Watergate.

Just as there has been no mass murderer in history – not Stalin, not Mao, not even Pol Pot – who has not found someone amongst the English-speaking intelligentsia to put in a good word for him, so the writer Gore Vidal was quoted telling *The Oklahoman* newspaper of McVeigh, 'He's very intelligent. The boy's got a sense of justice.' At a book festival he later compared McVeigh to the American Revolutionary hero Paul Revere. This was moral equivalence being taken to disgraceful levels. Yet even Mr Vidal did not suggest that Bill Clinton invite Timothy McVeigh to the White House and allow him to raise money.

A modern Great Power that today constantly works against the unity and amity of the English-speaking peoples is Hollywood. Just as Hollywood has been responsible, post-Watergate, for feeding the sense of betrayal and paranoia about the American political and military Establishment, and has presented a fantastically skewed portrait of the American mission in Vietnam, so it also churns out movies that are determinedly Anglophobic. Hollywood is thus institutionally racist in the way that it portrays the United Kingdom, both in her past and present.

Movies such as *Michael Collins*, *Rob Roy*, *Patriot* and *Braveheart* are as anti-British today as the Alexander Korda propaganda movies were pro-British during the Second World War. Hollywood political correctness has fastened on the Brit, especially the imperialist Brit, as a safe target for sustained abuse and misrepresentation. Outrageously factually inaccurate, these films do have an effect on the way the American public views Britain and the British. For example, Mel Gibson's *Patriot* (2000) was the story of an American general in the War of Independence, Francis Marion, who fights a brilliant guerrilla war against the evil British invaders. When the movie's historians discovered that in real life Marion raped his slaves and hunted Red Indians for sport, they changed the hero's name to Benjamin Martin, but one thing stayed the same: the movie's villains are as usual the treacherous, cowardly, evil, sadistic Brits.

In October 1996, *Michael Collins* was released, a badly skewed biopic of the Irish republican leader. In it a British armoured car is shown firing on an audience in a sports stadium (which never happened), car bombs were portrayed decades before that weapon was invented, and the film depicted the torture and murder by the British in 1922 of an informer – Ned Broy – who in fact had died peacefully in his bed half a century later, having drawn a British pension for many years. When the Irish director, Neil Jordan, himself a history graduate, was told that Irish historians had pinpointed these and very many more such falsifications, he simply answered, 'Well, fuck them.'

A week later, *Some Mother's Son* was released, about the IRA hunger-strikers in the Maze Prison in 1981, written and directed by Terry George, who, far from being an objective witness to the events, had served three years in prison in Northern Ireland for possession of a gun with intent to endanger life. When it was screened at the prestigious Hamptons Film Festival in 1996, these were some of the remarks made afterwards by ordinary Americans leaving the cinema: 'Those bloody British. I do hate them a lot.' 'God, I hate Thatcher.' 'The way they speak, the way they act – I hate the British.' Both consciously and subconsciously, Hollywood Anglophobia has a severely deleterious effect on the way that one part of the English-speaking peoples views another.

The following March saw a film entitled *The Devil's Own* open in American cinemas in which the hero, played by Brad Pitt, was an IRA terrorist and the villain was a British Intelligence officer, Harry Sloane, who kills Irishmen in cold blood and who *The Times* film critic described as 'the sort of sadist Hollywood once dressed in Nazi jackboots'.[33] Michael Medved, the Manhattan film critic, wrote of 'the movie's devilish attempt to rationalise, and, ultimately, to glamorise the most deadly sort of political violence'. Columbia Pictures were embarrassed by Mr Pitt's own denunciation of the movie as 'the most irresponsible bit of film-making – if you can call it that – that I have ever seen'.[34] The *New York Post* described the film as 'an eloquent apology for murderous terrorism', and the Irish former foreign minister, Conor Cruise O'Brien, explained how such a distorted film could have been released: 'In the structure of the American movie industry there is a large Irish-American lobby which is basically pro-IRA. Any film which depicted the IRA overall unfavourably would run into trouble at the box office.'

British films such as *The Ploughman's Lunch, Hidden Agenda* and *Defence of the Realm*, in the sage opinion of the British critic Bryan Appleyard,

> established in the audience's mind the idea of the British Establishment as a uniquely and intrinsically corrupt organism. Later came *In the Name of the Father*, with its laughably distorted account of the [wrongly convicted] Birmingham Six, designed to convince the Americans . . . that our legal system was an irredeemably vicious servant of Imperialism. But never mind the truth – the Irish-American market is much more important than the English domestic market.[35]

Since 9/11, it has proved more difficult for Hollywood to glamorise terrorism, but still the bad guys are the Brits.

The list of films depicting Britons as villains is now so long as to amount to a virtual declaration of war on the United Kingdom by a geographically small but globally incredibly powerful suburb of Los Angeles. Charles Dance in *Michael Collins*, Tim Roth in *Rob Roy*, Jeremy Irons and Alan Rickman in the *Die Hard* movies, Ben Kingsley playing the evil 'Hood' in *Thunderbirds*, Anthony Hopkins as the cannibalistic serial killer Hannibal Lecter in *The Silence of the Lambs* – all are either Britons speaking in British accents or Americans playing Britons but also using British accents. The subtext is clear: a British accent is shorthand for villainy. The careers of Christopher Plummer, Richard E. Grant, Brian Cox, Tim Curry, Jonathan Pryce, Christopher Lee and many others have been boosted by their ability to denote evil simply by their deployment of strangulated English vowels. As one critic has recently pointed out, actors 'now make a decent Hollywood living out of being snooty, murderous and psychotic in particularly English ways'.[36]

The toffee-nosed Brits in *Gandhi*, *The Last of the Mohicans* and *Pocohontas* are mere caricatures, as is Rickman's Sheriff of Nottingham in *Prince of Thieves*. But the British officers in *Titanic* are portrayed as battening down the steerage-class hatches, thereby deliberately condemning to death the happy, jig-dancing Irish working-class folk below. This never happened. In *The Messenger: The Story of Joan of Arc* (1999), starring Milla Jovovich, Dustin Hoffman, Faye Dunaway and John Malkovich, it was perhaps inevitable that the English should be the villains, considering the context, but was it really necessary to have an Englishman rape the corpse of the heroine's elder sister? Would any other racial grouping but Britons be depicted as stooping to necrophilia in politically correct modern Hollywood?

The Australian-born actor Mel Gibson has made a successful Hollywood career out of libelling Britons in many of his movies. *Gallipoli* (1981) showed moronic but heartless British officers cheerfully sacrificing brave young Anzacs; *The Patriot* (2000) depicted an entirely fictitious massacre of innocents in a church by British redcoats; *Braveheart* (1995) portrayed the English as little better than Nazis; the *Die Hard* movies often have British villains; and the only surprise of his biblical movie *The Passion of the Christ* was that Pontius Pilate spoke in Latin and the Pharisees in Aramaic, rather than in cut-glass, upper-class English accents.

As the English author A.N. Wilson wrote of a character in his novel *Hearing Voices*, 'He usually played unscrupulous conmen or the cold-hearted "brains" in criminal gangs. His villainy was made apparent to cinema audiences by his English accent.' Even in Disney cartoons such as *The Lion King*, the role of the treacherous, murderous lion Scar is played by Jeremy Irons with an upper-class English accent, as is George Sanders' Shere Khan the tiger in *The Jungle Book*. *Pocohontas* is also very anti-British. Admittedly, Cruella De Ville in *101 Dalmatians* and Captain Hook in *Peter Pan* were British in the books, in the way that lions and tigers are not, but many other Disney animal villains are recognisably English also.

Of course history movies have long embroidered upon or simply ignored the truth, conflating events, altering time-sequences, eliding characters, omitting difficult facts and occasionally simply making things up altogether, although as the old clerihew went:

> Cecil B. de Mille,
> Rather against his will,
> Was persuaded to leave Moses
> Out of *The Wars of the Roses*.

Distortion has been going on for decades. Errol Flynn's *Objective Burma* (1945) so ignored the British contribution to the Burmese campaign that it was not shown in Britain until 1952; *Saving Private Ryan* omitted the British and Canadian contributions to D-Day; *U-571* (2000) portrayed the Americans capturing an Ultra machine, which was actually done by the Royal Navy, and so on. As the film critic Philip French has written, 'Nothing gets people going like a massacre', and the massacre of Indians at Amritsar as shown in the 1982 movie *Gandhi* was a case in point.[37] The Richard Attenborough movie portrayed the British rulers of India such as the Viceroy, Lord Irwin, played by John Gielgud, as snobbish incompetents and fools, whereas they were the successful colonisers for two centuries of a vast subcontinent, where they made up a fraction of 1% of the population, and Irwin himself was a Fellow of All Souls who in reality twice put his career on the line in order to meet and negotiate with Gandhi.

'The villains used to be the Germans, the Japanese or the Russians,' Larry Mark, producer of *Jerry Maguire*, has explained, 'but they protested. If the English get a bad rap they can take it.' In a counter-intuitive way, Britons have almost taken it as a compliment that Hollywood bothers to cast them as villains. It shows they still matter in the world; after all, no-one tries negatively to stereotype the Finns, Norwegians or Thais. Yet there is a point where the incessant negative portrayals must affect the way that ordinary Americans view their closest and most dependable ally. As Appleyard concluded, 'Certainly, we can survive the abuse until the next stereotype comes along, but much more is involved here. For the bad Englishman amounts to a statement about the Atlantic relationship that will prove more lasting than any mere film.'

In mid-July 1995, the Bosnian Serbs pushed aside the Dutch force in the Muslim enclave of Srebenica in eastern Bosnia-Herzegovina, even though it had been designated a 'safe haven' by the United Nations. 'A belated pin-prick air strike, which was all that the UN bureaucracy would authorize,' records an historian of these terrible events, 'made no impact at all.'[38] Although tens of thousands of Bosnian Muslim women and children refugees were forced northwest to Tuzla, back in Srebenica over 7,000 Bosnian Muslim men were massacred and thrown into mass graves. It was by far the worst war crime to have been committed in Europe since the end of the Second World War, on a par with the Katyn massacre of 1940, except that unlike Katyn it was widely reported at the time. Yet it had no effect on the policy of the Major Government, which was to persist in seeing the struggle as a civil war in a country – Yugoslavia – that no longer existed, to enforce an arms embargo on the Bosnian Government and to discourage the United States from pursuing even a limited military response.

Nor was it just the Dutch who were humiliated at Srebenica; British SAS teams in the enclave also stood aside as the massacres took place, ordered not to act.[39] For years the Foreign Office had argued that any attempt by NATO to attack the Bosnian Serbs would result in the eastern enclaves falling, so nothing was done militarily. As it was they fell anyhow, with great loss of life. Mass graves were still being discovered ten years later, in 2005. Yet Bosnia saw by far the largest UN deployment in history, with no fewer than 39,922 men of the United Nations Protection Force (UNPROFOR) deployed there from September 1994.

Three days after Srebenica fell, the Bosnian Serbs attacked another UN-designated 'safe haven', Zepa, which fell on 25 July, and then the UN-designated 'safe haven' of Bihać in north-western Bosnia-Herzegovina, whose defence cost much loss of life. The crisis finally led the US Senate to pass a bill on 26 July enabling the US unilaterally to lift the embargo on arms supplies to Bosnian forces. It was passed by the House of Representatives on 1 August, finally opening up the prospect of Bosnia being able to defend her people against murderous Bosnian Serb incursions. Yet President Clinton vetoed the bill on 11 August. On 28 August, Serb troops in Bosnia-Herzegovina mortared a market place in Sarajevo, killing thirty-seven people. Only then did NATO aircraft begin their assault on Serb positions in Bosnia, flying 800 sorties by 13 September. Within days the Bosnian Serbs were on the retreat.

The leaders of the English-speaking peoples – particularly President Clinton, John Major and their foreign policy advisors – must bear a heavy responsibility for their failure to act sooner when genocide was being committed on such a scale on the European continent. Even worse, perhaps, was their constant refusal to allow the Bosnian Government to buy weapons to defend itself. 'Douglas, Douglas,' Margaret Thatcher is said to have told Major's Foreign Secretary, Douglas Hurd, one of the architects of Britain's Bosnian policy, 'you would make Neville Chamberlain look like a war-monger.'[40]

The reference to Chamberlain is instructive. 'The appeasement of the Serbs', considered Ivo Daalder, who was soon to join the US National Security Council staff, 'really hurt the image of Britain.' The British position was driven by, in the succinct words of one commentator, 'the profoundly Serbophile key-decision makers in the Government, the Foreign and Commonwealth Office swimming in a murky sea of sentimentality and bogus history about Tito and the Second World War, ... appalling anti-Americanism, and Orientalizing ideology about the Balkans'.[41] It was a lethal brew.

From the beginning Margaret Thatcher had advocated a tough military response against the Bosnian Serbs and their allies. 'We could have stopped this,' she said in December 1992.

We could still do so. ... But for the most part we in the West have actually given comfort to the aggressor. ... We have repeatedly stated in public that we will not intervene militarily, so removing even a nagging uncertainty from the minds of the generals in Belgrade. We have accepted the flouting of successive Security Council resolutions by Serbia, whose aircraft are still free to drop cluster bombs on children.

The following April, she described Bosnia as becoming 'a killing field the like of which I thought we would never see in Europe again', a situation that she said was 'not worthy of Europe, not worthy of the West and not worthy of the United States. ... This is happening in the heart of Europe and we have not done any more to stop it. It is in Europe's sphere of influence. It should be in Europe's sphere of conscience. ... We are little more than an accomplice to massacre.'[42]

Echoing the title of the best book about Europe's moral, political, diplomatic and military failings in the Balkans between 1991 and 1998, this was Britain's post-war 'unfinest hour'.[43] Especially since, as *The Times'* Central Europe correspondent Adam Le Bor has pointed out, 'Had air strikes taken place in 1991 or 1992, rather than 1995, many more Bosnians, of all denominations, would still be alive.'[44]

The air strikes against Bosnian Serb positions – so long delayed by London's refusal to countenance them – eventually took place under American insistence in Operations Storm and Deliberate Force. On 30 August, NATO aircraft flew 300 sorties in the first twelve hours, and by 13 September they had completed over 800 missions. Under their cover, Bosnian forces were able to launch an offensive in western and central Bosnia-Herzegovina that reduced Serbian-controlled territory from three-quarters down to half of the country. These joint attacks were extremely successful and quickly proved that the Serbs were no longer the doughty fighters as described by Second World War-era writers such as Fitzroy Maclean and Rebecca West, but merely genocidal murderers who fled when faced with overwhelming force from the skies. By 1 November, peace talks were being held between all the major parties at the Wright-Paterson air-force base near Dayton, Ohio, which ended in the formal signing of the peace plan at the Elysée Palace in Paris on 14 December. Within two months of the air strikes that the British had held up for years, the Serbs were at the negotiating table, and within three they had agreed to a peace deal. Air power had once again proved its efficacy.

The Major Government's consistent opposition to air strikes and its discouragement of direct American intervention created the worst Anglo-American relations since Suez. The Ambassador to Washington, Sir Robert Renwick, mentioned Suez in his (superbly ambiguously entitled) memoir *Fighting with Allies,* and the British permanent representative on the UN Security Council,

Sir David Hannay, agreed that it was the most 'sustained and damaging rift' for forty years (i.e. since Suez). In retrospect, it seems incredible that the British, in the words of one commentator, 'were so blinkered they were prepared to threaten the Special Relationship with the US for the sake of the appeasement of Serbia'.[45]

'Western prestige and alliance solidarity suffered,' commented Sir Percy Cradock, Major's foreign policy advisor. 'Yugoslavia must have been one of the first instances of Britain siding with Europe against the United States in a major international crisis. ... The wrangling between London and Washington over Yugoslavia would be costly and would contribute significantly to the decline in the Special Relationship in the later Major years.'[46]

If those were the views of Major's own side, the Americans went much further. Richard Holbrooke, the US Assistant Secretary of State who brokered the Dayton Accords that brought peace to the region in 1995, argues that it was worse than Suez, 'because Suez came at the height of the Cold War, the strain then was containable. Bosnia, however, had defined the first phase of the post-Cold War relationship between Europe and the United States, and seriously damaged the Atlantic relationship.' The Anglophile American Ambassador in London, Raymond Seitz, said he 'could sometimes smell a whiff of Suez', and Tony Lake, US National Security Advisor from 1993, described it as 'potentially a crisis worse than Suez'.

This was all because, as Bernard Simms put it in his fine philippic *Unfinest Hour*,

> Britain consistently refused to go to the military aid of an embattled member of the United Nations. She tenaciously obstructed all attempts by other countries – especially the United States – to provide such help. Indeed, thanks to her determined advocacy of the international arms embargo, Britain would not even allow the Sarajevo government to defend itself. ... Above all, Britain's policy on Bosnia led to a sustained fight on the Security Council and in NATO with her most important ally, the United States. ... What proportion of the tens of thousands of murdered civilians and millions of refugees should be attributed to this mistaken policy is unknowable, but certainly substantial.[47]

For all the talk of wishing to protect British lives by not imposing the West's will on the Bosnian Serbs, in fact although eighteen British servicemen lost their lives in the three-year non-confrontational, humanitarian strategy adopted by the Major Government, once the long-delayed actual shooting war against the Bosnian Serbs started there were no British fatalities at all. Similarly, only one US serviceman lost his life in Bosnia from hostile action, and none in Kosovo.[48] Furthermore, although there were 54,000 international troops in Bosnia in 1996, the level was down to 19,000 by 2003, with US forces of 16,500 in 1996 down to 4,250 in 2003. Once again, a combination of overwhelming air power and a willingness to take direct action delivered both strategic success and low casualties.

On Sunday, 13 November 1995, there was a bomb blast at the United States' military complex in Riyadh, Saudi Arabia. Seven people were killed on that occasion by Islamic fundamentalist terrorism, yet still the English-speaking peoples did not recognise that war was being made upon them.

On Thursday, 1 May 1997, Tony Blair's Labour Party won a landslide victory over John Major's Conservatives, winning 418 seats to their 165, and the Liberal Democrats' 46. The Conservative Party's eighteen-year domination of British politics – the longest period one Party had continuously held power in Britain since the Napoleonic Wars – ended with the worst electoral defeat they had suffered in over ninety years. Even in 1906 Arthur Balfour had won 43.6% of the vote, compared with John Major's 30.7% in 1997.

With a booming economy, no great popular demand for European social legislation or stronger unions, and with unemployment falling every month for the previous two years – in stark contrast to much of the rest of the European Union – the Conservatives were nonetheless punished severely at the polls. John Major's Party had been riven by splits over Europe ever since Margaret Thatcher had been overthrown in the internal Party coup in November 1990, and his by turns weak and then suddenly obstinate leadership had added greatly to its travails. Under pressure from powerful pro-European figures in his Cabinet – including the Chancellor of the Exchequer Kenneth Clarke, the Deputy Prime Minister Michael Heseltine and the Foreign Secretary Douglas Hurd – Major was forced into policy positions over European integration that a significant minority of Conservative Party members could not in conscience support.

By hinting to European audiences that Britain might under certain circumstances join a single currency, while simultaneously assuring his domestic supporters that of course in fact he never really would, Major was forced to adopt contradictory positions and tortuous language which sounded both pusillanimous and shifty. This was the wording of the compromise hammered out on 23 January 1997, for example: 'Upon the information available to us at present, we reached the conclusion as that it was very unlikely, though not impossible, that countries' performances against the criteria would be sufficiently clear and stable for it to proceed safely on 1 January 1999.' It was hardly a simple, rousing rallying cry to use on the nation's doorsteps in the forthcoming general election.

The result was that Major looked weak and vacillating over the most important issue facing the country. This was highlighted by the sadly pathetic name he gave to his policy: 'Wait and see.' However tough he might have been on education or health issues, Major was doubted and distrusted on the issue over which many Britons cared deeply. Instead of fighting against much of his Party, he ought to have tilted away from the small cabal of pro-European colleagues in his Cabinet. He did not have a strong enough personality for that, however, something else that the British people divined by the time of the 1997 election.

The electorate is accused of having a memory deficit disorder; however it was the recollection of Britain's forcible self-ejection from the Exchange Rate Mechanism (ERM) on 'Black Wednesday', 16 September 1992, which still rankled five years later. By disbelieving the Government's protestations that sterling would stay in the ERM, the international financier George Soros made a profit of over £1 billion in one day. Much more seriously, the centrepiece of the Government's economic strategy lay in ruins; interest rates at one point soared to 15% yet no minister resigned or even apologised. It was not a black day for the British economy – the ERM had cost a million jobs and destroyed 100,000 businesses – but it was a dark one for parliamentary democracy. The pro-euro new Chancellor, Kenneth Clarke, then made matters worse by breezily talking of one day re-entering the ERM, which infuriated the tens of thousands of people whose homes were repossessed during the period of its operation.

After 'Black Wednesday' the Government's popularity, which had touched 50% during the Gulf War eighteen months earlier, collapsed to 30%, from whence it never recovered, at times falling as low as 23%. Opinion polls are of course notoriously fallible, but a four-and-a-half-year continuously disastrous showing without any significant blips indicated a settled conviction on behalf of the British people that John Major had forfeited their confidence.

Since the fall of Margaret Thatcher in November 1990, it was inevitable that British politics would go through something of a post-heroic phase, but John Major's manifest failure to grow into the role of prime minister was remarkable, indeed almost unprecedented. His mangling of the English language was perhaps excusable in a man of little formal education, but was nevertheless sad to hear in a British prime minister. Other premiers have acquired at least a patina of charisma after seven years in power, but not him. Lord Curzon called Stanley Baldwin, who became prime minister in 1923, 'a man of the utmost insignificance'. It was not true of Baldwin, who turned out to have formidable reserves of hidden strength, but it was true of Mr Major.

Major only became prime minister because, after the fall of Mrs Thatcher, he was neither the ultra-liberal Foreign Secretary, Douglas Hurd, nor her political assassin, Michael Heseltine. Thatcher, who had wildly over-promoted Major to Chancellor of the Exchequer, wrongly believed him to be the heir to her ideological legacy. Very soon after securing him victory, the Thatcherites discovered their mistake. Major spoke of wanting Britain to be 'at the heart of Europe', without explaining what in practice this meant. Later, he was tape-recorded calling the three euro-sceptics in his Cabinet 'bastards' and ruined his nice-neighbour image by being caught on tape saying, 'I'm going to fucking crucify the Right.' In one sentence he thus managed to swear, blaspheme, split an infinitive and make a promise he could not keep.

With only the limited vision of a Party *apparatchik* – he was a Party Whip in the House of Commons before becoming a minister – Major was unable to win the support of even two-thirds of his Parliamentary Party when his Cabinet colleague John Redwood stood against him for the Party leadership in the summer of 1995. Redwood adopted the slogan 'No Change, No Chance', which was proved to be prescient by the 1997 election. Over issues such as the citizen's charter; a hotline to complain about motorway cones; surrenders over qualified majority voting in Europe; the EU working time directive; and much else, especially over Bosnia, Major was shown to be a figure of pathos.

One area where Major was thought to be entirely personally innocent of the disasters which struck his ministry was over 'sleaze'. The resignations of no fewer than seventeen of his ministers in twenty-two months over sexual and financial misdemeanours could not have been foreseen. In some cases, as with his friend David Mellor, Major certainly held on to the errant politicians for too long, but overall he cannot be blamed for his consistent run of appalling luck. Of course had anyone known that Major had earlier been conducting an affair with one of his fellow ministers, Edwina Currie, (fortuitously) while his wife was away in his Huntingdonshire constituency, he would have been laughed out of office.

Major weakened himself in November 1994 when he withdrew the Party Whip from eight Conservative MPs over the European issue, something that Neville Chamberlain never did to opponents of appeasement in the Thirties and which also never happened to the Suez rebels of 1956. By this gross act of intolerance, against patriots whose only concern was the protection of British sovereignty, he showed how at heart he was a Conservative hack politician rather than a Tory statesman, and essentially unfit for high office, let alone the premiership of the United Kingdom.

On 5 March 1496, King Henry VII granted a Charter authorising a citizen of Venice of Genoese extraction called John Cabot the right 'to hoist the English flag on shores hitherto unknown to Christian people'. Cabot had been a pilot and navigator in the Eastern Seas before coming to live with his family in Bristol, and on St John's Day, 24 June 1497, he and his Bristolians in a ship called the *Matthew* stepped ashore at Newfoundland, unwittingly founding the British Empire in the process. Half a millennium later to the very week, on 30 June 1997, Prince Charles and the last British Governor of Hong Kong,

Chris Patten, sailed away from Hong Kong on the royal yacht *Britannia*, putting the final full-stop to Britain's imperial story.

At a summit meeting in Madrid on 8 July 1997, the leaders of NATO formally invited Hungary, Poland and the Czech Republic to join the alliance in 1999. It was a massive alteration to the 1944 'percentages agreement' agreed between Churchill and Stalin and then ratified at Yalta, moving the borders of the alliance hundreds of miles to the east, and very far from the North Atlantic after which the organisation was originally named in 1949.

European anti-Americanism after 2001 has often been attributed to the actions of George W. Bush, rather than to the innate resentment always directed against the world's strongest power, yet in 1997 America was blamed for hubristically provoking Russian revanchism by extending NATO. That year, the French Foreign Minister Hubert Vedrine complained of American 'hyper-power'; US capital punishment was held to prove America's 'cultural inferiority', and the Clinton Administration's rejection of an anti-landmine treaty was held up as evidence of American unilateralism.[49] It was in Clinton's 1990s, not in the era of George W. Bush and Donald Rumsfeld, that the German weekly *Der Spiegel* wrote of how 'Americans are acting, in the absence of limits put on them by anybody or anything, as if they own a blank cheque in their McWorld.' (Notice the neat slipping of the obligatory sneering reference to fast food into the criticism of the USA as unilateralist.) Those anti-Americans who routinely claim that they are 'not anti-American, just anti-Bush' were usually, on examination, anti-American long before the forty-third President ever arrived on the scene.

The death of Diana, Princess of Wales, in the early morning of Sunday, 31 August 1997, in a car crash in the Pont de l'Alma underpass in Paris, provided another opportunity for wild conspiracy theories to abound, especially over the internet. The facts that her chauffeur, Henri Paul, had been drinking and was on prescription drugs, the Princess had not worn her safety-belt, and the car was going far too fast due to efforts of the world's paparazzi to photograph her with her boyfriend Dodi Fayed, were not accepted as the reason for the crash by a large number of people across the world brought up on post-Watergate Hollywood conspiracy movies.

September 1997 seemed like a weird sociological moment for many Britons who thought theirs a buttoned-up people little accustomed to extravagant displays of emotion, at least for someone the vast majority of them had never met personally. No less than £25 million was spent on 1.3 million bouquets of flowers, and within days vast moats of cellophane lapped around the various royal palaces in London. In Tim Rice's inspired lyrics from *Evita*, ordinary people worldwide were 'falling over themselves to get all of the misery right' in September 1997. Yet in fact there was plenty of precedent for such out-pourings, as the weeping admirals at Nelson's funeral and the emotional scenes at Wellington's and Churchill's show.

As one of the Queen's biographers, Ben Pimlott, sagely pointed out, the reaction to the Princess's death could 'be seen as a powerful expression of the continuing grip of the idea of royalty on the popular imagination'. It is difficult to envisage two billion people worldwide bothering to watch the funeral service of the ex-daughter-in-law of an elected head of a British republic, however glamorous, elegant and philanthropic. The monarchy stayed popular in post-Diana Britain, and not just because of the Adonic good looks of her eldest son. By the morning of Tuesday, 9 April 2002, some 200,000 people of all ages and conditions of life had filed past the catafalque in Westminster Hall containing the body of Queen Elizabeth the Queen Mother.

In May 1998, the ABC reporter John Miller interviewed Osama bin Laden at his mountaintop camp in southern Afghanistan. For all Miller's over-polite questions and statements – 'Do you have a message for the American people?', 'You are like the Middle East version of Teddy Roosevelt' – bin Laden's answers elicited a fascinating insight into his concept of international relations, especially with regard to America's lack of willpower. Ten years earlier to the month, the USSR had begun withdrawing her troops from Afghanistan after her eight-and-a-half-year occupation of that country. Bin Laden believed that Islam had humiliated one superpower and he was clearly excited at the prospect of breaking the will of the other.

'We do not care what the Americans believe,' bin Laden told Miller when he was described as the world's most wanted man. 'What we care for is to please Allah.' He went on to invoke Hiroshima and Nagasaki to claim that 'America has no religion that can deter her from exterminating whole peoples. ... We do not have to differentiate between military or civilian.' Of the bomber of the World Trade Center, Ramzi Yousef, he said, 'He acted with zeal.' He went on, 'The Soviet Union entered Afghanistan late in December of '79. The flag of the Soviet Union was folded once and for all on the 25th of December ten years later. It was thrown in the wastepaper basket. Gone was the Soviet Union forever. We are certain that we shall – with the grace of Allah – prevail over the Americans and the Jews.'

Bin Laden subsequently predicted the break-up of America into separate states and the 'wiping out' of the Saudi royal family, before explaining how after the Russian defeat in Afghanistan,

The legend of the invincibility of the superpowers vanished. Our boys no longer viewed America as a superpower. ... Our boys were shocked by the low morale of the American soldier [in Somalia] and they realised that the American soldier was just a paper tiger. He was unable to endure the strikes that were dealt to his army, so he fled, and America had to stop all its bragging ... America stopped calling itself world leader and master of the new world order, and its politicians realised that those titles were too big for them and that they were unworthy of them. I was in Sudan when this happened. I was happy to learn of that great defeat that America suffered.

In fact, the Americans had only suffered eighteen deaths and seventy-one people wounded in Somalia, but clearly President Clinton's decision to withdraw US troops from that country gave Al-Queda a morale boost of astonishing proportions.

Prestige is a tangible currency in the Middle East, as British imperialists knew when they swore to avenge General Gordon's murder in Khartoum in 1885. Even though it took them thirteen years to achieve that, at the battle of Omdurman in 1898, few in the region were left in any doubt that the war of vengeance would eventually come. By contrast, the Clinton Administration missed what in retrospect was the opportunity of the decade, to capture bin Laden before he left Sudan.

Elsewhere in his interview, bin Laden stated that unless the American people elected 'an American patriotic government that caters to their interests and not to the interests of the Jews', Al-Queda would 'inevitably move the battle to American soil, just as Ramzi Yousef and others have done'. There could hardly be a clearer message that the World Trade Center would once more come under attack.

On Friday, 7 August 1998, two lorry bombs exploded outside the US Embassies in Nairobi in Kenya and Dar es Salaam in Tanzania. Three hundred and thirty-one people were killed on that occasion by Islamic fundamentalist terrorism, yet the English-speaking peoples slept on. Clinton, in the words of the writer Hazhir Teimourian, 'went on blithely oblivious of the danger, settling for a token attack with cruise missiles on two training camps. The result was September 11.'

Edward Luttwak, senior fellow at the Center for Strategic and International Studies, identified US generals as well as Clinton Administration politicians as being partly responsible for the lack of a tough response. 'In 1998,' he has written,

when Osama bin Laden and his Al-Qaeda training camps were identified as a serious threat, plans were drawn up to attack both them and him. At that time Afghanistan had no air defence perimeter, so that any aircraft could fly in and out unmolested, and it had no ground patrols along its borders. The US was maintaining a vast panoply of Special Operations forces – some 29,000 men. ... Each time an operation was proposed, the chiefs of staff demanded impossibly detailed 'actionable' Intelligence and exhaustive feasibility studies. They imposed the most restrictive preconditions, demanding assurances that no casualties would be suffered – or inflicted – because of extreme concern about collateral damage.[50]

Once again the English-speaking peoples' fundamental decency was allowed to compromise their safety.

One of the best opportunities to kill or capture Osama bin Laden was missed in February 1999 when members of the United Arab Emirates' ruling family flew to Kahandaron for a hunting trip, where bin Laden set up camp next to them and visited them. Any American attack on the (well-monitored) camp was opposed by Richard Clarke, the White House's counter-terrorism advisor, who had recently visited the Emirates and had received fulsome promises of co-operation.[51] (Yet it was the same Mr Clarke who five years later was to publish a book highly critical of President Bush's failure to act against Al-Queda.)

On 20 December 1999, CIA chiefs and high-ranking US Service Chiefs were gathered in Washington to discuss, and then give the authority for, another operation to kill Osama bin Laden. The CIA man in charge, 'Mike', gave a final report in which he mentioned that bin Laden's location was close to a mosque. According to the Iranian author and Intelligence expert Amir Taheri, 'The revelation caused a commotion and led to the cancellation of the operation.'[52] Testifying before the 9/11 Commission, 'Mike' explained that the committee was concerned that 'shrapnel might hit the mosque and offend Muslims', before going on to assert, 'This was our last chance to kill bin Laden before 9/11.'[53] Of two other US attempts on bin Laden's life, one was cancelled because of President Clinton's concern about collateral damage and the other because of fears for a sheikh of the United Arab Emirates who had seen fit to visit bin Laden. At one point, the Clinton Administration asked the Afghan mujahadeen leader, Ahmad Shah Massoud, to capture bin Laden, but to ensure that he was not 'maltreated' in the process.[54] 'You guys must be crazy,' was Massoud's reply, 'you never change!'[55]

On Good Friday 1998, an agreement was made which was to lead to the formal suspension on 28 July 2005 of what Sinn Fein-IRA called 'the armed struggle', the campaign of terror that it had waged since 1969 in order to try to force the six counties of Ulster to join the twenty-six counties of the South so as to form one nation comprising the whole island of Ireland. It is thought finally to have 'put beyond use' a large amount of its weapons – at least to the satisfaction of the monitor, the Canadian General John de Chastelain – some time in the autumn of 2005. (Although Sinn Fein did not in fact support the Good Friday agreement in 1998, it later claimed authorship of it against its original creators – the Ulster Unionists, their allies and the Social Democratic

The eastern German city of
Dresden before the
(devastating but justified)
Allied bombing raids of
13–15 February 1945

And afterwards ...

The second great assault on the English-speaking peoples – Axis Fascism – defeated: US servicemen celebrate in New York on the morning of 1 May 1945

Facing page
(Above) Alexander Fleming, whose discovery of penicillin revolutionised medicine
(Middle) John Logie Baird, whose invention of television is a less clear-cut boon
(Below) Dr Jonas Salk, who pioneered the vaccine for polio

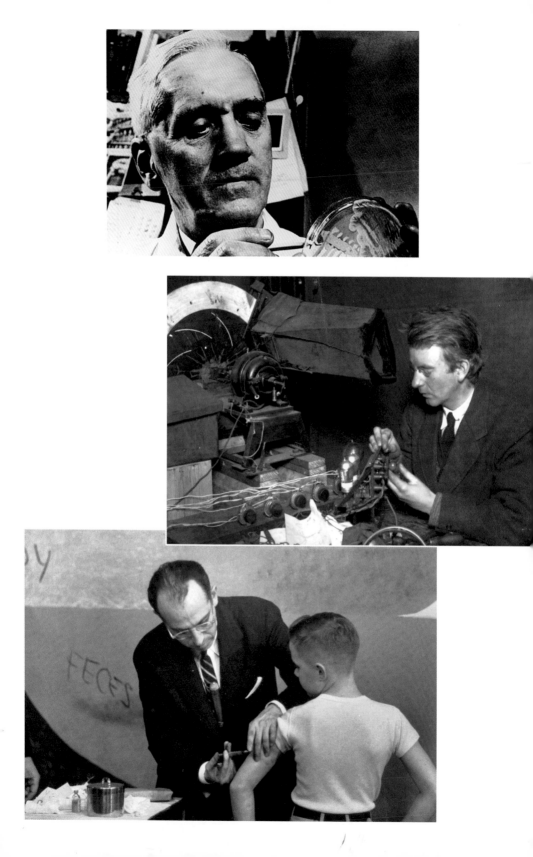

Irish resentment: Queen Victoria being removed from their parliament building, Leinster House, in 1948 (She was taken to a much more appreciative Sydney)

The remains of the Nelson Pillar in O'Connell Street, Dublin, on the morning of 9 March 1966, after Irish nationalists blew up the statue of the admiral who protected their country from invasion

Edmund Hillary, a 33-year-old New Zealander, who with the Nepalese Tenzing Norgay, conquered Mount Everest in May 1953. Shown here with his open circuit oxygen apparatus

Sixteen years after the English-speaking peoples planted one of their flags on the world's highest mountain, they planted another on the Moon. Astronaut Edwin E. 'Buzz' Aldrin on the lunar surface on 20 July 1969

Helicopter humiliations of the Seventies: Richard M. Nixon somewhat
unaccountably gives the 'V' for victory sign as he departs the White House
after his resignation on 9 August 1974

A US marine helicopter removes embassy staff and civilians from the American
Embassy in Saigon on 30 April 1975

The third great assault on the English-speaking peoples – Soviet Communism – defeated. A statue of V. I. Lenin being removed from the Moscow Barracks in 1991

Saddam Hussein's statue gets the same treatment in Baghdad on 9 April 2003

The fourth great assault on the English-speaking peoples – Islamicist
Terrorism – brings mass-murder to the American continent on
11 September 2001, and the 'Long War' begins

and Labour Party – which was a deft piece of political opportunism greatly assisted by the British and Irish Governments.)

By the time the IRA ordered all its units 'to dump arms' and try to achieve a 'united' Ireland via 'purely political and democratic programmes through exclusively peaceful means' in July 2005, the total death toll of 'the Troubles' stood at 3,637, with an estimated 45,000 further people injured. The IRA and its sister republican paramilitary organisations, as well as the Protestant loyalist paramilitary organisations that had sprung up to counter them, let off 15,300 bombs, perpetrated 36,000 shootings and had 30,000 convictions for terrorism-related offences. In all, no fewer than 300,000 British troops had to be deployed in the province over three decades, and the estimated cost to the Northern Irish economy was over £100 billion.[56]

By agreeing to end their campaign without their declared aim being won, the IRA in effect admitted defeat, although obviously neither they nor the British Government put it like that. Although the IRA had managed to murder so many innocent people over the course of thirty-four years, the British people – in particular the Loyalist community of Ulster – stayed firm in their insistence that only the Northern Irish people themselves should decide which country they should belong to, and in election after election the majority voted for candidates who wanted the province to remain British. The British people thus emerged from a sustained, incredibly bloody, long-term insurgency campaign with their principles intact, even though as part of the price for peace they had to endure the disgusting sight of convicted murderers – including several pathological homicidal maniacs – being freed from prison under the terms of the agreement.

It is too early to predict whether the peace will hold, but if it does it will represent a clear-cut victory for Unionism over republicanism in Northern Ireland. Despite the republican nail-bombings, shootings, tortures, 'disappearances', knee-cappings and 'punishment'-beatings over more than three decades, Queen Elizabeth II is still sovereign of Northern Ireland and the democratic will of the majority there has been protected.

One unexpected positive by-product of the horrors of the long campaign in Ulster has been the way in which the British people were able to take the Al-Queda attacks in London of 7 July 2005 almost in their stride. Had they not had the experience of IRA attacks over so very many years, it is possible that they might have panicked when they came under such sudden and violent assault. Instead, they showed a resigned disgust that had become their natural default position when dealing with such outrages.

The defeat of the Major Government in 1997 left the way open for a far more vigorous prosecution of the Balkan problem personified by Slobodan Milosevic. Early March 1998 had seen the Serbian leader send troops into the southern province of Kosovo to kill hundreds of men, women and children and crush the nine-year campaign by its Albanian majority to regain their autonomy. On 9 March, Milosevic was given ten days by NATO and Russia to withdraw troops before economic sanctions were imposed. Yet on 25 March, they agreed to delay imposing these sanctions for one month. On 29 April, the countries froze Yugoslavia's foreign assets, to little avail since by 29 July, after four days of fighting, Serbian forces had routed the Kosovo Liberation Army, displacing over 100,000 Albanians.

In May 1999, NATO air attacks were successful in driving the Serbs out of Kosovo, without any Western casualties whatsoever. These were undertaken in defiance of the United Nations Security Council, which, as with the invasion of Iraq in 2003, did not condone the operations. Nonetheless, NATO's BLU-114/B 'graphite bomb' deployed against Serbian forces disabled 70% of its power grid. This 'soft bomb' exploded a cloud of hundreds of ultra-fine carbon fibre wires over electrical installations, which caused the systems to short-circuit. It was a devastatingly hi-tech way to prosecute war, against which the Serbs had no response.

In 1999, the US air force had no fewer than 4,413 aircraft, including 179 bombers and 1,666 fighters and attack aircraft. By 2002, its arsenal included the Boeing B-2 Spirit, which for all that it cost $1.3 billion is the most advanced long-range, multi-role bomber ever produced, with enough stealth characteristics to allow it to penetrate enemy defences unobserved virtually anywhere in the world. Companies like the Lockheed Martin Corporation, which had global sales totalling $25.5 billion in 1999, had long produced the machines that bring victory to the cause of the English-speaking peoples.

The Albanian refugees were allowed to return to their country, unlike the Bosnian Muslims who had been expelled from places like Banja Luka, Zvornik and Srebenica.[57] Now it was time for the 200,000-strong Serbian population to quit Kosovo bag and baggage, which one-third of them had already done by early July 1998. With the Kosovan War, Britain finally regained some of the honour and prestige she had lost over Bosnia in the first half of the Nineties. The defeat in Kosovo finally led, in October 2000, to the overthrow of Milosevic, a victory for the English-speaking peoples working in harness that could never have been achieved by John Major and Douglas Hurd's policies of the 'level playing field'. Milosevic was to die in prison unlamented.

On 22 April 1999, Tony Blair delivered a speech to the Chicago Economic Club that was virtually ignored at the time, since it was swamped by the news of Kosovo, but which should have been studied carefully around the world – not least in Baghdad – because it contained some revolutionary thoughts about the doctrine of pre-emptive intervention in the internal affairs of other sovereign states. What Blair did in Chicago was to expand the specific case of

Kosovo into a general right to intervene, if five criteria were met. The Foreign Office's lawyers were deeply concerned about the speech – which they had not been sent in advance – since it seemed to cut international law and the United Nations out of the equation. As neither had achieved much against totalitarian and rogue states, this was partly what the new 'Blair Doctrine' was intended to achieve.[58]

Blair's speech in Chicago mentioned the Iraqi dictator Saddam Hussein by name as a possible future candidate for removal from office, a full eighteen months before George W. Bush had even arrived at the White House, and nearly four years before the war against Iraq. To present Blair as a mere poodle of the Americans, therefore, represents a profound misunderstanding of the dynamics of the Special Relationship.

On Thursday, 12 October 2000, suicide bombers rammed a dinghy packed with explosives into the hull of an American warship, the *USS Cole*, anchored in Aden. Seventeen sailors were killed on that occasion and seventy injured, once again by Islamic fundamentalist terrorism, yet the English-speaking peoples still slept on. As the Middle East expert David Pryce-Jones was to write about this and other attacks, 'Through the 1990s, President Clinton and his Administration had not taken Al-Queda seriously, absorbing the harm and the killing it did with an ineffective cruise missile here or there.'[59]

In his 957-page autobiography, Clinton devoted a total of two paragraphs to the *Cole* attack, saying that, 'We all thought it was the work of bin Laden and Al-Queda, but we couldn't be sure,' and that although 'We came close to launching another missile strike at him in October,' he didn't, but nonetheless, 'I was very frustrated, and I hoped that before I left office we would locate bin Laden for a missile strike.'[60] This was deeply disingenuous. There had been several such 'locations' identified over the years, which the Clinton Administration failed to follow up with effective attacks.

The final figures for the 7 November 2000 presidential election in Florida were: Bush/Cheney: 2,912,790 versus Gore/Lieberman: 2,912,253.[61] The voting was thus freakishly close, and since the electoral college votes stood at Gore 266 Bush 246, everything hinged on the twenty-five electoral votes from Florida. The problem was that the difference of a few hundred votes out of approximately six million cast in Florida was itself 'considerably smaller than the margin of error in counting the votes'.[62] The result was five weeks of legal challenge and counter-challenge, culminating in the 5-4 ruling by the US Supreme Court on 12 December for Bush, halting the manual recount ordered by the Florida Supreme Court. For anyone fascinated by the details of those exciting five weeks of the imbroglio, within six months several highly readable books were published, including James W. Caesar and Andrew E. Busch's *The*

Perfect Tie, David von Drehle's *Deadlock*, the *New York Times' Thirty-Six Days* and E.J. Dionne Jnr and William Kristol's *Bush v Gore*.

Although any number of (usually Democrat-supporting) law professors accused the Supreme Court of behaving in a politically partisan manner – the five justices who found for Bush, namely Chief Justice Rehnquist, Sandra Day O'Connor, Antonin Scalia, Anthony Kennedy and Clarence Thomas, were Republican appointees – close analysis of their actual arguments supports their judgment. As Peter Berkowitz of George Mason University Law School has put it, the Supreme Court

> held that the state-wide manual recount of undervotes (undamaged ballots on which the [voting] machines detected no vote for President) ordered by the Florida Supreme Court suffered from a variety of infirmities – absence of a uniform standard for determining the intention of the voter on similar ballots, arbitrary exclusion of overvotes (undamaged ballots on which machines detected a vote for more than one candidate for president), arbitrary inclusion of the results of partial or unfinished country recounts, and use of untrained and unsupervised personnel to conduct the hand recounts – that taken together constituted a violation of the Equal Protection Clause of the Fourteenth Amendment and must end.[63]

Although Bush won the electoral college vote, Gore took a higher proportion of the votes cast; although unusual, it was the fifth time such a discrepancy had occurred since 1824.

A referendum held on the question of an Australian republic saw a 54.4% 'No' vote on 6 November 1999. For all that Australia's republicans had long argued that it was offensive to compel newly immigrated Australians to swear allegiance 'to an elderly Englishwoman, for the most part resident in Berkshire', in fact many Greek, Italian and Vietnamese-born Australians were not only perfectly happy to do so, but voted in large numbers to retain the Queen as their head of state, rightly seeing her sovereignty as a guarantee of the political stability that they badly wanted in their new home.[64]

In the great discussion about Australia's true identity the answer was, in the view of the Australian-born historian Dr John Adamson, 'so blindingly obvious – that Australia is a fundamentally British culture, enriched and ornamented by non-British influences into an idiosyncratic synthesis all its own – that [if generally accepted] it would destroy the main national topic of conversation, other than sporting results'.[65] Nonetheless, in 2004, Dr Germaine Greer wrote a book entitled *Whitefella Jump Up: The Shortest Way to Nationhood*, in which the veteran controversialist argued that Australia should discard her British heritage and adopt Aboriginal language, customs and mythologies, partly because 'an Aboriginal republic would be a lot sexier'.

Greer further argued that Australia should become 'a hunter-gatherer society', which prompted the moral philosopher Anthony Daniels to retort that, 'No concrete suggestion is forthcoming as to how the five million Sydney-siders, for example, are to transform themselves into a bow-and-arrow brigade, living on assorted roots, grubs and game. Of course, like all great conurbations, Sydney already has its hunter-gatherers: they're called burglars, but I don't suppose that's what she meant.'[66]

According to Greer, Australians needed to rediscover what she calls their 'aboriginality', because it was the spirituality of Aborigines that allowed her – as she sat 'on my mattress under the river gums' – to 'feel all around me a new kind of consciousness in which the self was subordinate to *awelye*, the interrelationship of everything, skin, earth, language'. Before this kind of guff is entirely dismissed, however, it needs to be pointed out, as Daniels does, that Greer was indulging in 'a kind of moral exhibitionism, a claim of superiority to all those who haven't communed with the Aboriginal Brahman on mattresses under river gums and who have made lives for themselves in Sydney or Melbourne'. The reason that intellectuals such as Greer are often infuriated by Australia is largely because that country simply has no need for them, as it is, as Daniels goes on to state, 'about as good as modern, large-scale human societies get'.[67]

It is safe to assume that Australians will not adopt Dr Greer's preferred solution, which was to

> Declare their country and themselves Aboriginal, down would come the Blue Ensign and the Southern Cross and up would go the emblem of the black sky, the red earth and the golden sun. ... To accept Aboriginality would be to deny the validity of the annexation of the continent by the British monarch. The planting of Union Jacks on tiny bits of it would be seen from the Aboriginal point of view and would be understood to have been entirely insignificant. ... In this version of events colonisation was attempted and failed.

One of the purposes of this book is to explain how English-speaking colonisation, principally of course of the United States but also crucially of Canada, Australia, New Zealand and the Caribbean – though not Ireland, which has a very different identity – has succeeded triumphantly, and that those states represent the last, best hope for Mankind. Far from being 'entirely insignificant', the spread of the English-speaking peoples' political culture has been the most significant historical development since the invention of gunpowder and the printing press.

In January 2001, John Howard, the Australian Prime Minister, visited Britain at the head of a large and distinguished delegation of serving and former Australian statesmen, including four ex-premiers, to commemorate the centenary of the creation of the Commonwealth of Australia. Rarely in history has the journey from colony to nation been so smooth as in the cases of Australia and New Zealand. Britain can take enormous credit from the fact that these 'settler' colonies have become democratic, stable, economically advanced and happy countries which today stand in the vanguard of world civilisation. Each combines the best aspects of sovereign independence with the advantages that come from their deep historical, linguistic, cultural and often familial ties with what used to be termed without self-consciousness 'the Mother Country'.

SEVENTEEN

The Fourth Assault: Islamicist Terrorism and its *De Facto* Allies

11 September 2001 – 15 December 2005

Al-Queda attacks the continental USA – Tony Blair stands 'shoulder-to-shoulder' with America – The motivation and background of Al-Queda operatives – Anti-Americanism – The limits of Intelligence-gathering – The 'Coalition of the Willing' – Prestige and realpolitik – The invasion of Afghanistan – The oil-for-food scandal – No WMD are found – Saddam's 'useful idiots' – General Franks' deception operations – Saddam captured alive – Al-Queda attacks Madrid – Abu Ghraib and Guantanamo Bay – John Howard, George Bush and Tony Blair convincingly re-elected – Al-Queda attacks London – US and UK military losses – Democratic elections in Iraq

'People will endure their tyrants for years, but they tear their deliverers to pieces if a millennium is not created immediately.'

Woodrow Wilson on board USS *George Washington*, December 1918

'By God's leave, we call on every Muslim who believes in God and hopes for reward to obey God's command to kill the Americans and plunder their possessions wherever he finds them and whenever he can.'

Osama bin Laden, 2001[1]

'The present Iraqi regime has shown the power of tyranny to spread violence and discord in the Middle East. A liberated Iraq can show the power of freedom to transform that vital region, by bringing hope and progress into the lives of millions. America's interests in security, and America's belief in liberty, both lead in the same direction: to a free and peaceful Iraq.'

President Bush's speech to the American Enterprise Institute,
26 February 2003

'We must make sure that its work is fruitful, that it is a reality and not a sham, that it is a force for action and not merely a frothing of words, and that it is a

true temple of peace in which the shields of many nations can some day be hung up, and not merely a cockpit in a Tower of Babel.'
Churchill speaking about the United Nations at Fulton, Missouri,
5 March 1946

'Surprise happens so often that it's surprising that we're still surprised by it.'
Paul Wolfowitz, West Point Commencement Address, 2 June 2001

'If a suicide bomber targeted and killed civilians in Oxford Street he would be called a "terrorist"; at a bus stop in Tel Aviv, a "militant"; in Baghdad, an "insurgent". Where is Orwell?'   Letter to *The Times*, November 2004[2]

'We've never been a colonial power. Any nation that begins in a revolt against taxation without representation is going to be reluctant to embark on enterprises that involve ruling without representation.'
Donald Rumsfeld, May 2004[3]

'The Americans behave like a kind but strict uncle in a pith helmet.'
Vladimir Putin, December 2004[4]

'If only the French would cease to occupy themselves with politics, they would be the most attractive people in the world.'   Oliver Wendell Holmes

At 08.46 and then seventeen minutes later at 09.03 on Tuesday, 11 September 2001, the Twin Towers of the World Trade Center in Manhattan were hit by hijacked aeroplanes; they collapsed at 09.59 and 10.29 respectively, killing 2,749 people. Meanwhile, a third hijacked plane hit the Pentagon in Washington DC, killing a further 180 people. The sublimely brave passengers of a fourth plane, led by Americans Todd Beamer, Jeremy Glick, Thomas Burnett and Mark Bingham, rushed its hijackers, and in the course of trying to overpower them the plane crashed in Pennsylvania killing all those on board, but saving either the Capitol or the White House from a fate similar to the Twin Towers and the Pentagon. Mr Beamer's call to his compatriots just before they stormed the cockpit – 'Are you ready, guys? Let's roll!' – today ranks as one of the great rallying cries of the English-speaking peoples in combat.

Almost 3,000 people were killed by Al-Queda, Osama bin Laden's Islamic fundamentalist terrorist organisation, on 9/11, including sixty-seven Britons. It was by far the worst terrorist atrocity in modern history. Finally the English-speaking peoples woke up. Not since Pearl Harbor had there been a direct attack on such a scale on American territory, and not since the British burned the White House in 1814 had there been such an attack on continental USA.[5]

That terrible day the American people had painfully to re-learn the lesson that President Roosevelt had taught them in his fourth Inaugural Address in 1945, that 'We have learned that we must live as men, not as ostriches, nor as dogs in the manger.' For over a decade since the fall of the Berlin Wall, successive presidents and CIA directors had treated the threat of Islamo-fascist fundamentalist terrorism with too little appreciation of the true threat it posed.

The world did not change on 11 September, but the English-speaking peoples' understanding of it did. As Donald Rumsfeld put it in his testimony before the Senate Armed Services Committee on 9 July 2003, 'We acted because we saw the existing evidence in a new light, through the prism of our experience on September the eleventh.'[6] For in fact Islamic terrorists had been waging a war against the United States for twenty years, a conflict in which the attacks on the US Marines in Beirut in 1982, against the Twin Towers in February 1993, against the US troops in Mogadishu in October 1993, against two US bases in Saudi Arabia in November 1995 and June 1996, against the American Embassies in East Africa in August 1998 and against the USS *Cole* in October 2000 were only the most high-profile manifestations. Those who accuse Messrs Bush and Blair of exacerbating Islamicist terrorism through their invasions of Afghanistan and especially Iraq fail to appreciate that murderous and pitiless war-making was already well under way long before 2003. If anything, the War against Terror was a very belated response. If those invasions had taken place far earlier than 2003, perhaps in 1999 under President Clinton's watch, once the evidence of Al-Queda's terrorist activities and Saddam Hussein's malicious disruption of the work of the UN inspectors was beyond doubt, the victories in Afghanistan and Iraq would have been far quicker and easier than was subsequently the case.

Before 9/11, successive Administrations of both political complexions had decided to treat these assaults as terrorist-criminal acts rather than acts of asymmetric warfare, despite Osama bin Laden's very specific periodic declarations of war against the United States. Only after 9/11 were the English-speaking peoples finally prepared to fight the struggle properly and employ every element of national power to form a coherent and strong response. 'We learned about an enemy who is sophisticated, patient, disciplined, and lethal,' reported the 9/11 Commission, set up by President Bush to inquire into the events of that dreadful day. 'The enemy rallies broad support in the Arab and Muslim world by demanding redress of political grievances, but its hostility toward us and our values is limitless. Its purpose is to rid the world of religious and political pluralism, the plebiscite, and equal rights for women. It makes no distinction between military and civilian targets. "Collateral damage" is not in its lexicon.'[7]

The public statements of Osama bin Laden as transmitted to the world via the Arab television station Al-Jazeera soon made it clear that the demands of Al-Queda were so extravagant that no Western nation could ever accept them. They included the re-creation of the caliphate across the Arab crescent from Pakistan to southern Spain and the universal implantation of *Sharia* law. This was fortunate, because were it possible to appease Al-Queda, history suggests that there would have been voices raised in the West – especially in Western Europe – in favour of doing just that. Even as it was, after a series of bombings and attempted bombings by two Al-Queda cells in London in July 2005, the attacks were blamed by some on Britain's involvement in the invasion of Iraq, despite the fact that Al-Queda's campaign against the liberal democracies had long pre-dated that. As earlier chapters have attempted to show, since 1900 there have always been those amongst the English-speaking peoples prepared to appease, apologise for and even on occasion to laud and aid their mortal enemies.

Tony Blair was working on his speech to the Trades Union Congress in the Fitzherbert suite of the Grand Hotel in Brighton at 1.48 p.m. on 11 September when an aide told him that a plane had crashed into the World Trade Center. Assuming it to be a freak accident, he worked on. When the second plane hit, he watched on television the scenes in New York. The Prime Minister's reactions were those of most of the rest of Mankind: 'horror and disbelief'.[8] He was watching when the third plane hit the Pentagon at 2.43 p.m. (British time) and then put in a short appearance at the conference, where, visibly shaken, he told the delegates: 'There have been the most terrible, shocking events in the United States of America in the last hours. I am afraid we can only imagine the terror and carnage there and the many, many innocent people who have lost their lives. This mass terrorism is the new evil in our world today.' Later he recalled, in an interview on Boston television, 'Sometimes things happen in politics, an event so cataclysmic that, in a curious way, all the doubt is removed. From the outset, I really felt very certain as to what had to be said and done.' He stayed certain. In his conversations with Jacques Chirac of France, Gerhard Schröder of Germany and Vladimir Putin of Russia, he was delighted that all three leaders seemed to be 'totally on board, right from the outset'.[9]

Verbal expressions of support and sympathy were one thing; swift and decisive Anglo-American action to avert panic-selling of dollars immediately after the attacks was another. The high and mutual regard between the Governor of the Bank of England, Professor Mervyn King, and the Federal Reserve Deputy Chairman, Roger Ferguson, ensured that one short conversation on 9/11 was enough to open a $30 billion line of dollar credits, which kept the US currency in the United Kingdom stable and averted the danger of a global financial crisis following on from the national security one. It was a fine example of how Britain instinctively stood 'shoulder-to-shoulder' with America in her moment of peril. This reaction in turn gave Britain a say in – though of course not a veto over – what was decided in Washington. As Henry Kissinger has written, Anglo-American relations are 'so matter-of-factly intimate that it was psychologically impossible to ignore British views'.

President Bush's first conversation with a foreign leader – at 7.30 a.m. East Coast Time on Wednesday, 12 September – was with Tony Blair, another indication of the enduring importance of the Special Relationship to the Americans. Blair found Bush 'very calm'. They discussed the United States' response. 'We are not interested in simply pounding sand for the sake of demonstrating we are going to do something,' Bush said, adding that this would be a 'mission for a Presidency', thus proving that the lessons of the Clinton years had finally been learned. Blair then wrote out by hand a five-page memorandum, which was faxed to the White House and which concluded that, in the words of one report of it, 'the cancer was not confined to Afghanistan, or indeed Al-Qaeda, and they had to make plans to act against all who financed, supported or sponsored terrorism, wherever they existed in the world'.[10] The Left's characterisation of Tony Blair as being Bush's poodle is thus no more accurate than that of Thatcher being Reagan's or of Macmillan being Kennedy's, or indeed of Churchill being Roosevelt's, although those accusations have each been made in their time.

On 20 September, Blair flew to New York to attend a memorial service for those who died. It was there that the British Ambassador read out a message from the Queen in which was contained the phrase that 'Grief is the price we pay for love.' The Prime Minister then flew on to Washington, where, standing by the window in the Blue Room of the White House, which looks out towards the Washington Monument, he was 'delighted' to be told by Bush that the President was going to announce to a joint session of Congress that evening: 'Either you are with us or you are with the terrorists.'

Blair watched that speech, which was interrupted thirty-one times with ovations, from the Senate gallery. At one point Bush said, 'I'm so honoured the British prime minister has crossed the ocean to show his unity with America. ... Thank you for coming, friend.' Although it is true that every post-war prime minister except Edward Heath and John Major has set a high value on the Special Relationship, only Churchill and Thatcher had brought it to such a fine pitch as did Blair. This was underlined soon afterwards in his powerful address to the Labour Party Conference, in which he said of the American people, 'We were with you at the first. We will stay with you till the last.' In this he was as good as his word.

If America's part in the Second World War started in 1941 with the counter-intuitive but nonetheless hard-headed analysis that it was necessary to fight Germany first, then her War against Terror began similarly. Although Saddam

Hussein had not been implicated in the attacks of 9/11, Iraq was the world's leading state-sponsor of terrorism and an openly and oft-declared foe of the English-speaking peoples, who had led the coalition that had foiled his attempt to dominate the Middle East in 1990–1. Tony Blair, giving evidence in the House of Commons before 21 January 2003, two months before the invasion of Iraq, categorically accepted: 'Whenever I am asked about the linkage between al-Qaeda and Iraq, the truth is that there is no information I have that directly links Iraq to September 11. . . . I think that the justification for what we are doing in respect of Iraq has got to be made separately from any potential link with al-Qaeda.'[11] Saddam's support for non-Al-Queda terrorism was central to that.

Just as the Roosevelt Administration and Churchill Government had agreed to destroy Hitler first, even though Japan was the immediate enemy that had struck America, so the Bush Administration and Blair Government correctly identified the importance of removing the core problem in the Middle East – Saddam Hussein – even though the immediate enemy that had struck the United States had been Al-Queda. Tony Blair's part in formulating the Allies' post-9/11 military strategy was powerfully reminiscent of Churchill's role in encouraging the concept of 'Germany First' in December 1941. What might seem illogical at the time can often look clear-sighted much later in light of the wider struggle. Although the successful attack on Afghanistan, which expelled Al-Queda and its Taliban protectors from all the country's key areas, was much more than a sideshow, it was never going to be the main event if the English-speaking peoples and their wide alliance were going to engage their major Middle Eastern tormentor.

Some Western commentators have argued that Al-Queda attacked the United States on 9/11 largely because of her support for Israel, and that this could have been averted if only successive American Administrations had tried harder to solve the Israel-Palestinian problem. This is utterly to misinterpret the true nature of Al-Queda. None of the nineteen hijackers was Palestinian, and Osama bin Laden's primary goal was to drive the Americans out of Saudi Arabia. The annihilation of Israel would only come later, once US power had been expelled from the Middle East. As Richard Beeston of The Times has succinctly put it, 'The notion of 11 September being called off because of a fresh bout of US diplomacy in the Levant is ridiculous.'[12] American support for Israel has always been a noble response to, not a provocative cause of, fanatical Islamicist anti-Semitic terrorism.

An exhaustive study undertaken by Dr Marc Sageman of the University of Pennsylvania into the life histories of 400 Al-Queda members and their close allies shows that traditional motives ascribed to terrorists – poverty, desperation and ignorance – also do not generally apply. Instead, 17.6% came from the upper class of their societies and 54.9% from the middle class. Of those whose educational records were available, 28.8% had some college education, 33.3% had a college degree and 9% had a postgraduate degree. Ahmed Omar Sheikh, the Briton who murdered the American journalist Daniel Pearl, attended the London School of Economics. Far from being brainwashed in madrassa religious schools, 90.6% had had a secular education. In their career paths, 42.5% were professionally employed as lawyers, teachers, doctors and so on, and only 24.6% had unskilled jobs. For those whose marital status was known, 73% were married and most of those had children.[13] Poverty, alienation and ignorance were thus emphatically not the primary motivations for Al-Queda activity. (The killing of Mr Pearl inaugurated a new and particularly vile method of Al-Queda murder: the videoing of a hostage's throat being slashed or head being chopped off.)

A useful tool in analysing the mentality of the 9/11 suicide pilots and the suicide bombers who have followed them is the 150-page 1951 bestseller *The True Believer* by Eric Hoffer. The author, an autodidact and former New York docks longshoreman, made a precise study of the similarities between the fanaticism of several mass movements including first-century AD Christianity, early sixteenth-century Protestantism, Jacobinism, Nazism, communism and Muslim fundamentalism, finding that, 'There is a certain uniformity in all types of dedication, of faith, of pursuit of power, of unity and of self-sacrifice.'[14]

Al-Queda have variously defined their aims as the recreation of the caliph-ate, the complete expulsion of Western influences from the land of Islam, and the conversion of the world to the Muslim faith and *Sharia* law, none of which have any chance of being fulfilled, especially not through the terrorist route chosen, yet, as Hoffer argued, that if anything strengthens rather than weakens its adherents' fanaticism. As the Israeli Ambassador to London, Zvi Heifetz, pointed out in October 2005,

> The word *jihad* may be literally translated as "striving". It is an important clue because, in the distorted perspective of the global jihadists, waging war against the West is not a means to an end but the end in itself. Political objectives secured in the course of the struggle may be a welcome bonus but they are not the spiritual or intellectual point.[15]

The unimaginative, bourgeois, earth-bound English-speaking peoples refuse to dream dreams, see visions and follow fanatics and demagogues, from whom they are protected by their liberal constitutions, free press, rationalist philosophy and representative institutions. They are temperamentally less inclined towards fanaticism, high-flown rhetoric and Bonapartism than many other peoples in History. They respect what is tangible and, in politics at least, suspect what is not. But as Hoffer recognised in fanatical movements long before Al-Queda, 'In all ages men have fought more desperately for beautiful cities yet to be built and gardens yet to be planted. ... Dreams, visions and wild hopes are mighty weapons and realistic tools.'

Hoffer recognised how a conception of the past – or at least a highly idealised view of it – is an indispensable political weapon for a fanatical mass movement, since 'It develops a vivid awareness, often specious, of a distant glorious past ... to show up the present as a mere interlude between past and future', both of which were glorious. To that end, Hitler lauded Arminius, who defeated the Romans in 9 AD, and the Jacobins harked back to the pre-historical era of the 'Noble Savage'. How much more powerful a motivation, then, when the past was not only glorious but relatively recent.

Yet Al-Queda needed no specious, idealised view of the House of Osman that ruled the Ottoman Empire for 470 years, comprising at different stages parts of Spain, the North African littoral, Egypt, Greece, the Arabian Pen-insula, Mesopotamia, Syria and Lebanon, much of south-eastern Europe up to the gates of Vienna and, of course, Turkey. By any standards of History or Civilisation, the Empire that was ruled by thirty-six sultans of that House until 1922 was indeed impressive, and at times glorious. Bin Laden's reference in a videotape message after 9/11 to the abolition of the caliphate in March 1924 – the action of a Muslim, Kemal Atatürk – shows how acutely Al-Queda regrets the decline of the secular power and influence of Islam.

As well as an unappeasable desire for revenge for everything that has befallen the Muslim world since it stood at the gates of Vienna in 1683, Al-Queda acts out of the same sense of envious rage that has always actuated peoples who view the world's hegemonic power, whatever that power is or has been and however benign it might be. (To appreciate quite how long ago it was since the Ottomans were in the ascendant – and thus the length of the fundamentalists' legacy of resentment against the West – 1683 in Europe saw the Rye House Plot against King Charles II and in the New World it was the year that William Penn published *A General Description of Pennsylvania*.) There are good reasons why the United States should spend the billions it does relieving AIDS distress in Africa, tsunami victims in Asia, providing debt-relief throughout the Third World, and so on, but the hope of winning popularity should not be one of them.

Once again there had been a painful defeat in the opening engagement of a conflict. The sinking of the USS *Maine* (however it might have happened), the Boer invasion of Cape Colony, the retreat from Mons in 1914, the evacu-ation from Dunkirk, the attack on Pearl Harbor, the fall of Seoul, the Gulf of Tonkin incident, the capture of Port Stanley, the invasion of Kuwait, then 9/11: all fit into a long-established pattern of reverses that have befallen the English-speaking peoples in the opening stages of almost every war they had fought over the previous century – or beyond, if one also includes the nineteenth-century battles of the Alamo, Little Big Horn, Isandhlwana and Maiwand. Yet after every single one of those reverses and defeats, the English-speaking peoples were awoken as to what it would take to fight the war, and in all but Vietnam they went on to taste victory.

Furthermore, it is often small nations, rather than other Great Powers, which have tested the resolve of the English-speaking peoples. The Boers, Filipinos, North Koreans, Egyptians, North Vietnamese, Argentinians and Iraqis – for all that some of them might have been backed by Great Powers – were not particularly powerful in themselves, but they presented challenges no less important for the fact that they were not Wilhelm II's Germany or Hirohito's Japan. The end of Great Power status is often signalled by a successful challenge from a much lesser adversity, as Austria-Hungary found with Serbia, France at Dien-Bien-Phu in Indo-China, Britain at Suez and the USSR in Afghanistan. The United States could simply not afford to allow either the Taliban's Afghanistan or Saddam Hussein's Iraq to continue to mock her after 9/11. The worst bloodshed in history tends to arise when nations make an unwarranted bid for world-primacy; no potential successor could be left in any doubt that the United States was still a potent superpower more than capable of swatting a self-appointed irritant such as Saddam's Iraq.

The 9/11 attack brought out the virulence of anti-Americanism in all its ugliness. Palestinians danced in the streets of Gaza, many others in the Middle East celebrated less publicly, and the French philosopher Jean Baudrillard wrote of his own and his countrymen's 'prodigious jubilation in seeing this global superpower destroyed', saying of the terrorists responsible, 'Ultimately they were the ones who did it, but we were the ones who wanted it,' and that 'everyone without exception had dreamt' of such a cataclysm hitting America.[16] Gallingly for anti-Americans as rabid as him, America was soon to prove that far from being 'destroyed' by 9/11, she was galvanised in the same way that she had been by the *Lusitania* sinking and the assault on Pearl Harbor.

In Britain, intellectuals such as the author William Boyd denounced the Special Relationship as 'this faltering, gimcrack, unequal relationship', arguing in the pages of the *Times Literary Supplement:* 'We have had to live with the Churchillian myth of a special relationship with the US ever since the Second World War – and we continue to pay the price.' (Boyd's article was replete with factual errors, but that did not detract from the passion of his thesis.) Either denouncing the Relationship or denying that it even existed, many British commentators – especially of the Left – hoped to sever Britain's intimate and long-standing links with her closest ally.

Military Intelligence is necessarily an inexact science. To gain human intelligence on Saddam's Iraq involved having people who were willing to risk torture and execution not only on their own behalf, but also upon that of their families and colleagues as well. To expect, as so many armchair Intelligence experts since have, that all information on Iraqi capabilities could be supported by more than one source was simply to ask too much of any Intelligence service. Some defectors from Saddam's regime did speak to Western security services, telling them what everyone assumed was the case: that the dictator had weapons of mass destruction (WMD). (In 2001, for example, Adnan Ihsan Saeed al-Haideri, a civil engineer, said that he had visited twenty secret facilities for chemical, biological and nuclear weapons. He supported his claims with copies of Iraqi government contracts, complete with technical specifications.)[17] Saddam had used WMD in the past, had admitted to having them as recently as 1995, and had nothing in his previous behaviour that suggested that he might have destroyed them in the meantime. Indeed, why *should* a dictator whose power was entirely based on his ability to terrorise, voluntarily destroy weapons designed to achieve this? That central question still lies unanswered at the time of writing in January 2006.

'Iraq was the only country in the world that had recently used weapons of mass destruction,' Senator John McCain told me in November 2004. 'It had them in 1991, and every intelligence agency in the world believed that it still had them. We viewed Iraq as the greatest threat.'[18] The English-speaking peoples' experiences at Pearl Harbor in 1941, Dieppe in 1942, the Tet offensive in 1968, the Falkland Islands in 1982 and the Gulf War in 1991 all suggest that Intelligence is only part – and often by no means the most important part – of the story. Throughout the history of the English-speaking peoples since 1900, Military Intelligence has been patchy at best, with the almost sole (but vital) exception of the decryption of German codes during the First and Second World Wars. Yet that does not absolve Western leaders from the duty of taking decisions based on the best analysis available, which is what George W. Bush and Tony Blair had to do with regard to Iraq after 9/11. In the murky world of secret Intelligence, there is no counsel of perfection. As the CIA Director George Tenet told Georgetown University in February 2004, 'By definition, Intelligence deals with the unclear, the unknown, the deliberately hidden. What the enemies of the United States hope to deny, we work to reveal. In the Intelligence business, you are almost never completely right or completely wrong.' Over WMD, however, Tenet went badly wrong.

Appearing before the United Nations Security Council on 5 February 2003, the US Secretary of State Colin Powell held up a vial of white powder to represent Iraq's stocks of anthrax. 'My colleagues,' he said, 'every statement I make today is backed up by sources, solid sources. These are not assertions. What we're giving you are facts and conclusions based on solid intelligence.'[19] As a former four-star general, Chairman of the Joint Chiefs of Staff and National Security Advisor, Powell's words carried enormous weight. Although his estimation turned out to be wrong, there can be absolutely no doubt that this man of unimpeachable integrity believed them implicitly, because they were based on the very best Intelligence that the US and her allies could call upon at the time. That it was wrong was the fault of the Intelligence agencies, not the politicians who had to take decisions based upon it. (As well as the CIA and MI6, the Intelligence services of Russia, Israel, Germany, France and China all also took it for granted that Saddam had WMD.)

The first piece of recorded military intelligence in history is contained in a papyrus sent to Thebes 4,000 years ago, which reports: 'We have found the track of 32 men and three donkeys', evidence of a raiding party or the advance guard of an invasion force. Since then the espionage industry has become far more sophisticated technologically, but nothing has proved more valuable than human Intelligence ('humint'), which can only be gleaned from winning the trust of an opponent. Since the Al-Queda higher leadership largely coalesced over twenty years before 9/11 in the mujahadeen struggle against the USSR, that has proved impossible. Similarly, many of the people closest to Saddam had been with him since his 1968 coup. The leaders of the English-speaking peoples had to extrapolate what they could from what military Intelligence they had, as well as their knowledge of Saddam's track record. Their conclusions were the same that any reasonable, intelligent, objective person would have also come to at the time: that the War against Terror could not be won unless Saddam Hussein was overthrown, and that it was too much of a risk for the English-speaking peoples *not* to topple his regime.

Furthermore, it was not just the CIA and MI6 that provided Intelligence: in April 1995, the United Nations Special Commission (UNSCOM) weapons experts reported to the UN Security Council that 'Iraq had concealed its biological weapons program and had failed to account for three tons of growth material for biological agents.' After the defection that year of a senior official, Iraq herself admitted to making weapons from thousands of litres of anthrax, botulinim toxin and aflatoxin for use with Scud warheads, aerial bombs and aircraft. By September 2002, UNSCOM had concluded that 'Iraq's declarations on biological agents vastly understated the extent of its program, and that Iraq actually produced two to four times the amount of most agents, including Anthrax and Botulinim toxin, than it had declared.'

UNSCOM also reported in September 2002 that 'Iraqi accounting and current production capabilities strongly suggest that Iraq maintains stockpiles of chemical agents, probably VX, Sarin, Cyclosarin and Mustard agent.' Furthermore, 'Iraq has not accounted for hundreds of tons of chemical precursors and tens of thousands of unfilled munitions, including Scud variant missile warheads', let alone 'at least fifteen thousand artillery rockets that in the past were its preferred vehicle for delivering nerve agents, nor has it accounted for about 550 artillery shells filled with mustard agent'. It would have been a gross dereliction of duty on behalf of the leaders of the English-speaking peoples to have overlooked what Saddam's behaviour seemed to suggest, even though we now believe that he was misleading them. The United Nations' own inspectors said, 'There was a strong presumption that Saddam had ten thousand litres of anthrax, which could have been contained in a single petrol tanker,' yet the CIA and MI6 have been demonised for merely agreeing with them.[20]

In a dossier released by the British Government on 24 September 2002, a great deal of accurate information was given about Saddam Hussein's regime, capabilities and likely intentions. However, there was also a claim that 'some of the WMD' – without specifying whether these would be a short- or long-range – could be ready 'within 45 minutes of an order to use them'. Elsewhere it was reported that Iraq was attempting to construct a ballistic missile capable of hitting Cyprus, where there were large British military bases. Although battlefield chemical weapons could indeed be used within forty-five minutes, Saddam had no ballistic missiles with which to hit Cyprus at that time, and neither the dossier nor the Prime Minister ever claimed he yet had. As the MI6 source of the claim, Lieutenant-Colonel al-Dabbagh of Iraqi air defence, told Saddam's biographer, Con Coughlin of the *Sunday Telegraph*, 'We could have fired these within half an hour,' yet, as he also pointed out, they were only for battlefields in Iraq and Kuwait.

The forty-five-minute claim represented only one sentence on page 17 of the dossier's main text, albeit repeated twice within its internal summaries and once in Tony Blair's foreword. The Prime Minister then mentioned it once on presenting the dossier to Parliament that day. It was picked up by *The Sun* newspaper with the headline, 'Brits 45 Mins From Doom', and by some other papers, but was otherwise largely ignored. Crucially the Government's public relations experts ('spin-doctors') did nothing to disabuse *The Sun* or anyone else of the lurid interpretation that the newspaper had placed on two separate pieces of information that had been conflated. (It is not the Government's duty, or within its capacity, to correct every inaccurate Press story.)

The forty-five-minute claim then lay buried, at least until after the war broke out six months later. Of some 45,000 questions that were asked in Parliament between the publication of the dossier and the outbreak of war, only two referred to it. Mr Blair did not refer to it in his speech preparing the country for war in March 2003, nor did anyone raise it with him. The subsequent claims made by the anti-war movement, therefore, that it played a central role in the Government's case for war, are quite untrue.[21]

It was anyhow not enough that Iraq should not possess WMDs; UN resolutions made it incumbent on that country to prove that it did not, and Saddam's behaviour in expelling UN weapons inspectors in 1998 strongly suggested that he should not have been given the benefit of any doubt. The Al-Dawrah 'Foot and Mouth Disease Vaccine Facility' was one of two known top-level bio-containment facilities in Iraq that had an extensive air-handling and filtering system. Iraq had already admitted that it had been a biological weapons facility in the past. In 2001, Iraq announced that she would begin renovating the plant without UN approval, ostensibly to produce vaccines that she could more easily and quickly import through the UN. Any rational person would conclude, knowing what was already known of the Ba'athist regime, that WMDs would soon be produced there.

Partly as a result of the culture of distrust of the Establishment that had been built up in the three decades since Watergate, many in the West have assumed that there was a conspiracy between politicians and the security services to take the English-speaking peoples to war full in the knowledge that Iraq had no WMDs. Despite two sober, hugely in-depth investigations in Britain, carried out by men of the highest personal and professional probity, namely the law lord Lord Hutton and the former Cabinet Secretary Lord Butler, the media saw fit to denounce both as 'whitewashes', which they were patently not. Both in-depth inquiries probed very hard into the circumstances surrounding the outbreak of the Iraq War and both concluded that the Government had acted in good faith, although other criticisms were made.

It says much about how far post-Watergate paranoia about the motivation and honesty of public servants had gone that very many people genuinely believed that an American Administration and a British Government deliberately lied about the level of threat they believed Saddam posed in order to send US and British troops to fight and die in Iraq. Any such conspiracy would have had to have involved large numbers of utterly unprincipled people in the very highest reaches of government, the security services and armed forces. In fact it was a foul slur completely unsubstantiated by the facts. Although Bush and Blair have been widely denounced as liars by anti-war groups, by infantile political-comedians such as Michael Moore and Al Franken, and even on occasion by their Democratic and Conservative Party oppositions, to say what you devoutly believe to be the truth at the time – but which later turns out untrue – is not a 'lie' under any generally accepted construction of the word.

'I apologise', said Tony Blair in October 2004, 'for any information given in good faith that turned out to be wrong.' This was the central issue – good faith – and the electorates in Australia, America and Britain all had to decide between October 2004 and May 2005 whether the information truly was given in good faith. In all three countries they re-elected their leaders with very good majorities, suggesting that for all the conspiracy theorists and anti-war propagandists, most of the English-speaking peoples accepted that the incorrect information had nonetheless been given honestly.

Since Saddam had been the only leader to use biological weapons since Mussolini in Abyssinia, against the Iranians, the Marsh Arabs and the Kurds, there was a good deal of circumstantial evidence of his ruthlessness. Furthermore, the US Commander-in-Chief of the 'Coalition of the Willing' in Iraq, General Tommy Franks, was informed by both King Abdullah of Jordan and President Mubarak of Egypt that they had been told by Saddam that he would use WMD against the Americans.

In the same month as Blair's apology, Charles Duelfer, leader of the Iraq Survey Group, presented to Congress his *Comprehensive Report* on the issue of WMD. Though under-reported at the time, because it failed to fit in with the media's conspiracy theory preconceptions, this explained that Saddam's illegal military procurement budget ran at $500 million per annum between 1996 and 2003, with illicit oil contracts providing the funding. Duelfer further proved that Iraq had maintained weapons programmes that placed her in material breach of, amongst others, the key US Security Council Resolution 1441. He also surmised, as most other objective people would have, that Saddam intended to resume WMD production the moment that UN sanctions were lifted, while spending millions in bribes to individuals in China, France and Russia who were involved in the decision-making process. Duelfer also uncovered one Iraqi Intelligence report saying that French politicians had assured Saddam in writing that France would veto any second UN resolution, which it sure enough threatened to do in March 2003.

Since the terminal demise of the principles of the Washington Address, the American people have appreciated that their vital interests have lain far beyond her borders. The rest of the English-speaking peoples have known how far-flung their interests have been for far longer. It is therefore not enough to state, as various anti-war propagandists have, that simply because Saddam Hussein was not an immediate threat to US or British servicemen he should not have been overthrown. He threatened Western friends and allies in the region, harboured those who had murdered US servicemen and civilians, and occasionally Iraqi rockets were fired at RAF and USAF planes patrolling over the no-fly zones agreed in 1991. As Churchill said after the assassination of King Feisal and Prime Minister Nuri-es-Said of Iraq in 1958,

> The Middle East is one of the hardest-hearted areas in the world. It has always been fought over, and peace has only reigned when a major power has established firm influence and shown that it would maintain its will. Your friends must be supported with every vigour and if necessary they must be avenged. Force, or perhaps force and bribery, are the only things that will be respected. It is very sad, but we had all better recognise it. At present our friendship is not valued and our enmity is not feared.[22]

The Iraq War should not be seen as some kind of brand new military engagement in the Middle East, so much as the culmination of hitherto-unfinished business left over at the time of the Gulf War twelve years before. Quite apart from WMD, the British and American Governments also concentrated their case for war on other unanswerable humanitarian and terrorism-related factors that have survived the failure to discover WMD after the invasion. These also helped to justify the invasion of Iraq, but such was the influence of the anti-war movement – and the media's concentration on the WMD issue – that they were partly drowned out. Yet as Alex Van der Stoel, the UN special rapporteur on human rights for Iraq, had reported, the abuses there were 'so grave that it has few parallels in the years that have passed since the Second World War'.[23]

Earlier chapters have established how important prestige has always been in the *realpolitik* that governs international relations. For the English-speaking peoples after 9/11 to have permitted Saddam to continue to mock their power; attempt to shoot down RAF and USAF planes over the no-fly zones; profit from the Oil-for-Food scandal while Iraqi children starved to death; pay $25,000 to the families of each Palestinian suicide-murderer; threaten his peaceful pro-Western Arab neighbours; ignore and jeer at sixteen UN resolutions passed over nine years; and summarily expel UN weapons inspectors, would have made a War against Terror that did not involve toppling Saddam not worth the name. In 1993, the Iraqi Intelligence Service attempted to assassinate President George Bush Snr and the Emir of Kuwait with a powerful car bomb. Iraq also sheltered the Mujahadeen-e-Khalq Organisation (which had killed US soldiers and civilians), the Palestine Liberation Front, Abu Abbas (who murdered the US citizen Leon Klinghoffer on the cruise ship *Achille Lauro*), the Abu Nidal organisation (responsible for the deaths or wounding of 900 people in twenty countries), Abdul Rahman Yassin (who mixed the chemicals for the 1993 World Trade Center bombing) and such other notorious terrorists. There were therefore plenty of sound reasons for overthrowing Saddam quite separate either from WMDs or his monstrous domestic human rights record. Nor was time on the English-speaking peoples' side. Saddam had two vicious, sadistic sons, one of whom – Uday – was a rapist and mass murderer, whom he was grooming to succeed him.

'Looking at the what-ifs seems to me to be extremely important,' said the British Foreign Secretary Jack Straw on 26 January 2004. 'If we'd walked away, Saddam would have been re-emboldened, a destabilising force in the whole of the Middle East. The authority of the UN and the security of the Middle East would have been further undermined.' Straw – one of the best British Foreign Secretaries of the post-war era – also pointed out that even after high penetration of the Provisional IRA over thirty years, the British Army still didn't know the whereabouts of their weapon stockpiles, and that Ulster was a fraction of the size of Iraq.

By 2002, Iraq was declared in breach of almost every single one of the obligations set out during nearly a decade of binding United Nations Security Council resolutions (UNSCRs), all designed to protect the rest of the Middle East from Saddam. Between 29 November 1990 and 17 December 1999, there were no fewer than sixteen of these, namely 678, 686, 687, 688, 707, 715, 949, 1051, 1060, 1115, 1134, 1137, 1154, 1194, 1205 and 1284. Under their terms, Saddam Hussein was required to, among other things: 'destroy all of his ballistic missiles with a range greater than 150 kilometers; stop support for terrorism and prevent terrorist organisations from operating within Iraq; help account for missing Kuwaitis and other individuals; return stolen Kuwaiti property and bear financial liability for damage from the Gulf War; and . . . end his repression of the Iraqi people'.[24] He did none of these. Although Saddam probably had no more WMD by 1998, he certainly acted precisely as though he had.

On 8 November 2002, the Security Council voted unanimously in favour of Resolution 1441, which threatened that 'serious consequences' would follow further material breaches, yet those on the Left who had spent decades trumpeting the superior morality of the UN over the governments of the English-speaking peoples, preferred to see the Security Council's resolutions continue to be scorned rather than have the United States' case for war strengthened.

Saddam could have complied with Resolution 1441, albeit with huge loss of face, but he chose not to. Unfortunately the British Government – under extreme pressure from its Labour Party backbenchers – made a fetish of attempting to secure a second UN resolution specifically authorising war, thereby wasting further precious months.

The English-speaking peoples and their allies had the perfect moral right to invade Iraq whether she had flouted numerous UNSCRs or not; their freedom of manoeuvre and that of NATO could not be allowed to be circumscribed by the United Nations, an organisation whose interests are fundamentally different from – and occasionally opposed to – theirs. One of the most serious ramifications of the Iraq War was that a significant proportion of the English-speaking peoples seem to have believed that military action could only be legitimate if specifically authorised by the United Nations.

Bruno Tertrais, until 2001 the special assistant to the Director of Strategic Affairs in the French Defence Ministry, and certainly no friend of the Bush Doctrine, was forced to admit in his recent book *War Without End* that, 'The worldwide coalition against terrorism is in fact the widest in history: 134 countries offered their assistance to the United States after September 11, and ninety took part in one way or another in Operation Enduring Freedom (twenty-seven of them inside Afghanistan itself).'[25] No fewer than twenty-one nations – including Estonia, Poland and even Mongolia – also took part in the war against Saddam.

The countries that had 'boots on the ground' in Iraq in October 2003 were so many and varied that it made a mockery of the accusation that the United States was acting 'unilaterally' there. Taken alphabetically there were Albanians peace-keeping in northern Iraq; Azerbaijanis protecting religious and historic monuments; 7,400 Britons with more on the way; Bulgarians patrolling Karbala, south of Baghdad; Central American and Dominican Republic troops in south-central Iraq; Czech military police; Danish light infantry units; a battalion of Dutch Marines; Estonian mine-divers and cargo-handlers; Georgian sappers and medics; a Hungarian transportation contingent; 3,000 Italians; Moldovan de-mining experts; New Zealand and Norwegian army engineers; soldiers and police from the Philippines; no fewer than 2,400 Poles; Portuguese policemen; 800 Romanians; Slovakian military engineers; some South Koreans; 1,300 Spaniards; Thais assigned to humanitarian operations; over 1,600 Ukrainians from a mechanised unit; as well as troops from El Salvador, Honduras, Kazakhstan, Latvia, Lithuania, Macedonia and Nicaragua. This was hardly the United States 'going it alone', as the domestic and foreign opponents of the war constantly alleged.

Tertrais went on to acknowledge what he correctly analysed as 'the unwavering nature of the Bush-Blair partnership', adding,

> The United Kingdom certainly had good reasons to get directly involved in the war: its past of colonial involvement in the area, its status as America's unfailing partner in all the operations against Iraq between 1991 and 2003, its experience in the struggle against terrorism, as well perhaps as its indirect responsibilities in the development of Islamism in Europe (the country having long been a haven of tolerance for extremism).[26]

The major difference between Woodrow Wilson's attempt to spread self-determination after the Great War and George W. Bush's attempt to spread democracy after 9/11 is that the vehicle Wilson chose to use, the League of Nations, was fundamentally flawed as soon as America failed to join, whereas 'the Coalition of the Willing' was driven principally by the military might of the English-speaking peoples. As well as the USA and Britain, Canada provided troops for the liberation of Afghanistan and Australia provided them for Operation Enduring Freedom. 'In a world where the only alternative is the moral posturing of arthritic international organisations such as the EU or the UN' read an editorial in the largest-selling British broadsheet newspaper on the sixtieth anniversary of D-Day, 'the transatlantic partnership is the only force that can still offer freedom to distant lands. 'Then, as now, the Atlantic alliance in arms is an awesome thing.'[27]

Nor was it true that George W. Bush had somehow invented a doctrine of 'the pre-emptive strike', as has been alleged. If the threat to their interests was serious enough, the English-speaking peoples have long been willing to strike first. In 1801, George Canning did not wait for the Danish Navy to be used against Britain by Napoleon, but ordered Admiral Parker and his second-in-command Vice-Admiral Nelson to attack it at Copenhagen.[28] The Germans did not directly attack the English-speaking peoples in either 1914 or 1939, but both times Britain declared war against them first. Churchill pre-emptively bombarded the Outer Dardanelles Forts in 1914 two days before Britain declared war against the Ottoman Empire. Similarly, France had been Britain's ally until her armistice in mid-June 1940 and was not an enemy belligerent after it, but in early July Churchill ordered the sinking of the French Fleet at Oran. He thought it safer to shoot first and answer Prime Minister's Questions later, and the House of Commons rose as one man to cheer him for pre-emptively keeping French capital ships like the *Richelieu* and the *Jean Bart* out of the hands of Admiral Raeder.

A political leader of the English-speaking peoples in the perilous twenty-first century has higher responsibilities than to outdated precepts based on obsolete concepts of strategy. Since the Treaty of Westphalia of 1648, nation states have been the basic entities of the international system, but modern terrorism respects no borders. Today, it is better for the English-speaking peoples to be safe than to be ethically superior with regard to international law (although Article 51 of the UN Charter *does* anyhow allow the right of pre-emptive self-defence under certain circumstances, codifying the customary law that had been in being since the Canadian Caroline case of 1837). If a pre-emptive attack on Al-Queda bases in Afghanistan under the Clinton Presidency would have prevented the 9/11 outrage, it would have been justifiable under a precept that is greater than the whole panoply of international law – the basic right to self-protection. As Enoch Powell pointed out during the Falklands crisis, that right was 'inherent in us' and it existed 'long before the United Nations was ever thought of'. (Indeed, long before international law was ever thought of either, for that matter or the Treaty of Westphalia.)

Enemy powers have not been deterred from attacking Pearl Harbor, South Korea, the Falkland Islands or Kuwait because of international law; all that such rules have done is to hamstring the English-speaking peoples, but never their unscrupulous foes. As far back as 1996, Margaret Thatcher warned that America and her allies would have to deal with 'the proliferation of weapons of mass destruction ... by pre-emptive means', and she was right. Although Saddam did not turn out to have WMD, the English-speaking peoples must at the very least be absolutely certain he could never acquire them. Just as generals tend always to be ready to fight the last war rather than the next one, so international law covers the exigencies of the

Cold War, rather than the nihilistic, high-tech, stateless terrorism that characterises the present one.

The invasion of Afghanistan was undertaken full in the knowledge that the country's terrain made it legendarily difficult to govern. Foreigners had attempted it since the reign of Alexander the Great in the fourth century BC. 'Even Alexander's hold had been fleeting,' records the historian Ben McIntyre. 'Macedonian, Mogul, Persian, Russian, British and Soviet armies had all tried, and failed, to control the Afghan tribes.'[29] What made it any more likely that the English-speaking peoples' expeditionary force – including contingents from America, Britain and Canada – would succeed where so many others had failed? 'Of the five royal descendants of Dost Mohammed Khan's tribe to rule Afghanistan in the twentieth century,' relates McIntyre, 'three were assassinated and two were forced into exile.' The last was Zahir Shah, who had become king aged eighteen after he had witnessed his father's assassination in 1933. (He ruled wisely and introduced freedom of speech and voting rights for women, before being ousted in 1973 when he was on holiday in Italy.)

As notorious as Afghanistan's political instability was the viciousness of her power struggles. When the Soviet Union had been forced by the US-backed mujahadeen to quit Afghanistan in 1990 – after 50,000 Russians and one million Afghans had been killed – their puppet ruler Mohammed Najibullah unwisely stayed on in Kabul to continue to fight. After taking sanctuary in the United Nations' compound as the enemy closed in on the capital in 1995, he was captured, castrated, and his body was dragged around the city behind a truck and then exhibited upside down in the Kabul bazaar.

The Stars and Stripes had flown over part of Afghanistan once before in history, in 1839 when the Chester County, Pennsylvania-born Josiah Harlan had unfurled it at the start of his short-lived personal rule there. 'Relying on an alloy of brass neck and steely self-confidence,' the Quaker-born adventurer braved bandits, quicksand and sixteen-foot crocodiles to carve out an impressive fiefdom there. He put his success down to his nationality. 'Over the principal tent, a few feet above the apex,' Harlan recalled many years later,

> the American flag displayed its stars and stripes, flickering in the quietly drifting breeze. ... In the midst of that wild landscape, the flag of America seemed a dreamy illusion of the imagination, but it was the harbinger of enterprise which distance, space and time had not appalled, for the undaunted sons of Columbia are second to no people in the pursuit of adventure wherever the world is trodden by man.[30]

The 2001 campaign in Afghanistan was successful; American, British and Australian special forces, aided by dominant American air-power and the enlistment of the anti-Taliban Northern Alliance of Afghans, quickly overthrew the Kabul Government and expelled Al-Queda from their terrorist bases and training camps in that country.[31] It was an impressive victory by the English-speaking peoples and their allies in some of the toughest terrain in the world. However, Osama bin Laden managed to escape, most probably into Northern Pakistan. Nonetheless, the continued failure to capture him did at least concentrate Western minds on the fact that the War against Terror was far from over. In May 2006, the British soldier Lieutenant-General David Richards took command of the international force in Afghanistan, in charge of significant numbers of American troops, thereby exploding another myth about US insistence on exercising military control at all times.

The first Muslim Middle Eastern country in history to replace its government through a free election was Turkey in 1950. Unfortunately, it was also one of the last. Yet on Sunday, 18 September 2005, millions of Afghans braved Taliban threats in order to vote in the country's first parliamentary elections in over thirty years. The polling for provincial councils as well as the Wolesi Jirga (lower house) in Kabul was hailed by President Hamid Karzai, who said, 'We are proud of this day; we are proud of our people,' even though the election strengthened the opposition parties.

Although twenty-two people were killed by the Taliban in the forty-eight hours prior to the elections, turn-out was high. As Ahmed Rashid, the author of the book *Taliban,* wrote the next day, 'Stories of electoral heroism are as moving as the sacrifices made by the Afghans while fighting the Soviet Union and the Taliban. Hundreds of women defied custom to stand and campaign in a predominantly male environment.'[32] No fewer than 5,800 women put themselves forward for the Wolesi Jirga, a quarter of the seats of which were reserved for them. The return of democracy to Afghanistan after three decades was a fine achievement of the English-speaking peoples, protecting that country from Al-Queda's re-infestation.

By late August 2002, there were enough US forces stationed on the Kuwaiti border with Iraq to effect a successful invasion once the order was given. Yet it took another seven months for that to happen, since the Bush Administration rashly decided to exhaust every possible avenue in order to give Saddam a chance to back down, and hopefully to leave Iraq. In order to help Tony Blair politically, placate international opposition to the coming war and perhaps also to avoid the conflict altogether, the US pursued a policy that in fact only had the effect of expanding the peace movement, emboldening French, German and Russian opposition to the war and allowing time for Saddam to put in place elaborate plans for insurgency operations once the initial stage of the campaign was lost. Money and arms were stockpiled during those months that were to prove invaluable later.

The United Nations was not merely ineffective, as the League of Nations had been before the Second World War, but downright obstructive and – like many other unaccountable bureaucracies in history – grossly corrupt. 'With the demeaning behaviour demanded of the United Nations to try to get the Iraq resolution through in early 2003 by trying to outbid the French to get the vote of Cameroon on the UN Security Council,' concluded Professor Deepak Lal, 'no self-respecting power and certainly not one as powerful as the United States should, or is likely to, put up with this remnant of the old international order.'

An organisation that permitted totalitarian Libya to chair its Human Rights Commission and the UNSCOM-banning Iraq to chair its Disarmament Commission had clearly gone beyond parody and could not be permitted to circumscribe the foreign and defence policies of the English-speaking peoples. Nor could small undemocratic states such as Cameroon and Guinea, as well as other autocracies and kleptocracies on the Security Council, be allowed to prevent the extension of representative institutions to Iraq. The United Nations is based, as Lal points out, upon 'the anthropomorphic identification of states as persons, and the presumption of an essential harmony of interests between these equal world citizens', which is so at variance with the reality of international relations as to make the organisation almost redundant in crises, as was proved all too regularly in Bosnia, Rwanda, Somalia, Kosovo and latterly Darfur.[33] Indeed, it is possible that the United Nations actually makes such situations worse by giving the impression that something is being done when it often is not, thereby taking the pressure off the Great Powers to act. As Lord Salisbury once put it, a balcony that appears to be safe but is not is far more dangerous than having no balcony at all.

As well as giving Saddam much-needed time, the corruption of the United Nations' Oil-for-Food programme had provided him with equally essential Western currency. The Security Council handled around $64 billion from the programme's inception in 1996 until it was wound up after the 2003 war. Medicines and other supplies intended for the Iraqi people were routinely exported out of the country and sold on the international black market, while the genuine sufferings of the Iraqi people were blamed on the UN and US sanctions by the Ba'athist regime. Somehow during that time, the Iraqi regime managed to skim off over $1.8 billion in illegal revenues, just over half of it from smuggling outside the UN scheme.[34]

In effect, through internal UN corruption, Saddam was able to use the United Nations as a giant money-laundering scheme. Long after the pro-gramme came to an end, the General Accounting Office, the investigative arm of the US Congress, estimated that Saddam made vast amounts on kickbacks from international companies working within the scheme, despite it being run by UN officials and monitored by a Security Council sub-committee. US officials discovered that Iraq charged an illegal 'surcharge' of between ten and thirty-five cents on every barrel of oil that it sold within the UN scheme. It also demanded a 10% 'after-sale service fee' from firms selling humanitarian goods to the country under the programme.[35]

Furthermore, at least $1.1 billion was paid directly to people at the UN to cover the costs of administering the scheme, a 2.2% commission approved by the Security Council, for which no reliable audits were carried out nor accounts submitted. Claude Hankes-Drielsma, an advisor to the Iraqi governing council, testified to the House of Representatives Committee on Government Reform in April 2004 that tracking that money had been 'key' to untangling the corruption scandal, and that the programme 'provided Saddam Hussein with a convenient vehicle through which he bought support internationally by bribing'. Files in the Oil Ministry in Baghdad contained 'memorandums of understanding' that suggested that Saddam could decide which UN officials operated within Iraq. The person who was in overall charge of organising the Iraqi end of the operation was his Foreign Minister, Tariq Aziz.

In a separate abuse, Iraq's suppliers overvalued goods shipped into the country and then paid kickbacks to the Iraqi regime, providing it with hard currency. 'Thousands of tons of food delivered under the UN programme were later revealed to be rotten, and many of the medicines – particularly those imported from Russia – were found to be out of date.'[36] The man appointed to head the official investigation, Paul Volcker, the former Chairman of the Federal Reserve Bank, had the Security Council's backing but no powers to compel witnesses to testify; he also had to rely on the co-operation of foreign governments, UN staff and former Saddam regime members, which was not always forthcoming. Nonetheless, his 623-page report was damning and found that of the 4,758 companies involved in the programme, kickbacks were paid in connection with humanitarian aid contracts of 2,265 of them and oil surcharges were paid in connection with the contracts of at least 148. This was larceny on a vast scale.

Vouchers were also given by the Saddam regime to prominent people outside Iraq entitling them to purchase quantities of Iraqi crude oil; these vouchers were themselves tradeable. Recipients included the French former Interior Minister, Charles Pasqua; the head of the Liberal Democrat Party in Russia, Vladimir Zhironovsky; several Middle Eastern politicians; Russian Communist Party officials; even a Swiss Catholic priest, who put the profits in his Vatican bank account. Roberto Formigoni, president of the Lombardy region of Italy, received oil rights over twenty-seven million barrels, recorded as 'special requests for Italy', the Volcker Report stated. French anti-war campaigners also received allocations. 'The abuses were widespread,' reported the *Sydney Morning Herald* when the Report was published in October 2005. 'Kickbacks on humanitarian goods were traced to companies or individuals from sixty-six countries, while payments of surcharges were made by entities from forty countries.'[37]

It was true that in the 1980s the West did much to arm and aid Saddam's Iraq, when he was seen as a useful buffer against the ambitions of Iran. The laws of *realpolitik*, which have governed international relations since the Treaty of Westphalia, require countries to conform to the dictum that 'my enemy's enemy is my friend'. Even the generally severely anti-communist Winston Churchill embraced the USSR the moment Hitler invaded her in 1941. The laws of Nature decree that all living entities alter, adapt, develop, mature, collapse and die over time, and relations between states are no different. 'Men are very apt to run into extremes; hatred to England may carry some into excessive confidence in France,' George Washington wrote to Henry Laurens, President of the Continental Congress in November 1778. 'I am heartily disposed to entertain the most favourable sentiments of our new ally and to cherish them in others to a reasonable degree; but it is a maxim founded on the universal experience of mankind, that no nation is to be trusted farther than it is bound by its interest; and no prudent statesman or politician will venture to depart from it.'[38] These words of Washington's have continuing relevance because they covered unchanging principles, unlike his Farewell Address.

When Ayatollah Khomeini's Iran was the principal enemy in the Middle East, it made perfect sense to support her mortal enemy Iraq. When Saddam dropped his pro-Western stance for an aggressively anti-Western one, however, it made just as much sense to end that support. A similar case can be made for the decision of the Carter and Reagan Administrations to arm the mujahadeen guerrillas in Afghanistan with Stinger missiles after the Soviet invasion there. 'We have no eternal allies and we have no perpetual enemies,' Lord Palmerston told the House of Commons in March 1848. 'Our interests are eternal and perpetual, and those interests it is our duty to follow.' Just as Stalin was the West's ally in 1945 but its antagonist by 1948, so Saddam's Iraq stepped firmly into the enemy camp with his invasion of Kuwait in 1990. It was, as so often with the English-speaking peoples' enemies in the past, entirely his choice. Yet for the Left to claim that because Iraq was once an ally, it was somehow illegitimate of the English-speaking peoples to invade her years later under totally different circumstances showed staggering naïveté.

Equally naïve was the argument that Saddam and the Taliban should not have been overthrown because the West was not also willing to go to war against other dictatorships, such as those of Burma, Zimbabwe and North Korea. That democracies cannot be installed across all the globe by force did not make it illegitimate to install two in the Middle East, especially once 9/11 had focused the American public's attention on the threat emanating from that region. As Tony Blair told the former editor of *The Times* Peter Stothard on 13 March 2003, 'What amazes me is how many people are happy for Saddam to stay. They ask why we don't get rid of Mugabe, why not the Burmese lot? . . . I don't because I can't, but when you can, you should.'[39]

Saddam could have taken comfort on 15 February 2003, when huge demonstrations against the forthcoming war were held in London and across Europe. The demonstration in Rome of three million people is believed by the *Guinness Book of World Records* to be the largest political gathering in history. By thus demonstrating to Saddam the deep divisions in the West over military action, these marches and speeches made it correspondingly less likely that he might back down at the eleventh hour. This great parade of European conscience therefore, incredibly self-indulgently, made war even more likely than it already was.

Just as the Korean and Vietnam Wars had seen Western apologists for the North Korean and North Vietnamese Governments, so in January 1994 the then Labour MP George Galloway had visited Saddam Hussein in Baghdad and told him, on Iraqi television, 'I salute your courage, your strength, your indefatigability. And I want you to know that we are with you' – adding, in Arabic – 'until victory, until victory, until Jerusalem.' He also told his 'Excellency' that there were Palestinian families 'who [named] their newborn sons Saddam'.[40] As earlier chapters have shown in the cases of Beatrice Webb, Wilfred Burchett and Jane Fonda in earlier conflicts, the English-speaking peoples have always produced individuals willing to propagandise for totalitarian dictatorships.

The moment that President Bush came to authorise the invasion of Iraq was conducted with the seemly behaviour expected of such a serious event. As Donald Rumsfeld's deputy Paul Wolfowitz reminisced two years later,

> I think someone once said that decision making is usually trying to choose the least crappy of the various alternatives. I really admire people like President Bush who are good at it. I was in the Oval Office the day he signed the executive order to invade Iraq and I know how painful that was. He actually went out in the Rose Garden to be alone for a little while. It's hard to imagine how hard that was.[41]

In their cynicism and ideological opposition to America's wars, anti-Bush propagandists such as Michael Moore will simply not acknowledge that decisions such as Lyndon Johnson's to escalate the Vietnam War or Richard Nixon's to bomb Cambodia or Ronald Reagan's to invade Grenada do weigh heavily with presidents, and Bush's decision over Iraq was no different. Presidents who genuinely admire the military – and none did so more than Johnson, Nixon, Reagan and Bush – are the least likely to order soldiers into mortal combat. Similarly, the more God-fearing the president, the more conscious he is likely to be of eventual judgment before a far more august tribunal than simply the US Congress or even the bar of History.

'The duty of a politician', said the British nineteenth-century historian Bishop Mandell Creighton, 'is to educate the people, not to obey them.' In the debate on military action in the House of Commons on 18 March 2003, Tony Blair said that terrorism represented 'a fundamental assault on our way of life'.[42] He spent relatively little time justifying the forthcoming war on humanitarian grounds, concentrating instead on Saddam Hussein's repeated violations of the UN Security Council resolutions, and won the vote by 396 to 217 votes, a majority of 179 in a house of 659 seats, despite 139 members of his Labour Party voting for a rebel amendment. The large majority helped to remind many Americans that when the stakes are high and allies are needed for a major and dangerous operation, the United States cannot count on any friend more stalwart than the other nations of the English-speaking peoples, particularly Great Britain and Australia.

(Interestingly, formal war was not declared against Iraq in 2003, any more than it had been against Argentina in 1982, Egypt in 1956 [which was always designated a 'police action'] or North Korea in 1950. A state of war brings formal obligations on both sides, and the last time that Britain declared war was against Japan's ally, the Kingdom of Siam, in 1942.)

As in the Gulf conflict – another non-'war' – dire predictions were made about the disasters that were about to befall the coalition forces during the initial stages of the conflict. (There had also been incorrect estimations that hundreds of thousands of Afghan civilians would die in the winter of 2001/2 as a result of being caught in between the coalition forces and Taliban guerrillas fighting to the last in the hills above Kabul.) The British journalist Robert Fisk compared the defences he witnessed being made in Baghdad to those of Stalingrad, whereas in fact, as *The Times* reporter Richard Beeston has put it, 'The Iraqi capital fell in the short time that it took the first American armoured division column to drive into the city from the airport.' The Iraqi army in the field was routed in twenty-one days. Once again in the history of the English-speaking peoples, air power had been central to victory. Within days of the coalition attack not a single Iraqi aircraft was to be seen in the air.

The coalition commander was US General Tommy Franks. In his auto-biography, the use of strategic deception was revealed to have yet again been employed as a key element to victory. Rather like Operations Fortitude North and Fortitude South before D-Day, the intention was to persuade the enemy that the main thrust of the attack was due to take place hundreds of miles to the north, thereby forcing him to keep significant forces far from the place where it was really intended. It worked perfectly. Saddam was lulled by a double agent codenamed April Fool into believing that the coalition was 'planning to build up only a portion of its ground force in Kuwait, while preparing a major airborne assault into northern Iraq from above Tikrit to the oilfields above the city of Kirkuk. Helicopter-borne air-assault forces would then reinforce the paratroopers. Then, once several airstrips were secured, C-17 transports would deliver tanks and Bradley fighting vehicles to join them.' In the event, reconnaissance imagery showed how, 'Despite our sizeable build-up of forces in Kuwait to the south, Saddam's Republican Guard and regular army divisions had not moved significantly from their northerly position – no doubt waiting for an assault that would never come.'[43] Post-war interrogations of Iraqis confirmed that this was indeed the case.

The way in which the English-speaking fighting men respected their adversaries, from 'the Fuzzy Wuzzy' in the Sudan, to 'Johnny Boer', to 'Fritz the Hun', to the Argentine pilots in the Falklands, was also echoed by General Franks, who described Osama bin Laden as not merely 'a deadly adversary' but also 'a worthy, bold commander of dedicated and capable forces'.

One of the major criticisms of the US administrator Paul Bremer's running of Iraq after the fall of Saddam was that he disbanded the Iraqi army, elements of which then sought re-employment as *fedayeen* militiamen in the insurgency. But as Jonathan Foreman, a *New York Post* journalist embedded with the 4th battalion of the 64th Armored Regiment of the US Army, pointed out,

> Anyone who was there in April 2003 (and who wasn't doing their reporting from a hotel bar) could tell you, there was no Iraqi army for Bremer to disband. The Iraqi army had disbanded itself. It had ceased to exist. . . . Most of the Iraqis had simply doffed their uniforms and gone home between 21 March and 15 April. The truth is that when Bremer ordered the disbanding of the old Iraqi army on 23 May, he was merely formalising a state of affairs that already existed.[44]

Foreman went on to argue that co-opting the Iraqi army to police the liberated cities 'would have risked disaster on every level', since former Ba'athist officers made up much of the resistance. 'Any use by the Coalition of Saddam's armed forces – the forces that put three hundred thousand Iraqi civilians into mass graves – would instantly have alienated both the Shia and the Kurds. . . . Indeed if you're going to employ Saddam's savage, brutal, coercive machinery to maintain order in Iraq, then why overthrow the regime at all?' That same month, thousands of those bereaved by Saddam started to uncover the mass graves of their relatives murdered by his regime.

These graves, containing the corpses of Saddam' s victims over three decades, continued to be discovered at regular intervals, and are still being uncovered at the time of writing.

Although plenty of mass graves were discovered after Iraq's liberation, no Weapons of Mass Destruction were. When US Army historians had the opportunity to question Saddam's senior generals, Ba'ath party officials and advisors about what had happened, a situation rich in irony was uncovered. As the *Economist* reported on 18 March 2006,

> Some of the ruling circle never stopped believing, even after the war, that Iraq had WMD, even though Saddam himself knew otherwise. When he revealed the truth to members of his Revolutionary Command Council not long before the war, their morale slumped. But he refused a suggestion to make the truth clear to the wider world on the ground that his presumed possession of WMD was a form of deterrence.

Of course, far from being a form of deterrence, the Americans' genuine belief that Saddam possessed and might use WMD, and was busily creating more and yet deadlier ones, was one of the reasons they decided to overthrow him.

On Thursday, 2 May, President George W. Bush landed onto the deck of the aircraft carrier USS *Abraham Lincoln* in the co-pilot's seat of a Navy S-3B Viking turbofan jet. The aircraft made a 'tailhook' landing at 150 mph, coming to a complete stop in less than 400 feet, emphasising yet again the undoubted superiority of US aero-technology. The President had taken a turn at the controls during the flight, which had to be made by plane since the carrier was too far from land for helicopters.

In declaring the end of the major combat operations phase of the war, Bush was filmed with a large sign featuring the Stars and Stripes and the words 'Mission Accomplished' behind him. 'The banner was a Navy idea,' explained its spokesman Commander Conrad Chun. 'It signified the successful completion of the ship's deployment.' (The *Lincoln* had been deployed for 290 days during the Afghanistan and Iraq campaigns, longer than any other nuclear-powered aircraft carrier in history.)

The President made it very clear that he did not believe the mission of 'the Coalition of the Willing' – as opposed to that of the *Lincoln* – had yet been accomplished, however, saying,

> We have difficult work to do in Iraq. We're bringing order to parts of that country that remain dangerous. We're pursuing and finding leaders of the old regime, who will be held to account for their crimes. ... We're helping to rebuild Iraq, where the dictator built palaces for himself, instead of hospitals and schools. And we will stand with the new leaders of Iraq as they establish a government of, by, and for the Iraqi people. The transition from dictatorship to democracy will take time, but it is worth every effort. Our coalition will stay until our work is done. Then we will leave, and we will leave behind a free Iraq.

Brave words, and all the braver because it had already become clear that various anti-US and anti-democratic forces had coalesced to fight an insurgency war designed to confound his hopes for 'a free Iraq'.

*

On 13 December 2003, Saddam Hussein was captured alive, hauled out of a hole in the ground in which he had been hiding since the defeat of his armies. Those who had predicted that he would disappear in the same way that Osama bin Laden had were shown to be wrong. Although bin Laden himself has evaded capture – at least up until the time of writing – so too had Paul Kruger and Kaiser Wilhelm II in earlier wars, but neither made the eventual victories over them any less complete.

'If we were a true empire,' said US Vice-President Dick Cheney in January 2004, 'we would currently preside over a much greater piece of the earth's surface than we currently do. That's not the way we operate.' In his State of the Union speech the same month, President Bush agreed: 'We have no desire to dominate, no ambitions of empire.' Instead, by that month the United States had spent over $100 billion rebuilding Iraq, a fantastically high figure and testament both to that country's generosity and her sense of international responsibility, but also to the fact that in the modern world only the English-speaking peoples have the necessary wealth – let alone the will – to rid countries of their tyrants. Vast sums were spent in the past on the Hoover Moratorium, Lend-Lease, the Marshall Plan, the Berlin airlift, re-supplying Israel during the Yom Kippur War and any number of other fantastically expensive US initiatives; bringing representative institutions to Iraq would now be no different.

The cost to the United States of fighting the Iraq War was approximately $48 billion, which seems like a significant amount, yet when one takes into account the $13 billion per annum it was already costing to confront Saddam, it represented only four years' containment costs. Furthermore, the money allocated by the Bush Administration to the occupation and reconstruction of both Afghanistan and Iraq represented a mere 0.8% of US GDP.[45] The figure is so low partly because the GDP of the United States is so astonishingly high; in 2002, America accounted for no less than 31% of the entire global output. The American economy was two-and-a-half times the size of Japan's, eight-and-a-half times China's and thirty times larger than Russia's.

The Iraq War was also one of the cheapest engagements of its kind in the past century for the United Kingdom. By late September 2005, the entire conflict had only cost the British taxpayer £3.1 billion, less than 10% of British defence spending in the single year 2004, at a time when defence spending had fallen to 2.3% of GDP from a Cold War figure of 5%.[46] With total government spending at over £200 billion per annum, intervention in Iraq cost only £910 million for 2004, less than half of 1% of the total. Rarely can the British taxpayer have received such excellent value for money in the public services.

*

Bombings carried out in Madrid by Al-Queda on 11 March 2004 killed 192 people, injured 1,500 and resulted, after an incompetent response by the Spanish Government which initially blamed the Basque terrorist group ETA, in a disastrous change of ministry at the elections. The incoming socialist government announced that it would withdraw Spain's troops from Iraq. The terrorists' response to this attempted appeasement was merely to plant a 22 lb bomb on the railway track between Madrid and Seville, which was fortunately discovered on 2 April when part of the 430-foot cable was spotted. The nihilism inherent in Al-Queda's programme was evident from the statement it made at the time of the Madrid bombings: 'We choose death while you choose life.'

The publication of photographs of piles of naked prisoners simulating sex, hooded men with electrode clips attached to their arms, and grinning American servicemen and women at Abu Ghraib prison revealed serious abuses there, although nothing like the murder and torture common in any number of contemporaneous Middle Eastern political gaols. There followed no fewer than four official reports into what had taken place at Abu Ghraib, all of which concluded that the sadism demonstrated by the military policemen was not condoned by either any US Army doctrine or any orders from superiors; indeed, they went against everything that was in the interrogation rule-book. The incredibly extensive official documentation accompanying the reports was in itself, as the political commentator Alasdair Palmer has pointed out, 'astonishing testament to the legalistic nature of the American Government and its willingness to open itself to public scrutiny, and to that extent it is good evidence that the Bush Administration has not sunk into the kind of lawless dictatorship that some of its more hysterical opponents claim'.

One such might be the Senate Minority Whip Richard J. Durbin, who likened some US troops' misbehaviour at Abu Ghraib to the Nazis, the Soviet gulag and Pol Pot's Cambodian killing fields.[47] Speaking on the record, a senior French minister called the American President a 'serial killer' and a German minister compared the American leader to Adolf Hitler. The Australian journalist John Pilger told the readers of Britain's *Daily Mirror* in January 2003 that, 'The current American elite is the Third Reich of our times', and elsewhere claimed that, 'The Americans view Iraqis as *Untermenschen,* a term that Hitler used in *Mein Kampf* to describe Jews, Romanies and Slavs as subhumans'.[48] Nelson Mandela meanwhile accused President Bush of 'wanting to plunge the world into a Holocaust'. Not to be outdone, the British actor Corin Redgrave has suggested that the President might even be worse than Hitler, as 'even the Nazis allowed the Red Cross to visit their prisoners'.[49] (In fact, the International Red Cross has full access to detainees at all times at Guantanamo Bay in Cuba, and even has an office there. By contrast, the

United Nations refused an invitation to visit, yet nonetheless published a report claiming that torture took place there. It was the first time that force-feeding designed to save hunger-strikers' lives has been designated as 'torture'.)

The collapse of discipline at Abu Ghraib was a result of chronic manpower shortage due to the unexpectedly strong post-war Saddamite insurgency, and the fact that some of the military policemen involved were clearly little better than Appalachian mountain-cretins, but that does not justify comparing the scandal to the My Lai massacre in Vietnam of March 1968, as some anti-war commentators attempted to do. Neither did it justify attempting to blame Donald Rumsfeld, Richard Cheney or even President Bush for what went on there, as the veteran American journalist Seymour M. Hersh also tried to do. There has never been a war in history that has not had a seamy underside of abuse.

In January 2005, the ringleader of the Abu Ghraib abuses, Charles Grainer, was sentenced to ten years' imprisonment, having failed to produce any evidence to suggest that he was acting under orders from above. Janis Karpinski, the US Army Reserve general whose military police unit was in charge of the prison during the scandal, was demoted on a broad charge of dereliction of duty and relieved of her command. Of course that was not enough for the conspiracy theorists, who attempted to connect the highest reaches of the Pentagon and White House to the scandal, but then nothing possibly ever could be.

The detention without trial at the US naval base of Guantanamo of suspected Taliban and Al-Queda fighters captured in Afghanistan and Iraq provided the Left with a brand new opportunity for spurious moral equivalence. The British left-wing weekly, the *New Statesman*, billed as 'Exclusive', for example, an article entitled 'America's Gulag: Bush's secret torture network of prisons and planes', featuring an American flag on a Soviet-style concentration camp watch-tower and Bush wearing the lapels of a Soviet camp-guard's uniform.[50] As seems obligatory in articles of this sort, the capital 'R' in America was reversed. 'Just like Solzhenitsyn's system, the American archipelago operates as a secret network that remains largely unseen by the world,' the article stated. 'Guantanamo is the Gulag of our time,' agreed the general secretary of Amnesty International, Irene Khan, which if true proves how much better our time is than any earlier ones, since the Soviet gulag was responsible for six million deaths, whereas no-one was killed at Guantanamo.

Capturing and detaining enemy combatants has been the practice of the United States, Great Britain and their allies in every modern war. Under the law of war, there is no requirement that a detaining power charge enemy combatants with crimes or give them access to lawyers. The English-speaking peoples certainly did not do so in the First or Second World Wars. Under American law, the authority to detain enemy combatants exists independently of the judicial or criminal law system. It is rather a function of the President's role as Commander-in-Chief under the Constitution. Since Al-Queda is a terrorist organisation rather than a state, and therefore neither a signatory nor covered by the Geneva Conventions, their members are not entitled to POW status. And even if they were covered by the Conventions, they would still not be considered POWs, since they do not carry weapons openly, wear uniforms, follow responsible command or comply with the laws of war, as required under Article 4.

Detainees at Guantanamo Bay are provided with shelter, clothing, the means to send and receive mail, reading materials, three meals a day that meet cultural dietary requirements, medical care, prayer beads and rugs and copies of the Koran. Over twenty senators, 110 representatives, 150 congressional staffers and more than 1,000 American and international journalists have visited the prison, which was certainly not allowed in previous wars. Furthermore, 180 detainees have been released in the period to February 2006, at least twelve of whom returned to the fight against 'the Great Satan' America. Around 300 remain there, including self-confessed enemy combatants, terrorist trainers, recruiters, bomb-makers, would-be suicide bombers and terrorist financiers. America is right to keep them there.

Earlier presidents have resorted to extra-constitutional means when the Republic was under attack. Abraham Lincoln's policy of arresting secessionists in Maryland without trial in April 1861 forced the Supreme Court's Chief Justice, Roger B. Taney, to remind the chief executive of his presidential oath to 'take care that the laws be faithfully executed' and warned that his actions in denial of *habeas corpus* would mean that 'the people of the United States are no longer living under a government of laws'. Lincoln simply ignored Taney and kept secessionists such as John Merryman, the lieutenant of a pro-Confederate drill company, locked up in Fort McHenry in Baltimore Harbour.[51] History has forgiven Lincoln for his actions. Similarly, on 19 February 1942 the liberals' hero Franklin D. Roosevelt ordered the internment of 120,000 Americans of Japanese heritage, a policy administered by Earl Warren, later Chief Justice of the United States. By contrast, President Bush has not needed to resort to unlawful means to prosecute the War against Terror, despite the greatest of provocations.

It is quite untrue that the American neo-conservatives who initiated the Iraq War refused to accept that any mistakes were made. On 15 September 2004, Paul Wolfowitz, the former Deputy Defense Secretary, told the author Mark Bowden that in his view an Iraqi provisional government should have been established in Baghdad 'the day we got there', instead of having the US labelled as being 'an occupation authority'. As a result Al-Jazeera was able to draw (entirely spurious) parallels with Israel's post-1967 occupation of Palestinian territories. It had been the State Department that had opposed the recommendations of the Pentagon to recognise a provisional government.

Wolfowitz was straightforward in admitting with regard to the Saddamite insurgency that, 'I think most people underestimated how tough these bastards are. ... The heart of the problem is that 35 years of raping and murdering and torturing created a hard core that is incredibly brutal and a population that it incredibly scared: one relatively easy to intimidate.' Although during the pre-war build-up, 'We also had report after report of Iraqi brigade and division commanders who were promising to bring their units over to our side, I don't think there was a single such event that actually took place.'

With Saddam still describing himself as president of Iraq, and 'his cronies' having access to millions of dollars in Syrian, Lebanese and Jordanian bank accounts, Wolfowitz drew a telling comparison with 1945, sayings, 'It's as though the Nazis, after their defeat, still controlled Nuremberg and had bank accounts and sanctuary in Switzerland and co-operation from some other country like Iran.' When asked whether he had believed that Saddam had WMD stockpiles, he answered:

> What really bothered me was biological weapons and we know they made them. They were given a chance to come clean under 1441. We caught them lying on the declarations on not insignificant things – mostly on the missiles they were working on and the UAVs [unmanned aerial vehicles]. And there was lots of evidence of obstructing inspectors and hiding things. You had a very dangerous character who played with terrorists, who had regularly declared hostile intentions towards us and toward our allies in the Persian Gulf, who definitely had a capacity to make these weapons and was extremely dangerous, and much more dangerous in the light of September 11 than before. And that's where September 11 changed the calculation. I think it would have been irresponsible to leave him alone.[52]

On 28 October 2004, Wolfowitz also showed how the now-standard accusation that the United States ought to have flooded Iraq with troops is also not one that stands up to much analysis. As Wolfowitz pointed out, it was 'actionable intelligence' that was the problem, not numbers of coalition troops. The supposed lack of available troops was the criticism that was made during the Boer War, Gallipoli and Vietnam, even though in fact each place was awash with troops for much of those conflicts and the real problems were quite different. Moreover, as Wolfowitz admitted, 'If you have more troops, that creates a new set of problems. You have a heavier American footprint, which means alienating more people.' No war in history has been fought perfectly, and the counter-insurgency operation in Iraq has been fought no worse than many. Certainly, great courage has been shown by troops on the ground.

*

The Australian election results of October 2004 saw a landslide for John Howard's Liberal-National Party coalition, defeating Mark Latham's Labor Party and winning eighty-six seats in the House against Labor's sixty. Although the Government's strong record on the economy was the most important domestic issue, Iraq also played an important role. Whereas Howard – very ably assisted by his Foreign Minister Alexander Downer – had spoken in defence of President Bush and the war, Latham had promised to withdraw Australian troops by Christmas 2004 if elected. It was thus a timely help to President Bush, who three weeks before his own election would have been badly damaged if an English-speaking country had announced that it was pulling out of the coalition. As the British journalist Charles Moore wrote of the subsequent treble victories of Bush, Howard and Blair, 'Anglo-Saxon political culture still has enough self-confidence not to fear leadership in war, but to see it as a necessary attribute of a robust democracy. Which is a good thing.'

The re-election of President Bush in November 2004 was unsurprising in the light of the fact that no sitting president had ever been defeated in an election during a major war. To cashier a commander-in-chief mid-struggle would give succour to the enemy, which is something the American people had a patriotic reluctance to do. Given the widespread domestic opposition to the Iraq War, the scale of Bush's victory was remarkable. For the first time in US history, all the Southern states voted Republican. Bush was re-elected with 62.04 million votes against the Democrat contender John Kerry's 59.03 million, more than the entire 59.9 million population of France. Bush became the first US president since 1988 to win over 50% of the popular vote, on a turn-out of 60.3% of those eligible to vote – the highest since 1968.

On Thursday, 5 May 2005, the Labour Government in Britain was re-elected in an unprecedented third landslide victory. The two pro-war parties, Labour and the Conservatives, polled nearly 70% of the total votes cast between them.

Two months later, on Thursday, 7 July 2005, four suicide-murderers exploded devices at underground stations around London and on a No. 30 bus in Tavistock Square that killed fifty-two innocents as well as the bombers themselves. 'I can tell you now that you will fail in your long-term objectives to destroy our free society,' were the defiant words of Ken Livingstone, the left-wing Mayor of London and no ally of Tony Blair in the War against Terror. 'In the days that follow, look at our airports and seaports, and even after your cowardly attacks, you will still see people from around the world coming here to achieve their dreams. Whatever you do, however many you kill, you will fail.' There was no mass panic. The terrorists responsible for the attacks, who called themselves the Secret Group of Al-Queda's Jihad in Europe, boasted that, 'Here is Britain now burning with fear and terror.' Anyone who was present in the capital at the time knows that to be utterly untrue. The city that survived the Blitz and the V-weapons campaign showed disgusted resignation and mourned its dead, but did not consider bowing, any more than it had in earlier conflicts.

By late January 2006, the United States had lost 2,237 soldiers killed in Iraq, less than 4% of those who died in either Korea or Vietnam.[53] Great Britain had lost 100 dead, of whom more than a quarter had died in traffic accidents or training. 'The number [of British soldiers] killed in combat over the past year has been twelve,' reported the *Spectator* in February 2006, 'far lower than even the quietest years in Northern Ireland.' Meanwhile, fewer US troops had died in Afghanistan in the twelve months to February 2006 than in motorbike crashes in the continental USA. Seen in their historical perspective, therefore, the casualty figures were astonishingly low. Single engagements like the battle of Belleau Wood in the Great War or taking Tarawa Island in the Second World War had cost the US more fatalities than the entire Iraq War to date.

Furthermore, as a proportion of the total number of Americans, only 0.008% died bringing democracy to important parts of the Middle East in 2003-5. So, for all the sadness and tragedy of each American and British life lost, in the wider context Iraq ought to be seen as another very significant victory of the English-speaking peoples over yet another variety of fascism. Similarly, the number of Iraqis killed, variously estimated at around 25,000 to 30,000, needs to be seen in the context of the report of the Iraq Human Rights Centre in Kadhimiya, which calculated in 2004 that 'more than seventy thousand people would have died in the last year if Saddam had still been in charge'. United Nations figures show how wars in the second half of the twentieth century averaged 30,000 deaths globally. The death toll in Iraq was therefore below average up to January 2006, however much the media might have done its best to imply otherwise.

Al-Queda was wrong to assume from the experiences of Beirut in 1983 and Somalia in 1993 that Americans would refuse to tolerate substantial levels of casualties in Afghanistan and Iraq. As Michael Barone, co-author of *The Almanac of American Politics,* has pointed out, 'Americans will tolerate very high levels of military casualties if they believe that their leaders are on the road to victory. They tolerated them in Vietnam from 1965 to 1968, and ceased to do so only when their leaders seemed no longer to be seeking to win. Polls show that some of Eugene McCarthy's voters in New Hampshire wanted the war waged more vigorously, not less.' The two years that saw the highest numbers of casualties in American history – 1864 and 1944 – also witnessed the incumbent commanders-in-chief re-elected. As Barone extrapolated, 'After

Sherman marched from Atlanta and the GIs landed in Normandy, voters saw that American forces were headed for victory.'

In the Iraqi elections of December 2005, full democracy – rather than merely representative institutions – finally arrived in Iraq. Ten million Iraqis braved threats from the undiminished insurgency to record a 70% turn-out. The English-speaking peoples had written the latest chapter in their long history of bringing liberty to places which had previously known fascism of one form or another, but hopefully not the last. As Tony Blair had told a meeting of the Parliamentary Labour Party in February 2003, 'People say you are doing this because the Americans are telling you to do it. I keep telling them that it's worse than that. I believe in it.'[54]

Conclusion

'We might have been a free and great people together.'

Thomas Jefferson, 1776[1]

'I am here to tell you that, whatever form your system of world security may take, however the nations are grouped and ranged, whatever derogations are made from national sovereignty for the sake of the larger synthesis, nothing will work soundly or for long without the united effort of the British and American peoples. If we are together nothing is impossible. If we are divided all will fail. I therefore preach continually the doctrine of fraternal association of our two peoples ... for the sake of service to Mankind and for the honour that comes to those who faithfully serve great causes.'

Winston Churchill, Harvard University, 6 September 1943

'In today's wars, there are no morals, and it is clear that Mankind has descended to the lowest degrees of decadence and oppression.'

Osama bin Laden, May 1998

'The descendants of the 17th-century commonwealth, the mostly Protestant diaspora of English-speaking peoples, will always see the world through particular eyes.' Sir Simon Jenkins, *The Times*, March 2004

'September the eleventh was for me a wake-up call. Do you know what I think the problem is? That a lot of the world woke up for a short time and then turned over and went back to sleep.' Tony Blair, July 2005[2]

The Italians are rightly proud of the Cæsars and preserve the memory and relics of the Roman Empire with diligence and love. The Greeks venerate Periclean Athens as much as the Macedonians do the achievements of Alexander the Great. France's moment of *la Gloire* under Napoleon is today burnished even by French republicans, just as the greatness of King Philip II is admired by Spaniards. The palaces of Peter the Great and Catherine the

Great are kept pristine by Russians. Egyptians still feel proud of the New Kingdom's Pharaohs of the eighteenth, nineteenth and twentieth dynasties. Recollection of the reign of Gustavus Adolphus is uplifting for Swedes, and the highest decoration in Uzbekistan is the Order of Temur, named after the conqueror known to Westerners as Tamerlaine. The Portuguese esteem Prince Henry the Navigator and the Austrians their great Hapsburg Emperor, Charles V. A toast to 'The Great Khan' (Genghis) will still – despite decades of official disapproval – have Mongolians leaping to their feet. Indeed, there is no country, race or linguistic grouping that is expected – indeed required – to feel shame about the golden moment when they occupied the limelight of World History. Except, of course, the English-speaking peoples.

The fact that first the British and then the American hegemonies have held global sway since the Industrial Revolution is perceived as the source of profound, self-evident and permanent guilt. Ever since the 1960s, academics, the Left-liberal intelligentsias, and the social and political establishments of both countries have been united in the belief that English-speaking imperialism was evil. This is bad enough for Britain, whose time in the sun has been over for half a century, but the politics of the pre-emptive cringe is even worse for modern America, which is still enjoying her moment of world primacy, yet is being enjoined on all sides to apologise for it already, long before it is even over.

It was the Athenian historian Thucydides who first thought of uniting the four distinct but successive and related conflicts between Athens and Sparta from 431 BC to 404 BC into one great Peloponnesian War, the subject of his classic narrative composition. Similarly, the four distinct but successive attacks on the security of the English-speaking peoples, by Wilhelmine Germany, the Axis powers, Soviet communism and now Islamic fundamentalism ought to be seen as one overall century-long struggle between the English-speaking peoples' democratic pluralism and fascist intolerance of different varieties.

Historians will long continue to debate precisely when the baton of world leadership passed from one great branch of the English-speaking peoples, the British Empire and Commonwealth, to the other, the American Republic, but it certainly took place some time between the launch of Operation Torch in November 1942 and D-Day in June 1944. It wasn't handed over in any formal or official sense, of course, but the leadership of the Free World that lay in Churchill's hands before Pearl Harbor was certainly held by Roosevelt three years later. The baton was not passed easily, as in a relay race, but neither was it forcibly snatched, as on most other occasions in history when one nation supplants another in the sun.

The way that the Suez crisis of 1956 italicised a power-shift that had already taken place raised an ire in Britain that has still not fully abated, yet it is naïve to hope that a world power will act against its own perceived best interests out of linguistic solidarity or a feeling of *auld lang syne* for a shared wartime past. The fact that in retrospect it was clearly in America's long-term interests to permit Britain and France to swat the nascent Arab nationalism personified by Colonel Nasser is ironic, but immaterial. The fact nonetheless remains that of all the peoples of the world who could have supplanted her, the British, Australian, Canadian, New Zealand, West Indian and Irish peoples were immensely fortunate that it was the Americans who did. The surprising phenomenon is not that the United States acted in her own perceived national interest immediately after the Second World War and at the time of Suez – any Great Power would have done the same thing – but how often over the century the genuine national interests of the English-speaking peoples have coincided; and never more so than today.

'Collaboration of the English-speaking peoples threatens no one,' wrote Churchill in 1938. 'It might safeguard all.' He was quite wrong, of course, both then and now. The collaboration of the English-speaking peoples threatened plenty of people, and still does. Just as it threatened the Axis' ambitions and subsequently the Soviets', today in very different ways Middle Eastern tyrants, Islamic fundamentalist terrorists, rogue states, world-government uni-globers, Chinese hegemonists and European federalists have every right to feel threatened by what that collaboration might still achieve in the future.

The English-speaking peoples did not invent the ideas that nonetheless made them great: the Romans invented the concept of Law, the Greeks one-freeman-one-vote democracy, the Dutch modern capitalism, the Germans Protestantism, and the French can lay some claim to the Enlightenment (albeit alongside the Scots). Added to those invaluable ideas, however, the English-speaking peoples have produced the fine practical theories behind constitutional monarchy, the Church-State divide, free speech and the separation of powers. They have managed to harness foreign modes of thought for the enormous benefit of their societies, whilst keeping their native genius for scientific, technological, labour-saving and especially military inventions.

It is emphatically not that the English-speaking peoples are inherently better or superior people that accounts for their success, therefore, but that they have perfected better systems of government, ones that have tended to increase representation and accountability while minimising jobbery, nepotism and corruption. These in turn have allowed them to achieve their full potential, while some other peoples on the planet have remained mired in authori-tarianism, totalitarianism and institutionalised larceny. The English-speaking peoples are unromantic and literal-minded, and do not dream of future utopias like French or Russian revolutionaries; instead, they root their hopes in what is tangible and tested. 'I confess myself to be a great admirer of tradition,'

Churchill told the House of Commons in March 1944. 'The longer you can look back, the farther you can look forward.'

Many – indeed most – of the English-speaking peoples' theories of government, such as the First Amendment of the US Constitution that guarantees freedom of speech and thus the ability of the media to expose corruption, or the Northcote-Trevelyan reforms that ended institutionalised corruption in the British Civil Service, were in place before 1900. Part of their genius has been rigidly to abide by the general principles of 1776 in the United States and of the 1688 Glorious Revolution in the case of most of the rest of the English-speaking peoples. That is ultimately why today their economies account for more than one-third of global GDP, despite their combined population of 335.7 million making up only 7.5% of the world's population.[3]

In the two political (though not military) defeats of the English-speaking peoples since 1900 – Britain's at Suez and America's in Vietnam – the operational side of events went relatively well from the start. Otherwise their wars tend to begin very badly indeed. In both cases the initial provocations came from abroad, with the sudden nationalisation of the Suez Canal in July 1956 and the North Vietnamese attack on USS *Maddox* in the Gulf of Tonkin on 2 August 1964. Taken together with the Spanish declaration of war against America in 1898, the Boers' declaration of war on Britain in 1899, Germany's attack on France through Belgium in 1914, the threat to America contained in the Zimmermann Telegram in 1917, Hitler's invasion of Poland in 1939, Japan's attack on Pearl Harbor and Hong Kong in 1941, the Berlin blockade of 1948, North Korea's assault on South Korea in 1950, Argentina's grabbing of the Falkland Islands in 1982, Saddam Hussein's invasion of Kuwait in 1990, and latterly the Al-Queda attacks of 9/11, an identifiable pattern emerges: that of the essentially pacific English-speaking peoples and their allies coming under sudden, unprovoked and usually lethal attack from an aggressive foe whose assaults must be militarily avenged if honour and prestige are to be secured.

The reason that prestige is so important in international affairs is not because of pride or self-importance, but because it is a tangible currency in the *realpolitik* that governs relations between states. Because the most costly wars in modern history have arisen whenever there is confusion about which is the world's pre-eminent power, anything that emphasises the true situation is good for security and stability. Today, fortunately for themselves but also for most of the rest of the world, the English-speaking peoples occupy that hegemonic place.

As the devoutly Anglican British Prime Minister, Lord Salisbury, pointed out, international affairs cannot be conducted according to the Lord's Prayer or the Sermon on the Mount. The harsh truth of *realpolitik* is that if you turn the other cheek or forgive those who trespass against you, disaster often strikes. The world is at its most peaceful when Great Powers are under no illusions as to where they stand in the global pecking order. By taking such an aggressive stance over the War against Terror since 9/11 – and especially by overthrowing the Taliban and Ba'athist regimes in 2002 and 2003 – the English-speaking peoples unmistakably demonstrated to the rest of the world that they still enjoy global hegemony. They have thus made less likely the type of clash that historically has cost the most lives in the period since 1900: a struggle between the Great Powers.

For all the evident unpopularity of the Iraq War in some circles, it has reminded the world that although the English-speaking peoples put up with a good deal of insolence and defiance from Saddam over twelve years, they would not be mocked indefinitely. The speed and ease with which Saddam Hussein's army of well over half-a-million men was defeated in a matter of three weeks by the smaller forces of the coalition in March 2003 was an object lesson in courage, professionalism and superior technology.

The coalition's willingness to stay in Iraq and fight against the post-Saddam insurgency there – while re-electing the American, Australian and British leaders in the process – was further proof to the world that it was serious about allowing Iraqis to decide their own government for the first time in over thirty years. When over ten million Iraqis voted in their general election of December 2005 – at 70% a far higher turn-out than in most Western countries, despite the threats – it was shown that democracy is as popular a concept in the Middle East as it is rare. Far from being an aberration, the foreign policy pursued by the USA, Great Britain, Australia and other countries of the English-speaking peoples since 9/11 derives from the mainstream of their historical tradition.

The English-speaking peoples are constantly berated by the Left and by churches over the levels of debt they are owed by Third World countries. One reviewer in the *Times Literary Supplement* has described such debt as 'the newest version of empire – the novel American method of maintaining world dominance by keeping the old colonies massively, permanently and irre-deemably in debt, and demanding payment in strong dollars. As an exercise in raw power, this makes even the Spanish looting of Latin America seem sophisticated.'[4] In fact, of course, the amount America receives in debt-service payments from the Third World is a minute proportion of her GDP; all loans were voluntary and therefore not a form of imperialisn. The fact that the borrowers have often wasted their money on corruption and white-elephant prestige projects can hardly be blamed on America; it is commercial banks rather than the USA herself which do the lending in most cases, but Wash-ington does provide huge amounts of debt relief each year through the Highly

Indebted Poor Countries Initiative, which it does not have to. Finally, if a country borrows in dollars – which haven't always been strong and certainly aren't at the time of writing in 2006 – it must expect to repay in either that or another currency acceptable to the lender. As so often, this critique of the USA, for all its sarcasm and aggression, fails to stand up to close examination.

Both absolute poverty and the gap between rich and poor in the United States is also often held against the country by anti-Americans, but the fact remains that the poor there are a good deal better off than the poor almost anywhere else in the world. Over 46% of America's poor – as defined by the US Government's Census Bureau – own their own homes, 72% have washing machines, 60% own microwave ovens, 92% have colour TV sets, 76% have air-conditioning and 66% own one or more cars. Two-thirds of poor households have an average of two rooms per person, and the average poor American has more living space than the average individual in Paris, London, Vienna or Athens.[5] Obesity, rather than hunger or malnutrition, is the danger for the children of America's poor, who nonetheless are growing up to be an average of one inch taller and ten pounds heavier than the GIs who stormed the Normandy beaches in the Second World War. According to *The Progress Paradox* by Gregg Easterbrook, the editor of the *New Republic* magazine, if one strips out immigration, for the nine out of ten Americans who are native-born inequality is declining, due in part to the rising affluence of African-Americans.

The hackneyed line that 'When America sneezes, the rest of the world catches a cold' also has its obverse side, that when virtuous phenomena take place in America, the rest of the world benefits. When American doctors find the cure for various diseases – as they do more than any other nation – all can celebrate. The fact that America has won far more Nobel Prizes – 270 between 1907 and 2004 – than any other country is a reflection of the English-speaking peoples' thirst for new knowledge. In 1900, only 382 PhDs were conferred in the entire United States, yet between 1900 and 1950 the number of PhDs awarded in the fields of science, medicine and technology increased 16.2 times faster than the population.[6]

The success of the Anglo-Saxon model in higher education is mirrored by its success as the best of the many forms of capitalism. The incredible regenerative power of American capitalism was underlined in January 2006 when the Dow Jones hit 11,000 for the first time since 9/11. Even a global War against Terror had not doused American optimism for long. The Promethean power of free markets to provide material benefits has enriched the world. The extension of representative institutions since the early 1940s first to Western Europe and Japan, then to the Indian sub-continent, then Palestine, then to parts of Asia, then to Latin America, then to Eastern Europe and Russia, then to much of Africa, and recently to Afghanistan and Iraq, is also in great part down to

America's willingness – when not under direct threat herself – to extend her birthright across the globe.

When the threat from Marxism-Leninism was mortal during parts of the Cold War, the democratisation process that is America's default position had to take second place to stability and anti-communist tyrants unfortunately had to be tolerated. As with Stalin in 1941, or with Saddam in the Seventies when Iran was the greater threat, *realpolitik* dictated that 'my enemy's enemy is my friend'. It is perfectly true, therefore, that there were monstrous human rights abuses committed by US allies such as in Guatemala during the Cold War, but it takes a particular kind of anti-American to blame these on the United States rather than on the Guatemalans themselves. To have undermined pro-American regimes over their human rights abuses during a period when the most likely alternative was an anti-American Marxist-Leninist regime, would have been the height of irresponsibility. Quite apart from the geo-strategic implications, it would not have led to an improved human rights situation either, as the 94.4 million people killed by communism since 1917 bear witness.

It is not out of sentimentality or naïve utopianism that the English-speaking peoples actively support the extension of representative institutions throughout the world, but out of hard-headed self-interest. The so-called 'neo-conservative' drive to export liberal democracy actuated British statesmen such as George Canning and Lord Palmerston in the nineteenth century, just as the concept of pre-emptive warfare was practised by the Royal Navy in the Napoleonic Wars and since, including against the Vichy Fleet at Oran in 1940. George W. Bush has not invented a new doctrine therefore; he has simply adapted an old one to new and equally terrifying circumstances. In that sense, the fact that 9/11 was not a chemical, biological or nuclear attack was a god-send, in that it finally woke the English-speaking peoples up to the fact that war was being waged against them, but in a way that did not leave hundreds of acres of downtown Manhattan as a sea of radio-active, cancer-inducing rubble.

When freed from the isolationist impulse, the desire to liberate from tyranny runs deep in the English-speaking peoples' psyche; it was they who first came up with the then-unusual notion of first impeding and then abolishing Slavery by force of arms. In many ways they are still carrying out the task, as the women of Afghanistan and the majority of Iraqis can attest. Yet in countries too feudal, theocratic, tribal or obscurantist for an experiment in representative institutions to result in genuine pluralism, democracy must sadly wait, especially if the likely result would be governments elected that were violently opposed to the West. The stable Cold War conditions are already being seen by some as a golden age, which they were certainly not. Old hatreds have produced new terrors in new guises. In the wars of the future, germs will be more dangerous than Germans. Nor are the wars getting shorter; indeed, they seem to be elongating exponentially: the Great War took four years, the Second World War six, Vietnam eleven and the Cold War forty-three. No-one can tell how long the war between Western democratic pluralism and Islamic fundamentalist terrorism might take, but it will certainly not be of short duration. It is already correctly being dubbed 'The Long War' in the Pentagon.

In trying to understand why the English-speaking peoples have been successful in exporting their political culture in the period since 1900, the fact that they have not suffered the trauma, humiliation, expense and fear involved in being invaded, unlike all their major geopolitical rivals – principally France, Russia, Germany, Japan and China – played a major part. In many ways, the 'broad sunlit uplands' that Churchill promised future generations in the darkest days of 1940 are where the English-speaking peoples abide today. For when last has there been a period of six decades with no major war between any of the European Great Powers? When has every continent (except Africa) advanced materially every decade for over half a century? When have scientific and technological innovation, and the free market that delivers their fruits, been so vibrant?

As Churchill said in his 1943 Harvard speech (the Ur-text of this book),

> Law, language, literature – these are considerable factors. Common conceptions of what is right and decent, a marked regard for fair play, especially to the weak and poor, a stern sentiment of impartial justice, and above all a love of personal freedom . . . these are the common conceptions on both sides of the ocean among the English-speaking peoples.

They connect the peoples of the United States, Great Britain, Canada, Australia, New Zealand, the British West Indies and – more often than not – Eire. Instead of distancing themselves from the heritage of the rest of the English-speaking peoples, as some siren voices in each of those places suggest, all of them should take pride in it. National identity is all the stronger for it.

There have been a number of sins and errors committed by the English-speaking peoples since 1900, as was inevitable in the course of human affairs. Amongst their crimes, follies and misdemeanours have been: underestimating the capabilities of the Turks at Gallipoli and the Japanese before Pearl Harbor; the failure to dismember Germany in 1919; not doing more to try to strangle Bolshevism in its cradle in 1918-20; Woodrow Wilson's mismanagement of the Senate in 1919 and the subsequent refusal of the United States to join the League of Nations in 1920; Britain treating France rather than Germany as the more likely enemy in the 1920s; not opposing Hitler's re-militarisation of the Rhineland in 1936; allowing too few visas to Jews wanting to escape Nazi Germany; doing too little to publicise the Holocaust once the true facts were known for certain; transporting non-Soviet citizens to Stalin after Yalta;

botching the 1947 transfer of power in India; the fervent support of the State Department for closer European integration after the Second World War; allowing Nasser to nationalise the Suez Canal; encouraging the Hungarians to rise in 1956; misleading the Commonwealth about the true implications of Britain joining the EEC; waiting for a century after Lincoln's Emancipation Address genuinely to emancipate Black Americans; fighting only for stalemate in Vietnam; the Carter Administration pursuing *Détente* long after its initial purposes were exhausted; appeasing the Serbs for so long after the collapse of Yugoslavia; failing to overthrow Saddam Hussein after the Gulf War; encouraging the Kurds and Shias to rise against him while allowing Iraq the use of helicopter gun-ships; trusting the United Nations to operate the Oil-for-Food programme honestly; relying too much on Intelligence-led WMD arguments to justify the Iraq War; waiting so long for a second UN resolution before attacking Iraq, and subsequently not turning the administration of the country over to a provisional Iraqi government immediately upon Saddam's fall. It is a long and at times shameful catalogue of myopic and failed states-manship, but most other Powers would have done worse, and a century is a very long time in politics. Most of these oversights and errors were made out of good intentions.

Plenty of doom-sayers have predicted disaster for America's *imperium* in the twenty-first century. Many factors have been adduced for why this is inevitable, in a genre known as 'declinist literature'. A useful check-list was provided by the distinguished historian Walter Lacquer in February 2003 in a *Times Literary Supplement* review of a profoundly pessimistic book entitled *The End of the American Era*, by Charles A. Kupchan, Professor of International Relations at Georgetown University:

> Unilateralism on one hand; arrogance and lack of patience to cooperate with allies, as well as isolationism, on the other; adding up to an unwillingness to pay the price for empire. The American economy will simply not be strong enough to sustain the country's role as the globe's strategic guardian. Among other sources of weakness, the author sees the false promises of globalization, American dependence on foreign capital, the weakness and vulnerability of American industry, the destructive consequences of the digital revolution, economic and social inequality among nations and within societies. Kupchan disapproves of the fact that younger Americans watch too much television and sport, and SUVs [sports utility vehicles] are clogging American highways and city streets even though the owners only get thirteen miles to the gallon. He complains about the lagging performance of American institutions of governance and the penetration of politics by corporate money.[7]

It was quite a list – except the one about SUVs, which sounds like a personal gripe – but in order for the American Era to end, another nation must take its place. There was plenty of 'lagging performance of institutions of governance' in Ancient Rome, let alone 'economic and social inequalities', but until Attila the Hun arrived, Rome was the dominant power for over six centuries.

Furthermore, we have been here before; the 1980s also saw a spate of declinist books, such as Paul Kennedy's *Rise and Fall of the Great Powers*, which predicted in 1988 that America's imperial overreach would produce bankruptcy as a result of its irresponsible arms race with the Soviet Union. As Lacquer wrote of Lupchan, who predicts that the European Union would replace America as hegemonic superpower, 'The temptation to draw far-reaching political conclusions concerning the future from present economic trends is always there and should always be resisted. Most truly important issues in the life of nations cannot be quantified, and are not found in the *Statistical Abstract of the United States* and similar works of reference.'[8] Trees never grow to the sky. Even though China replaced Britain as the world's fourth-largest power in terms of GDP in 2006, and is set to overtake Germany in 2008, she nonetheless has severe political, social and environmental problems to overcome before she can threaten the United States (at least economically).

The American economy – despite the War against Terror – is still the power-house of the world, as it has been for over three-quarters of a century. In 2003, America's industries and workers produced almost $500 billion more goods and services than in 2002. That means that America added to the size of her economy an amount equal to a Brazil, or an India, or over one-and-a-half Russias. Of the world's ten largest businesses, measured by market capitalisation, eight were in the US (and the other two – BP and HSBC Holdings – were British). Americans bought over sixteen million cars and light trucks and some two million fridges that year. What Henry A. Wallace in 1942 described as 'the century of the common man' and others have dubbed 'the American Century' has in fact been the English-speaking peoples' century, and it is far from over.

'Sometimes it takes a foreigner to open your eyes,' recalled a recent British contributor to the *Spectator*. 'A Norwegian diplomat told me long ago that he was taught at school, as British kids aren't, that Britain gave the world industrialisation, democracy and football – its economic system, its political system and its fun.'[9] There are plenty of causes for hope amongst the English-speaking peoples: Gonville and Caius College, Cambridge, until recently had more Nobel Prize-winners than France; the most recognised word on the planet is not the name of a dictator or political theorist but of a refreshing fizzy drink, 'Coca-Cola'; on Christmas Day 2004, more than one million phone calls were made between Britain and America; more people – 750 million – speak English as a second language than as a first one; Canada has taken part in more United Nations peacekeeping operations – fifty-five by 2004 – than any other country in the world except Fiji (a country with the Union Jack in its flag). Best of all, most Americans post-9/11 now view George Washington's isolationist Farewell Address in its proper historical context as an obsolete policy stance that has been comprehensively overtaken by events, rather like the Founding Fathers' compromise over slavery.

Yet the phrase 'Anglo-centric' is still a term of disapprobation, at least among the English-speaking peoples themselves. A recent work has even criticised Dr Samuel Johnson's dictionary, complaining that it transmitted 'an image of English and Englishness which is not just predominantly middle-class, but also backward-looking, Anglocentric, and male'.[10] Considering that the (male) Dr Johnson was compiling a book entitled *A Dictionary of the English Language* on necessarily backward-looking historical principles, using citations from authors who in those days were overwhelmingly male, one wonders how the great work could have been anything much different? Many citizens of the English-speaking peoples resemble the Jacobin in George Canning's rhyme, who was, 'A steady patriot of the world alone, / The friend of every country but his own.'

When a British pro-American left-winger, Jonathan Freedland, published a book in 1998 – on the Fourth of July, no less – which was subtitled *How Britain Can Live the American Dream,* he recalled how,

> The Leftie response was unsurprising. How could anyone admire a country that gorged itself on junk food, still executed criminals and wouldn't treat the sick till they produced a credit card? What was there to emulate in a land of Bible-bashing, gun-wielding simpletons, trapped in a sclerotic political system warped by cash and painfully ignorant of the rest of the world?[11]

His spirited reply, that America was in fact 'a vigorous democracy and an engaged civil society, still captivated by the dream of self-government – a dream made manifest by a degree of volunteerism, philanthropy and local autonomy that put Britain to shame,' was commendable, but what was more interesting was the sheer fury of the left-wing reaction, and at a time when Bill Clinton was President. 'Anti-Americanism is now written into the European psyche,' believes the writer Leo McKinstry, 'the last acceptable prejudice in a culture that makes a fetish of racial equality.'[12]

Instead of creating an outpouring of thanks and affection for the United States, the demise of communism, ironically enough, made Europe safe for anti-Americanism once again. As one writer has put it, 'The threat from a common enemy during the Cold War helped to put anti-American attitudes on hold. The common disdain ... for American civilisation – its vulgar materialism, its rootless cosmopolitanism, its shallow optimism, its lack of the tragic sense – emerged once again when the common enemy disappeared.'[13]

As George Kennan observed in his famous *Foreign Affairs* article in 1947, anti-Americanism is sometimes simply unappeasable because, like those Irish republicans who cannot accept that Roger Casement was a promiscuous homosexual, for some people these things become 'essentially theological, in the end a matter of faith rather than reason'. As Jonathan Swift said, it is useless to try to reason a man out of something he was not reasoned into. Some of the rants of anti-Americans – especially since the Iraq War – more closely resemble attacks of Tourette's Syndrome than rational criticism.

One of the most common criticisms of the United States is that her citizens do not travel abroad; only about 18% of adult Americans hold passports. Yet the astonishing geographical variety to be found in the United States makes it far less necessary for Americans to leave their continent than Europeans. Living on a land mass that comprises San Francisco, the Great Lakes, the Rocky Mountains, the Shenandoah Valley, Philadelphia, the Grand Canyon, Chicago, Californian wineries, New England villages, the Niagara Falls, the Appalachian Mountains, the Capitol, Colorado ski resorts, the Nevada Desert, New York City, Hawaiian beaches, the Mid-Western prairies, Southern swamps, everglades and bayous, the Yosemite National Park, wonders of the natural world and almost every conceivable type of flora and fauna, as well as extremes of temperature and climate, all girt by the globe's greatest two oceans, Americans have less reason to own passports than any other people on earth.

The Mississippi River is over 4,000 miles long and pours a billion cubic feet of water into the Gulf of Mexico every week. Yellowstone Park is half the size of England's largest county, Yorkshire; another American National Park can boast sixty glaciers. If the entire British Isles were dropped into the Great Lakes, there would be room for a further 9,000 square miles. 'Vast is America,' wrote H.L. Gee in 1943, 'a modern world in itself.' The very best of Western civilisation's painting, music, sculpture and culture can be enjoyed in the great American museums, art galleries and concert halls. With European countries such as Luxembourg and Liechtenstein so small that, in Woody Allen's gag, 'they could carpet them', cross-border travel is an absolute necessity for many Europeans in a way it simply is not in the continental United States. The relatively small number of Americans who own passports should not be such a cause for European derision.

The contradictions inherent in anti-Americanism were pinpointed by a senior broadcaster named Henri Astier recently, who wrote of how,

> We are happy to view American society as both utterly materialistic and insufferably religious; it is predominantly racist and absurdly politically correct; Americans are both boring conformists and reckless individualists; US corporations can do whatever they want and are stifled by asinine liability laws. Furthermore, in the same breath the United States is accused of 'unilateralism' but also of

shirking its international responsibilities. America is blamed for intervening every-where, and expected to save Mexico from default, protect Taiwan from China, mediate between India and Pakistan ... get the two Koreas talking, etc.[14]

The explanation for all this double-speak given by the French philosopher Jean-François Revel is the correct one: anti-Americanism 'can only be explained in psychological terms. Anti-American recriminations stroke a society's collective ego by drawing attention away from its own failures.' Thus the highly censored Arab media alleges that the War against Terror has muzzled freedom of speech in America; the Organisation of African States calls for 'a Marshall Plan for Africa' despite having enjoyed the equivalent of four such cash injections in four decades; Europeans 'find a reassuring explanation for the Continent's catastrophic loss of status' by blaming American hegemony, rather than attributing it to their own two continental suicide attempts within thirty years during the twentieth century.'[15]

Only the English-speaking peoples need not indulge in this kind of self-indulgence, because through our Special Relationship – whose relevance has never been more powerfully tangible in the entire post-war period than since 9/11 – we are part of the hegemonic power that the Arabs, Africans and Europeans so self-referentially loathe. For all that the English-speaking peoples might hold different views over carbon emissions or steel tariffs, in the great world-historical struggle, as Tony Blair put it so perfectly, our shared interests dictate that we stand 'shoulder-to-shoulder' with our cousins, allies and co-linguists.

Churchill was right in his Harvard speech when he declared, 'If we are together, nothing is impossible.' In the last century, the Union Jack has flown on Everest and the Stars and Stripes on the Moon, and together the English-speaking peoples have brought down tyrannies across four continents, cured disease after disease, delivered unheard-of prosperity to hundreds of millions, made their tongue the global *lingua franca*, won by far more Nobel Prizes than anyone else in both absolute and per capita terms, and smoothly passed the baton of global leadership from one of their constituent parts to another, right in the middle of a debilitating war. Their only possible limiting factor seems to have been a recurring, inexplicable, undeserved form of anguished intro-spection that makes them doubt their own abilities and moral worth.

Back in 1900, any number of rivals might have snatched hegemony from the English-speaking peoples. The British Empire was overstretched and had no army to speak of, at least not one that could have engaged a Great Power on equal terms; the United States had neither a significant army nor navy and was only beginning to discover a global ambition. By contrast, the economically formidable Imperial Germany was flexing her *weltpolitik* from China to Venezuela to Samoa and building a world-class High Seas Fleet, France had a huge global empire and a thirst for *revanche* against Britain over her humiliation at Fashoda only three years previously; Russia was industrialising successfully, heavily armed and carefully eyeing British India; the Ottoman Empire, Austria-Hungary, Italy and Japan looked relatively weak but could certainly not be written off entirely, especially the last.

A little over a century later the landscape could not be more different. Not one but two lunatic attempts to force geopolitical matters through military rather than commercial means have left Germany a pacifist husk and wrecked French power as much as her own; Russia suddenly capitulated in her long struggle to impose communism on the rest of the world and is now the weakest she has been since the 1905 Revolution. All of those countries, as well as Austria-Hungary, Italy and Japan, have been invaded and occupied at least once, most of them twice, with all the dislocation and demoralisation that that entails.

The English-speaking peoples, by total contrast, today know no rival in might, wealth or prestige. The most likely future challenger on the far horizon is China – not a contender in 1900 – which still has very far to go before she can threaten to supplant them. A few fanatical malcontents from the former Ottoman Empire have proven their ability to strike a painful blow to the heart of the greatest city of the English-speaking peoples, it is true, but their fury is a mark of their enemies' primacy rather than a serious threat to it. Even were terrorists to strike a further, perhaps chemical, biological or nuclear blow against one of the English-speaking peoples' principal cities, it would not destroy that primacy. As George Will has observed, 'Al-Queda has no rival model about how to run a modern society. Al-Queda has a howl of rage against the idea of modernity.'[16]

At the closing stage of the battle of Waterloo, once the Emperor Napoleon's Imperial Guard had been defeated in its final great assault on the Anglo-Allied lines, the Duke of Wellington raised his cocked hat and gave the order: 'Go forward and complete your victory.' With Soviet communism now lying in the dust, and with representative institutions, free enterprise, the English language, military superiority and the rule of law their talismans as of old, it is clear that the English-speaking peoples have done just that.

On 26 January each year, the Roman Empire celebrated the festival of Feria Latina, commemorating the origins of the Latin-speaking peoples, held at Alba Longa, once their principal city. (As *Pontifex Maximus*, Julius Cæsar officiated at it seven weeks before his assassination.) The English-speaking peoples are far too self-deprecating to copy such a celebration of themselves, but perhaps they should, because today they are the last, best hope for Mankind. It is in the nature of human affairs that, in the words of the hymn, 'Earth's proud empires pass away', and so too one day will the long hegemony of the English-speaking peoples. When they finally come to render up the report of their global stewardship to History, there will be much of which to boast. Only when another power – such as China – holds global sway, will the human race come to mourn the passing of this most decent, honest, generous, fair-minded and self-sacrificing *imperium*.

NOTES

EHR refers to *English Historical Review*
NA refers to the British National Archives at Kew
TLS refers to *The Times Literary Supplement*

INTRODUCTION

1 NA CAB 195/2
2 Lal, *In Praise of Empires*, p. 45
3 *Daily Telegraph*, 1 January 2000, p. 7
4 Gordon-Duff, *It Was Different Then*,
  p. 1
5 Winston Churchill in the *News of the
  World*, 22 May 1938
6 *The Times*, 1 January 1901
7 Roberts, *Salisbury*, p. 810
8 Pagden, *Peoples and Empires*, p. 159
9 *The Times*, 1 January 1901
10 Roberts, *op. cit.*, p. 50
11 Beard and Beard, *The Rise of American
  Civilisation*, p. 377
12 *The Times*, 14 May 1901
13 Ronaldshay, *The Life of Lord Curzon*, vol.
  I, p. 254
14 Williams, *Chronology of the Modern
  World 1763–1992* (hereafter *Chronology*),
  p. 396
15 Charles Wheeler in the *Literary Reviews*,
  March 2002, p. 11
16 Pagden, *op. cit.*, p. 28
17 Alfred Lee Papers, Federation ephemera
  album
18 Seddon Papers, series 1, file 3
19 Powell, *My American Journey*, p. 22
20 *United Irishman*, 31 March 1900
21 Pašeta, 'Nationalist Responses to Two
  Royal Visits to Ireland, 1900 and 1903',
  pp. 488–504

ONE: 1900–4

1 Belfield, *The Boer War*, p. xxiii
2 Morris, *Theodore Rex*, p. 313
3 Brands, *T.R.: The Last Romantic*, p. 84
4 *Ibid.*, p. 73
5 O'Gara, *Theodore Roosevelt and the Rise
  of the Modern Navy*, pp. 3–12
6 Edmund Morris letter to the *TLS*, 8
  March 2002, p. 17
7 John Vincent in the *Spectator*, 15 June
  2002, p. 39
8 Edmund Morris letter to the *TLS*, 8
  March 2002, p. 17
9 Francisco E. Gonzalez letter to the *TLS*,
  19 April 2002, p. 17
10 Ernest R. May in the *TLS*, 1 February
  2002, p. 9
11 Charles Wheeler in the *Literary Review*,
  March 2002, p. 11
12 ed. Wilson, *The International Impact of
  the Boer War*, pp. 107–22
13 Beard and Beard, *The Rise of American
  Civilisation*, vol. 2, p. 373
14 Reyes, *A Legislative History of America's
  Economic Policy Toward the Philippines*,
  p. 192
15 Ellis, *His Excellency*, p. 235
16 Adler, *The Isolationist Impulse*, pp. 25–6
17 Nelson, *The Philippines*, pp. 48ff
18 Fernández, *The Philippine Republic*,
  pp. 173–4
19 Keesing, *The Philippines*, p. 43

20 Beard and Beard, *op. cit.*, p. 485

21 Blount, *The American Occupation of the Philippines*, p. 456

22 Keesing, *op. cit.*, pp. 43ff

23 McKinley Papers, series 1, reel 14, 17 January 1901

24 Taft Papers, series 3, reel 72

25 Forbes, *The Philippine Islands*, pp. 72–3

26 Day, *The Philippines*, p. 108

27 eds Brown and Lewis, *The Oxford History of the British Empire*, vol. IV, p. 232

28 Rosenthal, *Stars and Stripes in Africa*, p. 142

29 Judd and Surridge, *The Boer War*, pp. 229–32

30 ed. Maurice, *The History of the War in South Africa 1899–1902*, vol. IV, p. 64

31 Judd and Surridge, *op. cit.*, p. 195

32 Wilson, *Attitudes*, pp. 113–15

33 Dimbleby and Reynolds, *An Ocean Apart*, p. 48

34 Ferguson, *American Diplomacy and the Boer War*, p. 208

35 Mulanax, *The Boer War and American Politics and Diplomacy*, p. 83

36 Anthony Browne in the *Spectator*, 23 July 2005, p. 10

37 Penlington, *Canada and Imperialism 1896–1899*, p. 217

38 ed. Lycett, *Rudyard Kipling: Selected Poems*, p. 85

39 eds Omissi and Thompson, *The Impact of the South African War*, pp. 233–50

40 Penlington, *op. cit.*, pp. 30ff

41 *Ibid.*, p. 45

42 *Ibid.*, p. 53

43 eds Omissi and Thompson, *op. cit.*, p. 234

44 Micklethwait and Wooldridge, *The Company*, p. 2

45 *Ibid.*

46 *Ibid.*, p. 9

47 Williams, *Chronology*, p. 409

48 *Annual Register*, 1901, II, p. 5

49 Strouse, *Morgan*, p. 4

50 eds Milner, O'Connor and Sandweiss, *The Oxford History of the American West*, p. 491

51 *Ibid.*, p. 495

52 *The Argus*, 4 January 1901, p. 4

53 *Sunday Telegraph* review, 25 July 2004, p. 12

54 eds Omissi and Thompson, *op. cit.*, p. 257

55 Knightley, *Australia*, pp. 50–1

56 *Ibid.*, p. 54

57 Grimshaw, 'Federation as a Turning Point in Australian History', p. 26

58 Jason Groves letter to *The Times*, 12 June 2004

59 Barton Papers, A6/1 1901/364

60 Knightley, *op. cit.*, p. 48

61 *Ibid.*, p. 46

62 *Sunday Star Times*, 30 October 2005, p. 2

63 Bull, 'The Formation of the United Irish League', pp. 404–5

64 *Ibid.*, p. 410

65 McGarry, *Eion O'Duffy*, p. 29

66 McLaughlin, 'The British in the Air', p. 80

67 Mackinder, *The Scope and Methods of Geography and the Geographical Pivot of History*, pp. 3–9

68 *Ibid.*, p. 10

69 Landes, *The Wealth and Poverty of Nations*, p. 311

70 Colm Tóibín in the *TLS*, 30 September 1999, p. 37

71 Keogh, *Jews in Twentieth-Century Ireland*, p. 51

TWO: 1905–14

1 Morris, 'The Murder of H. St G. Galt', pp. 1–15

2 *Ibid.*, p. 6

3 Forward, *You Have Been Allocated Uganda*, p. 47

4 David Blair in the *Daily Telegraph*, 30 April 2005, p. 18

5 Richard Beeston in *The Times*, 3 August 2004, p. 14

6 Cocker, *Rivers of Blood, Rivers of Gold*, pp. 345–6

7 Anderson, *Histories of the Hanged*, and Elkins, *Britain's Gulag*

8 Owen, *Lord Cromer*, p. 394

9 M.E. Yapp in the *TLS*, 26 January 2001, p. 9

10 Fromkin, *Europe's Last Summer*, p. 296

11 Zuber, *Inventing the Schlieffen Plan*, *passim*

12 Raymond Carr in the *Spectator*, 9 June 2001, p. 39

13 Donald Cameron Watt in the *TLS*, 17 May 2002

14 Williams, *Chronology*, p. 405

15 ed. Smart, *The Diaries and Letters of Robert Bernays*, p. 214

16 Richard Hamblyn in the *Sunday Times* review, 9 October 2005, p. 50; Eugen Weber in the *TLS*, 17 February 2006, p. 12

17 Winchester, *A Crack in the Edge of the World*, *passim*; and Fradkin, *The Great Earthquake and Firestorms of 1906*, *passim*

18 Segar, 'The Struggle for Foreign Trade', pp. 523–4

19 *Ibid.*

20 Beefsteak Club betting book

21 King, 'The Institutions of Monetary Policy', p. 2

22 eds Gere and Sparrow, *Geoffrey Madan's Notebooks*, p. 87

23 Zimmermann, *First Great Triumph*, p. 7

24 Conway, *Conway's All the World's Fighting Ships* (hereafter *All the World's*), p. 137

25 Zimmermann, *op. cit.*, p. 6

26 Judd, *The Quest for C, passim*

27 *Ibid.*

28 *New Statesman*, 6 June 2005, p. 40

29 Diane Atkinson in the *New Statesman*, 6 June 2005, pp. 38–41

30 Grunberger, *A Social History of the Third Reich*, p. 321

31 Richard Overy in the *TLS*, 11 December 1998, p. 6

32 Williams, *The Hutchinson Chronology of World History*, vol. IV (hereafter *Hutchinson*), p. 332

33 Hendrik, *The Life and Letters of Walter Hines Page*, vol. I, p. 144

34 *Ibid.*, pp. 282–3

35 Ian Buruma in *The Times*, 3 August 2004, p. 16

36 Lownie, *John Buchan*, p. 120

37 Fromkin, *op. cit.*, p. 295

38 *Ibid.*, p. 296

39 John Keegan in the *TLS*, 12 April 2002, p. 25

40 Elton, *Imperial Commonwealth*, p. 473

41 *Hansard*, vol. 65, cols 1808–27

42 *Ibid.*

43 *Ibid.*, col. 1827

THREE: 1914–17

1 eds Gere and Sparrow, *Geoffrey Madan's Notebooks*, p. 116

2 Gregory, 'War Enthusiasm in 1914', p. 69

3 *Ibid.*, p. 72

4 Burleigh, *Earthly Powers*, p. 448

5 John Keegan in the *TLS*, 12 April 2002, p. 25; Horne and Kramer, *German Atrocities, passim*

6 John Horne in the *TLS*, 6 February 2004, p. 4

7 Pershing *My Experiences in the World War*, vol. I, pp. 7–8

8 Raymond Carr in the *Spectator*, 9 June 2001

9 *The Times*, 25 August 1914, p. 7

10 *Ibid.*

11 Gregory, *op. cit.*, p. 80

12 James, *Imperial Warrior*, p. 71

13 ed. Seymour, *The Intimate Papers of Colonel House*, vol. I, p. 383

14 Knight, 'Fighting on the Beaches', p. 28

15 *RUSI Journal*, 1901, p. 1322

16 ed. Gilbert, *The Straits of War*, p. 27

17 *Ibid.*, p. 10

18 *Ibid.*, p. 136

19 Knight, *op. cit.*, p. 30; Chasseaud and Doyle, *Grasping Gallipoli*, pp. 265–9

20 Alexander Turnbull Library, NLNZ MS-5583-3

21 Blainey, *A Shorter History of Australia*, p. 155

22 Burton, 'Spy Fever', pp. 37–9

23 ed. Norwich, *The Duff Cooper Diaries*, pp. 8–10

24 *Ibid.*, pp. 8–9

25 Charmley, *Duff Cooper*, pp. 23–4

26 ed. Seymour, *op. cit.*, pp. 448–51

27 Pollock, *The League of Nations*, pp. 71–3

28 Ellis, *The Social History of the Machine Gun*, pp. 113–14

29 Woodward, *Great Britain and the War of 1914–18*, p. 35 n.2

30 Ellis, *op. cit.*, p. 130

31 *Ibid.*, p. 86

32 Jackson, *Private 12768*, p. 7

33 *Ibid.*, p. 8

34 NA CO 137/709/25738

35 Howe, *Race, War and Nationalism*, p. 41

36 Burns, *The History of the British West Indies*, p. 701

37 NA CO 551/81/48100

38 Howe, *op. cit.*, pp. 30–1

39 *Ibid.*, p. 34

40 *Ibid.*, p. 38

41 Beckett, *A Nation in Arms*, pp. 13–14

42 Howe, *op. cit.*, pp. 100–1

43 *EHR*, cxix 480 (February 2004), p. 254

44 Elton, *Imperial Commonwealth*, p. 475. For a different view, see Ferguson, *The Pity of War*, pp. 143–73

45 Gliddon, *The Aristocracy and the Great War*, p. ix

46 *Ibid.*, p. xvii

47 *Salisbury Review*, Autumn 2004, p. 44

48 Roy Foster in the *TLS*, 21 October 2005, p. 3

49 McCormack, *Roger Casement in Death*, *passim;* Keith Jeffrey in the *TLS*, 15 November 2002, p. 24

50 Roy Foster in the *TLS*, 21 October 2005, p. 4

51 Harrison, *Ireland and the British Empire*, p. 26

52 Ørvik, *The Decline of Neutrality*, p. 43

53 Ludendorff, *Ludendorff's Memoirs 1914–1918*, pp. 102–3

54 Pershing, *op. cit.*, p. 7

55 *Ibid.*, pp. 8–9

56 Coleman, *The Last Exquisite*, p. 2

57 Todman, *The Great War*, p. 105

58 Neillands, *The Great War Generals of the Western Front 1914–1918*, *passim*

59 Holmes, *Tommy*, *passim*

60 Corrigan, *Mud, Blood and Poppycock*, p. 10

61 Gee, *American England*, p. 163

62 Cecil, *Lansdowne*, p. 26

63 Lloyd George, *War Memoirs of David Lloyd George*, vol. I, p. 515

64 *Ibid.*, p. 519

65 eds Sheffield and Bourne, *Douglas Haig: War Diaries and Letters 1914–1918*, p. 259

66 Lloyd George, *op. cit.*, p. 521

67 Cecil, *op. cit.*, p. 27

68 Grigg, *Lloyd George*, p. 14

69 Pershing, *op. cit.*, p. 8

70 John Vincent in the *Spectator*, 15 June 2002; Charles Wheeler in the *Literary Review*, May 2000, p. 4

71 Black, *The British Seaborne Empire*, p. 269

72 Wills Papers, 27 December 1902

73 Tuchman, *The Zimmermann Telegram*, p. 10

74 James, 'Room 40', p. 50; Jones, 'Alfred Ewing and "Room 40"', p. 66

75 Ewing, *The Man of Room 40*, p. 175

76 *Ibid.*

77 Beesley, *Room 40*, p. 171

78 Andrew, 'Codebreakers at King's', p. 2

79 Jones, *op. cit.*, p. 87; Andrew, *ibid.*

80 Jones, *ibid.*, p. 88

81 *Ibid.*, p. 72

82 James, *op. cit.*, p. 51

83 Toye, *For What We Have Received, passim*

84 Jones, *op. cit.*, pp. 86–7

85 James *op. cit.*, p. 87

86 Tansill, *America Goes to War*, p. 635

87 Wills Papers, May-July 1979, *passim*

88 *Ibid.*, July 1979, *passim;* Jones, *op. cit.*, pp. 86–7

89 Williams, *Pétain*, pp. 158–9

90 Rose, *King George V*, p. 174

91 Pipes, *Three Whys of the Russian Revolution*, p. 43

92 Kinross, *Atatürk*, p. 118

93 Dictionary of New Zealand Biography website: www.dnzb.nz also www.nzedge.com

94 John Campbell in *ibid.*

FOUR: 1918–19

1 McCormick, *The Mask of Merlin*, p. 144
2 *Ibid.*
3 Robert Blatchford in the *Illustrated Sunday Herald*, 13 October 1918
4 McCormick, *op. cit.*, pp. 146–7
5 *Sunday Telegraph* review, 27 March 2005, p. 15
6 Dupuy and Dupuy, *The Collins Encyclopedia of Military History from 3,500 BC to the Present* (hereafter *Encyclopedia*), p. 1075
7 *Ibid.*, p. 1083
8 Lovat, *March Past*, p. 55
9 McKinstry, *Rosebery*, p. 523
10 Bigelow Papers, 34A, 9 March 1927 and 18 October 1927
11 Macmillan, *Peacemakers*, p. 500
12 *The Times*, 27 December 1918
13 *Royalty Digest*, summer 2004, p. 57
14 Mitchell, 'Woodrow Wilson as "World Saviour"', pp. 8-9
15 *The Times*, 15 February 1929, p. 15
16 Fromkin, *Europe's Last Summer*, p. 278
17 Clark, *The Tories*, p. 3
18 Powell and Maude, *Biography of a Nation*, p. 183
19 Root Papers, box 92
20 Nicolson, *Peacemaking 1919*, p. 207
21 Theodore Roosevelt Papers, series 2, reel 412, 28 December 1918
22 Bond, 'Amritsar 1919', p. 666
23 *Ibid.*, p. 667–9
24 Collet, *The Butcher of Amritsar*, p. 283
25 Bond, *op. cit.*, p. 669
26 Denis Judd in *BBC History*, May 2005, p. 54; Nicholas Fearn in the *Independent on Sunday*, 11 May 2005, p. 31; Tony Gould in the *Spectator*, 16 April 2005, p. 45; Frank Fairfield in the *Literary Review*, April 2005, p. 16
27 Bond, *op. cit.*, p. 676
28 Herman, *To Rule the Waves*, p. 517
29 Toye, *For What We Have Received*, p. 196
30 Knight, *The Pursuit of Victory*, p. 520
31 T.G. Otte in the *TLS*, 26 October 2001, p. 8

32 *Daily Telegraph*, 1 September 2005, p. 16
33 McCormick, *op. cit.*, p. 314
34 *Ibid.*, 12 January 2005, p. 13
35 Churchill, *The Aftermath*, p. 206
36 Wilson Papers, series 6, reel 462
37 Mantoux, *The Carthaginian Peace*, p. 5
38 *Ibid.*
39 *Congressional Record*, vol. 59, part 3, pp. 2696ff
40 Millin, *General Smuts*, vol. II, p. 174
41 Mantoux, *op. cit.*, p. 3
42 Keynes, *The Economic Consequences of the Peace*, p. 59
43 Lloyd George, *The Truth about the Peace Treaties*, vol. I, p. 223
44 Mantoux, *op. cit.*, p. 11
45 Wilson Papers, series 7B, reel 480
46 Bryce Papers, USA 7, fol. 192ff.
47 *Ibid.*, fol. 169ff
48 *Ibid.*, fol. 193ff

FIVE: 1920–9

1 Kissinger, in *American Foreign Policy in Washington DC*, 1969, p. 93
2 *New Statesman*, 1 July 1939
3 Hart, 'The Protestant Experience of Partition in Southern Ireland', pp. 81–99
4 Wheatcroft, *The Strange Death of Tory England*, p. 140
5 Harold Perkin in the *TLS*, 9 July 2004, p. 10
6 ed. Cross, *A Century of Icons, passim*
7 Perkin, *op. cit.*
8 Williams, *Chronology*, p. 483
9 ed. Soames, *Speaking for Themselves*, p. 229
10 *News of the World*, 15 May 1938
11 G. Calvin Mackenzie in the *TLS*, 13 October 2000, p. 13
12 Figes, *A People's Tragedy*, p. 646
13 Sol Sanders letter to the *TLS*, 6 July 2001
14 ed. Pottle, *Champion Redoubtable, Daring to Hope, passim*
15 Roberts, 'Northern Territory Colonization Schemes', p. 420
16 *News of the World*, 22 May 1938

17 Kershaw, *Hitler: Hubris*, pp. 248–9
18 Toland, *Adolf Hitler*, p. 199; Kershaw, *ibid.*, p. 677 n.148 and n.149
19 Mackinder, 'The Round World and the Winning of the Peace', pp. 595–605
20 Walton, 'Feeling for the Jugular', p. 32
21 *Ibid.*, p. 22
22 *Ibid.*, p. 27
23 *Ibid.*, p. 30
24 Vincent Crapanzano in the *TLS*, 15 June 2001, p. 11
25 see Foner, *From Ellis Island to JFK*, for these and many more statistics
26 *TLS*, 24 December 2004, p. 9
27 *Ibid.*
28 Murray, *Human Accomplishment*, p. 282
29 Library of Congress, *Index to the Calvin Coolidge Papers*, p.v
30 Coolidge Papers, series 1, box 295
31 Egremont, *Balfour*, p. 333
32 *Hansard*, vol. 65, col. 286
33 Baldwin Papers, box 129, f.17
34 Donald Cameron Watt letter to the *TLS*, 12 December 2003
35 Pilpel, *Churchill in America*, pp. 34–56
36 Gilbert and Churchill, *Winston S. Churchill*, vol. 1, pp. 434–5
37 *Ibid.*, vol. v, p. 301
38 *Ibid.*, p. 308
39 Jeremy Noakes in the *TLS*, 5 December 2005, p. 23
40 ed. Weinberg, *Hitler's Second Book*, p. 107
41 *Ibid.*
42 *Ibid.*, p. 113
43 Ackermann, *Cool Comfort, passim*
44 Andrew Ballantyne in the *TLS*, 24 January 2003, p. 36
45 Alexander Masters in the *TLS*, 10 January 2003, p. 6
46 Kamm and Baird, *John Logie Baird, passim*
47 Anthony Browne in the *Spectator*, 23 July 2005, p. 11
48 *Ibid.*
49 Richard Cavendish in *History Today*, March 2005, p. 59
50 DNB 1951–60, p. 361
51 *Ibid.*, p. 362

SIX: 1929–31

1 Hayek, *The Constitution of Liberty*, p. 520 n.1
2 Bergreen, *Capone*, p. 20
3 J. Castellar-Gassol, *Dali, passim*
4 ed. Ramsden, *George Lyttelton's Commonplace Book* (hereafter *Lyttelton*), p. 71
5 *Ibid.*
6 *Daily Telegraph*, 9 September 2005
7 Mary Kenny in the *Literary Review*, March 2005, p. 12
8 *The Times*, 4 August 1998, p. 16
9 James Bowman in the *New Criterion*, November 2005, pp. 54–5
10 Micklethwait and Wooldridge, *The Right Nation*, p. 310
11 Conway, *All the World's*, p. 137
12 *Quarterly Review*, vol. 112, October 1862, p. 535
13 Williams, *Chronology*, p. 523
14 Meltzer, *A History of the Federal Reserve*, vol. I, p. 271
15 Barone, *Hard America, Soft America*, p. 28
16 Meltzer, *op. cit.*
17 *Ibid.*, vol. I, p. x
18 Friedman and Friedman, *Free to Choose*, p. 71
19 *Ibid.*
20 *Ibid.*, p. 74
21 *Ibid.*, p. 81
22 Lipset and Marks, *It Didn't Happen Here, passim*
23 Felipe Fernandez-Armesto in the *Literary Review*, August 2000, p. 19
24 Williams, *Chronology*, p. 535
25 Caplan, 'The Failure of Canadian Socialism', p. 93
26 *Ibid.*
27 *Ibid.*
28 Niall Ferguson in the *Literary Review*, September 2000, p. 24
29 *Ibid.*
30 Martin Filler in the *TLS*, 30 March 2001, p. 3
31 *Ibid.*
32 Williamson, *Stanley Baldwin*, p. 41
33 Jenkins, *Baldwin*, p. 119

34 Chisholm and Davie, *Beaverbrook*, p. 305

SEVEN: 1931–9

1 ed. Ramsden, *Lyttelton*, p. 87
2 Montefiore, *Stalin*, pp. 218–26
3 *Ibid.*, p. 23
4 *Ibid.*, pp. 32–7
5 Niall Ferguson in the *Sunday Telegraph*, 11 December 2005, p. 19
6 *New Criterion*, March 2005, p. 60
7 *Tourists of the Revolution, The People's Flag*, First Circle Films, 14 December 1999
8 Amis, *Koba the Dread*, p. 21n
9 Webb and Webb, *Soviet Communism*, pp. 431, 432, 433, 435
10 Alexander Papers, 5/1/3
11 Steffens, *The Autobiography of Lincoln Steffens*, p. 799
12 *Ibid.*, p. 797
13 *Tribune*, 19 November 1937
14 *Tourists of the Revolution, The People's Flag, op. cit.*
15 Amis, *op. cit., passim*
16 Keith Windschuttle in the *New Criterion*, May 2003, pp. 1–10
17 Farrell, *Mussolini*, p. 225
18 ed. Ramsden, *op. cit.*, p. 72
19 Attlee Papers, Mss Eng c.4792 fol. 60
20 *Ibid.*, fol. 65
21 Howard Gotlieb Archives, Boston University
22 Steiner, *The Lights That Failed*, p. 692
23 *TLS*, 9 November 2001, p. 10
24 Powell, *FDR's Folly, passim*
25 Black, *Franklin Delano Roosevelt*, p. 352
26 Bork, *Coercing Virtue*, p. 3
27 Jenkins, *Franklin Delano Roosevelt*, p. 94
28 Cadbury, *Seven Wonders of the Industrial World*, p. 329
29 *Ibid.*, p. 294
30 *Ibid.*, p. 297
31 G. Calvin Mackenzie in the *TLS*, 13 October 2000, p. 13
32 Juliet Townsend in the *Spectator*, 18/25 December 2004, p. 90
33 Lycett, *Rudyard Kipling*, pp. 584–5
34 ed. Ramsden, *op. cit.*, pp. 98–9

35 ed. Eberle, *The Hitler Book*, pp. 17–18
36 McCallum, *Public Opinion and the Last Peace*, p. 166
37 *Ibid.*
38 Williams, *The People's King, passim*
39 ed. Self, *The Neville Chamberlain Diary Letters*, p. 226
40 Elliott, *A Dictionary of Politics*, p. 237
41 Will, 'Insurrection and the Development of Political Institutions', p. 10
42 *Ibid.*, p. 75, n.5
43 *Ibid.*, p. 12
44 *Ibid.*, p. 17
45 *Ibid.*
46 Williams, *Chronology*, pp. 561 and 567
47 Lord Mountbatten speech to Winston S. Churchill, Society of Edmonton, Alberta, in 1966, *Finest Hour*, no. 127, summer 2005, p. 18
48 ed. Self, *op. cit.*, p. 348
49 *Ibid.*, 17 December 1937, p. 294
50 Parker, *Churchill and Appeasement, passim*
51 Stewart, *Burying Caesar*, p. 4
52 Ian Sayer Papers, 28 March 1938
53 Attlee Papers, Mss Eng c.4792 fol. 85
54 *Ibid.*, fol. 88
55 Załuski, *The Third Estate*, p. 97
56 A. J. Sherman in the *TLS*, 6 July 2001, p. 29
57 Herman, *To Rule the Waves, passim*
58 ed. Smart, *The Diaries and Letters of Robert Bernays 1932–1939*, p. 219
59 *American Historical Review*, vol. 110, no. 1, February 2005
60 Sarna, *American Judaism, passim*
61 Geoffrey Wheatcroft in the *Spectator*, 22 May 2004, p. 48
62 Murray, *Human Accomplishment*, p. 283
63 ed. Pickersgill, *The Mackenzie King Record, vol. 1 1939–1944*, pp. 238–9
64 *The Press*, 17 September 1938
65 ed. Norwich, *The Duff Cooper Diaries*, p. 260
66 *Ibid.*, p. 261
67 Hogg, *The Left Was Never Right*, p. 200
68 National Archives of Australia, A 981/4, CZE 18 part I, part II and A2937 185CZE

69 *Ibid.*, AA 1972/341/2, 7 October 1938
70 Wodehouse, *The Code of the Woosters*, p. 143
71 Jones, *Mosley*, p. 90
72 *Ibid.*, p. 93
73 *The Press*, 7 May 1938, p. 8
74 Jonathan Mirsky in the *Spectator*, 8 October 2005, p. 47
75 My thanks to Michael Barone, 12 December 2005
76 Butler and Butler, *British Political Facts 1900–1994*, p. 157
77 *Ibid.*, p. 157
78 ed. Norwich, *op. cit.*, p. 260
79 *The Press*, 1 October 1938, p. 16
80 Mantoux, *The Carthaginian Peace*, p. 7
81 Christopher York Papers, 29 March 1939
82 Eden, *Portrait of Churchill*, p. 11

EIGHT: 1939–41

1 Attlee Papers, Mss Dep. 25 fol. 9
2 Menzies, *Afternoon Light*, p. 16
3 *Ibid.*
4 *The Press*, 4 September 1939, p. 8
5 Savage Papers, box 1361, fol. 1B
6 Dwyer, *De Valera*, p. 234
7 O'Halpin essay in ed. Wylie, *European Neutrals and Non-Belligerents during the Second World War*; Geoffrey Best in the *TLS*, 17 May 2002, p. 30
8 Richard Woods in the *Sunday Times*, 28 April 1996
9 Ó Drisceoil, *Censorship in Ireland 1939–1945, passim*
10 *Hansard*, 4 February 1938
11 Inglis, *Neutrality*, pp. 216–17
12 ed. Parrish, *The Simon and Schuster Encylopaedia of World War II* (hereafter *Encyclopaedia*), p. 362
13 Hull, *The Memoirs of Cordell Hull*, vol. 1, p. 697
14 Horne, *The Terrible Year*, p. 101
15 Attlee Papers, Mss Eng c.4793 fol. 16
16 Coward, *The Lyrics of Noël Coward*, pp. 269–70
17 Morley, *A Talent to Amuse*, p. 222
18 Wheatcroft, *The Strange Death of Tory England*, p. 144
19 Hull, *Irish Secrets*, p. 60
20 *Ibid.*, p. 134
21 *Documents on German Foreign Policy*, vol. 8, no. 473
22 Colm Tóibín in the *TLS*, 30 September 1999, p. 38
23 Hull, *op. cit.*, pp. 36 and 190–1
24 *Ibid.*, p. 192
25 ed. Pottle, *Daring to Hope*, pp. 317–18
26 Savage Papers, box 1361 fol. 1B
27 Christopher York Papers, 12 October 1943
28 Edwards, 'R.G. Menzies's Appeals to the United States May-June 1940', p. 66
29 Lukacs, *Churchill*, p. 173
30 Collins and Lapierre, *Freedom at Midnight*, p. 58
31 *Ibid.*
32 Churchill Papers, 20/13
33 Burns, *The History of the British West Indies*, pp. 702–3
34 Stewart, *Burying Cæsar*, p. 430
35 Acton, *Memoirs of an Aesthete*, p. 369
36 Ousby, *Occupation*, p. 109
37 Dannreuther, *Somerville's Force H*, p. 31
38 ed. Hardy, *Isaiah Berlin*, p. 337
39 ed. Parrish, *Encyclopaedia*, p. 155
40 Roberts, *Eminent Churchillians*, p. 177
41 ed. Hardy, *op. cit.*, p. 326
42 Lang Papers, 176 ff. 348
43 Ian Sayers Papers, 5 March 1961
44 *Ibid.*, 6 March 1961
45 Barker, *Children of the Benares*, p. 114
46 *National Geographic*, vol. lxxviii, no. 6, December 1940, p. 822
47 Colton, 'Aviation in Commerce and Defence', p. 685
48 ed. Parrish, *Encyclopaedia*, pp. 362–3
49 Goodhart Papers, Eng Mss c.2925
50 Norman Longmate in the *Literary Review*, October 2005, p. 36
51 *Ibid.*
52 Klimmer, 'Everyday Life in Wartime England', p. 497
53 *Ibid.*, p. 533
54 Mortimer, *The Longest Night, passim*
55 Fraser Papers, series 1, file 8
56 Fraser Papers, series 1, file 3
57 ed. Butler, *My Dear Mr Stalin*, p. 4

58 Beevor, *Berlin, passim*
59 Lindbergh Papers, Container 36
60 Montagu Curzon in the *Spectator*, 20 August 2005, p. 42
61 Ross, *The Last Hero*, p. 312–13
62 Black, *Franklin Delano Roosevelt*, pp. 687–8
63 ed. Weinberg, *Hitler's Second Book*, *passim*
64 ed. Trevor-Roper, *Hitler's Table Talk 1941–1944*, p. 196
65 eds Eberle and Uhl, *The Hitler Book*, p. 79
66 *Ibid.*

NINE: 1942–4

1 In conversation with the author in September 1989
2 Gilbert, *Churchill and America, passim*
3 Addison, *Churchill*, p. 199
4 ed. Parrish, *Encyclopaedia*, p. 393
5 eds Danchev and Todman, *War Diaries 1939–1945*, pp. 281–2
6 *Ibid., passim*
7 *Ibid.*, p. 650
8 *Ibid.*, p. 249
9 *Ibid.*, p. 473
10 David Dilks, 'The Role of Sir Winston Churchill and the Contribution made by the Commonwealth Countries to the Second World War', p. 24
11 Hasluck, *The Evolution of Australian Foreign Policy 1938–1965*, p. 39
12 *Sydney Morning Herald*, 29 December 1941, p. 6; *Canberra Times*, 29 December 1941, p. 2
13 Black, *Franklin Delano Roosevelt*, p. 712
14 Dupuy and Dupuy, *Encyclopedia*, p. 1309
15 Black, *op. cit.*, p. 1063
16 Sumiko Higashi in the *TLS*, 2 April 2004
17 Mankiw, *Principles of Economics*, p. 746
18 Nicholas, 'Overlord, Over-Ruled and Over There', p. 46
19 Roseman, *The Villa, the Lake, the Meeting, passim*
20 Zoe Waxman in the *English Historical Review*, cxix, 484, November 2004, p. 1466

21 Davis, *Late Victorian Holocausts, passim*; David Arnold in the *TLS*, 23 March 2001, p. 30
22 Simplich, 'Behind the News in Singapore', pp. 83
23 Bayly and Harper, *Forgotten Armies*, pp. 106–55
24 Anthony Milner in the *TLS*, 7 January 2005, p. 8
25 Birks Papers, MS 1413/56
26 www.naa.gov.au
27 Curtin Papers, RC00810 no. 7/1/2
28 Curtin Papers, A5954/69
29 Fraser Papers, series 1, file 8
30 Heiferman, *World War II*, p. 208
31 Keegan, *Intelligence in War*, pp. 240–8
32 eds Danchev and Todman, *op. cit.*, pp. 335–6
33 Klimmer, 'Everyday Life in Wartime England', p. 522
34 Gary Sheffield in the *TLS*, 31 March 2003, p. 31
35 eds Heiber and Glantz, *Hitler and His Generals*, pp. 615–16
36 Noble Frankland in the *Spectator*, 17 July 2004, p. 33
37 Jackson, *The British Empire in the Second World War, passim*
38 Christopher Lee in the *Literary Review*, September 2004, p. 11
39 *Hansard*, vol. 386
40 *Evening Standard*, 8 February 1943
41 Hinchingbrooke, *Full Speed Ahead!*, p. 22
42 Butler and Butler, *British Political Facts*, p. 287
43 Joanna Bourke in the *TLS*, 8 June 2001, p. 8
44 Noble Frankland letter to the *TLS*, 6 June 2001, p. 17
45 Royle, *Patton*, pp. 17–18
46 *Ibid.*, p. 23
47 *Ibid.*, p. 34
48 Reynolds, *Monty and Patton*, p. 285
49 *Ibid.*, p. 146
50 Whiting, *The Field Marshal's Revenge*, pp. 217–19
51 eds Eberle and Uhl, *The Hitler Book*, p. 175n

52 ed. James, *Churchill Speaks 1897–1963*, pp. 815–17
53 Black, *op. cit.*, p. 881
54 Williams, *Hutchinson*, pp. 328–61
55 Drea, *MacArthur's ULTRA*, p. 226
56 From conversations with James Woolsey of the CIA, Sir John Scarlett of the JIC, Sir Stephen Lander of MI5 and another Intelligence chief
57 Drea, *op. cit.*, pp. 227–8
58 ed. Hardy, *Isaiah Berlin*, p. 486

TEN: 1944–5
1 Gilbert, *D-Day*, p. 127
2 *Ibid.*, p. 57
3 Small, *The Forgotten Dead, passim*
4 Lacouture, *De Gaulle*, pp. 520–7
5 *Ibid.*, p. 521
6 *Ibid.*, p. 523
7 *Ibid.*, pp. 525–6
8 Gilbert, *op. cit.*, p. 61n
9 Interview with Paul Woodadge, 22-23 July 2005
10 D'Este, *Eisenhower*, p. 527
11 Interview with Paul Woodadge, 22–23 July 2005
12 Gilbert, *op. cit.*, p. 86
13 Interview with Paul Woodadge, 22–23 July 2005
14 *Ibid.*
15 Hastings, *Overlord*, pp. 109–10
16 Percy, *A Bearskin's Crimea*, p. 199
17 Gilbert, *op. cit.*, pp. 40–1
18 *Ibid.*, pp. 178–9
19 Interview with Paul Woodadge, 22-23 July 2005
20 Hastings, *Armageddon*, p. 105
21 Marshall Papers, reel 9, p. 252
22 *Ibid.*
23 *Ibid.*, p. 304
24 Overy, *Why the Allies Won*, p. 225
25 Hughes-Wilson, *Military Intelligence Blunders*, pp. 2–3
26 Gilbert, *Churchill*, p. 796
27 NA FO 800/302
28 Gunther, *Inside USA*, chart
29 *Ibid.*
30 *Ibid.*

31 ed. Butler, *My Dear Mr Stalin*, p. xiii
32 ed. Hardy, *Isaiah Berlin*, p. 546
33 Doenecke and Stoler, *Debating Franklin D. Roosevelt's Foreign Policies*, pp. 214–15
34 Martin Gilbert's speech to International Churchill Conference, Lansdowne, Virginia, October 2002
35 Taylor, *Dresden*, Appendix B
36 *Ibid.*, p. 405
37 Joanna Bourke in the *TLS*, 8 June 2001, p. 28
38 Taylor, *op. cit.*, p. 413
39 *Ibid.*, Appendix B
40 Ian Sayer Papers
41 Max Hastings in the *Mail on Sunday*, 9 January 2005, p. 73
42 Ben Fenton in the *Daily Telegraph*, 28 February 2005
43 *Ibid.*
44 Justin Wintle in the *Literary Review*, November 2000, p. 30
45 Dwyer, *De Valera*, p. 283
46 Ørvik, *The Decline of Neutrality*, p. 271
47 Astley, *The Inner Circle*, p. 206
48 Attlee Papers, Mss Eng c.4793 fol. 32
49 *Ibid.*, fol. 12
50 Fry, 'A Reconsideration of the British General Election of 1935 and the Electoral Revolution of 1945', *passim*
51 *Military Illustrated*, no. 200, January 2005, p. 25
52 ed. Bowman, *Chronicle of the Twentieth Century*, p. 134
53 *Financial Times* magazine, 6–7 August 2005, p. 27
54 Bourke, *The Second World War*, p. 223
55 Kagan, 'Why America Dropped the Bomb', pp. 17–23
56 Murray Sayle in the *New Yorker*, 31 July 1995; Hiroshima Special Report, *Newsweek*, 24 July 1995
57 *TLS*, 3 May 2002, p. 17
58 Kagan, *op. cit.*, p. 19
59 ed. Parrish, *Encyclopaedia*, p. 458
60 Kagan, *op. cit.*, p. 21
61 Ibid., p. 22
62 Allan Massie, *Daily Telegraph* books review, 19 February 2005, p. 11

63 ed. Parrish, *op. cit.*, p. 632

64 Maddox, *Weapons for Victory, passim*

65 Kagan, *op. cit.*, p. 23

66 *Daily Telegraph* books review, 19 February 2005, p. 11

67 DeGroot, *The Bomb*, p. 99

ELEVEN: 1945–8

1 Browne, *Long Sunset*, p. 41

2 Williams, *Chronology*, p. 597

3 John Horne in the *TLS*, 6 February 2004, p. 4

4 Andrew Kenny in the *Spectator*, 13 August 2005, p. 29

5 Jeffrey Hart in the *New Criterion*, November 2005, p. 29

6 Lind, *Vietnam*, p. 222; *The Times*, 26 January 2006, p. 43

7 ed. Hardy, *Isaiah Berlin*, p. 622

8 Henri Astier in the *TLS*, 10 January 2003, p. 3

9 Donald Cameron Watt in the *TLS*, 17 May 2002

10 Rohwer and Monakov, *Stalin's Ocean-Going Fleet, passim*

11 *BBC History*, January 2005, p. 7

12 Lewis, *Changing Direction*, p. 179

13 *Ibid.*, p. 229

14 *Ibid.*, p. 232

15 Ramsey Papers, 37 ff. 364–70, 15 July 1963

16 British Movietone Newsreel, 15 October 1945

17 Williams, *Hutchinson*, p. 355

18 ed. Pottle, *Champion Redoubtable*, p. 291

19 Dobson, *US Wartime Aid to Britain 1940–1946*, p. 224

20 Bullock, *Ernest Bevin*, pp. 443 and 464 n.2

21 *Ibid.*, p. 476

22 Penney, 'John Douglas Cockcroft 1897–1967', p. 175

23 Reynolds, *Australia's Bid for the Atomic Bomb, passim*

24 ed. James, *Churchill Speaks*, p. 909

25 *Spectator*, 5 November 2005, p. 61

26 Roberts, *Eminent Churchillians*, pp. 126–31

27 Attlee Papers, Mss Eng c.4793 fol. 66

28 Gupta, *Delhi Between Two Empires 1803–1931*, p. 179

29 Attlee Papers, Mss Eng c.4793 fol. 59

30 Graham Macklin in *BBC History*, January 2005, p. 8

31 Isaacson and Thomas, *The Wise Men*, p. 455

32 *Ibid.*, p. 458

33 C.V. Glines on www.indianamilitary.org

34 Tanenhaus, *Whittaker Chambers*, p. 218

35 Herman, *Joseph McCarthy*, pp. 84ff

36 *Ibid.*, pp. 84ff

37 ed. Swan, *Alger Hiss, Whittaker Chambers, and the Schism in the American Soul*, pp. 306–7

38 Tanenhaus, *op. cit.*, p. 169

39 *Ibid.*, pp. 518–20

40 Reeves, *The Life and Times of Joe McCarthy*, p. 674

41 *Ibid.*

42 David Caute in the *Spectator*, 28 October 2000, p. 65

43 Leach, *Europe*, p. 223

44 Grondana, *Commonwealth Stocktaking*, p. xii

45 Gedmin and Kennedy, 'Selling America – Short', p. 71

TWELVE: The 1950s

1 Frank Johnson in the *Spectator*, 13 October 2001, p. 40

2 Donovan, *Tumultuous Years*, pp. 307–10

3 Attlee Papers, Mss Dep. 114 fol. 149

4 *Ibid.*, fol. 153

5 Garrett, *Prisoner of War*, p. 207

6 *Ibid.*, p. 211

7 *Ibid.*, p. 218

8 *New York Times*, 16 November 1998; Weathersby, 'Deceiving the Deceivers', *passim*

9 Simpson, *Human Rights and the End of Empire*, pp. 728–9

10 Bernard Porter in the *TLS*, 8 February 2002, p. 12

11 *TLS*, 18 April 2004, p. 19

12 Saunders, *Who Paid the Piper, passim*

13 Simon Heffer in the *Literary Review*, September 2003, p. 37

14 Murray, *Human Accomplishment*, p. 262

15 Conquest, *The Dragons of Expectation*, p. 158

16 Simon Heffer in the *Literary Review*, September 2003, p. 38

17 William D. Rubinstein in the *TLS*, 26 October 2001, p. 13

18 Deedes, *Brief Lives*, p. 114

19 Torsten Meissner in the *TLS*, 28 June 2002, p. 11

20 Robinson, *The Man Who Deciphered Linear* B, p. 16

21 *Ibid.*, pp. 102–3

22 ed. Willams, *Chronology*, p. 342

23 British Movietone News, 18 April 1955

24 Gaddis, *The Cold War, passim*

25 *Ibid.*, fol. 2

26 Moran, *Winston Churchill*, p. 237

27 James, *Anthony Eden*, p. 453

28 ed. Williams, *The Diary of Hugh Gaitskell 1945–1956*, pp. 551–2

29 ed. Boyle, *The Eden-Eisenhower Correspondence 1955–1957*, pp. 155–6

30 *Ibid.*

31 *Ibid.*

32 Clark Papers, 160 fol. 2

33 Thorpe, *Eden, passim*

34 Martin, *Australian Prime Ministers*, p. 200

35 James, *op. cit.*, p. 557

36 *Spectator*, 13 October 2001, p. 40

37 Diane Kunz in the *Spectator*, 3 November 1990, pp. 25–6

38 *The Times*, 22 November 2004

39 Gerald Frost in the *New Criterion*, November 2004, p. 74

40 John Ramsden in the *TLS*, 10 May 2002, p. 13

41 *Ibid.*

42 Goldsworthy, *Losing the Blanket, passim*

43 Carl Bridge in the *TLS*, 21 February 2003, p. 31

44 *Canadian Historical Review*, vol. 86, no. 1, March 2005, pp. 142–4

45 *International Herald Tribune*, 1 November 2005, p. 6

46 Roberts, *The Holy Fox*, p. 72

47 Kosek, 'Richard Gregg, Mohandas Gandhi and the Strategy of Nonviolence', pp. 1318–22

48 *Ibid.*, p. 1336

49 ed. Coleman, *A History of Georgia*, pp. 362–3

50 *Ibid.*, p. 362

51 Lunghi and Conquest, *Soviet Imperialism*, p. 39

52 Martin Luther King Papers, Howard Gotlieb Research Center, Boston University

53 *Ibid.*

54 *Ibid.*

55 G. Calvin Mackenzie in the *TLS*, 13 October 2000, p. 13

56 Michael Burleigh in the *Literary Review*, December 2005, p. 24

57 Davies, *Mission to Moscow*, pp. 222–30

58 Aitken, *Richard Nixon*, pp. 259–60

59 de Botton, *Status Anxiety*, p. 33

60 Aitken, *op. cit.*, p. 261

61 de Botton, *op. cit.*, p. 34

62 ed. Trevor-Roper, *Hitler's Table Talk*, p. 22

THIRTEEN: The 1960s

1 Kissinger, *American Foreign Policy*, p. 56

2 eds Milner, O'Connor and Sandweiss, *The Oxford History of the American West*, p. 491

3 *Ibid.*

4 Andrew Rotter in the *TLS*, 25 May 2001, p. 12

5 Freedman, *Kennedy's Wars*, pp. ix–xii, 415–19

6 Andrew Rotter in the *TLS*, 25 May 2001, p. 12

7 *Ibid.*

8 Caute, *The Dancer Defects*, p. 484

9 *Ibid.*, p. 485

10 *Ibid.*, p. 487

11 Dimbleby and Reynolds, *An Ocean Apart*, p. 238

12 *Ibid.*, p. 239

13 National Archives PREM 11 fol. 3665

14 www.digitalhistory.uh.edu

15 ed. Asmal, *Nelson Mandela*, pp. 12–13

16 Ramsey Papers, 66 ff. 222, 6 May 1964

17 Simon C. Smith in the *EHR*, cxix 480, (February 2004), p. 271
18 Philip Mandler in the *TLS*, 23 July 2004, p. 13
19 Beschloss, *Taking Charge*, p. 411 n.1
20 Bennett, *The Anglosphere Challenge*, p. 279
21 Perlstein, *Before the Storm, passim;* letters to the *TLS*, 28 June and 19 July 2002, p. 17
22 Browne, *Long Sunset*, p. 327
23 Douglas Johnson in the *TLS*, 24 December 2001
24 Miller, *The Cold War*, p. 386
25 *The Times*, 8 March and 9 March 1966, p. 10
26 *The Times*, 27 April 2004, p. 36
27 Sugden, *Nelson*, p. 3
28 *The Times*, 27 April 2004, p. 36
29 John A. Barnes in the *Opinion Journal, Wall Street Journal*, 24 August 2005
30 Chang and Halliday, *Mao*, p. 537
31 Windschuttle, 'Mao and the Maoists', p. 9
32 Gaddis, *We Now Know*, p. 63
33 Windschuttle, *op. cit.*, pp. 9–11
34 *Guardian*, 9 October 2004
35 Windschuttle, *The Killing of History*, p. 6
36 *Daily Telegraph*, 10 October 2004
37 *Daily Telegraph*, obituary of Derrida, 10 October 2004
38 Lind, *Vietnam*, pp. 254–5
39 *Wall Street Journal*, 3 August 1995
40 Roger Kimball in the *New Criterion*, November 2004, p. 6
41 *The Times*, 23 December 1854, p. 9
42 Michael Burleigh in the *Literary Review*, December 2005, p. 24
43 Frum, *How We Got Here*, p. 345
44 Fraser, *The Hollywood History of the World*, p. 244
45 Lind, *op. cit.*, p. 256
46 Dupuy and Dupuy, *Encyclopedia*, p. 1333
47 Johnson, *A History of the American People*, p. 896
48 Lind, *op. cit.*, p. 261
49 Blainey, *This Land is all Horizons*, p. 58

50 *Ibid.*
51 Sidney Mintz in the *TLS*, 14 September 2001, p. 7
52 *Sunday Star Times* (New Zealand), 30 October 2005, p. 2
53 Pagden, *Peoples and Empires*, p. 133
54 Sinclair McKay in the *Daily Telegraph* Arts, 30 April 2005, p. 6
55 Williams, *Chronology*, p. 662
56 *Literary Review*, September 2005, p. 35

FOURTEEN: The 1970s
1 NA CAB 164/988
2 *Spectator*, 19 May 2001, p. 45
3 William H. Pritchard in the *TLS*, 14 January 2005, p. 19
4 Garrett, *Prisoner of War*, pp. 219–31
5 *Ibid.*, p. 224
6 Isaacson, *Henry Kissinger*, p. 256
7 Kissinger, *American Foreign Policy*, p. 57
8 US Congress House Committee on Internal Security, Travel to Hostile Areas HR 16742, 19–25 September 1972, p. 761
9 *Wall Street Journal*, 2 August 1975
10 *The Times*, 2 April 2005, p. 41
11 *Conservative Weekly News Letter*, vol. 18, no. 37, 15 September 1962, p. 1
12 Tony Bennett in the *TLS*, 9 March 2001, p. 32
13 *The European Journal*, vol. 10, no. 7, May-June 2003, pp. 2–3
14 Leach, *Europe*, p. 223
15 Kissinger, *op cit.*, p. 74
16 Leach, *op. cit.*, p. 223
17 Lal, *In Praise of Empires*, p. 77
18 Bennett, *The Anglosphere Challenge*, p. 279
19 *Sunday Telegraph*, 28 November 2004, p. 12
20 Klaus Dodds in *BBC History*, December 2004, p. 46
21 *Guardian*, 8 November 2002
22 *Daily Telegraph*, 3 August 2004
23 Reeves, *President Nixon*, p. 606
24 Kissinger, *Years of Upheaval*, p. 538
25 *Daily Telegraph*, 26 November 2005, p. 24; *Weekly Standard*, 21 November 2005, p. 19

26 Butler and Butler, *British Political Facts*, pp. 68–9
27 Paul Johnson in the *Spectator*, 18 June 2005, p. 10
28 Walters, *The Mighty and the Meek*, p. 31
29 Johnson, *op. cit.*, p. 920
30 Reeves, *op. cit.*, p. 499
31 Johnson, *op. cit.*, p. 921
32 Nixon, *In the Arena*, p. 35
33 Carroll, *Australia's Governors-General*, p. 159
34 *Ibid.*, p. 162
35 Blainey, *A Shorter History of Australia*, p. 222
36 Foreman, *The Pocket Book of Patriotism*, p. 58
37 Kissinger, *Years of Renewal*, p. 541
38 *Spectator*, 21 November 1998, pp. 25–7
39 Dickie, *The New Mandarins*, p. 86
40 Miller, *The Cold War*, p. 127
41 *Ibid.*, p. 366
42 Lind, *op. cit.*, p. 258
43 William Shawcross in correspondence with the author, 30 November 2005; Niall Ferguson in the *Sunday Telegraph*, 11 December 2005, p. 19
44 Frum, *op. cit.*, p. 308
45 Norman Stone in the *Literary Review*, December 2005, p. 16
46 Frum, *op. cit.*, p. 310
47 *Ibid.*

FIFTEEN: The 1980s

1 Robert Skidelsky in the *Evening Standard*, 30 May 1995, p. 29
2 Federal Reserve website; Cannon, *President Reagan*, p. 275
3 *Guardian*, 11 June 2004
4 eds Skinner, Anderson and Anderson, *Reagan In His Own Hand*, *passim*
5 James Bowman in the *TLS*, 7 September 2001
6 eds Milner, O'Connor and Sandweiss, *The Oxford History of the American West*, p. 530
7 Freedman, *The Official History of the Falklands Campaign*, vol. 1, pp. 3–32
8 Nott, *Here Today, Gone Tomorrow*, p. 257
9 Freedman, *op. cit.*, p. 209
10 Nott, *op. cit.*, p. 258
11 Lawrence Freedman in the *TLS*, 8 June 2001, p. 28
12 Weinberger, *Fighting for Peace*, p. 206
13 Conversation with Caspar Weinberger, 13 October 2005
14 Weinberger, *op. cit.*, p. 205
15 Freedman, *op. cit.*, II, p. 381
16 *Ibid.*, pp. 383–4
17 *Ibid.*, p. 379
18 James Delingpole in the *Spectator*, 10 July 2004, p. 45
19 Ryan, 'Manning Clark', p. 22
20 *Ibid.*, p. 9
21 Partington, 'Australian Anglophobia', p. 14
22 Véliz, 'Bad History', p. 22
23 Partington, *op. cit.*
24 Peter Kelly letter to the *TLS*, 16 February 2001, p. 17
25 *TLS*, 16 February 2001, p. 17; 23 February 2001, p. 17; 9 March 2001, p. 17; 30 March 2001, p. 21; 10 August 2001, p. 15
26 Clark, *Meeting Soviet Man*, pp. 12 and 86
27 Wheatcroft, *The Strange Death of Tory England*, p. 169; Thatcher, *Downing Street Years*, p. 263
28 Cannon, *op. cit.*, p. 317
29 Volkogonov, *The Rise and Fall of the Soviet Empire*, p. 365
30 *Ibid.*, p. 366
31 Cannon, *op. cit.*, p. 440
32 *Washington Post*, 22 November 2005, p. A28
33 Miller, *The Cold War*, p. 30
34 Reagan, *An American Life*, p. 547
35 Weinberger, *op. cit.*, p. 291
36 Thatcher, op. cit., p. 326
37 *Ibid.*, p. 329
38 *Ibid.*, p. 326
39 Cannon, *op. cit.*, p. 448
40 *Ibid.*, p. 449
41 Piers Brendan review in the *Sunday Telegraph* review, 12 June 2005, p. 11
42 *Ibid.*
43 *Spectator*, 5 November 2005, p. 61

44 Padraic Kenney in the *TLS*, 14 January 2005, p. 10

45 Keith Joseph Memorial Lecture to Centre for Policy Studies, 11 January 1996, p. 5

46 *Auckland Star*, 6 March 1985, p. A8

47 *New Zealand Herald*, 4 March 1985, p. 11

48 *Auckland Star*, 8 March 1985, p. A6

49 *New Zealand Herald*, 11 March 1985, p. 8

50 Volkogonov, *op. cit.*, p. 492

51 *Ibid.*, p. 449

52 Biggs, 'The Q-Man', pp. 117–18

53 *Ibid.*, p. 118

54 Library of Congress, *The Iran-Contra Affair*, p. 25

55 Iran-Contra Affair Papers, item 02614

56 Library of Congress, *op. cit.*, p. 25

57 Stewart, *History of The Times*, p. 387

58 www.state.gov

59 *Daily Telegraph*, 28 January 2005, p. 23

60 *Ibid.*

61 *Spectator*, 3 December 2005, p. 36

62 David Pryce-Jones in the *Literary Review*, May 2002, p. 36

63 Pryce-Jones, *The War That Never Was*, p. 105

64 Robert Conquest in the *TLS*, 2 June 1995, p. 5

65 Sharansky, *The Case for Democracy*, pp. 137–8

66 *Daily Telegraph*, 25 February 2005, p. 16

67 Pryce-Jones, *op. cit.*, p. 161

68 *Sunday Telegraph* review, 7 May 1995, p. 9

SIXTEEN: 1990–11 September 2001

1 *Melbourne Age*, 24 October 2005, p. 13

2 Stewart, *History of The Times*, p. 386

3 *Atlantic Monthly*, September 1990

4 Schwarzkopf, *It Doesn't Take A Hero*, p. 316

5 Robert Harris in the *Daily Telegraph*, 10 May 2005

6 Dupuy and Dupuy, *Encyclopaedia*, pp. 1479–80

7 Justin Wintle in the *Literary Review*, November 2000, p. 30

8 *Washington Post*, 21 November 2005, p. A15

9 *Ibid.*

10 Carlton, *The West's Road to 9/11*, p. 230

11 *Ibid.*, p. 232

12 *Spectator*, 19 January 2002, p. 40

13 I.M. Lewis in the *TLS*, 8 June 2001, p. 4

14 *Ibid.*

15 Paul Johnson in the *Sunday Telegraph* review, 16 January 2005, p. 11

16 Law No. 94-665 of 4 August 1994 Relative to the Use of the French Language

17 Tom Ladner in *Global Network*, 10 December 1996, p. 2

18 Davidson, *Voltaire in Exile*, pp. 30n and 63n

19 *The Times*, 31 January 2005, p. 31

20 *Ibid.*

21 Claiborne, *The Life and Times of the English Language*, p. 289

22. *Ibid.*, p. viii

23 Bragg, *The Adventure of English*, pp. 297–312

24 *Ibid.*, p. 305

25 *Spectator*, 13 November 2004, p. 40

26 Diarmid Ó Muirithe in the *Literary Review*, February 2005, p. 42

27 Ostler, *Empires of the Word*, pp. 541–9

28 *Sunday Telegraph* review, 27 February 2005, p. 13

29 Alan Taylor in *The Author*, summer 2005, p. 79

30 G. Calvin Mackenzie in the *TLS*, 13 October 2000, p. 13

31 *Ibid.*, p. 14

32 *TLS*, 15 June 2001, p. 18

33 *The Times*, 27 March 1997

34 *Ibid.*

35 Bryan Appleyard, *Sunday Times*, 17 November 1996

36 *Ibid.*

37 *BBC History*, January 2005, pp. 36–9

38 Simms, *Unfinest Hour*, p. 316

39 *Ibid.*, p. 317

40 *Ibid.*, p. 50

41 James Pettifer in the *TLS*, 10 May 2002, p. 29

42 Simms, *op. cit.*, p. 50

43 *Ibid.*, *passim*

44 James Pettifer in the *TLS*, 10 May 2002, p. 29

45 *Ibid.*

46 Cradock, *In Pursuit of British Interests*, p. 191

47 Simms, *op. cit.*, p. 52

48 Mark Mazower in the *TLS*, 14 February 2003, p. 6

49 Gedmin and Kennedy, 'Selling America – Short', p. 72

50 *Sunday Telegraph*, 28 March 2004

51 *Ibid.*

52 Amir Taheri in the *Sunday Telegraph* review, 22 August 2004, p. 11

53 *Ibid.*

54 *Ibid.*

55 Burke, *Al-Queda*, *passim*; Hazhir Teimourian in the *Literary Review*, September 2003, p. 36

56 *Daily Telegraph*, 29 July 2005, p. 1

57 Simms, *op. cit.*, p. 345

58 Seldon, *Blair*, p. 399

59 *Spectator*, 27 November 2004, p. 48

60 Clinton, *My Life*, p. 925

61 Florida State website

62 Peter Berkowitz in the *TLS*, 27 July 2001, p. 27

63 *Ibid.*

64 John Adamson in the *Sunday Telegraph* review, 25 July 2004

65 *Ibid.*

66 Anthony Daniels in the *Spectator*, 3 July 2004, p. 35

67 *Ibid.*

SEVENTEEN: 11 September 2001–15 December 2005

1 *Washington Post*, 22 November 2005, p. A28

2 Letter from Robert Eschbach to *The Times*, 12 November 2004

3 *National Post*, 8 May 2004

4 *Washington Post*, 12 December 2004, p. A24

5 Steven Beller in the *TLS*, 15 March 2002, p. 14; Shawcross, *Allies*, p. 13

6 Tertrais, *War Without End*, p. 46

7 The 9/11 Commission, *Final Report . . .*, p. xvi

8 Seldon, *Blair*, p. 484

9 *Ibid.*

10 *Ibid.*, p. 491

11 Tyrie, 'Mr Blair's Poodle Goes to War', p. 9 n.9

12 Richard Beeston in the *Spectator*, 22 October 2005, p. 52

13 *Spectator*, 2 April 2005, p. 15

14 Hoffer, *The True Believer*, p. ix

15 *Spectator*, 15 October 2005, p. 30

16 George Walden in the *Sunday Telegraph* review, 27 March 2005, p. 11; Henri Astier in the *TLS*, 10 January 2003, p. 3

17 White House Background Paper on Iraq, 12 September 2002

18 Senator John McCain in conversation with the author, 30 November 2004

19 *International Herald Tribune*, 16 November 2004, p. 8

20 Jack Straw, *Today programme*, 26 January 2004

21 Melanie Phillips in the *Spectator*, 27 September 2003, p. 18; George Tenet at Georgetown University, 5 February 2004; *The Times*, 7 February 2004, p. 2; *Hansard*, 24 September 2002, col. 4

22 Browne, *Long Sunset*, p. 166

23 *Spectator*, 24 April 2004

24 White House Background Paper on Iraq, 12 September 2002

25 Tertrais, *op. cit.*, p. 43

26 *Ibid.*, p. 44

27 *Daily Telegraph*, 6 June 2004

28 ed. Stelzer, *Neoconservatism*, pp. 271–2

29 McIntyre, *Josiah the Great*, p. 308

30 *Ibid.*, p. 73

31 Keegan, *The Iraq War*, p. 100

32 *Daily Telegraph*, 19 September 2005, p. 12

33 Lal, *In Defence of Empires*, p. 57

34 *The Times*, 22 April 2004, p. 5; *Sydney Morning Herald*, 29–30 October 2005, p. 17

35 *The Times*, ibid.

36 *Sunday Telegraph*, 25 April 2004, p.24

37 *Sydney Morning Herald*, 29–30 October 2005, p. 17

38 Ellis, *His Excellency*, p. 123

39 Stothard, *Thirty Days*, p. 42

40 *Weekly Standard*, 17 November 2005, p. 17

41 *Sunday Times* review, 19 June 2005, p. 9

42 *Hansard*, 18 March 2003, col. 772

43 *Sunday Times*, 1 August 2004, p. 7; Franks, *American Soldier*, ch. 10, 'The Plan'

44 *New York Post*, 26 April 2004

45 *The Times*, 24 April 2004, p. 11

46 *Daily Telegraph*, 29 September 2005, p. 2

47 *Washington Times*, 21 November 2005, p. A6

48 John Pilger in the *Daily Mirror*, 29 January 2003

49 Brendon O'Neill in the *Spectator*, 21 January 2006, p. 23

50 *New Statesman*, 17 May 2004

51 Donald, *Lincoln*, p. 299

52 *Sunday Times* review, 19 June 2005, p. 9

53 By the time of writing in December 2005, 2,151 US servicemen had died in Iraq; the figures for earlier wars' dead are taken from Dupuy and Dupuy, *Encyclopaedia*, pp. 1083, 1309, 1365 and 1333

54 Riddell, *Hug Them Close*, p. 1

CONCLUSION

1 McCullough, *John Adams*, p. 135

2 *The Times* T2, 27 July 2005, p. 2

3 Anthony Browne in the *Spectator*, 23 July 2005, p. 11; United Nations Population Division figures

4 Landeg White in the *TLS*, 26 October 2001, p. 11

5 www.heritage.org/research/welfare/bg1713.cfm

6 Murray, *Human Accomplishment*, p. 433

7 Walter Lacquer in the *TLS*, 21 February 2003, pp. 9-10

8 *Ibid.*

9 Anthony Browne in the *Spectator*, 23 July 2005, p. 10

10 Hitchings, *Dr Johnson's Dictionary*, p. 138

11 Jonathan Freedland in the *Spectator*, 7 August 2004, p. 14

12 *Sunday Telegraph*, 30 January 2005, p. 24

13 Ian Buruma in the *Financial Times* magazine, 20 August 2005, p. 24

14 Henri Astier in the *TLS*, 10 January 2003, p. 3

15 *Ibid.*

16 *Finest Hour*, winter 2005–6, no. 129, p. 22

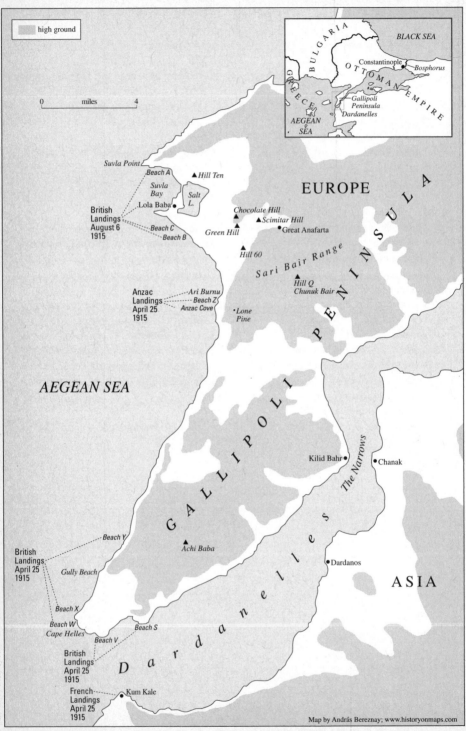

high ground

0   miles   4

BULGARIA

BLACK SEA

GREECE

OTTOMAN EMPIRE

Constantinople   Bosphorus

Gallipoli
Peninsula
Dardanelles

AEGEAN
SEA

Suvla Point

Beach A

Suvla
Bay

▲ Hill Ten

EUROPE

Salt
L.

Lola Baba ●

British
Landings
August 6
1915

Beach C

Beach B

Green Hill

Chocolate Hill
▲
▲ ▲ Scimitar Hill
▲ ● Great Anafarta

▲
Hill 60

Sari Bair Range

Anzac
Landings
April 25
1915

Ari Burnu
Beach Z
Anzac Cove

● Hill Q
Chunuk Bair

● Lone
Pine

GALLIPOLI PENINSULA

AEGEAN SEA

Kilid Bahr ●

● Chanak

The Narrows

British
Landings
April 25
1915

Beach Y

▲
Achi Baba

● Dardanos

Gully Beach

ASIA

Beach X

Beach W
Cape Helles

Beach S

Beach V

Dardanelles

British
Landings
April 25
1915

French
Landings
April 25
1915

● Kum Kale

Map by András Bereznay; www.historyonmaps.com

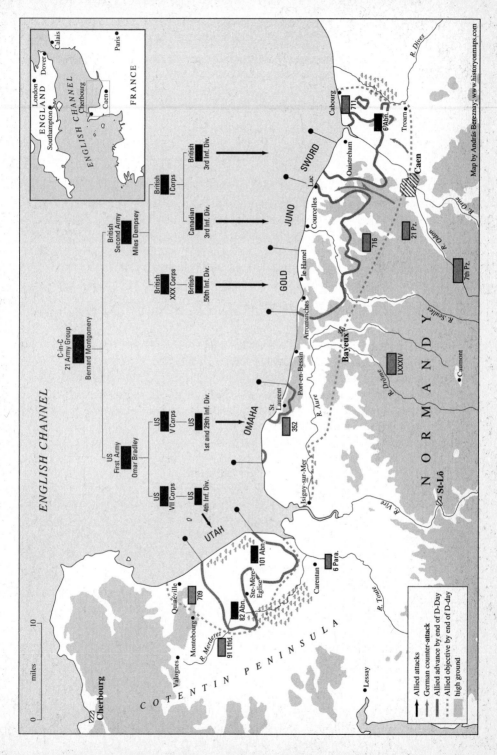

Map by András Bereznay; www.historyonmaps.com

ENGLISH CHANNEL

Cherbourg

COTENTIN PENINSULA

miles
0   10

Valognes

Montebourg

R. Merderet

Quineville

91 Lftd.

709

82 Abn.

Ste-Mère-Eglise

101 Abn.

6 Para.

Carentan

Lessay

R. Taute

UTAH

US
VII Corps

US
4th Inf. Div.

US
First Army

Omar Bradley

US
V Corps

US
1st and 29th Inf. Div.

OMAHA

St Laurent

Isigny-sur-Mer

R. Vire

St-Lô

R. Aure

352

Port-en-Bessin

Arromanches

Caumont

LXXXIV

R. Drôme

R. Seulles

Bayeux

GOLD

le Hamel

British
XXX Corps

British
50th Inf. Div.

British
Second Army

Miles Dempsey

Courcelles

JUNO

Luc

Canadian
3rd Inf. Div.

716

7th Pz.

R. Odon

21 Pz.

Caen

R. Orne

Ouistreham

SWORD

British
I Corps

British
3rd Inf. Div.

Cabourg

711

6 Abn.

Troarn

R. Dives

N O R M A N D Y

C-in-C
21 Army Group

Bernard Montgomery

INSET MAP:

ENGLAND

London
Dover
Calais
Southampton
Cherbourg
Caen

ENGLISH CHANNEL

Paris

FRANCE

LEGEND:

Allied attacks

German counter-attack

Allied advance by end of D-Day

Allied objective by end of D-day

high ground

BIBLIOGRAPHY

Since any attempt to establish a comprehensive bibliography for a subject so vast as the history of the English-speaking peoples in the twentieth century would be absurd, I merely list those archives I have visited and the books, newspapers and learned articles that I have consulted, quoted from or referred to in the text. All books were published in London unless otherwise stated. Due to microfilming and the internet, several manuscript collections can be accessed from different institutions. I have listed only those I visited.

ARCHIVE AND MANUSCRIPT COLLECTIONS

George Bliss Agnew (New York Public Library)
A.V. Alexander (Churchill Archives Centre, Cambridge)
Robert Jackson Alexander (New York Public Library)
Dean Alfange (New York Public Library)
H.H. Asquith (Bodleian Library, Oxford)
Clement Attlee (Bodleian Library, Oxford)
Newton Diehl Baker (New York Public Library)
Stanley Baldwin (Cambridge University Library)
Sir Edmund Barton (National Archives of Australia, Canberra)
The Beefsteak Club (by kind permission of the Club Secretary)
Professor J.D. Bernal (Cambridge University Library)
Poultney Bigelow (New York Public Library)
Thomas Lawrence Birks (Auckland War Memorial Museum Library)
Sol Bloom (New York Public Library)
Lord Brand (Bodleian Library, Oxford)
Brooks's Club (by kind permission of the Club Secretary)
Viscount Bryce (Bodleian Library, Oxford)
Lord Cadogan (Churchill Archives Centre, Cambridge)
Ashley T. Cole (New York Public Library)
Sir Joseph Cook (National Archives of Australia, Canberra)
Sir Winston Churchill (Churchill Archives Centre, Cambridge)
William Clark (Bodleian Library, Oxford)
Gordon Coates (Alexander Turnbull Library, National Library of New Zealand)
William Bourke Cochran (New York Public Library)
Calvin Coolidge (Manuscript Division, Library of Congress)
John Curtin (National Archives of Australia, Canberra)
D.H. Davis (Auckland War Memorial Museum Library)

Alfred Deakin (National Archives of Australia, Canberra)

The Marquess of Crewe (Cambridge University Library)

Archbishop Randall Thomas Davidson (Lambeth Palace Library)

Sir Alfred Ewing (by kind permission of Mr David Wills)

Lord Fisher (Churchill Archives Centre, Cambridge)

Archbishop Geoffrey Fisher (Lambeth Palace Library)

George William Forbes (Alexander Turnbull Library, National Library of New Zealand)

Henry Atherton Forster (New York Public Library)

Peter Fraser (Archives of New Zealand)

William Gerhardie (Cambridge University Library)

Charles Goodell (New York Public Library)

Arthur Lehman Goodhart (Bodleian Library, Oxford)

Lord Gore-Booth (Bodleian Library, Oxford)

William Hall-Jones (Alexander Turnbull Library, National Library of New Zealand)

Lord Hankey (Churchill Archives Centre)

Warren G. Harding (Manuscript Division, Library of Congress)

Gertrude Heyman (New York Public Library)

Sidney Holland (Alexander Turnbull Library, National Library of New Zealand)

Keith Holyoake (Alexander Turnbull Library, National Library of New Zealand)

Sir Archibald Hurd (Churchill Archives Centre)

Harold L. Ickes (Manuscript Division, Library of Congress)

Iran-Contra Affair (Manuscript Division, Library of Congress)

Martin Luther King (Boston University Library)

Archbishop Cosmo Gordon Lang (Lambeth Palace Library)

Alfred Lee (New South Wales State Library, Sydney)

Alan Lennon-Boyd (Bodleian Library, Oxford)

Charles A. Lindbergh (Manuscript Division, Library of Congress)

Lord Lloyd (Churchill Archives Centre)

Vito Marcantonio (New York Public Library)

George C. Marshall (Manuscript Division, Library of Congress)

H.E. McVeagh (Auckland War Memorial Museum Library)

William McKinley (Manuscript Division, Library of Congress)

Sir Robert Menzies (National Archives of Australia, Canberra)

Joseph Molloy (Auckland War Memorial Museum Library)

Walter Nash (Alexander Turnbull Library, National Library of New Zealand)

John J. Pershing (Manuscript Division, Library of Congress)

Archbishop Arthur Michael Ramsey (Lambeth Palace Library)

Sir George Houston Reid (National Archives of Australia, Canberra)

Theodore Roosevelt (Manuscript Division, Library of Congress, and Boston University Library)

Elihu Root (Manuscript Division, Library of Congress)

Cameron Sadler (Auckland War Memorial Museum Library)

Joseph Savage (Alexander Turnbull Library, National Library of New Zealand)

Ian Sayer (by kind permission of Mr Ian Sayer)

Lady Schonland (Cambridge University Library)

Sir George Schuster (Bodleian Library, Oxford)

Richard Seddon (Archives of New Zealand)

Sir Louis Spears (Churchill Archives Centre, Cambridge)
William Howard Taft (Manuscript Division, Library of Congress)
Archbishop William Temple (Lambeth Palace Library)
Viscount Templewood (Cambridge University Library)
Lord Vansittart (Churchill Archives Centre)
John Wallace (Auckland War Memorial Museum Library)
John Christian Watson (National Archives of Australia, Canberra)
Sir Frank Whittle (Churchill Archives Centre)
Stuart Wilson (Auckland War Memorial Museum Library)
Woodrow Wilson (Manuscript Division, Library of Congress)
Christoper York (private possession)
Sir Alfred Zimmern (Bodleian Library)

NEWSPAPERS AND PERIODICALS

The Age (Melbourne)
The American Historical Review
The American Spectator
The Argus (Australia)
Atlantic Monthly *Auckland Star*
The Australian
Australian Association for the Advancement of Science
Australian Historical Studies
Australian Journal of Politics and History
Australian Outlook
Australian Quarterly
BBC History Magazine
Barbados Museum and Historical Society Journal
Canadian Historical Review
Canberra Times
Chronicles
Commentary
Contemporary British History
Daily Telegraph
Dominion Post (New Zealand)
Economist
Encounter
The European Journal
Finest Hour
Foreign Affairs
History Today
International Affairs
Irish Historical Studies
Journal of American History

Journal of Imperial and Commonwealth History
Quadrant
Quarterly Journal of Military History
Literary Review
Melbourne Herald-Standard
National Post (Toronto)
New York City Round
New Yorker
National Geographic
National Interest
The National Observer (Australia)
New Criterion
New York Review of Books
New York Times
New Zealand Herald
New Zealand Institute
New Zealand Journal of History
The Observer
Picture Post
Proceedings of the American Philosophical Society
Quadrant (Australia)
Quarterly Review
Royalty Digest
Royal United Services Institute Journal
Salisbury Review
South China Morning Post
Spectator
Star Times (New Zealand)
Stand To!: The Journal of the Western Front Association

Sydney Morning Herald

The Press (New Zealand)

The Times

Times Higher Educational Supplement

Times Literary Supplement

Uganda Journal

Wall Street Journal

Washington Post

Washington Times

Weekly Standard

BOOKS

Ackerman, Marsha E., *Cool Comfort: America's Romance with Air-Conditioning*, (Washington DC) 2002

Adams, R.J.Q., *Bonar Law*, 1999

Addison, Paul, *Churchill: The Unexpected Hero*, 2005

Adler, Selig, *The Isolationist Impulse: Its Twentieth Century Reaction*, 1957

Aitken, Jonathan, *Richard Nixon*, 1993

Alperovitz, Gar, *Atomic Diplomacy: Hiroshima and Potsdam*, 1965

Amis, Martin, *Koba the Dread*, 2002

Anderson, David, *Histories of the Hanged: Britain's Dirty War and the End of the Empire*, 2005

Andrew, Christopher, *Théophile Delcassé and the Making of the Entente Cordiale*, 1968

Andrew, Christopher, and Mitrokhin, Vasili, *The Mitrokhin Archive: The KGB in Europe and the West*, 1999

Angier, F.R., and others, *The Making of the West Indies*, 1961

The Annual Register, 1901

Armitage, David, *The Ideological Origins of the British Empire*, 2001

Askin, *Gallipoli: A Turning Point*, (undated)

ed. Asmal, Kader, *Nelson Mandela: In His Own Words*, 2004

Astley, Joan Bright, *The Inner Circle: A View of War at the Top*, 1971

Atkinson, Neill, *Adventures in Democracy: A History of the Vote in New Zealand*, (Dunedin) 2003

Ayearst, Morley, *The British West Indies: The Search for Self-Government*, 1960

Babbin, Jed, *Inside the Asylum: Why the United Nations and Old Europe are Worse Than You Think*, (Washington DC) 2004

Baker, Anne Pimlott, *The Pilgrims Society of Great Britain: A Centennial History*, 2002

Balfour, Patrick, *Society Racket*, 1933

Baly, Lindsay, *Horsemen, Pass By: The Australian Light Horse in World War I*, 2004

Barker, Ralph, *Children of the Benares*, 1987

Barnett, Correlli, *The Swordbearers: Studies in Supreme Command in the First World War*, 1963

Barnett, Correlli, *The Audit of War*, 1986

Barnett, Correlli, *The Lost Victory*, 1995

Barnett, Correlli, *The Verdict of Peace*, 2001

Barone, Michael, *Hard America Soft America*, (New York) 2004

Barone, Michael, and Cohen, Richard, *The Almanac of American Politics*, (Chicago) 2004

Bartholomew, James, *The Welfare State We're In*, 2004

Bayly, Christopher, and Harper, Tim, *Forgotten Armies: The Fall of British Asia 1941-1945*, 2004

Beale, Howard K., *Theodore Roosevelt and the Rise of America to World Power*, (Baltimore) 1956

Beard, Charles A., and Beard, Mary R., *The Rise of American Civilisation*, vol. 2, 1930

Beckett, Ian F.W., *A Nation in Arms*, 1985

Beesley, Patrick, *Room 40: British Naval Intelligence 1914-18*, 1982

Beevor, Antony, *Stalingrad*, 1998

Beevor, Antony, *Berlin: The Downfall*, 2002

Belfield, Eversley, *The Boer War*, 1975

Belich, James, *Paradise Reforged: A History of the New Zealanders from the 1880s to the Year 2000*, 2002

Bell, David V., *The Roots of Disunity: A Study of Canadian Political Culture*, (Toronto) 1992

Bennett, James C., *The Anglosphere Challenge*, 2004

Bercuson, David, *One Christmas in Washington: Roosevelt and Churchill Forge the Grand Alliance*, 2005

Bergreen, Laurence, *Capone: The Man and His Era*, (New York) 1994

Bernhardi, F. von, *Germany and the Next War*, 1911

Bernstein, George L., *The Myth of Decline: The Rise of Britain Since 1945*, 2004

Beschloss, Michael R., *Taking Charge: The Johnson White House Tapes 1963-64*, (New York) 1998

Best, Geoffrey, *Churchill: A Study in Greatness*, 2001

Best, Geoffrey, *Churchill and War*, 2005

Bhagwati, Jagdish, *In Defence of Globalisation*, 2004

Bigelow, Poultney, *Prussian Memories 1864-1914*, 1916

Bigelow, Poultney, *Seventy Summers*, 1925

Black, Conrad, *Franklin Delano Roosevelt: Champion of Freedom*, 2003

Black, Jeremy, *The British Seaborne Empire*, 2004

Blainey, Geoffrey, *The Tyranny of Distance*, 1968

Blainey, Geoffrey, A *Shorter History of Australia*, 2000

Blainey, Geoffrey, *This Land is all Horizons: Australian Fears and Visions*, (Sydney) 2001

Blount, James H., *The American Occupation of the Philippines 1898-1912*, 1912

Booker, Christopher, *A Looking-Glass Tragedy: The Controversy over the Repatriations from Austria in 1945*, 1997

Bork, Robert H., *Coercing Virtue: The Worldwide Rule of Judges*, (Canada) 2002

Bourke, Joanna, *The Second World War: A People's History*, 2001

ed. Bowman, John S., *Chronicle of the Twentieth Century*, 2004

ed. Boyer, Paul S., *The Oxford Companion to United States History*, 2001

ed. Boyle, Peter G., *The Eden-Eisenhower Correspondence 1955-1957*, (North Carolina) 2005

Bozo, Frédéric, *Two Strategies for Europe: De Gaulle, the United States and the Atlantic Alliance*, 2001

Bradley, James, *Flags of Our Fathers*, 2000

Bragg, Melvyn, *The Adventure of English: The Biography of a Language*, 2003

Braithwaite, Rodric, *Across the Moscow River: The World Turned Upside Down*, 2002

Brands, H.W., *T.R.: The Last Romantic*, 1997

ed. Braybon, Gail, *Evidence, History and the Great War: Historians and the Impact of 1914-18*, 2003

Brendan, Piers, *Ike: The Life and Times of Dwight D. Eisenhower*, 1987

Brogan, Hugh, *The Penguin History of the United States of America*, 1999

eds Brown, Judith, and Louis, William Roger, *The Oxford History of the British Empire*, vol. IV, 1999

Brown, Kevin, *Alexander Fleming and the Antibiotics Revolution*, 2004

Browne, Anthony, *The Sleep of Reason*, 2005

Browning, Christopher R., *Ordinary Men: Reserve Police Battalion 101 and the Final Solution in Poland*, 1998

ed. Bruce, Lawrence, *Messages to the World: The Statements of Osama bin Laden*, 2005

Buchan, John, *The Thirty-Nine Steps*, 2000

Buchanan, Patrick J., *A Republic, Not an Empire: Reclaiming America's Destiny*, 1999

Bullock, Alan, *Hitler and Stalin: Parallel Lives*, 1991

Bullock, Alan, *Ernest Bevin: A Biography*, 2002

Bunting, Sir John, *R. G. Menzies: A Portrait*, 1988

Burdon, R.M., *The New Dominion: A Social and Political History of New Zealand 1918-1939*, 1965

Burgess, Simon, *Stafford Cripps: A Political Life*, 1999

Burke, Jason, *Al-Queda: Casting a Shadow of Terror*, 2003

Burleigh, Michael, *Earthly Powers: Religion and Politics from the Enlightenment to the Great War*, 2005

Burns, Sir Alan, *The History of the British West Indies*, 1965

ed. Busch, Briton C., *Canada and the Great War*, (Montreal) 2003

Butler, David, and Butler, Gareth, *British Political Facts 1900-1994*, 1994

ed. Butler, Susan, *Mr Dear Mr Stalin: The Complete Correspondence of Franklin D. Roosevelt and Joseph V. Stalin*, (New York) 2005

Butow, Robert J.C., *Japan's Decision to Surrender*, (Stanford) 1954

Cadbury, Deborah, *Seven Wonders of the Industrial World*, 2004

Caesar, James W., and Busch, Andrew E., *The Perfect Tie: The True Story of the 2000 Presidential Election*, 2001

Cameron Watt, Donald, *Succeeding John Bull*, 1984

Campbell, Charles S., *From Revolution to Rapprochement: The United States and Great Britain 1783-1900*, (New York) 1974

Campbell, John, *Edward Heath*, 1993

Cannon, Lou, *President Reagan: The Role of a Lifetime*, 1991

ed. Careless, J.M.S., *The Canadians 1867-1967*, 1968

Carlton, David, *The West's Road to 9/11: Resisting, Appeasing and Encouraging Terrorism Since 1970*, 2005

Carroll, Brian, *Australia's Governors-General: From Hopetoun to Jeffery*, (Sydney) 2004

Castellar-Gassol, J., *Dalí: A Perverse Life*, 1984

Catchpole, Brian, *The Korean War: 1950-53*, 2000

Caute, David, *The Dancer Defects: The Struggle for Cultural Supremacy During the Cold War*, 2003

Cecil, Algernon, *British Foreign Secretaries*, 1927

Cecil, Algernon, *Queen Victoria and her Prime Ministers*, 1953

Chambers, James, *Palmerston*, 2004

ed. Chambers, John Whiteclay, *The Oxford Companion to American Military History*, 1999

Chambers, Whittaker, *Witness*, 1953

Chang, Jung, and Halliday, Jon, *Mao: The Untold Story*, 2005

Charmley, John, *Duff Cooper*, 1986

Chasseaud, Peter, and Doyle, Peter, *Grasping Gallipoli*, 2005

Chaudhuri, Nirad, *Thy Hand, Great Anarch!*, 1987

Chisholm, Anne, and Davie, Michael, *Beaverbrook*, 1992

Churchill, Winston, *The Aftermath*, 1929

Churchill, Winston, *A History of the English-Speaking Peoples*, vol. 4, 2002

Claiborne, Robert, *The Life and Times of the English Language*, 1990

Clark, Alan, *The Tories: Conservatives and the Nation State 1922-97*, 1999

Clark, Charles Manning, *Meeting Soviet Man*, 1960

Clark, Charles Manning, *The Quest for Grace*, 1991

Clark, Jonathan, *Our Shadowed Present: Modernism, Postmodernism and History*, 2003

Clarke, Richard A., *Against All Enemies: Inside America's War Against Terror*, 2004

Clinton, Bill, *My Life*, 2004

ed. Clowes, William Laird, *The Royal Navy: A History*, vol. VII, 1901

ed. Coates, Tim, *War in the Falklands 1982*, 2001

Cocker, Mark, *Rivers of Blood, Rivers of Gold: Europe's Conflict with Tribal Peoples*, 1998

ed. Coleman, Kenneth, *A History of Georgia*, 1991

Coleman, Verna, *The Last Exquisite: A Portrait of Frederic Manning*, (Melbourne) 1990

Collett, Nigel, *The Butcher of Amritsar: General Reginald Dyer*, 2005

Collins, Larry, and Lapierre, Dominique, *Freedom at Midnight*, 1976

Colvin, John, *Twice Around the World*, 1991

Condiliffe, J.B., *New Zealand in the Making*, 1930

Condiliffe, J.B., and Willis, T.G. Airey, *A Short History of New Zealand*, 1935

Connelly, Mark, *Reaching for the Stars: A New History of Bomber Command in World War Two*, 2001

Conquest, Robert, *The Great Terror*, 1968

Conquest, Robert, *Reflections on a Ravaged Century*, 1999

Conquest, Robert, *The Dragons of Expectation*, 2005

Conway Maritime Press, *Conway's All the World's Fighting Ships 1860-1905*, 1979

Coogan, Tim Pat, *De Valera: Long Fellow, Long Shadow*, 1996

Corbin, Jane, *The Base: In Search of Al-Queda*, 2002

Cornwell, John, *Hitler's Scientists: Science, War and the Devil's Pact*, 2003

Correspondents of the *New York Times*, *Thirty-Six Days: The Complete Chronicle of the 2000 Presidential Election Crisis*, (New York) 2001

Corrigan, Gordon, *Mud, Blood and Poppycock: Britain and the Great War*, 2004

Corrigan, Gordon, *Blood, Sweat and Arrogance and the Myths of Churchill's War*, 2006

Coughlin, Con, *Saddam: The Secret Life*, 2005

Coward, Noël, *The Lyrics of Noël Coward*, 1983

ed. Cowley, Robert, *The Great War: Perspectives on the First World War*, 2004

Cradock, Sir Percy, *In Pursuit of British Interests: Reflections on Foreign Policy under Margaret Thatcher and John Major*, 1997

Cray, Ed, *General of the Army: George C. Marshall*, 1990

Critchell, J.T., and Raymond, J., *History of the Frozen Meat Industry*, 1912

ed. Cross, Mary, *A Century of Icons: 100 Products and Slogans from the Twentieth Century Consumer Culture*, (Westport, CT) 2004

Crouch, Tom, and Jakab, Peter, *The Wright Brothers and the Invention of the Aerial Age*, (New York) 2003

Crystal, David, *English as a Global Language*, 2003

Dallas, Gregor, *1918: War and Peace*, 2000

Danchev, Alex, *On Specialness: Essays in Anglo-American Relations*, 2000

eds Danchev, Alex, and Todman, Daniel, *War Diairies 1939-1945: Field Marshal Lord Alanbrooke*, 2001

Danner, Mark, *Torture and Truth: Abu Ghraib and America in Iraq*, 2005

Dannreuther, Raymond, *Somerville's Force H: The Royal Navy's Gibraltar-based Fleet, June 1940 to March 1942*, 2005

Davidson, Ian, *Voltaire in Exile: The Last Years 1753-78*, 2004

Davie, Michael, *Anglo-Australian Attitudes*, 2000

Davies, David Twiston, *The Daily Telegraph Book of Military Obituaries*, 2003

Davies, Joseph E., *Mission to Moscow*, 1942

Davis, Kenneth Sydney, *FDR: The War President 1940-1943*, 2001

Davis, Mike, *Late Victorian Holocausts: El Nino Famines and the Making of the Third World*, 2000

Dawson, Robert MacGregor, *The Development of Dominion Status 1900-1936*, 1937

Day, Beth, *The Philippines: Shattered Showcase of Democracy in Asia*, (New York) 1974

Day, David, *Menzies and Churchill at War*, 1986

Dean, John, *Worse Than Watergate*, 2004

de Botton, Alain, *Status Anxiety*, 2004

DeConde, Alexander, *The American Secretary of State: An Interpretation*, 1962

Deedes, William, *Brief Lives*, 2004

DeGroot, Gerard, *The Bomb: A History of Hell on Earth*, 2004

Dell, Edmund, *A Strange Eventful History: Democratic Socialism in Britain*, 2000

Denton, J.A., *When Hell Was in Session*, 1976

D'Este, Carlos, *Eisenhower: Allied Supreme Commander*, 2002

Dickie, John, *The New Mandarins: How British Foreign Policy Works*, 2004

Dilks, David, *The Great Dominion: Winston Churchill in Canada 1900-1954*, 2005

Dimbleby, David, and Reynolds, David, *An Ocean Apart: The Relationship between Britain and America in the Twentieth Century*, 1988

Dionne, E.J., and Kristol, William, *Bush v Gore: The Court Cases and the Commentary*, (Washington DC) 2001

Dobson, Alan P., *US Wartime Aid to Britain 1940-1946*, 1986

Dodd, Clement H., *Discord on Cyprus: The UN Plan and After*, 2004

Dodson, Alan P., *US Wartime Aid to Britain 1940-1946*, 1986

Doenecke, Justus D., *Storm on the Horizon: The Challenge to American Intervention 1939-1941*, (New York) 2000

Doenecke, Justus D., and Stoler, Mark A., *Debating Franklin D. Roosevelt's Foreign Policies 1933-1945*, 2005

ed. Doenecke, Justus D., *In Danger Undaunted: The Anti-Interventionist Movement of 1940-41 as Revealed in the Papers of the America First Committee*, (Stanford) 1990

Donald, David Herbert, *Lincoln*, 1995

Donovan, Robert J., *Conflict and Crisis: The Presidency of Harry S. Truman 1945-48*, 1977

Donovan, Robert J., *Tumultuous Years: The Presidency of Harry S. Truman 1949-53*, 1982

Douglas, Sir Arthur P., *The Dominion of New Zealand*, 1909

Dow, Mark, *American Gulag: Inside US Immigration Prisons*, (California) 2004

Drea, Edward J., *MacArthur's ULTRA: Codebreaking and the War against Japan 1942-45,* (University of Kansas) 1992

Drehle, David von, *Deadlock: The Inside Story of America's Closest Election,* 2001

Drummond, J., *The Life and Times of Richard John Seddon,* (Christchurch, *NZ)* 1906

Duggan, John P., *Neutral Ireland and the Third Reich,* (Dublin) 1989

Dupuy, R. Ernest, and Dupuy, Trevor N., *The Collins Encyclopaedia of Military History from 3,500BC to the Present,* 1993

Dutton, David, *Anthony Eden: A Life and Reputation,* 1997

Dwyer, T. Ryle, *Irish Neutrality and the USA,* 1977

Dwyer, T. Ryle, *De Valera: The Man and the Myths,* 1995

eds Eberle, Henrik, and Uhl, Matthias, *The Hitler Book: The Secret Dossier Prepared for Stalin,* 2005

Eden, Guy, *Portrait of Churchill,* (undated)

Egremont, Max, *Balfour: A Life of Arthur James Balfour,* 1980

Egremont, Max, *Under Two Flags: The Life of Major General Sir Edward Spears,* 1997

Elkins, Caroline, *Britain's Gulag: The Brutal End of Empire in Kenya,* 2005

Elliott, Florence, *A Dictionary of Politics,* 1973

Ellis, John, *The Social History of the Machine Gun,* 1987

Ellis, Joseph J., *Founding Brothers,* 2000

Ellis, Joseph J., *His Excellency: George Washington,* 2004

Elton, Lord, *Imperial Commonwealth,* 1945

Emmott, Bill, *20:21 Vision: The Lessons of the Twentieth Century for the 21st,* 2003

Evans, E.W., *The British Yoke: Reflections on the Colonial Empire,* 1949

Evans, Harold, *Men in the Tropics: A Colonial Anthology,* 1949

Evans, Harold, *The American Century,* 1998

Evans, Harold, *They Made America,* 2004

Evans, Luther Harris, *The Virgin Islands: From Naval Base to New Deal,* (Ann Arbor) 1945

Ewing, A.W., *The Man of Room 40: The Life of Sir Alfred Ewing,* 1939

Farrell, Nicholas, *Mussolini: A New Life,* 2003

Ferguson, John H., *American Diplomacy and the Boer War,* (Philadelphia) 1939

Ferguson, Niall, *The Pity of War,* 1998

Ferguson, Niall, *Colossus: The Rise and Fall of the American Empire,* 2004

Fernández, Leandrotti, *The Philippine Republic,* 1926

Fernández-Armesto, Felipe, *The Americas: The History of a Hemisphere,* 2003

Fischer, Fritz, *Germany's Aims in the First World War,* 1967

Fishman, Jack, *If I Lived My Life Again,* 1974

Fisk, Robert, *The Great War for Civilisation,* 2005

ed. Fleming, Laurence, *Last Children of the Raj: British Childhoods in India 1919-1950,* 2 vols, 2004

Fleming, Thomas, *The Illusion of Victory: America in World War One,* (New York) 2003

Foner, Nancy, *From Ellis Island to JFK: New York's Two Great Waves of Immigration,* (New York) 2001

Forbes, W. Cameron, *The Philippine Islands,* (Harvard) 1945

Foreman, Jonathan, *The Pocket Book of Patriotism,* 2005

Forward, Alan, *You Have Been Allocated Uganda,* 1999

Foster, R.F., *Modern Ireland 1600-1972,* 1988

Fox, Annette Baker, *Freedom and Welfare in the Caribbean: A Colonial Dilemma*, (New York) 1949

Fradkin, Philip L., *The Great Earthquake and Firestorms of 1906*, (Berkeley) 2006

Franken, Al, *Lies and the Lying Liars Who Tell Them*, 2003

Franks, General Tommy, *American Soldier*, 2004

Fraser, General Sir David, *Alanbrooke*, 1982

Fraser, George Macdonald, *The Hollywood History of the World*, 1988

Fraser, George Macdonald, *Quartered Safe Out Here*, 2000

Fraser, Steve, *Wall Street: A Cultural History*, 2005

Freedland, Jonathan, *Bring Home the Revolution: How Britain Can Live the American Dream*, 1998

Freedman, Lawrence, *Kennedy's Wars*, 2001

Freedman, Lawrence, *The Official History of the Falklands Campaign*, 2 vols, 2005

French, Patrick, *Liberty or Death: India's Journey to Independence and Division*, 1997

Freytag-Loringhoven, Baron von, *Deductions from the World War*, 1918

Friedman, Milton, and Friedman, Rose, *Free to Choose*, 1979

Friedman, Walter A., *Birth of a Salesman: The Transformation of Selling in America*, (Harvard) 2004

Fromkin, David, *Europe's Last Summer: Why the World Went to War in 1914*, 2004

Frum, David, *How We Got Here: The Seventies*, 2000

Frum, David, *The Right Man: The Surprise Presidency of George W. Bush*, 2003

Fuller, J.F.C., *Empire Unity and Defence*, 1934

Fussell, Paul, *The Boys' Crusade: American GIs in Europe*, 2004

Gaddis, John Lewis, *We Now Know: Rethinking Cold War History*, 1997

Gaddis, John Lewis, *The Cold War*, 2006

Garraty, John A., *The American Nation: A History of the United States*, (New York) 1991

Garrett, Richard, *Prisoner of War: The Uncivil Face of War*, 1981

Gaskin, M.J., *Blitz: The Story of 29th December 1940*, 2005

Gee, H.L., *American England*, 1943

eds Gere, J.A., and Sparrow, John, *Geoffrey Madan's Notebooks*, 1981

Gerolymatos, André, *The Balkan Wars*, 2005

Gilbert, Sir Martin, *Churchill: A Life*, 1991

Gilbert, Sir Martin, *D-Day*, 2004

Gilbert, Sir Martin, *Churchill and America*, 2005

Gilbert, Sir Martin, and Churchill, Randolph, *Winston S. Churchill*, 8 vols, 1966-88

ed. Gilbert, Sir Martin, *The Straits of War: Gallipoli Remembered*, 2000

Gilmour, David, *The Long Recessional: The Imperial Life of Rudyard Kipling*, 2002

Gilmour, David, *The Ruling Caste: Imperial Lives in the Victorian Raj*, 2005

Glazebrook, G.P. de T., *A Short History of Canada*, 1950

Glees, Anthony, and Davies, Philip H., *Spinning the Spies: Intelligence, Open Government and the Hutton Inquiry*, 2004

Gliddon, Gerald, *The Aristocracy and the Great War*, (privately published) 2002

Goldsworthy, David, *Losing the Blanket: Australia and the End of Britain's Empire*, (Melbourne) 2002

Gombrich, E.H., *A Little History of the World*, 2005

Gordon-Duff, John, *It Was Different Then*, (privately published) 1976

Granatstein, J.L., *Twentieth-Century Canada*, (Toronto) 1986

Graubard, Stephen, *The Presidents: The Transformation of the American Presidency from Theodore Roosevelt to George W. Bush*, 2005

Gray, Herbert Branston, and Turner, Samuel, *Eclipse or Empire?*, 1916

Grayling, A.C., *Among the Dead Cities: Was the Allied Bombing of Civilians in World War II a Necessity or a Crime?*, 2006

Greer, Germaine, *Whitefella Jump Up*, 2004

Griffin, David Ray, *The New Pearl Harbor: Disturbing Questions about the Bush Administration and 9/11*, 2004

Griffiths, Sir Percival, *Empire Into Commonwealth*, 1969

Griffiths, Richard, *Patriotism Perverted: Captain Ramsay, the Right Club and British Anti-Semitism 1939-40*, 1998

Grigg, John, *Lloyd George: War Leader 1916-1918*, 2002

Grondana, L. St Clare, *Commonwealth Stocktaking*, 1953

Groot, Gerard De, *The Bomb: A Life*, 2004

Grunberger, Richard, *A Social History of the Third Reich*, 2005

Gullace, Nicoletta F., '*The Blood of our Sons*': *Men, Women and the Renegotiation of British Citizenship During the Great War*, 2002

Gunther, John, *Inside USA*, 1948

Gupta, Narayani, *Delhi Between Two Empires 1803-1931: Society, Government and Urban Growth*, 1981

Halevy, Efraim, *Man in the Shadows: Inside the Middle East Crisis With a Man Who Led the Mossad*, 2006

Hamshere, Cyril, *The British in the Caribbean*, 1972

Hanson, Victor Davis, *Why the West Has Won: Carnage and Culture from Salamis to Vietnam*, 2002

ed. Hardy, Henry, *Isaiah Berlin: Flourishing, Letters 1928-1946*, 2005

Harris, Kenneth, *Attlee*, 1995

Harrison, Henry, *Ireland and the British Empire*, 1937

Harrison, Ian, *The Book of Firsts: The Stories Behind the Amazing Breakthroughs of the Modern World*, 2003

Hart, Robert A., *The Great White Fleet*, (Boston) 1965

Hasluck, P., *The Evolution of Australian Foreign Policy 1938-1965*, (Canberra) 1970

Hastings, Max, *Overlord: D-Day and the Battle for Normandy*, 1984

Hastings, Max, *The Korean War*, 1988

Hastings, Max, *Armageddon: The Battle for Germany 1944-45*, 2004

Hawke, G.R., *The Making of New Zealand: An Economic History*, 1985

Hayek, F.A., *The Constitution of Liberty*, 1999

Hays, Constance, *Pop: Truth and Power at the Coca-Cola Company*, 2004

eds Heiber, Helmut, and Glantz, David M., *Hitler and His Generals: Military Conferences 1942-1945*, 2002

Heiferman, Ronald, *World War II*, 1973

Hendrik, Burton J., *The Life and Letters of Walter Hines Page*, 3 vols, 1922

Hennessy, Peter, *The British Prime Minister*, 2000

Herman, Arthur, *Joseph McCarthy*, 2000

Herman, Arther, *To Rule the Waves: How the British Navy Shaped the Modern World*, (New York) 2004

Hersh, Seymour, *Chain of Command: The Road from 9/11 to Abu Ghraib*, 2004

Hinchingbrooke, Viscount, *Full Speed Ahead! Essays in Tory Reform*, 1943

Hirst, John, *The Sentimental Nation: The Making of the Australian Commonwealth*, 2002

Hitchens, Christopher, *The Trial of Henry Kissinger*, 2001

Hitchens, Peter, *The Abolition of Liberty*, 2004

Hitchings, Henry, *Dr Johnson's Dictionary*, 2005

Hoffer, Eric, *The True Believer*, (New York) 1966

Hogg, Quintin, *The Left Was Never Right*, 1945

Holland, Richard, *Augustus: Godfather of Europe*, 2004

Holmes, Richard, *Tommy: The British Soldier on the Western Front 1914-1918*, 2005

Holmes, Richard, *In the Footsteps of Churchill*, 2005

Holt, Thaddeus, *The Deceivers: Allied Military Deception in the Second World War*, 2004

Horne, Alistair, *The Terrible Year: The Paris Commune 1871*, 2004

Horne, John, and Kramer, Alan, *German Atrocities, 1914*, 2000

Howarth, David, *The Dreadnoughts*, 1980

Howe, Glenford Deroy, *Race, War and Nationalism: A Social History of West Indians in the First World War*, 2002

Hughes-Wilson, John, *Military Intelligence Blunders*, 1999

Hull, Cordell, *The Memoirs of Cordell Hull*, 2 vols, 1948

Hull, Mark M., *Irish Secrets: German Espionage in Wartime Ireland 1939-1945*, 2003

Huntingdon, Samuel P., *Who Are We?: America's Great Debate*, 2004

Hurd, Douglas, *Memoirs*, 2003

Hyam, Ronald, and Henshaw, Peter, *The Lion and the Springbok: Britain and South Africa since the Boer War*, 2003

Ignatieff, Michael, *Human Rights as Politics and Idolatry*, (Princeton) 2001

Inglis, Brian, *The Story of Ireland*, 1965

Isaacson, Walter, *Henry Kissinger*, 1992

Isaacson, Walter, *Benjamin Franklin: An American Life*, 2003

Isaacson, Walter, and Thomas, Evan, *The Wise Men: Six Friends and the World They Made*, 1986

Iwan-Müller, E.B., *Lord Milner and South Africa*, 1902

Jackson, Ashley, *The British Empire and the Second World War*, 2006

Jackson, John, *Private 12768: Memoir of a Tommy*, 2004

Jackson, Julian, *Charles De Gaulle*, 2003

Jackson, Keith, and Harré, John, *New Zealand*, 1969

James, C.L.R., *The Case for West Indian Self-Government*, 1933

James, Lawrence, *Imperial Warrior: The Life and Times of Field Marshal Viscount Allenby 1861-1936*, 1993

James, Lawrence, *Warrior Race: A History of the British at War*, 2002

James, Robert Rhodes, *Anthony Eden*, 1986

ed. James, Robert Rhodes, *Churchill Speaks 1897-1963: Collected Speeches in Peace and War*, 1981

ed. Jane, Fred T., *Fighting Ships, 1914*

Jenkins, Roy, *Baldwin*, 1987

Jenkins, Roy, *Churchill*, 2001

Jenkins, Roy, *Franklin Delano Roosevelt*, 2003

ed. Johnson, Frank, *Conversations with Hugh Trevor-Roper*, (unpublished) 2005

Johnson, Paul, *A History of the American People*, 2000

Johnston, Mark, and Stanley, Peter, *Alamein: The Australian Story*, 2002

Jones, Nigel, *Rupert Brooke: Life, Death & Myth*, 1999

Jones, Nigel, *Mosley*, 2004

Judd, Alan, *The Quest for C: Sir Mansfield Cumming and the Founding of the British Secret Service*, 1999

Judd, Denis, and Surridge, Keith, *The Boer War*, 2002

Kagan, Robert, *Paradise and Power*, 2003

Kamm, Antony, and Baird, Malcolm, *John Logie Baird*, 2002

Kamm, Oliver, *Anti-Totalitarianism: The Left-Wing Case for a Neoconservative Foreign Policy*, 2005

Kampfner, John, *Blair's Wars*, 2003

Karabell, Zachary, *Parting the Desert: The Creation of the Suez Canal*, 2003

Kavanagh, Dennis, and Seldon, Anthony, *The Powers Behind the Prime Minister: The Hidden Influence of Number Ten*, 1999

Keegan, John, *Intelligence in War: Knowledge of the Enemy from Napoleon to Al-Qaeda*, 2003

Keegan, John, *The Iraq War*, 2004

Keesing, Felix M., *The Philippines: A Nation in the Making*, 1937

Kemp, Paul, *The Admiralty Regrets: British Warship Losses of the Twentieth Century*, 1999

Kennan, George, *Memoirs 1925-50*, (New York) 1988

Kennedy, Paul, *The Rise and Fall of the Great Powers*, 1988

Kennedy, Robert F., *Thirteen Days: The Cuban Missile Crisis 1962*, 1968

Keogh, Dermot, *Jews in Twentieth-Century Ireland: Refugees, Anti-Semitism and the Holocaust*, (Cork) 1999

Kershaw, Ian, *Hitler: Hubris 1889–1936*, 1998

Kershaw, Ian, *Hitler: Nemesis 1936-1945*, 2000

Keynes, John Maynard, *The Economic Consequences of the Peace*, 1920

Kinne, Derek, *The Wooden Boxes*, 1955

Kinross, Patrick, *Atatürk: The Rebirth of a Nation*, 1993

Kissinger, Henry A., *Nuclear Weapons and Foreign Policy*, 1957

Kissinger, Henry A., *A World Restored: Castlereagh, Metternich and the Restoration of Peace 1812-1822*, 1957

Kissinger, Henry A., *The Troubled Partnership: A Reappraisal of the Atlantic Alliance*, 1965

Kissinger, Henry A., *American Foreign Policy: Three Essays*, 1969

Kissinger, Henry A., *The White House Years*, 1979

Kissinger, Henry A., *Years of Upheaval 1973-77*, 1982

Kissinger, Henry A., *United States Policy in Central America*, 1984

Kissinger, Henry A., *Diplomacy: The History of Diplomacy and the Balance of Power*, 1994

Kissinger, Henry A., *Years of Renewal*, 1999

Kissinger, Henry A., *Does America Need a Foreign Policy?: Towards a Diplomacy for the 21st Century*, 2001

Kissinger, Henry A., *Ending the Vietnam War*, 2003

Klimmer, Harvey, *They'll Never Quit*, 1941

eds Knight, Franklin W., and Palmer, Colin A., *The Modern Caribbean*, 1989

Knight, Roger, *The Pursuit of Victory: The Life and Achievement of Horatio Nelson*, 2005

Knightley, Phillip, *Australia: A Biography of a Nation*, 2000

Kristol, Irving, *Neo-Conservatism: Selected Essays 1949-1995*, 1995

Kupchan, Charles A., *The End of the American Era: US Foreign Policy and the Geopolitics of the Twenty-First Century*, (New York) 2005

Lacour-Guyet, *A History of South Africa*, 1977

Lacouture, Jean, *De Gaulle: The Rebel 1890-1944*, 1990

Lal, Deepak, *In Praise of Empires: Globalization and Order*, 2004

Lamb, Richard, *Mussolini and the British*, 1997

Landes, David S., The *Wealth and Poverty of Nations*, 1998

eds Lane, Ann, and Temperley, Harold, *The Rise and Fall of the Grand Alliance 1941-45*, 1995

Leach, Rodney, *Europe*, 2004

Lee, John, *The Warlords: Hindenburg and Ludendorff*, 2005

Lefever, Dr Robert, *The Diary of a Private Doctor*, 1988

Leitz, Christian, *Nazi Germany and Neutral Europe During the Second World War*, 2000

Lentin, Antony, *Lloyd George and the Lost Peace: From Versailles to Hitler 1919-1940*, 2001

Le Queux, William, *Britain's Deadly Peril*, 1915

Levin, Bernard, *The Pendulum Years*, 1970

Lewis, Bernard, *The Crisis of Islam*, 2003

Lewis, Julian, *Changing Direction: British Military Planning for Post-War Strategic Defence 1942-1947*, 1988

Lind, Michael, *Vietnam: The Necessary War*, 1999

Lipset, Martin, and Marks, Gary, *It Didn't Happen Here: Why Socialism Failed in the United States*, 2000

Lloyd George, David, *War Memoirs of David Lloyd George*, 2 vols, 1938

Lloyd George, David, *The Truth About the Peace Treaties*, vol. I, 1938

Longden, Sean, *Hitler's British Slaves: British and Commonwealth PoWs in German Industry 1939-1945*, 2005

ed. Louis, William Roger, *Yet More Adventures with Britannia: Personalities, Politics and Culture in Britain*, 2005

Lovat, Lord, *March Past*, 1978

Lownie, Andrew, *John Buchan: The Presbyterian Cavalier*, 1995

Ludendorff, General, *Ludendorff's Memoirs 1914-1918*, 1933

Lukacs, John, *Churchill: Visionary, Statesman, Historian*, 2002

Lukacs, John, *Remembered Past*, (Delaware) 2005

Lukacs, John, *Democracy and Populism: Fear and Hatred*, 2005

Lunghi, Hugh, and Conquest, Robert, *Soviet Imperialism*, 1962

'L.W.', *Fascism: Its History and Significance*, 1924

Lycett, Andrew, *Rudyard Kipling*, 1999

ed. Lycett, Andrew, *Rudyard Kipling: Selected Poems*, 2004

Lyons, F. S.L., *The Irish Parliamentary Party 1890-1910*, vol. 4, 1951

MacArthur, Brian, *Surviving the Sword: Prisoners of the Japanese 1942-45*, 2005

eds MacDonagh, Oliver, and Mandle, I.W.F., *Ireland and Irish-Australia*, 1986

MacDonald, Robert, *The Fifth Wind: New Zealand and the Legacy of a Turbulent Past*, 1989

Mackinder, Sir Halford J., *Democratic Ideals and Reality*, 1919

Mackinder, Sir Halford J., *The Scope and Methods of Geography and the Geographical Pivot of History*, 1951

Macmillan, Margaret, *Peacemakers: The Paris Conference of 1919*, 2001

Maddox, Robert, *Weapons for Victory: The Hiroshima Decision Fifty Years Later*, (Missouri)
    1995
Major, John, *The Autobiography*, 1999
Mankiw, N. Gregory, *Principles of Economics*, (New York) 2003
Mann, Jessica, *Out of Harm's Way: The Wartime Evacuation of Children from Britain*,
    2005
Mantoux, Étienne, *The Carthaginian Peace, or The Economic Consequences of Mr Keynes*,
    1952
Márai, Sándor, *Embers*, 2003
Martin, Allan, *Australian Prime Ministers*, 2000
ed. Maurice, Frederick, *The History of the War in South Africa 1899-1902*, 4 vols, 1908
M'Carthy, Michael J.F., *Five Years in India 1895-1900*, 1901
McCallum, Ronald, *Public Opinion and the Last Peace*, 1944
McConville, Michael, *Ascendancy to Oblivion: The Story of the Anglo-Irish*, 1986
McCormack, W.J., *Roger Casement in Death, or Haunting the Free State*, (Dublin) 2002
McCormick, Donald, *The Mask of Merlin: A Critical Study of David Lloyd George*, 1963
McCullough, David, *John Adams*, 2001
McGarry, Fearghal, *Eion O'Duffy: A Self-Made Hero*, 2005
ed. McGibbon, Ian, *The Oxford Companion to New Zealand Military History*, 2000
McGrath, John, *Prisoner of War: Six Years in Hanoi*, 1975
McIntyre, Ben, *Josiah the Great*, 2004
McKercher, Brian C.J., *The Second Baldwin Government and the United States 1924-1929:
    Attitudes and Diplomacy*, 1984
McKinstry, Leo, *Rosebery: Statesman in Turmoil*, 2005
Meacham, John, *Franklin and Winston: A Portrait of a Friendship*, 2004
Mead, Gary, *The Doughboys: America and the Great War*, 2000
Medawar, Jean, and Pyke, David, *Hitler's Gift: Scientists Who Fled Nazi Germany*, 2001
Meikle, Louis S., *The Confederation of the British West Indies versus Annexation to the United
    States of America*, 1912
Meltzer, Allan H., *A History of the Federal Reserve, vol. I 1913-1951*, 2003
Menzies, Sir Robert, *Afternoon Light*, 1967
Meredith, Martin, *The State of Africa: A History of Fifty Years of Independence*, 2005
Messenger, Charles, *Call To Arms: The British Army 1914-18*, 2005
Micklethwait, John, and Wooldridge, Adrian, *The Company: A Short History of a
    Revolutionary Idea*, 2003
Micklethwait, John, and Wooldridge, Adrian, *The Right Nation: Why America is Different*,
    2004
Miller, David, *The Cold War: A Military History*, 1998
Miller, Nathan, *The United States Navy*, 1977
Millin, S.G., *General Smuts*, 2 vols, 1936
eds Milner, Clyde A., O'Connor, Carol A., and Sandweiss, Martha A., *The Oxford History
    of the American West*, 1994
Mombauer, Annika, *Helmuth von Moltke and the Origins of the First World War*, 2001
Montague Browne, Sir Anthony, *Long Sunset*, 1995
Montefiore, Simon Sebag, *Stalin: The Court of the Red Tsar*, 2003
Moore, Sara, *How Hitler Came to Power*, 2006
Moran, Lord, *Winston Churchill: The Struggle for Survival 1940-1965*, 1966

Morefield, Jean, *Covenants Without Swords: Idealist Liberalism and the Spirit of Empire*, 2004

Morley, Sheridan, *A Talent to Amuse: A Biography of Noël Coward*, 1986

Morris, Edmund, *Theodore Rex*, 2002

Morrison, S.A., *Middle East Survey*, 1954

Mortimer, Gavin, *The Longest Night: Voices from the London Blitz*, 2005

Mulanax, Richard B., *The Boer War and American Politics and Diplomacy*, (Maryland) 1994

Munro, Dana G., *Intervention and Dollar Diplomacy in the Caribbean*, (Princeton) 1964

Murphy, James H., *Abject Loyalty: Nationalism and Monarchy during the Reign of Queen Victoria*, (Cork) 2001

Murray, Charles, *Human Accomplishment: The Pursuit of Excellence in the Arts and Sciences 800 BC to 1950*, 2003

Murray, Douglas, *Neoconservatism: Why We Need It*, 2005

Naughtie, James, *The Accidental American: Tony Blair and the Presidency*, 2004

Neillands, Robin, *A Fighting Retreat: The British Empire 1947-97*, 1996

Neillands, Robin, *The Great War Generals of the Western Front 1914-1918*, 1998

Nelson, Raymond, *The Philippines*, 1968

Nicolson, Harold, *Peacemaking 1919*, 1933

Nicolson, Nigel, *Long Life*, 1997

Nixon, Richard, *In the Arena: A Memoir of Victory, Defeat and Renewal*, 1990

ed. Norwich, John Julius, *The Duff Cooper Diaries*, 2005

Nott, John, *Here Today, Gone Tomorrow: Recollections of an Errant Politician*, 2002

ed. O'Day, Alan, *Reactions to Irish Nationalism*, 1987

Ó Drisceoil, Donal, *Censorship in Ireland 1939-1945*, 1996

Odo, Franklin, *No Sword to Bury: Japanese Americans in Hawai'i during World War II*, (Philadelphia) 2004

O'Gara, Gordon C., *Theodore Roosevelt and the Rise of the Modern Navy*, (Princeton) 1943

O'Halpin, Eunan, *The Decline of the Union: British Government in Ireland 1892-1920*, 1981

O'Hegarty, Patrick, *A History of Ireland under the Union 1801-1922*, 1952

Oliver, W.H., *The Story of New Zealand*, 1960

ed. Oliver, W.H., and Williams, B.R., *The Oxford History of New Zealand*, 1981

eds Omissi, David, and Thompson, Andrew S., *The Impact of the South African War*, 2002

Ørvik, Nils, *The Decline of Neutrality 1914-1945*, 1953

Oshinsky, David, *Polio: An American Story*, 2005

Ostler, Nicholas, *Empires of the Word: A Language History of the World*, 2005

Ousby, Ian, *Occupation: The Ordeal of France 1940-1944*, 1997

Overy, Richard, *Why the Allies Won*, 1995

Overy, Richard, *Bomber Command 1939-1945*, 1997

Overy, Richard, *Interrogation: The Nazi Elite in Allied Hands, 1945*, 2001

Owen, Roger, *Lord Cromer: Victorian Imperialist, Edwardian Proconsul*, 2004

Ozment, Steven, *A Mighty Fortress: A New History of the German People*, 2004

Packer, Herbert L., *Ex-Communist Witnesses: Four Studies in Fact Finding*, 1962

Pagden, Anthony, *Peoples and Empires*, 2001

Page, Max, *The Creative Destruction of Manhattan 1900-1940*, (Chicago) 2001

Parker, R.A.C., *Churchill and Appeasement*, 1995

Parker, R.A.C., *The Second World War*, 2001

ed. Parrish, Thomas, *The Simon and Schuster Encyclopaedia of World War II*, (New York) 1978

Paxman, Jeremy, *The English: A Portrait of a People*, 2002

Pelling, H., *America and the British Left*, 1956

Penlington, Norman, *Canada and Imperialism 1896-1899*, (Toronto) 1965

Percy, Algernon, *A Bearskin's Crimea*, 2005

Perkins, Bradford, *The Great Rapprochement: US.-British 1895-1914*, (New York) 1968

Perle, Richard, and Frum, David, *An End to Evil*, 2004

Perlstein, Rick, *Before the Storm: Barry Goldwater and the Unmaking of the American Consensus*, 2002

Pershing, John J., *My Experiences in the World War*, 2 vols, 1931

ed. Pickersgill, J.W., *The Mackenzie King Record, vol. 1 1939-1944*, (Toronto) 1960

Pilpel, Robert H., *Churchill in America 1895-1961*, 1976

Pipes, Richard, *Three Whys of the Russian Revolution*, (Toronto) 1995

Pollock, Frederick, *The League of Nations*, 1920

Pollock, John, *Kitchener*, 1998

Ponsonby, Sir Frederick, *Recollections of Three Reigns*, 1957

Porch, Douglas, *Hitler's Mediterranean Gamble*, 2004

ed. Pottle, Mark, *Champion Redoubtable: The Diaries and Letters of Violet Bonham Carter 1914-45*, 1998

ed. Pottle, Mark, *Daring to Hope: The Diaries and Letters of Violet Bonham Carter 1946-1969*, 2000

Powell, Colin, *My American Journey*, 1995

Powell, J. Enoch, and Maude, Angus, *Biography of a Nation*, 1955

Powell, Jim, *FDR's Folly: How Roosevelt and his New Deal Prolonged the Great Depression*, 2004

Prados, John, *Hoodwinked: The Documents that Reveal how Bush Sold Us a War*, 2004

Prochaska, Frank, *The Republic of Britain: 1760-2000*

Pugsley, Christopher, *The Anzac Experience: New Zealand, Australia and the Empire in the First World War*, (Auckland) 2004

Pryce-Jones, David, *The War That Never Was: The Fall of the Soviet Empire 1985-1991*, 1995

ed. Ramsden, James, *George Lyttelton's Commonplace Book*, 2002

Read, David Herbert, *Lincoln*, 1995

Reagan, Ronald, *An American Life*, 1990

Rees, Laurence, *The Nazis: A Warning from History*, 1997

Rees, Laurence, *Auschwitz: The Nazis and the 'Final Solution'*, 2005

Reeves, Richard, *President Nixon: Alone in the White House*, 2002

Reeves, Thomas C., *The Life and Times of Joe McCarthy*, 1982

Reeves, William Pember, *The Long White Cloud*, 1924

Reid, T.R., *The United States of Europe: The New Superpower and the End of American Supremacy*, (New York) 2004

Rennell, Tony, *Last Days of Glory: The Death of Queen Victoria*, 2000

Renwick, Robin, *Fighting with Allies: America and Britain in Peace and War*, 1996

Reyes, A.M. José S., *A Legislative History of America's Economic Policy Toward the Philippines*, (New York) 1923

Reynolds, Michael, *Monty and Patton: Two Paths to Victory*, 2005

Reynolds, P.A., *British Foreign Policy in the Inter-War Years*, 1954

Reynolds, Wayne, *Australia's Bid for the Atomic Bomb*, (Melbourne) 2000

Riddell, Peter, *Hug Them Close: Blair, Clinton, Bush and the 'Special Relationship'*, 2003

Ring, Jim, *How the English Made the Alps*, 2000

Ritter, Gerhard, *The Schlieffen Plan*, 1958

Roberts, Andrew, *The Holy Fox: A Life of Lord Halifax*, 1991

Roberts, Andrew, *Eminent Churchillians*, 1994

Roberts, Andrew, *Salisbury: Victorian Titan*, 1999

Robertson, Patrick, *Shell Book of Firsts*, 1974

Robinson, Andrew, *The Man Who Deciphered Linear B: The Story of Michael Ventris*, 2002

Rodger, N.A.M., *The Safeguard of the Sea*, 1997

Roger, Philippe, *The American Enemy: The History of French Anti-Americanism*, 2005

Rohwer, Jürgen, and Monakov, Mikhail S., *Stalin's Ocean-Going Fleet: Soviet Naval Strategy and Shipbuilding 1935-1953*, 2002

Rolo, P.J.V., *The Entente Cordiale*, 1969

Ronaldshay, Earl of, *The Life of Lord Curzon*, 2 vols, 1927

Rose, Kenneth, *Superior Person: A Portrait of Curzon and his Circle in Late Victorian England*, 1969

Rose, Kenneth, *King George V*, 1983

Rose, Norman, *The Cliveden Set: Portrait of an Exclusive Fraternity*, 2000

Roseman, Mark, *The Villa, the Lake, the Meeting: Wannsee and the Final Solution*, 2002

Rosenthal, Eric, *Stars and Stripes in Africa*, (Cape Town) 1968

Roskill, Stephen, *Hankey: Man of Secrets*, vol. 2, 1972

Ross, Angus, *New Zealand's Record in the Pacific Islands in the Twentieth Century*, (Auckland) 1969

Ross, Walter Sanford, *The Last Hero: Charles A. Lindbergh*, 1976

Roth, Joseph, *Hotel Savoy*, 2000

Rovere, Richard H., *Senator Joe McCarthy*, 1959

Royle, Trevor, *Patton: Old Blood and Guts*, 2005

Sainsbury, Keith, *Churchill and Roosevelt at War*, 1994

Sarna, Jonathan D., *American Judaism: A History*, (Yale) 2004

Saroop, Narindar, *The Last Indian*, 2004

Saunders, Frances Stonor, *Who Paid the Piper? The CIA and the Cultural Cold War*, 1999

Sayer, Ian, and Botting, Douglas, *Nazi Gold*, 1998

Schmitz, Oscar, *The Land Without Music*, (undated)

Schneer, Jonathan, *London 1900: The Imperial Metropolis*, 1999

Scholefield, G.H., *New Zealand in Evolution*, 1909

Schwarzkopf, General H. Norman, *It Doesn't Take A Hero*, 1992

Searle, G.R., *A New England?: Peace and War 1886-1918*, 2004

Seldon, Anthony, *Blair*, 2004

ed. Self, Robert, *The Neville Chamberlain Diary Letters: The Downing Street Years 1934-1940*, 2005

Seth, Ronald, *The Sleeping Truth: The Hiss-Chambers Affair*, 1968

ed. Seymour, Charles, *The Intimate Papers of Colonel House*, 4 vols, 1926

Sharansky, Natan, *The Case for Democracy*, 2004.

Shawcross, William, *Allies: The United States, Britain, Europe and the War in Iraq*, 2003 eds Sheffield, Gary, and Bourne, John, *Douglas Haig: War Diaries and Letters 1914-1918*, 2005

Sherman, A.J., *Island Refuge: Britain and Refugees from the Third Reich 1933-39*, 1973

Shiletto, Carl, *Devils and Eagles in Normandy 1944: American, British and Canadian Airborne Forces in Normandy*, 2004

Simms, Brendan, *Unfinest Hour: Britain and the Destruction of Bosnia*, 2001

Simpson, A.W. Brian, *Human Rights and the End of Empire: Britain and the Genesis of the European Convention*, 2001

Simpson, John, *The Wars Against Saddam*, 2004

Sinclair, Keith, *William Pember Reeves: New Zealand Fabian*, 1965

Sinclair, Keith, *A Destiny Apart: New Zealand's Search for National Identity*, (Wellington) 1986

Skidelsky, Robert, *John Maynard Keynes: Hopes Betrayed 1883-1920*, 1983

Skidelsky, Robert, *John Maynard Keynes: Fighting for Britain 1937-1946*, 2000

eds Skinner, Kiron K., Anderson, Annelise, and Anderson, Martin, *Reagan In His Own Hand*, 2001

Small, Ken, *The Forgotten Dead: Why 946 American Servicemen Died off the Coast of Devon in 1944*, 1989

ed. Smart, Nick, *The Diaries and Letters of Robert Bernays 1932-1939*, 1996

Snowman, Daniel, *The Hitler Émigrés: The Cultural Impact on Britain of Refugees from Nazism*, 2002

ed. Soames, Mary, *Speaking for Themselves: The Personal Letters of Winston and Clementine Churchill*, 1998

Solow, Barbara Lewis, *The Land Question and the Irish Economy 1870-1903*, (Harvard) 1971

Somerville, Christopher, *Our War: How the British Commonwealth Fought the Second World War*, 1998

Soto, Hernando de, *The Mystery of Capital: Why Capitalism Triumphs in the West and Fails Everywhere Else*, 2000

Sprout, Harold, and Sprout, Margaret, *The Rise of American Naval Power 1776-1918*, (Princeton) 1966

Stacey, C.P., *Canada in the Age of Conflict: A History of Canadian External Politics, vol. II 1921-1948*, (Toronto) 1981

Stafford, David, *Churchill and the Secret Service*, 1997

Steffens, Lincoln, *The Autobiography of Lincoln Steffens*, (New York) 1931

Steiner, Zara, *The Lights That Failed: European International History 1919-1933*, 2005

ed. Stelzer, Irwin, *Neoconservatism*, 2004

Stephan, Alexander, *'Communazis': FBI Surveillance of German Émigré Writers*, 2000

Stevenson, David, *1914-1918: The History of the First World War*, 2004

Stewart, Graham, *Burying Caesar: Churchill, Chamberlain and the Battle for the Tory Party*, 1999

Stewart, Graham, *The History of The Times: The Murdoch Years*, 2005

Stothard, Peter, *Thirty Days: A Month at the Heart of Blair's War*, 2003

Strachan, Hew, *The First World War*, 2003

Strong, Kenneth, *Intelligence at the Top*, 1968

Strouse, Jean, *Morgan: American Financier*, 1999

Sugden, John, *Nelson: A Dream of Glory*, 2004

ed. Swan, Patrick, *Alger Hiss, Whittaker Chambers, and the Schism in the American Soul,* (Delaware) 2003

ed. Talbott, Strobe, *Khruschcev Remembers,* 1970

Tamarkin, M., *Cecil Rhodes and the Cape Afrikaners,* 1996

Tanenhaus, Sam, *Whittaker Chambers,* 1997

Tanner, Stephen, *The Wars of the Bushes,* 2004

Tansill, Charles Callin, *America Goes to War,* (Boston) 1938

Taylor, Frederick, *Dresden: Tuesday 13 February 1945,* 2004

Tertrais, Bruno, *War Without End,* 2005

Thatcher, Margaret, *The Downing Street Years,* 1993

The 9/11 Commission, *Final Report of the National Commission on Terrorist Attacks Upon the United States,* 2002

Thorne, Christopher, *Allies of a Kind: The United States, Britain, and the War Against Japan,* 1978

Thorpe, D.R., *Eden: The Life and Times of Anthony Eden, First Earl of Avon, 1897-1977,* 2003

Todman, Dan, *The Great War: Myth and Memory,* 2005

Toland, John, *Adolf Hitler,* 1976

Tolstoy, Nikolai, *Victims of Yalta,* 1977

Tolstoy, Nikolai, *Stalin's Secret War,* 1981

Tolstoy, Nikolai, *The Minister and the Massacres,* 1986

Tombs, Robert, and Tombs, Isabelle, *That Sweet Enemy: The French and the British from the Sun King to the Present,* 2006

Townshend, Charles, *Easter 1916: The Irish Rebellion,* 2005

Toye, Francis, *For What We Have Received,* (New York) 1948

ed. Trevor-Roper, Hugh, *Hitler's Table Talk 1941-1944,* 2000

Truman, Harry S., *Year of Decisions,* 1955

Truman, Harry S., *Years of Trial and Hope,* 1956

Tuchman, Barbara, *The Zimmerman Telegram,* 1971

ed. Twiston Davies, David, *The Daily Telegraph Book of Naval Obituaries,* 2004

Urban, Mark, *Ten British Commanders Who Shaped the World,* 2005

Vat, Dale Van Der, *Grand Scuttle,* 1982

Véliz, Claudio, *The New World Order of the Gothic Fox,* (California) 1994

Vile, M.J.C., *Politics in the USA,* 1976

Volkogonov, Dmitri, *The Rise and Fall of the Soviet Empire,* 1998

ed. Waites, Neville H., *Troubled Neighbours: Franco-British Relations in the Twentieth Century,* 1971

Walters, Vernon A., *The Mighty and the Meek: Dispatches from the Front Line of Diplomacy,* 2001

Ward, Stuart, *Australia and the British Embrace,* (Melbourne) 2002

Watson, Peter, *A Terrible Beauty: A History of the People and Ideas that Shaped the Modern Mind,* 2001

Webb, Sidney, and Webb, Beatrice, *Soviet Communism: A New Civilisation?,* 1935

ed. Weinberg, Gerhard L., *Hitler's Second Book: The Unpublished Sequel to Mein Kampf,* 2003

Weinberger, Caspar W., *Fighting for Peace,* 1990

eds Weinreb, Ben, and Hibbert, Christopher, *The London Encyclopaedia,* 1983

Weinstein, Allen, *Perjury: The Hiss-Chambers Case*, (New York) 1978

Welsh, Frank, *Great Southern Land: A New History of Australia*, 2004

Wheatcroft, Geoffrey, *The Strange Death of Tory England*, 2005

Whiting, Charles, *The Field Marshal's Revenge: The Breakdown of a Special Relationship*, 2004

Whittle, Sir Frank, *Jet*, 1953

Wilcox, Craig, *Australia's Boer War*, 2002

Will, Henry A., *Constitutional Change in the British West Indies 1880-1903*, 1970

Williams, Andrew, *D-Day to Berlin*, 2004

Williams, Charles, *Pétain*, 2005

Williams, Hywel, *Cassell's Chronology of World History*, 2005

Williams, Neville, *Chronology of the Modern World 1763-1992*, 1994

Williams, Neville, *The Hutchinson Chronology of World History*, vol. IV, 1999

ed. Williams, Philip M., *The Diary of Hugh Gaitskell 1945-1956*, 1983

Williams, Susan, *The People's King: The True Story of the Abdication*, 2003

Williamson, Philip, *Stanley Baldwin*, 1999

Wilson, A.C., *New Zealand and the Soviet Union: A Brittle Relationship*, (Wellington) 2004

Wilson, A.N., *After the Victorians* 2005

ed. Wilson, Keith, *The International Impact of the Boer War*, 2001

Winchester, Simon, *Outposts: Journeys to the Surviving Relics of the British Empire*, 2003

Winchester, Simon, *A Crack in the Edge of the World: The Great American Earthquake of 1906*, 2005

Windschuttle, Keith, *The Killing of History: How Literary Critics and Social Theorists are Murdering Our Past*, 1996

Wodehouse, P.G., *The Code of the Woosters*, 2000

Woodcock, George, *Canada and the Canadians*, 1973

Woodman, Richard, *The Real Cruel Sea: The Merchant Navy in the Battle of the Atlantic 1939-1943*, 2004

Woodward, Bob, *Plan of Attack*, 2004

Woodward, Sir Llewellyn, *Great Britain and the War of 1914-18*, 1967

Woodward, Sir Llewellyn, *British Foreign Policy in the Second War*, 5 vols, 1970-6

ed. Wylie, Neville, *European Neutrals and Non-Belligerents during the Second World War*, 2001

Zakaria, Fareed, *From Wealth to Power: The Unusual Origins of America's World Role*, 1998

Załuski, Andrzej, *The Third Estate*, 2003

Zimmermann, Warren, *First Great Triumph: How Five Americans Made Their Country a Great Power*, (New York) 2002

Zuber, Terence, *Inventing the Schlieffen Plan: German War Planning 1871-1914*, 2002

Zuckerman, Larry, *The Rape of Belgium: The Untold Story of World War I*, 2004

ARTICLES

Anderson, Stuart, 'Racial Anglo-Saxonism and the American Response to the Boer War', *Diplomatic History*, vol. 2, no. 3, summer 1978

Andrew, Christopher, 'Codebreakers at King's', *King's Parade magazine*, King's College, Cambridge (undated)

Bellamy, Christopher, 'Jean de Bloch', *RUSI Journal*, April 1992

Black, Conrad, 'Britain's Final Choice: Europe or America?', *Centre for Policy Studies*, 1998

Bond, Brian, 'Amritsar 1919', *History Today*, October 1963

Briggs, Joe Bob, 'The Q-Man' *The National Interest*, no. 74, winter 2003/4

Bull, Philip, 'The Formation of the United Irish League, 1898-1900: The Dynamics of Irish Agrarian Agitation', *Irish Historical Studies*, vol. 33, no. 132, November 2003

Burton, Sarah, 'Spy Fever', *BBC History*, May 2005

Caplan, Gerald L., 'The Failure of Canadian Socialism: The Ontario Experience 1932-1945', *The Canadian Historical Review*, vol. 44, 1963

Cecil, Hugh, 'Lord Lansdowne: From the Entente Cordiale of 1904 to the "Peace Letter" of 1917: A European Statesman Assessed', Foreign & Commonwealth Office, 2004

Chomsky, Noam, 'The Carter Administration: Myth and Reality', *The Australian Quarterly*, vol. 50, no. 1, April 1978

Colton, F. Barrows, 'Aviation in Commerce and Defence', *National Geographic*, vol. Lxxviii, no. 6, December 1940

Dilks, David, 'Collective Security, 1919 and Now', University of Hull, 1993

Dilks, David, 'Great Britain, The Commonwealth and the Wider World 1939-45', University of Hull, 1998

Dilks, David, 'The Solitary Pilgrimage": Churchill and the Russians 1951-1955', Churchill Society for the Advancement of Parliamentary Democracy, 1999

Dilks, David, 'The Role of Sir Winston Churchill and the Contribution made by Commonwealth Countries to the Second World War', Royal Borough of Kensington and Chelsea, November 2005

Edwards, P.J., 'R.G. Menzies's Appeals to the United States May-June 1940', *Australian Outlook*, vol. 28, no. 1, April 1974

Fry, Geoffrey, 'A Reconsideration of the British General Election of 1935 and the Electoral Revolution of 1945', *History Today*, February 1991

Gedmin, Jeffrey, and Kennedy, Craig, 'Selling America – Short', *The National Interest*, winter 2003/4

Gowing, Margaret, 'The Origins of Britain's Status as a Nuclear Power', Oxford Project for Peace Studies Paper, no. 11, 4 November 1987

Gregory, Adrian, 'British War Enthusiasm in 1914: A Reassessment', in ed. Braybon, Gail, *Evidence, History and the Great War*, 2003

Grimshaw, Patricia, 'Federation as a Turning Point in Australian History', *Australian Historical Studies*, vol. 33, no. 118, 2002

Hart, Peter, 'The Protestant Experience of Partition in Southern Ireland', in eds English, Richard, and Walker, Graham, *Unionism in Modern Ireland*, 1996

James, Sir William, 'Room 40', *University of Edinburgh Journal*, vol. xxii, no. 1, spring 1965

Jones, R.V., 'Alfred Ewing and "Room 40"', *Notes and Records of the Royal Society of London*, vol. 34, no.1, July 1979

Kagan, Robert, 'Why America Dropped the Bomb', *Commentary*, vol. 100, no. 3, September 1995

Keegan, Sir John, 'The Self-Made Scot', *New Criterion*, vol. 23, no. 2, October 2004

King, Prof. Mervyn, 'The Institutions of Monetary Policy', Richard T. Ely Lecture, *The American Economic Review*, vol. 94, no. 2, May 2004

Klimmer, Harvey, 'Everyday Life in Wartime England', *National Geographic*, vol. lxxix, no. 4, April 1941

Knight, Ian, 'Fighting on the Beaches', *Military Illustrated*, no. 193

Kosek, Joseph Kip, 'Richard Gregg, Mohandas Gandhi, and the Strategy of Nonviolence', *The Journal of American History*, vol. 91, no. 4, March 2005

Mackinder, Sir Halford, 'The Round World and the Winning of the Peace', *Foreign Affairs*, no. 21, 1943

Mansfield, Harvey, 'The Manliness of Theodore Roosevelt', *New Criterion*, vol. 23, no. 7, March 2005

McLaughlin, Terence, 'The British in the Air', *History Today*, February 1978

Mitchell, David, 'Woodrow Wilson as "World Saviour": Peacemaking in 1919', *History Today*, January 1976

Morris, H.F., 'The Murder of H. St G. Galt', *The Uganda Journal*, vol. 24, no. 1, March 1960

Nasson, Bill, 'Delville Wood and South African Great War Commemoration', *English Historical Review*, February 2004

Nicholas, David, 'Overlord, Over-Ruled and Over There', *History Today*, April 2005

Packer, Richard, 'The Good is Oft Interred', *The Salisbury Review*, autumn 2004

Partington, Geoffrey, 'Australian Anglophobia: Manning Clark at Oxford', *The Salisbury Review*, autumn 2004

Pašeta, Senia, 'Nationalist Responses to Two Royal Visits to Ireland, 1900 and 1903', *Irish Historical Studies*, no. 31, 1998-9

Penney, Lord, 'John Douglas Cockcroft 1897-1967', *Biographical Memoirs of the Fellows of the Royal Society*, vol. 14, November 1968

Penney, Lord, 'The Nuclear Explosive Yields at Hiroshima and Nagasaki', *Philosophical Transactions of the Royal Society of London*, vol. 266, no. 1177, 11 June 1970

Pickles, Katie, 'A link in "The Great Chain of Empire friendship": The Victoria League in New Zealand', *Journal of Imperial and Commonwealth History*, vol. 33, no. 1, January 2005

Report of the Presidential Commission of Inquiry on the Intelligence Capabilities of the United States regarding Weapons of Mass Destruction, 2004

Roberts, Stephen H., 'Northern Territory Colonization Schemes', *Australasian Association for the Advancement of Science*, August 1924

Ryan, Peter, 'Manning Clark', *Quadrant*, September 1993

Segar, H.W., 'The Struggle for Foreign Trade', *Transactions of the New Zealand Institute*, vol. 40, 1907

Shawcross, William, 'The Cynicism of the Defeatists', *Spectator*, 24 April 2004

Simplich, Frederick, 'Behind the News in Singapore', *National Geographic*, vol. lxxviii, no. 1, July 1940

Tyrie, Andrew, 'Mr Blair's Poodle Goes to War: The House of Commons, Congress and Iraq', *Centre for Policy Studies*, 2004

Véliz, Claudio, 'Bad History', *Quadrant*, May 1982

Walton, R.D., 'Feeling for the Jugular: Japanese Espionage at Newcastle 1919-1926', *The Australian Journal of Politics and History*, vol. 32, no.1, 1986

Weathersby, Kathryn, 'Deceiving the Deceivers: Moscow, Beijing, Pyongyang and the Allegations of Bacteriological Weapons Use in Korea', *Cold War International History Project*, Woodrow Wilson Center, Washington

Will, W. Marvin, 'Insurrection and the Development of Political Institutions: The 1937

Rebellion and the Birth of Labour Parties and Labour Unions in Barbados', *Journal of the Barbados Museum and Historical Society*, vol. 39, 1991

Windschuttle, Keith, 'The Hypocrisy of Noam Chomsky', *New Criterion*, May 2003

Windschuttle, Keith, 'The Journalism of War', *New Criterion*, June 2005

Windschuttle, Keith, 'Mao and the Maoists', *New Criterion*, October 2005

Windsor, Philip, 'The Occupation of Germany', *History Today*, February 1963

Wright, Sir Paul, 'The Festival of Britain: Some Memories', *RSA Journal*, vol. 143, no. 5459, May 1995

Zuber, Terence, 'The Schlieffen Plan: Fantasy or Catastrophe?', *History Today*, September 2002